XXIst International Symposium on the Jahn-Teller Effect 2012

Journal of Physics: Conference Series Volume 428

Tsukuba, Japan
26-31 August 2012

ISBN: 978-1-62748-305-6
ISSN: 1742-6588

Printed from e-media with permission by:

Curran Associates, Inc.
57 Morehouse Lane
Red Hook, NY 12571

Some format issues inherent in the e-media version may also appear in this print version.

Printed by Curran Associates, Inc. (2013)

For permission requests, please contact the Institute of Physics
at the address below.

Institute of Physics
Dirac House, Temple Back
Bristol BS1 6BE UK

Phone: 44 1 17 929 7481
Fax: 44 1 17 920 0979

techtracking@iop.org

Additional copies of this publication are available from:

Curran Associates, Inc.
57 Morehouse Lane
Red Hook, NY 12571 USA
Phone: 845-758-0400
Fax: 845-758-2634
Email: curran@proceedings.com
Web: www.proceedings.com

TABLE OF CONTENTS

C_{60} AND ORGANIC COMPOUNDS

SYMMETRY, VIBRONIC COUPLING, ETC

DEFECTS, INTERCALATION, LIGAND FIELDS

FERROELECTRICITY AND MULTIFERROICITY

MAGNETISM

SUPERCONDUCTIVITY

With the support of

University of Tsukuba

University of Tsukuba
Center for Computational Sciences

The Institute of Natural Sciences, College of Humanities & Sciences, Nihon University

Kyoto University Foundation

The Physical Society of Japan

The Chemical Society of Japan

Izumi Science and Technology Foundation

Tsukuba Expo`85 Memorial Foundation

Tsukuba Tourist and Convention Association

ScienceX Corporation

Scientific and Social Programs

Sunday, 26 August 2012

14:00-18:00 Registration Desk

14:00-16:00 Tea Ceremony

16:00-17:00 Opening Ceremony

Ceremony for the Late Prof. C. Bates

Welcome by H. Koizumi

17:00-18:30 Welcome Reception

Monday, 27 August 2012

9:00-17:00 Registration Desk

Chair R. Kato (RIKEN)

9:30-10:05 S. Saito (Ookayama, Japan) "Electron-phonon interaction, electron-electron interaction, and superconductivity in C_{60} fullerides and diamond"

10:05-10:40 J. Dunn (Nottingham, UK) "Theoretical modeling of Jahn-Teller distorted C_{60} anions on a surface"

10:40-10:55 Coffee Break

Chair K. Yabana (U of Tsukuba)

10:55-11:20 K. Kamaras (Budapest, Hungary) "Dynamic Jahn–Teller effect in the insulating state of correlated fulleride superconductors"

11:20-11:45 T. Akasaka (Tsukuba, Japan) "New vistas in fullerene endohedrals"

11:45-12:10 S. Mahapatra (Hyderabad, India) "Photophysics of fluorinated benzene"

12:10-12:35 H. Goto (Tsukuba, Japan) "Asymmetric synthesis produces optical rotation, circular dichroism, and interference color controllable polymers"

12:35-13:30 Lunch Break

13:30-15:00 Poster Session I

Chair T. Hotta (Tokyo Metropolitan U)

15:00-15:45 D. Khomskii (Cologne, Germany) "Orbital effects in insulator-metal transitions: vanadium oxides": canceled due to sudden sickness

15:45-16:10 T. Takami (Toyonaka, Japan) "Correlation between the Jahn-Teller effect and the spin state in thermoelectric cobalt oxides"

16:10-16:25 Coffee Break

Chair N. Kojima (U of Tokyo)

16:25-16:50 M. Kaplan (Boston, USA) "Vibronic orbital magnetism"

16:50-17:15 A. Stroppa (L'Aquila, Italy) "Polar and magneto-electric properties of anti-ferrodistortive ordered Jahn-Teller distortions in a metal-organic multiferroic"

17:15-17:40 V. Hyzynakov (Tartu, Estonia) "Optical Jahn-Teller effect: contribution of phonons"

17:40-18:05 B. Tsukerblat (Be'er Sheva, Israel) "Electronic and vibronic problems in nanosized mixed valence clusters: advances and challenges"

18:05-19:00 Poster Session I

Tuesday, 28 August 2012

9:00-17:00 Registration Desk

9:00- 9:30 Poster Session I

Chair H. Nakai (Waseda U)

9:30-10:05 A. Bohm (Austin, USA) "The uses of groups and semigroups in quantum theory"

10:05-10:40 W. Domcke (Munich, Germany) "Jahn-Teller theory beyond the standard model"

10:40-11:15 D. Yarkony (Baltimore, USA) "Computational Studies of the Role of Conical Intersections in the Spectra of Jahn-Teller molecules and their substitutional isomers"

11:15-11:30 Coffee Break

Chair J. Zwanziger (Dalhausie U)

11:30-12:05 W. Ernst (Graz, Austria) "Homo- and heteronuclear alkali trimers-conical intersections and spin-orbit coupling"

12:05-12:40 C. A. Mead (Savannah, USA) "Symmetry and degeneracy manifolds in Jahn-Teller molecules"

12:40-13:40 Lunch

Chair Y. Hatsugai (U of Tsukuba)

13:40-14:15 J. Zwanziger (Halifax, Canada) "Polarization and magnetization in solids: the geometric connection"

14:15-14:50 Q. Niu (Austin, USA) "Geometric phase effects on electronic properties"

14:50-15:25 T. Kondo (Tsukuba, Japan) "Observation of Landau levels in potassium-intercalated graphite under zero magnetic field"

15:25-15:40 Coffee Break

Chair C. Satoko (Nihon U)

15:40-16:15 C. Daul (Friburg, Switzerland) "Ligand field theory: A eighty years old young theory"

16:15-16:50 N. Kojima (Komaba, Japan) "Progress of extended spin-crossover phenomena based on ligand field theory"

16:50-17:05 Coffee Break

Chair C. Daul (U of Fribourg)

17:05-17:30 M. Moreno (Santander, Spain) "Puzzling questions in ligand-field theory"

17:30-17:55 G. Lucovsky (Raleigh, USA) "Tanabe-Sugano diagrams applied to nearest-neighbor Si/Ge and transition metal atom dangling bonds in vacated O-atom defects in Si/GeO"

17:55-18:20 S. Sugano (Nagano, Japan) "Optical studies of transitions between multiplet terms with same orbital configuration" (read by H. Koizumi)

18:20-19:00 Poster Session I

Wednesday, 29 August 2012

9:00-12:00 Registration Desk

9:00- 9:30 Poster Session I

Chair W. Ernst (Tech U of Graz)

9:30-10:05 H. Wörner (Zurich, Switzerland) "Conical intersection dynamics probed by high-harmonic spectroscopy"

10:05-10:40 H. Nakai (Waseda, Japan) "Symmetry rules for electronic excitations between degenerate orbitals in high-symmetry systems"

10:40-10:55 Coffee Break

Chair W. Domcke (Tech U of Munich)

10:55-11:20 S. Matsika (Philadelphia, USA) "Nuclear dynamics of model systems with two and three-state conical intersections."

11:20-11:45 S. Adhikari (Kolkata, India) "Diabatic Hamiltonian from molecular symmetry adapted non-adiabatic coupling terms"

11:45-12:10 Y. Liu (Harbin, China) "Pseudo Jahn-Teller origin of nonplanarity and rectangular-ring structure of tetrafluorocyclobutdiene"

12:25 Departure for Mt.Tsukuba

13:00 Lunch at Edoya

16:00 Return from Edoya to Tsukuba

17:15-18:00 Concert (Koto)

18:00-20:30 Banquet

Thursday, 30 August 2012

9:00-12:00 Registration Desk

9:00- 9:30 Poster Session II

Chair M. Kaplan (Simmons College)

9:30-10:15 I. Bersuker (Austin, USA) "A local approach to solid state problems: pseudo Jahn-Teller origin of ferroelectricity, multiferroicity, and magnetic-ferroelectric crossover"

10:15-10:40 S. Kojima (Tsukuba, Japan) "Dynamic Polar Nanoregions and Broken Local Symmetry in Relaxor Ferroelectrics Proved by Inelastic Light Scattering "

10:40-10:55 Coffee Break

Chair Y. Moritomo (U of Tsukuba)

10:55-11:30 Y. Murakami (Tsukuba, Japan) "Impurity effect on orbital

ordering in layered manganite"
11:30-11:55 V. Polinger (Seattle, USA) "Ferroelectric phase transitions in cubic perovskites"
11:55-12:20 S. Streltsov (Ekaterinburg, Russia) "Elongated vs. compressed octahedra: Jahn-Teller effect in the transition metal compounds with partially filled eg-subshell"
12:20-13:30 Lunch
13:30-14:30 Poster Session II
Chair V. Polinger (U of Washington)
14:30-14:55 T. Sato (Kyoto, Japan) "Vibronic coupling density and related concepts"
14:55-15:20 T. Hotta (Hachioji, Japan) "Fermi-surface topology and superconductivity induced by Jahn-Teller phonons"
15:20-15:45 S. Shamoto (Tokai, Japan) "Lattice breathing by the electron round scattering"
15:45-16:10 S. Ishihara (Sendai, Japan) "Dynamical Jahn-Teller effect and spin liquid in an orbital degenerate magnet"
16:10-16:25 Coffee Break
Chair Y. Murakami (KEK)
16:25-16:50 V. V. Gudkov (Ekaterinburg, Russia) "Resonance Anomalies in Interaction of Ultrasound with Jahn-Teller Centers in ZnSe:Cr in magnetic fields"
16:50-17:15 J. A. Aramburu (Santander, Spain) "What is the origin of the different structure of $CuCl_4(NH_3)_2$- and $CuCl_4(H_2O)_2$- units ? "
17:15-17:40 E. Gindensperger (Strasbourg, France) "Vibronic and spin-orbit interactions in the photophysics of Re(I) Complexes"
17:40-18:05 T. Mizokawa (Kashiwa, Japan) "Jahn-Teller effects in transition-metal compounds with small charge-transfer energy"
18:05-19:00 Poster Session II

Friday, 31 August 2012
9:00- 9:30 Poster Session II

Chair T. Mizokawa (U of Tokyo)

9:30-10:15 H. Oyanagi (Tsukuba, Japan) " Local lattice distortion in superconducting cuprates studied by XAS"

10:15-10:50 A. Fujimori (Hongo, Japan) "Electron-phonon coupling in multilayer cuprate superconductor"

10:50-11:05 Coffee Break

Chair K. Shiraishi (U of Tsukuba)

11:05-11:40 H. Kamimura (Kagurazaka, Japan) "On the important role of the anti-Jahn-Teller effect in cuprate superconductors"

11:40-12:25 A. Bussmann-Holder (Stuttgart, Germany) "Polaron formation, coherence and interaction with strain in high temperature superconducting copper oxides."

12:25-13:25 Lunch Break

Chair H. Oyanagi (AIST)

13:25-14:10 K. A. Müllller (Zurich, Switzerland) "Some unique superconductive properties of cuprates" (video + teleconference)

14:10-14:45 M. Fujita (Sendai, Japan) "Doping-induced hourglass spin excitations in cuprate oxides"

14:45-15:20 H. Koizumi (Tsukuba, Japan), "What is the phase variable in superconductors ? "

15:20-15:30 Closing Ceremony

第２１回ヤーン・テラー効果に関する国際会議：
物理と化学における対称性の破れ

平成２４年8月２６日－３１日

筑波大学大学会館

ホームページ　http://www.ccs.tsukuba.ac.jp/CCS/jt2012/

特別招待講演者

K. A. Muller (1987 Nobel Prize 受賞)

C. A. Mead (2012 Wigner Medal 受賞)

特別ゲスト

田辺行人 (東京大学名誉教授)

菅野　暁 (東京大学名誉教授)

上村　洸 (東京理科大学教授、
東京大学名誉教授)

JT 2012

Jahn-Teller Effect

トピックス

1. 銅酸化物超伝導体と鉄砒素系超伝導体
2. C60 超伝導体と有機超伝導体
3. 強誘電体とマルチフェロイック化合物
4. 協力的ヤーン・テラー効果と軌道秩序
5. バイブロニック系の光学的、分光学的性質
6. 円錐型交差と化学反応
7. 数学的、数値的計算法の進歩
8. ヤーン・テラー効果と磁性

特別セッション

A. 配位子場理論

B. 分子系、固体系におけるゲージ場の出現

共催

筑波大学

筑波大学計算科学研究センター

日本物理学会

日本化学会

後援

泉科学技術振興財団

京都大学財団

つくば万博記念財団

All papers published in this volume of *Journal of Physics: Conference Series* have been peer reviewed through processes administered by the proceedings Editors. Reviews were conducted by expert referees to the professional and scientific standards expected of a proceedings journal published by IOP Publishing.

Theoretical modelling of Jahn-Teller distorted C_{60} anions on a surface

A J Lakin, H S Alqannas and J L Dunn

School of Physics and Astronomy, University of Nottingham, University Park, Nottingham, NG7 2RD, UK

E-mail: janette.dunn@nottingham.ac.uk

Abstract. Through the use of scanning tunnelling microscopy (STM), it is possible to directly observe the molecular orbitals associated with a particular molecule. For the charged ions of the C_{60} fullerene, the interpretation of these experimental images has an additional complication due to the inherent presence of the Jahn-Teller (JT) effect. In this work, the influence of the JT effect on STM images is examined. We also include interactions between the molecule and both the surface substrate on which it rests, and, when present within a monolayer, the nearest neighbours. Simple symmetry arguments are used to explain the effects of these external interactions on the energy of the different JT wells relating to molecular distortions of D_{3d}, D_{5d}, and D_{2h} symmetry. We first investigate the C_{60}^- monoanion, and then, through the construction of multi-electron states, move on to consider anions with higher charges. It is found that for high symmetry orientations of the molecule on the surface, the wells that are degenerate in the absence of external interactions split into equal energy subgroups, with the grouping dependent on the orientation of the molecule. Hückel molecular orbital theory is then used to investigate the effect this has on STM images. We show that when dynamic JT effects are considered, the images are always formed from some linear combination of the squares of the individual single electron molecular orbitals that make up the lowest unoccupied molecular orbital of the neutral molecule.

1. Introduction

Through the use of scanning tunnelling microscopy (STM), molecules which undergo a Jahn-Teller (JT) distortion can be directly imaged, providing useful information on the nature of the symmetry reduction this effect causes. The ions of the C_{60} fullerene molecule are ideal candidates for analysing this because the large size of the molecule makes it easily discernible via STM. High degeneracies in the electronic and vibrational states are lifted by the JT effect, producing splittings in the electronic states. A signature of this splitting is apparent in observed STM images.

For isolated anions of C_{60}, the JT effect is expected to reduce the symmetry of the molecule from I_h to either D_{5d}, D_{3d}, D_{2h}, or C_{2h} [1, 2, 3], producing a selection of isoenergetic wells in the lowest adiabatic potential energy surface (APES). When quadratic JT effects are strong, the JT effect is static; the energy barrier between each of these wells is high, causing the system to become locked into a single distortion. The effect this would have on the STM images of the C_{60}^- ion has been discussed previously [4]. Alternatively, if the quadratic JT effects are weaker, the JT effect will be dynamic. The energy barrier between minimum energy wells will be sufficiently small to allow the molecular distortion to pseudorotate between the equivalent lowest energy distortions. If the barriers in the APES are relatively high, the system is expected to spend negligible time in intermediate distorted configurations, and the distortion will 'hop' between wells. If the barrier is smaller, the time spent at intermediate positions will become significant giving a hindered pseudorotation. In the limit of a negligible barrier height, a free pseudorotation occurs, where equal time is spent in all possible distortions. It is expected that at realistic coupling strengths, the pseudorotation occurs on a timescale that is much quicker than that of STM measurements, and therefore all lowest energy distortions will be imaged simultaneously.

In constructing STM images, we will use Hückel molecular orbital (HMO) theory, rather than the more commonly used density functional theory (DFT), due to its speed and simplicity. As π orbitals contribute the overwhelming proportion of the bonding in C_{60}, little error is introduced when constructing molecular orbitals by neglecting σ bonding, and it is this assumption that forms the basis of HMO theory. Indeed, there is excellent agreement between the molecular orbitals obtained for the neutral C_{60} molecule using both DFT and HMO theory [5, 6, 7, 8].

In this work, interactions between C_{60} ions and both the surface substrate and neighbouring ions will be treated using simple symmetry arguments to model the effect on the molecular distortion. The effect these interactions have on the C_{60}^- anion will then be investigated, and a comparison made between the theoretical method and experimentally obtained images. We will then extend the method to the higher charged anions by constructing multi-electron states that are functions of the single electron orbitals that make up the lowest unoccupied molecular orbital (LUMO). A comparison will be made with experimental images of the unoccupied states of the C_{60}^{4-} anion.

2. Modelling the Jahn-Teller effect and external interactions in C_{60}^-

We will assume that C_{60} molecules are adsorbed onto a surface, either as an isolated molecule or as part of a monolayer. In this section, we will consider the case in which STM probes the states of the singly-charged anion, C_{60}^-. This could represent the case where C_{60}^- ions are intrinsically present on a surface. Alternatively, the STM measurement process could result in the C_{60}^- states being probed. A brief overview will be given of how the JT effect is modelled in the monoanion, and of the way in which an external interaction is treated. The two interactions will then be combined to see the effect of different molecular orientations on the surface adsorbed, JT-active anion.

2.1. Jahn-Teller interactions

The LUMO of C_{60} is three fold degenerate and is described by the T_{1u} irrep. When the molecule is ionised to form C_{60}^-, the additional electron occupies this T_{1u} state, and couples to eight h_g modes to give a $T_{1u}\otimes 8h_g$ JT effect. Each mode will correspond to different nuclear displacements. However, these displacements are too small to be discernible in an STM experiment. As only the splitting of the electronic state is required, the only consideration is of what symmetry the molecule reduces to, and for this, a single effective mode can be considered.

The JT interaction can be expressed by a linear coupling Hamiltonian and two (due to the non-simple reducibility of the $H\otimes H$ product) quadratic coupling Hamiltonians. We will follow [1] and write the Hamiltonians using two-fold molecular axes $\{X, Y, Z\}$ which align with double bonds of the molecule. When a C_{60} molecule is adsorbed on a surface, we will later also use axes $\{x, y, z\}$ where z refers to the direction perpendicular to the surface.

The full JT Hamiltonian can be written as:

$$H_{JT} = V_1 H_1 + V_2 H_2 + V_3 H_3 + \frac{1}{2}\mu\omega^2 \sum_{\gamma} Q_{\gamma}^2. \tag{1}$$

Journal of Physics: Conference Series **428** (2013) 012001 doi:10.1088/1742-6596/428/1/012001

Here, V_1, V_2 and V_3 are the linear and two quadratic coupling constants respectively, μ is the mass and ω is the frequency of the (effective) h_g mode, and γ sums over all 5 of the vibrational coordinates Q_γ of the vibrational h_g mode. These Qs are denoted $\{Q_\theta, Q_\epsilon, Q_4, Q_5, Q_6\}$ which transform according to the linear combination of the d-type orbitals

$$\{\sqrt{\frac{3}{8}}d_{3Z^2-R^2} + \sqrt{\frac{5}{8}}d_{X^2-Y^2}, \sqrt{\frac{3}{8}}d_{X^2-Y^2} - \sqrt{\frac{5}{8}}d_{3Z^2-R^2}, d_{YZ}, d_{ZX}, d_{XY}\}. \tag{2}$$

The forms of the matrices representing these Hamiltonians can be found in ref. [1].

Diagonalising the Hamiltonian results in an energy expression that is a function of the coupling constants and the Qs. For given values of the coupling constants, the lowest energy eigenvalue can be minimised with respect to the Qs revealing the symmetry of the distorted molecule, which, for C_{60}^- is predominantly to either D_{3d} or D_{5d} symmetry. Following the convention set out in ref. [1], the six D_{5d} wells are labelled A to F and the ten D_{3d} wells are labelled a to j. For some regions of coupling values it is also possible that the molecule distort to D_{2h} symmetry, which has 15 wells which will be labelled A to O as in [4]. Distortion to C_{2h} symmetry is also possible for a very small range of coupling constants, but this will not be considered in this paper.

2.2. External interactions

We will assume that the effect on a C_{60} molecule of interactions with a surface substrate is to perturb the atoms closest to the surface in some way such that the symmetry is reduced. For many common surfaces, such as Ag or Au, it will be appropriate to treat the surface as being homogeneous in the plane of the surface. The effect on the C_{60} molecule is therefore to remove the equivalence between the z axis (perpendicular to the surface) and the x and y axes (in the plane of the surface). For a C_{60} oriented with a pentagon prone to the surface, this can be visualised in terms of the symmetry reduction depicted in figure 1. The fivefold rotation about the (z) axis through the centre of the uppermost pentagon remains as a symmetry axis. However, fivefold rotations about axes through the centres of the other pentagons are no longer symmetry operations. Similarly, reflections in various vertical planes are allowed but not in other planes that would be equivalent in the undistorted molecule. In this case, the reduction in symmetry induced by adsorption on the surface is to the C_{5v} point group. Similarly, when the molecule is adsorbed with a hexagon or double bond uppermost, the symmetry is reduced to C_{3v} or C_{2v} respectively. It is important to note that figure 1 is given as a guide to represent the reduction in symmetry and we are not suggesting the molecule distorts in exactly this manner.

Characters associated with each symmetry operation for each component of the icosahedral T_{1u} LUMO can be found, and matched with the character table of the lower-symmetry point group to reveal any degeneracies. The result of this is that, with respect to a basis $\{\psi_x, \psi_y, \psi_z\}$ defined such that the z component is perpendicular to the surface, the ψ_x and ψ_y form a doublet and the ψ_z is a singlet when the molecule is adsorbed with a pentagonal or hexagonal face prone to the surface. For all other adsorption orientations, three singlets are expected. It is therefore straightforward to model the effect this has on the molecule by constructing a simple diagonal Hamiltonian.

To remain consistent with the JT Hamiltonian in equation (1) the surface interaction is defined as:

$$\mathcal{H}_{sz} = -\frac{V_1^2}{\mu\omega^2}\begin{bmatrix} 0 & 0 & 0 \\ 0 & \Delta_2 & 0 \\ 0 & 0 & \Delta_1 \end{bmatrix}, \tag{3}$$

where the ψ_x component is treated as the zero in energy, and Δ_2 and Δ_1 are dimensionless parameters representing the energy difference between the ψ_x state and the ψ_y and ψ_z states

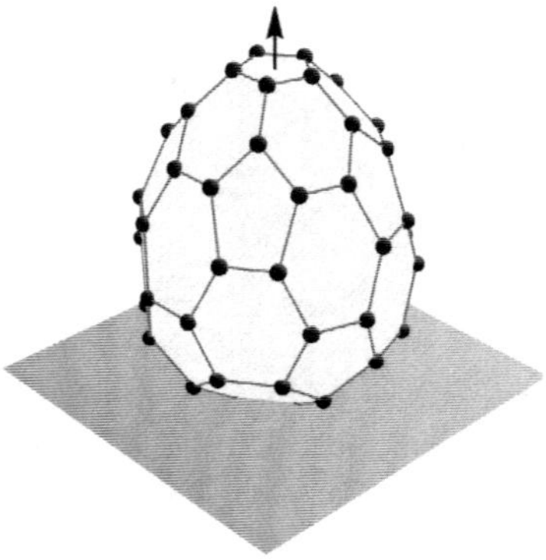

Figure 1. (Colour online) Representation of the reduction in symmetry introduced by a surface interaction on the pentagon prone molecule. Note that this is a visual aid to the symmetry reduction experienced by the molecule, and is not intended to be a real representation of the way the molecule distorts.

respectively. Thus, for orientations where a doublet remains, Δ_2 will be equal to zero. By using this definition it is assumed that the strength of the surface interaction does not affect the basis functions at all, and merely changes the difference in energy between the different states. Comparing with DFT calculations undertaken by Pascual et al. [5] shows the Hückel method produces images that can't be distinguished from those from DFT.

The surface interaction Hamiltonian in equation (3) acts in a basis where the ψ_z component of the LUMO relates to the perpendicular to the surface. In order to combine this Hamiltonian with the JT Hamiltonian which is written in terms of a 2-fold Z-axis, a simple transformation is needed to change to a common basis.

Intermolecular interactions can be modelled via a splitting of the components of the LUMO in a similar manner to that described above for the surface interaction. When present as a monolayer, C_{60} molecules form a hexagonal lattice, and so, by working out what symmetry operations remain when a molecule is placed at each lattice point, the splitting of the orbitals can be found. It is assumed throughout that each molecule is oriented identically on the surface. While this is not always the case in practice, experimental data obtained on C_{60} monolayers indicates that this does occur in a lot of instances. It is therefore a good starting point for investigating the effects of intermolecular interactions. The orientation of molecules in a hexagonal lattice determines which of the symmetry operations associated with this group remain.

The resultant point group symmetries for different molecular orientations are shown in table 1. We note that aligning a reflectional plane of the molecule with a reflectional plane of the lattice results in a higher symmetry. Using this information it is again possible to construct a diagonal Hamiltonian to represent the induced splitting. However, for the instances where more than one T_{1u} component is described by the same irrep., the resultant splitting in the electronic states can't be modelled using group theoretical ideas alone, as it is not known what linear combination of states represent the true singlets.

As both the intermolecular and surface interaction Hamiltonians are diagonal and act on the same basis, it is sensible to combine them to form one Hamiltonian representing the external interaction. As the differences in energy between the molecular orbitals are not known and are varied as part of the investigation, the magnitudes of the splitting can be reduced to variables that represent the joint effect of the splitting induced by both the surface and intermolecular interactions.

Journal of Physics: Conference Series **428** (2013) 012001 doi:10.1088/1742-6596/428/1/012001

Table 1. The reduction in symmetry of C_{60} molecules situated on a hexagonal lattice and the splitting of the T_{1u} states $\{\psi_x, \psi_y, \psi_z\}$. Results are given for different molecular orientations. The column headed 'Reflection' indicates whether a reflection plane of the molecule is aligned with a reflection plane of a hexagonal lattice.

Orientation	Reflection	Symmetry	Splitting
Hexagon prone	Yes	D_{3d}	$\psi_x + \psi_y \to E_u$ $\psi_z \to A_{2u}$
	No	S_6	$\psi_x + \psi_y \to E_u$ $\psi_z \to A_u$
Double Bond prone	Yes	D_{2h}	$\psi_x \to B_{3u}$ $\psi_y \to B_{2u}$ $\psi_z \to B_{1u}$
	No	C_{2h}	$\psi_x \to B_u$ $\psi_y \to B_u$ $\psi_z \to A_u$
All other orientations	Yes	C_{2h}	$\psi_x \to B_u$ $\psi_y \to A_u$ $\psi_z \to B_u$
	No	C_i	All components transform as A_u

2.3. Results

The combined JT and external interaction Hamiltonians are a function of the two quadratic coupling constants V_2 and V_3, the vibrational coordinates Q_γ, and the external interaction parameters Δ_1 and Δ_2. Diagonalising this Hamiltonian therefore provides a function in terms of these variables. Before looking at the effect on the STM current (and hence the STM image) the various parameters have, it is worthwhile considering the effect the external interaction has on the energy of the C_{60}^- JT wells of different symmetry.

For a single molecule (i.e. where no intermolecular interaction is present) only the surface interaction needs to be considered. For the hexagon and pentagon prone molecular orientations, ψ_x and ψ_y form a doublet, which means that Δ_2 can be set to zero, so it is possible to look at the effects on the individual wells of Δ_1 alone. The results of this are shown in table 2, where it can be seen that the JT wells form isoenergetic groups dependent on the orientation of the molecule. For the double-bond prone orientation, ψ_x and ψ_y are assumed to have negligible splitting in this table, although the calculation method also allows us to investigate the more general case where this is non-zero.

The grouping of the wells has consequences when considering the resultant STM images as the distortion will tunnel between isoenergetic wells in finite JT coupling. Figure 2 represents the electronic states $c_x\psi_x + c_y\psi_y + c_z\psi_z$ for each JT well as points with coordinates $\{c_x, c_y, c_z\}$ for (a), D_{5d}, and (b), D_{3d} type distortions, labelled with the molecular C_2 Z-axis. The changes in the electronic states as the molecular distortion pseudorotates on a direct path between adjacent wells are shown as solid lines. As the pseudorotational process is expected to occur at a much faster timescale than that of an STM experiment, STM images will result from a superposition of all electronic states sampled. We will assume that the system spends most of its time in the wells and not consider the intermediate configurations, which will be a reasonable approximation for moderate coupling strengths (with appreciable barriers between wells). In fact, including intermediate configurations has been shown to have very little effect on the predicted images [10].

Table 2. Groupings of equivalent D_{5d}, D_{3d} and D_{2h} JT wells for pentagon, hexagon and double-bond prone molecular orientations. Groups on the same row between braces are related to each other by reflections in the x–z plane. The groups that can be global minima in our model when $\Delta_1 > 0$ are marked with a $^+$ and when $\Delta_1 < 0$ with a *, assuming that $\Delta_2 = 0$ for the double-bond prone orientation.

	D_{5d}	D_{3d}	D_{2h}
Pentagon	C$^+$	a,b,c g,i$^+$	*A,C,D,E,H*
	A,B,D,E,F*	d,e,f,h,j*	*B,F,K,L,O*
			G,I,J,M,N
Hexagon	C,E,F$^+$	c$^+$	*A,M,N*
	A,B,D*	{a,f,j},{b,e,h}*	{*B,D,J*},{*C,I,O*}
		d,g,i	*E,F,H*
			G,L,K
Double-bond	C,D$^+$	a,b$^+$	*A*
	E,F*	{e},{f}*	*F*
	A,B	c,d	*G*
		{g,j},{h,i}	{*E,I*},{*H,J*}
			{*B,N*},{*M,O*}
			{*C,K*},{*D,L*}

To model the tunnelling current, a simple Tersoff-Hamann approach is used [9], which means that in our problem (for an s-type tip) the current will be proportional to the sum of the squares of the wavefunctions of each degenerate minimum-energy molecular orbital function ψ_i, i.e.:

$$I \propto \sum_i \psi_i^2. \tag{4}$$

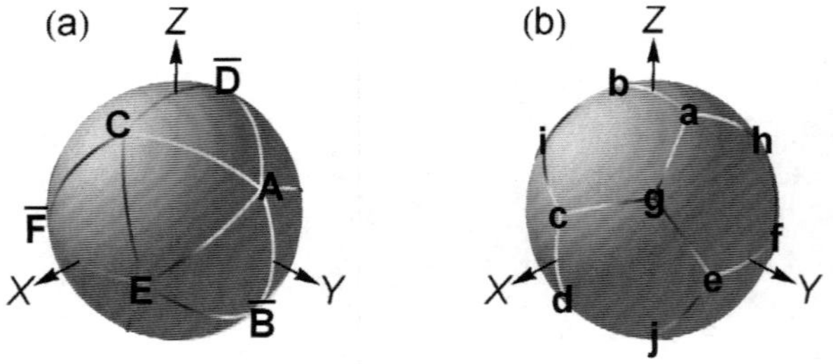

Figure 2. (Colour online) Representation of the electronic states $c_x\psi_x + c_y\psi_y + c_z\psi_z$ for each JT well as points with coordinates $\{c_x, c_y, c_z\}$. (a) is for JT parameters resulting in a D_{5d} distortion, and (b) is for parameters giving a D_{3d} distortion. The solid lines represent changes in the electronic states when the system pseudorotates on a direct path between nearest-neighbour wells.

Comparing the results in table 2, with the electronic states for the wells as shown in figure 2, it can be seen that for a positive value of Δ_1 (i.e. a lowering of energy of the ψ_z state), it is the wells closest to the poles (with respect to a 2-fold Z-axis) which are favoured. For a negative value of Δ_1, it is the wells closest to the equatorial $X - Y$ plane that are favoured.

For the highly symmetric orientations considered in table 2, it is found that summing over the squares of the lowest energy electronic states results in an STM image that is always constructed from some linear combination of the form $a_x\psi_x^2 + a_y\psi_y^2 + a_z\psi_z^2$. This occurs because the wells are symmetrically distributed around the molecular Z axis, and hence when the current is found from the sum of the degenerate states, any cross terms will always cancel.

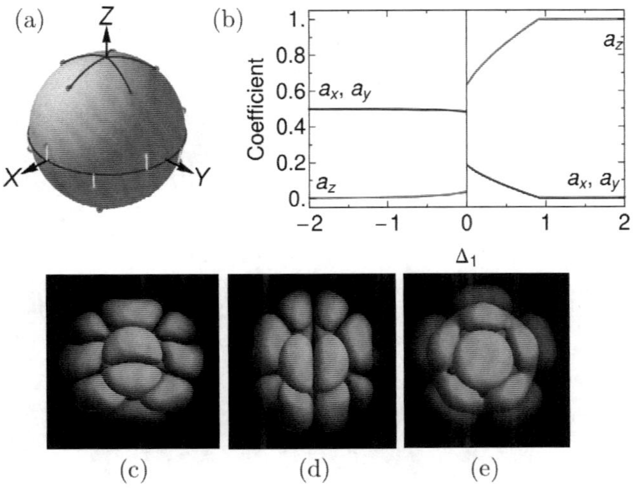

Figure 3. (Colour online) (a) Movement of the 'coordinates' $\{c_x, c_y, c_z\}$ specifying the components of the electronic states as the strength of the external interaction Δ_1 changes. The positions of the states evolve from those of the JT wells to either the poles or the equatorial plane, (b) variation of the coefficients $\{a_x, a_y, a_z\}$ in the expression for the tunnelling current I with the strength of the external interaction, theoretical STM scans of (c) ψ_z^2, (d) ψ_y^2 and (e) ψ_x^2. All results are for the pentagon prone molecule with JT parameters that favour a D_{3d} distortion.

An example of how the magnitude of the surface interaction affects the observed current is shown in figure 3, where the C_{60} molecule is oriented with a pentagonal face prone to the surface, and the JT coupling constants are chosen to favour a D_{3d} distortion. Also shown are theoretical STM simulations of ψ_x^2, ψ_y^2 and ψ_z^2 (calculated using HMO theory). The resultant STM image for any strength of surface interaction will result from a superposition of the appropriate contributions from each component. Although the exact form of such images must be determined by direct calculation, a rough indication of the expected results can be deduced from a visual examination of the results for the separate components.

The crossover to a different solution when $\Delta_1 = 0$ indicates that a different group of wells become lowest in energy when the sign of the surface interaction changes. As indicated in table 2, wells $\{a, b, c, g, i\}$ are favoured for a positive surface interaction, and wells $\{d, e, f, h, j\}$ are favoured when the surface interaction is negative. This effect can be visualised by plotting the coefficients of the electronic state for each well, and seeing how the coefficients alter as the surface interaction changes, as is shown in fig. 3(a). It can be seen that choosing $\Delta_1 > 0$ such that the ψ_z component is lowered in energy relative to ψ_x and ψ_y causes the coordinates to shift towards the z axis of the sphere. Choosing $\Delta_1 < 0$ so that the energy of ψ_x and ψ_y are lowered causes the coordinates to shift towards the equator.

It should be noted that the JT coupling constants are chosen to favour a certain symmetry for the isolated molecule, and do not represent the (lower) symmetry of the distorted molecule once the external interaction is considered. The introduction of the surface interaction generally reduces the symmetry to C_{2h}.

2.4. Comparison with experiment

We will now consider published experimental results that may show a signature of the JT effect. The work presented in [11] shows images of a C_{60} monolayer on an alkylthiol buffer layer. While in [11], the molecules are assumed to be charge neutral, charge transfer between the molecule and substrate, or, as is particularly relevant when a buffer layer is present, between the molecule and tip, may result in images relating to the states of a charged ion, and as such the observation of images relating to the JT-distorted molecule should still be considered as a possibility.

We start by seeking values for the coefficients $\{a_x, a_y, a_z\}$ that result in an image that matches the left-hand image in figure 1 of [11], reproduced here in figure 4(a). We find that the best match comes when we take the C_{60} molecule to be tilted in the y–z plane such that a molecular C_2 axis is around 0.1 radians from the perpendicular to the surface and with $\{a_x, a_y, a_z\} \approx \{0, 0.13, 0.87\}$. The image resulting from this match is shown in figure 4(b). This is different to the match of the authors of [11], which is shown as an inset to figure 4(a). As our match is only a qualitative match by eye, determination of the coefficients is not exact; a_y can vary between around 0.075 and 0.2 with corresponding changes in a_z between 0.925 and 0.8 [10].

(a) 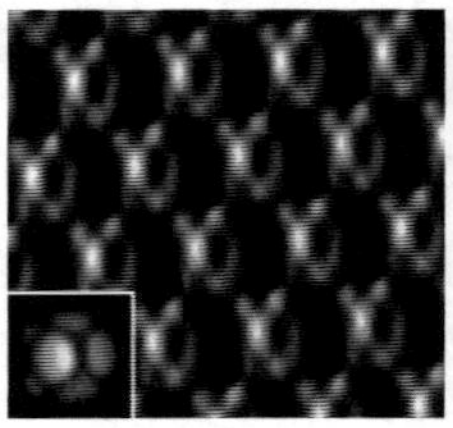 (b) 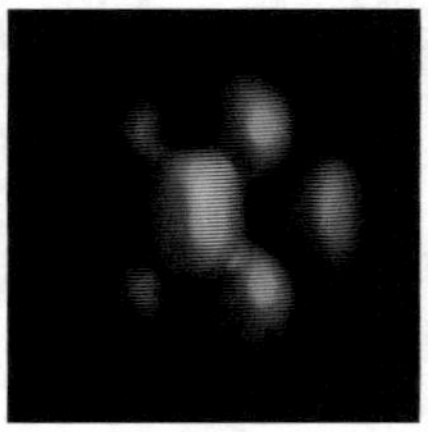

Figure 4. (Colour online) Comparison between (a) an experimental image obtained in ref. [11] (with theoretical interpretation from this reference inset), and (b) the theoretical STM image obtained using the method outlined here. (a) reprinted from figure 1 of LF Yuan et al. J. Am. Chem. Soc. **125**, 169 (2003) [11]. Copyright 2003 American Chemical Society, Copyright (2003).

The next step in determining a theoretical match to the experimental image in [11] is to determine whether it is possible to find combinations of the JT and external interaction parameters that result in the required form for the current I. It is found that this is indeed possible for a range of JT parameters that favour D_{3d} wells a and b, alongside external interaction parameters that set the ψ_z state of a double bond prone molecule lowest in energy. Also, it is important to note that we can't get our match without using a JT model, as it is only when we superimpose results due to different distortions that unwanted cross-terms in the expression for I cancel. Hence our match represents a different physical situation, and isn't simply an alternative explanation of the same case.

It is worth noting that the tilt in the y–z plane needed to match the experimental image reduces the symmetry associated with both the surface interaction and the intermolecular interaction (as one of the reflection axes of the molecule is lost, which previously aligned with one of the lattice). This introduces the problem where two of the components are described by the same irrep., and as such the correct combination of states needed to describe the splitting can't be determined from group theory alone. However, as the deviation from the double bond prone case is small, the assumption is made that the components split in the same way as for the double bond prone molecule. The fact that an accurate match with experiment is found shows that this assumption appears valid.

Journal of Physics: Conference Series **428** (2013) 012001 doi:10.1088/1742-6596/428/1/012001

3. Higher charged states

So far, only the C_{60}^- monoanion has been considered. However, experimental images are available for C_{60}^{3-}, C_{60}^{4-} [12], and C_{60}^{5-} [13] ions as well. Investigating the higher charge states is inherently more difficult due to the Coulomb interaction between the electrons, which affects the electron distribution around the molecule and hence the observed current during STM. Due to electron-hole equivalence, the JT effect in C_{60}^{5-} is the same as in the monoanion. As such this doesn't need further consideration. This work will concentrate on the C_{60}^{4-} anion, which is modelled as two interacting holes, and is considered by looking at the $T_{1u} \otimes T_{1u}$ interaction, which results in $A_g \oplus H_g$ antisymmetric states and a T_{1g} symmetric state [2]. The coupling between the T_{1g} state and the H_g vibrational state is the same as for the monoanion and so only the coupling with the antisymmetric state will be considered here. The C_{60}^{3-} ion may also be modelled by considering the product $T_{1u} \otimes T_{1u} \otimes T_{1u}$, resulting in a JT effect between the h_g vibration, and the $T_{1u} \oplus H_u$ electronic states [14]. However, this will not be considered here.

It is found that for the hexagon and pentagon molecule, a component of the H_g state transforming as the spherical harmonic $d_{3Z^2-R^2}$ is a singlet, with components transforming as $d_{X^2-Y^2}$ and d_{XY} forming one doublet and the components transforming as d_{YZ} and d_{ZX} forming a second doublet. For all other orientations, five singlets transforming as the above d-orbitals are formed. Note that this basis is defined differently to the JT basis, which uses linear combinations $H_{g\theta}$ and $H_{g\epsilon}$ of $d_{3Z^2-R^2}$ and $d_{X^2-Y^2}$ as defined in equation (2) and in [15]. Therefore an additional transformation needs to be made to align the external interaction and JT Hamiltonian bases. The states transforming as d_{yz}, d_{zx} and d_{xy} will be called H_{g4}, H_{g5} and H_{g6} respectively.

Clearly additional parameters are introduced due to the combined Coulomb and external interactions compared to C_{60}^- due to the larger dimension of the electronic basis, as each distinct state needs to be set at a different energy. Also, where there are repeated representations in the symmetry reduction, we do not know from group theory alone what basis states to take for the external interaction Hamiltonian. However, a unique form can be written down from knowledge that the single-hole states ψ_x, ψ_y and ψ_z will be at different energies to each other. We note that [16] gives values for the term splitting between the A and H states due to the Coulomb interaction. Using an estimate for the linear coupling constant in [17] to convert the values to the units used here leads to a value for the term splitting of 2.4. We will use this value to illustrate our results below. However, appropriate values for the real C_{60} system could vary somewhat from this.

In order to obtain expressions for the current that would be observed during STM, we express the two hole basis states in terms of the single hole molecular orbital functions. The hole interaction transforms as $T_{1u} \otimes T_{1u}$, thus, by using the Clebsch-Gordon coefficients in [15], the correct combinations of the two hole states can be constructed. This can then be combined with the correct spin state to produce the appropriate two-hole states.

To obtain an expression for the observed STM current, the square of some combination(s) of the basis states needs to be taken. This introduces problems due to the inclusion of the electron spin in the wavefunctions. To solve this, it is noted that the distribution of one electron around the molecule is dependent on the position of the second electron. It is therefore possible to integrate over all space for one of the electrons to obtain the current in terms of a single electron distribution function. The orthonormality of the states ensures that the integrals that arise from the products of the various states always go to one or zero, resulting in a simple expression that is made up of the single electron molecular orbitals only.

With a form obtainable for the effective current, it is then possible to investigate the effect of the surface and Coulomb interactions on observed STM images. In much the same way as was obtained for the monoanion, groupings of JT wells occur, and, despite the fact that the basis is now formed from a two hole combination, the images are again constructed from some linear combination of the squares of the single hole (electron) molecular orbitals for the high symmetry orientations. However, a particular well will provide a different contribution to the current when compared with the monoanion, and as such the resultant images are different.

Figure 5 shows an example where the molecule is situated with a pentagon prone to the surface, with JT coupling constants chosen to prefer a D_{3d} molecular distortion. In the figure, the energy of the A state has been set to 2.4 relative to that of the $H_{g\theta}$ state, the splitting between $H_{g\theta}$ and the H_{g4}, H_{g5} doublet, denoted Δ_2, is taken to be 0, and the splitting between $H_{g\theta}$ and the $H_{g\epsilon}, H_{g6}$ doublet, denoted Δ_1, is varied. It can be seen that in the asymptotic limit of a strong surface interaction, the current is still formed from all of the molecular orbital functions, with none tending to zero. This is because the two hole states H_{g4}, H_{g5} and H_{g6} are formed from states where the two holes are never in the same single-hole orbital. As such, any state involving these two-hole states will always contain contributions from ψ_x^2, ψ_y^2 and ψ_z^2. This means that observed STM images are in general likely to be less distinctive than those from the monoanion. However, the current is still made up from contributions from ψ_x^2, ψ_y^2 and ψ_z^2 with no cross-terms.

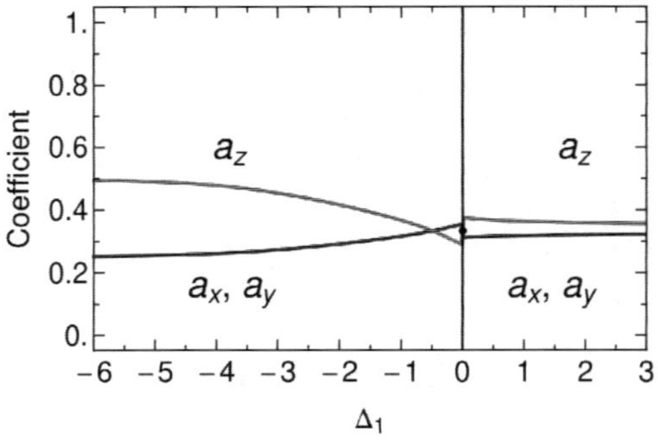

Figure 5. (Colour online) Graphical representation of the coefficients of the square of the single electron molecular orbitals, with varying strength of surface interaction parameter Δ_1 for the pentagon prone molecule with JT parameters to favour a D_{3d} distortion.

Using this idea, it is possible to obtain a match with the experimental image obtained by Wachowiak et al. [12]. The best match, shown in figure 6, arises from a molecule in the double-bond prone orientation and with a current $I \approx 0.015\psi_x^2 + 0.985\psi_y^2$.

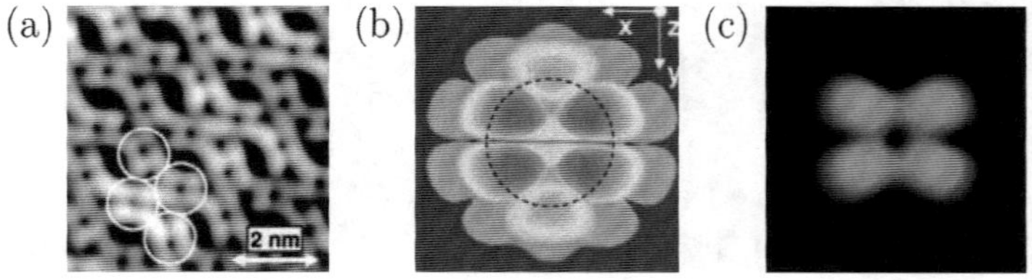

Figure 6. (Colour online) Comparison between an experimentally obtained image and the theoretical interpretation in [12] (a and b), and the best-fit image obtained by matching the current to $I \approx 0.015\psi_x^2 + 0.985\psi_y^2$ (c). (a) and (b) are reprinted from parts of figures 3 and 4 of [12] with permission from AAAS.

The next step in the identification of the origins of the image is to determine what values of the JT and surface interaction parameters could produce the required current. Examination of our graphs of the contributions from ψ_x^2, ψ_y^2 and ψ_z^2 shows that the only situation that can result in the required current is if the JT parameters prefer a D_{3d} distortion, with the external interaction picking out wells e and f. The magnitudes of the coefficients of ψ_x and ψ_y are the same for these two wells, but the signs are different. We can therefore conclude that the required combination of ψ_x^2 and ψ_y^2 occurs if wells e and f are taken to have electronic states $0.122\psi_x \pm 0.992\psi_y$ respectively. It is found that a wide variety of the external interaction parameters and quadratic coupling constants result in this state.

It is noted that in order to obtain a match which contains only the ψ_x and ψ_y single electron states, the states can't involve H_{g4}, H_{g5} or H_{g6} (for the reasons mentioned above), which means that these states must be higher in energy than the $H_{g\theta}$ and $H_{g\epsilon}$ states. As the external interaction for the double bond prone molecule splits the H_g state into 5 singlets, the assumption can be made that the external interaction is such that these are suitably higher in energy and so don't contribute to the problem.

Our match to the image in [12] is obtained from JT parameters preferring a D_{3d} distortion, but once the surface and Coulomb interactions are included, the result is a dynamic distortion of C_{2h} symmetry. This is different to the interpretation in [12], where the image was explained by a static JT distortion of D_{2h} symmetry. This corresponds to a sole contribution from the ψ_y component in our notation, and corresponds to an image with a nodal plane (seen as in the horizontal division of the image in figure (6)(b) into two halves and the similar vertical division in figure 3(d)) which is not present in the experimental image. However, the evidence that our match is more appropriate than that given in [12] is not conclusive as factors such as experimental blurring could produce a similar effect to the inclusion of a small ψ_x component in the states. We note that previous work on the $p^2 \otimes h$ JT system that describes this anion suggests that isolated C_{60} molecules should not favour D_{2h} distortions, and instead distort to either D_{3d} or D_{5d} [2]. Studies of isolated ions in a storage ring indicate that the distortion is most likely to be to D_{3d} symmetry [18], as did an analysis of spectra from electrochemically-generated C_{60} ions in solution [17]. However, the evidence is again not conclusive.

4. Conclusion

In this work, a method has been introduced to model the STM images produced for the JT active ions of C_{60} when present on a surface, either individually, or as part of a monolayer. This method utilises group theoretical techniques to model the reduction in symmetry of the molecule due to both interactions with the surface and from interactions with nearest neighbours (when present). It is found that for C_{60}^-, C_{60}^{2-} and C_{60}^{4-} ions, the resultant STM current I is always formed from a linear combination of ψ_x^2, ψ_y^2 and ψ_z^2 when the molecule is situated with a C_2, C_3, or C_5 symmetry axis prone to the surface. Matches with experimental images have been obtained for both C_{60}^- and C_{60}^{4-} anions. In both cases, our interpretations are different to those in the original works.

Clearly, the work given here is a starting point for future investigations. Only a small selection of orientations and JT parameters have been looked at, and particularly in the case of C_{60}^{4-}, the effect of a number of different external interaction parameters still need to be explored. However, the method can easily be extended to look at these cases. In addition, the C_{60}^{3-} case, which has not been looked at here, also needs to be investigated, although incorporating the external interactions in this case is more complicated due to the presence of repeated irreps. Experimental data also exists for this anion [12].

A hindrance in expanding on this work is the lack of experimental images within the literature. It is hoped that as experimental techniques continue to improve, additional data will soon become available to firstly verify the findings here, and secondly to examine the effect of different orientations on the images. In doing so, it is hoped that clear indications of the JT effect will be observable through interpretation of the experimental images.

Acknowledgments
AJL gratefully acknowledges the support of EPSRC (UK) for funding this work under a Doctoral Training Award.

References
[1] Dunn J L and Bates C A 1995 *Phys. Rev. B* **52** 5996
[2] Sindi L M, Hands I D, Dunn J L and Bates C A 2007 *J. Mol. Struct.* **838** 78
[3] Lakin A J, Hands I D, Bates C A and Dunn J L 2012 The quadratic $p^3 \otimes h$ Jahn-Teller system as a model for the C_{60}^{3-} anion *Vibronic Interactions and the Jahn-Teller effect: Theory and Applications* (*Progress in Theoretical Chemistry and Physics* vol 23) ed Atanasov M and Daul C (Dordrecht: Springer) pp 231–243
[4] Hands I D, Dunn J L and Bates C A 2010 *Phys. Rev. B* **82** 155425
[5] Pascual J I, Gómez-Herrero J, Rogero C, Baró A M, Sánchez-Portal D, Artacho E, Ordejón P and Soler J M 2000 *Chem. Phys. Lett.* **321** 78
[6] Casarin M, Forrer D, Orzali T, Petukhov M, Sambi M, Tondello E and Vittadini A 2007 *J. Phys. Chem. C* **111** 9365
[7] Hands I D, Dunn J L and Bates C A 2010 *Phys. Rev. B* **81** 205440
[8] Heinrich B W, Rastei M V, Choi D J, Frederiksen T and Limot L 2011 *Phys. Rev. Lett.* **107** 246801
[9] Tersoff J and Hamann D R 1985 *Phys. Rev. B* **31** 805
[10] Dunn J L, Lakin A J and Hands I D, 2012 *New Journal of Physics* In press.
[11] Yuan L F, Yang J, Wang H, Zeng C, Li Q, Wang B, Hou J G, Zhu Q and Chen D M 2003 *J. Am. Chem. Soc.* **125** 169–172
[12] Wachowiak A, Yamachika R, Khoo K H, Wang Y, Grobis M, Lee D H, Louie S G and Crommie M F 2005 *Science* **310** 468
[13] Wang Y, Yamachika R, Wachowiak A, Grobis M, Khoo K, Lee D H, Louie S G and Crommie M F 2007 *Phys. Rev. Lett.* **99** 086402
[14] Chancey C C and O'Brien M C M 1997 *The Jahn-Teller Effect in C_{60} and other Icosahedral Complexes* (Princeton: Princeton University Press)
[15] Fowler P W and Ceulemans A 1985 *Mol. Phys.* **54** 767
[16] Nikolaev A V, Bodrenko I V and Tkalya E V 2008 *Phys. Rev. A* 012503
[17] Hands I D, Dunn J L, Bates C A, Hope M J, Meech S R and Andrews D L 2008 *Phys. Rev. B* **77** 115445
[18] Tomita S, Andersen J U, Bonderup E, Hvelplund P, Liu B, Nielsen S B, Pedersen U V, Rangama J, Hansen K and Echt O 2005 *Phys. Rev. Lett.* **94** 053002

Mott localization in the correlated superconductor Cs_3C_{60} resulting from the molecular Jahn-Teller effect

Katalin Kamarás[1], Gyöngyi Klupp[1], Péter Matus[1], Alexey Y. Ganin[2], Alec McLennan[2], Matthew J. Rosseinsky[2], Yasuhiro Takabayashi[3], Martin T. McDonald[3], Kosmas Prassides[3]

[1] Institute for Solid State Physics and Optics, Wigner Research Centre for Physics, Hungarian Academy of Sciences, P.O. Box 49, H-1525 Budapest, Hungary

[2] Department of Chemistry, University of Liverpool, Liverpool L69 7ZD, UK

[3] Department of Chemistry, Durham University, Durham DH1 3LE, UK

E-mail: kamaras.katalin@wigner.mta.hu

Abstract.
Cs_3C_{60} is a correlated superconductor under pressure, but an insulator under ambient conditions. The mechanism causing this insulating behavior is the combination of Mott localization and the dynamic Jahn-Teller effect. We show evidence from infrared spectroscopy for the dynamic Jahn-Teller distortion. The continuous change with temperature of the splitting of infrared lines is typical Jahn-Teller behavior, reflecting the change in population of solid-state conformers. We conclude that the electronic and magnetic solid-state properties of the insulating state are controlled by molecular phenomena. We estimate the time scale of the dynamic Jahn-Teller effect to be above 10^{-11} s and the energy difference between the conformers less than 20 cm^{-1}.

1. Introduction

Trivalent fulleride salts belong to the family of correlated superconductors, similar to high temperature superconducting cuprates [1, 2]. They show a dome-shaped T_c (superconducting transition temperature) versus lattice constant phase diagram [3] with an antiferromagnetic insulating state above a critical lattice constant [1]. The transition from metallic/superconducting to insulating behavior is generally attributed to Mott localization. Fullerides showing similar behavior have been known for a long time. Metallic fcc (face centered cubic) mixed alkali $Cs_xA_{(3-x)}C_{60}$ (A=K,Rb) phases have been prepared, which were in the lattice constant range where T_c decreases upon expansion [4]. Guest molecules, like cubane [5] or ammonia [6] were also used to expand the lattice. This approach leads in most cases to a change in crystal structure, inducing a crystal field with lower symmetry which distorts the fulleride molecular ions. Metal-insulator transitions in these compounds have been confirmed [7, 8], but since these occur in a low-symmetry crystal field, the external and internal contributions could not be unambiguously distinguished. Cs_3C_{60} is in this respect a key compound. It is an expanded structure, shows a metal-to-insulator transition upon application of pressure [3, 9] and retains its cubic structure in the whole pressure range.

Cs_3C_{60} has two cubic polymorphs, an ordered A15 [3] and a merohedrally disordered fcc one [10]. At ambient pressure they are insulators and magnetic with S=1/2, but turn into metals upon application of pressure [11], exhibiting superconducting transition temperatures T_c up to 38 K [3]. The metal-insulator transition upon increasing distance between the fulleride ions and the localized magnetic moments point to strong electron-electron correlation and consequently, Mott localization.

The localization mechanism was suggested to involve a Jahn-Teller distortion [12], which is difficult to prove experimentally. Atomic displacements in the distorted molecular ions have been estimated to fall in the range of 0.05 ångstroms, hard to detect even by sensitive neutron scattering experiments. However, since the distortion involves a symmetry change, vibrational spectroscopy can be a sensitive indicator. The $T_{1u}(4)$ infrared-active mode at 1429 cm^{-1} is most widely used to detect both charge and symmetry changes in C_{60} [13]. From symmetry considerations for an I_h point group object placed in a cubic environment, it follows that this mode will not show splitting caused by the environment. Therefore, all splittings found can be directly related to the symmetry reduction following from the molecular Jahn-Teller effect.

2. The Jahn–Teller effect in solid fullerides

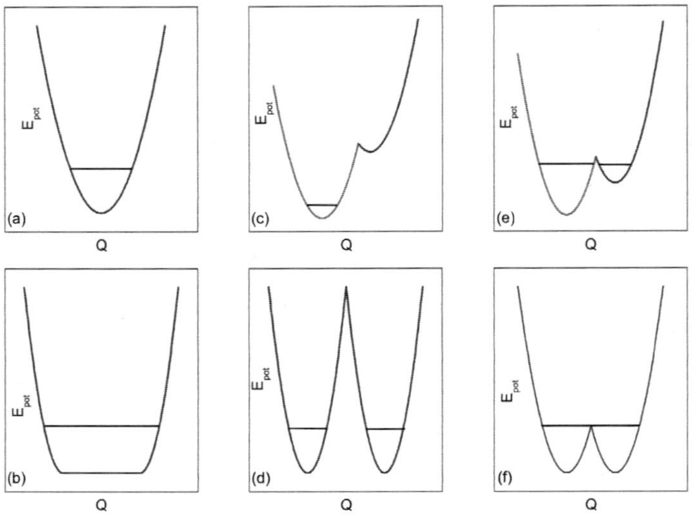

Figure 1. Possible potential energy surfaces of the C_{60}^{3-} molecular ion in Cs_3C_{60}. (a) The possible D_{2h} distortions in all directions have the same energy. (b) No energy barrier between differently directed distortions causes an average I_h symmetry. (c, e, f) Two inequivalent directions of differing energy separated by a small energy barrier. (c) Static distortion in one of the directions. (d) Two inequivalent directions of similar energy separated by a large energy barrier. A small energy difference between the two wells is possible. Static distortion. (e) Dynamic distortion with two solid-state conformers with differing population. (f) Hindered pseudorotation between differently directed distortions.

According to theoretical calculations [14] the symmetry of the Jahn–Teller distortion of C_{60}^{3-} is D_{2h}. The potential energy surface of a single distortion is schematically illustrated in Fig. 1 (a). The possible combinations of fifteen equivalent twofold axes among the symmetry elements of icosahedral C_{60} define thirty equivalent distortions among which the molecule can move. The corresponding motion is called pseudorotation (see the movie in the Supplementary Information). The molecule performing pseudorotation is dynamically distorted. If there is no barrier between the different D_{2h} distortions (free pseudorotation), the molecular symmetry will be restored to I_h in time average [15] (see Fig. 1 (b)). Introducing a barrier between the D_{2h} distortions results in hindered pseudorotation [15] (see Fig. 1 (e) and (f)). If the time scale of a spectroscopic measurement is shorter than that of pseudorotation, the D_{2h} distortion can be detected. The time scale of infrared excitations being of the order of 10^{-11} s, spectra corresponding to distorted states correspond to pseudorotations slower than this value. The last possibility is that with a barrier high enough to cause static distortion along one special axis [15] (see Fig. 1 (c) and (d)).

In the solid one has to take into account the effect of the crystal field determined by the potential created by the neighboring Cs^+ ions. The undistorted site symmetry is T_h in these cubic crystals, so the crystal field alone would not lower the symmetry of the fulleride ions below that point group.[16]. The number of twofold axes along which Jahn-Teller distortion can occur is reduced to three, and if one considers the C_{60}^{3-} ion in a T_h point group there will be just two inequivalent directions. This is why two kinds of distortions appear in Fig. 1 (c)-(f).

Thermal expansion results in weaker crystal field acting towards equalizing the two kinds of distortions. In addition, at higher temperature the higher energy states become increasingly populated. Thus, for example, heating will lead from the state in Fig. 1 (c) to (e), and further to (f). Raising the temperature will also lower the barriers between potential energy wells (best seen in Fig. 1 (d)) and they can be more easily overcome.

3. Experimental

The Cs_3C_{60} samples used in this study were prepared by solution chemistry routes as described elsewhere [3, 10]. Sample composition and structure were determined by x-ray diffraction. Infrared measurements were done on both fcc-rich and A15-rich materials, in KBr pellets with 0.25 cm^{-1} resolution [16].

4. Signatures of the Mott insulating state

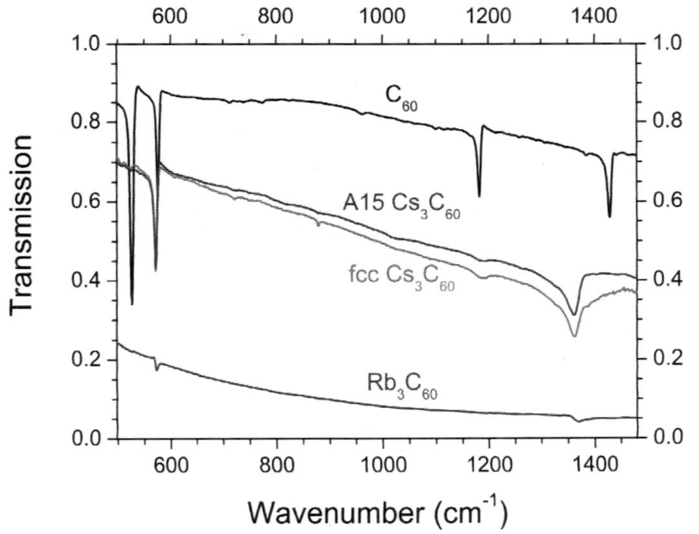

Figure 2. Infrared spectrum of A15 and fcc Cs_3C_{60} in the range of intramolecular vibrations at room temperature. The room temperature spectra of insulating C_{60} and metallic Rb_3C_{60} are included for comparison. The metallic character influences both the background and the lineshape.

The measured infrared spectra of both polymorphs together with that of C_{60} and of metallic Rb_3C_{60} are shown in Fig. 2. The spectrum of pure C_{60} consists of four lines, a consequence of the high symmetry of the fullerene ball [17]. In accordance with previous data [18], the highest frequency mode softens due to the charge added on the fullerene, showing approximately the same shift in Cs_3C_{60} and in Rb_3C_{60}. The lineshape of this mode, however, is different for Rb_3C_{60} and for Cs_3C_{60}. Rb_3C_{60} has a Fano lineshape [19] as a consequence of coupling between vibrational states and the electronic continuum of the same energy. The non-Fano lineshape in both polymorphs of Cs_3C_{60} signals the absence of metallic electrons, therefore these materials are insulators, in agreement with previous measurements [1]. Metallic electrons also cause a continuous background absorption, which is indeed present in Rb_3C_{60}, but not in either of the two Cs_3C_{60} polymorphs.

Another signature of the insulating state is that it facilitates the appearance of the Jahn–Teller effect: molecular ions can suffer a Jahn–Teller distortion only if their electrons are localized. The asymmetrically broadened lineshape in both Cs_3C_{60} compounds (Fig. 2) indicates splitting of the vibrational lines, a consequence of the symmetry reduction owing to the Jahn–Teller distortion [16]. To further study the temperature and the polymorph dependence of the splitting, temperature dependent measurement were performed on both Cs_3C_{60} polymorphs.

5. Jahn–Teller effect in Cs_3C_{60}

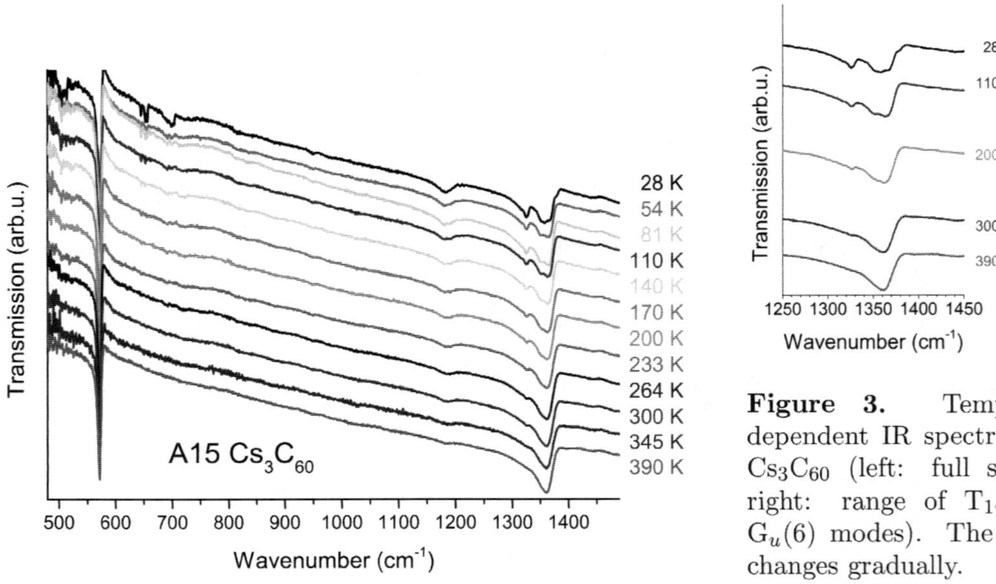

Figure 3. Temperature-dependent IR spectra of A15 Cs_3C_{60} (left: full spectrum, right: range of $T_{1u}(4)$ and $G_u(6)$ modes). The splitting changes gradually.

Figures 3 and 4 illustrate the gradual changes in the splitting, characteristic of both polymorphs (although not exactly identical). Newly activated low intensity modes between 600-800 cm^{-1} can also be observed at low temperature. These changes indicate that the symmetry lowering is different in the two Cs_3C_{60} polymorphs. The weak new modes disappear at differing temperatures underlining the smooth nature of the transition. As we will see, this is a sign of temperature-dependent solid-state conformers, implying a dynamical Jahn–Teller effect [15].

The low-temperature multiple peak structure and its dependence on crystal structure is a result of the interplay between crystal field and Jahn–Teller distortion. The Cs^+ ions that surround the C_{60}^{3-} ion constitute a cubic environment and do not cause splitting in the T_{1u} modes. However, the energies of molecular distortions, that are equivalent in a free molecular ion, can differ slightly in a crystal when pointed at different crystallographic directions. The observation of the signatures of the Jahn–Teller distortion in the vibrational spectra rules out the case of Fig. 1 (b) (free pseudorotation). Statically ordered distortions in the lattice with the same energy (Fig. 1 (a)) would lead to a temperature-independent spectrum consisting of one set of lines (maximum 3+4, for triply degenerate T_{1u} and quadruply degenerate G_u, respectively) and is also contrary to observation. The number of possible allowed IR lines (in principle 1* or 2*(3+4)) corresponds to the number of populated potential energy wells (1 in Fig. 1 (a-c) and 2 in Fig. 1 (d-f)). According to the experiments, this number continously decreases on heating. This finding can be qualitatively explained by a model where the population of the wells changes with temperature.

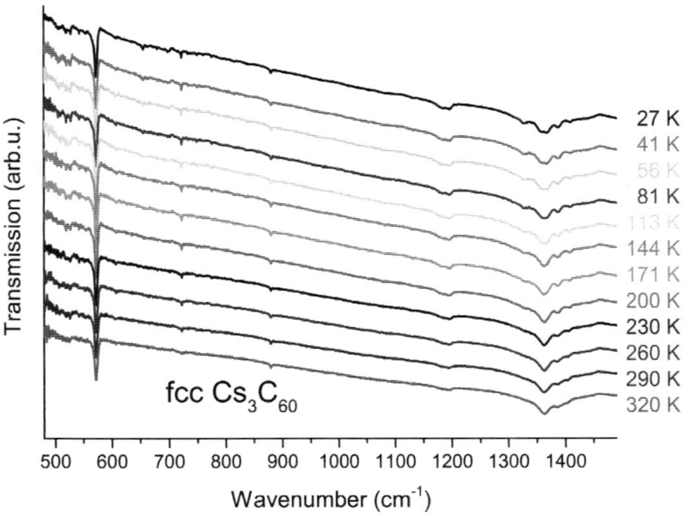

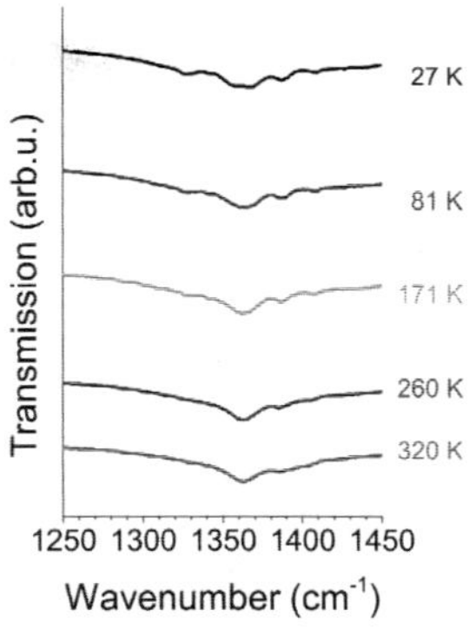

Figure 4. Temperature dependent IR spectra of fcc Cs_3C_{60}. The splitting changes gradually.

The simplest of these models is that of the static-to-dynamic transition (from Fig. 1 (d) to (f)) on heating. At low temperature the differently directed distortions would be frozen in with static and disordered distortion. Thermal expansion lowers the barrier and the increased thermal energy from pseudorotation helps to overcome it. In this case the transition would be located in a considerably narrower temperature range, but we observe a continuous change over the whole temperature range we measured. The model based on Fig. 1 (d), where the barrier is too high for a transition even at high temperature, can also be rejected based on the measured temperature-dependent infrared data. A transition could also happen from the static distortion shown in Fig. 1 (c) to the dynamic ones Fig. 1 (e) and (f). The number of lines should first increase on heating, then start to decrease. As we cannot prove such a behaviour, this case can also be ruled out.

This leaves us with the dynamic distortions Fig. 1 (e) and (f), the latter developing from the former on increasing temperature. This situation corresponds to different solid-state conformers with different occupational probability [16]. On heating, both their energy difference and their populational difference gradually disappears [15], as observed in the spectra. Fewer kinds of distortions mean fewer lines in the infrared spectra. The two equilibrium conformations with a low barrier (Fig. 1e) are connected by hindered pseudorotation [15]. The fact that we observe the distorted states puts the lower limit of the time scale of this pseudorotation to 10^{-11} s, and the lowest temperature being 28 K can be reconciled with an upper limit of 20 cm^{-1} for the barrier between the two conformers.

6. Conclusions

We studied the temperature-dependent infrared spectra of both polymorphs of the correlated superconductor Cs_3C_{60} in their normal state. The vibrational signatures reveal the insulating state of these materials, and establish the interplay of Mott localization and the Jahn-Teller effect in forming this state. In addition, the crystal field influences the vibrational levels making the spectra temperature- and polymorph-dependent. The presence of temperature-dependent solid-state conformers validates the proof of the dynamic Jahn–Teller effect.

Acknowledgments

Funding for this research was provided by the Hungarian National Research Fund (OTKA) (Grant No. 75813), the Engineering and Physical Sciences Research Council(Grant No.

EP/G037132 and EP/G037949), and the EU-Japan project LEMSUPER (Grant No. NMP3-SL-2011-283214).
Supplementary Information accompanies this paper.

References

[1] Takabayashi Y, Ganin A Y, Jeglič P, Arčon D, Takano T, Iwasa Y, Ohishi Y, Takata M, Takeshita N, Prassides K and Rosseinsky M J 2009 *Science* **323** 1585
[2] Tosatti E 2009 *Science* **323** 1570
[3] Ganin A Y, Takabayashi Y, Khimyak Y Z, Margadonna S, Tamai A, Rosseinsky M J and Prassides K 2008 *Nature Mater.* **7** 367
[4] Dahlke P, Denning M S, Henry P F and Rosseinsky M J 2000 *J. Am. Chem. Soc.* **122** 12352
[5] Pekker S, Kováts E, Oszlányi G, Bényei G, Klupp G, Bortel G, Jalsovszky I, Jakab E, Borondics F, Kamarás K, Bokor M, Kriza G, Tompa K and Faigel G 2005 *Nature Mater.* **4** 1468
[6] Rosseinsky M J, Murphy D W, Fleming R M and Zhou O 1993 *Nature* **364** 425
[7] Allen K M, Heyes S J and Rosseinsky M J 1996 *J. Mater. Chem.* **6** 1445
[8] Iwasa Y, Shimoda H, aned Y Maniwa T T M P, Zhou O and Mitani T 1996 *Phys. Rev. B* **53** R8836
[9] Ihara Y, Alloul H, Wzietek P, Pontiroli D, Mazzani M and Riccò M 2010 *Phys. Rev. Lett.* **104** 256402
[10] Ganin A Y, Takabayashi Y, Jeglič P, Arčon D, Potočnik A, Baker P J, Ohishi Y, McDonald M T, Tzirakis M D, McLennan A, Darling G R, Takata M, Rosseinsky M J and Prassides K 2010 *Nature* **466** 221
[11] Ihara Y, Alloul H, Wzietek P, Pontiroli D, Mazzani M and Riccò M 2011 *Europhys. Lett.* **94** 37007
[12] Capone M, Fabrizio M, Castellani C and Tosatti E 2009 *Rev. Mod. Phys.* **81** 943
[13] Kamarás K, Forró L and Iwasa Y 1997 *Phys. Rev. B* **55** 10999
[14] Auerbach A, Manini N and Tosatti E 1994 *Phys. Rev. B* **49** 12998
[15] Bersuker I B 2006 *The Jahn-Teller Effect* (Cambridge University Press)
[16] Klupp G, Matus P, Kamarás K, Ganin A Y, McLennan A, Rosseinsky M J, Takabayashi Y, McDonald M T and Prassides K 2012 *Nat. Commun.* **3** 912
[17] Krätschmer W, Fostiropoulos K and Huffman D R 1990 *Chem. Phys. Lett.* **170** 167
[18] Pichler T, Winkler R and Kuzmany H 1994 *Phys. Rev. B* **49** 15879
[19] Iwasa Y, Watanabe S, Kaneyasu T, Yasuda T, Koda T, Nagata M and Mizutani N 1993 *J. Phys. Chem. Solids* **54** 1795

Journal of Physics: Conference Series **428** (2013) 012003 doi:10.1088/1742-6596/428/1/012003

Vibronic couplings in cycloadditions to fullerenes

Naoki Haruta, Tohru Sato, Naoya Iwahara and Kazuyoshi Tanaka

Department of Molecular Engineering, Graduate School of Engineering, Kyoto University, Nishikyo-ku, Kyoto 615-8510, Japan

E-mail: tsato@moleng.kyoto-u.ac.jp

Abstract.
It is known to be difficult to predict the regioselectivity of cycloadditions to fullerenes based on the frontier orbital theory. To discuss the chemical reactivity of C_{60}, vibronic coupling constants for localized stretching vibrational modes of C_{60} anion which are the hypothetical reaction modes of cycloadditions are calculated. Furthermore, the product ratio of C_{60} bisadduct ($C_{60}X_2$) is evaluated in the same way by considering the electronic state of C_{60} monoadduct anion ($C_{60}X$ anion). It is found that there is a good correlation between vibronic coupling constants for the localized modes and experimental regioselectivity.

1. Introduction

Chemical reaction involves the formation and breaking of chemical bonds which connect atoms in molecules. Bond formation and breaking are caused by nuclear motions (molecular vibrations) with the rearrangement of electrons. The frontier orbital theory[1, 2], which is the most successful theory on the chemical reactivity, discusses molecular/bimolecular stabilization by considering only the rearrangement of electrons, that is, electron delocalization (charge-transfer) interactions via frontier orbitals. In addition to the frontier orbital interactions, vibronic (electron-vibration) couplings[3, 4] can give rise to stabilization via molecular distortion in the early stage of a reaction. The vibronic coupling can be therefore a key concept to understand chemical reactions beyond the frontier orbital theory.

The frontier orbital theory has difficulty in predicting the chemical reactivity in some cases. For example, large molecules with delocalized frontier orbitals and molecules with degenerate frontier orbitals are such a case. The delocalization or degeneracy tends to smear a reactive site. The typical example with these problems is the cycloadditions to C_{60} fullerene (Figure 1). Nucleophilic cycloadditions to C_{60} have been studied widely in material chemistry[5, 6]. The C_{60} cage has two types of bonds. One is the bond between adjacent hexagon rings (6:6 bonds), and another is the bond between a hexagon ring and a pentagon ring (6:5 bonds). Cycloadditions are experimentally known to occur on 6:6 bonds[5, 6]. According to the frontier orbital theory, the regioselectivity of nucleophilic cycloadditions is indicated by the distributions of t_{1u} LUMOs of C_{60}. However, since the averaged LUMO density has the totally symmetric distribution which is equivalent at each carbon atom, the regioselectivity cannot be predicted by the frontier orbital theory.

Recently, we have proposed that the concept of vibronic coupling density is effective to predict the chemical reactivity[7, 8, 9]. The vibronic coupling density gives a local picture of the strength of linear vibronic interaction. The vibronic coupling density for the effective vibrational mode of

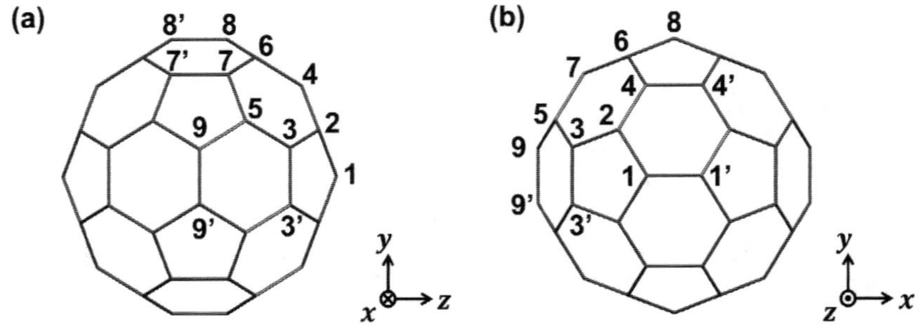

Figure 1. C_{60} fullerene and the labels of its atoms.

C_{60} anion illustrates the reactive bonds of C_{60} clearly[9]. We have found that the C_{60} cage has ethylene moieties as a functional group by comparing the distribution of the vibronic coupling density of C_{60} anion with ethylene anion.

Furthermore, we have recently calculated vibronic coupling constants for localized stretching vibrational modes of C_{70} anion which reproduce the experimental reactivity of each bond of C_{70}[10]. The localized stretching vibrational modes are the hypothetical reaction modes of cycloadditions.

In this study, in order to estimate the quantitative chemical reactivity of each bond of C_{60}, we calculate vibronic coupling constants for localized modes of its anion. In addition, we evaluate the product ratio of C_{60} bisadduct ($C_{60}X_2$, Figure 2) in the same way by considering the electronic state of C_{60} monoadduct anion ($C_{60}X$ anion) based on that of C_{60} anion.

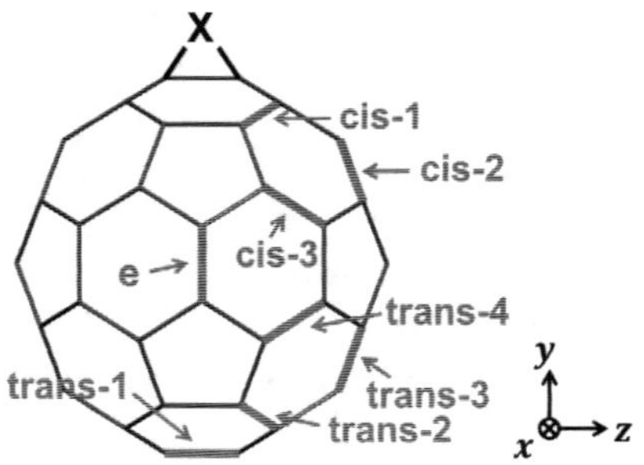

Figure 2. The names of C_{60} bisadduct isomers. X is diene or diene-like reactant. The red bonds with arrows indicate the bonds added by another reactant.

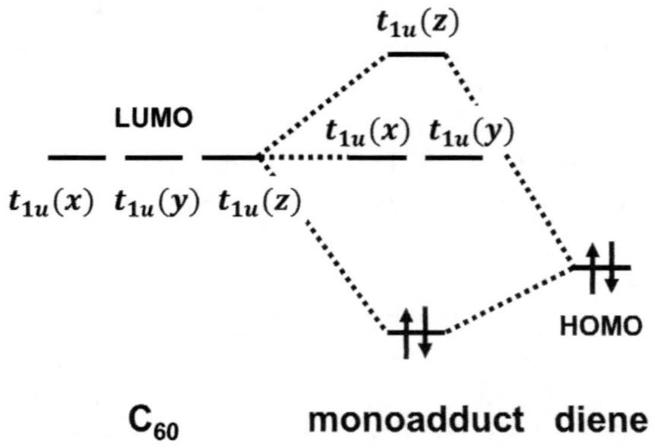

Figure 3. The diagram of orbital levels of C_{60}, C_{60} monoadduct and diene.

2. Theory

We discuss vibronic couplings in chemical reactions based on the Parr and Yang's conceptual density functional theory[11]. The ground electronic energy E of an isolated reactant is the functional of the number of electrons N and the potential $u(\mathbf{r})$ attracting a single electron: $E = E[N, u]$. They have introduced an electronic chemical potential μ as $(\partial E/\partial N)_u$. Since μ

is also the functional of N and $u(\mathbf{r})$: $\mu = \mu[N, u]$, its differential is given by

$$d\mu = 2\eta dN + \int f(\mathbf{r})du(\mathbf{r})d^3\mathbf{r}, \tag{1}$$

where an absolute hardness η is

$$\eta := \frac{1}{2}\left(\frac{\partial \mu}{\partial N}\right)_u = \frac{1}{2}\left(\frac{\partial^2 E}{\partial N^2}\right)_u, \tag{2}$$

and a Fukui function $f(\mathbf{r})$ is

$$f(\mathbf{r}) := \left(\frac{\delta \mu}{\delta u}\right)_N = \left(\frac{\partial \rho}{\partial N}\right)_u. \tag{3}$$

$\rho(\mathbf{r})$ is a total electron density. Based on equation (1), Parr and Yang have proposed the principle of chemical reactions that a chemical reaction occurs such that the initial $|d\mu|$ is the maximum. Since $f(\mathbf{r})$ is approximately the frontier orbital density, their principle in their opinion leads to the frontier orbital theory.

Our theory is also based on equation (1). We consider the case of an electron acceptor such as C_{60}. We assume that $f(\mathbf{r})$ can be approximated by the electron density difference $\Delta\rho(\mathbf{r})$ between its neutral state and its anionic state. We also assume that $du(\mathbf{r})$ can be expressed by a reaction coordinate s:

$$du(\mathbf{r}) = v_s(\mathbf{r})ds, \tag{4}$$

where a potential derivative $v_s(\mathbf{r})$ is defined as

$$v_s(\mathbf{r}) := \left(\frac{\partial u}{\partial s}\right)_{\mathbf{R}_0}. \tag{5}$$

$\mathbf{R}_0$ is molecular reference geometry. As a result, equation (1) can be rewritten as

$$d\mu = 2\eta dN + \int \Delta\rho(\mathbf{r})v_s(\mathbf{r})d^3\mathbf{r}ds = 2\eta dN + \int \eta_s(\mathbf{r})d^3\mathbf{r}ds, \tag{6}$$

where vibronic coupling density $\eta_s(\mathbf{r})$ for the reaction mode s[7, 8] is defined as

$$\eta_s(\mathbf{r}) := \Delta\rho(\mathbf{r}) \times v_s(\mathbf{r}). \tag{7}$$

The integration of the vibronic coupling density $\eta_s(\mathbf{r})$ is known to be exactly equal to a linear vibronic coupling constant V_s for the reaction mode s:

$$V_s := \left\langle \Psi^- \left| \left(\frac{\partial \hat{H}}{\partial s}\right)_{\mathbf{R}_0} \right| \Psi^- \right\rangle = \int \eta_s(\mathbf{r})d^3\mathbf{r}, \tag{8}$$

where $\hat{H}$ denotes a molecular Hamiltonian and Ψ^- denotes an anionic state. In general, the reaction mode s is a mixture of totally symmetric modes, Jahn–Teller active modes and the other modes. Note that the vibronic coupling constant defined here is different from the Jahn–Teller coupling constant in the $T \otimes h$ Jahn–Teller Hamiltonian[12] by the Clebsch–Gordan coefficients even if mode s coincides with some Jahn-Teller modes in C_{60}.

The integration of $\eta_s(\mathbf{r})$ is also equal to the gradient of the anionic potential surface due to the Hellmann-Feynman theorem[13, 14]:

$$V_s = \int \eta_s(\mathbf{r})d^3\mathbf{r} = \left(\frac{\partial\left(E^- + U_{nn}\right)}{\partial s}\right)_{\mathbf{R}_0}, \tag{9}$$

where E^- denotes an electronic energy of the anionic state, and U_{nn} the sum of nuclear-nuclear repulsive potentials. The anionic state is stabilized in the direction such that V_s is negative. Therefore, we have proposed that chemical reaction occurs such that V_s is a minimum, and its local picture is given by $\eta_s(\mathbf{r})$[10].

In our previous work, we have regarded an effective vibrational mode as a reaction mode s[7, 8, 9]. The effective vibrational mode is the steepest descent of the anionic state, that is constructed by only totally symmetric modes and Jahn–Teller active modes. As for cycloadditions, some localized stretching vibrational modes can be regarded as a reaction mode rather than the effective mode[10]. Each localized mode is a linear combination of some intramolecular vibrational modes including totally symmetric modes, Jahn–Teller active modes and the others. All the localized modes construct the effective mode. The localized stretching vibrational vector $\mathbf{v}_{\mathrm{loc}(A,B)}$ of the bond between atom A and atom B is defined as

$$\mathbf{v}_{\mathrm{loc}(A,B)} := \sum_{\xi=x,y,z} \frac{M_B}{\sqrt{M_A^2+M_B^2}} \frac{R_{A\xi}-R_{B\xi}}{\|\mathbf{R}_A-\mathbf{R}_B\|}\mathbf{e}_{A\xi} - \sum_{\xi=x,y,z} \frac{M_A}{\sqrt{M_A^2+M_B^2}} \frac{R_{A\xi}-R_{B\xi}}{\|\mathbf{R}_A-\mathbf{R}_B\|}\mathbf{e}_{B\xi}, \quad (10)$$

where M_A denotes the mass of nucleus A, $R_{A\xi}$ is the Ath ξ-component of the molecular geometry $\mathbf{R}$ constructed with N atoms represented by the $3N$-dimensional orthonormal vectors $\{\mathbf{e}_{A\xi}\}$:

$$\mathbf{R} = \sum_A \sum_{\xi=x,y,z} R_{A\xi}\mathbf{e}_{A\xi}, \quad (11)$$

and $\mathbf{R}_A$ is defined as

$$\mathbf{R}_A := \sum_{\xi=x,y,z} R_{A\xi}\mathbf{e}_{A\xi}. \quad (12)$$

The directions of the localized stretching modes are chosen such that the corresponding bonds lengthen. In particular, for the case of a diatomic molecule, the localized mode is exactly equal to its normal mode. All the localized modes are not a reaction mode. However, we introduce some of them as hypothetical reaction modes to evaluate the reactivity of each bond.

3. Computational Method

In our previous study, we have calculated the T_{1u} states of C_{60} anion with I_h symmetry employing B3LYP/6-311G(d)[15]. In order to evaluate the reactivity of each bond, we calculated vibronic coupling constants for the localized stretching vibrational modes in the $T_{1u}(x)$, $T_{1u}(y)$ $T_{1u}(z)$ states of C_{60} anion employing our previous results. In addition, we calculated the vibronic coupling densities for some localized stretching vibrational modes.

Next, we discuss the product ratio of C_{60} bisadduct ($C_{60}X_2$). For this purpose, we require the electronic state of C_{60} monoadduct anion ($C_{60}X$ anion). In this study, we regard two states of three-fold degenerate states of C_{60} anion with I_h symmetry as C_{60} monoadduct anion since $t_{1u}(x)$ and $t_{1u}(y)$ LUMOs of C_{60} correspond to the LUMOs of C_{60} monoadduct as shown in Figure 3. We ignore the slight energy splitting between pseudo-degenerate LUMOs of C_{60} monoadduct. Such simple method is effective to discuss the chemical reactivity of C_{60} multiadduct ($C_{60}X_n$). In order to evaluate the reactivity of C_{60} monoadduct, we calculated vibronic coupling constants for the localized stretching vibrational modes in the $T_{1u}(x)$ and $T_{1u}(y)$ states of C_{60} anion.

All the calculations of vibronic coupling constants and vibronic coupling densities are performed by using our codes.

4. Results and discussion

In Table 1, we show vibronic coupling constants for the localized stretching vibrational modes in the $T_{1u}(z)$ state of C_{60} anion. The negative values of the vibronic coupling constants indicate that the anionic state is stabilized by molecular distortion along the corresponding bond-elongated modes. On the other hand, the positive values indicate the destabilization. Table 1 shows that the elongation of $\mathbf{8}-\mathbf{8}'$ bond gives the maximum stabilization, and the elongation of $\mathbf{3}-\mathbf{3}'$ bond gives the maximum destabilization. Since a reactive bond will be longer in cycloadditions

Table 1. Vibronic coupling constants for the localized stretching vibrational modes in the $T_{1u}(z)$ state of C_{60} anion. The bond notation is defined in Figure 1.

bond	vibronic coupling constant in 10^{-4} a.u.
$\mathbf{1}-\mathbf{1}'$	0.14
$\mathbf{1}-\mathbf{2}$	-0.06
$\mathbf{2}-\mathbf{3}$	0.22
$\mathbf{2}-\mathbf{4}$	-0.06
$\mathbf{3}-\mathbf{3}'$	0.80
$\mathbf{3}-\mathbf{5}$	-0.86
$\mathbf{4}-\mathbf{4}'$	0.24
$\mathbf{4}-\mathbf{6}$	-0.24
$\mathbf{5}-\mathbf{7}$	0.69
$\mathbf{5}-\mathbf{9}$	0.04
$\mathbf{6}-\mathbf{7}$	-0.38
$\mathbf{6}-\mathbf{8}$	0.69
$\mathbf{7}-\mathbf{7}'$	-0.35
$\mathbf{8}-\mathbf{8}'$	-1.33
$\mathbf{9}-\mathbf{9}'$	-0.09

to C_{60}, $\mathbf{8}-\mathbf{8}'$ bond can be regarded as the most reactive bond, and $\mathbf{3}-\mathbf{3}'$ bond the most unreactive bond. Vibronic coupling densities for localized modes at $\mathbf{8}-\mathbf{8}'$ and $\mathbf{3}-\mathbf{3}'$ bonds are shown in Figure 4. Since vibronic coupling density for the localized mode at $\mathbf{8}-\mathbf{8}'$ bond has the similar distribution to ethylene anion, $\mathbf{8}-\mathbf{8}'$ bond can be regarded as the functional group of ethylene moiety. It should be noted that $\mathbf{8}-\mathbf{8}'$ bond is a reactive bond between adjacent hexagon rings (6:6 bond) and $\mathbf{3}-\mathbf{3}'$ bond is an unreactive bond between a hexagon ring and a pentagon ring (6:5 bond).

In addition, in order to consider the degeneracy of anionic states, we averaged the vibronic coupling constants for localized modes over the T_{1u} three-fold degenerate states:

$$\tilde{V}_{\mathrm{loc}(A,B)} = \frac{1}{3}\left(V_{\mathrm{loc}(A,B)}\left[T_{1u}(x)\right] + V_{\mathrm{loc}(A,B)}\left[T_{1u}(y)\right] + V_{\mathrm{loc}(A,B)}\left[T_{1u}(z)\right]\right), \tag{13}$$

where $V_{\mathrm{loc}(A,B)}\left[T_{1u}(\xi)\right]$ denotes a vibronic coupling constant for a localized mode in the $T_{1u}(\xi)$ anionic state. We obtain two types of the averaged vibronic coupling constants since all the 6:6 bonds and all the 6:5 bonds become equivalent by this average, respectively. The averaged vibronic coupling constants at the 6:6 bonds are -0.43×10^{-4} a.u., and those at the 6:5 bonds are 0.22×10^{-4} a.u. Since the number of 6:6 bonds and 6:5 bonds are 30 and 60 in the C_{60} cage, we summed up these averaged vibronic coupling constants to be -12.82×10^{-4} a.u. and 13.44×10^{-4} a.u., respectively. This result shows that 6:6 bonds are much more reactive than 6:5 bonds, that is consistent with experimental findings[5, 6].

Next, we discuss the chemical reactivity of a C_{60} monoadduct in the same way. C_{60} bisadduct isomers are named as shown in Figure 2. We calculated only the 6:6 bonds which is the reactive bonds of C_{60}. We calculated vibronic coupling constants for the localized modes, and averaged

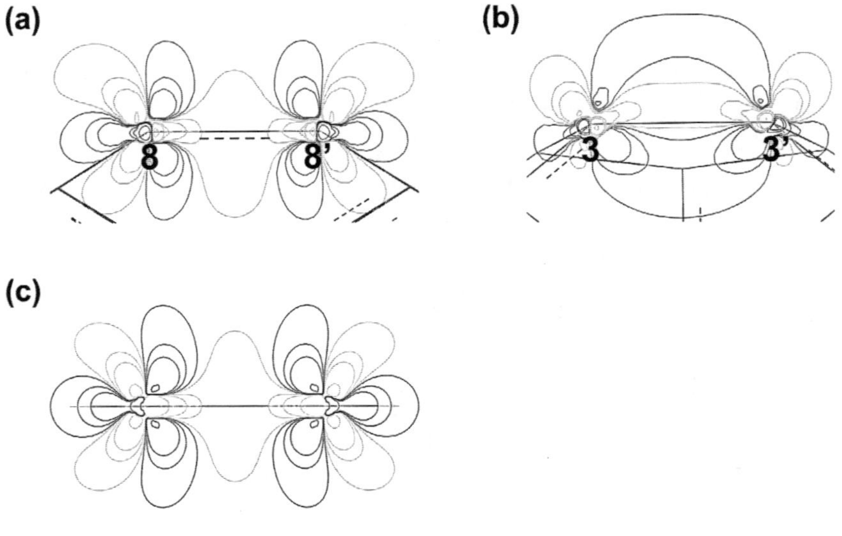

Figure 4. Vibronic coupling densities for localized modes at (a) $\mathbf{8} - \mathbf{8}'$ and (b) $\mathbf{3} - \mathbf{3}'$ bonds of C_{60} anion, and (c) vibronic coupling density for an effective mode of ethylene anion. The violet lines indicate positive and the cyan lines indicate negative.

the vibronic coupling constants for the localized modes over the $T_{1u}(x)$ and $T_{1u}(y)$ anionic states of C_{60}, which can be regarded as the anionic states of C_{60} monoadduct:

$$\tilde{V}_{\mathrm{loc}(A,B)} = \frac{1}{2}\left(V_{\mathrm{loc}(A,B)}\left[T_{1u}(x)\right] + V_{\mathrm{loc}(A,B)}\left[T_{1u}(y)\right]\right). \tag{14}$$

In addition, the averaged vibronic coupling constants for the localized modes which give the same C_{60} bisadduct isomer are summed up (Figure 5). According to experimental findings[16],

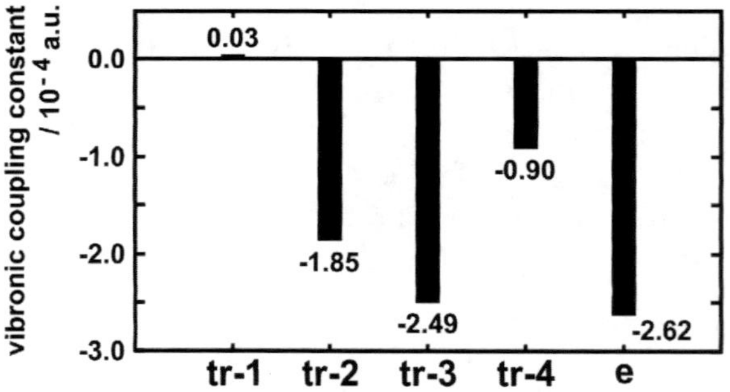

Figure 5. Sum of the averaged vibronic coupling constants for the localized modes which give the same C_{60} bisadduct isomer. In this figure, **tr** indicates **trans**. The notation of **trans−1** to **trans−4** and **e** is defined in Figure 2.

the order of the yields is **e** > **trans−3** > **trans−2** > **trans−4** > **trans−1** on the whole. The experimental facts are consistent with our results. **Cis**−products are also obtained experimentally much less than **trans**−products. However, the product ratio of **cis**−products are not reproduced by our method since we ignored steric effects so that the reactivities of **cis−1**, **cis−2** and **cis−3** are equal to **trans−2**, **trans−3** and **trans−4**, respectively.

Journal of Physics: Conference Series **428** (2013) 012003 doi:10.1088/1742-6596/428/1/012003

5. Conclusion

In terms of vibronic coupling, we discussed the chemical reactivity of C_{60} and its monoadduct ($C_{60}X$). We found that vibronic coupling constants for the localized stretching vibrational modes correlate with the experimental product ratios of cycloadditions to C_{60} and its monoadduct. Our simple method used to evaluate the reactivity of C_{60} monoadduct is also effective to investigate the reactivities of all multiadducts ($C_{60}X_n$) without much computational costs. We believe that our theory is helpful for the prediction of chemical reactivities of higher fullerenes, endohedral fullerenes and the other attractive reactants.

References

[1] Fukui K, Yonezawa T and Shingu H 1952 *J. Chem. Phys.* **20** 722–725
[2] Fukui K 1971 *Acc. Chem. Res.* **4** 57–64
[3] Fischer G 1984 *Vibronic Coupling: The Interaction between the Electronic and Nuclear Motions* (London: Academic Press)
[4] Bersuker I B and Polinger V Z 1989 *Vibronic Interactions in Molecules and Crystals* (New York: Springer-Verlag)
[5] Yurovskaya M A and Trushkov I V 2002 *Russ. Chem. Bull. Int. Ed.* **51** 367–443
[6] Thilgen C and Diederich F 2006 *Chem. Rev.* **106** 5049–5135
[7] Sato T, Tokunaga K and Tanaka K 2008 *J. Phys. Chem. A* **112** 758–767
[8] Sato T, Tokunaga K, Iwahara N, Shizu K and Tanaka K 2009 *The Jahn–Teller Effect: Fundamentals and Implications for Physics and Chemistry* (*Springer Series in Chemical Physics* vol 97) ed Köppel H, Yarkony D R and Barentzen H (Berlin: Springer–Verlag) pp 99–129
[9] Sato T, Iwahara N, Haruta N and Tanaka K 2012 *Chem. Phys. Lett.* **531** 257–260
[10] Haruta N, Sato T and Tanaka K 2012 *J. Org. Chem.* **77** 9702–9706
[11] Parr R G and Yang W 1984 *J. Am. Chem. Soc.* **106** 4049–4050
[12] Chancey C C and O'Brien M C M 1997 *The Jahn–Teller Effect in* C_{60} *and Other Icosahedral Complexes* (Princeton: Princeton University Press)
[13] Hellmann H 1937 *Einführung in die Quantenchemie* (Leipzig: Deuticke and Company)
[14] Feynman R P 1939 *Phys. Rev.* **56** 340–343
[15] Iwahara N, Sato T, Tanaka K and Chibotaru L F 2010 *Phys. Rev. B* **82** 245409 1–10
[16] Nakamura Y, Takano N, Nishimura T, Yashima E, Sato M, Kudo T and Nishimura J 2001 *Org. Lett.* **3** 1193–1196

Vibronic bands in the HOMO-LUMO excitation of linear polyyne molecules

Tomonari Wakabayashi and Yoriko Wada
Department of Chemistry, Kinki University, Higashi-Osaka 577-8502, Japan
E-mail: wakaba@chem.kindai.ac.jp

Naoya Iwahara and Tohru Sato
Department of Molecular Engineering, Kyoto University, Kyoto 615-8510, Japan
E-mail: tsato@scl.kyoto-u.ac.jp

Abstract. Hydrogen-capped linear carbon chain molecules, namely polyynes $H(C{\equiv}C)_nH$ ($n{\geq}2$), give rise to three excited states in the HOMO-LUMO excitation. Electric dipole transition from the ground state is fully allowed to one of the three excited states, while forbidden for the other two low-lying excited states. In addition to the strong absorption bands in the UV for the allowed transition, the molecules exhibit weak absorption and emission bands in the near UV and visible wavelength regions. The weak features are the vibronic bands in the forbidden transition. In this article, symmetry considerations are presented for the optical transitions in the centrosymmetric linear polyyne molecule. The argument includes Herzberg-Teller expansion for the state mixing induced by nuclear displacements along the normal coordinate of the molecule, intensity borrowing from fully allowed transitions, and inducing vibrational modes excited in the vibronic transition. The vibronic coupling considered here includes off-diagonal matrix elements for second derivatives along the normal coordinate. The vibronic selection rule for the forbidden transition is derived and associated with the transition moment with respect to the molecular axis. Experimental approaches are proposed for the assignment of the observed vibronic bands.

1. Introduction

Vibronic interaction creates finest fingerprints in the electronic spectra of polyatomic molecules. A symmetry forbidden transition becomes weakly allowed by the mixing of states due to the vibronic coupling, resulting in the intensity borrowing from fully allowed transitions. A noticeable example is found for the $\tilde{A}{\leftarrow}\tilde{X}$ transition in benzene, for which some absorption bands are observed around 260 nm [1].

Here, we discuss a series of linear hydrocarbon molecules consisting of a conjugated sp-carbon chain terminated by a hydrogen atom at each end. The centrosymmetric linear molecules, namely polyynes $H(C{\equiv}C)_nH$ ($n{\geq}2$), can be produced experimentally by laser ablation of carbon particles in organic solvents [2, 3]. The series of polyyne molecules exhibits a systematic trend in the UV absorption spectra, shifting to longer wavelengths according to the molecular size n [4]. In addition to the strong absorption bands in the UV, weak absorption features have been identified in the near UV region [5]. These features are associated with intrinsically forbidden, but vibronically allowed transitions in the molecule [6].

2. The HOMO-LUMO excitation in linear polyyne molecules

The ground state electron configuration in the polyyne molecule is such that doubly degenerate π orbitals constitute a ladder of orbital energy levels in which the *gerade* and *ungerade* orbitals appear alternately above and below the HOMO-LUMO gap, e.g., $\ldots(\pi_u)^4(\pi_g)^4(\pi_u)^4(\pi_g)^0(\pi_u)^0\ldots$ for $C_{10}H_2$ and $\ldots(\pi_g)^4(\pi_u)^4(\pi_g)^4(\pi_u)^0(\pi_g)^0\ldots$ for $C_{12}H_2$. Non-degenerate σ_g and σ_u orbitals locate upper and lower sections of the π orbitals. Since the neutral polyyne molecule has a closed shell electronic structure, its ground state is totally symmetric, Σ_g^+. The HOMO-LUMO excitation gives rise to three excited states, $\pi_u \otimes \pi_g = \Sigma_u^+ \oplus \Sigma_u^- \oplus \Delta_u$, for both series of $C_{4m+2}H_2$ (π_u for HOMO) and $C_{4m}H_2$ (π_g for HOMO). Electric dipole transition from the ground state is fully allowed to the Σ_u^+ state, while forbidden to the low-lying Σ_u^- or Δ_u state.

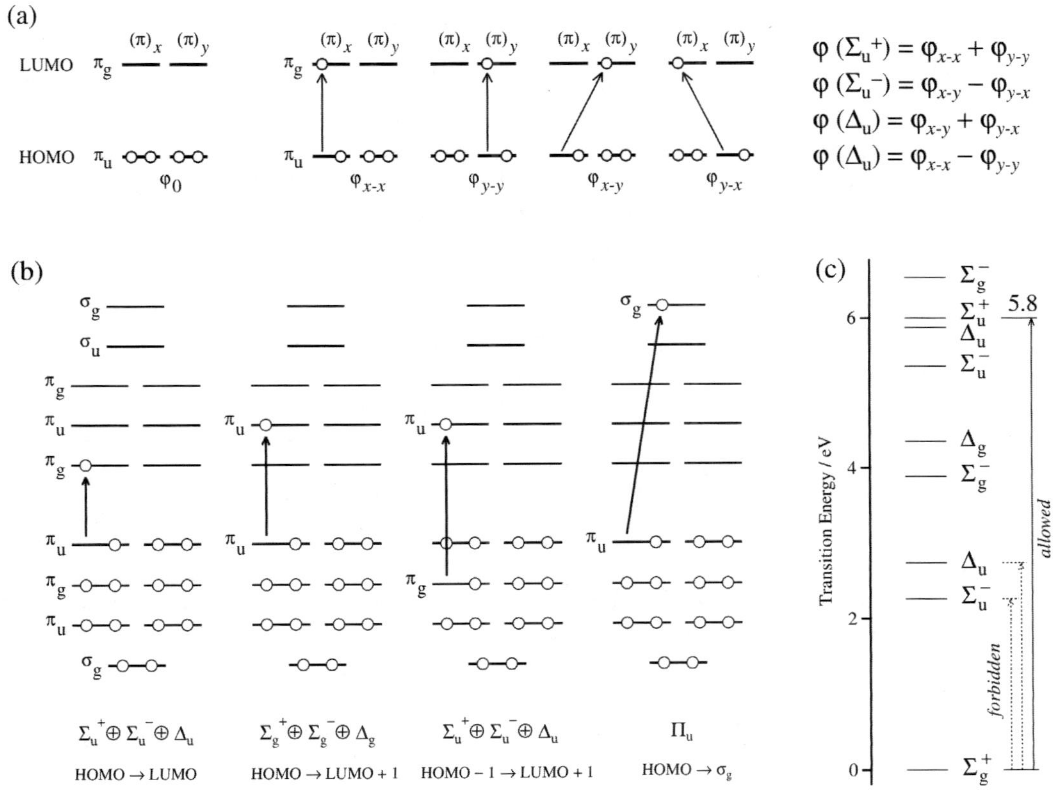

Figure 1. (*a*) Electron configurations for the ground state (left panel) and excited states stemming from the HOMO-LUMO excitation (right panel) in the linear polyyne molecule, $C_{4m+2}H_2$, in $D_{\infty h}$ point group symmetry. Four singly excited electron configurations in the π_u–π_g transition, φ_{x-x}, φ_{y-y}, φ_{x-y}, and φ_{y-x}, constitute an orthogonal set of linear combinations for three symmetry species, $\pi_u \otimes \pi_g = \Sigma_u^+ \oplus \Sigma_u^- \oplus \Delta_u$. (*b*) Configurations for single-electron excitations providing various excited states as represented for C_6H_2. Among the states from a π_u–π_g excitation for the HOMO-LUMO excitation, Σ_u^- and Δ_u come close in energy, while Σ_u^+ occupies always the upper most energy level. (*c*) Calculated transition energies for $C_{10}H_2$ (TDDFT). The first allowed transition, $\Sigma_u^+ \leftrightarrow \Sigma_g^+$, has an oscillator strength of f_z=5.8.

Figure 1(*a*) illustrates the electron configuration for the ground state (left panel) and those for singly excited states stemming from the HOMO-LUMO excitation. Since HOMO and LUMO are doubly degenerate, there are four possible electron configurations within the spin-singlet manifold. In the $D_{\infty h}$ point group symmetry, the four configurations are classified into three symmetry species, Σ_u^+, Σ_u^-, and Δ_u. Each of these states are represented by a linear combination of the singly excited configurations. The Σ_u^+ state contains excitations of an electron between π orbitals in the same direction, while the Σ_u^- state contains excitations between π orbitals perpendicular to each other. The doubly degenerate Δ_u state includes both the parallel and perpendicular excitations.

Figure 1(b) depicts some single electron excitations, using a schematic orbital energy diagram for C_6H_2. The left column illustrates the HOMO-LUMO excitation, generating three symmetry species, Σ_u^+, Σ_u^-, and Δ_u. The second column shows the electron excitation from HOMO to LUMO+1. Since the two orbitals have the same parity, u, the resulting three excited states have *gerade* symmetry, Σ_g^+, Σ_g^-, and Δ_g. The excitation from a lower occupied level is also possible, e.g., from HOMO−1 to LUMO+1 as in the third column. With this higher energy excitation, another set of excited states including the same symmetry species as the HOMO-LUMO excitation is generated. In the fourth column in Fig. 1(b), the excitation from HOMO (π_u) to an upper lying σ_g orbital gives rise to a Π_u state.

Figure 1(c) plots the transition energy for singly excited states in $C_{10}H_2$ (TDDFT). The spin-singlet excited states are the consequence of the state mixing within the same symmetry species. The three excited states consisting mostly of the HOMO-LUMO excitation are the lowest two excited states, Σ_u^- and Δ_u, and an upper lying excited state, Σ_u^+. The Σ_u^+ state is the lowest energy excited state among those to which the optical transition from the ground state is fully allowed, namely the first allowed transition. The Π_u state to which the electric dipole transition is also allowed from the ground state (see the fourth column in Fig. 2(b)), has much higher transition energy (~14 eV by the TDDFT calculation and not shown in Fig. 1(c)) than that for Σ_u^+ (~6 eV by TDDFT). In Fig. 2(c), it is noticeable that multiple states are locating in-between the lowest two excited states, Σ_u^- and Δ_u, and the upper Σ_u^+ state. These states stem from the electron configurations with higher transition energies (see the second and third column in Fig. 1(b)). Among the excited states stemming from the excitation between a pair of degenerate orbitals, the state for the optically allowed transition, in this case the Σ_u^+ state, is always the highest in its transition energy [7]. Therefore, several states from higher energy excitations are located below the Σ_u^+ state.

The electric dipoles, ex, ey, and ez, have irreducible representations, π_u, π_u, and σ_u^+, respectively. Therefore, the transition moment for $\Sigma_u^+ \leftrightarrow \Sigma_g^+$ is parallel to the molecular axis z, while the transition moments in (x, y) for $\Pi_u \leftrightarrow \Sigma_g^+$ are perpendicular to the molecular axis z.

3. Observations and conjectures

In a linear molecule in $D_{\infty h}$, the optical transition from the ground state, Σ_g^+, to an excited state other than Σ_u^+ or Π_u is forbidden by the orbital symmetry, e.g., $\Delta_u \leftarrow / \rightarrow \Sigma_g^+$ or $\Sigma_u^- \leftarrow / \rightarrow \Sigma_g^+$. This selection rule strictly applies to the vibrational 0–0 band. The origin band is not observed for the forbidden transition. The missing origin in the optical spectra poses a difficulty for the assignment of the spectral features. However, in some cases, transitions are weakly allowed by the electric dipole mechanism when accompanied by activation or deactivation in a specific vibrational mode.

Haink and Jungen analyzed UV absorption spectra for diacetylene, C_4H_2, and triacetylene, C_6H_2, in the gas phase. The missing origin was located at 258 cm^{-1} below the prominent band in the electronic transitions, $\Delta_u \leftarrow \Sigma_g^+$ and $\Sigma_u^- \leftarrow \Sigma_g^+$. The observed band was associated with an excitation of the π_g vibrational mode [8]. Ding et al. observed resonant two photon ionization spectra for the series of polyynes, $C_{2n}H_2$ of n=3–7, in the gas phase. The vibronic selection rule for the forbidden transitions, $\Delta_u \leftarrow \Sigma_g^+$ and $\Sigma_u^- \leftarrow \Sigma_g^+$, was mentioned based on the

Herzberg-Teller coupling [6]. During Raman spectroscopy for the polyyne molecules $C_{2n}H_2$ of n=5–9 in solutions [9], new optical emission band systems were observed in the near UV and visible wavelength regions. The emission spectra systematically shifted to longer wavelengths with the increasing molecular size n [10].

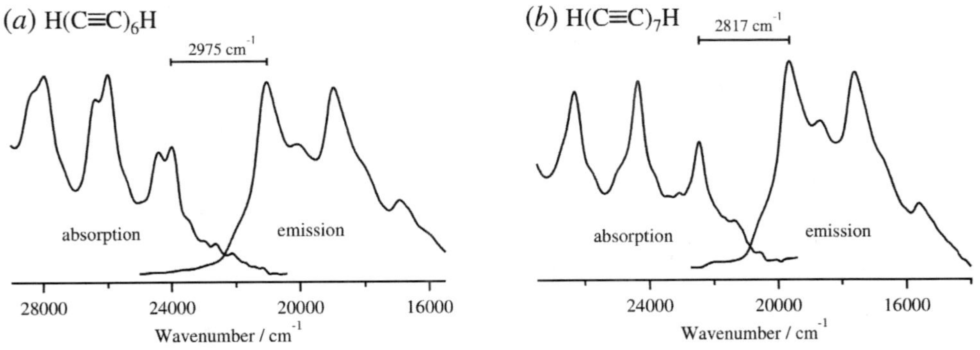

Figure 2. Vibronic bands in the absorption and emission spectra for the forbidden transition, $\Delta_u \leftrightarrow \Sigma_g^+$, in the polyyne molecules (*a*) $C_{12}H_2$ and (*b*) $C_{14}H_2$ in hexane. For the emission spectra, the excitation wavelength was tuned for the allowed transition, $\Sigma_u^+ \leftrightarrow \Sigma_g^+$, in the UV [10].

Figure 2 shows the absorption and laser induced optical emission spectra for (*a*) $C_{12}H_2$ and (*b*) $C_{14}H_2$ in hexane [10]. The vibrational progression is noted in the absorption and emission spectra, showing a characteristic vibrational frequency, 1800–2100 cm^{-1}, for the totally symmetric σ_g stretching vibrational mode of sp-carbon chains. The absorption and emission spectra constitute a mirror image. The missing origin between the absorption and emission features is a promise of the forbidden transition in the molecule. The lowest energy band in the absorption and the highest energy band in the emission are largely separated by 2975 and 2817 cm^{-1} for these polyynes. If the pair of the absorption and emission bands belong to the same electronic transition, and if the spectral shift due to stabilization by the solvent molecules, so called a Stokes shift, is not so large, the missing origin should locate in the middle of the two bands. In this case, such a low frequency π_g mode at ω_e=258 cm^{-1} can not explain for the location of the missing origin.

4. Vibronic interactions in the symmetry forbidden transition

The aim of this article is to provide symmetry considerations on the vibronic selection rule for the forbidden transition in the linear polyyne molecule. In the following sections, the theory of vibronic transitions is summarized and applied to the polyyne molecule in $D_{\infty h}$ point group symmetry. Vibronic transition mechanisms are based on the state mixing induced by symmetry lowering due to nuclear displacements along the normal coordinate for a vibrational mode. Intensity borrowing is expected for the forbidden transition when an excited state connected to the initial or the final state by an allowed transition is involved in the mixed states. Vibrational modes that activate the optical transition, namely inducing modes, depend on the state symmetries. The number of necessary vibrational quanta for an inducing mode is associated with the transition moment of allowed transitions in the intensity borrowing. The vibronic selection rule is deduced for a specific case of the $\Delta_u \leftrightarrow \Sigma_g^+$ transition.

Journal of Physics: Conference Series **428** (2013) 012004 doi:10.1088/1742-6596/428/1/012004

5. The theory of vibronic transitions

5.1. Vibronic interactions in the Hamiltonian

Nuclear displacements in a molecule from their equilibrium geometry at a high symmetry cause perturbation, H', in the Hamiltonian, H. The unperturbed Hamiltonian, H_0, is constructed for the equilibrium geometry [11]. The perturbation under the distortion, H', can be expressed by a series of Taylor expansion around the equilibrium geometry by the displacements along the normal coordinates, Q_k and Q_l, for the kth and lth vibrational modes, respectively.

$$H = H_0 + H', \tag{1}$$

$$H' = \sum_k \left(\frac{\partial H}{\partial Q_k}\right)_0 Q_k + \sum_{k,l} \frac{1}{2}\left(\frac{\partial^2 H}{\partial Q_k \partial Q_l}\right)_0 Q_k Q_l + H.O. \tag{2}$$

Here, we take both the linear and quadratic terms in order to consider all possible allowed transitions in the intensity borrowing within the first order perturbation.

5.2. The Herzberg-Teller expansion

Using the linear and quadratic terms in Eqn. (2), the wavefunction for an electronic state is represented within the framework of the first order perturbation theory [12].

$$\phi_n^{(1)}(q,Q) = \phi_n^{(0)}(q,0) + \sum_{m\neq n}\left[\sum_k \frac{\left\langle \phi_m^{(0)} \middle| \left(\frac{\partial H}{\partial Q_k}\right)_0 Q_k \middle| \phi_n^{(0)} \right\rangle}{E_n - E_m} + \sum_{k,l} \frac{\left\langle \phi_m^{(0)} \middle| \frac{1}{2}\left(\frac{\partial^2 H}{\partial Q_k \partial Q_l}\right)_0 Q_k Q_l \middle| \phi_n^{(0)} \right\rangle}{E_n - E_m}\right] \phi_m^{(0)}(q,0). \tag{3}$$

The electronic wavefunction, $\phi_n^{(1)}(q,Q)$, as a function of the electron coordinates, q, under the nuclear displacements, Q, is represented by a linear combination of unperturbed wavefunctions, $\phi_m^{(0)}(q,0)$, at the equilibrium geometry, Q=0. The wavefunction, $\phi_m^{(0)}$, is intrinsically a complex number thus contains a phase factor. For our purpose of symmetry considerations, only the real part is presented in the following discussion.

5.3. The Born-Oppenheimer approximation

The vibronic wavefunction, $\psi_n(q,Q)$, in the nth electronic state is represented by a product of the electronic wavefunction, $\phi_n^{(1)}(q,Q)$ in Eqn. (3), and the vibrational wavefunction, $\chi_v(Q)$, with vibrational quantum numbers, v [13, 14, 15].

$$\psi_n(q,Q) = \phi_n^{(1)}(q,Q)\chi_v(Q). \tag{4}$$

For an excited state, e, with vibrational quantum numbers, v', the vibronic wavefunction is

$$\psi_e = \phi_e^{(0)}\chi_{v'} + \sum_{m\neq e}\left[\sum_k \frac{\left\langle \phi_m^{(0)} \middle| \left(\frac{\partial H}{\partial Q_k}\right)_0 \middle| \phi_e^{(0)} \right\rangle}{E_e - E_m} Q_k + \sum_{k,l} \frac{\left\langle \phi_m^{(0)} \middle| \frac{1}{2}\left(\frac{\partial^2 H}{\partial Q_k \partial Q_l}\right)_0 \middle| \phi_e^{(0)} \right\rangle}{E_e - E_m} Q_k Q_l\right] \phi_m^{(0)}\chi_{v'}, \tag{5}$$

and for the ground state, g, with vibrational quantum numbers, v'', it is

$$\psi_g = \phi_g^{(0)}\chi_{v''} + \sum_{m\neq g}\left[\sum_k \frac{\left\langle \phi_m^{(0)} \middle| \left(\frac{\partial H}{\partial Q_k}\right)_0 \middle| \phi_g^{(0)} \right\rangle}{E_g - E_m} Q_k + \sum_{k,l} \frac{\left\langle \phi_m^{(0)} \middle| \frac{1}{2}\left(\frac{\partial^2 H}{\partial Q_k \partial Q_l}\right)_0 \middle| \phi_g^{(0)} \right\rangle}{E_g - E_m} Q_k Q_l\right] \phi_m^{(0)}\chi_{v''}, \tag{6}$$

where the nuclear displacements, Q_k and Q_l, are eligible for escaping the integral over the electron coordinates.

5.4. Transition moments

For the electric dipole, $\mu_\rho = e\rho$ $(\rho = x, y, z)$, $\mu = ex + ey + ez$, the transition moment between the ground state and an excited state is represented as

$$\langle \mu_\rho \rangle = \langle \psi_g | e\rho | \psi_e \rangle \quad , \quad \langle \mu \rangle = \langle \mu_x \rangle + \langle \mu_y \rangle + \langle \mu_z \rangle. \tag{7}$$

The integral is taken over all the electron coordinates, q, and the nuclear coordinates, Q. The oscillator strength, f, is proportional to the square of the transition moment and the transition energy, $f \propto \Delta E |\langle \mu \rangle|^2$.

5.5. Allowed and forbidden transitions

Using the wavefunctions for the ground and excited states in Eqn. (5) and (6), the matrix element for the transition moment, μ, is calculated as

$$\langle \mu \rangle = \langle \psi_g | \mu | \psi_e \rangle \tag{8}$$

$$\begin{aligned}
&= \langle \phi_g^{(0)} | \mu | \phi_e^{(0)} \rangle \langle \chi_{v''} | \chi_{v'} \rangle \\
&+ \sum_{m \neq e} \langle \phi_g^{(0)} | \mu | \phi_m^{(0)} \rangle \left[\sum_k \frac{\left\langle \phi_m^{(0)} \middle| \left(\frac{\partial H}{\partial Q_k} \right)_0 \middle| \phi_e^{(0)} \right\rangle}{E_e - E_m} \langle \chi_{v''} | Q_k | \chi_{v'} \rangle \right. \\
&+ \left. \sum_{k,l} \frac{\left\langle \phi_m^{(0)} \middle| \frac{1}{2} \left(\frac{\partial^2 H}{\partial Q_k \partial Q_l} \right)_0 \middle| \phi_e^{(0)} \right\rangle}{E_e - E_m} \langle \chi_{v''} | Q_k Q_l | \chi_{v'} \rangle \right] \\
&+ \sum_{m \neq g} \langle \phi_m^{(0)} | \mu | \phi_e^{(0)} \rangle \left[\sum_k \frac{\left\langle \phi_g^{(0)} \middle| \left(\frac{\partial H}{\partial Q_k} \right)_0 \middle| \phi_m^{(0)} \right\rangle}{E_m - E_g} \langle \chi_{v''} | Q_k | \chi_{v'} \rangle \right. \\
&+ \left. \sum_{k,l} \frac{\left\langle \phi_g^{(0)} \middle| \frac{1}{2} \left(\frac{\partial^2 H}{\partial Q_k \partial Q_l} \right)_0 \middle| \phi_m^{(0)} \right\rangle}{E_m - E_g} \langle \chi_{v''} | Q_k Q_l | \chi_{v'} \rangle \right].
\end{aligned} \tag{9}$$

When the 0th order integral, $\langle \phi_g^{(0)} | \mu | \phi_e^{(0)} \rangle$, is not zero, the first term predominates the transition moment, $\langle \mu \rangle$. The relevant transition is designated as *allowed* or a *dipole allowed* transition. On the other hand, when the integral, $\langle \phi_g^{(0)} | \mu | \phi_e^{(0)} \rangle$, vanishes, the relevant electronic transition is designated as *forbidden* or a *symmetry forbidden* transition. In this case, however, having non-vanishing matrix elements in the latter four terms in Eqn. (9), transitions between particular pairs of vibronic states can have non-zero contributions to the transition moment. The selection rule for the vibronic bands in the forbidden transition can be obtained by looking for such contributions.

5.6. Franck-Condon integrals

For the allowed transition in the first term in Eqn. (9), the integral for vibrational wavefunctions, $\langle \chi_{v''} | \chi_{v'} \rangle$, is represented explicitly as

$$\langle \chi_{v''} | \chi_{v'} \rangle = \prod_j^{3N-5} \langle v_j'' | v_j' \rangle. \tag{10}$$

The magnitude of this integral is approximated using a harmonic oscillator for each vibrational mode in the lower and upper electronic states. For the jth mode, the harmonic oscillator in the lower state, $|v_j''\rangle$, is located at the equilibrium geometry, Q=0, while that in the upper state,

Journal of Physics: Conference Series **428** (2013) 012004 doi:10.1088/1742-6596/428/1/012004

$|v'_j\rangle$, at the potential minimum in the excited state. The Franck-Condon integral, $\langle v''_j|v'_j\rangle$, can be calculated separately for each vibrational mode.

In the allowed transition, vibronic patterns are totally governed by the Franck-Condon factors, $|\langle v''_j|v'_j\rangle|^2$. Among all the possible combinations of vibrational energy levels, some contribute significantly to the vibronic pattern. For an extreme case in which the potential surfaces in the upper and lower electronic states are similar and both harmonic, the Franck-Condon factors, $|\langle v''_j|v'_j\rangle|^2$, are unity only for the case of $v''_j = v'_j$, providing contributions only from 0–0, 1–1, ... transitions in all the vibrational modes.

Concerning the vibronic bands in the forbidden transition, four vibronic terms in Eqn. (9) are relevant, in which the nuclear displacements, Q_k and Q_l, are involved in the integral over the nuclear coordinates, Q, as

$$\langle \chi_{v''}|Q_k|\chi_{v'}\rangle = \langle v''_k|Q_k|v'_k\rangle \prod_{j\neq k} \langle v''_j|v'_j\rangle, \tag{11}$$

$$\langle \chi_{v''}|Q_kQ_l|\chi_{v'}\rangle = \langle v''_k|Q_k|v'_k\rangle\langle v''_l|Q_l|v'_l\rangle \prod_{j\neq k,l} \langle v''_j|v'_j\rangle, \tag{12}$$

where the subscripts, k, l, and j, specify the normal modes. Note that the quadratic term in Eqn. (12) includes two cases, the diagonal term, $k = l$, for the overtone and the off-diagonal term, $k \neq l$, for the combination.

$$\langle \chi_{v''}|Q_k^2|\chi_{v'}\rangle = \underbrace{\langle v''_k|Q_k^2|v'_k\rangle}_{\text{overtone}} \prod_{j\neq k} \langle v''_j|v'_j\rangle,$$

$$\langle \chi_{v''}|Q_kQ_l|\chi_{v'}\rangle = \underbrace{\langle v''_k|Q_k|v'_k\rangle\langle v''_l|Q_l|v'_l\rangle}_{\text{combination}} \prod_{\substack{j\neq k,l\\ k\neq l}} \langle v''_j|v'_j\rangle.$$

The number of vibrational modes in the centrosymmetric linear polyyne molecule, $H(C{\equiv}C)_nH$, is $3N-5 = 6n+1$. They are classified into four symmetry species, σ_g, σ_u, π_g, and π_u. For σ_g (σ_u), only σ_g^+ (σ_u^+) modes exist and there is no σ_g^- (σ_u^-) mode. The σ_g (σ_u) modes correspond to the symmetric (anti-symmetric) stretching vibrations with nuclear displacements along the molecular axis, z. The π_g (π_u) modes correspond to the trans-bending (cis-bending) vibrations with nuclear displacements perpendicular to the molecular axis, (x, y). The numbers of vibrational modes are $n+1$, n, n, and n for the σ_g, σ_u, π_g, and π_u modes, respectively. Note that each of the π_g and π_u modes is doubly degenerate according to the degeneracy in the coordinates perpendicular to the molecular axis, (x, y).

Concerning the trans-bending mode of π_g symmetry in $D_{\infty h}$, the degeneracy is lifted by the nuclear displacements along the π_g normal coordinate to lower the symmetry into a_g and b_g in C_{2h}. For the a_g and b_g modes, the planes of nuclear motions are perpendicular to each other. This applies also to the cis-bending π_u mode. For a doubly degenerate π_g mode, the quadratic vibronic coupling in Eqn. (12) is associated with an overtone, π_g^2, for the case of a_g^2 or $b_g{}^2$, or a combination, $\pi_g\pi'_g$, for the case of a_gb_g. This is important for considerations on the vibronic coupling in the $\Sigma_u^-{\leftrightarrow}\Sigma_g^+$ and $\Delta_u{\leftrightarrow}\Sigma_g^+$ forbidden transitions.

6. Symmetry considerations on the transition moment

6.1. The intensity borrowing

For large contributions to the transition moments in Eqn. (9), $\langle\phi_g^{(0)}|\mu|\phi_m^{(0)}\rangle$ and $\langle\phi_m^{(0)}|\mu|\phi_e^{(0)}\rangle$, must be large. The former is associated with the dipole allowed transition from the ground state such as $\Sigma_u^+{\leftrightarrow}\Sigma_g^+$ and $\Pi_u{\leftrightarrow}\Sigma_g^+$. The latter for the transition between excited states also contributes significantly. The vibronic transition becomes allowed with the aid of the transition moment in the allowed transitions, thus the mechanism is called *intensity borrowing*.

Journal of Physics: Conference Series **428** (2013) 012004 doi:10.1088/1742-6596/428/1/012004

In the following discussion, we restrict our attention to a specific case for the $\Delta_u \leftrightarrow \Sigma_g^+$ transition in the linear polyyne molecule. The allowed transitions to be considered are those connecting either the initial or final state to another. Possible intermediate states are Π_u and Σ_u^+ for the allowed transitions connected with the ground state, Σ_g^+, while Π_g and Δ_g for the allowed transitions connected with the excited state, Δ_u. The latter two are rationalized by the direct products, $\Pi_g \otimes \Delta_u = \Pi_u \oplus \Phi_u$ and $\Delta_g \otimes \Delta_u = \Sigma_u^+ \oplus \Sigma_u^- \oplus \Gamma_u$. Each of the direct products includes, at least, one representation corresponding to the electric dipoles, Π_u for (x, y) and Σ_u^+ for z. These four transitions are classified by their transition moments, (x, y) and z, and associated with the linear and quadratic terms in the vibronic coupling, respectively. The important vibronic terms in the $\Delta_u \leftrightarrow \Sigma_g^+$ transition are represented explicitly as

$$
\begin{aligned}
\langle \mu \rangle &= \langle \Sigma_g^+ | \langle \chi_{v''} | \mu | \chi_{v'} \rangle | \Delta_u \rangle \qquad (13) \\
&= \sum_k \left\langle \Sigma_g^+ \middle| ex + ey \middle| \Pi_u \right\rangle \frac{\left\langle \Pi_u \middle| \left(\frac{\partial H}{\partial Q_k} \right)_0 \middle| \Delta_u \right\rangle}{E(\Delta_u) - E(\Pi_u)} \langle v_k'' | Q_k | v_k' \rangle \prod_{j \neq k} \langle v_j'' | v_j' \rangle \\
&+ \sum_k \left\langle \Pi_g \middle| ex + ey \middle| \Delta_u \right\rangle \frac{\left\langle \Sigma_g^+ \middle| \left(\frac{\partial H}{\partial Q_k} \right)_0 \middle| \Pi_g \right\rangle}{E(\Pi_g) - E(\Sigma_g^+)} \langle v_k'' | Q_k | v_k' \rangle \prod_{j \neq k} \langle v_j'' | v_j' \rangle \\
&+ \sum_{k,l} \left\langle \Sigma_g^+ \middle| ez \middle| \Sigma_u^+ \right\rangle \frac{\left\langle \Sigma_u^+ \middle| \frac{1}{2} \left(\frac{\partial H}{\partial Q_k \partial Q_l} \right)_0 \middle| \Delta_u \right\rangle}{E(\Delta_u) - E(\Sigma_u^+)} \langle v_k'' | Q_k | v_k' \rangle \langle v_l'' | Q_l | v_l' \rangle \prod_{j \neq k,l} \langle v_j'' | v_j' \rangle \\
&+ \sum_{k,l} \left\langle \Delta_g \middle| ez \middle| \Delta_u \right\rangle \frac{\left\langle \Sigma_g^+ \middle| \frac{1}{2} \left(\frac{\partial H}{\partial Q_k \partial Q_l} \right)_0 \middle| \Delta_g \right\rangle}{E(\Delta_g) - E(\Sigma_g^+)} \langle v_k'' | Q_k | v_k' \rangle \langle v_l'' | Q_l | v_l' \rangle \prod_{j \neq k,l} \langle v_j'' | v_j' \rangle . \qquad (14)
\end{aligned}
$$

6.2. Inducing modes

Once the intermediate state symmetry is determined from considerations on the intensity borrowing, the next step is to find vibrational modes, Q_k and Q_l, which provide non-zero contributions to the vibronic coupling, $\langle \Pi_u | (\partial H / \partial Q_k)_0 | \Delta_u \rangle$ and $\langle \Sigma_u^+ | \frac{1}{2} (\partial H / \partial Q_k \partial Q_l)_0 | \Delta_u \rangle$. Since the unperturbed Hamiltonian, H_0, is totally symmetric, its derivatives with respect to Q_k and $Q_k Q_l$ have the same symmetry as the relevant vibrational mode symmetry, $\Gamma(Q_k)$ and $\Gamma(Q_k) \otimes \Gamma(Q_l)$. For non-vanishing contributions, the direct product of the vibrational species and the two electronic species must contain a totally symmetric species. In other words, the symmetry species of the derivative must be the same as one of the symmetry species in the direct product of the two electronic states.

The linear molecule in $D_{\infty h}$ has four fundamental species in the vibrational modes, σ_g^+, σ_u^+, π_g, and π_u. For the vibronic coupling, $\langle \Pi_u | (\partial H / \partial Q_k)_0 | \Delta_u \rangle$, the symmetry species for the electronic part is $\Pi_u \otimes \Delta_u = \Pi_g \oplus \Phi_g$. Then, the vibrational mode of interest should be π_g. Concerning the quadratic coupling, $\langle \Sigma_u^+ | \frac{1}{2} (\partial H / \partial Q_k \partial Q_l)_0 | \Delta_u \rangle$, the symmetry species for the electronic part is $\Sigma_u^+ \otimes \Delta_u = \Delta_g$. The vibrational δ_g species is produced by an overtone or a combination of π_g modes, $\pi_g \otimes \pi_g = \sigma_g^+ (\oplus \sigma_g^-) \oplus \delta_g$. Also, by an overtone or a combination of π_u modes, $\pi_u \otimes \pi_u$, the vibrational δ_g species can be produced.

6.3. The vibronic selection rule in the $\Delta_u \leftrightarrow \Sigma_g^+$ transition

Suppose that a single π_g mode among the n π_g modes is the *inducing mode* which is responsible for the main spectral features in the vibronic transition. In the summation over all the vibrational modes, Q_k and Q_l, in Eqn. (14), most terms other than those for the inducing mode, $Q_k = Q_l = \pi_g$, are eliminated for their null contribution.

$$\begin{aligned}\langle\mu\rangle &= \left\langle\Sigma_g^+\middle|ex+ey\middle|\Pi_u\right\rangle\frac{\left\langle\Pi_u\middle|\left(\frac{\partial H}{\partial Q_{\pi_g}}\right)_0\middle|\Delta_u\right\rangle}{E(\Delta_u)-E(\Pi_u)}\langle v''_{\pi_g}|Q_{\pi_g}|v'_{\pi_g}\rangle\prod_{j\neq\pi_g}\langle v''_j|v'_j\rangle \\ &+ \left\langle\Pi_g\middle|ex+ey\middle|\Delta_u\right\rangle\frac{\left\langle\Sigma_g^+\middle|\left(\frac{\partial H}{\partial Q_{\pi_g}}\right)_0\middle|\Pi_g\right\rangle}{E(\Pi_g)-E(\Sigma_g^+)}\langle v''_{\pi_g}|Q_{\pi_g}|v'_{\pi_g}\rangle\prod_{j\neq\pi_g}\langle v''_j|v'_j\rangle \\ &+ \left\langle\Sigma_g^+\middle|ez\middle|\Sigma_u^+\right\rangle\frac{\left\langle\Sigma_u^+\middle|\frac{1}{2}\left(\frac{\partial^2 H}{\partial Q_{\pi_g}^2}\right)_0\middle|\Delta_u\right\rangle}{E(\Delta_u)-E(\Sigma_u^+)}\langle v''_{\pi_g}|Q_{\pi_g}^2|v'_{\pi_g}\rangle\prod_{j\neq\pi_g}\langle v''_j|v'_j\rangle \\ &+ \left\langle\Delta_g\middle|ez\middle|\Delta_u\right\rangle\frac{\left\langle\Sigma_g^+\middle|\frac{1}{2}\left(\frac{\partial^2 H}{\partial Q_{\pi_g}^2}\right)_0\middle|\Delta_g\right\rangle}{E(\Delta_g)-E(\Sigma_g^+)}\langle v''_{\pi_g}|Q_{\pi_g}^2|v'_{\pi_g}\rangle\prod_{j\neq\pi_g}\langle v''_j|v'_j\rangle. \end{aligned}\tag{15}$$

We further focus on a specific case of the emission from the vibrationally ground state, $v'_j = 0$ for all the $3N-5$ modes, in the upper electronic state, Δ_u, to multiple vibrational levels, v''_j, in the lower electronic ground state, Σ_g^+. Among the considerable number of Franck-Condon integrals, most are zero except for transitions to the $v''_j = 0$ level. An exceptional case for the HOMO-LUMO excitation in the polyyne molecule is a series of transitions for the CC stretching σ_g mode, providing the vibrational progression in v''_{σ_g}.

Harmonic oscillator considerations on the inducing π_g mode provide the vibronic selection rule, $\Delta v_{\pi_g} = \pm 1$ for the linear coupling associated with the transition moment in (x, y) and $\Delta v_{\pi_g} = \pm 2$ for the quadratic coupling associated with the transition moment in z. In the case of the emission from the vibrational $v' = 0$ level in the excited Δ_u state, the formula in Eqn. (15) is simplified as follows, including the vibrational progression for $v''_{\sigma_g} = 0-4$.

$$\mu_{x,y} = \left[\left\langle\Sigma_g^+\middle|\begin{matrix}ex\\ey\end{matrix}\middle|\Pi_u\right\rangle\frac{\left\langle\Pi_u\middle|\left(\frac{\partial H}{\partial Q_{\pi_g}}\right)_0\middle|\Delta_u\right\rangle}{E(\Delta_u)-E(\Pi_u)} + \left\langle\Pi_g\middle|\begin{matrix}ex\\ey\end{matrix}\middle|\Delta_u\right\rangle\frac{\left\langle\Sigma_g^+\middle|\left(\frac{\partial H}{\partial Q_{\pi_g}}\right)_0\middle|\Pi_g\right\rangle}{E(\Pi_g)-E(\Sigma_g^+)}\right]\langle 1|Q_{\pi_g}|0\rangle\langle v''_{\sigma_g}|0\rangle, \tag{16}$$

$$\mu_z = \left[\left\langle\Sigma_g^+\middle|ez\middle|\Sigma_u^+\right\rangle\frac{\left\langle\Sigma_u^+\middle|\frac{1}{2}\left(\frac{\partial^2 H}{\partial Q_{\pi_g}^2}\right)_0\middle|\Delta_u\right\rangle}{E(\Delta_u)-E(\Sigma_u^+)} + \left\langle\Delta_g\middle|ez\middle|\Delta_u\right\rangle\frac{\left\langle\Sigma_g^+\middle|\frac{1}{2}\left(\frac{\partial^2 H}{\partial Q_{\pi_g}^2}\right)_0\middle|\Delta_g\right\rangle}{E(\Delta_g)-E(\Sigma_g^+)}\right]\langle 2|Q_{\pi_g}^2|0\rangle\langle v''_{\sigma_g}|0\rangle. \tag{17}$$

7. Determining factors for the transition intensity

Figure 3 summarizes the mechanisms for the vibronic bands in the optical (*a*) absorption and (*b*) emission in the forbidden transition, $\Delta_u \leftrightarrow \Sigma_g^+$, in the centrosymmetric polyyne molecule. Each of the four intermediate states is connected both to the initial and final states by a dipole allowed transition in one side and by a vibronic coupling in the other. In the vibronic coupling, one (two) vibrational quantum (quanta) is (are) excited in the inducing mode. The appearance of spectral features is a matter of relative amplitudes between the terms in Eqn. (15). The factors which determine the magnitude of the vibronic term in Eqn. (15) are examined in the following.

7.1. Strengths in the vibronic coupling

The primarily important factor is the vibronic coupling in the form of $\langle\phi_m^{(0)}|(\partial H/\partial Q_k)_0|\phi_n^{(0)}\rangle$ or $\langle\phi_m^{(0)}|\frac{1}{2}(\partial H/\partial Q_k\partial Q_l)_0|\phi_n^{(0)}\rangle$. The former is an off-diagonal matrix element for the first derivative. Similar coupling appears in considerations on the pseudo Jahn-Teller effects [16]. The latter

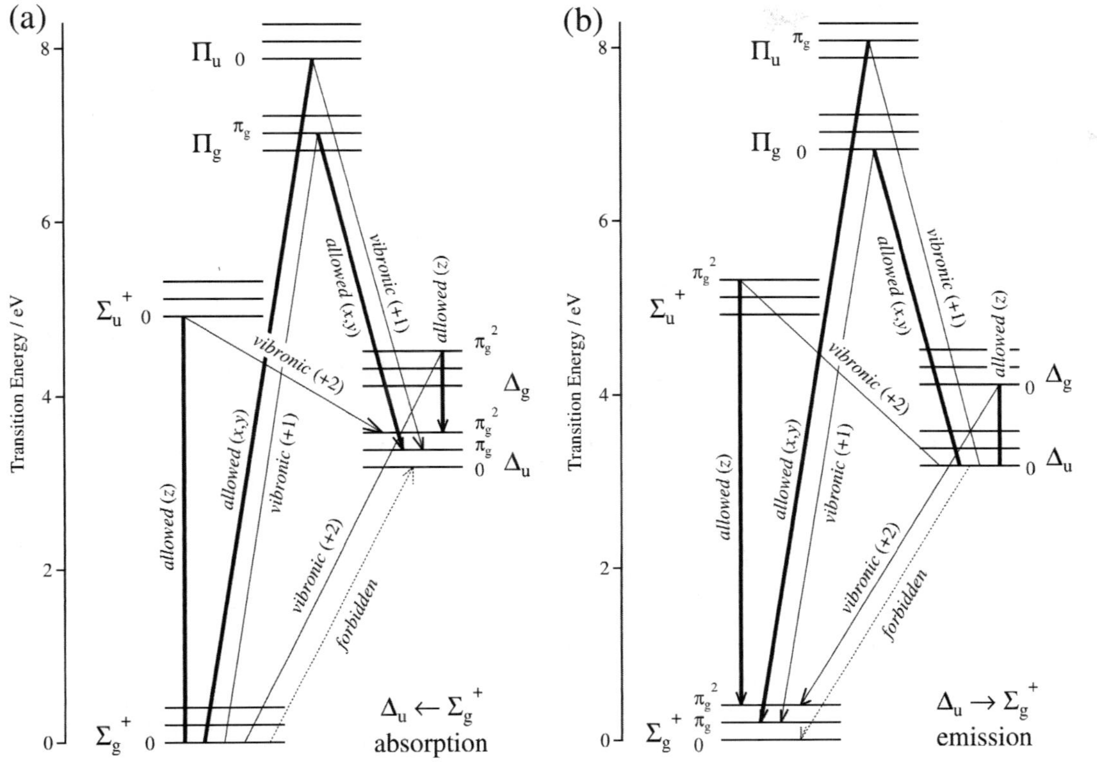

Figure 3. Schematic drawing for the mechanism of vibronic bands in the optical (*a*) absorption and (*b*) emission spectra for the $\Delta_u \leftrightarrow \Sigma_g^+$ transition in the centrosymmetric polyyne molecule, $C_{10}H_2$. For the forbidden transition, the intensity borrowing is necessary from the fully allowed transition, $\Pi_u \leftrightarrow \Sigma_g^+$, $\Sigma_u^+ \leftrightarrow \Sigma_g^+$, $\Pi_g \leftrightarrow \Delta_u$ or $\Delta_g \leftrightarrow \Delta_u$, as depicted by a thick line with *allowed* (ρ). The symbol, $\rho = x, y$, or z, in parenthesis represents the direction in the transition moment with respect to the molecular axis, z. The vibronic coupling is depicted by a thin line with *vibronic* (Δv_{π_g}). The number, $\Delta v_{\pi_g} = +1$ or $+2$, in parenthesis represents the number of vibrational quanta necessary for activating the transition by the excitation of the inducing vibrational mode, π_g. Note that the 0–0 band in the $\Delta_u \leftrightarrow \Sigma_g^+$ transition is strictly forbidden by the mechanism, thus the signal is missing both in the absorption and emission spectra.

is an off-diagonal matrix element for the second derivative. This is beyond the conventional matrix elements in considerations on symmetry breaking systems. Calculations of these vibronic coupling matrix elements are the subject in theoretical developments in the future. It is our belief that the magnitude of the quadratic coupling, $\langle \phi_m^{(0)} | \frac{1}{2}(\partial^2 H/\partial Q_k^2)_0 | \phi_n^{(0)} \rangle \langle v_k'' | Q_k^2 | v_k' \rangle$, can be comparable to the linear coupling, $\langle \phi_m^{(0)} | (\partial H/\partial Q_k)_0 | \phi_n^{(0)} \rangle \langle v_k'' | Q_k | v_k' \rangle$.

7.2. Transition moments in the allowed transition

For the intensity borrowing, transition moments in the allowed transition are important. In the polyyne molecule, oscillator strengths for $\Sigma_u^+ \leftrightarrow \Sigma_g^+$ transitions are concentrated to the first allowed transition in the UV, ~5 eV. This is regarded as the plasmon excitation with $f_z = 5.8$ (TDDFT for $C_{10}H_2$) and with an absorption coefficient $\epsilon \sim 2 \times 10^5$ L mol^{-1} cm^{-1} [4]. The $\Pi_u \leftrightarrow \Sigma_g^+$ transition has been suggested to locate in the vacuum UV region, ~8 eV, close to the ionization threshold [5], with $f_x + f_y = 2.3$ (TDDFT for $C_{10}H_2$). Although any allowed transition involving the Δ_u state has not been observed, their oscillator strengths should be comparable to the other allowed transitions. Note that the relation between the transition moment and the oscillator strength is $|\mu_{mn}|^2 \propto f_{mn}/\Delta E_{mn}$, where ΔE_{mn} denotes the transition energy.

7.3. Energy denominators
The perturbation coefficient includes a denominator by the energy difference between electronic states connected by the vibronic coupling. As is seen in Figure 3, the magnitude of the denominator for the quadratic term, $|E(\Delta_u) - E(\Sigma_u^+)|$, is small. Thus, substantial contributions are expected by the intensity borrowing from the $\Sigma_u^+ \leftrightarrow \Sigma_g^+$ transition. Terms with denominators of $|E(\Delta_g) - E(\Sigma_g^+)|$ and $|E(\Delta_u) - E(\Pi_u)|$ may also contribute to some extent. The magnitude of the denominator, $|E(\Pi_g) - E(\Sigma_g^+)|$, is the largest, thus probably having a less contribution. Excited states for which the transition energies have been experimentally determined are the Σ_u^+ and Δ_u states. Therefore, the comparison between the four denominators remains an estimate.

7.4. Vibrational off-diagonal elements associated with the inducing mode
Equation (15) contains three types of integrals for the vibrational wavefunctions in the forms of $\langle v_k''|Q_k|v_k'\rangle$, $\langle v_k''|{Q_k}^2|v_k'\rangle$, and $\langle v_j''|v_j'\rangle$, where k represents the inducing mode and j the other modes. The former two are associated with the linear and quadratic vibronic couplings. When the scaling factors in Q_k and Q_k^2 are normalized or cancelled by the factors in the vibronic coupling derivatives, all the vibrational integrals are counted as order unity. With a simple estimation based on the harmonic oscillator approximation, the integrals with unitary amplitudes are those for $\Delta v_k = \pm 1$ and ± 2 for the linear and quadratic integrals, respectively. For polyyne molecules, the inducing mode is a trans-bending π_g mode as discussed in Section 6.2 and 6.3 [5, 6, 8, 10, 17]. The importance of the quadratic term has been emphasized first with the observation of the pair of absorption and emission spectra in the forbidden transition [10].

7.5. Franck-Condon factors – the vibrational progression
Among the three types of vibrational integrals, the latter one, $\langle v_j''|v_j'\rangle$, is the Franck-Condon integral. In the harmonic oscillator approximation, the integrals for $\Delta v_j = 0$ have an unitary amplitude. However, it is the electronic transition, thus the difference in the molecular geometry between the upper and lower electronic states causes the vibrational progression in some vibrational modes. Actually, for the $\Sigma_u^+ \leftarrow \Sigma_g^+$ allowed transition, the vibrational progression is conspicuous for the symmetric stretching σ_g mode with an increment of $\sim$2000 cm^{-1} [4, 9]. Also, for the vibronic spectra in the $\Delta_u \leftrightarrow \Sigma_g^+$ forbidden transition in Fig. 2(*a*) and (*b*) [6, 8, 10], the vibrational progression in the same σ_g mode is discernible.

7.6. Possible interference effects between the terms
Finally, it should be emphasized again that the transition intensity is proportional to the square of the transition moment and that the phase factor has not been considered in the present article. Interference effects may occur when the squared sum of the elementary transition moments are calculated explicitly. Some vibronic features can be pronounced and some can be diminished. Here, we just mention about the possibility for the interference effects. This may force some modifications to our description on the intensity for the vibronic bands in the forbidden transition.

8. The selection rule and the transition moment
Our considerations on the vibronic transition in the centrosymmetric linear polyyne molecule complete when both the linear and quadratic vibronic couplings are included. The linear term is

associated with the transition of the single quantum excitation in the inducing vibrational mode with an intensity borrowing from the allowed transitions with a transition moment perpendicular to the molecular axis, (x, y). On the other hand, the quadratic term is associated with the transition of the double quanta excitation in the inducing vibrational mode with an intensity borrowing from the allowed transitions with a transition moment in parallel with the molecular axis, z. From the symmetry considerations on the inducing mode in Section 6, the selection rule for the vibronic bands in the $\Delta_u \leftarrow \Sigma_g^+$ transition is established.

The remaining questions are (1) which π_g mode among the n π_g modes is really the inducing mode, (2) where the missing origin is located, and (3) which vibronic term is dominating in the observed vibronic bands. For answering these questions, the vibronic considerations presented in this article constitute a milestone casting a light for the direction in the future. First, the direct comparison in the absorption and emission spectra provides information on the vibrational frequency for the inducing mode as well as for the location of the missing origin [10]. Second, the direction of the transition moment is a key feature in the vibronic transition. Since the transition moment is related to the vibronic selection rule, the correlation between the transition intensity and the polarization of a linearly polarized light should be an interesting issue for understanding the mechanisms in the vibronic transition.

9. The experimental approach

Experimental approaches based on the idea presented above are under way in our group. Molecular crystals containing a trace amount of highly oriented polyyne molecules are the subject for absorption spectroscopy using a linearly polarized light. The allowed transition, $\Sigma_u^+ \leftarrow \Sigma_g^+$, is known to have the transition moment in parallel with the molecular axis, thus the absorption should be strong when polarization of the incident light gets along the line of the molecular axis in the crystal. If the absorption intensity for the vibronic bands in the forbidden transition follows the intensity for those in the allowed transition upon rotation of the polarization of light, the vibronic bands must have the transition moment in the same direction as the allowed transition. In this case, the vibronic bands should be induced by the excitation of an overtone in the π_g mode. Therefore, the missing origin should be separated from the observed vibronic band by twice the frequency of the inducing π_g mode. If the absorption intensities in the allowed and forbidden transitions show anti-correlation upon rotation of the polarization of light, the vibronic bands in the forbidden transition must be associated with the single quantum excitation in the inducing π_g mode with the transition moment perpendicular to the molecular axis.

One of the critical issues for such experiments is to prepare an appropriate single crystal in which the linear polyyne molecules are highly oriented. Transparency down to shorter wavelengths, ~220 nm, is also a necessary condition. We are trying to prepare such a crystalline form of molecular solids in which the size-selected polyyne molecules are embedded. Preparation of the size-selected polyynes as well as cyanopolyynes is in progress in our laboratory [18]. Very recently, we found that polyyne molecules form a relatively stable molecular complex with iodine molecules in non-polar solvents. For the polyyne-iodine complex, the vibronic bands in the forbidden transition were dramatically intensified in the absorption spectra [19, 20]. With the experiments in solutions and solids as well as in the gas phase, information is available for further understanding of the mechanisms for the vibronic transition in the linear polyyne molecules.

10. Conclusions

Based on the Herzberg-Teller coupling for the intensity borrowing, the optical selection rule was derived for the vibronic bands in the symmetry forbidden transition, $\Delta_u \leftrightarrow \Sigma_g^+$, in the linear polyyne molecules, $C_{2n}H_2$, in $D_{\infty h}$ point group symmetry. In addition to the linear vibronic coupling by the first derivative, the quadratic vibronic coupling by the second derivative was taken into account. The quadratic coupling is associated with the vibronic transition having the transition moment in parallel with the molecular axis in the linear polyyne molecule. It is induced by the excitation of an overtone in the vibrational π_g mode, for which the intensity borrowing from the fully allowed transition, $\Sigma_u^+ \leftrightarrow \Sigma_g^+$, is expected.

References

[1] Steinfeld J I 1985 *Molecules and Radiation: An Introduction to Modern Molecular Spectroscopy, 2nd Edition* (The MIT Press) chapter 9 pp 272–280
[2] Tsuji M, Tsuji T, Kuboyama S, Yoon S-H, Korai Y, Tsujimoto T, Kubo K, Mori A and Mochida I 2002 *Chem. Phys. Lett.* **355** 101
[3] Tabata H, Fujii M, Hayashi S, Doi T and Wakabayashi T 2006 *Carbon* **44** 3168
[4] Eastmond R, Johnson T R and Walton D R M 1972 *Tetrahedron* **28** 4601
[5] Kloster-Jensen E, Haink H-J and Christen H 1974 *Helv. Chim. Acta* **57** 1731
[6] Ding H, Schmidt T W, Pino T, Güthe F and Maier J P 2003 *Phys. Chem. Chem. Phys.* **5** 4772
[7] Nakai H 2009 *Rules for excited states of degenerate systems: Interpretation by frozen orbital analysis* (Advances in The Theory of Atomic and Molecular Systems: Conceptual and Computational Quantum Chemistry, Progress in Theoretical Chemistry and Physics) eds Piecuch P, Maruani J, Delgado-Barrio G and Wilson S (Springer, Dordrecht-Heidelberg-London-New York) pp 363–395
[8] Haink H-J and Jungen M 1979 *Chem. Phys. Lett.* **61** 319
[9] Wakabayashi T, Tabata H, Doi T, Nagayama H, Okuda K, Umeda R, Hisaki I, Sonoda M, Tobe Y, Minematsu T, Hashimoto K and Hayashi S 2007 *Chem. Phys. Lett.* **433** 296
[10] Wakabayashi T, Nagayama H, Daigoku K, Kiyooka Y and Hashimoto K 2007 *Chem. Phys. Lett.* **446** 65
[11] Sato T, Tokunaga K and Tanaka K 2006 *J. Chem. Phys.* **124** 024314
[12] Eyring H, Walter J and Kimball G E 1944 *Quantum Chemistry* (John Wiley & Sons, Inc.) chapter 7 pp 92–99
[13] Orlandi G and Siebrand W 1973 *J. Chem. Phys.* **58** 4513
[14] Lin S H and Eyring H 1974 *Proc. Natl. Acad. Sci. USA* **71** 3415
[15] Lin S H and Eyring H 1974 *Proc. Natl. Acad. Sci. USA* **71** 3802
[16] Bersuker I B 2006 *The Jahn-Teller Effect* (Cambridge University Press) chapter 4 pp 110–161
[17] Turowski M, Crépin C, Gronowski M, Guillemin J-C, Coupeaud A, Couturier-Tamburelli I, Piétri N and Kolos R 2010 *J. Chem. Phys.* **133** 074310
[18] Wakabayashi T, Saikawa M, Wada Y and Minematsu T 2012 *Carbon* **50** 47
[19] Wada Y, Wakabayashi T and Kato T 2011 *J. Phys. Chem.* B **115** 8439
[20] Wada Y, Morisawa Y and Wakabayashi T 2012 *Chem. Phys. Lett.* **541** 54

Journal of Physics: Conference Series **428** (2013) 012005 doi:10.1088/1742-6596/428/1/012005

C_{60}^{+} and B_{80}^{+}: A Comparative Study of the Jahn-Teller Effect

H Ramanantoanina[a], J T Muya[b], A Ceulemans[b] and C Daul[a]

[a] Department of Chemistry, University of Fribourg
Chemin du musée 9, 1700 Fribourg, Switzerland
[b] Department of Chemistry and Institute for Nanoscale Physics and Chemistry, Katholieke Universiteit Leuven
Celestijnenlaan 200F, B-3001 Leuven, Belgium

E-mail: claude.daul@unifr.ch

Abstract. The ground state wave function of the neutral icosahedral C_{60} and B_{80} belong to the totally symmetric representation, where the HOMOs are fivefold degenerate and form the basis of the H_u representation of the I_h point group. Hence both C_{60}^{+}, and B_{80}^{+} are prone to a molecular distortion of the Jahn-Teller type. Density Functional Theory calculation is applied and revealed that a minimum energy configuration in D_{5d} point group is obtained for C_{60}^{+}; whereas a slight S_6 distortion of a D_{3d} nuclear configuration is obtained for B_{80}^{+}. Thus the vibronic coupling between the ^2H_u electronic states of both systems with the degenerate normal modes in the I_h point group are analysed and presented here in a comparative point of view. Moreover, a simple and efficient procedure, which is fully non-empirical, based on the harmonic approximation, is presented in order to calculate the Jahn-Teller parameters and the first order vibronic coupling coefficient.

1. Introduction

The properties and the structure of non-planar boron allotrope were the subject of enormous work, [1-6] since the prediction of the existence of the boron buckyball B_{80}. [7] The B_{80} molecule is of significant interest due to the fact that B_{80} and the analogue carbon buckyball C_{60} are isoelectronic. The necessity to characterize the B_{80} molecule also emerged from some physical properties, such as electrical conductivity [8] and hydrogen storage ability, [9] which appear to be superior comparing to those properties obtained in C_{60}.

The distinguishing structural feature of the boron buckyball, as compared to the carbon buckyball, is the presence of twenty extra boron atoms, capping the twenty six-membered ring boron atoms of the buckyball (figure 1). Thus the neutral B_{80}, like the case of the neutral C_{60}, might exhibit the highest possible finite point group in nature, I_h. Quite a number of theoretical investigations have been devoted to the structure of the neutral B_{80} molecule. Although Szwacki and Tymczak [10] postulated the existence of B_{80} with a nuclear configuration in the I_h point group, using high level wave function theory calculation, Gopakumar et al., [11] together with Gunashinge et al., [12] predicted that this icosahedral

Journal of Physics: Conference Series **428** (2013) 012005 doi:10.1088/1742-6596/428/1/012005

geometry is instable leading to the formation of two stable structures in the T_h point group, using Density Functional Theory (DFT) calculation. The corresponding descent in symmetry has been recently analysed by Muya et al., [13] and has been identified as the consequence of a weak pseudo-Jahn-Teller mechanism.

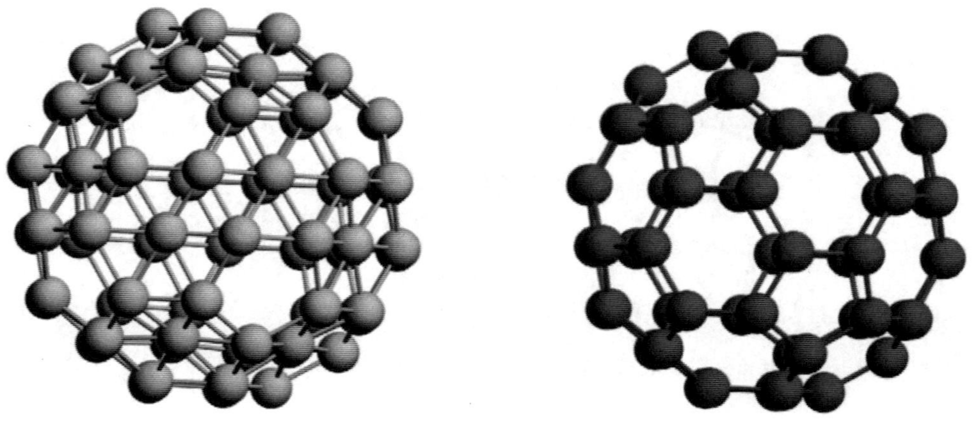
(a) (b)

Figure 1. Structures of the neutral B_{80} (a) and the neutral C_{60} (b) in their icosahedral nuclear configurations.

Nevertheless, the structures of neutral B_{80} and C_{60} in their high symmetry point groups suppose that either the anionic or the cationic species are characterized by degenerate electronic states, giving rise to a Jahn-Teller instability. Moreover, this instability for the case of B_{80}^+ or B_{80}^- will be superimposed on the pseudo-Jahn-Teller instability of the neutral species, leading to a complicated distortion mechanism, which has no counterpart in the case of C_{60}^+ or C_{60}^-. Hence the analysis of the structural distortion from the high symmetry arrangement to the global minimum of B_{80}^+ and C_{60}^+ presents a challenge because of the superposition of effect produced by many different normal modes.
Herein, we report the analysis of the Jahn-Teller distortion, which is occurring in the icosahedral B_{80}^+ and compare the results of that we obtained for C_{60}^+. In order to tackle the multimode problem, the Intrinsic Distortion Path (IDP) model, [14] in which the distortion is represented as a linear combination of all totally symmetric normal modes of the low symmetry minimum energy configuration, is performed. Within this model, it is possible to directly separate the contribution of the different normal modes to the Jahn-Teller distortion, their energy contribution to the Jahn-Teller stabilization energy, and the forces at the high symmetry point, giving further insight into the first order vibronic coupling coefficient in both B_{80}^+ and C_{60}^+ Jahn-Teller effect.

Journal of Physics: Conference Series **428** (2013) 012005 doi:10.1088/1742-6596/428/1/012005

2. Methodology

The DFT calculations reported in this work have been carried out using the Gaussian program package, Gaussian2009. [15] The Local Density Approximation (LDA) characterized by the Vosko-Wilk-Nussair (VWN) parametrization, [16] the Generalized Gradient Approximation (GGA) characterized by the Perdew-Burke-Ernzehof (PBE) functional, [17] and the Hybrid functional characterized by the B3LYP [18] parameterization were used for the analysis of the Adiabatical Potential Energy Surface (APES) of B_{80}^{+}. The 6-31G(D) basis set was used to represent the boron and the carbon atoms. Analytical harmonic frequencies were calculated.
To facilitate the understanding of our proposed model, we first outline a very basic part of the former achievements relevant to our analysis of the Jahn-Teller effect. Within the multideterminental approach, [19] we are able to evaluate the Jahn-Teller parameters (e.g. Jahn-Teller Energy) using direct DFT calculation. Then the approach has been successfully applied to a series of typical Jahn-Teller active molecules. [20-21] The approach itself can be found elsewhere, [19] and requires the DFT calculation of the high symmetry (HS) and low symmetry (LS) geometries and energies of the species. Although, the geometry and energy of the LS species can be found in a quite straightforward way, those corresponding to the HS species need further discussion. We always represented the geometry of the HS species as the structure exhibited in the so-called Average of Configuration (AOC) type of calculation. [19] This procedure prevents us to consider the HS geometry as the geometry obtained for the neutral species and turns out to be the appropriate way to calculate the DFT Jahn-Teller energy. [19] [22]
Here, HS and LS denote the geometry obtained from the LDA calculation of the non-Jahn-Teller parent molecule, in this case the neutral species and the global minimum of the cationic species, respectively. It is noteworthy to point out that the choice of the neutral parent molecule as the HS structure does not bias our calculation since the Jahn-Teller parameters we are looking for, are extracted from the harmonic approximation without fitting to the DFT results.
The distortion vector is regarded as the difference between the mass-weighted Cartesian coordinate corresponding to the HS and the LS nuclear configurations, equation (1).

$$\vec{R}_{JT} = \vec{R}_{HS} - \vec{R}_{LS} = \sum_{k=1}^{3N-6} w_{LS,k} \vec{Q}_{LS,k} \tag{1}$$

where $\vec{R}_{HS}$ and $\vec{R}_{LS}$ are column vectors, which contain 3N elements (N being the total number of atoms in a molecule) of the mass-weighted Cartesian coordinates, corresponding to the optimized LDA geometry of the HS and the LS nuclear configurations, respectively.

Furthermore $\vec{R}_{JT}$ can be represented in terms of the normal modes at the LS nuclear configuration, equation (1), where the weighting factor $w_{LS,k}$ is a scalar, which represents the linear combination coefficient of the kth mass-weighted normal coordinate $\vec{Q}_{LS,k}$ of the LS nuclear configuration obtained from the LDA frequency analysis. Since our approach deals with the first order Jahn-Teller distortion, only totally symmetric modes in the LS point group might have a non-zero $w_{LS,k}$.
While the LS nuclear configuration is taken as reference point, within the harmonic approximation, the contribution to the stabilization energy in terms of the normal modes in the LS conformation is calculated in a simple analytical form, equation (2).

$$E_{LS,k} = \frac{1}{2} w_{LS,k} \left| \vec{Q}_{LS,k} \right|^2 \lambda_{LS,k} \tag{2}$$

where $\lambda_{LS,k}$ represents the eigenvalue of the Hessian obtained from the LDA frequency calculation on the LS global minimum energy.
Therefore, using the same approach, the forces, which correspond to each normal mode through the distortion, can be obtained as:

$$\vec{F}_{IDP} = \sum_{k=1}^{3N-6} w_{LS,k} \lambda_{LS,k} \mathbf{M}^{\frac{1}{2}} \vec{Q}_{LS,k} \tag{3}$$

where **M** is a diagonal matrix with atomic masses in triplicate form as elements. $\vec{F}_{IDP}$ can be projected in either LS or HS normal modes using the vector properties of the (3N-6) normal modes orthogonal basis.
Recently we proposed a simple and efficient approach for the determination of the Jahn-Teller parameters. [22] As the JT magnitude $q_{HS,k}$ and the forces $F_{HS,k}$ are obtained in a straightforward way; the energy contributions to the Jahn-Teller stabilization energy of the normal modes in the HS nuclear configuration require a further refinement.
We presented in reference [22] that two sets of mass-weighted normal coordinates can be correlated as following:

$$\mathbf{Q}_{LS}^{T} \cdot \mathbf{Q}_{HS} = \mathbf{C} \tag{4}$$

where $\mathbf{Q}_{LS} = \left[\vec{Q}_{LS,1} \ \vec{Q}_{LS,2} \ \cdots \ \vec{Q}_{LS,3N-6} \right]$ and $\mathbf{Q}_{HS} = \left[\vec{Q}_{HS,1} \ \vec{Q}_{HS,2} \ \cdots \ \vec{Q}_{HS,3N-6} \right]$ are matrices with 3N x (3N-6) elements, which contain the kth mass-weighted normal coordinates of the LS and HS nuclear configuration, respectively. **C** is a (3N-6) x (3N-6) squared correlation matrix and the element $\mathbf{C}_{i,j}^2$ of the matrix **C** gives the overlap between the two sets of normal coordinates.
Therefore the energy contribution of each HS normal mode to the Jahn-Teller stabilization energy can be easily obtained from equation (2) as:

$$E_{HS,k} = \sum_{i=1}^{3N-6} \left(C_{k,i} \right)^2 E_{LS,i} \tag{5}$$

To summarize, using the scheme presented above, we can calculate:

1. The Jahn-Teller magnitude $q_{HS,k}$, considering:

$$\vec{R}_{JT} = \sum_{k=1}^{3N-6} v_k \vec{Q}_{HS,k} \tag{6}$$

one can obtain the Jahn-Teller magnitude relative to the HS normal modes, in the present case, as:

$$q_{HS,k} = \left| v_k \vec{Q}_{HS,k} \right| / \sqrt{m} \tag{7}$$

Journal of Physics: Conference Series **428** (2013) 012005 doi:10.1088/1742-6596/428/1/012005

where m stands for the atomic mass of either boron atom or carbon atom, i.e., 11.0093 a.u. and 12.0000 a.u., respectively.

2. The energy contribution of each normal mode at the HS nuclear configuration $E_{HS,k}$ to the Jahn-Teller stabilization energy as given in equation (5). Therefore the Jahn-Teller stabilization energy is formulated as:

$$E_{JT} = \sum_{k=1}^{3N-6} E_{HS,k} \tag{8}$$

3. The forces $F_{HS,k}$, which exert each normal mode at the HS nuclear configuration to the distortion as:

$$\vec{F}_{IDP} = \sum_{k=1}^{3N-6} \left|F_{HS,k}\right| \vec{Q}_{HS,k} \tag{9}$$

3. Results and Discussion

The ground state of B_{80}^+

We have analysed the APES of B_{80}^+ by means of DFT calculation, using the LDA, the GGA and the hybrid functional, table 1. Without going into much detail, because some of these results have been already published, [23] it can be seen from table 1, that according to the DFT calculations, the ^2A_u conformation in S_6 symmetry is the global minimum. This is confirmed by the absence of any imaginary frequency, while vibrational analysis was performed. In the other case, similar analysis of the APES of C_{60}^+ always predicted a global minimum in D_{5d} symmetry, [22] in agreement with several reported calculations. [24-25]

Table 1. Total electronic energies (Hartrees) of epikernels D_{5d}, D_{3d}, D_{2h} and S_6 of B_{80}^+ with three different DFT methods; relative energies to the global minimum are given in parenthesis (cm^{-1}), taken from reference [23].

Symmetry	Γ_{el}[a]	VWN/6-31G(D)	PBE/6-31G(D)	B3LYP/6-31G(D)	# img. frq. (Γ_{vib})[b]
D_{5d}	$^2A_{1u}$	-1974.37750 (515.77)	-1984.80333 (602.50)	-1987.31611 (1095.62)	7 ($e_{1g}+e_{2g}+e_{2u}+a_{2u}$)
D_{2h}	$^2B_{1u}$	-1974.37928 (125.10)	-1984.80492 (392.86)	-1987.31956 (338.43)	3 ($b_{1g}+b_{2g}+b_{3g}$)
D_{3d}	$^2A_{1u}$	-1974.37945 (87.79)	-1984.80607 (140.46)	-1987.32036 (162.85)	1 (a_{2g})
S_6	^2A_u	-1974.37985 (0.00)	-1984.80671 (0.00)	-1987.32110 (0.00)	0 (-)

[a] Electronic state
[b] number of imaginary frequencies obtained from the DFT frequency calculation and their symmetry in parenthesis.

The D_{5d}, D_{2h} and D_{3d} epikernel symmetries of B_{80}^+ exhibit imaginary frequencies (table 1), which indicate a further instability of the parent geometry. The choice of the LS point group of the Jahn-Teller distortion is mostly based on the epikernel principle, [26] which states that extremum point on the APES are more favourable to maximal epikernels than to the lower ranking ones. It is noteworthy to highlight that at least, the icosahedral B_{80}^+ distorts in a trigonal route, unless the icosahedral C_{60}^+ prefers a pentagonal distortion. Nevertheless, the D_{3d} structure of B_{80}^+ is characterized by a single imaginary frequency of a_{2g} symmetry. This mode, which is anti-symmetric with respect to the twofold axes and the dihedral symmetry planes, reduces the D_{3d} geometry to the S_6 point group. This difference between the ground state symmetry of C_{60}^+ and B_{80}^+ is then attributed to the presence of persistent pseudo-Jahn-Teller forces, present in the neutral B_{80}. [23]
The hybrid functional B3LYP overestimates the relative energies, whereas the GGA and LDA functionals give a relatively reasonable range of energy.

Multimode Jahn-Teller effect of B_{80}^+

B_{80}^+ and C_{60}^+ have a ^2H_u electronic ground state in their icosahedral nuclear configurations, I_h point group. Nine electrons (one hole) fill the fivefold degenerate HOMO. Hence B_{80}^+ and C_{60}^+ are Jahn-Teller active molecules and are subjected to a lowering of the symmetry of their nuclear configuration. According to the DFT analysis of the ground state of both cations, the descent in symmetry goes to S_6 and D_{5d}, respectively for the case of B_{80}^+ and C_{60}^+. In D_{5d} point group, the electronic state splits into $^2A_{1u}$, $^2E_{1u}$ and $^2E_{2u}$, while ^2A_u and two times ^2E_u are obtained in the S_6 point group. B_{80} in I_h symmetry has 234 vibrational modes, where seven among them were calculated with an imaginary frequency forming the basis of g_g and t_{2u} representation. There are three a_g, four t_{1g}, five t_{2g}, eight g_g and eleven h_g vibrational modes, which at least one component of the degenerate modes becomes totally symmetric under the descent in symmetry to S_6 (except the case of g_g modes, where two components become totally symmetric). Thus in the S_6 point group, there are thirty-nine totally symmetric normal modes, which can mix and contribute to our analysis of the Jahn-Teller distortion. Out of the thirty-nine totally symmetric normal modes in S_6 point group, with our model, we are able to identify the two most important vibrations with respect to the Jahn-Teller distortion. These particular vibrations are illustrated in figure 2, mainly correspond to the first g_g vibration in the I_h configuration of B_{80}^+, which has an imaginary frequency and contribute almost 86 % to the Jahn-Teller distortion.
C_{60} in I_h nuclear configuration has 174 vibrational modes, where only one component of each of the eight h_g modes becomes totally symmetric under the descent in symmetry to D_{5d}, together with the two icosahedral a_g modes. Therefore, there are ten totally symmetric normal modes in D_{5d} point group, which contribute to the distortion of the I_h C_{60}^+. Out of those ten totally symmetric normal modes, the distortion is mainly due to the lowest frequency ones, which is a part of the squashing h_g modes of I_h C_{60}. [22]

Journal of Physics: Conference Series **428** (2013) 012005 doi:10.1088/1742-6596/428/1/012005

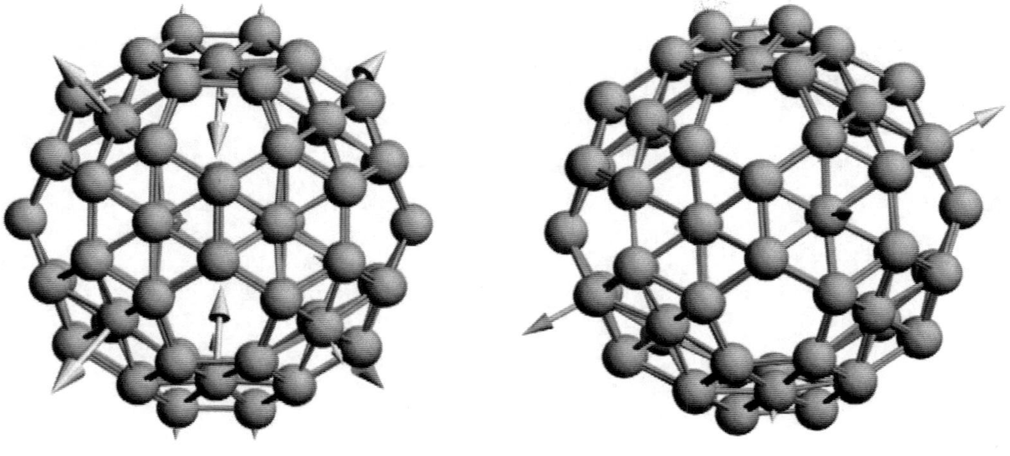

(a) (b)

Figure 2. Illustration of the two a_g normal modes of the S_6 ground state of B_{80}^+, involved in the Jahn-Teller distortion: a_g #02 (165 cm^{-1}) (a) and a_g #03 (190 cm^{-1}) (b).

In order to complete the study of the Jahn-Teller distortion in B_{80}^+ and C_{60}^+, we calculated the Jahn-Teller magnitude $q_{HS,k}$, the Jahn-Teller energy $E_{HS,k}$ and the forces $F_{HS,k}$, according to the equations (7), (5) and (9), respectively. The results of the calculations using the original formulation of the harmonic approximation are tabulated in table 2 and table 3, respectively, for B_{80}^+ and C_{60}^+ Jahn-Teller problems. According to the group theory prediction, the destabilization of the ^2H_u degenerate state of B_{80}^+ and C_{60}^+ involves the distortion force representative of g_g and h_g normal modes ($H_u \otimes H_u = \{t_{1g} \oplus t_{2g} \oplus g_g\} \oplus [a_g \oplus g_g \oplus 2\ h_g]$). This is reproduced in table 2 and table 3, since the t_{1g} and t_{2g} normal modes of the I_h B_{80}^+ participate in a relatively weak contribution (table 2), either to the Jahn-Teller distortion or to the energy stabilization. Thus the calculated forces to those normal modes are almost negligible. However, the Jahn-Teller radius of the distortion of the I_h B_{80}^+ (0.406 Å) is 2.7 times larger than the radius corresponding to the distortion of the geometry of the I_h C_{60}^+ (0.151 Å). Then we have used the distortion vector $\vec{R}_{JT}$ of B_{80}^+ to separate the eighty boron atoms in two different fragments. The first fragment grouped the sixty boron atoms that form the buckyball, while the second fragment consisted to the remaining 20 boron caps. Therefore 18 % (i.e. 0.173 Å) of the distortion is occurring on this first group of boron atoms, while 82 % (i.e. 0.367 Å) of the distortion is due to the twenty boron caps. Thus the relaxation of the geometry of B_{80}^+ takes place mainly at the twenty boron caps and turns out to be a complex combination of Jahn-Teller and pseudo-Jahn-Teller mechanisms. A detailed analysis of these results will be found elsewhere. [23] Moreover, the Jahn-Teller stabilization energy of B_{80}^+ (1289 cm^{-1}) is 2.3 times higher than that obtained for the stabilization of C_{60}^+ (575 cm^{-1}), while the total force corresponding to the B_{80}^+ problem is found two times weaker in comparison to the total force corresponding to the C_{60}^+ problem (table 2 and table 3).

Table 2. Analysis of 1st order Jahn-Teller distortion of the I_h B_{80}^+: The Jahn-Teller magnitude $q_{HS,k}$ [Å], the stabilization energy $E_{HS,k}$ [cm^{-1}] and the forces $F_{HS,k}$ [mdyne].

Mode	$q_{HS,k}$	$E_{HS,k}$	$F_{HS,k}$ [a]
a_g #01	0.102	98	0.073
a_g #02	0.073	53	0.061
a_g #03	0.007	2	0.043
t_{1g} #01	0.002	0	0.006
t_{1g} #02	0.002	0	0.005
t_{1g} #03	0.000	0	0.001
t_{1g} #04	0.001	0	0.003
t_{2g} #01	0.001	1	0.002
t_{2g} #02	0.001	0	0.004
t_{2g} #03	0.000	0	0.004
t_{2g} #04	0.001	1	0.002
t_{2g} #05	0.000	0	0.003
img[b] g_g #01	0.379	489	0.090
g_g #02	0.052	40	0.032
g_g #03	0.003	1	0.004
g_g #04	0.010	1	0.011
g_g #05	0.005	2	0.014
g_g #06	0.002	24	0.068
g_g #07	0.007	9	0.065
g_g #08	0.005	8	0.039
h_g #01	0.025	358	0.017
h_g #02	0.001	63	0.010
h_g #03	0.033	64	0.015
h_g #04	0.017	16	0.032
h_g #05	0.004	17	0.041
h_g #06	0.001	2	0.012
h_g #07	0.005	8	0.035
h_g #08	0.012	6	0.033
h_g #09	0.001	6	0.036
h_g #10	0.007	18	0.071
h_g #11	0.002	1	0.014
Total	0.406	1289	0.208

[a] 1 mdyne represents 10^{-8} N
[b] img stands for imaginary

Journal of Physics: Conference Series **428** (2013) 012005 doi:10.1088/1742-6596/428/1/012005

Table 3. Analysis of the 1st order Jahn-Teller distortion of the I_h C_{60}^+: The Jahn-Teller magnitude $q_{HS,k}$ [Å], the stabilization energy $E_{HS,k}$ [cm^{-1}] and the forces $F_{HS,k}$ [mdyne].

Mode	$q_{HS,k}$	$E_{HS,k}$	$F_{HS,k}$ [a]
a_g #01	0.004	0	0.010
a_g #02	0.006	30	0.103
h_g #01	0.143	246	0.069
h_g #02	0.036	39	0.046
h_g #03	0.001	0	0.003
h_g #04	0.021	50	0.087
h_g #05	0.002	2	0.012
h_g #06	0.010	23	0.122
h_g #07	0.016	100	0.253
h_g #08	0.014	86	0.263
Total	0.151	575	0.416

[a] 1 mdyne represents 10^{-8} N

The calculated forces of the normal modes of HS nuclear configurations give an information about the linear vibronic coupling coefficient in B_{80}^+ and C_{60}^+ problems. Considering the forces obtained for the distortion of the I_h C_{60}^+, the two hardest anti-squashing h_g#07 and h_g #08 modes have the largest vibronic coupling coefficient, as previously determined by other theoretical studies. [22, 24] Although, the squashing h_g #01 mode has a smaller force, its contribution to the Jahn-Teller energy is the most important. For the case of B_{80}^+, the coupling is found stronger among the g_g modes than the h_g modes, although the energy stabilizations by h_g and g_g modes are almost similar.

4. Conclusion

In this paper, our general approach to analyse the multimode Jahn-Teller distortion is applied to study the Jahn-Teller effect of the cationic form of the icosahedral boron and carbon buckyball (C_{60}^+ and B_{80}^+). It is shown how a simple model based on harmonic approximation can be successfully applied in order to determine the Jahn-Teller parameters, i.e. the Jahn-Teller magnitude, the Jahn-Teller energy and the forces at the HS reference point. The origin of our approach states that the Jahn-Teller distortion may be expressed as a linear combination of all totally symmetric normal modes in the LS global minimum energy conformation. This specific formulation based on the Hessian of the LS structure has several advantages. The LS structure is true minima on the APES and the potential energy expression has a simple analytical form in the harmonic approximation. Therefore while the Jahn-Teller magnitude and the forces are obtained in a straightforward way, due to the vector properties of the normal modes, the Jahn-Teller stabilization energy with respect to the HS normal modes can be obtained using the correlation matrix between the normal modes in the HS and LS nuclear configurations.

Hence we presented here the static part of the multimode problem in B_{80}^+ and compared the results to that obtained for C_{60}^+. The ground state geometry of B_{80}^+ distorts slightly from the D_{3d} to the S_6 symmetry, while the ground state geometry of C_{60}^+ has a D_{5d} symmetry. Thus these two inorganic cations present the only two possible solutions of the solving of the standard Jahn-Teller problem corresponding to the quintet H ⊗ (g ⊕ 2h) Jahn-Teller instability. [27]

5. Acknowledgments

This work is supported by the Swiss National Science Foundation and the KULeuven Research Council (GOA, ID and DOB programs). The authors are grateful to Professor Maja Gruden-Pavlovic and Doctor Matija Zlatar from the university of Belgrade (Serbia) for invaluable discussion about the general methodology. Professor Minh Tho Nguyen from the university of Leuven (Belgium) is also acknowledged for his constant advices and interest.

6. References

[1] Yan Q B, Shen X L, Zheng Q R, Zhang L Z and Su G 2008 *Phys. Rev. B* **78**, 201401
[2] Baruah T, Pederson M R and Zope R R 2008 *Phys. Rev. B* **78**, 045408
[3] Prasad D L V K and Jemmis E D 2008 *Phys. Rev. Lett.* **100**, 165504
[4] Pochet P, Genovese L, De S, Goedecker S, Caliste D, Ghasemi S A, Bao K and Deutch T 2011 *Phys. Rev. B* **83**, 081403
[5] Zope R R and Baruah T 2011 *Phys. Rev. B* **80**, 033410
[6] Li J L and Yang G W 2009 *J. Phys. Chem. C* **113**, 18292
[7] Szwacki N G, Sadrzadeh A and Yakobson B I 2007 *Phys. Rev. Lett.* **98**, 165
[8] (a) He H, Pandey R, Boustani I and Karna S P 2010 *J. Phys. Chem. C* **114**, 4149; (b) Bean D E, Muya J T, Fowler P W, Nguyen M T and Ceulemans A 2011 *Phys. Chem. Chem. Phys.* **13**, 20855
[9] Olguim M, Baruah T and Zope R R 2011 *Chem. Phys. Lett.* **514**, 66
[10] Szacki N G and Tymczak C J 2010 *Chem. Phys. Lett.* **494**, 80
[11] Gopakumar G, Nguyen M T and Ceulemans A 2008 *Chem. Phys. Lett.* **450**, 175
[12] Gunashinge R N, Kah C B, Quarles K D and Wang X-Q 2011 *Appl. Phys. Lett.* **98**, 261906
[13] Muya J T, Sato T, Nguyen M T and Ceulemans A 2012 *Chem. Phys. Lett.* **543**, 111
[14] Zlatar M, Gruden-Pavlovic M, Schläpfer C-W and Daul C 2010 *J. Mol. Struct. THEOCHEM* **954**, 80
[15] Gaussian09, Revision B.01, Frisch M J et al., Gaussian, Inc., Wallindford CT, 2009
[16] Vosko S, Wilk L and Nussair M 1980 *Can. J. Phys.* **58**, 1200
[17] Perdew J P, Burke K and Ernzerhof M 1996 *Phys. Rev. Lett.* **77**, 3865
[18] Stephens P J, Devlin F J, Chabalowski C F and Frisch M J 1994 *J. Phys. Chem.* **98**, 11623
[19] Bruyndonckx R, Daul C, Manoharan P T and Deiss E 1997 *Inorg. Chem.* **36**, 4251
[20] Gruden-Pavlovic M, Garcia-Fernandez P, Andelkovic L, Daul C and Zlatar M 2011 *J. Chem. Phys. A* **115**, 10801
[21] Ramanantoanina H, Gruden-Pavlovic M, Zlatar M and Daul C 2012 *Int. J. Quantum Chem.* DO1: 10.1002/qua.24080
[22] Ramanantoanina H, Zlatar M, Garcia-Fernandez P, Daul C and Gruden-Pavlovic M 2013, *Phys. Chem. Chem. Phys.* **15**, 1252
[23] Muya J T, Ramanantoanina H, Daul C, Nguyen M T, Gopakumar G and Ceulemans A 2012, *Phys. Chem. Chem. Phys.* DOI: 10.1039/c2cp43171h
[24] Manini N, Dal Corso A, Fabrizio M and Tosatti E 2001 *Phyl. Mag. B* **81**, 793
[25] Luders M, Manini N, Gattari P and Tosatti E 2003 *Eur. Phys. J. B* **35**, 57
[26] Ceulemans A and Vanquickenborne L G 1989 *Struct. Bond.* **71**, 125
[27] Ceulemans A and Fowler P W 1990 *J. Chem. Phys.* **93**, 1221

A possibility for generation of two species of charge carriers along main-chain and side-chains for a π-conjugated polymer

Yuki Kudo, Kohsuke Kawabata, Hiromasa Goto*

Division of Materials Science, Faculty of Pure and Applied Sciences, University of Tsukuba, Tsukuba, Ibaraki 305-8573, Japan

Correspondence to H. Goto, e-mail: gotoh@ims.tsukuba.ac.jp

Abstract. Iodide doping produces charge carriers in π-conjugated polymers. Solitons can be generated in the case of polyacetylene, and polarons in the case of aromatic-type conjugated polymers. We synthesized a conjugated main-chain/side-chain polymer, which consists of polyene in the main-chain and aromatic-type conjugated units in the side-chains. Based on the SSH (Su, Schrieffer, Heeger) theoretical model of solitons in one-dimensional conjugated polymers, we experimentally carried out chemical doping to the main-chain/side-chains conjugated polymer. Generation of the charge carriers was examined by electron spin resonance spectroscopy. This study may lead to realization of a dual doping system of solitons and polarons in π-conjugation expanded to two-dimensional directions in polymers.

1. Introduction

π-Conjugated polymers consist of alternation of single and double bonds. The main-chains have delocalised π-electrons. Polyacetylene is a typical conjugated polymer. It shows metallic refraction because of its highly developed conjugated system, and displays quite high electrical conductivity comparable to metals such as copper, after vapor phase doping by iodine (chemical doping), accompanied by insulator-metal transition [1]. In this case, the charge species of polyacetylene can be characterized as solitons [2].

Aromatic type conjugated polymers also show conductivity upon chemical doping. In the case of aromatic type conjugated polymers, the charge carriers are referred to as polarons and bipolarons [3–4].

In the present study, we synthesize a π-conjugated polymer having a conjugated system in both the main-chain and the side-chains. The side-chains are directed perpendicular to the main-chain. After the doping by iodine, charge carriers are generated in both side-chains and main-chain, producing solitons in the main-chain and polarons or bipolarons in the side-chains. This can be referred to as a double doping process in conjugated polymers.

We have prepared a polyacetylene derivative with aromatic-type side-chains and doping effect for the polymer. Then, we experimentally consider charge carriers along the π-conjugated system.

The SSH model in a one-dimensional system was proposed for solitons of polyacetylene [2]. A ground state of polyacetylene described by the model as follows,

$$H_{\mathrm{SSH}} = H_{\pi} + H_{\pi-\mathrm{ph}} + H_{\mathrm{ph}} \tag{1}$$

$$H_\pi = -t_0 \sum_{n,s}(c^\dagger_{n+1,s}\, c_{n+1,s} + c^\dagger_{n,s} c_{n+1,s}) \tag{2}$$

$$H_{\pi\text{-ph}} = \alpha \sum_{n,s} (u_{n+1} - u_n)(\, c^\dagger_{n+1,s} c_{n,s} + c^\dagger_{n,s} c_{n+1,s}) \tag{3}$$

$$H_{\text{ph}} = \sum p_n^2/2M + K/2 \sum (u_{n+1} - u_n)^2. \tag{4}$$

Here, the dimerization coordinate and the lattice constant are defined by u_n and α, respectively. To clamp u_n at twofold ground state $E\,(u_0) = E\,(-u_0)$, $(-1)^n u$, the system supports nonlinear excitations.

$$\varphi_\text{n} = (-1)^n u_\text{n}. \tag{5}$$

So that,

$$\varphi_n = u_0\,, E(u_0), \tag{6}$$

$$\varphi_n = -u_0\,, E(-u_0). \tag{7}$$

If φ_n changes very slowly from $-u_0$ to u_0, there will be large range surrounding $n = 0$ where condensation energy is reduced, and it rises again. Width of the solitons is expressed by ξ.

$$\varphi_\text{n} \cong u_0 \tanh[(n - n_0)\alpha/\xi]. \tag{8}$$

And if there is a solution of E_0, because the spectrum of H_{el} is symmetric about E_0, then H_{el} satisfies

$$\sum \langle \text{n}|H_{el}|\text{m}\rangle \psi_0 = 0. \tag{9}$$

This model has been proposed by Su, Schrieffer, and Heeger for polyacetylene [2]. In the case of aromatic conjugated polymers [3–5], chemical doping allows generation of polarons along the main-chains [6]. Figure 1 shows an example of polarons along poly(*p*-phenylene).

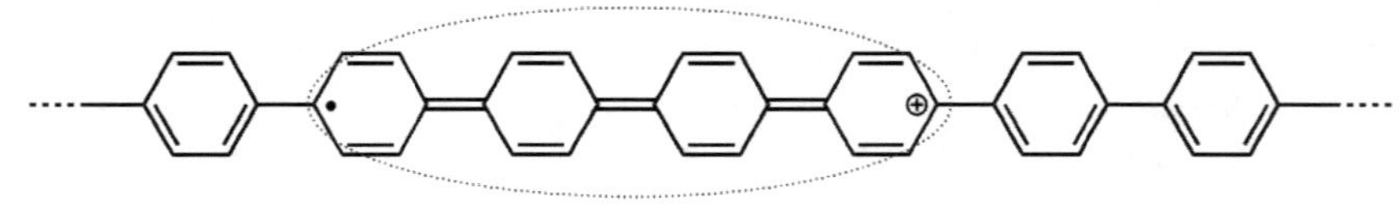

Figure 1. Radical cations (polarons) of poly(*p*-phenylene)

2. Results and discussion

2.1 Synthesis polymer

Firstly, the disubstituted acetylene monomer was synthesized with the Sonogashira coupling reaction by using Pd(0) and CuI catalysts in the base condition. Next, the monomer was polymerized with $TaCl_5$ catalyst in toluene solution via metathesis mechanism. The resultant polymer was washed with a large amount of methanol, the product was collected by reprecipitation followed by drying in vacuum to yield a yellow polymer with fluorescence.

2.2 Chemical doping for polyacetylene derivatives

Polyacetylene shows high electrical conductivity upon vapor phase doping by iodine as an electron accepter for generation of charge carriers.

The chemical structure of di-substituted polyacetylene is shown in Figure 2.

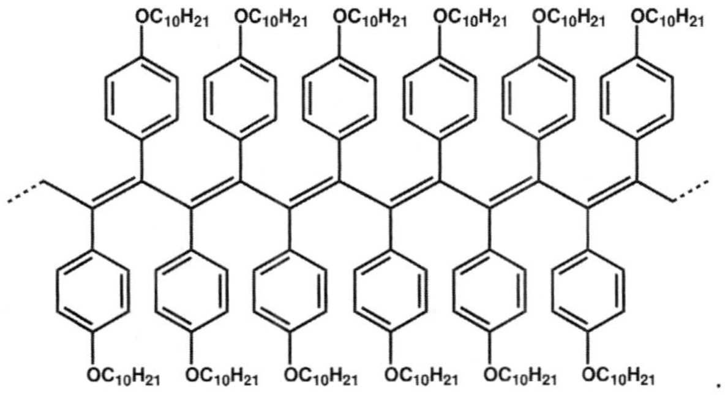

Figure 2. Neutral state of polymers

Chemical vapor phase doping by iodine produces charge carriers along the main-chain as follows,

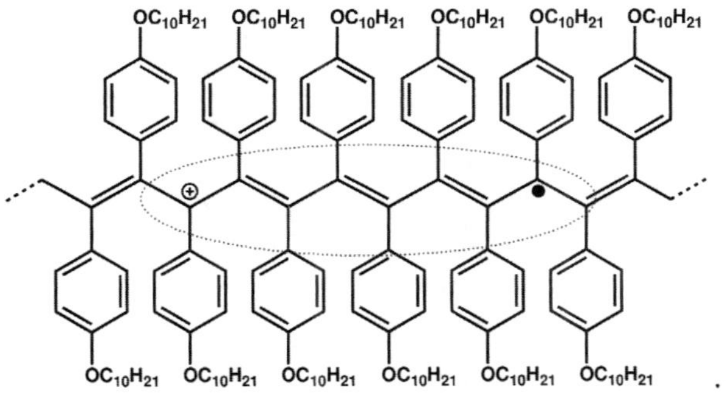

Figure 3. Radical cations in the main-chain

In the case of the polymers having conjugated substituent located perpendicular to main-chain, the side-chains can also be doped with an electron accepter, resulting in formation of polarons in the aromatic group in the side-chains.

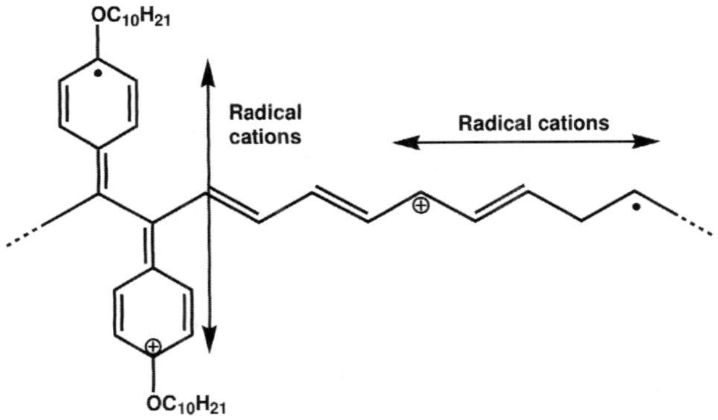

Figure 4. Radical cations in the polymer

Furthermore, main-chain and side-chains doping can be performed. Figure 4 shows a plausible structure of generation of charge carriers along both the main-chain and the side-chains upon iodine doping.

Electron spin resonance (ESR) of the di-substituted polymer with in-situ iodine doping was carried out. In the undoped state the polymer shows no ESR signal. The Lorenz line shape of the ESR signal appeared with iodine doping. Increase of spin concentration and line width were observed. The *g*-value increased slightly to 2.00393 (60 min doping) from g = 2.00378 (10 min doping) with heavy doping. Line width of the peak-to-peak (ΔH_{pp}) increases to 0.29 mT (60 min) from 0.19 mT (10 min). A small shoulder appeared in the spectral form in the heavy doped state. Heavy doping may allow formation of solitons and polarons in both the main-chain and the side-chains, but evidence of solitons and polarons cannot be separated in the spectroscopy measurement. Furthermore, the side-chains may play a role of stabilizer for charge carriers.

Electrical conductivity upon iodine doping of the polymer increased to ~4.9 $\times 10^{-5}$ S/cm (obtained with four-point probe method at room temperature) from insulator range, indicating occurrence of electrical conduction by generation of the charge carriers.

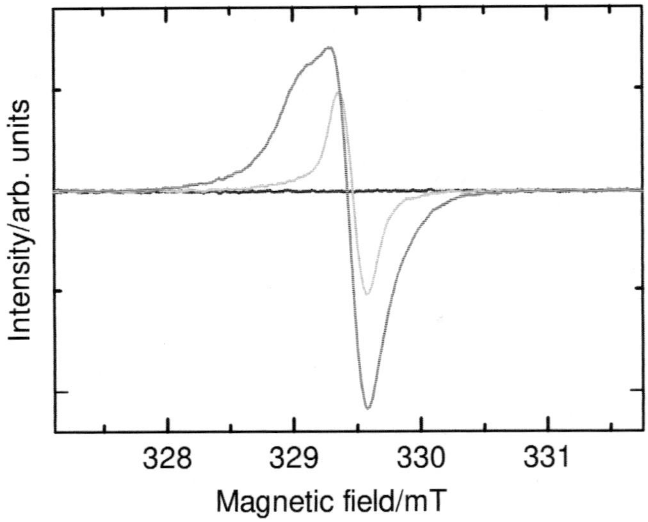

Figure 5. Electron spin resonance (ESR) spectroscopy results of in-situ iodine doping of the polymer. Blue line: 0 min of doping time; green line: 10 min; red line: 60 min.

Increase of charge carrier indicates increase of radicals of polymer. In other words, the ESR results suggest generation of charge carrier in entire structure of the di-substituted polyacetylene. Increase of conductivity and existence of paramagnetic species for the polymer implies generation of solitons (radical cations along the polyene) and polarons (radical cations along the aromatic units) by iodine doping.

Temperature dependence on inverse of resistance ($1/R$) for the heavy-doped sample with iodine vapor was examined (Figure 6). $1/R$ ($\propto\sigma$) values of the doped polymer decreased with decrease of temperature. This is a typical behavior of semi-conducting polymers. The $1/R$ vs. $1/T$ plots for the sample clearly shows an inflection point at 258 K.

2.3 Molecular form

Computer modeling evaluated the ideal form of the polymer, indicating steric repulsion between substituents. The distortion introduces deformation of linear shape in the main-chain, resulting in

Journal of Physics: Conference Series **428** (2013) 012006 doi:10.1088/1742-6596/428/1/012006

limitation of expansion of the π-conjugated system. Also, the perturbation might restrict free movement of charge carriers along the main-chain.

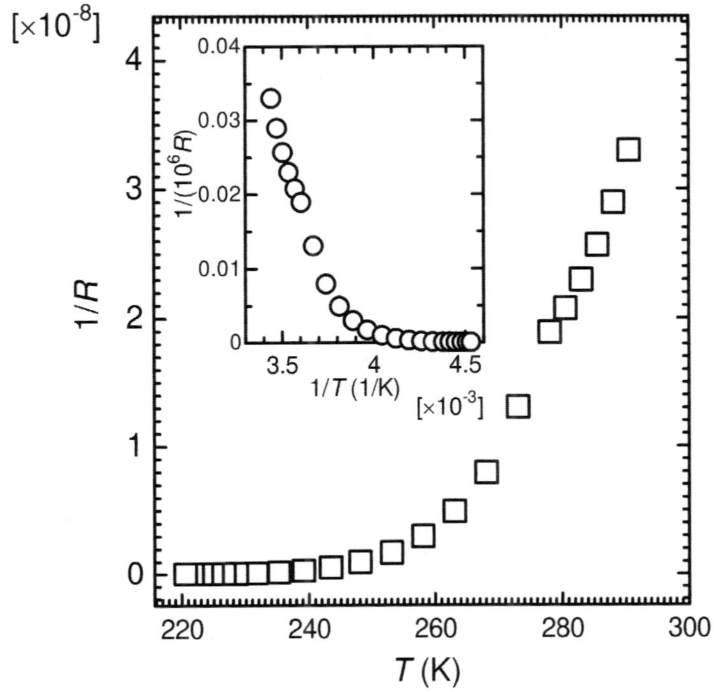

Figure 6. Inverse of resistance ($1/R$) as a function of T for the heavy doped sample with iodine. Inset shows $1/R$ vs. $1/T$ plots.

However, the computer simulation demonstrated that the entire molecular shape is linear, and the main-chain (inside of substituents described by the balls in Figure 7) forms helical structure. The charge carrier on the main-chain can move along the helical architecture.

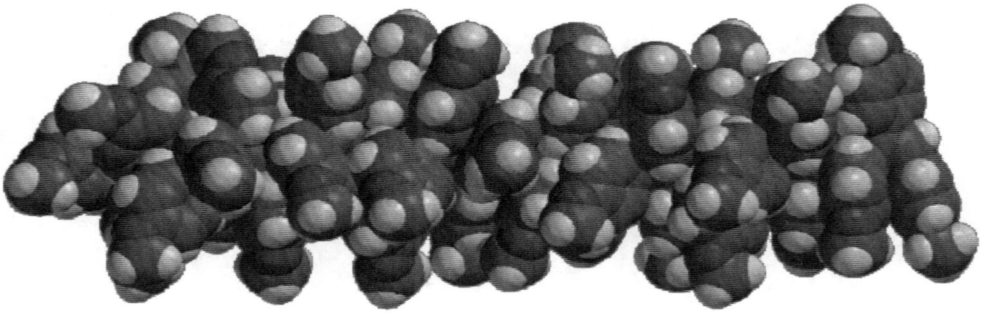

Figure 7. Computer simulation of the polymer described by balls model. Geometries of this polymer have been optimized at the MMFF (Merck Molecular Force Field method, developed at Merck).

Conclusion

A possibility of generation of solitons along the di-substituted polyacetylene derivative is discussed. Restriction of mobility of the charge carrier along the main-chain due to deformation of linear shape of the main-chain may occur because of steric hindrance between side-chains. ESR results strongly

Journal of Physics: Conference Series **428** (2013) 012006 doi:10.1088/1742-6596/428/1/012006

suggest generation of charge carriers. This can be related with double doping models for polyphenylenevinylenes [7].

Techniques

Samples were synthesized in an inert gas with Schlenke tubes. ESR measurements of the polymer were carried out with a JEOL JES TE-200 spectrometer with 100 kHz modulations. Electrical conductivities and resistance for the sample were obtained with Mitsubishi MCP-T610 and Hioki SM8213. $1/R$ vs. T plots were obtained with two-probe method.

References

[1] Ito T and Shirakawa H 1975 *Journal of Polymer Science*: *Polymer Chemistry* **13** 1943
[2] Heeger A J, Kivelson S, Schriffer J R and Su W-P 1988 *Reviews of Modern Physics* **60** 781
[3] Kaneto K, Hayashi S, Ura S and Yoshino K 1985 *Journal of the Physical Society of Japan* **54** 1146
[4] Hayashi S, Nakajima S, Kaneto Kand Yoshino K 1986 *Journal of the Physical Society of Japnan* **55** 3995
[5] Chien J C W, 1984 *Polyacetylene: Chemistry, Physics, and Materials Science* Academic press London 470
[6] Campbel D K, Bishop A R, 1981 *Physical Review B* **24** 4859
[7] Kawashima H, Kawabata K and Goto H 2012 *Journal of Polymer Science Part A, Polymer Chemistry* **50** 1530

Symmetry and degeneracy manifolds in Jahn-Teller molecules

C A Mead[1,2]

[1] 92 Bartram Road, Savannah, GA 31411, USA

[2] Professor Emeritus of Chemistry and Chemical Physics, University of Minnesota

E-mail: aldykar@att.net

Abstract: We consider problems in dealing with molecular systems of n identical nuclei. One problem is that of finding suitable internal coordinates. For $n \leq 4$, these can be simply the internuclear distances. For $n > 4$, it is shown that, with perhaps one exception, there is no internal coordinate system that treats all nuclei equivalently. We also consder the properties of conical intersections between two Born-Oppenheimer electronic energy surfaces, in particular the problem of identifying the two coordinates that remove the degeneracy to first order in the near neighborhoods of symmetry manifolds.

1. Introduction

In this article, we consider the description of Born-Oppenheimer electronic energies and wave functions in molecular systems containing n identical nuclei. Section 2 considers the problem of finding an internal coordinate system (dimension $3n - 6$, apart from overall translation and rotation) that, as much as possible, treats the nuclei equivalently. It will be shown that, apart from one possible exception, no completely satisfactory such solution exists for $n > 4$. Section 3 considers the behavior of conical intersections between pairs of energy surfaces, concentrating on the case of symmetry-determined (C_{nv}) degeneracy, and for this case, out of the $(3n - 7)$ coordinates that break the degeneracy, identifying the two that make the intersection truly conical, i.e., that break the degeneracy in first order. There is also a brief discussion of other symmetries. There is a brief discussion in Section 4. We wish to emphasize that we are *not* claiming to give an exhaustive review of the literature, only mentioning a few relevant references, while acknowledging that there are also others. We are also presenting little that is actually new, much of the article being a sort of review, hopefully with new insights into known problems, and emphasizing aspects that are not considered in most treatments.

2. Internal coordinate system

Apart from center-of-mass location and orientation in space, there are $3n - 6$ “internal” coordinates for a system of n identical nuclei, and naturally it is desirable to choose these coordinates in such a way that all n nuclei are treated equivalently. This is easily solved if n is 3 or 4, since for these numbers, and these only, the number of internuclear distances equals the number of desired coordinates [1]:

$$n(n-1)/2 = 3n - 6 \text{ for } n = 3, 4 \quad (1)$$

Journal of Physics: Conference Series **428** (2013) 012007 doi:10.1088/1742-6596/428/1/012007

Moreover, since permutations of identical nuclei also permute the internuclear distances, the coordinates can be decomposed into subsets belonging to irreducible representations of the permutation goups S_3 and S_4, as follows:

n = 3:

# # #	(3)	dim = 1
# # #	(2,1)	dim = 2

n = 4:

# # # #	(4)	dim = 1
# # # #	(3,1)	dim = 3
# # # #	(2,2)	dim = 2

In the above table, the Young diagrams for the representations are shown symbolically with # standing for a box, the numbers in parentheses are the numbers of boxes in the rows of the diagrams, and "dim" of course stands for the dimension of the representation.

For more general *n*, we can ask whether there might be a set of symmetric functions of 2 or more nuclei that might serve as coordinates, e.g. *f(a,b)*, a symmetric function of the positions of nuclei *a* and *b* (such as the internuclear distance), or *f(a,b,c)*, a symmetric function of nuclei *a,* b, and *c* (such as the area of the triangle formed by the three). Perhaps other types of coordinates may be imagined, but the author has been unable to think of any, so we will restrict ourselves to potential coordinates of this type. For $n > 4$, as we know, the number of such functions is not equal to the number of desired coordinates, $d = 3n - 6$, but we may hope that, when the functions are broken up into combinations belonging to irreducible representations of S_n, some one of these, or some combination, might do the trick.

To consider this question, we first observe that there must be some combination of the internal coordinates that is totally symmetric in the nuclei and that represents the overall size of the system, i.e. that belongs to the irreducible representation (n). So our task is to find other irreducible representations whose dimensions add up to $d - 1 = 3n - 7$. This is also the number of symmetry-breaking coordinates for C_{nv} and other symmetries in which all nuclei are equivalent.

The irreducible representations for symmetric functions of ν nuclei are:

$$(n), (n\text{-}1,1), (n\text{-}2,2), ..., (n\text{-}\nu,\nu).$$

Table 1 shows relevant numbers for symmetric functions of 2 and 3 nuclei, and for *n* up to eight.

n	*f(a,b)*	*f(a,b,c)*	*d-1*	*(n-1,1)*	*(n-2,2)*	*(n-3,3)*
gen	n(n-1)/2	n(n-1)x (n-2)/6	3n-7	n-1	n(n-3)/2	n(n-1)x (n-5)/6
3	3	1	2	2	-	-
4	6	4	5	3	2	-
5	10	10	8	4	5	-
6	15	20	11	5	9	5
7	21	35	14	6	14	14
8	28	56	17	7	20	28

Table 1: Data for symmetric functions decomposed into irreducible representations of the permutation groups. Columns 2 and 3 show the number of symmetric functions of the two types, column 4 gives the desired dimension $d - 1$, the last three give the dimensions of the three included irreducible representations. Row 2 gives general formulas, the other rows the results for each *n*.

Perusal of Table 1 shows that, in general, the dimensions of the irreducible representations for $n > 4$ do not add up. A possible exception might be for $n = 5$ using two different, but independent, symmetric functions, and choosing (4,1), dimension 4, for both. But any such function must be a function only of the *relative* positions of the two particles, and there does not seem to be any such function that is independent of the internuclear distance.

Much more interesting is the case of $n = 7$, for which both (5,2) and (4,3) have dimension 14, exactly what is desired! So for this case it would seem to be feasible to define such a coordinate system, made up of symmetric functions of two or three nuclei, and with all treated equivalently. This would seem to be worthy of further investigation.

Beyond 7, the dimensions of the irreducible representations rapidly become too large, and the same holds if one considers symmetric functions of more than three nuclei. We conclude that $n = 7$ is the only reasonably promising possibility.

Journal of Physics: Conference Series **428** (2013) 012007 doi:10.1088/1742-6596/428/1/012007

3. Intersections of Born-Oppenheimer electronic energy surfaces

3a. General considerations

The properties of intersections between Born-Oppenheimer electronic energy surfaces, especially conical intersections, have been the subject of many studies, going back to the classic paper by Jahn and Teller [2] and including those of Longuet-Higgins et al ("LHOPS") [3] and Herzberg and Longuet-Higgins [4]. Mead and Truhlar [5] have further studied the matter, which turns out to be closely related to the so-called geometric, or Berry, phase [6]. The author has published a review article [7], and more recently there has been one by Wittig [8]. An elegant purely group-theoretical analysis of some aspects of the problem has been given by Ceulemans and Linjen [9].

Throughout this paper, we assume that time-reversal invariance holds, i.e. that there is no external magnetic field, and throughout most of this section we also assume that the total electronic spin is integer, which is equivalent to the possibility of choosing electronic wave functions to be real [10]. A final subsection will present some brief comments on the half-odd integer spin case.

For a system of n nuclei, the number of internal degrees of freedom (apart from overall translation and rotation) is $d = 3n - 6$. As is well known [11], degeneracy of two electronic energy levels requires that two conditions must be satisfied, from which it follows that a degeneracy manifold (the submanifold of the d-dimensional internal configuration space where degeneracy holds) will be of dimension $d - 2 = 3n - 8$, and there will be two coordinates (degeneracy-removing coordinates) that describe the distance of a point in the d-dimensional space from the degeneracy manifold. For configurations with symmetry, there may be symmetry-determined degeneracy, but the symmetry manifold is in nearly all cases a lower-dimensional submanifold of the degeneracy manifold. For example, for C_{nv} and regular polyhedron symmetry, the only symmetry-preserving coordinate is the totally symmetrical breathing one, so the dimension of the symmetry manifold is just one.

An intersection is *conical* if the difference in energy $\Delta E = E_1 - E_2$ between the two states depends linearly on the degeneracy-removing coordinates in the infinitesimal neighborhood of the degeneracy; otherwise it is a *glancing* intersection. For a glancing intersection, not only must the two degeneracy conditions be satisfied, but also all d components of the gradient of ΔE must vanish, a total of $d + 2$ conditions. Since the number of conditions for a glancing intersection exceeds the number of available coordinates, this means that intersections must necessarily be conical, except possibly within a symmetry submanifold, where perhaps the gradient may be required by symmetry to vanish. From this point of view, the classic result of Jahn and Teller [2] may be regarded as simply removing this last hope, except for linear configurations.

3b. The Herzberg-Longuet-Higgins sign change

Consider two electronic states, φ_1 and φ_2 which are degenerate at a given point (or manifold) in nuclear configuration space, and define coordinates along the gradients at the degeneracy point of $(H_{11} - H_{22})/2$ and H_{12}:

$$x = r\cos\theta\,;\; y = r\sin\theta\,. \tag{2}$$

In the near neighborhood of the intersection, the Hamiltonian is proportional to

$$\hat{H} \propto r\begin{pmatrix} \cos\theta & \sin\theta \\ \sin\theta & -\cos\theta \end{pmatrix} \tag{3}$$

with eigenvalues $\pm r$ and eigenfunctions

$$\begin{pmatrix} \cos\theta/2 \\ \sin\theta/2 \end{pmatrix}, \begin{pmatrix} -\sin\theta/2 \\ \cos\theta/2 \end{pmatrix}. \tag{4}$$

Thus, it is seen that the eigenfunctions undergo a sign change (the Herzberg-Lounguet-Higgins, HLH sign change [4]) when the coordinates traverse a full circuit around the intersection, meaning that the real-valued electronic wave functions are not single-valued. The consequences of the HLH sign change have been the subject of many studies [5,6,7].

However, in the case of degeneracies required by symmetry, which are the ones most often studied, one has to be careful about what symmetry-breaking coordinates to use, since not all symmetry-breaking coordinates break the degeneracy in first order, and hence not all generate an HLH sign change. Moreover, the angle θ appearing in eq. (1) is not always the same as the natural angle in the coordinate system, and though there may be more than one pair of coordinates with the proper symmetry to remove the degeneracy, only one such pair actually does so.

A simple example is shown in Figure 1.

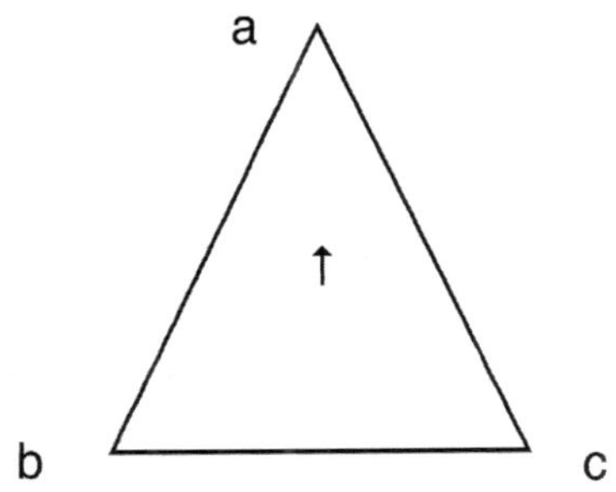

Figure 1: Triangular pyramid, viewed from above, with the apex atom displaced toward corner *a*, breaking the C_{3v} symmetry.

In fig. 1, we define the *x* direction as pointing from the baseline *bc* to corner *a*, the *y* direction as from corner *c* to corner *b*, and *z* perpendicular to the base. The degenerate states are taken to be atomic *p* states centered at the center of the base triangle, $|x>$ and $|y>$. With the apex moved toward a as in the figure, there is stll a plane of symmetry xz, so the eigenfunctions will be just $|x>$ and $|y>$, and the effective hamiltonian will be proportional to

$$\hat{h}_a = |x\rangle\langle x| - |y\rangle\langle y| . \tag{5}$$

Similarly, we can define kets rotated through an arbitrary angle φ:

$$|x_\varphi\rangle = |x\rangle\cos\varphi + |y\rangle\sin\varphi \,;\, |y_\varphi\rangle = -|x\rangle\sin\varphi + |y\rangle\cos\varphi \tag{6}$$

and corresponding operators $\hat{h}_\varphi$, also $\hat{h}_b$, $\hat{h}_c$ for kets rotated toward corners b, c, i.e. through angles $\pm 2\pi/3$.

Now, the degeneracy-removing coordinates for this case must belong to irreducible representation (irrep) E of C_{3v} , but there are two pairs of coordinates that have this symmetry: the pseudorotation of the atoms at the base, and the x, y coordinates describing the displacement of the apex. Assuming for the present that the base remains equilateral, consider the following possibilities for the apex displacement: First, if the base is very large and the p functions concentrated in a small place, then the wave functions will barely notice the base atoms, and will mainly just follow the apex displacement, so if the apex is displaced at an angle θ from the x direction, the effective hamiltonian will be proportional to

$$\hat{h}_\theta = \begin{pmatrix} \cos 2\theta & \sin 2\theta \\ \sin 2\theta & -\cos 2\theta \end{pmatrix} , \tag{7}$$

so the wave function is rotated through half of 2θ, or just θ. There is no HLH sign change for the apex displacement, the intersection is glancing with respect to apex displacement, and the degeneracy is removed in first order just through the base pseudorotation.

On the other hand, an equally plausible Hamiltonian for the apex rotation, showing a preference for the nearest base corner to the displaced apex, might be:

$$\begin{aligned} \hat{h} &= \hat{h}_a \cos\theta + \hat{h}_b \cos\left(\theta - {}^{2\pi}\!/_{3}\right) + \hat{h}_c \cos\left(\theta + {}^{2\pi}\!/_{3}\right) \\ &= \frac{3}{2}\begin{pmatrix} \cos\theta & -\sin\theta \\ -\sin\theta & -\cos\theta \end{pmatrix}. \end{aligned} \tag{8}$$

In this case, the wave function is rotated through *minus* one half θ, and there is an HLH sign change. Since base pseudorotation also presumably removes the degeneracy, the true degeneracy-removing coordinate pair is a reinforcing combination of pseudorotation and apex rotation, while they cancel each other in the other combination.

The moral of this is that, when there are two or more coordinate pairs that have the correct symmetry for degeneracy removal, finding the right combination will nearly always require calculation; there will be no purely group-theoretical way to find it.

We now take up the symmetry- and degeneracy-removing combinations for the near neighborhoods of C_{nv} configurations.

3c. Systems of n identical nuclei near C_{nv} configurations

In the this subsection, we consider systems of *n* nuclei distorted slightly from C_{nv} symmetry. Higher symmetries, such as O_h, have subgroups C_{kv}, so these are really the only groups that need to be considered. The next subsection will contain a brief discussion of other possible symmetries. We emphasize that there is no need to assume that any C_{nv} configuration is rigid or stable, but only that such configurations exist in the configuration space, which they obviously do. The dimension of the C_{nv} manifold is just unity. Of the *(3n - 7)* symmetry-breaking coordinates, we seek to identify the two that break the degeneracy in first order. For this purpose, we need the character tables for the C_{nv} groups, which are given in Tables 2 and 3:

$C_{2v+1,v}$	E	$2C_{2v+1}$	$2C^2_{2v+1}$	...	$2C^v_{2v+1}$	$(2v+1)\sigma$
A_1	1	1	1	...	1	1
A_2	1	1	1	...	1	-1
E_1	2	2 cos τ	2 cos 2τ	...	2 cos vτ	0
E_2	2	2 cos 2τ	2 cos 4τ	...	2 cos 2vτ	0
...	...	...	...	...	...	...
E_j	2	2 cos jτ	2 cos 2jτ	...	2 cosvjτ	0
...	...	...	...	...	...	...
E_v	2	2 cos vτ	2 cos 2vτ	...	2 cos v^2τ	0

doi:10.1088/1742-6596/428/1/012007

Table 2: Character table for $C_{n\nu}$ with n odd, $2\nu + 1$. In the table, $\tau = 2\pi/(2\nu + 1)$.

$C_{2\nu,\nu}$	E	$2C_{2\nu}$	...	$2C_{2\nu}^{\nu-1}$	C_2	$\nu\sigma_v$	$\nu\sigma_d$
A_1	1	1	...	1	1	1	1
A_2	1	1	...	1	1	-1	-1
B_1	1	-1	...	$(-1)^{\nu-1}$	$(-1)^{\nu}$	1	-1
B_2	1	-1	...	$(-1)^{\nu-1}$	$(-1)^{\nu}$	-1	1
E_1	2	$2\cos\rho$	...	$2\cos(\nu-1)\rho$	-2	0	0
...	...	...	...	...	...	...	...
$E_{\nu-1}$	2	$2\cos(\nu-1)\rho$	...	$2\cos(\nu-1)^2\rho$	$2(-1)^{\nu-1}$	0	0

Table 3: Character Table for $C_{n\nu}$ with n even, 2ν. In the table, $\rho = \pi/\nu$.

For most cases, it suffices to restrict ourselves to prime values of $(2\nu + 1)$, since for two primes p and q, $C_{pq\nu}$ can be distorted into $C_{p\nu}$ or $C_{q\nu}$ and degeneracy will remain. An exception occurs, e.g. for E_{nq} when distorted to $C_{p\nu}$. Here the character for the element C_p becomes

$$2\cos\frac{2\pi nqp}{qp} = 2$$

and the degeneracy splits into A_1 and A_2.

For degenerate electronic states with symmetry E_j in $C_{n\nu}$, the nuclear displacement coordinates that can have nonvanishing matrix elements are those contained in $E_j \times E_j$, for which the characters are:

$$\chi(E) = 4\,;\, \chi\left(C_n^k\right) = 2 + 2\cos\frac{2\pi(2kj)}{n}\,;\, \chi(\sigma) = 0\,. \tag{9}$$

This decomposes into $A_1 + A_2 + E_{2j}$. A_1 is simply a scalar, shifting both energies equally; A_2 is the antisymmetric "cross product" $x_a y_b - x_b y_a$, which vanishes for components of the same vector. The potentially degeneracy-removing coordinates in first order are therefore those of E_{2j}. But now we have to determine exactly what "E_{2j}" means in this context.

For $n = 2v + 1$, ν even: For j = 1, 2, 3, ..., ν/2, 2j of course is just 2, 4, 6, ..., ν. For j = 1 + ν/2, 2j is ν + 2 = (2ν + 1) - (ν - 1), so the effective "2j" is (ν - 1). For successive j after this, we get ν - 3, ν - 5, etc.

For $n = 2v + 1$, ν odd: Now we get 2, 4, 6, ... (ν-1), then (ν+1) = (2ν + 1) - ν, so the next ones are ν, ν - 2, etc.

For $n = 2v$, ν even, we get 2j = 2, 4, 6, ..., (ν-2). Next is "E_v", which is actually $B_1 + B_2$. Then the same ones again: ν - 2, ν - 4, etc.

For $n = 2v$, ν odd, we obtain 2, 4, 6, ..., (ν-1), then ν + 1 = 2ν - (ν-1), so we again get the same ones over again, ν - 1, ν - 3, etc.

Now, for each degeneracy, we know the symmetry of the potentially first-order degeneracy-removing coordinates. We must now determine which coordinates have this symmetry, and if there are more of these than are needed, which ones actually do lift the degeneracy in first order.

The coordinates for small deviations from C_{nv} can be considered to be of three types: Out-of-plane (OP), and in-plane stretch (ST), directly away from or toward the center, and rotation (RO), in plane perpendicular to ST. We first consider $n = 2v + 1$.

For both OP and ST, the characters are:

$$\chi(E) = 2v + 1\,;\ \chi(C) = 0\,;\ \chi(\sigma) = 1$$
$$\rightarrow A_1 + E_1 + E_2 + \ldots + E_v$$

For RO, it's the same except that $\chi(\sigma) = -1$, so its representations are

$$A_2 + E_1 + E_2 + \ldots + E_v\ ,$$

and for all the coordinates we have

$$2A_1 + A_2 + 3\left(E_1 + E_2 + \ldots + E_v\right).$$

Of these, the OP A_1 is translation, the OP E_1 is rotation, and in plane the A_2 is rotation and one E_1 is translation, so the remaining true internal coordinates belong to the representations

$$A_1 + E_1 + 3\left(E_2 + E_3 + \ldots + E_v\right).$$

For $n = 2v$, a similar analysis gives for the internal coordinates

$$A_1 + B_1 + 2B_2 + E_1 + 3\left(E_2 + E_3 + \ldots + E_{v-1}\right).$$

The A_1 coordinates are symmetry-preserving "breathing" coordinates. Potentially degeneracy-removing are all the E coordinates for odd n, the E_{2k} for even n, and also B_1 and B_2 when n is a multiple of 4. For odd $n = 2\nu + 1$, the E_1 coordinates are uniquely identified as degeneracy-removing for electronic states degenerate with E_ν symmetry, but for all other E symmetries there are three apparent contenders.

The situation is helped somewhat in that we can eliminate the OP E coordinates. These are "folding" coordinates, where the n-gon is being folded along a diagonal line. A positive fold as viewed from above is the same as a negative one viewed from below, so the lower-energy state is the same for both directions and the intersection is a glancing one with respect to these coordinates.

This still leaves two E_j coordinate pairs, $j > 1$, for each symmetry, and there is no general way to decide which combination is the degeneracy-removing one. For any particular pair of degenerate wave functions, one combination of the two pairs will cancel each other out, the other will remove the degeneracy. Unfortunately, this must be calculated.

We have discussed the determination of degeneracy-removing coordinates in first order , the coordinates that make the intersection truly conical. Relatively little can be said about the global nature of the degeneracy manifolds, but still something can be ascertained. There are $(n-1)!/2$ different C_{nv} configurations, differing by permutations of the nuclei around the ring. Each has its own degeneracy manifold, which we have determined in principle in the near neighborhood of the C_{nv} configuration. Further away, by continuity, it can be stated that the coordinates of each degeneracy manifold will transform in the same way under C_{nv}, now considered as a subgroup of the permutation group.

3d. Other symmetries

Of course, symmetries other than C_{nv} are possible. These include octahedral and icosahedral symmetries for appropriate n, and for general n, for example, there is C_{kv}, $k < n$, for which the dimensionality is $1 + n - k$, with k nuclei forming a regular k-gon and the others located on the axis through the center of the polygon. The polygon has a breathing degree of freedom, and each of the $n - k$ other nuclei is allowed to move up and down the axis. For $n = 2\nu$, for example, there is also the prism-shaped $C_{\nu v}$, which has dimension 3, since both ν-gons can breath independently and the height of the prism can change.

3e. Note on half-odd integral spin

In the case where time-reversal invariance holds and the electronic spin is half-odd integer, there is always the two-fold Kramers degeneracy. For an intersection of two Kramers doublets, five conditions are required [10], so here there are cases, such as T_d symmetry for $n = 4$, in which the dimensions of symmetry and degeneracy manifolds are identical. The analog of the HLH sign change is a unitary transformation in the space of a doublet upon being taken along a closed path [7].

4. **Discussion**

We have discussed two problems connected with systems of n identical nuclei: The problem of constructing an internal coordinate system treating the nuclei equivalently, as nearly as possible, and that of determining the coordinates that lift a symmetry-determined degeneracy to first order, thus making the intersection truly conical. The results are not of a sensational nature, but still we hope that our treatment will be somewhat useful.

References

[1] Keating S P and Mead C A 1985 *J Chem Phys* **82** 5102
[2] Jahn H A and Teller E 1937 *Proc Roy Soc London Ser A* **161** 220
[3] Longuet-Higgins H C *et al* 1958 *Proc Roy Soc London Ser A* **244** 1
[4] Herzberg G and Longuet-Higgins H C 1963 *Discuss Faraday Soc* **35** 77
[5] Mead C A and Truhlar D G 1979 *J Chem Phys* **70** 2284
[6] Berry M V 1984 *Proc R Soc London Ser A* **392** 45
[7] Mead C A 1992 *Revs Modern Phys* **64** 51
[8] Wittig C 2012 *Phys Chem Chem Phys* **14** 6409
[9] Ceulemans A and Linjen E 2008 in *The Jahn-Teller Effect,* Springer Series in Chemical Physics, Vol. 97 (Köppel H, Yarkony D R, and Barentzen H, eds). p. 25
[10] Wigner E P 1959 *Group Theory*, translated by J J Griffin (Academic, New York), Chapter 26
[11] Landau L D and Lifshitz E M 1965 *Quantum Mechanics, Non-Relativistic Theory, 2'nd edition,* translated by J B Sykes and J S Bell (Pergamon, New York) §79

The molecular symmetry adapted non - adiabatic coupling terms and diabatic Hamiltonian matrix

Saikat Mukherjee[1], Sudip Bandyopadhyay[2], Amit Kumar Paul[1] and Satrajit Adhikari[1*]

[1]Department of Physical Chemistry and Raman Center for Atom, Molecule and Optical Sciences
Indian Association for the Cultivation of Science
Jadavpur, Kolkata -700 032 INDIA
[2] Department of Chemistry, Maulana Azad College, Kolkata - 700 013, INDIA

E-mail: pcsa@iacs.res.in

Abstract.
We calculate the adiabatic Potential Energy Surfaces (PESs) and the Non - Adiabatic Coupling Terms (NACTs) for the excited electronic states ($2^2E'$ and $1^2A_1'$) of Na_3 cluster at the MRCI level by using *ab initio* quantum chemistry package (MOLPRO), where the NACTs are adapted with Molecular Symmetry (MS) by employing appropriate Irreducible Representations (IREPs). Such terms are incorporated into the Adiabatic to Diabatic Transformation (ADT) equations to obtain the ADT angles to construct the continuous, single - valued, symmetric and smooth 3×3 diabatic Hamiltonian matrix.

1. Introduction

Since the non - adiabatic coupling terms (NACTs) have important roles[1–12] in the description of scattering and radiation - less processes in molecular system, it is a matter of contemporary research how gracefully one can handle the NACTs instead of neglecting them forcibly[13–20]. As the adiabatic potential energy surfaces (PESs) and the NACTs are physically indicative and the removable components of those NACTs are usually very sharp functions of nuclear coordinates with singularities[21,22] in the configuration space (CS), one needs to perform a unitary transformation to obtain the diabatic representation, where couplings among the electronic states are smooth functions of nuclear coordinates and thereby, the dynamical calculations on the diabatic PESs would be numerically accurate and stable. Such a transformation from adiabatic to diabatic representation of SEs for a given sub - Hilbert space is guaranteed only when the vector fields created by the NACTs satisfy the so called Curl Conditions,[17,23] *i.e.* the non - removable components of NACTs are approximately zero.

Baer and Alijah[24] predicted the existence of quantization of NACTs for a given sub - Hilbert space to construct continuous and single valued diabatic potential matrix. Such quantization was confirmed by Yarkony *et al.*[25] for removable component of NACTs with singularity in the CS. Sarkar *et al.*[7–12] has generalized the Born - Oppenheimer (BO) treatment of any three/four coupled electronic states with detailed analysis of Curl Conditions in terms of electronic basis functions/ADT angles, where the explicit form of NACTs, their Curl - Divergence equations,

Journal of Physics: Conference Series **428** (2013) 012008 doi:10.1088/1742-6596/428/1/012008

Curl Condition and diabatic PESs in terms of Adiabatic to Diabatic Transformation (ADT) angles were formulated. This approach provides a practical way to handle the NACTs with singularity at certain point(s)/seam in the CS, and help to construct continuous, single valued, smooth and symmetric diabatic PESs. Finally, since molecular symmetry (MS) imposes certain constraints on the CIs and NACTs with respect to the symmetry adapted nuclear coordinates,[26] it is necessary to incorporate MS in the *ab initio* calculated NACTs to calculate ADT angles to construct diabatic PESs and thereby, one may expect to observe "correct" features of nuclear dynamics and spectral profiles.

The spectroscopy of alkali metal clusters is an intense area of research with convenient one - electron - valence configuration and few vibrational degrees of freedom (DOF) for easy analysis both by experimental measurements and theoretical calculations. L. Wöste *et al.*[27–29] and W. E. Ernst *et al.*[30–32] have done extensive work to generate the absorption spectra of Na_3, where those trimers are also important for exhibiting JT interaction in D_{3h} symmetry between the excited $2^2E'$ states. In Li_3 system, R. G. Sadygov and D. R. Yarkony showed[33] unusual CIs of C_{2v} symmetry along with that of D_{3h} symmetry. H. Koizumi and I. B. Bersuker modified[34] the existing PJT model Hamiltonian[35] for excited states ($2^2E'$ and $1^2A'_1$) of Na_3 cluster in such a way that the corresponding adiabatic PESs for $2^2E'$ states show four CIs. We had also demonstrated the *ab initio* calculated PESs and NACTs of $2^2E'$ and $1^2A'_1$ states of Na_3 cluster at the SA-CASSCF level, where four CIs were reported[36–38] between the states of $2^2E'$ symmetry.

In this article, we recalculate *ab initio* PESs and NACTs of $2^2E'$ and $1^2A'_1$ states of Na_3 cluster at the MRCI level, incorporate sign correction of the NACTs according to the proper IREPs of MS group [D_{3h}(M)], solve the ADT equations with those MS adapted NACTs to obtain the ADT angles and construct the diabatic PESs as functions of nuclear coordinates to demonstrate the workability of our beyond BO theory.

2. Theory

2.1. Beyond Born - Oppenheimer theory for three - state sub - Hilbert space

We assume a three - state electronic sub - Hilbert space (i.e., a complete space for the present case) with conical intersections (CIs) among the states anywhere in the nuclear configuration space (CS). The Born - Oppenheimer[39,40] (BO) expansion of the molecular wavefunction for this subspace of the Hilbert space and the total electron - nuclei Hamiltonian in the adiabatic representation are:

$$\begin{aligned}
\Psi(s_e, s_n) &= \sum_{i=1}^{3} \psi_i(s_n)\xi_i(s_e; s_n), \\
\hat{H} &= \hat{T}_{s_n} + \hat{H}_e(s_e, s_n), \\
\hat{T}_{s_n} &= -\frac{\hbar^2}{2}\sum_i \left(\frac{\nabla^2_{s_n,i}}{m_i}\right), \\
\hat{H}_e(s_e, s_n)\xi_i(s_e; s_n) &= u_i(s_n)\xi_i(s_e; s_n) \qquad (1)
\end{aligned}$$

where the eigenfunctions $[\xi_i(s_e; s_n)]$ of the electronic Hamiltonian, $\hat{H}_e(s_e, s_n)$, are defined by the sets of nuclear (s_n) and electronic (s_e) coordinates with nuclear coordinate dependent eigenvalues, $u_i(s_n)$. $\hat{T}_{s_n}$ is the nuclear kinetic energy (KE) operator and the expansion coefficient, $\psi_i(s_n)$, are actually the nuclear wavefunctions.

When the total electron - nuclear Hamiltonian and the BO expansion of the molecular wavefunction [Eq. (1)] for a sub - Hilbert space are substituted in the time independent SE $[\hat{H}\Psi(s_e, s_n) = E\Psi(s_e, s_n)]$, we can arrive at the following compact form of kinetically coupled nuclear equation:

Journal of Physics: Conference Series **428** (2013) 012008 doi:10.1088/1742-6596/428/1/012008

$$-\frac{\hbar^2}{2m}\left(\vec{\nabla}+\vec{\tau}\right)^2\psi+(U-E)\,\psi=0, \tag{2}$$

where the non - adiabatic coupling and adiabatic PES matrices have the following form:

$$\boldsymbol{\vec{\tau}}=\begin{pmatrix} 0 & \vec{\tau}_{12} & \vec{\tau}_{13} \\ -\vec{\tau}_{12} & 0 & \vec{\tau}_{23} \\ -\vec{\tau}_{13} & -\vec{\tau}_{23} & 0 \end{pmatrix} \quad \text{and} \quad U_{ij}=u_i\delta_{ij}, \tag{3}$$

respectively. The non - adiabatic coupling matrix elements are defined as:

$$\vec{\tau}_{ij} \;=\; \langle\xi_i(s_e;s_n)|\vec{\nabla}\xi_j(s_e;s_n)\rangle. \tag{4}$$

If the three states constitute a sub - Hilbert space, it is possible to transform ($\boldsymbol{\psi}=\mathbf{A}\boldsymbol{\psi}^d$) the adiabatic nuclear SE [Eq. (2)] to the diabatic one:

$$-\frac{\hbar^2}{2m}\nabla^2\psi^d+(W-E)\psi^d=0, \qquad W=A^{\dagger}UA \tag{5}$$

under the condition:

$$\vec{\nabla}\mathbf{A}+\boldsymbol{\vec{\tau}}\mathbf{A}=0. \tag{6}$$

This equation is known as the Adiabatic - Diabatic Transformation (ADT) condition[13]. In order to find a meaningful solution of Eq. (6), we have to ensure that the chosen form of **A** matrix must (a) be orthogonal at any point in CS ; (b) have elements that are cyclic functions with respect to a parameter.

As the model form of (3×3) **A** has to be an orthogonal matrix and the ortho - normality conditions demand the fulfillment of six relations, three independent variables, *viz.*, Euler like angles of rotation [$\theta_{12}(s_n)$, $\theta_{23}(s_n)$ and $\theta_{13}(s_n)$], commonly called ADT angles, are the natural requirement to construct the three - state **A** matrix by taking the product of three (3×3) rotation matrices, $\mathbf{A}_{12}(\theta_{12})$, $\mathbf{A}_{23}(\theta_{23})$, and $\mathbf{A}_{13}(\theta_{13})$ as shown below:

$$\mathbf{A}(\boldsymbol{\theta_{12}},\boldsymbol{\theta_{23}},\boldsymbol{\theta_{13}}) \;=\; \mathbf{A}_{12}(\theta_{12})\cdot\mathbf{A}_{23}(\theta_{23})\cdot\mathbf{A}_{13}(\theta_{13}). \tag{7}$$

The above model form of **A** matrix [Eq. (7)] and the anti - symmetric form of $\boldsymbol{\tau}$ matrix [Eq. (3)] are substituted in Eq. (6), where the simple manipulation as performed by Top and Baer[41] and Alijah and Baer[24] leads to the following equations for ADT angles:

$$\vec{\nabla}\theta_{12} \;=\; -\vec{\tau}^{12}+\tan\theta_{23}(\vec{\tau}^{13}\cos\theta_{12}-\vec{\tau}^{23}\sin\theta_{12}), \tag{8a}$$

$$\vec{\nabla}\theta_{23} \;=\; -(\vec{\tau}^{13}\sin\theta_{12}+\vec{\tau}^{23}\cos\theta_{12}), \tag{8b}$$

$$\vec{\nabla}\theta_{13} \;=\; -\frac{1}{\cos\theta_{23}}(\vec{\tau}^{13}\cos\theta_{12}-\vec{\tau}^{23}\sin\theta_{12}), \tag{8c}$$

which in turn brings the explicit form[7] of τ matrix elements in terms of ADT angles:

$$\vec{\tau}_{12} \;=\; -\vec{\nabla}\theta_{12}-\sin\theta_{23}\vec{\nabla}\theta_{13}, \tag{9a}$$

$$\vec{\tau}_{23} \;=\; \sin\theta_{12}\cos\theta_{23}\vec{\nabla}\theta_{13}-\cos\theta_{12}\vec{\nabla}\theta_{23}, \tag{9b}$$

$$\vec{\tau}_{13} \;=\; -\cos\theta_{12}\cos\theta_{23}\vec{\nabla}\theta_{13}-\sin\theta_{12}\vec{\nabla}\theta_{23}. \tag{9c}$$

The components of *ab initio* calculated NACTs are substituted in Eq. (8) and those coupled stiff differential equations are solved by using “Backward Differentiation Formula (BDF)” to calculate the ADT angles. When we substitute those ADT angles in Eq. (10), the diabatic potential matrix elements are obtained as given below:

$$\begin{aligned} W_{11} &= u_2\cos^2\theta_{23}\sin^2\theta_{12} + u_3(\cos\theta_{12}\sin\theta_{13} + \cos\theta_{13}\sin\theta_{12}\sin\theta_{23})^2 \\ &+ u_1(\cos\theta_{12}\cos\theta_{13} - \sin\theta_{12}\sin\theta_{13}\sin\theta_{23})^2, \end{aligned} \tag{10a}$$

$$\begin{aligned} W_{22} &= u_2\cos^2\theta_{12}\cos^2\theta_{23} + u_3(-\sin\theta_{12}\sin\theta_{13} + \cos\theta_{12}\cos\theta_{13}\sin\theta_{23})^2 \\ &+ u_1(-\cos\theta_{13}\sin\theta_{12} - \cos\theta_{12}\sin\theta_{13}\sin\theta_{23})^2, \end{aligned} \tag{10b}$$

$$W_{33} = u_3\cos^2\theta_{13}\cos^2\theta_{23} + u_1\cos^2\theta_{23}\sin^2\theta_{13} + u_2\sin^2\theta_{23}, \tag{10c}$$

$$\begin{aligned} W_{12} &= W_{21} = u_2\cos\theta_{12}\cos^2\theta_{23}\sin\theta_{12} + u_3[(-\sin\theta_{12}\sin\theta_{13} + \cos\theta_{12}\cos\theta_{13}\sin\theta_{23}) \\ &\times (\cos\theta_{12}\sin\theta_{13} + \cos\theta_{13}\sin\theta_{12}\sin\theta_{23})] - u_1[(\cos\theta_{13}\sin\theta_{12} + \cos\theta_{12}\sin\theta_{13}\sin\theta_{23}) \\ &\times (\cos\theta_{12}\cos\theta_{13} - \sin\theta_{12}\sin\theta_{13}\sin\theta_{23})], \end{aligned} \tag{10d}$$

$$\begin{aligned} W_{13} &= W_{31} = \cos\theta_{23}[-(u_1 - u_3)\cos\theta_{12}\cos\theta_{13}\sin\theta_{13} \\ &+ \sin\theta_{12}(-u_2 + u_3\cos^2\theta_{13} + u_1\sin^2\theta_{13})\sin\theta_{23}], \end{aligned} \tag{10e}$$

$$\begin{aligned} W_{23} &= W_{32} = -\cos\theta_{23}[-(u_1 - u_3)\cos\theta_{13}\sin\theta_{12}\sin\theta_{13} \\ &- \cos\theta_{12}(u_2 - u_1\sin^2\theta_{13} - u_3\cos^2\theta_{13})\sin\theta_{23}]. \end{aligned} \tag{10f}$$

One can use Eqs. (9) to formulate the explicit form of Curl equation[7] for each NACT in terms of ADT angles:

$$\text{Curl } \tau_{pq}^{12} = C_{12} = Z_{12} = -\cos\theta_{23}(\nabla_q\theta_{23}\nabla_p\theta_{13} - \nabla_p\theta_{23}\nabla_q\theta_{13}), \tag{11a}$$

$$\begin{aligned} \text{Curl } \tau_{pq}^{23} = C_{23} = Z_{23} &= \cos\theta_{12}\cos\theta_{23}(\nabla_q\theta_{12}\nabla_p\theta_{13} - \nabla_p\theta_{12}\nabla_q\theta_{13}) \\ &- \sin\theta_{12}\sin\theta_{23}(\nabla_q\theta_{23}\nabla_p\theta_{13} - \nabla_p\theta_{23}\nabla_q\theta_{13}) \\ &+ \sin\theta_{12}(\nabla_q\theta_{12}\nabla_p\theta_{23} - \nabla_p\theta_{12}\nabla_q\theta_{23}), \end{aligned} \tag{11b}$$

$$\begin{aligned} \text{Curl } \tau_{pq}^{13} = C_{13} = Z_{13} &= \sin\theta_{12}\cos\theta_{23}(\nabla_q\theta_{12}\nabla_p\theta_{13} - \nabla_p\theta_{12}\nabla_q\theta_{13}) \\ &+ \cos\theta_{12}\sin\theta_{23}(\nabla_q\theta_{23}\nabla_p\theta_{13} - \nabla_p\theta_{23}\nabla_q\theta_{13}) \\ &- \cos\theta_{12}(\nabla_q\theta_{12}\nabla_p\theta_{23} - \nabla_p\theta_{12}\nabla_q\theta_{23}), \end{aligned} \tag{11c}$$

where the Curl due to vector product of NACTs and analyticity of ADT matrix are defined as: $C_{ij} = (\boldsymbol{\tau}_q\boldsymbol{\tau}_p)_{ij} - (\boldsymbol{\tau}_p\boldsymbol{\tau}_q)_{ij}$ and $Z_{ij} = \frac{\partial}{\partial p}\tau_q^{ij} - \frac{\partial}{\partial q}\tau_p^{ij}$, respectively with $\nabla_p = \dfrac{\partial}{\partial p}$ and $\nabla_q = \dfrac{\partial}{\partial q}$. The *Cartesian coordinates* p and q denote nuclear DOFs.

2.2. Molecular symmetry adaptation to NACTs for $D_{3h}(M)$ group

The "true" symmetry group for the electron - nuclei Hamiltonian of molecules is called molecular symmetry (MS) group which, in general, contains R symmetry operations, $\hat{O}_i$, i=1,2,... ..,R, including the identity operation E, all feasible permutations P of the spatial as well as spin coordinates of the *equivalent* nuclei, the inversion $E^\star$ of *all* nuclear coordinates s_k and electronic coordinates s_e, and the inversion-permutation $P \cdot E^\star$. Each symmetry operation $\hat{O}$ of the MS group of a molecule transforms the vibronic modes, the three angles (Euler) to describe the orientations of a rigid body in a three dimensional Euclidean space and the nuclear spin, which enables us to express $\hat{O}$ as a product of three different operations,[42] *i.e.*, $\hat{O} = \hat{O}_a \cdot \hat{O}_b \cdot \hat{O}_c$; where

$\hat{O}_a$ produces change in vibronic degrees of freedom (DOF), $\hat{O}_b$ changes the Euler angles and $\hat{O}_c$ permutes the nuclear spin. At this point, we take the example of sodium trimer that belongs to the point group D_{3h}. If one locates the system-fixed (x,y,z) axes on a Na_3 cluster such that the xy plane be the molecular plane (plane of symmetry, σ_h) and the z axis be the threefold symmetry axis (C_3), the relations between the MS group [D_{3h}(M)] and the corresponding point group [D_{3h}] operations are as follows[42] (one representative element from each class of the MS group is shown):

$$E = E \cdot R^0 \cdot p_0 \tag{12a}$$
$$(123) = C_3^1 \cdot R^{\frac{2\pi}{3}} \cdot p_{123} \tag{12b}$$
$$(23) = C_{2x} \cdot R^{\pi_z} \cdot p_{23} \tag{12c}$$
$$E^\star = \sigma_{xy} \cdot R^{\pi_z} \cdot p_0 \tag{12d}$$
$$(132)^\star = S_3^{-1} \cdot R_z^{\frac{\pi}{3}} \cdot p_{132} \tag{12e}$$
$$(23)^\star = \sigma_{xz} \cdot R_y^{\pi} \cdot p_{23} \tag{12f}$$

where "R^ϕ"s are members of the corresponding molecular rotation group and "p"s denote nuclear permutation. There are two important theorems[26] on MS, which relate the IREPs with different NACTs.

(i) Theorem 1:
The first theorem states that when the IREP of a particular NACT, say τ_l^{ij}, is known for a specific symmetry adapted nuclear coordinate s_l, the IREP of another one, τ_k^{ij} between the same electronic states i, j for a different symmetry adapted nuclear coordinate, s_k, can be found out by the following relation:

$$\Gamma(\tau_k^{ij}) = \Gamma\left(\frac{\partial}{\partial s_k}\right) \times \Gamma\left(\frac{\partial}{\partial s_l}\right) \times \Gamma(\tau_l^{ij}), \tag{13}$$

where

$$\tau_k^{ij} = \langle \xi_i(s_e; s_n)| \frac{\partial}{\partial s_k} (\xi_j(s_e; s_n)) \rangle. \tag{14}$$

The possible combination of the IREPs of the NACTs can be determined by fixing each of them [τ_ks, (k=ρ, ϕ, x and y)] in succession to the totally symmetric representation (A_1'). If we choose $\tau_\rho = A_1'$, the IREPs of other three NACTs appear as $\tau_\phi = A_1''$, $\tau_x = \tau_y = E''$ from Eq. (13). Though there would be four such combination of the IREPs of the NACTs, only two possibilities survive as theorem 1 is valid only for those $\Gamma\left(\frac{\partial}{\partial s_l}\right)$ that belong to 1-d IREP (i.e., for l=ρ, ϕ). In other words, when we try to find out the combination of the IREPs of the τ_ks fixing τ_x or τ_y to A_1', the IREP for τ_y or τ_x turns into *reducible representation* contradicting the construction of Character Table as shown in Table (1).

Since we need to select the *correct* combination from the above two possible ones, we utilize the following quantization rule[17] [Eq. (15)] of the contour integrals over the NACTs evaluated along a closed loop L_g of nuclear coordinate s_n around a single conical intersection (CI):

$$\oint ds_n \cdot \tau(s_n|L_g) = \pm\pi. \tag{15}$$

It appears that the second combination is compatible with the quantization rule producing non-zero residue.

Table 1 Extended character table of molecular symmetry group $D_{3h}(M)$ with different NAC elements

D_{3h}(M)	E	(123) (132)	(12) (13) (23)	$E^\star$	(123)* (132)*	(12)* (13)* (23)*	coord.	deriv.	τ_k^{23} τ_k^{13}	τ_k^{12}
A_1'	1	1	1	1	1	1			τ_ϕ	
A_2'	1	1	-1	1	1	-1	ϕ	$\frac{\partial}{\partial\phi}$		τ_ϕ
E'	2	-1	0	2	-1	0	x,y	$\frac{\partial}{\partial x}, \frac{\partial}{\partial y}$	τ_x,τ_y	τ_x,τ_y
A_1''	1	1	1	-1	-1	-1			τ_ρ	
A_2''	1	1	-1	-1	-1	1	ρ	$\frac{\partial}{\partial\rho}$		τ_ρ
E''	2	-1	0	-2	1	0				

(ii) Theorem 2:

To state the second theorem [see Eq. (18)] , we need to consider a loop-type sequence of N molecular states with same spin multiplicity, say N doublet states, D_a, D_b, D_c,.....,D_y, $D_{z=a}$, for the Na_3 cluster. The corresponding NACTs are $\tau_k^{a,b}$, $\tau_k^{b,c}$,....., $\tau_k^{y,z=a}$, and their product is:

$$\tau_k^{a,b,c,...,y,z=a} = \tau_k^{a,b}\tau_k^{b,c}...\tau_k^{y,z=a}. \tag{16}$$

In terms of the IREPs, one can write:

$$\begin{aligned}\Gamma(\tau_k^{a,b,c,...,y,z=a}) &= \Gamma(\tau_k^{a,b})\times\Gamma(\tau_k^{b,c})\times...\times\Gamma(\tau_k^{y,z=a})\\ &= \Gamma(\xi_a(s_e;s_n))^2\times\Gamma(\xi_b(s_e;s_n))^2\\ &\times \quad ...\times\Gamma(\xi_y(s_e;s_n))^2\times\Gamma\left(\frac{\partial}{\partial s_k}\right)^N\end{aligned} \tag{17}$$

(a) For 1-d IREPs: [26]

$$\Gamma(\tau_k^{a,b,c,...,y,z=a}) = \begin{cases}\Gamma\left(\frac{\partial}{\partial s_k}\right) & \text{when } N \text{ is odd}\\ A_1' & \text{when } N \text{ is even}\end{cases} \tag{18}$$

If there is a CI between two electronic states i and $(i+1)$, the corresponding electronic eigenfunctions $\xi_i(s_e;s_n)$ and $\xi_{(i+1)}(s_e;s_n)$ will change their signs along the nuclear coordinate s_n at the time of crossing the CI. Therefore, while implementing the second theorem [Eq. (18)] on the NACTs, we need to interchange the signs of the NACTs $\tau_k^{i,j}$ and $\tau_k^{(i+1),j}$, i<j, at the point of CI with the same IREP *i.e.*

$$\Gamma(\tau_k^{i,j}) = \Gamma(\tau_k^{(i+1),j}). \tag{19}$$

(b) For 2-d IREPs:

Since Eq. (18) is only valid to find out the 1-d IREPs, we need to establish similar working equations for 2-d IREP cases as given below:

$$\begin{aligned}\Gamma(\tau_x^{23}) &= \Gamma(\langle\xi_2(s_e;s_n)|\frac{\partial}{\partial x}(\xi_3(s_e;s_n))\rangle)\\ &= \Gamma(\xi_2(s_e;s_n))\times\Gamma\left(\frac{\partial}{\partial x}\right)\times\Gamma(\xi_3(s_e;s_n))\end{aligned} \tag{20}$$

As $\Gamma\left(\dfrac{\partial}{\partial x}\right) = E'$ and $\Gamma(\tau_x^{23}) = E'$ [see Table (1)], it appears from Eq. (20):

$$\Gamma(\xi_2(s_e; s_n)) \times \Gamma(\xi_3(s_e; s_n)) = A_1', \tag{21a}$$

Similarly, for $\Gamma(\tau_x^{13}) = E'$, we have:

$$\Gamma(\xi_1(s_e; s_n)) \times \Gamma(\xi_3(s_e; s_n)) = A_1'. \tag{21b}$$

Thus,

$$\Gamma(\xi_1(s_e; s_n))^2 \times \Gamma(\xi_2(s_e; s_n))^2 \times \Gamma(\xi_3(s_e; s_n))^2 = A_1'. \tag{21c}$$

When we use Eq. (16) for the present case and substitute Eq. (21c), we obtain:

$$\begin{aligned}
\Gamma(\tau_x^{1,2,3,1}) &= \Gamma(\xi_1(s_e; s_n))^2 \times \Gamma(\xi_2(s_e; s_n))^2 \times \Gamma(\xi_3(s_e; s_n))^2 \times \Gamma\left(\frac{\partial}{\partial x}\right)^3 = (E')^3 \\
&= \Gamma(\tau_x^{12}) \times \Gamma(\tau_x^{23})^2 \\
&= \Gamma(\tau_x^{12}) \times (E')^2
\end{aligned} \tag{22}$$

Therefore, one gets $\Gamma(\tau_x^{12}) = E'$ and similarly, $\Gamma(\tau_y^{12}) = E'$.

3. *Ab initio* calculation for NACTs and its MS Adaptation

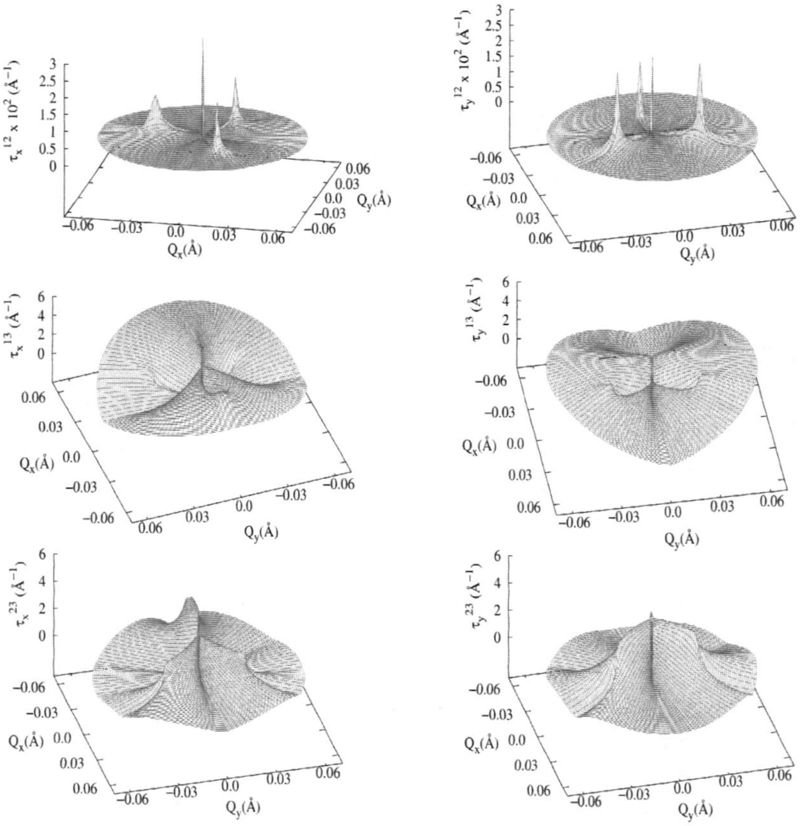

Fig. 1 The functional forms of the components of Non-Adiabatic Coupling (NAC) matrix elements as functions of Q_x and Q_y.

The vibrational modes involved in this calculation are: Symmetric stretching (Q_s), bending (Q_x) and antisymmetric stretching (Q_y). The "breathing mode", Q_s preserves the D_{3h} symmetry, whereas Q_x and Q_y distort the D_{3h} equilateral triangle. We have calculated the excited state PESs ($2^2E'$ and $1^2A_1'$) and NACTs by the MRCI level of theory with MOLPRO *ab initio* program.[43] The triple zeta or TZV basis functions augmented with a single set of *d* functions ($\zeta_d = 0.1$) are used to describe the electron correlation effects in a better way and two diffuse *s* functions ($\zeta_{4s} = 0.008736$, $\zeta_{5s} = 0.00345$)[35] are employed to improve the excited state description. Calculations at SA- CASSCF level were already presented[36–38] as functions of the normal modes Q_x and Q_y for three fixed Q_ss, namely 3.4, 3.7 and 4.0 Å, respectively. Since the 4th, 5th ($2^2E'$) and 6th ($1^2A_1'$) excited states of Na_3 cluster are the ones to be explored, the first 20 electronic states (generating 572 CSFs) are being involved at CASSCF reference to attain the closest possible variational minimum with respect to the exact energies and to ensure the formation of three state sub - Hilbert space within the range of CS of our investigation. Then, more accurate MRCI calculations are performed by incorporating the first six (6) states of the system obtained from the SA - CASSCF calculation and the finite numerical difference method (DDR) is applied to calculate NACTs among $2^2E'$ and $1^2A_1'$ states. The functional forms of the component of NACTs in normal mode coordinates (Q_x) and (Q_y) are shown in Fig. 1. All these calculations are carried out over the range ρ= 0.0 to 0.9 Å and $\phi = 0$ to 2π for a fixed $Q_s = 3.7$ Å. Fig. 1 displays there are four (4) CIs between the $2^2E'$ states of Na_3 cluster as reported[36–38] earlier, where one of them is being located at $(Q_x,Q_y) = (0,0)$ $[(\rho, \phi) = (0,0)]$ and the other three at $\rho \approx 0.03$ Å, $\phi=\pi/3$, π and $5\pi/3$ with $\rho = (Q_x^2+Q_y^2)^{1/2}$ and $\phi = \tan^{-1}(Q_y/Q_x)$.

The *ab initio* calculation can only provide the magnitudes of the NACTs, but their correct sign ("+" or "-") at each point of the CS are determined by employing molecular symmetry. We have incorporated MS adaptation scheme to NACTs through the following two steps: (i) The sign of each of the NACTs are fixed according to the nodal pattern of the corresponding IREPs as shown in panels (b, e) of Fig. 2, whereas the panels (a, d) depict the numerical magnitude of τ_ρ^{ij} and τ_ϕ^{ij} as obtained from MOLPRO package; (ii) The signs of τ_k^{13} and τ_k^{23} $[k=\rho,\phi]$ are interchanged at the point of CI to include crossing which is presented in panels (c, f) of Fig. 2.

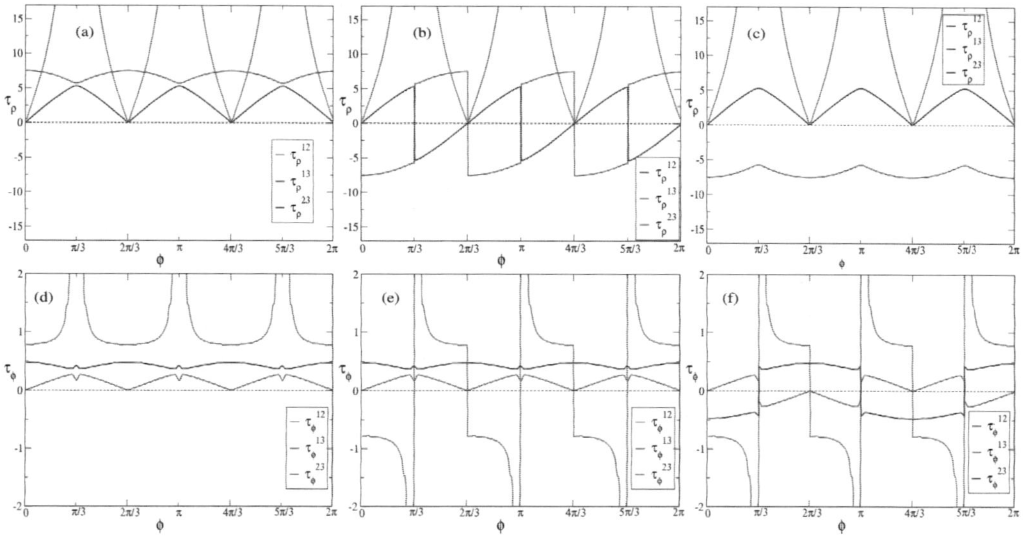

Fig. 2 Panels (a, d) show τ^{ij} vs. ϕ for a fixed ρ=0.03Å before MS adaptation. The IREP adapted τ^{ij} is displayed in panels (b, e), whereas panels (c, f) represent the same with appropriate crossing due to CIs.

4. ADT angles (θ_{ij}) and their residues ($\bar{\theta}^k_{ij}$)

The MS adapted NACTs are plugged into Eqs. (8) on a 2D grid of geometries in polar coordinates (ρ, ϕ) ranging from $\rho = 0$ to 0.9 Å and $\phi = 0$ to 2π to obtain the ADT angles. The Backward Differentiation Formula is used to solve such stiff differential equations with appropriate relative and absolute error tolerances to attain the required level of convergence. Fig. 3 describes the rectangular contour path - the differential equations for the ρ grid are integrated with positive increment from $\rho = 0$ to ρ_{max} for each positive step integration of the differential equations for the ϕ grid from 0 to 2π. It may be noted one can integrate two sets of coupled (ρ and ϕ) differential equations [Eqs. (8)] with different chosen paths, where each path will produce a set of ADT angles and such angles obtained from any choice of path are expected to show gauge invariance.

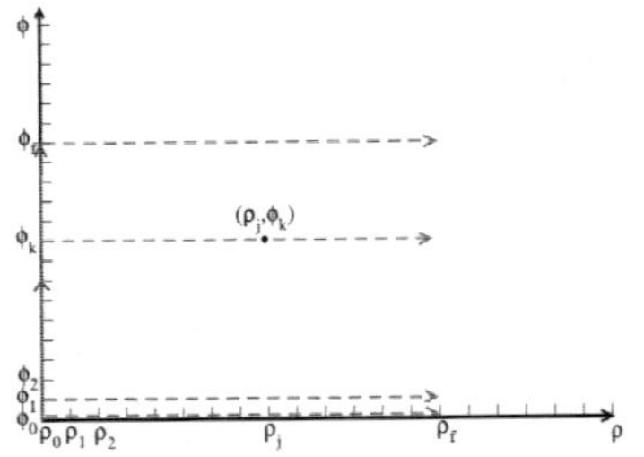

Fig. 3 The 2d rectangular paths along which the ADT equations are solved.

In panels (a - c) of Fig. 4, we display the three ADT angles $[\theta_{ij}(\rho, \phi)]$ as functions of ρ and ϕ, where in panels (d - f), the residues of the ADT angles $[\bar{\theta}^k_{ij} = \int_0^{2\pi} \theta_{ij}(\rho_k, \phi) \cdot d\phi]$ and their cumulative residues $[\bar{\theta}_{ij,l} = \sum_{k=1}^{l} \mid \bar{\theta}^{k+1}_{ij} - \bar{\theta}^k_{ij} \mid]$ are shown as function of ρ.

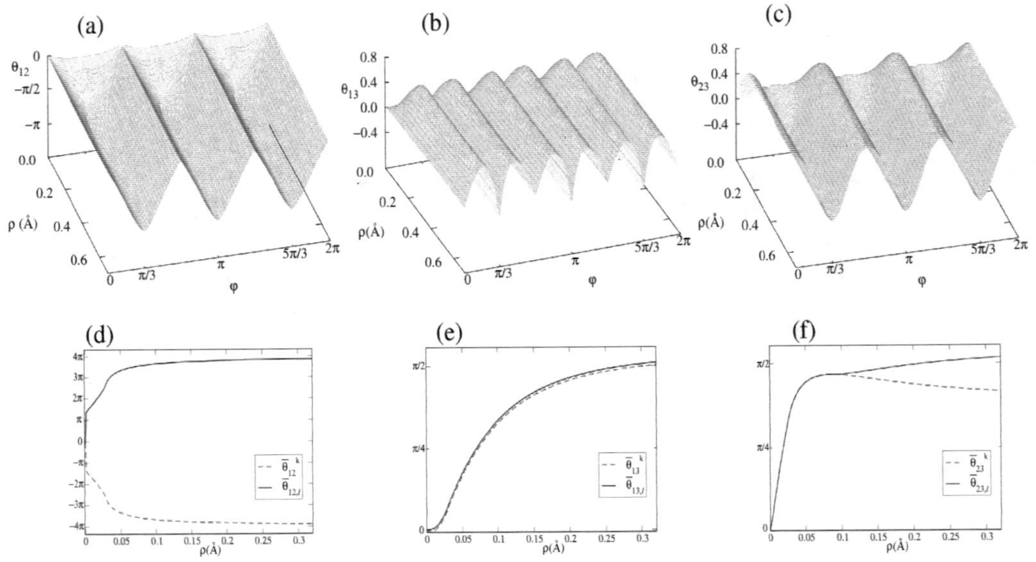

Fig. 4 Top panels (a - c): The functional form of the ADT angles solved on a 2d contour integral. Bottom panels (d - f): The residues of the corresponding ADT angles and the cumulative residues as function of ρ.

The numerical magnitude of $\bar{\theta}^k_{ij}$ essentially determines the number of CIs between a pair of states and depicts the gauge invariance for the chosen path of integration. When a system has multiple CIs, the contribution for each CI on residue ($\bar{\theta}^k_{ij}$) could be positive or negative and thereby, the cumulative residue ($\bar{\theta}_{ij,l}$) could be a better way of showing the total change of residue. The panel (d) of Fig. 4 reveals that the cumulative residue ($\bar{\theta}_{ij,l}$) for the ADT angle θ_{12} become close to 4π at asymptotic ρ showing four CIs between the $2^2E'$ states. On the other hand, the residue in the panel (e or f) of the Fig. 4 approaches to $\pi/2$ implying the gauge invariance of the path of contour integral.

We use the functional form of ADT angles (Fig. 4) to calculate the Curl condition by involving Eq. (11) to demonstrate the existence of sub - Hilbert space constituted by $2^2E'$ and $1^2A_1'$ states of Na_3 cluster. Fig. 5 displays the root mean square deviation (RMSD) between the mathematical (Z_{ij}) and ADT (C_{ij}) curl by employing the equation: $\chi_{\mathrm{Curl}\tau_{ij}}(\rho_k) = \frac{1}{N}\sqrt{\sum_l^N (C_{ij}(\rho_k,\phi_l) - Z_{ij}(\rho_k,\phi_l))^2}$, i,j = 1,2,3. Since the calculated RMSD values for Curl τ_{12} appears as absolutely zero, the other two RMSDs for Curl τ_{13} and Curl τ_{23} are presented (Fig. 5), where their negligibly small magnitude endorses the validity of three-state sub - Hilbert space.

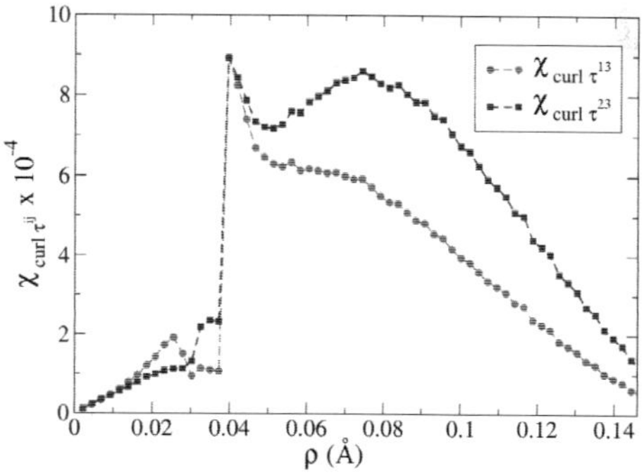

Fig. 5 The RMSDs between mathematical (Z_{ij}) and ADT (C_{ij}) Curls for τ_{13}, τ_{23} elements.

5. ADT and diabatic potential matrix

When we incorporate those ADT angles in Eq. (7), diagonal element(s) of the ADT matrix, A_{11}/A_{22} shows six sign changes in Fig. 6(a) and A_{33} displays no sign change in Fig. 6(b) as function of ϕ at asymptotic ρ.

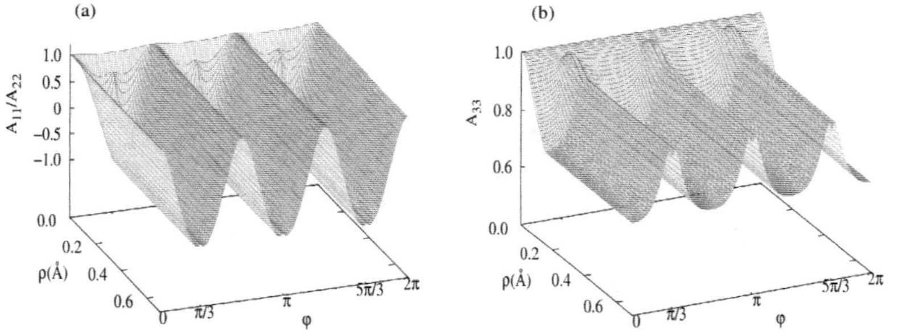

Fig. 6 The $(1,1)/(2,2)$ and $(3,3)$ elements of ADT matrix.

Though six sign changes in A_{11}/A_{22} elements of A matrix could be due to six CIs between the E' states, the residue / cumulative residue of the ADT angle θ_{12} indicates four CIs in the CS [see Fig. 4(d)]. One possibility could be due to the collapse of three central CIs on a single point at $\rho = 0$, but Longuet - Higgins' sign change technique can not be implemented here to locate the number of CIs at the central point ($\rho = 0$, $\phi = 0$).

Fig. 7 depicts the functional form of the diabatic matrix elements calculated by substituting the θ_{ij}s in Eqs. (10), where those ADT angles are obtained by integrating Eqs. (8) for the rectangular contour path of integration. The diabatic surfaces appear continuous, single valued, smooth and symmetric.

6. Summary

If few electronic states of a molecular system constitute a sub - Hilbert space over a range of CS, one can construct the diabatic Hamiltonian within the same range. We have calculated adiabatic PESs and NACTs for the states $2^2E'$ and $1^2A_1'$ of Na_3 system at MRCI level using quantum chemistry package MOLPRO, adapted MS on those NACTs, and solved the differential equations for the ADT angles (θ_{ij}s)[41] by invoking stiff equation solver technique [Backward Differentiation Formula (BDF)] over a 2D grid on normal modes, Q_x and Q_y ($\equiv \rho, \phi$) for a fixed Q_s. With

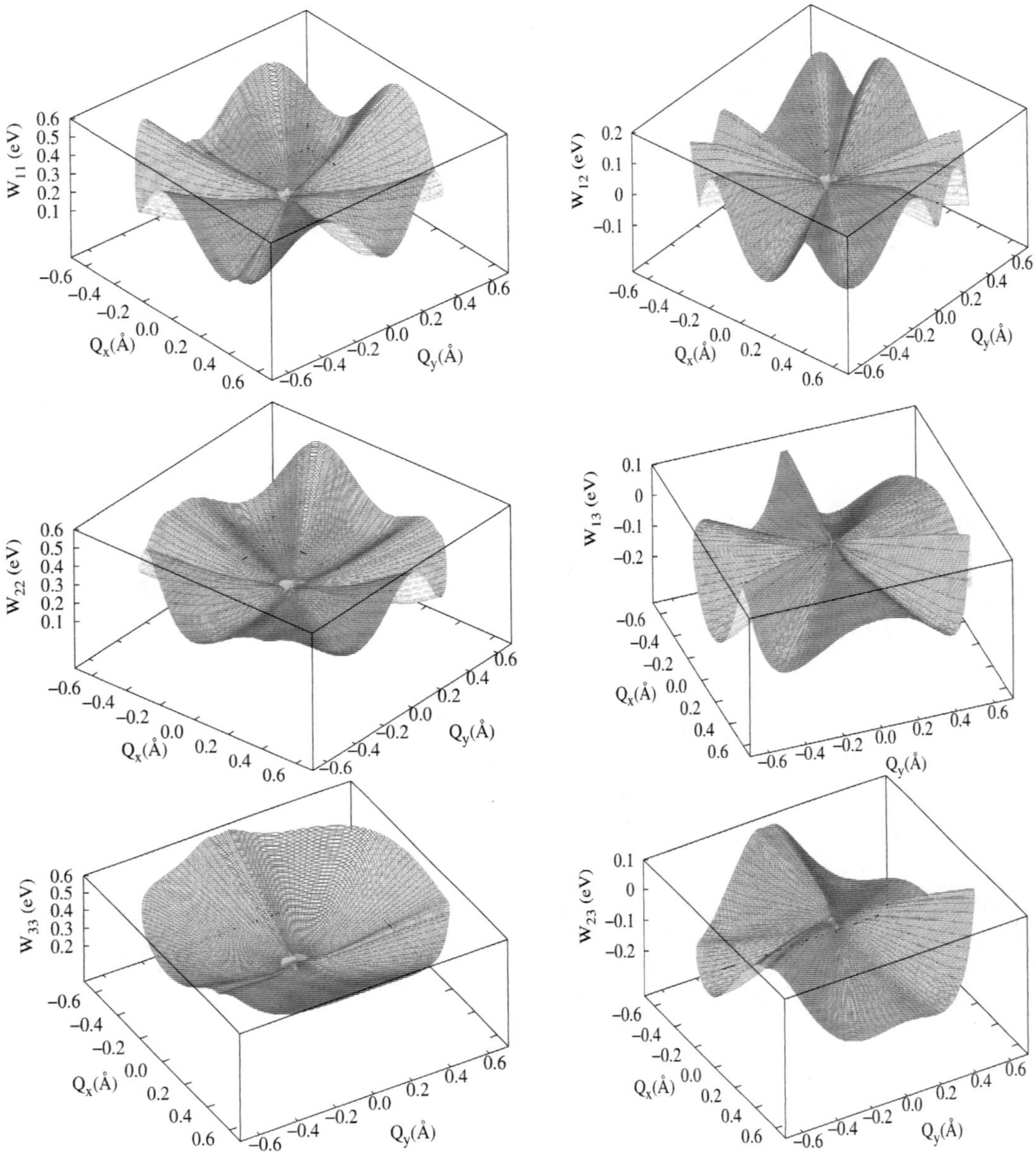

Fig. 7 The diabatic PES matrix elements (eV) as functions of the normal modes.

those ADT angles and the adiabatic PESs, we have constructed the diabatic potential energy matrix by employing the analytic expressions of its elements in terms of ADT angles [see Eq. (10)]. Calculated diabatic surfaces are not only symmetric and smooth functions but also show required number of crossings between its first two diagonal elements.

7. Acknowledgments

SM, AKP acknowledge DST, INDIA and CSIR, INDIA for research fellowship and SA acknowledges DST, INDIA through project no. SR/S1/PC-13/2008 for the research funding.

References

[1] Baer R, Charutz D, Kosloff R and Baer M 1996 *J. Chem. Phys.* **105** 9141

[2] Charutz D M, Baer R and Baer M 1997 *Chem. Phys. Letts.* **265** 629

[3] Adhikari S and Billing G D 1999 *J. Chem. Phys.* **111** 40

[4] Varandas A J C and Xu Z R 2000 *J. Chem. Phys.* **112** 2121

[5] Baer M, Lin S H, Alijah A, Adhikari S and Billing G D 2000 *Phys. Rev. A* **62** 32506

[6] Adhikari S, Billing G D, Alijah A, Lin S H and Baer M 2000 *Phys. Rev. A* **62** 32507

[7] Sarkar B and Adhikari S 2006 *J. Chem. Phys.* **124** 074101

[8] Sarkar B and Adhikari S 2007 *Indian J. Phys.* **81 (9)** 925

[9] Sarkar B and Adhikari S 2008 *J. Phys. Chem. A* **112** 9868

[10] Sarkar B and Adhikari S 2009 *Int. J. Quan. Chem.* **109** 650

[11] Paul A K, Sardar S, Sarkar B and Adhikari S 2009 *J. Chem. Phys.* **131** 124312

[12] Paul A K, Sarkar B and Adhikari S 2010 *Recent Advances in Spectroscopy* (Astrophysics and Space Science Proceedings, Springer - Verlag, Berlin, Heidelberg)

[13] Baer M 1975 *Chem. Phys. Lett.* **35** 112

[14] Mead C A and Truhlar D G 1982 *J. Chem. Phys.* **77** 6090

[15] Matsunaga N and Yarkony D R 1998 *Mol. Phys.* **93** 79

[16] Sadygov R G and Yarkony D R 1998 *J. Chem. Phys.* **109** 20

[17] Baer M 2002 *Phys. Rep.* **358** 75

[18] Kendrick B K, Mead C A and Truhlar D G 2002 *Chem. Phys.* **277** 31

[19] Abrol R and Kuppermann A 2002 *J. Chem. Phys.* **116** 1035

[20] Baer M, Vértesi T, Halász G J, Vibok A and Suhai S 2004 *Faraday Discuss* **127** 337

[21] Hellmann H 1937 *Einfuhrang in die Quantenchemie* (Franz Deutiche, Leipzig, Germany)

[22] Feynman R 1939 *Phys. Rev.* **56** 340

[23] Baer M 2006 *Beyond Born - Oppenheimer: Conical intersections and Electronic nonadiabatic coupling terms* (Wiley - Interscience)

[24] Alijah A and Baer M 2000 *J. Phys. Chem. A* **104** 389

[25] Yarkony D R 1996 *J. Chem. Phys.* **105** 10456

[26] Al-Jabour S, Baer M, Deeb O, Leibscher M, Manz J, Xu X and Zilberg S 2010 *J. Phys. Chem. A* **114** 2991

[27] Delacrétaz G and Wöste L 1985 *Surf. Sci.* **156** 770

[28] Delacrétaz G, Grant E R, Whetten R L, Wöste L and Zwanziger J W 1986 *Phys. Rev. Letts.* **56** 2598

[29] Rutz S, Kobe K, Kühling H, Schreiber E and Wöste L 1993 *Z. Phys. D.* **26** 276

[30] Ernst W E and Rakowsky S 1995 *Phys. Rev. Letts.* **74** 58

[31] Ohashi N, Tsuura M, Hougen J T, Ernst W E and Rakowsky S 1997 *J. Mol. Spec.* **184** 22

[32] Ernst W E and Golonzka O 2004 *Phys. Scrp.* **T112** 27

[33] Sadygov R G and Yarkony D R 1999 *J. Chem. Phys.* **110** 3639

[34] Koizumi H and Bersuker I B 1999 *Phys. Rev. Letts.* **83** 3009

[35] Cocchini F, Upton T H and Andreoni W 1988 *J. Chem. Phys.* **88** 6068

[36] Paul A K, Ray S, Mukhopadhyay D and Adhikari S 2011 *Chem. Phys. Letts.* **508** 300

[37] Paul A K, Ray S, Mukhopadhyay D and Adhikari S 2011 *J. Chem. Phys.* **135** 034107

[38] Paul A K, Ray S and Adhikari S 2012 *Vibronic Interactions and the Jahn - Teller Effect: Theory and Applications* (Progress in Theoretical Chemistry and Physics 23, Springer)

[39] Born M and Oppenheimer J R 1927 *Ann. Phys. (Leipzig)* **84** 457

[40] Born M and Huang K 1954 *Dynamical Theory of Crystal Lattices* (Oxford University Press, Oxford)

[41] Top Z H and Baer M 1977 *J. Chem. Phys.* **66** 1363

[42] Bunker P R and Jensen P 2006 *Molecular symmetry and spectroscopy: 2nd edition* (National Research Council of Canada)

[43] Werner H J, Knowles P J, Knizia G, Manby F R, Schütz M, Celani P, Korona T, Lindh R, Mitrushenkov A, Rauhut G, Shamasundar K R, Adler T B, Amos R D, Bernhardsson A, Berning A, Cooper D L, Deegan M J O, Dobbyn A J, Eckert F, Goll E, Hampel C, Hesselmann A, Hetzer G, Hrenar T, Jansen G, Köppl C, Liu Y, Lloyd A W, Mata R A, May A J, McNicholas S J, Meyer W, Mura M E, Nicklass A, O'Neill D P, Palmieri P, Peng D, Pflüger K, Pitzer R, Reiher M, Shiozaki T, Stoll H, Stone A J, Tarroni R, Thorsteinsson T and Wang M 2010 Molpro, version 2010.1, a package of ab initio programs see http://www.molpro.net

Journal of Physics: Conference Series **428** (2013) 012009 doi:10.1088/1742-6596/428/1/012009

Quasi-diabatic basis versus fixed electronic states in molecular systems

C A Mead[1,2]

[1] 92 Bartram Road, Savannah, GA 31411, USA

[2] Professor Emeritus of Chemistry and Chemical Physics, University of Minnesota

E-mail: aldykar@att.net

Abstract: We consider the use of quasi-diabatic bases to improve upon the Born-Oppenheimer approximation in molecular systems. In a nonrigorous, but still suggestive, manner, it is shown that, when the quasi-diabatic approximation is a good one, the use of "frozen" electronic states will often be equally good, or nearly so.

1. Introduction

A problem that has received much attention in molecular theory is the best way to handle the so-called nonadiabatic coupling terms (NACTs) between different Born-Oppenheimer electronic states. An approach that has been often used is to construct, as nearly as posssible, a so-called diabatic basis, a basis of electronic states which are linear combinations of a finite number of Born-Oppenheimer states, chosen in such a way that the NACTs vanish with respect to the new states. The transformation that achieves this is called the adiabatic to diabatic transformation, or ADT. The approach has been nicely described, with further references, by Baer [1].

However, it has also been shown [2,3] that an exact ADT does not, in general, exist, so that one must content oneself with trying to construct useful approximations. In this note, we present a nonrigorous analysis of when this is feasible, and in such cases, how the ADT compares with the simple use of frozen electonic states (FS), independent of the nuclear coordinates.

In Section 2, we introduce the notation and present the "curl condition" for the existence of an exact ADT. In Section 3, we show that this condition is the same as that for the exact use of FS. Section 4 contains a rough analysis of the two approaches when neither is exactly true. There is a brief discussion in Section 5. Much of our analysis is similar to what has been presented elsewhere [2,3] but from a somewhat different point of view. Throughout, we use atomic units.

2. Diabatic basis

We consider the subspace SS of the electronic Hilbert space HS, spanned by n Born-Oppenheimer electronic states $|\alpha(\boldsymbol{R})\rangle$, where $\boldsymbol{R}$ is the space of the nuclear coordinates. The NACT matrix $\tau_j(\boldsymbol{R})$ is defined by

$$\langle\alpha|\tau_j|\beta\rangle = \langle\alpha|\partial_j\beta\rangle \tag{1}$$

where j is any one of the nuclear coordinates.

A diabatic basis is a basis of kets $|\bar{\alpha}\rangle$ defined by

Journal of Physics: Conference Series **428** (2013) 012009 doi:10.1088/1742-6596/428/1/012009

$$|\bar{\alpha}\rangle = \sum_{\gamma} |\gamma\rangle A_{\gamma\alpha} \tag{2}$$

such that

$$\langle\alpha|\bar{\tau}_j|\beta\rangle \equiv \langle\bar{\alpha}|\partial_j\bar{\beta}\rangle = 0 \,. \tag{3}$$

It can easily be shown [1] that the matrix A must satisfy

$$\partial_j A = -\tau_j A \,. \tag{4}$$

Furthermore, if A is to satisfy (4) and be "locally single-valued", i.e. single-valued upon being taken around an infinitesimal closed curve not surrounding a singularity, we must have the "curl condition"

$$\partial_k \tau_j - \partial_j \tau_k + \tau_k \tau_j - \tau_j \tau_k = 0 \,. \tag{5}$$

Equation (5) is the necessary and sufficient condition for the existence of an exact ADT.

3. Consequences of curl condition

In this section, we assume that we have a basis of electronic kets over the entire HS, including SS and its reciprocal space RS, which are required to be locally single-valued. We define a matrix T over HS as

$$\langle\alpha|T_j|\beta\rangle = \langle\alpha|\partial_j\beta\rangle \,. \tag{6}$$

The necessary and sufficient condition for local single-valuedness is now

$$\partial_k T_j - \partial_j T_k + T_k T_j - T_j T_k = 0. \tag{7}$$

We can divide the matrix T into its parts referring to SS (τ), RS (S), and the parts coupling SS and RS (t) as follows:

$$T_j = \begin{pmatrix} \tau_j & t_j \\ -t_j^\dagger & S_j \end{pmatrix} . \tag{8}$$

The part of (7) that refers to SS is

$$\partial_k \tau_j - \partial_j \tau_k + \tau_k \tau_j - \tau_j \tau_k - t_k t_j^\dagger + t_j t_k^\dagger = 0 \,. \tag{9}$$

Journal of Physics: Conference Series **428** (2013) 012009 doi:10.1088/1742-6596/428/1/012009

Equations (7) and (9) are rigorous, depending only on the local single-valuedness of the basis kets. If there is to be an exact ADT, the curl condition (5) must also be satisfied. Subtracting (5) from (9), we obtain

$$t_j t_k^\dagger - t_k t_j^\dagger = 0\,. \tag{10}$$

Equation (10) is thus equivalent to the curl condition, and is necessary and sufficient for the existence of an exact ADT in SS.

There are basically two ways for (10) to be satisfied (or approximately satisfied), viz: (i) the elements of the *t*-matrices vanish (or are small); or (ii) the elements are neither zero nor small, but the combination in (10) cancels to zero (or a small number). Although possibility (ii) cannot be rigorously ruled out, it does require such a fortuitous cancellation that its appearanc in any actual molecular context must be extremely unlikely. Certainly any claim that we have (ii) in a real case would require proof. Hence, we will dismiss (ii) and concentrate on (i) as the only way in which an ADT in SS can exist.

For a diabatic basis to exist, then, we must have

$$t_j = 0\,, \tag{11}$$

i.e.

$$\langle \kappa | \partial_j \alpha \rangle = 0 \tag{12}$$

for any $|\kappa>$ in RS and $|\alpha>$ in SS. It is also easy to see that the projection operator on SS,

$$P_{SS} = \sum_{\alpha=1}^{n} |\alpha\rangle\langle\alpha| \tag{13}$$

must be independent of the nuclear coordinates, and thus so must SS itself. If $|\alpha>$ is a ket of a diabatic basis, however, we also have for any $|\beta>$ in SS

$$\langle \beta | \partial_j \alpha \rangle = 0\,. \tag{14}$$

The derivatives of the diabatic kets are thus orthogonal to all vectors in HS, and the diabatic kets are independent of the nuclear coordinats. They are frozen states, FS.

Thus, if a diabatic basis exist at all, it is unnecessary to try to integrate eq. (4) to find it. One just needs to find a basis in SS for some ***R***, and then use those frozen kets for all ***R***.

4. Approximate validity of ADT or FS

What, however, if the *t*-matrix elements are not exactly zero, but are small, say of order ε? This would mean in effect that the length scale for significant change in P_{SS} must be of order $1/\varepsilon$. A glance at eq. (9) then shows that the length scale appying to the τ-matrix elements must also be of order $1/\varepsilon$. This means that the electronic states of interest are insensitive to changes in nuclear coordinates of order of a few bohr or less, meaning that FS is also a good approximation for regions of dimensions no more that a few bohr. Such a condition is best satisfied for Rydberg states. For such states, the separation of nuclear energy levels is also small, and it is also unclear that the nonremovable part of the NACT will be small compared with the removable part.

Our rough analysis shows that, in many cases if not most when ADT delivers a good approximation, the ADT and FS approaches do not differ significantly from each other in results, as long as one considers a region in nuclear space of dimension no more than a few bohr, i.e. small compared with the length scale $1/\varepsilon$, and thus there is no need to make the much greater computational effort required by ADT. For larger regions, or order $1/\varepsilon$, the projection operator P_{SS} will change significantly, and the FS approach will fail.

5. Discussion

Our rough analysis indicates that there are many cases in which the existence of an accurate ADT also requires that the FC approach must be equally accurate in terms of order of magnitude. It must be emphasized, however, that this rough analysis is not rigorous for all cases, and also does not deal with possibility (ii) of Sec. 3. There may thus be real cases in which the ADT approach is most worthwhile.

References

[1] Baer M 2002 Phys. Rep. **358** 75
[2] McLachlan A D 1961 Mol. Phys. **4** 417
[3] Mead C A and Truhlar D G 1982 J. Chem. Phys. **77** 6090

Vibronic coupling density and related concepts

Tohru Sato[a,b], Motoyuki Uejima[a], Naoya Iwahara[a], Naoki Haruta[a], Katsuyuki Shizu[a], Kazuyoshi Tanaka[a]

[a]Department of Molecular Engineering, Graduate School of Engineering, Kyoto University, Kyoto 615-8510, Japan
[b]Unit of Elements Strategy Initiative for Catalysts & Batteries, Kyoto University, Nishikyo-ku, Kyoto 615-8510, Japan

E-mail: tsato@moleng.kyoto-u.ac.jp

Abstract. Vibronic coupling density is derived from a general point of view as a one-electron property density. Related concepts as well as their applications are presented. Linear and non-linear vibronic coupling density and related concepts, orbital vibronic coupling density, reduced vibronic coupling density, atomic vibronic coupling constant, and effective vibronic coupling density, illustrate the origin of vibronic couplings and enable us to design novel functional molecules or to elucidate chemical reactions. Transition dipole moment density is defined as an example of the one-electron property density. Vibronic coupling density and transition dipole moment density open a way to design light-emitting molecules with high efficiency.

1. Introduction

Vibronic couplings[1, 2, 3] give rise to Jahn-Teller effect in a degenerate state, reorganization energy in an ionized state, and non-radiative decay and vibrational relaxation in an excited state. Vibronic couplings have been investigated in density forms[4, 5]. The spatial integral of the density form, *vibronic coupling density* (VCD), yields a vibronic coupling constant (VCC). Thus, the distribution of a VCD provides an insight into the magnitude of a VCC. Moreover, the vibronic interaction can be analyzed on the basis of the electronic and vibrational structures by using the VCD concept, since the VCD can be decomposed as the product of an electron density/overlap density and a potential derivative.

The VCD concept has been applied for the designs of carrier-transporting molecules [6, 7, 8], analyses of inelastic tunneling spectroscopy[9], Jahn-Teller couplings in C_{60}[10], electron-vibration coupling in hydrocarbons [4, 11], and vibronic progression in the spectrum of an organic molecule[12].

In this article, we derive vibronic coupling densities from a general point of view, one-electron property density. We propose transition dipole moment density as other example of the one-electron property density. We summarize related concepts to VCD, orbital vibronic coupling density, reduced vibronic coupling density, atomic vibronic coupling constant, and effective vibronic coupling density. We present some illustrative examples of their applications. In Sec. 2, we define vibronic couplings, and they are expressed in density forms in Sec.3 as a one-electron property density. In Sec. 4, we define transition dipole moment density. We summarize related concepts to VCD in Sec. 5. Their applications are presented in Sec. 6. We summarize these results in Sec.7.

2. Vibronic Coupling Densities

We regard a stationary geometry $\mathbf{R}_0$ for an electronic state Ψ_0 as a reference system. A molecular Hamiltonian, $\hat{H} = \hat{T}_e + \hat{U}_{ee} + \hat{U}_{en} + U_{nn}$ where $\hat{T}_e$ denotes kinetic operator of electrons, $\hat{U}_{ee}$ electron-electron potential, $\hat{U}_{en}$ electron-nucleus potential, U_{nn} nucleus-nucleus potential, is expanded as the Hertzberg-Teller series in terms of a normal coordinate Q_α around the reference geometry $\mathbf{R}_0$:

$$\begin{aligned}
\hat{H}(\{Q_\alpha\}) &= \hat{H}(0) + \sum_\alpha \left(\frac{\partial \hat{H}}{\partial Q_\alpha}\right)_{\mathbf{R}_0} Q_\alpha + \frac{1}{2}\sum_{\alpha,\beta}\left(\frac{\partial^2 \hat{H}}{\partial Q_\alpha \partial Q_\beta}\right)_{\mathbf{R}_0} Q_\alpha Q_\beta \\
&\quad + \cdots + \frac{1}{n!}\sum_{\alpha_1,\cdots,\alpha_n}\left(\frac{\partial^n \hat{H}}{\partial Q_{\alpha_1}\cdots\partial Q_{\alpha_n}}\right)_{\mathbf{R}_0} Q_{\alpha_1}\cdots Q_{\alpha_n} + \cdots \qquad (1)\\
&= \hat{H}(0) + \sum_\alpha \hat{V}_\alpha Q_\alpha + \frac{1}{2}\sum_{\alpha,\beta}\hat{W}_{\alpha\beta}Q_\alpha Q_\beta + \cdots \\
&\quad + \frac{1}{n!}\sum_{\alpha_1,\cdots,\alpha_n}\hat{W}^{(n)}_{\alpha_1\cdots\alpha_n}Q_{\alpha_1}\cdots Q_{\alpha_n} + \cdots, \qquad (2)
\end{aligned}$$

where

$$\hat{V}_\alpha = \left(\frac{\partial \hat{H}}{\partial Q_\alpha}\right)_{\mathbf{R}_0} \qquad (3)$$

is the electronic part of a linear vibronic coupling,

$$\hat{W}_{\alpha\beta} = \left(\frac{\partial^2 \hat{H}}{\partial Q_\alpha \partial Q_\beta}\right)_{\mathbf{R}_0} \qquad (4)$$

is the electronic part of a quadratic vibronic coupling, and

$$\hat{W}^{(n)}_{\alpha_1\cdots\alpha_n} = \left(\frac{\partial^n \hat{H}}{\partial Q_{\alpha_1}\cdots\partial Q_{\alpha_n}}\right)_{\mathbf{R}_0} \qquad (5)$$

is the electronic part of an nth vibronic coupling. Note that $\hat{W}^{(1)}_\alpha = \hat{V}_\alpha$ and $\hat{W}^{(2)}_{\alpha\beta} = \hat{W}_{\alpha\beta}$.

The nuclear-electronic attractive potential $\hat{U}_{en}$ is written as

$$\hat{U}_{en} = \sum_i u(\mathbf{r}_i), \qquad (6)$$

where $\mathbf{r}_i$ is the position of electron i, and

$$u(\mathbf{r}_i) = -\sum_A \frac{Z_A e^2}{|\mathbf{r}_i - \mathbf{R}_A|} \qquad (7)$$

is the potential from nuclei A with the position $\mathbf{R}_A$ and the charge Z_A acting on a single electron. The nuclear-nuclear repulsive potential U_{nn} is written as

$$U_{nn} = \sum_{A,B}\frac{Z_A Z_B e^2}{|\mathbf{R}_B - \mathbf{R}_A|}. \qquad (8)$$

Journal of Physics: Conference Series **428** (2013) 012010 doi:10.1088/1742-6596/428/1/012010

The nth order derivative of potential u with respect to $Q_{\alpha_1}, \cdots, Q_{\alpha_n}$ is denoted by

$$\hat{w}^{(n)}_{\alpha_1,\cdots,\alpha_n}(\mathbf{r}) := \left(\frac{\partial^n u(\mathbf{r})}{\partial Q_{\alpha_1}\cdots\partial Q_{\alpha_n}}\right)_{\mathbf{R}_0}. \tag{9}$$

For the linear and quadratic derivative, we define

$$\hat{v}_\alpha(\mathbf{r}) := \hat{w}^{(1)}_\alpha(\mathbf{r}) = \left(\frac{\partial u(\mathbf{r})}{\partial Q_\alpha}\right)_{\mathbf{R}_0} \tag{10}$$

and

$$\hat{w}_{\alpha\beta}(\mathbf{r}) := \hat{w}^{(2)}_{\alpha\beta}(\mathbf{r}) = \left(\frac{\partial^2 u(\mathbf{r})}{\partial Q_\alpha \partial Q_\beta}\right)_{\mathbf{R}_0}, \tag{11}$$

respectively. The vibronic operators can be decomposed as follows:

$$\hat{V}_\alpha = \sum_i \hat{v}_\alpha(\mathbf{r}_i) + \left(\frac{\partial U_{nn}}{\partial Q_\alpha}\right)_{\mathbf{R}_0}, \tag{12}$$

$$\hat{W}_{\alpha\beta} = \sum_i \hat{w}_{\alpha\beta}(\mathbf{r}_i) + \left(\frac{\partial^2 U_{nn}}{\partial Q_\alpha \partial Q_\beta}\right)_{\mathbf{R}_0}, \tag{13}$$

$$\hat{W}^{(n)}_{\alpha_1,\cdots,\alpha_n} = \sum_i \hat{w}^{(n)}_{\alpha_1,\cdots,\alpha_n}(\mathbf{r}_i) + \left(\frac{\partial^n U_{nn}}{\partial Q_{\alpha_1}\cdots\partial Q_{\alpha_n}}\right)_{\mathbf{R}_0}. \tag{14}$$

It should be noted that the vibronic coupling operators consist of a sum of the one-electron operators, $\hat{v}_\alpha$, $\hat{w}_{\alpha\beta}$, and $\hat{w}^{(n)}_{\alpha_1\cdots\alpha_n}$, respectively.

The linear vibronic-coupling matrix is defined as

$$(\mathbf{V}_\alpha)_{lm} = \left\langle \Psi_l \left| \hat{V}_\alpha \right| \Psi_m \right\rangle, \tag{15}$$

where Ψ_l and Ψ_m are electronic wavefunctions at the reference geometry $\mathbf{R}_0$. We simply call a diagonal element of the linear vibronic-coupling matrix as a vibronic coupling constant (VCC) V_α hereafter. The quadratic vibronic-coupling matrix is defined as

$$(\mathbf{W}_{\alpha\beta})_{lm} = \left\langle \Psi_l \left| \hat{W}_{\alpha\beta} \right| \Psi_m \right\rangle, \tag{16}$$

and the nth order vibronic-coupling matrix is defined as

$$(\mathbf{W}^{(n)}_{\alpha_1\cdots\alpha_n})_{lm} = \left\langle \Psi_l \left| \hat{W}^{(n)}_{\alpha_1\cdots\alpha_n} \right| \Psi_m \right\rangle. \tag{17}$$

We simply call a diagonal element of the quadratic vibronic-coupling matrix as a quadratic vibronic-coupling constant (QVCC) $W_{\alpha\beta}$ hereafter.

The diagonal linear vibronic coupling density, which is simply called as a vibronic coupling density (VCD), is defined as

$$\eta_\alpha(\mathbf{r}) = \Delta\rho(\mathbf{r}) \times v_\alpha(\mathbf{r}), \tag{18}$$

where $\Delta\rho$ is an electron density difference between Ψ and the reference electronic state Ψ_0. The integral of the VCD yields the VCC:

$$V_\alpha = \int \eta_\alpha(\mathbf{r}) d^3\mathbf{r} = \int \Delta\rho(\mathbf{r}) \times v_\alpha(\mathbf{r}) d^3\mathbf{r}. \tag{19}$$

This equation is proved in the following Section. From this equation, the VCC can be analyzed based on electronic and vibrational states.

For the nth-order coupling, vibronic coupling density between Ψ_k and Ψ_l is defined as

$$\eta^{(n)}_{\alpha_1\cdots\alpha_n,kl}(\mathbf{r}) = \rho_{kl}(\mathbf{r}) \times \hat{w}^{(n)}_{\alpha_1\cdots\alpha_n}(\mathbf{r}). \tag{20}$$

$\rho_{kl}(\mathbf{r})$ is defined in the following Section. The diagonal matrix element of the nth-order vibronic-coupling matrix is written as

$$(\mathbf{W}^{(n)}_{\alpha_1\cdots\alpha_n})_{kk} = \int \rho_k(\mathbf{r}) \times \hat{w}^{(n)}_{\alpha_1\cdots\alpha_n}(\mathbf{r})d^3\mathbf{r} + \left(\frac{\partial^n U_{nn}}{\partial Q_{\alpha_1}\cdots\partial Q_{\alpha_n}}\right)_{\mathbf{R}_0}, \tag{21}$$

where $\rho_k = \rho_{kk}$. The off-diagonal element is written as

$$(\mathbf{W}^{(n)}_{\alpha_1\cdots\alpha_n})_{kl} = \int \rho_{kl}(\mathbf{r}) \times \hat{w}^{(n)}_{\alpha_1\cdots\alpha_n}(\mathbf{r})d^3\mathbf{r}. \tag{22}$$

For the quadratic coupling,

$$\eta^{(2)}_{\alpha\beta,kl}(\mathbf{r}) = \rho_{kl}(\mathbf{r}) \times \hat{w}_{\alpha\beta}(\mathbf{r}). \tag{23}$$

The QVCC is written as

$$(\mathbf{W}_{\alpha\beta})_{kk} = \int \rho_k(\mathbf{r}) \times \hat{w}_{\alpha\beta}(\mathbf{r})d^3\mathbf{r} + \left(\frac{\partial^2 U_{nn}}{\partial Q_\alpha \partial Q_\beta}\right)_{\mathbf{R}_0}, \tag{24}$$

and the off-diagonal QVCC

$$(\mathbf{W}_{\alpha\beta})_{kl} = \int \rho_{kl}(\mathbf{r}) \times \hat{w}_{\alpha\beta}(\mathbf{r})d^3\mathbf{r}. \tag{25}$$

3. Vibronic Coupling Density as a One-Electron Property Density

We prove Eqs. (19), (21), (22), (24), and (25) from a general point of view, since the vibronic coupling operators are a one-electron operator. The proof for VCDs including non-linear VCDs have been already published[8]. The present proof is an extension as a one-electron operator. The electron density of the N-electron state $\Psi_k(\mathbf{x}_1, \mathbf{x}_2, \cdots, \mathbf{x}_N)$ which can be either an ionized or an excited state is defined by

$$\rho_k(\mathbf{r}) := N \int \Psi^*_k(\mathbf{r}\, s_1, \mathbf{x}_2, \cdots, \mathbf{x}_N)\Psi_k(\mathbf{r}\, s_1, \mathbf{x}_2, \cdots, \mathbf{x}_N)ds_1\, d\mathbf{x}_2 \cdots d\mathbf{x}_N, \tag{26}$$

where $\mathbf{x}_i$ denotes space coordinate $\mathbf{r}_i$ and spin coordinate s_i[13].

The overlap density between the N-electron states, $\Psi_k(\mathbf{x}_1, \mathbf{x}_2, \cdots, \mathbf{x}_N)$ and $\Psi_l(\mathbf{x}_1, \mathbf{x}_2, \cdots, \mathbf{x}_N)$, is defined by

$$\rho_{kl}(\mathbf{r}) := N \int \Psi^*_k(\mathbf{r}\, s_1, \mathbf{x}_2, \cdots, \mathbf{x}_N)\Psi_l(\mathbf{r}\, s_1, \mathbf{x}_2, \cdots, \mathbf{x}_N)ds_1\, d\mathbf{x}_2 \cdots d\mathbf{x}_N, \tag{27}$$

$$\rho_{kk}(\mathbf{r}) = \rho_k(\mathbf{r}). \tag{28}$$

From the antisymmetric principle for Fermions:

$$\Psi_k(\mathbf{x}_{\sigma(1)}, \mathbf{x}_{\sigma(2)}, \cdots, \mathbf{r}s_{\sigma(i)=1}, \cdots, \mathbf{x}_{\sigma(N)}) = \text{sgn}(\sigma)\, \Psi_k(\mathbf{r}s_1, \mathbf{x}_2, \cdots, \mathbf{x}_N), \tag{29}$$

where $\sigma \in S_N$ (n^{th} order symmetric group). The product of two electronic wavefunctions satisfies

$$\begin{aligned} & \Psi_k^*(\mathbf{x}_{\sigma(1)}, \cdots, \mathbf{r}s_1, \cdots, \mathbf{x}_{\sigma(N)})\, \Psi_l(\mathbf{x}_{\sigma(1)} \cdots, \mathbf{r}s_1, \cdots, \mathbf{r}_{\sigma(N)}) \\ = & \ (\mathrm{sgn}(\sigma))^2\ \Psi_k^*(\mathbf{r}s_1, \mathbf{x}_2, \cdots, \mathbf{x}_N)\Psi_l(\mathbf{r}s_1, \mathbf{x}_2, \cdots, \mathbf{x}_N) \\ = & \ \Psi_k^*(\mathbf{r}s_1, \mathbf{x}_2, \cdots, \mathbf{x}_N)\Psi_l(\mathbf{r}s_1, \mathbf{x}_2, \cdots, \mathbf{x}_N). \end{aligned} \tag{30}$$

According to the antisymmetric principle of Fermions,

$$\begin{aligned} & \int \Psi_k^*(\mathbf{r}\, s_1, \mathbf{x}_2, \cdots, \mathbf{x}_N)\Psi_l(\mathbf{r}\, s_1, \mathbf{x}_2, \cdots, \mathbf{x}_N) ds_1\, d\mathbf{x}_2 \cdots d\mathbf{x}_N \\ = & \int \Psi_k^*(\mathbf{x}_1, \mathbf{r}\, s_2, \cdots, \mathbf{x}_N)\Psi_l(\mathbf{x}_1, \mathbf{r}\, s_2, \cdots, \mathbf{x}_N) d\mathbf{x}_1\, ds_2\, d\mathbf{x}_3 \cdots d\mathbf{x}_N \\ \cdots & \\ = & \int \Psi_k^*(\mathbf{x}_1, \mathbf{x}_2, \cdots, \mathbf{x}_{N-1}, \mathbf{r}\, s_N)\Psi_l(\mathbf{x}_1, \mathbf{x}_2, \cdots, \mathbf{x}_{N-1}, \mathbf{r}\, s_N) d\mathbf{x}_1 \cdots d\mathbf{x}_{N-1}\, ds_N \\ = & \ \frac{1}{N} \rho_{kl}(\mathbf{r}). \end{aligned} \tag{31}$$

This is also the case for the electron density $\rho_{kk}(\mathbf{r}) = \rho_k(\mathbf{r})$.

We consider a one-electron property $\hat{O}_1$ which is a sum of one-electron operators $\hat{o}_1$:

$$\hat{O}_1 = \sum_i \hat{o}_1(\mathbf{r}_i). \tag{32}$$

The vibronic coupling operators are such a one-electron property. An electric dipole moment operator is also a one-electron property operator. We will discuss a transition dipole moment which is regarded as an off-diagonal element of $\hat{\mu}$ as a density form in the following Section.

The matrix element of a one-electron operator $\hat{O}_1$ between Ψ_k and Ψ_l is written as

$$\begin{aligned} (\mathbf{O}_1)_{kl} = & \left\langle \Psi_k \middle| \hat{O}_1 \middle| \Psi_l \right\rangle \\ = & \int d\mathbf{x}_1\, d\mathbf{x}_2 \cdots d\mathbf{x}_N \Psi_k^*(\mathbf{x}_1, \mathbf{x}_2, \cdots, \mathbf{x}_N) \left(\sum_{i=1}^N \hat{o}_1(\mathbf{r}_i) \right) \Psi_l(\mathbf{x}_1, \mathbf{x}_2, \cdots, \mathbf{x}_N) \\ = & \int d\mathbf{r}_1\, \hat{o}_1(\mathbf{r}_1) \int ds_1\, d\mathbf{r}_2 \cdots d\mathbf{r}_N \Psi^*(\mathbf{r}_1\, s_1, \mathbf{x}_2, \cdots, \mathbf{x}_N)\Psi(\mathbf{r}_1\, s_1, \mathbf{x}_2, \cdots, \mathbf{x}_N) \\ & + \cdots + \\ & \int d\mathbf{r}_N\, \hat{o}_1(\mathbf{r}_N) \int d\mathbf{x}_1 \cdots d\mathbf{x}_{N-1}\, ds_N \Psi^*(\mathbf{x}_1, \mathbf{x}_2, \cdots, \mathbf{r}_N\, s_N)\Psi(\mathbf{x}_1, \mathbf{x}_2, \cdots, \mathbf{r}_N\, s_N) \\ = & \ \frac{1}{N} \int d\mathbf{r}_1\, \rho_{kl}(\mathbf{r}_1)\, \hat{o}_1(\mathbf{r}_1) + \cdots + \frac{1}{N} \int d\mathbf{r}_N\, \rho_{kl}(\mathbf{r}_N)\, \hat{o}_1(\mathbf{r}_N) \\ = & \int d\mathbf{r}\, \rho_{kl}(\mathbf{r})\, \hat{o}_1(\mathbf{r}) = \int d\mathbf{r}\, \Omega_{kl}(\mathbf{r}), \end{aligned} \tag{33}$$

where the one-electron property density $\Omega_{kl}(\mathbf{r})$ is defined as

$$\Omega_{kl}(\mathbf{r}) := \rho_{kl}(\mathbf{r}) \times \hat{o}_1(\mathbf{r}). \tag{34}$$

Therefore a one-electron property can be discussed in the density form.

For the vibronic coupling operators except for constants of the nuclear-nuclear potential derivatives, $\mathbf{O}_1 = \mathbf{W}^{(n)}_{\alpha_1 \cdots \alpha_n}$, $\hat{o}_1 = \hat{w}^{(n)}_{\alpha_1 \cdots \alpha_n}$, and $\Omega_{kl}(\mathbf{r}) = \eta^{(n)}_{\alpha_1 \cdots \alpha_n, kl}(\mathbf{r})$. Since the reference geometry $\mathbf{R}_0$ is taken at a stationary point of the reference electronic configuration,

$$\int \rho_0 \times \hat{v}_\alpha(\mathbf{r}) d^3\mathbf{r} + \left(\frac{\partial U_{nn}}{\partial Q_\alpha} \right)_{\mathbf{R}_0} = 0, \tag{35}$$

the first derivative of the nuclear-nuclear potential can be eliminated from the expression of the VCC:

$$V_\alpha = \int \rho_k(\mathbf{r}) \times \hat{v}_\alpha(\mathbf{r}) d^3\mathbf{r} - \int \rho_0(\mathbf{r}) \times \hat{v}_\alpha(\mathbf{r}) d^3\mathbf{r} = \int \Delta\rho_k(\mathbf{r}) \times \hat{v}_\alpha(\mathbf{r}) d^3\mathbf{r}, \tag{36}$$

where $\Delta\rho_k(\mathbf{r})$ is the electron density difference between Ψ_k and the reference state Ψ_0. Because of the stationary condition (35), the nuclear-nuclear part of the potential is eliminated from the expression for the linear coupling V_α (36).

For a Slater determinant Φ_0 constructed by orthonormal spin orbitals, $\{\chi(\mathbf{x}_i) = \psi(\mathbf{r}_i)\omega(s_i)\}$, a diagonal element of the one-electron property can be decomposed as[14]

$$\left\langle \Phi_0 \middle| \hat{O}_1 \middle| \Phi_0 \right\rangle = \sum_{a \in \mathrm{occ}} n_a \left\langle \psi_a(\mathbf{r}_i) \left| \hat{o}_1(\mathbf{r}_i) \right| \psi_a(\mathbf{r}_i) \right\rangle,$$

where n_a is the occupation number of ψ_a. For an off-diagonal element between Φ_0 and Φ_a^b with orbital ψ_a in Φ_0 replaced by ψ_b,

$$\left\langle \Phi_0 \middle| \hat{O}_1 \middle| \Phi_a^b \right\rangle = \left\langle \psi_a(\mathbf{r}) \left| \hat{o}_1(\mathbf{r}) \right| \psi_b(\mathbf{r}) \right\rangle$$

We define orbital one-electron property density $\omega_{ab}(\mathbf{r})$ as

$$\omega_{ab}(\mathbf{r}) := \psi_a^*(\mathbf{r})\psi_b(\mathbf{r})\hat{o}_1(\mathbf{r}) = \rho_{ab}(\mathbf{r}) \times \hat{o}_1(\mathbf{r}),$$

where $\rho_{ab}(\mathbf{r})$ denotes orbital overlap density (transition density). Orbital one-electron property can be decomposed in the density form:

$$\left\langle \psi_a(\mathbf{r}) \left| \hat{o}_1(\mathbf{r}) \right| \psi_b(\mathbf{r}) \right\rangle = \int \omega_{ab}(\mathbf{r}) d^3\mathbf{r}. \tag{37}$$

4. Other Example of One-Electron Property Density: Transition Dipole Moment Density

As an example of the one-electron property, we consider an electric dipole moment operator

$$\hat{\mu} = \sum_i e\mathbf{r}_i. \tag{38}$$

Therefore, $\hat{o}_1 = e\,\mathbf{r}_i$. We define an electric dipole-moment density as

$$\tau_{kl}(\mathbf{r}) = \rho_{kl}(\mathbf{r})\,(e\mathbf{r}). \tag{39}$$

For an electronic state Ψ_0, the diagonal element of $\hat{\mu}$ yields a permanent electric dipole moment of electrons:

$$\mu = \int \rho_0(\mathbf{r})\,(e\mathbf{r}) d^3\mathbf{r} = \int \tau_{00}(\mathbf{r}) d^3\mathbf{r}. \tag{40}$$

The integrand is called a permanent dipole-moment density which is the diagonal element of the electric dipole-moment density.

The off-diagonal element of $\hat{\mu}$ between Ψ_k and Ψ_l is a transition dipole moment.

$$\mu_{kl} = \int \rho_{kl}(\mathbf{r})\,(e\mathbf{r}) d^3\mathbf{r} = \int \tau_{kl}(\mathbf{r}) d^3\mathbf{r}, \tag{41}$$

where $\tau_{kl}(\mathbf{r})$ is called the transition dipole moment density (TDMD) which is the off-diagonal element of the electric dipole-moment density. Since absorptions, emissions, and scattering are dominated by the transition dipole moment, we can analyze and control optical processes via the electronic structures of molecules. For example, light-emitting or light-harvesting molecules with high efficiency can be designed based on the TDMD (39). We will investigate the transition dipole moments of anthracene and its derivatives later.

Journal of Physics: Conference Series **428** (2013) 012010 doi:10.1088/1742-6596/428/1/012010

5. Related Concepts to Vibronic Coupling Density

5.1. *Orbital Vibronic Coupling Density*

Vibronic coupling constant can be decomposed into orbital contributions. n^{th} order orbital vibronic coupling constant (OVCC) between orbitals ψ_a and ψ_b for modes $\alpha_1, \alpha_2, \cdots, \alpha_n$ is defined by

$$\left\langle \psi_a(\mathbf{r}) \left| \hat{w}^{(n)}_{\alpha_1 \cdots \alpha_n}(\mathbf{r}) \right| \psi_b(\mathbf{r}) \right\rangle. \tag{42}$$

The OVCC is also equal to the spatial integral of a density function $o^{(n)}_{\alpha_1 \cdots \alpha_n, ab}(\mathbf{r})$:

$$\left\langle \psi_a(\mathbf{r}) \left| \hat{w}^{(n)}_{\alpha_1 \cdots \alpha_n}(\mathbf{r}) \right| \psi_b(\mathbf{r}) \right\rangle = \int \rho_{ab}(\mathbf{r}) \times \hat{w}^{(n)}_{\alpha_1 \cdots \alpha_n}(\mathbf{r}) d^3\mathbf{r} = \int o^{(n)}_{\alpha_1 \cdots \alpha_n, ab}(\mathbf{r}) d^3\mathbf{r}, \tag{43}$$

where orbital vibronic coupling density (OVCD) is defined as

$$o^{(n)}_{\alpha_1 \cdots \alpha_n, ab}(\mathbf{r}) := \rho_{ab}(\mathbf{r}) \times \hat{w}^{(n)}_{\alpha_1 \cdots \alpha_n}(\mathbf{r}). \tag{44}$$

5.2. *Reduced Vibronic Coupling Density*

The potential derivatives $\hat{w}^{(n)}_{\alpha_1 \cdots \alpha_n}(\mathbf{r})$ have symmetric distribution near atoms. This gives rise to cancellation as we will see some examples later. We can eliminate such a zero contribution $\eta'_\alpha(\mathbf{r})$ from $\eta_\alpha(\mathbf{r})$ to define the reduced vibronic coupling density (RVCD) $\bar{\eta}_\alpha(\mathbf{r})$ which describes net contributions to the VCC [15]:

$$\bar{\eta}_\alpha(\mathbf{r}) = \eta_\alpha(\mathbf{r}) - \eta'_\alpha(\mathbf{r}), \tag{45}$$

and

$$\int \eta'_\alpha(\mathbf{r}) d^3\mathbf{r} = 0. \tag{46}$$

Therefore,

$$V_\alpha = \int \eta_\alpha(\mathbf{r}) d^3\mathbf{r} = \int \bar{\eta}_\alpha(\mathbf{r}) d^3\mathbf{r}. \tag{47}$$

5.3. *Atomic Vibronic Coupling Constant*

Nuclear-electronic and nuclear-nuclear potentials for nuclei A and B are written as

$$U_{en,A} = \sum_i -\frac{Z_A e^2}{|\mathbf{r}_i - \mathbf{R}_A|}, \tag{48}$$

and

$$U_{nn,A} = \sum_{B \neq A} \frac{Z_A Z_B e^2}{|\mathbf{R}_A - \mathbf{R}_B|}, \tag{49}$$

respectively, where $\mathbf{r}_i$ denotes the position of electron i, $\mathbf{R}_A$ that of nucleus A.

Nuclear-electronic attracting and nuclear-nuclear repulsive potentials are given by

$$U_{en} = \sum_A U_{en,A} = \sum_A \sum_i -\frac{Z_A e^2}{|\mathbf{r}_i - \mathbf{R}_A|}, \tag{50}$$

and

$$U_{nn} = \frac{1}{2} \sum_A U_{nn,A} = \frac{1}{2} \sum_A \sum_{B \neq A} \frac{Z_A Z_B e^2}{|\mathbf{R}_A - \mathbf{R}_B|}, \tag{51}$$

respectively.

Atomic vibronic coupling constant (AVCC) is defined as [16]

$$V_{mn,\alpha,A} = \left\langle \Psi_m \left| \left(\frac{\partial U_{en,A}}{\partial Q_\alpha} \right)_{\mathbf{R}_0} \right| \Psi_n \right\rangle + \delta_{mn} \left(\frac{\partial U_{nn,A}}{\partial Q_\alpha} \right)_{\mathbf{R}_0}. \tag{52}$$

The total sum of the AVCCs over all the nuclei is equal to the VCC:

$$V_\alpha = \sum_A V_{mn,\alpha,A}. \tag{53}$$

Therefore, the AVCC is an atomic contribution to the VCC.

5.4. *Effective Vibronic Coupling Density*

As we will see later, the VCD for a reaction mode can be regarded as a reactivity index[17]. An effective mode can be a candidate of the reaction mode. The effective mode s is defined by

$$s = \sum_\alpha \frac{V_\alpha}{\sqrt{\sum_{\alpha'} V_{\alpha'}^2}} Q_\alpha. \tag{54}$$

The VCD for the effective mode s which is called the effective VCD (EVCD) is defined as

$$\eta_s(\mathbf{r}) = \Delta\rho(\mathbf{r}) \times \left(\frac{\partial u(\mathbf{r})}{\partial s} \right)_{\mathbf{R}_0} = \Delta\rho(\mathbf{r}) \times v_s(\mathbf{r}). \tag{55}$$

6. Applications

In this Section, we present some illustrative examples. All the calculation were performed using the Gaussian09[18].

6.1. *Jahn-Teller Couplings in C_3H_3*

The VCD analysis is applied for a Jahn-Teller molecule, C_3H_3[19].

We have demonstrated a crucial role of orbital relaxation in vibronic couplings based on the VCD analysis. For non-degenerate π system, naphthalene anion, the VCCs are overestimated/underestimated without σ-type polarization taken into consideration[4]. For degenerate π system, C_{60} monoanion, the VCD analysis using a broken-symmetry wavefunction also suggests importance of orbital relaxation[10]. Calculations based on a single determinant give rise to a symmetry-broken wavefunction, though the energy splittings of degenerate levels are small in DFT calculations [20]. Nevertheless, for Jahn-Teller systems, a broken-symmetry wavefunction is not appropriate. A computational method in which both symmetry and orbital relaxation can be taken into account is the state-averaged CASSCF method. One should note that for the π systems CASSCF calculations including only π orbitals in the active space cannot reproduce the σ-type polarization, and the VCCs will not be reliable in such calculations. Therefore, the active space of the CASSCF calculations must include σ as well as π orbitals.

We evaluated the VCCs from the state-averaged CASSCF(7,8) wavefunction with the correct symmetry which includes the σ orbitals in the active space. The non-degenerate $C_3H_3^+$ is considered as the reference system. The optimization and frequency analysis for the reference system were performed within the RHF/STO-3G level of theory. The VCCs were calculated from the analytical gradients in C_3H_3.

In a single-determinant method, the Hartree-Fock and DFT method, the VCCs of Jahn-Teller modes calculated from wavefunctions with correct symmetries are equal to the OVCCs of the frontier orbitals due to the symmetry of the Clebsh-Gordan coefficients. In the

Journal of Physics: Conference Series **428** (2013) 012010 doi:10.1088/1742-6596/428/1/012010

CASSCF calculation, however, this is not the case, since it includes the orbital relaxations and configuration interactions. For comparison with the VCCs, the OVCCs were estimated from the numerical gradients of the LUMO levels.

Table 1 summarizes the calculated VCCs and OVCCs for the vibronically active modes, $2A_1^{'} \oplus 3E^{'}$, including the reorganization energies calculated from

$$E = \sum_{\alpha} \frac{V_\alpha^2}{2\omega_\alpha^2}. \tag{56}$$

The symmetric wavefunction yields the VCCs and OVCCs with the correct symmetry of the

Table 1. Frequencies, ν (cm^{-1}), vibronic coupling constants of C_3H_3, V_α (10^{-4} a.u.), the gradients of the LUMOs of $C_3H_3^+$, $\Delta\epsilon_L/\Delta Q_\alpha$ (10^{-4} a.u.), and the stabilization energies after A_1' modes (E_A) and E' modes (E_{JT}) (meV).

Mode	ν	V_α		$\Delta\epsilon_L/\Delta Q_\alpha$	
		Ψ_θ	Ψ_ϵ	ψ_θ	ψ_ϵ
$A_1'(1)$	1913.22	-9.826	-9.826	-2.436	-2.436
$A_1'(2)$	3757.99	-13.629	-13.629	-14.237	-14.237
$E'(1)\theta$	1059.77	-2.826	2.826	-3.943	3.938
$E'(1)\epsilon$	1059.77	-0.003	0.051		
$E'(2)\theta$	1480.32	-9.217	9.215	-11.228	11.248
$E'(2)\epsilon$	1480.32	-0.010	0.165		
$E'(3)\theta$	3670.56	-4.603	4.602	-6.485	6.430
$E'(3)\epsilon$	3670.56	-0.005	0.082		
E_A		259.08	259.08	104.69	104.69
E_{JT}		310.97	310.97	488.23	488.98

Clebsh-Gordan coefficients. Among the Jahn-Teller active modes, $E'(2)$ has the maximum coupling. The OVCCs with respect to $E^{'}$ Jahn-Teller active modes are larger than the corresponding VCCs. This difference between the VCC and the OVCC is explained on the basis of the VCD analysis. The AVCCs are tabulated in Table 2. For the $E'(2)$ mode, the main contribution comes from 1-C.

Table 2. Atomic vibronic coupling constants for C_3H_3 (10^{-4} a.u.)

ATOM	Mode				
	$a_1'(1)$	$a_1'(2)$	$e'(1)\theta$	$e'(2)\theta$	$e'(3)\theta$
1-C	-7.775	-3.393	-3.426	-7.193	-3.749
2-C	-2.741	-1.196	0.467	-1.304	0.535
3-C	-2.735	-1.193	0.466	-1.304	0.533
4-H	1.522	-3.489	0.543	1.205	-5.110
5-H	0.951	-2.180	-0.439	-0.310	1.595
6-H	0.950	-2.178	-0.438	-0.310	1.593
V_α	-9.826	-13.629	-2.826	-9.217	-4.603

Figure 1 shows the LUMO density and the electron density difference. Since the LUMO is a π orbital, the positive orbital density $\rho_{LU\theta}(\mathbf{r})$ is distributed outside the molecular plane. On

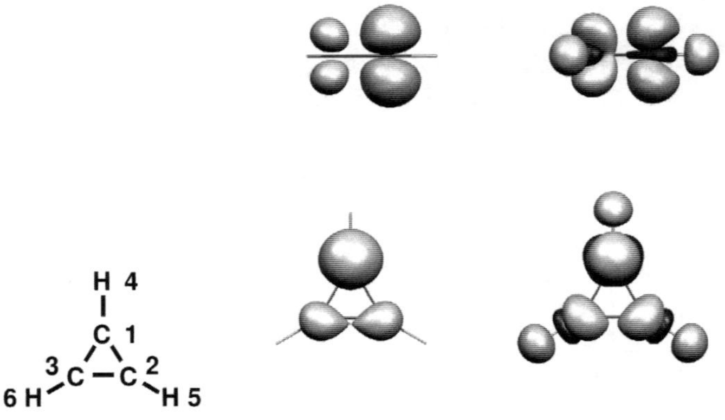

Figure 1. LUMO density (θ) $\rho_{\mathrm{LU}\theta}$ of $C_3H_3^+$ and $\Delta\rho$ of C_3H_3. The gray region is positive, and the blue region is negative. Threshold is 0.01 a.u.

the other hand, the electron density difference $\Delta\rho$ is different from the orbital density $\rho_{\mathrm{LU}\theta}(\mathbf{r})$. It should be noted that the negative contribution which originates from polarizations of doubly occupied σ orbitals appears in the molecular plane.

Figure 2 shows the VCD analysis for the $E'(2)$ mode. Since the displacements of the mode and the potential derivative (Figure 2(b)) are in the molecular plane, the σ polarization in $\Delta\rho$ yields large contributions in the VCDs. The negative VCD in the molecular plane gives rise to large contribution to the AVCC on 1-C (Figure 2, Table 2). This in-plane VCD originates from the σ density and will not appear without the orbital relaxation. The OVCC does not include the σ-polarization, and hence the OVCCs for the E' modes are larger than the VCCs. Thus, the OVCCs cannot be regarded as the VCCs. The VCD analysis suggests importance of the orbital relaxation in VCC calculations for both the non-Jahn–Teller and Jahn–Teller system.

(a) (b) (c)

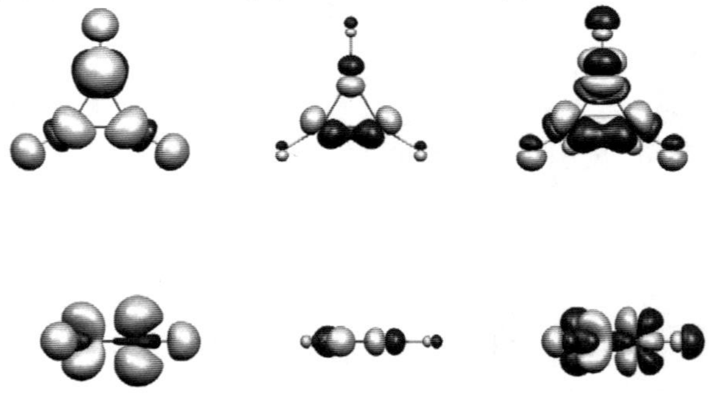

Figure 2. (a) Electron density difference $\Delta\rho$, (b) potential derivative $v_{E'(2)}$, and (c) vibronic coupling density $\eta_{E'(2)}$. The isosurface values of (c) is 5×10^{-4} a.u.

Journal of Physics: Conference Series **428** (2013) 012010 doi:10.1088/1742-6596/428/1/012010

6.2. Vibronic Couplings and Transition Dipole Moment in the S_1 state of Anthracene
Vibronic couplings in the S_1 state of anthracene are analyzed in terms of the VCD and transition dipole moments are analyzed in terms of the TDMD in order to enhance the radiative transition and suppress the radiationless transitions.

Diagonal VCCs give rise to vibrational relaxation in excited states. Figure 3 shows the diagonal VCCs in the Franck-Condon S_1 state. Mode 53 has the maximum coupling. Figure 4 shows the diagonal VCD analysis for the maximum coupling mode in the Franck-Condon S_1 state. The side edge bonds have the large contribution to the VCC. Figure 5 shows the AVCCs for the maximum coupling mode in the Franck-Condon S_1 state. The AVCCs on the side edge bonds also indicate the maximum values.

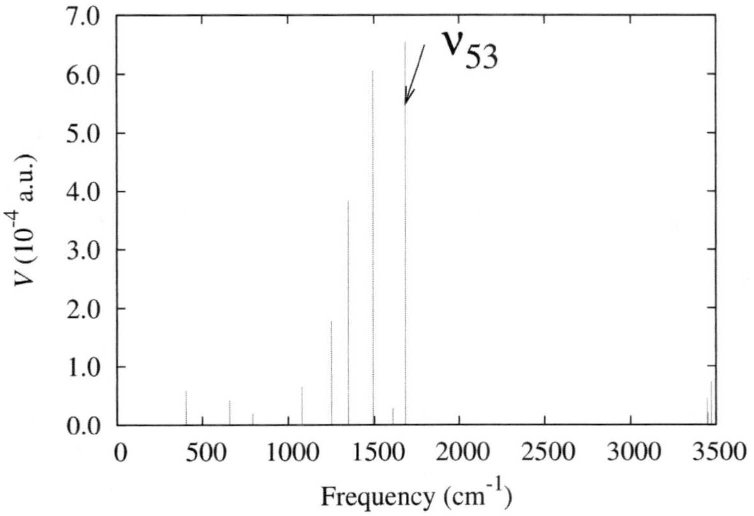

Figure 3. Vibronic coupling constants for anthracene in the Franck-Condon S_1 state.

To control the radiationless transitions, we calculated the off-diagonal VCCs between the S_1 and S_0 states for the adiabatic S_1 state. Figure 6 show the off-diagonal VCC between the S_1 and S_0 states. Figure 7 shows the off-diagonal VCD for the mode 43 between the S_1 and S_0 states. The large contribution originates from the central carbon atoms.

To enhance radiative transition, we calculated transition dipole moments for anthracene and its derivatives between the S_1 and S_0 states (Table 3). The TDMD can explain the reason why 9,10-dichloroanthracene has the largest dipole moment. We show the overlap density between the S_1 and S_0 states of anthracene and its derivatives in Figure 8. The overlap density on the chloride is the origin of the enhancement of transition dipole moment. The TDMDs are shown in Figure 9. The density on the chloride is strengthened because of the product of the z component of the position vector.

Table 3. Transition dipole moments between the S_1 and S_0 states in z direction μ_z.

	μ_z (a.u.)
Anthracene	-0.8827
9-chloroanthracene	-1.1047
9,10-dichloroanthracene	-1.2322

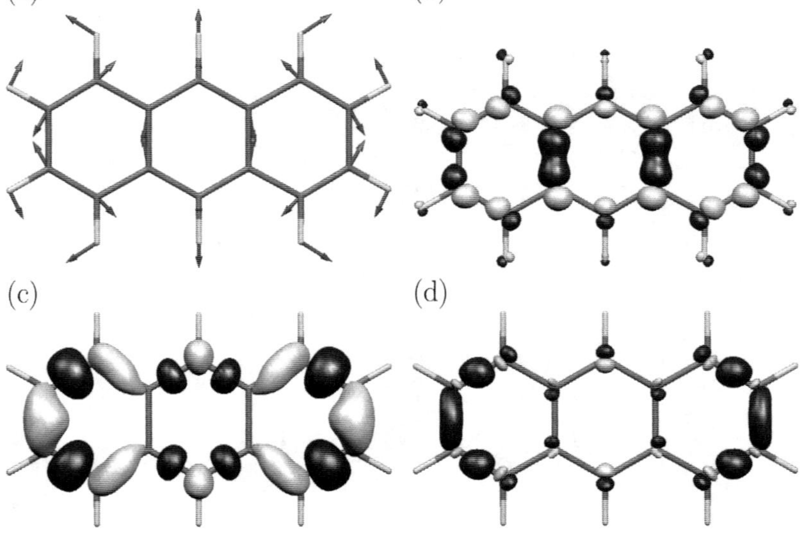

Figure 4. Anthracene in the Franck-Condon S_1 (B_{1u}) state: (a) maximum-coupling mode ν_{53} (ν_{53} = 1683.71 cm^{-1}); (b) derivative of the nuclear-electronic potential with respect to the maximum-coupling mode v_{53} ($|v_{53}| = 1 \times 10^{-2}$ a.u.); (c) electron-density difference $\Delta\rho$ ($|\Delta\rho| = 1 \times 10^{-3}$ a.u.); and (d) vibronic coupling density for the maximum-coupling mode η_{53} ($|\eta_{53}| = 1 \times 10^{-5}$ a.u.).

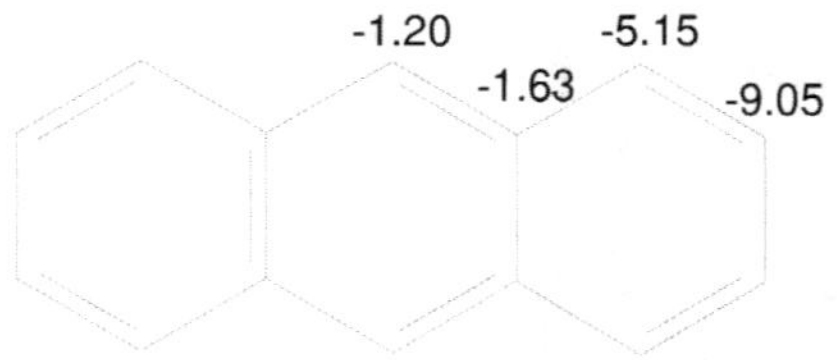

Figure 5. Atomic vibronic coupling constants for the maximum coupling mode of anthracene in the Franck-Condon S_1 state. Unit is 1×10^{-5} a.u.

The transition dipole moment can be assigned to each atoms. Transition dipole moment between k and l states is express as

$$\mu_{kl} = \langle \Psi_k | \hat{\mu} | \Psi_l \rangle = \sum_{\kappa} \sum_{\lambda} D^{kl}_{\kappa\lambda} \langle \phi_\kappa | e\mathbf{r} | \phi_\lambda \rangle, \tag{57}$$

where ϕ_κ indicates the basis function κ and $D^{mn}_{\kappa\lambda}$ indicates the density matrix. The basis function which belongs to atom A is denoted by ϕ_{κ_A}. Then, an atomic transition dipole moment μ^A_{kl} can be defined as

$$\mu^A_{kl} = \sum_{\kappa_A} \sum_{B} \sum_{\lambda_B} D^{kl}_{\kappa_A \lambda_B} \langle \phi_{\kappa_A} | e\mathbf{r} | \phi_{\lambda_B} \rangle. \tag{58}$$

The sum of the atomic transition dipole moment over all the atoms corresponds to the transition dipole moment. Figure 10 shows the atomic transition dipole moments. The atomic transition dipole moments represent the strength of TDMD in each atom.

The TDMD analyses illustrate that the chlorination enhances the radiative transitions.

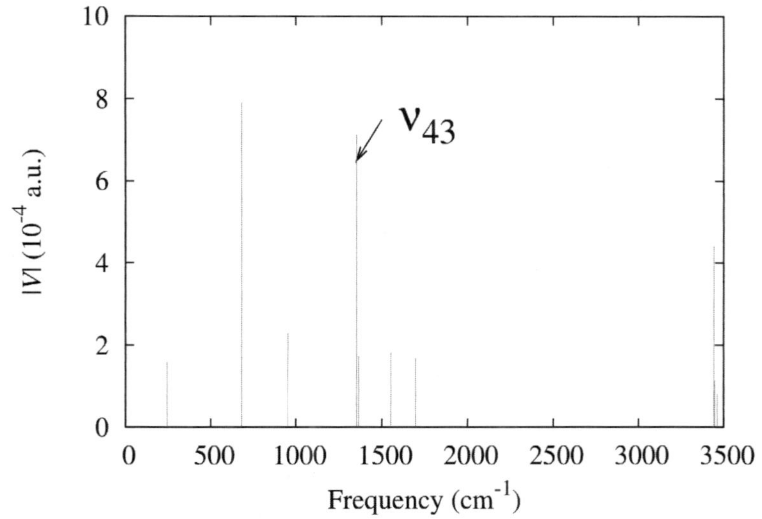

Figure 6. Off-diagonal vibronic coupling constants for anthracene between the S_1 and S_0 states.

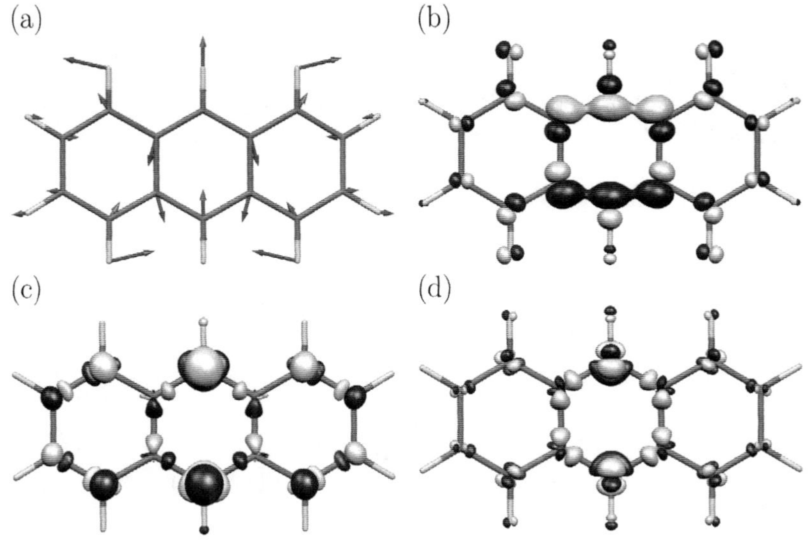

Figure 7. Anthracene in the adiabatic S_1 state: (a) the second largest coupling mode ν_{43} (ν_{43} = 1354.21cm^{-1}); (b) derivative of the nuclear-electronic potential with respect to the mode 43 v_{43} ($|v_{43}| = 1 \times 10^{-2}$ a.u.); (c) overlap density between the S_1 and the S_0 states ρ_{10} ($|\rho_{10}| = 5 \times 10^{-3}$ a.u.); and (d) vibronic coupling density for the mode 43 η_{43} ($|\eta_{43}| = 5 \times 10^{-5}$ a.u.).

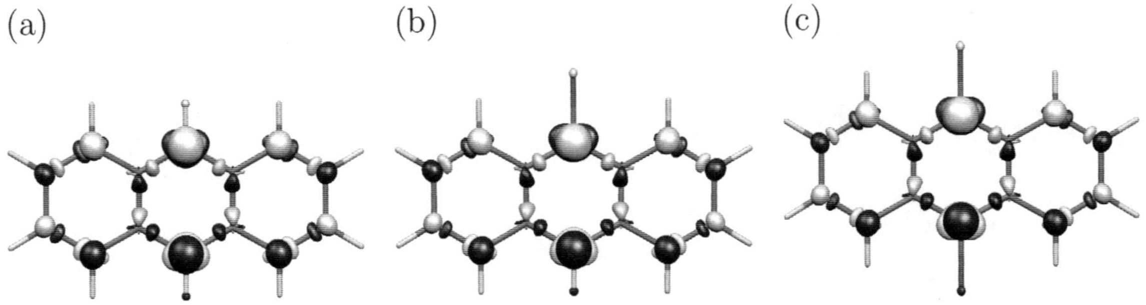

Figure 8. Overlap densities (transition densities) between the S_1 and S_0 states: (a) anthracene, (b) 9-chloroanthracene, and (c) 9,10-dichloroanthracene. Isovalue is 5×10^{-3} a.u. White regions are positive ; blue regions are negative.

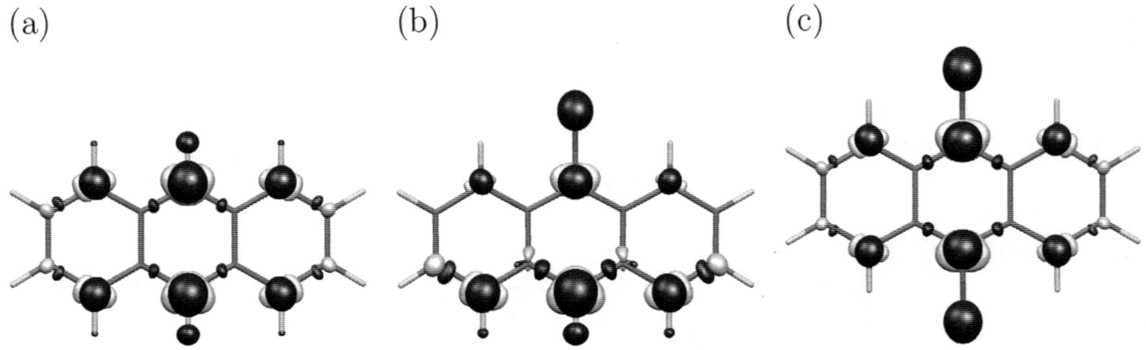

Figure 9. Transition dipole moment densities between the S_1 and S_0 states in z direction: (a) anthracene, (b) 9-chloroanthracene, and (c) 9,10-dichloroanthracene. Isovalue is 1×10^{-2} a.u. White regions are positive ; blue regions are negative.

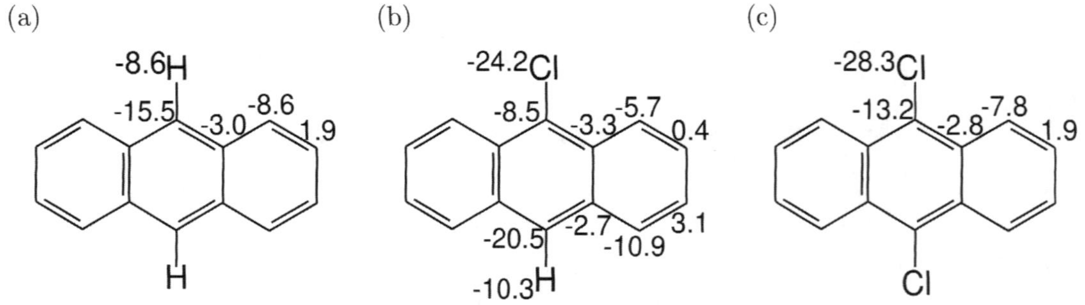

Figure 10. Atomic transition dipole moments in z-direction between the S_1 and S_0 states: (a) anthracene, (b) 9-chloroanthracene, and (c) 9,10-dichloroanthracene. Unit is 1×10^{-2} a.u.

Journal of Physics: Conference Series **428** (2013) 012010 doi:10.1088/1742-6596/428/1/012010

6.3. Pseudo Jahn-Teller Effect in NH_3

The pseudo Jahn–Teller Effect in NH_3 [21] is discussed in terms of QVCD and off-diagonal OVCD. Planar NH_3 is calculated using the B3LYP/STO-3G level of theory within the D_{3h} symmetry. The vibrational modes are

$$\Gamma_{\text{vib}} = a_1' \oplus a_2'' \oplus 2e'. \tag{59}$$

From the vibrational analysis, we obtained an imaginary a_2'' mode (902.2724i cm^{-1}). For the imaginary a_2'' mode, the QVCD and off-diagonal OVCD were calculated. The orbital levels are tabulated in Table 4. Since the HOMO is a_2'' and the LUMO is a_1', these levels couple with the

Table 4. Orbital levels using the B3LYP/STO-3G level of theory

No.	Irrep.	occupation	orbital energies /a.u.
1	a_1'	2	-14.048265
2	a_1'	2	-0.754146
3	e'	2	-0.406501
4	e'	2	-0.406501
5	a_2''	2	-0.108744
6	a_1'	0	0.350188
7	e'	0	0.505624
8	e'	0	0.505624

a_2'' mode:

$$a_2'' \times a_1' = a_2''. \tag{60}$$

The pseudo Jahn-Teller energy curve for the reference nuclear configuration with the D_{3h} ($Q_{a_2''} = 0$) is written as

$$E(Q_{a_2''}) \cong E(0) + \frac{1}{2}\left\{ \left\langle \Psi \left| \left(\frac{\partial^2 \hat{H}}{\partial {Q_{a_2''}}^2} \right)_0 \right| \Psi \right\rangle - 2 \frac{\left| \left\langle \psi_{a_2''} \left| \left(\frac{\partial \hat{H}}{\partial Q_{a_2''}} \right)_0 \right| \psi_{a_1'} \right\rangle \right|^2}{\epsilon_{a_1'} - \epsilon_{a_2''}} \right\} {Q_{a_2''}}^2,$$

where Ψ denotes the electronic wavefunction, $\psi_{a_2''}$ is the a_2'' HOMO, $\psi_{a_1'}$ is the a_1' LUMO, $\epsilon_{a_2''}$ is the orbital level of the HOMO, and $\epsilon_{a_1'}$ is the orbital level of the LUMO.

The QVCD is defined by the integrand of the first term of the right-hand side in

$$\left\langle \Psi \left| \left(\frac{\partial^2 \hat{H}}{\partial {Q_{a_2''}}^2} \right)_0 \right| \Psi \right\rangle = \int \rho(\mathbf{r}) \times w_{a_2''}(\mathbf{r}) d^3\mathbf{r} + \left(\frac{\partial^2 U_{nn}}{\partial {Q_{a_2''}}^2} \right)_0, \tag{61}$$

where $w_{a_2''}(\mathbf{r})$ denotes the quadratic potential derivative. The off-diagonal OVCD is the integrand of

$$\left\langle \psi_{a_2''} \left| \left(\frac{\partial \hat{H}}{\partial Q_{a_2''}} \right)_0 \right| \psi_{a_1'} \right\rangle = \int \rho_{\text{HO,LU}}(\mathbf{r}) \times v_{a_2''}(\mathbf{r}) d^3\mathbf{r}, \tag{62}$$

where $v_{a_2''}(\mathbf{r})$ is the potential derivative. The QVCD is shown in Figure 11. The electron density is totally symmetric, and mainly distributed on the nitrogen atom. However, since the quadratic potential derivative has small distribution on the nitrogen atom, the QVCD is almost equally distributed on the nitrogen and hydrogen atoms.

The OVCD is shown in Figure 13. The OVCD is localized on the nitrogen atoms. This means the pseudo Jahn-Teller instability originates from the lone pair of the nitrogen atoms.

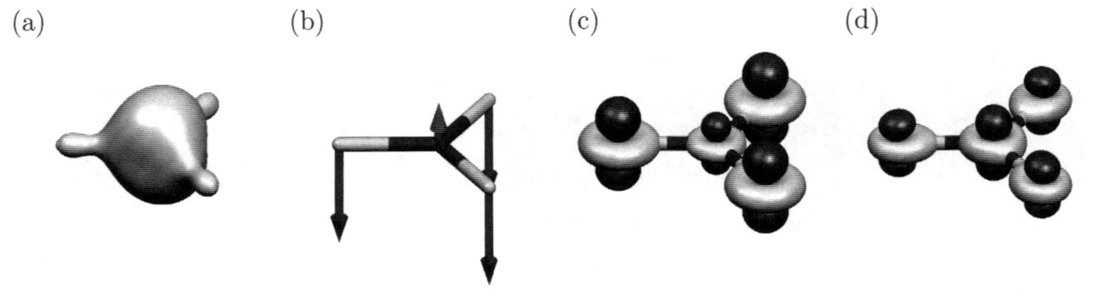

Figure 11. (a) Electron density $\rho(\mathbf{r})$ (isosurface value: 3.0×10^{-1} a.u.), (b) imaginary a_2'' mode, (c) quadratic potential derivative $w_{a_2''}(\mathbf{r})$ (isovalue: 3.0×10^{-4} a.u.), and (d) quadratic vibronic coupling density $\rho(\mathbf{r}) \times w_{a_2''}(\mathbf{r})$ (isovalue: 1.0×10^{-4} a.u.).

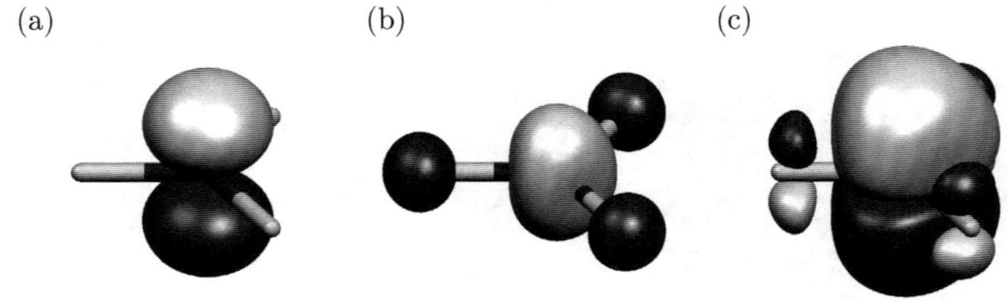

Figure 12. (a) a_2'' HOMO (isosurface value: 2.0×10^{-1} a.u.), (b) a_1'LUMO (isosurface value: 2.0×10^{-1} a.u.), and (c) their orbital overlap density $\rho_{\mathrm{HO,LU}}(\mathbf{r})$ (isosurface value: 5.0×10^{-3} a.u.).

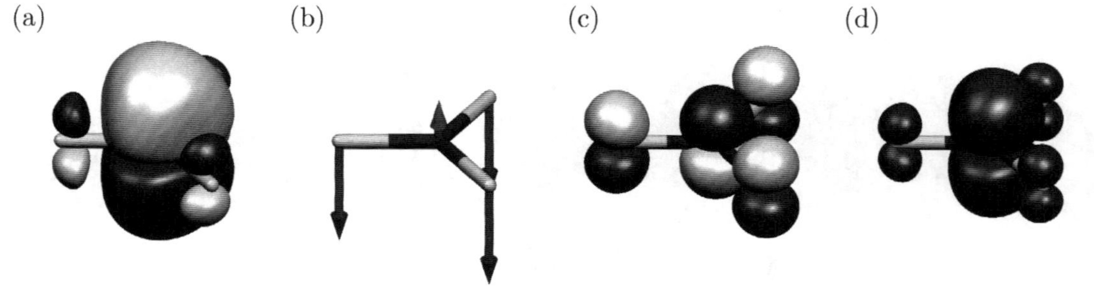

Figure 13. (a) orbital overlap density (transition density) $\rho_{\mathrm{HO,LU}}(\mathbf{r})$ (isosurface value: 5.0×10^{-3} a.u.), (b) a_2'' mode, (c) potential derivative $v_{a_2''}(\mathbf{r})$ (isosurface value: 1.0×10^{-2} a.u.), and (d) orbital vibronic coupling density $\rho_{\mathrm{HO,LU}}(\mathbf{r}) \times v_{a_2''}(\mathbf{r})$ (isosurface value: 1.0×10^{-4} a.u.)

6.4. Chemical Reactivity in Nucleophilic Cycloadditions

Regioselectivity for Diels-Alder reactions can be predicted on the basis of the frontier orbital theory [22, 23]. The regioselectivity of the Diels-Alder reaction to C_{60} is difficult to be predicted. It has been found that the EVCD is effective to predict the reactivity of C_{60}[17]. Here we discuss the regioselectivity in a Diels-Alder reaction to styrene (see Figure 14) using the EVCD. The reactive site of Diels-Alder reaction (b) is the C=C bond between α-C and β-C. All

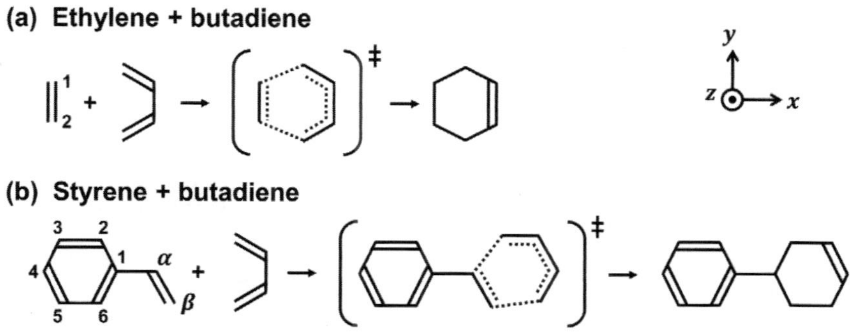

Figure 14. Diels-Alder reactions.

the calculations were performed at B3LYP/STO-3G level of theory. The LUMO of styrene is delocalized as shown in Figure 15. Figures 16 show the VCD analysis for the effective mode

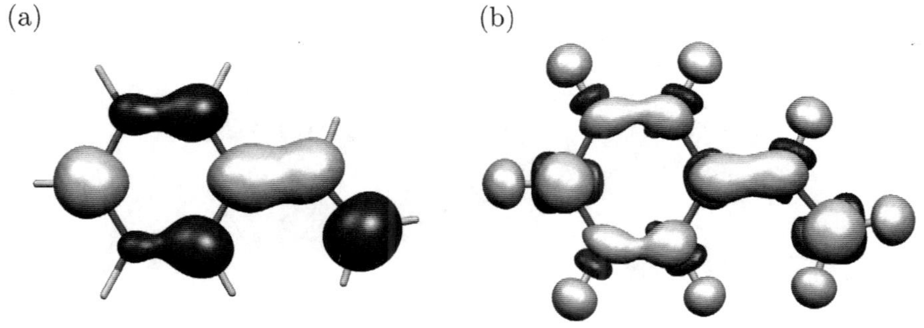

Figure 15. (a) LUMO (isosurface value: 5.0×10^{-2} a.u.) and (b) $\Delta\rho(\mathbf{r})$ (isovalue: 4.0×10^{-3} a.u.).

of the styrene anion. The EVCD distribution on α-C and β-C is almost the same as that of ethylene as shown in Figure 17. Figure 18 shows the cross section of the EVCD between α-C and β-C. Since the EVCD on the bond between α-C and β-C is similar to that of ethylene, it is predicted to be the reactive site based on the EVCD analysis.

7. Summary

Linear and nonlinear vibronic couplings are expressed in the density forms. VCDs illustrate the interactions between electronic and vibrational states as local pictures. With help of concepts of OVCD, RVCD, and AVCC, we can analyze and control the vibronic couplings. EVCD can be used as a reactivity index for chemical reactions. Vibronic couplings are one-electron property. We also define TDMD as a one-electron property density. The VCD and TDMD analyses enable

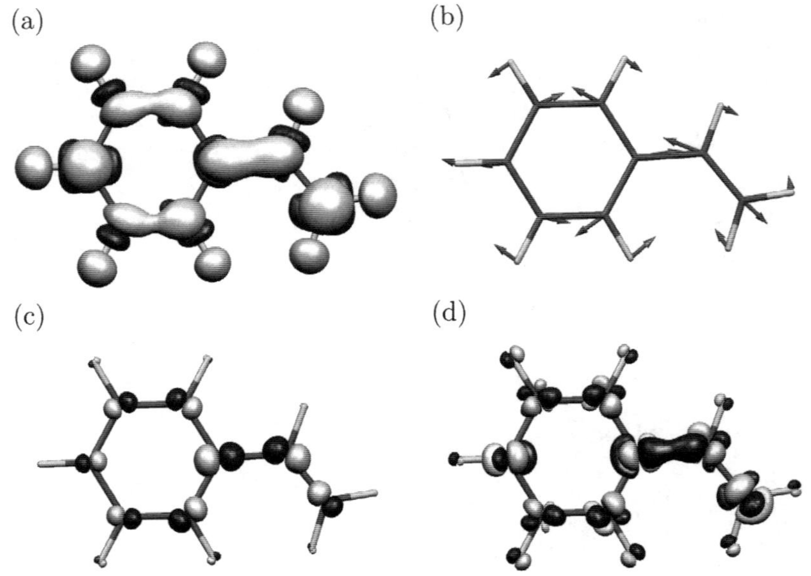

Figure 16. Vibronic coupling density analysis for the effective mode of styrene anion. (a) $\Delta\rho(\mathbf{r})$ (isosurface value: 4.0×10^{-3} a.u.), (b) effective mode, (c) $v_s(\mathbf{r})$ (isosurface value: 2.0×10^{-2} a.u.), and (d) $\eta_s(\mathbf{r})$ (isosurface value: 5.0×10^{-5} a.u.).

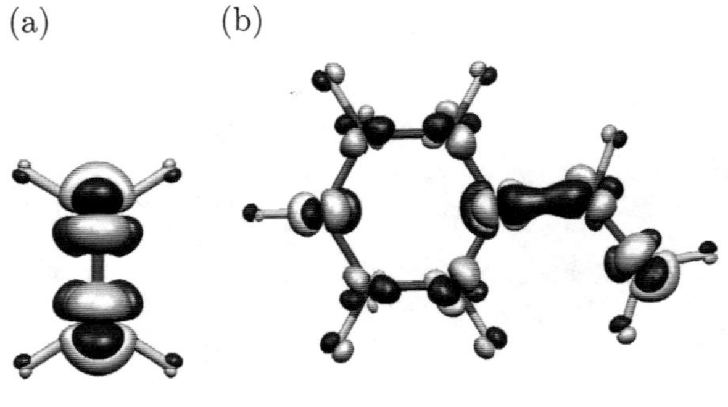

Figure 17. Comparison of the effective vibronic coupling density for styrene anion to that for ethylene anion. (a) ethylene anion $\eta_s(\mathbf{r})$ (isovalue: 2.0×10^{-4} a.u.) and (b) styrene anion $\eta_s(\mathbf{r})$ (isovalue: 5.0×10^{-5} a.u.).

us to suppress radiationless transitions and enhance radiative transitions with a molecular design. The VCD and TDMD open a way to design light-emitting molecules.

Acknowledgment
This work was supported by a Grant-in-Aid for Scientific Research (C) (24550210) from the Japan Society for the Promotion of Science (JSPS). Numerical calculations were performed partly at the Supercomputer Laboratory of Kyoto University and the Research Center for Computational Science, Okazaki, Japan.

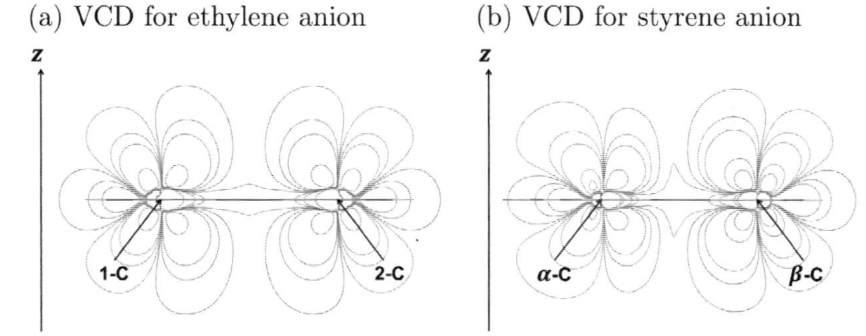

Figure 18. Effective vibronic coupling density for (a) ethylene anion and (b) styrene anion.

References

[1] Fischer G 1984 *Vibronic Coupling: The Interaction between the Electronic and Nuclear Motions* (London: Academic Press)
[2] Bersuker I and Polinger V 1989 *Vibronic Interactions in Molecules and Crystals* (Berlin: Springer)
[3] Bersuker I 2006 *The Jahn–Teller Effect* (Cambridge: Cambridge University Press)
[4] Sato T, Tokunaga K and Tanaka K 2008 *J. Phys. Chem. A* **112** 758–767
[5] Sato T, Tokunaga K, Iwahara N, Shizu K and Tanaka K 2009 *The Jahn–Teller Effect: Fundamentals and Implications for Physics and Chemistry* (*Springer Series in Chemical Physics* vol 97) ed Köppel H, Yarkony D R and Barentzen H (Berlin: Springer–Verlag) pp 99–129
[6] Shizu K, Sato T, Tanaka K and Kaji H 2010 *Appl. Phys. Lett.* **97** 142111 1–3
[7] Shizu K, Sato T, Ito A, Kaji H and Tanaka K 2011 *J. Mater. Chem.* **21** 6375–6382
[8] Sato T, Shizu K, Tanaka K and Kaji H 2012 *J. Photonics Energy* **2** 021201 1–10
[9] Shizu K, Sato T and Tanaka K 2010 *Nanoscale* **2** 2186–2194
[10] Iwahara N, Sato T and Tanaka K 2012 *J. Chem. Phys.* **136** 174315 1–6
[11] Sato T, Iwahara N and Tanaka K 2012 *Phys. Rev. B* **85** 161102(R) 1–5
[12] Shigemitsu Y, Uejima M, Sato T, Tanaka K and Tominaga Y 2012 *J. Phys. Chem.* **116** 9100–9106
[13] Parr R and Yang W 1989 *Density-Functional Theory of Atoms and Molecules* (Oxford: Oxford University Press)
[14] Szabo A and Ostlund N 1982 *Modern Quantum Chemistry: Introduction to Advanced Electronic Structure Theory* (New York: Macmillan)
[15] Shizu K, Sato T and Tanaka K 2010 *Chem. Phys. Lett.* **491** 65–71
[16] Shizu K, Sato T, Tanaka K and Kaji H 2010 *Chem. Phys. Lett.* **486** 130–136
[17] Sato T, Iwahara N, Haruta N and Tanaka K 2012 *Chem. Phys. Lett.* **531** 257–260
[18] Frisch M J, Trucks G W, Schlegel H B, Scuseria G E, Robb M A, Cheeseman J R, Scalmani G, Barone V, Mennucci B, Petersson G A, Nakatsuji H, Caricato M, Li X, Hratchian H P, Izmaylov A F, Bloino J, Zheng G, Sonnenberg J L, Hada M, Ehara M, Toyota K, Fukuda R, Hasegawa J, Ishida M, Nakajima T, Honda Y, Kitao O, Nakai H, Vreven T, Montgomery Jr J A, Peralta J E, Ogliaro F, Bearpark M, Heyd J J, Brothers E, Kudin K N, Staroverov V N, Kobayashi R, Normand J, Raghavachari K, Rendell A, Burant J C, Iyengar S S, Tomasi J, Cossi M, Rega N, Millam J M, Klene M, Knox J E, Cross J B, Bakken V, Adamo C, Jaramillo J, Gomperts R, Stratmann R E, Yazyev O, Austin A J, Cammi R, Pomelli C, Ochterski J W, Martin R L, Morokuma K, Zakrzewski V G, Voth G A, Salvador P, Dannenberg J J, Dapprich S, Daniels A D, Farkas O, Foresman J B, Ortiz J V, Cioslowski J and Fox D J 2009 *Gaussian 09 Revision B.01* Gaussian, Inc. Wallingford, CT
[19] Taekyn H, Graf F and Guenthard H H 1973 *J. Mol. Struct.* **15** 335–350
[20] Sato T, Tokunaga K and Tanaka K 2006 *J. Chem. Phys.* **124** 024314 1–12
[21] Bersuker I, Gorinchoi N and Polinger V 1984 *Theo. Chim. Acta* **66** 161–172
[22] Fukui K, Yonezawa T and Shingu H 1952 *J. Chem. Phys.* **20** 722–725
[23] Fukui K 1971 *Acc. Chem. Res.* **4** 57–64

Optical Jahn-Teller effect: contribution of phonons

Kaja Pae, V. Hizhnyakov

Institute of Physics, University of Tartu, Riia 142, 51014 Tartu, Estonia

Abstract

A theory of absorption and resonance Raman scattering of impurity centers in crystals with $E \times e$-type Jahn-Teller effect in the excited state is presented. The vibronic interaction with non-totally symmetric local modes and phonons is considered. It is shown that a number of specific quantum effects are apparent in the spectra; the vibronic interaction with phonons essentially determines the structure of the spectra.

Key words: absorption spectrum, resonance Raman scattering, Jahn-Teller effect, non-adiabaticity, phonons

1. Introduction

The goal of this communication is to study the manifestations of the non-adiabaticity of the dynamical Jahn-Teller (JT) effect [1-12] in the absorption and resonance Raman scattering (RRS) spectra of impurity centres in solids. An essential feature of the JT effect in this case is the participation of the ~*N* Avogadro phonons of the bulk [3, 8-11, 13-15]. However, the existing theories, including the recently developed multi-configurational time-dependent Hartree approximation [12, 16-24], allow one to consider the dynamical JT effect in the systems with finite (usually of the order of 10) contributing vibrations.

The dynamical JT effect with many frequencies has been also considered in earlier works [25, 26]. The author of Ref. [25] has considered only the states with the lowest energy by using the variational procedure. In Ref. [26] an approach, allowing one to consider the effect of large number N of modes within the frames of an efficient single JTE active mode, was proposed. O'Brien implies the fact that in harmonic approximation it is always possible to reduce the multimode problem to the case of vibronic interaction with single degenerated mode by adding an additional effective quadratic interaction of vibrations. She proposed a set of equations for the parameters of the effective quadratic interaction. However, it remains unclear how these equations can be solved.

Recently we proposed a method [27] allowing one to calculate the effective quadratic interaction supposing that contribution of single local mode to JT effect is stronger than that of phonons. Below we apply this method to the calculation of absorption spectrum and Raman excitation profiles (REPs) of JT systems. We are considering the trigonal centers in the crystals excited in resonance with the electronic transition from a non-degenerate to a two-fold degenerate electronic state of the *E*-representation interacting with the vibrations of *e*-representation (JT effect in this state is called the $E \otimes e$-problem). Our calculations are fully quantum mechanical, which is different from the calculations of REPs of JT systems in semi-classical approximation [28], performed earlier in [29-32].

2. Basic equations

In this communication we study absorption and RRS using the time-domain approach based on consideration of the Fourier transforms of the absorption and RRS amplitudes. In case of the dipole-allowed electronic transition the Fourier transform of the absorption spectrum is determined by the equation [32]

$$F_i(t) = e^{-\gamma_0 \tau} \langle i | \mu_\alpha e^{i\tau(H+V)} \mu_\alpha e^{-i\tau H} | i \rangle$$

analogously, the Fourier transform of the RRS amplitude (called as Raman Fourier amplitude) can be presented in the form [33] (1)

$$A_{\alpha\beta,if}(t) = \Theta(t) e^{-\gamma_0 \tau} \langle f | \mu_\beta e^{i\tau(H+V)} \mu_\alpha e^{-i\tau H} | i \rangle$$

Here H is the vibrational Hamiltonian in the ground electronic state, $H+V$ is the vibronic Hamiltonian of the excited state, γ_0 is the decay constant of the excited state, $\Theta(t)$ is the Heaviside step function. The amplitude of the RRS $a_{\alpha\beta,if}(\omega)$ is given by the Fourier transform of $A_{\alpha\beta,if}(t)$; the REP equals $|a_{\alpha\beta,if}(\omega)|^2$. Equations for the Fourier amplitudes of RRS for thermal equilibrium, see in Ref. [33].

We consider absorption and Raman scattering by trigonal centers excited in resonance with the transition from A_1 (ground) to E (excited) electronic states, representations. In harmonic approximation the vibrational Hamiltonian in the ground state reads

$$H = \sum_{\alpha, j \geq 0} \omega_j a^+_{\alpha j} a_{\alpha j}$$

(up to the unimportant here zero-point energy) with $\alpha = x, y$, $a_{\alpha j}$ and $a^+_{\alpha j}$ being the annihilation and creation operators; ω_j is the frequency of the normal mode j, index α stands for the rows of the symmetry representations. The electron(s) of the E-state interacts, besides totally-symmetric modes, with doubly-degenerate modes of e-representation (the corresponding vibronic problem is called $E \otimes e$-problem). If one takes two components of the excited state as $|-\rangle_e = (|x\rangle - i|y\rangle)/\sqrt{2}$ and $|+\rangle_e = (|x\rangle + i|y\rangle)/\sqrt{2}$, where $|x\rangle$ and $|y\rangle$ transform as components of the polar vector, respectively, then the linear (usually main) vibronic interaction with e vibrations gets the form [1-10] (2)

$$V = \kappa \sum_\alpha \sigma_\alpha Q_\alpha$$

where Q_α are the configurational coordinates of e-representation, σ_x and σ_z are the Pauli matrices, κ is the constant of vibronic coupling. In polar coordinates $V = kQ(\sigma_x \cos\varphi + \sigma_y \sin\varphi)$. The adiabatic potential (AP) curves of the ground and the excited state in the space of the configurational coordinates Q_x and Q_y is shown in Fig.1. The adiabatic potential of the $E \otimes e$ problem (the upper potential) has a shape of the Mexican hat.

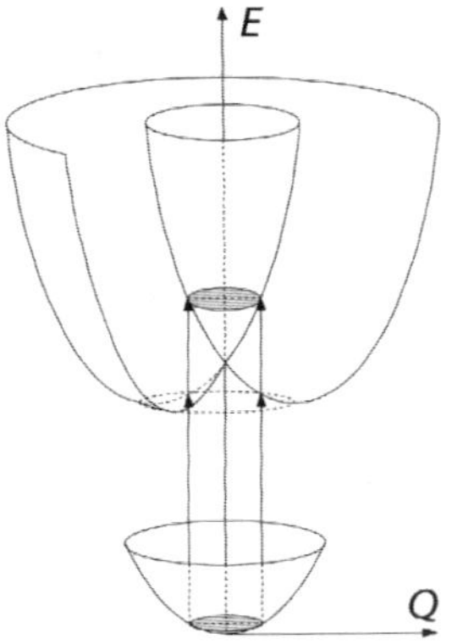

Fig.1. The scheme of the potentials and electronic transitions in the configurational coordinate space. The upper potential corresponds to the Jahn-Teller effect, $E \otimes e$-problem.

It is commonly accepted that in the case of a strong vibronic interaction, for describing optical spectra the semi-classical approximation can be used (see, e.g. [28]). In this approximation one neglects the commutator $[V,H]$. Then in the Condon approximation $\langle\alpha|\mu_\beta|g\rangle = M\delta_{\alpha\beta}$, $M = const$ we get (up to the factor M^2)

$$F(t) \simeq \langle 0|e^{itV}|0\rangle = 1 - \tau w(\tau/2)$$

$$A_{1,xy} = (2i/\tau)A_{2,xx} = \Theta(t)(i/2)\left(\tau + \left(2-\tau^2\right)w(\tau/2)\right)$$

where $\tau = t/\sqrt{2D}$, $D = \kappa^2\langle 0|Q^2|0\rangle$, $w(x) = \mathrm{e}^{-x^2}\int_0^x e^{y^2}dy$ is the Dawson function. The resulting spectra are given in Fig. 2.

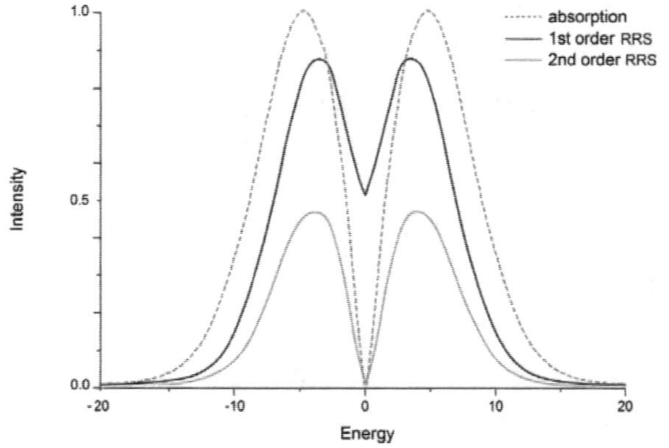

Fig. 2. Spectrum of absorption, first-order REP and second-order REPs for optical $E\times e$ problem in semi-classical approximation.

In the semi-classical approximation it is not important whether the contributing to V coordinates are normal coordinates or not. However in quantum theory it is of primary importance. Here we are considering the case when the coordinate Q of the given (any) row of the e-representation is the linear combination of a number of normal coordinates x_j (the subscript α is omitted for simplicity), but the main contribution to Q is given by one normal coordinate q_0 describing the local mode. In this case Q can be presented in the form

$$Q = \left(1+\lambda^2\right)^{-1/2}\left(x_0 + \lambda\sum_j e_j x_j\right) \tag{3}$$

where x_j ($j \geq 1$) are normal coordinates of phonons, e_{1j} are the normalized polarization vectors; $\sum_j e_j^2 = 1$, λ is a small dimensionless parameter. In Ref. [27] it was shown that in case of small λ the vibronic interaction with non-totally symmetric phonons (it is given by the $\propto\lambda$ term in the brackets of Eq. (3)) can be replaced by the following quadratic interaction of the main mode with phonons:

$$H' = \lambda\sum_{\alpha=x,y}\sum_{j\geq 1} \upsilon_j\left(a_{0\alpha}^+ + a_{0\alpha}\right)\left(a_{j\alpha} + a_{j\alpha}^+\right) \tag{4}$$

where

$$\upsilon_j = e_j\left(\omega_0^2 - \omega_j^2\right)/2\sqrt{\omega_0\omega_j}$$

If to neglect the mode-phonon interaction H' then the problem under consideration reduces to the JT problem in case of the single e-mode x_0. In this approximation vibronic interaction

Journal of Physics: Conference Series **428** (2013) 012011 doi:10.1088/1742-6596/428/1/012011

with the local mode and phonons of e-representation V is replaced by the vibronic interaction with the local mode of e-representation

$$V_0 = \kappa \sum_\alpha \sigma_\alpha x_{0\alpha}$$

and the vibrational Hamiltonian H is replaced by the Hamiltonian $H + H'$. This allows one to use, as the zero'th approximation the known numerical solutions of the latter problem and to take into account the interaction H' approximately by considering it as small perturbation.

Using the eigenstates $|n,m\rangle$ of the two-dimensional harmonic oscillator as the basis one can present the stationary of the JT $E \times e$ problem as follows [2, 34-38]

$$|v_\pm\rangle = \sum_n C_{2n,v} |2n, j-1/2\rangle |\pm\rangle_e + C_{2n+1,v} |2n+1, j+1/2\rangle |\mp\rangle_e \tag{5}$$

where the integer n determines the energy $\omega_0 n$ of the two-dimensional oscillator, the sign-alternative m determines the projection of the rotational momentum on z-axis ($|m| \le n$), v and j, determine the radial and the angular JT motion, respectively. At that the quantum number $j = m - 1/2 = \pm 1/2, \pm 3/2, \ldots$ is the projection of the full (electronic $s = 1/2$ and vibrational) rotational momentum on the z (C_3)- axis. The amplitudes $C_{l,v}$ and the energies E_v are the solutions of the given in Refs [2, 35-38] matrix equation with a three-diagonal matrix .The states $|v_+\rangle$ and $|v_-\rangle$ denote the components of the two-fold degenerate states (Kramers degeneracy) and give the same contribution to the spectra.

Here we are considering the case of zero temperature. In this case only the vibrational state with $n = m = 0$ in the initial electronic state is populated and only the vibronic levels with $j = \pm 1/2$ of the excited state give contributions to RRS. We use in Eq. (1), as the basis the solutions of the single e-mode $E \times e$ problem and apply the Dyson equation [39]. We get

$$e^{it(H+V)} \simeq e^{it(H+H'+V_0)} = \hat{T} \exp\left[i \int_0^t d\tau H'(\tau) \right] e^{it(H+V_0)}$$

where $H'(\tau) = e^{i\tau(H+V_0)} H' e^{-i\tau(H+V_0)}$. We find for $T = 0$

$$A_f(t) = \Theta(t) \sum_{vv'} C_{kv} C_{0v'} e^{itE_{v'}^+} \langle v_+ | \langle 0_{ph} | \hat{T} \exp\left[i \int_0^t d\tau H'(\tau) \right] | 0_{ph} \rangle | v'_+ \rangle \tag{6}$$

Applying then the cumulant expansion and taking into account first two cumulants we get (in full analogy with the calculation of $F(t)$ in Ref. [27])

$$A_f(t) \cong \Theta(t) \sum_v C_{fv} C_{0v} \exp[iE_v t + g_v(t)], \tag{7}$$

$$g_v(t) = \lambda^2 \sum_j \upsilon_j^2 \sum_{v'} |S_{vv'}|^2 \left(\frac{it}{E_v - E_{v'} - \omega_j} + \frac{e^{i(E_v - E_{v'} - \omega_j)} - 1}{(E_v - E_{v'} - \omega_j)^2} \right) \tag{8}$$

$S_{vv'} = 2 \sum_n C_{2n+1,v} \left(\sqrt{2n+1}\, C_{2n,v'} + \sqrt{2n+2}\, C_{2n+2,v} \right)$, υ_j is given above. Note that $e^{g_v(t)}$ are the same factors which describe the contribution of phonons in the Fourier transform of absorption [27]. At large time t these factors damp due to the decay of vibronic states v caused by creation of phonons.

3. Numerical

Below we present some results of the numerical calculations of absorption spectra and REPs for the $E\times e$ problem in the excited state, using the Debye-Van Hove model of phonons. In this model DOS of phonons equals $\rho(\omega)=(16/\pi)\omega^2\sqrt{1-\omega^2}$ and the interaction function $\upsilon^2(\omega)=\sum_j \upsilon_j^2\delta(\omega-\omega_j)$ has the form $\upsilon^2(\omega)=\rho(\omega)\omega\left(\omega_0^2-\omega^2\right)^2/2\pi\omega_0$. The sums over j in Eq. (8) were replaced by the integrals over ω; the sums over ν and n were calculated taking into account few hundreds of the levels ν and n. The results of the calculations of REPs of first and second order RRS for two different values of the vibronic interactions D with the main mode and two different values of the vibronic interaction with phonons λ^2 are given in Figs 3 and 4. We also performed calculations for very strong vibronic interaction with the main mode $D=1000$ and weak interaction with phonons (see Fig. 5).

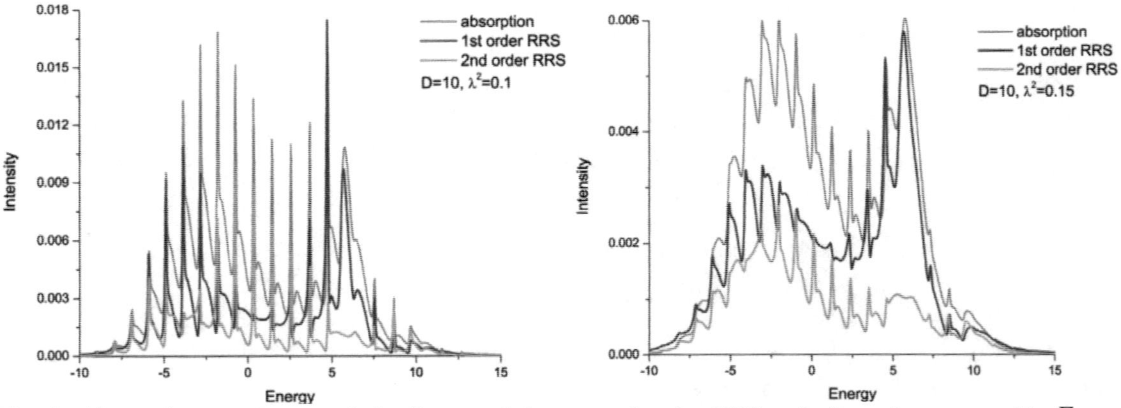

Fig. 3. Absorption spectrum and the first- and the second-order REPs of trigonal centers with $E\times e$-type Jahn-Teller effect in the excited state. $D=10$ is dimensionless energy of vibronic interaction with the main e-mode; λ is dimensionless vibronic interaction with non-totally symmetric phonons.

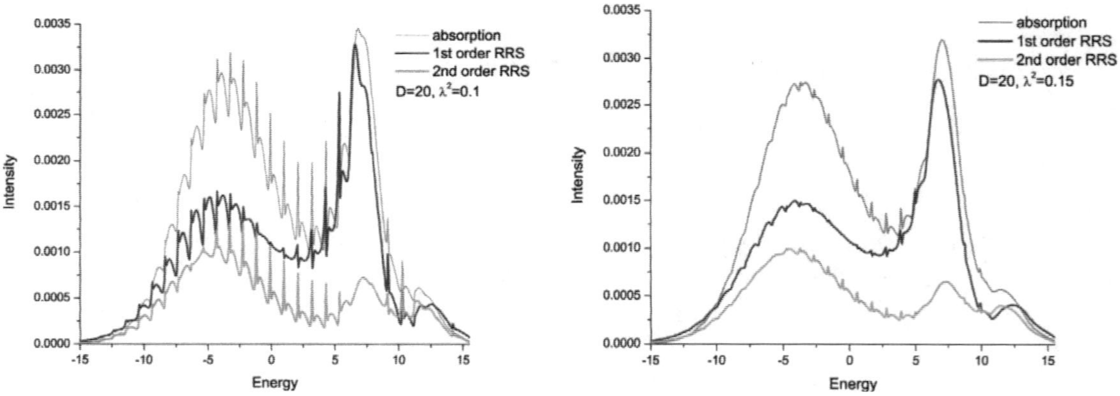

Fig. 4. The same as in Fig.3 but for $D=20$.

The absorption spectra and the REPs presented in Figs. 2-5 are given in arbitrary units, however, with correct relative intensities of the first- and the second-order RRS. The latter have a comparable intensity which is characteristic of RRS [41, 42]. The central parts of the spectra in Figs. 3-5 correspond to the excitation in the region of the conical intersection of AP in the excited state. The remarkable intensity of these parts in the spectra presented in these figures display the contribution of the non-adiabatic transitions between the sheets of AP. Somewhat smaller intensity of the high-energy part of the second-order REP as compared to absorption is due to larger size of the wave function of the finite (two-quantum) state as compared to the initial (zero-point) state. This makes the second-order RRS more probable when excitation occurs to the lower, broader part of the conical intersection of the $E \times e$ AP (see Fig. 1) than to the higher part of it.

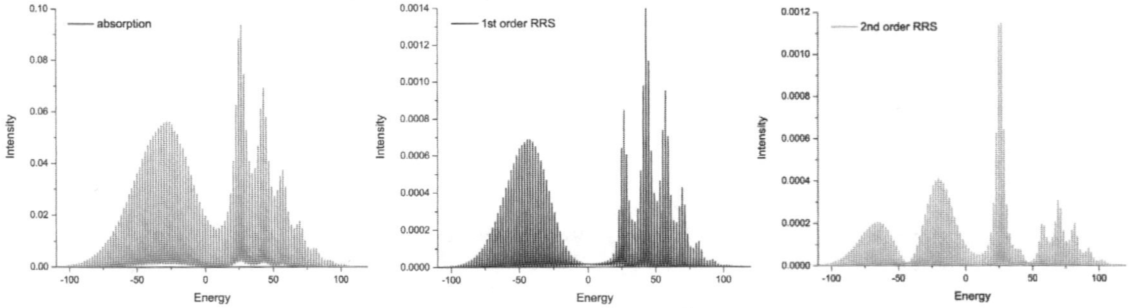

Fig. 5. The same as in Fig. 3, but for $D = 1000$ and $\lambda^2 < 0.01$. The presented spectra differ essentially from the given on Fig. 2 spectra calculated in semiclassical approximation.

Rather evident manifestation of the non-adiabaticity is displayed in the first-order REPs. We have in mind here the enhancement of the high-energy parts of them in Figs. 3, 4 as compared to absorption. Indeed, the first-order RRS is accompanied by a real vibronic transition in the conical intersection of the $E \times e$ AP. The coordinates of the upper part of the conical intersection are, on the average, smaller than that of the lower part of it. This results in higher probability of the vibronic transitions in the upper part of the conical intersection than in the lower part of it and in the enhancement of the higher energy part of the first-order REPs. Note that this effect works only for the moderate vibronic interaction D. However if D is very large, then the difference in the shape of the lower and upper parts of the conical intersection in the resonance region is small, due to what the intensities of the lower and higher energy parts of the first-order REP become comparable (see Fig. 5).

In the semi-classical approximation the virtual non-adiabatic transitions are not taken into account. This explains why the intensities of the central part of the absorption spectrum and of the second-order REP in Fig. 2 is small and it tends to zero if $E \to 0$. An essentially different situation, however, exists for the first-order RRS. This scattering is accompanied by the real transition between the sheets of the AP. The probability of this transition increases with $E \to 0$, compensating the decreasing of the density of states and causing the finite intensity of the semi-classical first-order REP at $E = 0$. However, in the quantum theory this intensity strongly diminishes (see Figs. 3-5). The reason of this diminishing is in the quantum interference. Here we take into account that according to the quantum theory the first-order RRS has two scattering channels [33]: a) the channel describing the resonances for incident light and b) the channel describing the resonances for scattered light. These two channels interfere destructively. It is this destructive interference which causes the decrease of the intensity of the first-order REP in the central part of the spectrum. Interference effects exist also for the second-order RRS. Here the two channels mentioned interfere constructively and they both interfere destructively with the scattering channel, having resonance for the intermediate frequency. This combined interference is responsible for the appearance of the two deepest minima in the second-order REP in Fig. 5.

Note also one more essential difference of the spectra in the quantum and the semi-classical theory: the series of the well-distinguished peaks in the envelopes of the high-energy parts of the spectra in Figs. 3-5 are absent in the spectra given in Fig. 2, obtained in semi-classical approximation. These peaks correspond to the excitation in resonance with the high-energy quantum levels of the upper part of the conical intersection, known as the Slonczewski resonances [42].

The presented results show also the importance of the interaction with phonons: already relatively moderate increase of the corresponding interaction parameter λ^2 in Figs. 3 and 4 causes a remarkable diminishing of the vibronic structure. The Fig. 5 demonstrates that the quantum theory of absorption and RRS does not coincide with the semi-classical theory even for rather strong vibronic interaction. However, as it follows from our calculations the remarkable interaction with phonons leads in case of large D to an essential diminishing of the difference of the spectra in the quantum and semi-classical.

Acknowledgments
The research was supported by the ETF grant TLOFY0145, by the project GLOFY7741 and by the European Union through the European Regional Development Fund (project TK114).

References

1. H.A. Jahn, E. Teller, Proc. R. Soc. Lond. Ser. A 161 (1937) 220.
2. H.C. Longuet-Higgins, U. ¨Opik, M.H. Pryce, R.A. Sack, Proc. R. Soc. A 244 (1958) 1.
3. M.D. Sturge, The Jahn-Teller Effect in Solids, Solid State Physics 20 (1967) pp.91-211.
4. R. Englman, The Jahn-Teller Effect in Molecules and Crystals (Wiley-Interscience, New York, 1972).
5. Yu.E. Perlin and B.S. Tsukerblat, Electron-vibrational Interaction in Optical Spectra of Impurity Paramagnetic ions (Shtintsa, Kishinev, 1974) (in Russian).
6. M.Z. Zgierski, M. Pawlikowski, J. Chem. Phys. 70 (1979) 3444.
7. G. Fischer, Vibronic Coupling (Academic, San Diego CA, 1984).
8. I.B. Bersuker, V.Z. Polinger, Vibronic Interactions in Molecules and Crystals (Springer, Berlin, 1989).
9. Y. Toyozawa, Optical Processes in Solids (Cambridge University Press, 2003).
10. W. Domcke, D.R. Yarkony, H. Köppel (eds.), Conical Intersections: Electronic Structure, Dynamics and Spectroscopy (Word Scientific, Singapore, 2004).
11. I.B. Bersuker, The Jahn–Teller Effect (Cambridge University Press, Cambridge, 2006).
12. Horst Köppel, David R. Yarkony, Heinz Barentzen (Eds), Fundamentals and Implications for Physics and Chemistry, Springer-Verlag Berlin Heidelberg 2009.
13. V.Z. Polinger, I.B. Bersuker,: Phys. Stat . Sol. (b) 95 (1979) 403.
14. Yu. B. Rosenfeld, A.V. Vaisleib, Sov. Phys. JETP 86 (1984) 1059.
15. Yu. E. Perlin, B. S. Tsukerblat. Optical Bands and Polarization Dichroism of Jahn-Teller Centers, in: Dynamical Jahn-Teller Effect in Localized Systems, Elsevier Publ. B. B. (1984) pp. 251-346, Amsterdam.
16. H.D. Meyer, U. Manthe, L.S. Cederbaum, Chem. Phys. Lett. 165 (1990) 73.
17. U. Manthe, H.D. Meyer, L.S. Cederbaum, J. Chem. Phys. 97 (1992) 3199.
18. M. Beck, A. Jäckle, G.A. Worth, H.D. Meyer, Phys. Rep. 324 (2000) 1.
19. H.D. Meyer, G.A. Worth, Theor. Chem. Acc. 109 (2003) 251.
20. Horst Köppel, David R. Yarkony, Heinz Barentzen (Eds.), Fundamentals and Implicationsfor Physics and Chemistry, Springer-Verlag Berlin Heidelberg 2009.
21. S. Faraji, H.D. Meyer, H. Köppel, J. Chem. Phys. 129 (2008) 074311.
22. M. Nooijen, J.G. Snijders, Int. J. Quantum Chem. S 26 (1992) 55.
23. J.F. Stanton, J. Gauss, J. Chem. Phys. 101 (1994) 8938.
24. H.D. Meyer, U. Manthe, L.S. Cederbaum, Chem. Phys. Lett. 165 (1990) 73.
25. J.R. Fletcher, J. Phys. C: Solid State Phys. 5 (1972) 852.
26. M.C.M. O'Brien, J. Phys. C: Solid State Phys. 5 (1972) 2045.
27. V. Hizhnyakov, K. Pae, T. Vaikjärv, Chemical Physics Letters, 525 (2012) 64.

28. Y. Toyozawa, W. Inoue, J. Phys. Soc. Japan 21 (1966) 1663.
29. V. Hizhyakov, I. Tehver, Proceedings of the Second International Conference on Light Scattering in Solids, Flammarion Science, pp. 57-61, 1972
30. S. Muramatsu, K. Nasu, M. Takahashi and K. Kaya: Chem. Phys. Letters 50 (1977) 284.
31. S. Muramatsu, K. Nasu, J. Phys. Soc. Jpn. 46 (1979) 189.
32. M. Lax, J. Chem. Phys. 20 (1953) 1752.
33. V. Hizhnyakov, I. Tehver, J. Raman Spectroscopy 24 (1993) 653.
34. W. Moffit, W. Thorson, Phys. Rev. 108 (1958) 1251.
35. V. Loorits, Izv. Akad. Nauk. ESSR, Fiz.-Mat. 29 (1980) 208.
36. M.C.M. O'Brien, S.N. Evangelou, Solid State Commun. 36 (1980) 29.
37. H. Köppel, E. Haller, L. S. Cederbaum and W. Domcke, Molecular Physics, 41 (1980) 669.
38. V. Hizhnyakov, N. Kristoffel, Jahn–Teller mercury-like impurities in ionic crystals, in: Yu E. Perlin, M. Wagner (Eds.), The Dynamical Jahn–Teller Effect in Localized Systems, Elsevier Science, 1984, p. 383.
39. F. Dyson, Phys. Rev. 75 (1949) 486, 1736.
40. R. Kubo, J. Phys. Soc. Jpn. 17(1962) 1100.
41. V. Hizhnyakov, I. Tehver, phys stat. sol. 21 (1967) 755.
42. J.C. Slonczewski, Phys. Rev. 131 (1963) 1596.

Manifestation of phonons in the optical spectra of pseudo Jahn-Teller systems

Taavi Vaikjärv

Institute of Physics, University of Tartu, Riia 142, 51014 Tartu, Estonia

Taavi.vaikjarv@ut.ee

Abstract. The system under consideration is a molecule or an impurity center in a crystal with a two-fold quasi-degenerate electronic state (pseudo Jahn-Teller system) in strong interaction with one main mode and in weak interaction with continuum of secondary modes of phonons. This system exhibits strong non-adiabatic couplings which have been demonstrated to be detectable in absorption and Raman excitation spectra. By choosing an appropriate axis of polarization the non-adiabatic transitions become directly observable.

1. MODEL AND FORMULA

Our task is to calculate the optical spectra of vibronic transitions between non-degenerate (ground) and two-fold quasi-degenerate electronic states. We consider how vibronic transitions between electronic states of close energy manifest themselves in the absorption spectrum and in the excitation spectra of the resonance Raman scattering (Raman excitation profiles). The Fourier transform of absorption and the Fourier amplitude of the resonance Raman scattering are [1,2]

$$F(t) = \sum_{\alpha} \langle i | \mu_\alpha e^{it(H+V)} \mu_\alpha e^{-itH} | i \rangle \tag{1}$$

$$\mathrm{A}_{\alpha\beta,\, if}(t) = \Theta(t) \langle f | \mu_\beta e^{it(H+V)} \mu_\alpha e^{-itH} | i \rangle , \tag{2}$$

where H is the vibrational Hamiltonian in the ground electronic state, $H+V$ is the vibronic Hamiltonian of the excited state and μ, the electric dipole operator. $|i\rangle$ and $|f\rangle$ denote the initial and final states respectively. Heaviside step-function, $\Theta(t)$, is used to takes into account the causality: the scattering photon appears after the incident one. The vibronic coupling V is taken in the linear approximation [3,4]

$$V = \kappa\sigma_x Q + \sigma_z \varepsilon \tag{3}$$

where Q is the odd configurational coordinate, ε is the half-splitting of the quasi-degenerate state, σ_x and σ_z are the Pauli matrices, κ is the constant of the vibronic coupling. The interaction with the totally-symmetric coordinates is described by unit matrix and gives additional factors in the equations for $F(t)$ and $A_{\alpha\beta,if}(t)$ which can be found by applying the standard theory.

The solving of the problem under consideration carried out in three steps[5]: First, the vibronic interaction with phonons in excited electronic state is transferred to the quadratic[6] interaction of phonons with the main mode. Secondly, the eigenvalue problem for the vibronic interaction with single mode in the excited state is solved numerically. Thirdly, the interaction of the main mode with phonons is taken into account using the second cumulant expansion.

After applying this technique we arrive at the following final equations for odd ($f = 2k+1$) and even order ($f = 2k$) Raman scattering excitation profiles. In one polarization of the incident light the equations read

$$A_{2k+1}(t) = \Theta(t)\sum_{\nu} C^{-}_{2k+1,\nu} C^{+}_{0,\nu} e^{itE^{+}_{\nu} + g^{+}_{\nu}(t)}, \tag{4}$$

$$A_{2k}(t) = \sum_{\nu} C^{+}_{2k,\nu} C^{+}_{0,\nu} e^{itE^{+}_{\nu} + g^{+}_{\nu}(t)}. \tag{5}$$

However if the incident light is polarized in perpendicular direction then in the previous equations the superscripts $+$ and $-$ should be transposed. $+$ and $-$ distinguish between the two quasi-degenerate electronic states. Here $E^{\pm}_{\nu}$ and $C^{\pm}_{\nu}$ are the eigenvalues and eigenstates of the following three-diagonal vibronic matrices[4]

$$\begin{pmatrix} \varepsilon & \sqrt{D} & 0 & \dots \\ \sqrt{D} & 1\mp\varepsilon & \sqrt{2D} & \dots \\ 0 & \sqrt{2D} & 2\mp\varepsilon & \dots \\ \dots & \dots & \dots & \dots \end{pmatrix} \begin{pmatrix} C^{\pm}_{0} \\ C^{\mp}_{1} \\ C^{\pm}_{2} \\ \dots\dots \end{pmatrix} = E^{\pm} \begin{pmatrix} C^{\pm}_{0} \\ C^{\mp}_{1} \\ C^{\pm}_{2} \\ \dots\dots \end{pmatrix}, \tag{6}$$

where $D = (\kappa/\omega_0)^2$.

The effects of phonons are described by the factor $\exp(g_{\nu}(t))$ [5], where

$$g^{+}_{\nu}(t) = \lambda^2 \sum_{j} \upsilon^2_j \sum_{\nu'} \left|S^{+}_{\nu\nu'}\right|^2 \left(\frac{it}{E^{+}_{\nu} - E^{-}_{\nu'} - \omega_j} + \frac{e^{i\left(E^{+}_{\nu} - E^{-}_{\nu'} - \omega_j\right)t} - 1}{\left(E^{+}_{\nu} - E^{-}_{\nu'} - \omega_j\right)^2} \right). \tag{7}$$

Here λ^2 is a small dimensionless parameter that determines the relative interaction strength with the phonon continuum, υ^2_j describes the phonon density of the states which is taken in the Debye-Van Hove model,

$$\upsilon^2(\omega) = 8\omega^3\left(\omega_0^2 - \omega^2\right)^2 \sqrt{\omega_M^2 - \omega^2} \Big/ \pi\omega_0\omega_M^6, \tag{8}$$

$$S^{+}_{\nu\nu'} = s^{+}_{\nu\nu'} + s^{-}_{\nu\nu'}; \quad s^{\pm}_{\nu\nu'} = \sum_{n} C^{\pm}_{2n+1,\nu}\left(\sqrt{2n+1}\, C^{\mp}_{2n,\nu'} + \sqrt{2n+2}\, C^{\mp}_{2n+2,\nu'}\right) \tag{9}$$

are the parameters of the vibronic states that characterize the vibronic correlation.

2. NUMERICAL CALCULATIONS

In the following the absorption, first and second order Raman scattering excitation spectra of the center with two-fold quasi-degenerate electronic state have been calculated. Non-totally symmetric modes of phonons are considered. The Raman scattering spectra are of special interest since they are, in general, more informative than absorption and luminescence spectra. Besides, using the resonant Raman scattering method the relatively weak spectra can be amplified many orders of magnitude in the experiment. The spectra are calculated for different values of ε at constant value of $k = 10$.

We choose polarization so that in the case of $\varepsilon = 0$ the Raman scattering appears only in the incident polarization. This corresponds to the incident light being polarized in the direction of the dimer. Consequently only one final electronic state is exited. Since, in case of $\varepsilon \neq 0$ the adiabatic interaction mixes the two states, Raman scattering appears also in perpendicular polarization. This means the adiabatic effects are directly observable in the perpendicular polarization.

3. RESULTS

The first contribution to the spectra by phonons is a band accompanying each spectral line which is most noticeable in the absorption spectra. It is worth pointing out that each line and band has its own shape. This is an implication that each vibronic state has a different intrinsic lifetime. This can be explained with great variance in the distance of neighboring energy levels ($E_{v}^{+} - E_{v'}^{-}$) for each vibronic energy state[7]. This interaction with the phonon continuum can be so strong that some lines are broadened to the extent that they disappear from the spectrum. It is also apparent, that the Raman spectra have a totally different shape from the absorption spectra as seen on Fig. 1. For example in the first order Raman spectra there is a gap at zero energy. This is caused by destructive wave function interference[2] between the excitation and scattering processes. Also two clear minima are seen in second order Raman spectra which is caused by an analogous process for the two-phonon scattering.

An interesting effect of nonadiabatic transitions becomes observable if the parameter ε is increased. As this can be viewed as a perturbation of the degenerate case, increase in this parameter leads to the increase in the probability of nonadiabatic transitions from one adiabatic energy surface to another. This can be seen in the increase of the intensities of the spectral lines in Raman scattering excitation profiles in the perpendicular polarization. One can also observe another more subtle effect. The intensity of the higher energy part of the spectrum for the perpendicular polarization is larger than that of the lower part as seen on Fig. 2. The spectrum in this polarization only appears due to vibronic transition between components with different polarization of the excited state. The reason behind this effect is the steep (narrow) shape of the upper energy surface given in Fig. 3. This means that smaller changes in the configural coordinates are needed for these transitions and this causes the higher probability of the transition. As a result the intensity of scattering in perpendicular polarization for the excitation above the central minimum of the adiabatic potential in the excited state is higher than for the excitation below the central maximum of the adiabatic potential.

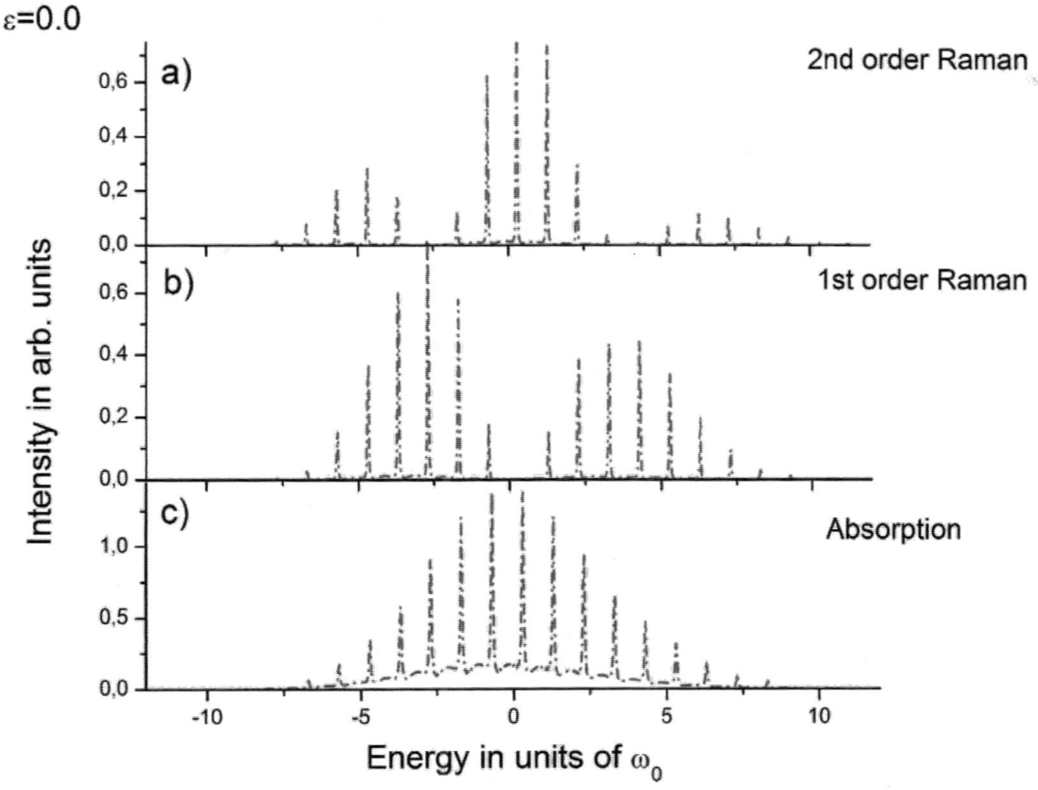

Figure 1. Absorption spectrum and Raman scattering excitation profiles in the case of $\varepsilon = 0$ and $k = 10$ where each line is accompanied by a phonon band. As a result of the quantum interference effects the Raman spectra have clear minima.

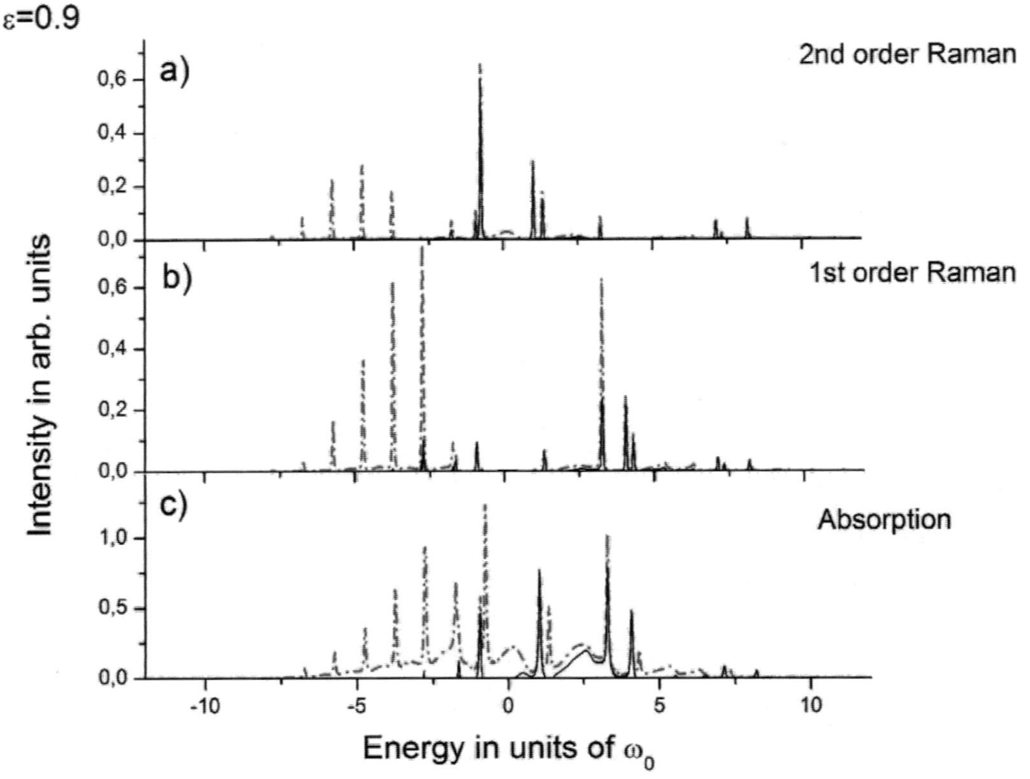

Figure 2. Spectra in the case of $\varepsilon = 0.9$ and $k = 10$. Black solid lines are the Raman scattering excitation profiles in the perpendicular polarization to the incident light which is a direct result of the phonon mixing of the states

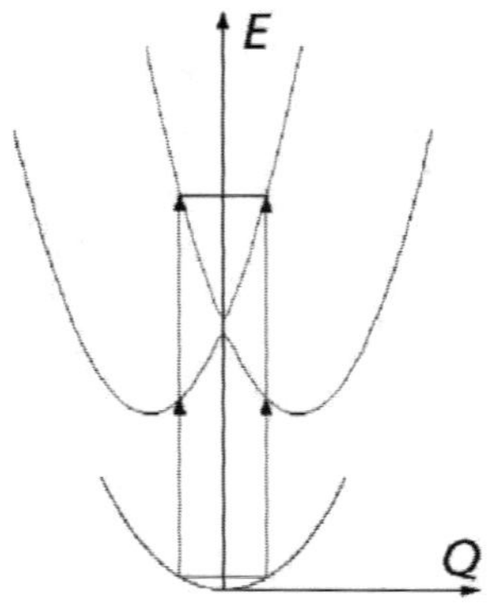

Figure 3. Qualitative shape of the adiabatic energy surface. Notice the difference in distance between the sheets of the adiabatic potential for the two electronic states near the central minimum

Acknowledgments
The research was supported by the ETF grant TLOFY0145, by the project GLOFY7741 and by the European Union through the European Regional Development Fund (project TK114).

References

[1] Lax M 1953 *J. Chem. Phys.* **20** 1752
[2] Hizhnyakov V and Tehver I 1993 *J. Raman Spectroscopy* **24** 653
[3] Fulton R and Gauterman M 1961 *J. Chem. Phys.* **35**, 1059 (1961), **41**, 2280 (1964).
[4] Loorits V 1979 Projavlenij rezonansa Fermi v opticheskih spektrah psevdo-effekta Jana-Tellera, *Akademija Nauk ESSP* Otdelenije fiziko-matematicheskih hauk, Preprint F-11
[5] Hizhnyakov V, Pae K and Vaikjärv T 2012 Optical Jahn-Teller effect in the case of local modes and phonons. *Chem. Phys. Lett.* **525–526** 64 - 68
[6] O'Brien M C M 1972 The dynamic jahn-teller effect with many frequencies: a simple approach to a complicated problem. *J. Phys. C: Solid State Phys.* **5** 2045.
[7] Hizhnyakov V, Pae K. and Vaikjärv T 2011 Vibronic Transitions to a State with Jahn-Teller Effect: Contribution of Phonons *Vibronic Interactions and the Jahn-Teller Effect: Theory and Applications* (Springer) 179 – 191

Kondo effect emerging from a spin-vibronic state

Takahiro Fuse[1] **and Takashi Hotta**[1,2]

[1] Department of Physics, Tokyo Metropolitan University, Tokyo 192-0397, Japan
[2] Advanced Science Research Center, Japan Atomic Energy Agency, Ibaraki 319-1195, Japan

E-mail: fuse-takahiro@tmu.ac.jp

Abstract. We discuss the Kondo effect in an electron system dynamically coupled with local Jahn-Teller phonons. For the purpose, we consider the two-orbital Anderson model including both Coulomb interactions and the coupling between electrons and Jahn-Teller phonons on an impurity site. Note that in this paper, we consider a quarter-filling case. First we analyze a local problem and confirm a spin-vibronic quartet ground state, characterized by the direct product of spin and vibronic degrees of freedom. Next we include the hybridization term and analyze the model by using a numerical renormalization group method. From the evaluation of entropy, specific heat, and several kinds of susceptibilities, it is found that spin and total angular momenta are simultaneously screened by conduction electrons, leading to the Kondo phenomenon due to the entropy release of $\ln 4$ of the local spin-vibronic state. A possibility of heavy-electron formation in the spin-vibronic Kondo effect is briefly discussed.

1. Introduction

Recently, the Kondo effect originating from phonon degrees of freedom have attracted renewed attention both in the experimental and theoretical research fields of condensed matter physics. In particular, in cage-structure materials such as filled skutterudites and clathrate compounds, local anharmonic oscillations of a guest atom in the cage, called *rattling*, have been considered to play crucial roles for the emergence of exotic magnetic phenomena [1]. A typical example of such interesting phenomena is heavy-electron state induced by rattling. In fact, in the filled-skutterudite compound $SmOs_4Sb_{12}$, it has been observed that the large electronic specific heat coefficient is almost independent of an applied magnetic field up to about 30 Tesla [2], in sharp contrast to the decrease of electronic specific heat coefficient due to the magnetic field in the conventional heavy-electron materials originating from the traditional spin Kondo effect. The mechanism of rattling-induced heavy-electron state has been theoretically investigated from various viewpoints [3, 4, 5, 6, 7, 8, 9, 10, 11, 12, 13, 14, 15].

Among several kinds of oscillation modes of the guest atom in the cage, here we focus on local Jahn-Teller mode. When such Jahn-Teller phonons are dynamically coupled with orbital degree of freedom of electrons, the local ground state is known to be a vibronic state characterized by clockwise and anti-clockwise rotational mode of Jahn-Teller phonons. Such geometric degree of freedom is screened by orbital moment of electrons and there occurs the release of an entropy $\ln 2$ of the vibronic state, leading to the Kondo-like phenomenon [16, 17, 18, 19]. Due to this type of Kondo effect, we expect the emergence of heavy-electron state, but this possibility was discussed with the use of the dynamical mean-field theory for the orbital-degenerate periodic Anderson model coupled with Jahn-Teller phonons [20]. When we explicitly consider the spin degree of freedom in the vibronic state, we expect a spin-vibronic quartet ground state, which should be given by the direct product of spin and vibronic degrees of freedom. We believe that it is an intriguing problem to pursue the Kondo effect emerging from the spin-vibronic state.

In this paper, we numerically analyze the two-orbital Anderson model including both Coulomb interactions and the coupling between electrons and Jahn-Teller phonons on an impurity site. For the convenience of the calculation, we introduce unitary transformations of electron and phonon operators, which naturally express the rotational mode of the Jahn-Teller phonons and remove the pair-hopping interaction term from the original Hamiltonian. By using a numerical renormalization method [21, 22], we evaluate entropy, specific heat, and several kinds of susceptibilities. It is found that the spin and total angular momenta are simultaneously screened by conduction electrons, leading to the Kondo phenomenon due to the entropy release of ln 4 of the local spin-vibronic state. Then, we point out a possibility of heavy-electron formation by the spin-vibronic Kondo effect. Throughout this paper, we use such units as $\hbar = k_{\rm B} = 1$.

2. Model Hamiltonian and Local Electron States

2.1. Model Hamiltonian

In this paper, we consider the two-orbital Anderson model including both Coulomb interactions and the coupling between electrons and Jahn-Teller phonons, expressed by

$$H = \sum_{\boldsymbol{k}\ell\sigma} \varepsilon_{\boldsymbol{k}} c^{\dagger}_{\boldsymbol{k}\ell\sigma} c_{\boldsymbol{k}\ell\sigma} + V \sum_{\boldsymbol{k}\ell\sigma} (c^{\dagger}_{\boldsymbol{k}\ell\sigma} d_{\ell\sigma} + \text{h.c.}) + H_{\rm loc}, \tag{1}$$

where $\varepsilon_{\boldsymbol{k}}$ is the dispersion of conduction electron, $c_{\boldsymbol{k}\ell\sigma}$ is the annihilation operator of conduction electron with momentum $\boldsymbol{k}$, orbital ℓ (=a and b), and spin σ (=$\uparrow$ and $\downarrow$), $d_{\ell\sigma}$ is the annihilation operator of impurity electron, and V denotes the hybridization between conduction and impurity electrons of the same orbitals. The final term $H_{\rm loc}$ indicates the local Hamiltonian, given by

$$\begin{aligned} H_{\rm loc} = {} & U \sum_{\ell} n_{\ell\uparrow} n_{\ell\downarrow} + U' n_a n_b + J \sum_{\sigma,\sigma'} d^{\dagger}_{a\sigma} d^{\dagger}_{b\sigma'} d_{a\sigma'} d_{b\sigma} + J' (d^{\dagger}_{a\uparrow} d^{\dagger}_{a\downarrow} d_{b\downarrow} d_{b\uparrow} + \text{h.c.}) \\ & + E_d n + \sqrt{\alpha}\omega [(a_2 + a_2^{\dagger})\tau_x + (a_3 + a_3^{\dagger})\tau_z] + \omega (a_2^{\dagger} a_2 + a_3^{\dagger} a_3 + 1), \end{aligned} \tag{2}$$

where U is intra-orbital Coulomb interaction, $n_{\ell\sigma} = d^{\dagger}_{\ell\sigma} d_{\ell\sigma}$, U' is inter-orbital Coulomb interaction, $n_{\ell} = n_{\ell\uparrow} + n_{\ell\downarrow}$, J and J' denote exchange and pair-hopping interaction, respectively, E_d denotes the level of localized electron, n is local electron number, given by $n = n_a + n_b$, α is non-dimensional electron-phonon coupling constant, a_2 and a_3 are annihilation operators for $(x^2 - y^2)$- and $(3z^2 - r^2)$-type Jahn-Teller phonons, respectively, $\tau_x = \sum_{\sigma} (d^{\dagger}_{a\sigma} d_{b\sigma} + d^{\dagger}_{b\sigma} d_{a\sigma})$, $\tau_z = \sum_{\sigma} (d^{\dagger}_{a\sigma} d_{a\sigma} - d^{\dagger}_{b\sigma} d_{b\sigma})$, and ω is the frequency of local Jahn-Teller phonons. Note that $U = U' + J + J'$ due to the rotational invariance in the orbital space, while $J = J'$ due to the fact that the electron wavefunction is real for the evaluation of Coulomb integrals.

Although it is possible to perform the numerical calculations even in the present form of the Hamiltonian, it is convenient to introduce the following unitary transformations for both electron and phonon operators [23]:

$$c_{\boldsymbol{k}\pm\sigma} = (c_{\boldsymbol{k}a\sigma} \mp i c_{\boldsymbol{k}b\sigma})/\sqrt{2}, \; d_{\pm\sigma} = (d_{a\sigma} \mp i d_{b\sigma})/\sqrt{2}, \; a_{\pm} = (a_3 \mp i a_2)/\sqrt{2}. \tag{3}$$

In the following, we use the index γ to distinguish the pseudo-orbital, + and −. From the above transformations, eqs. (1) and (2) are, respectively, rewritten as

$$H = \sum_{\boldsymbol{k}\gamma\sigma} \varepsilon_{\boldsymbol{k}} c^{\dagger}_{\boldsymbol{k}\gamma\sigma} c_{\boldsymbol{k}\gamma\sigma} + V \sum_{\boldsymbol{k}\gamma\sigma} (c^{\dagger}_{\boldsymbol{k}\gamma\sigma} d_{\gamma\sigma} + \text{h.c.}) + H_{\rm loc}, \tag{4}$$

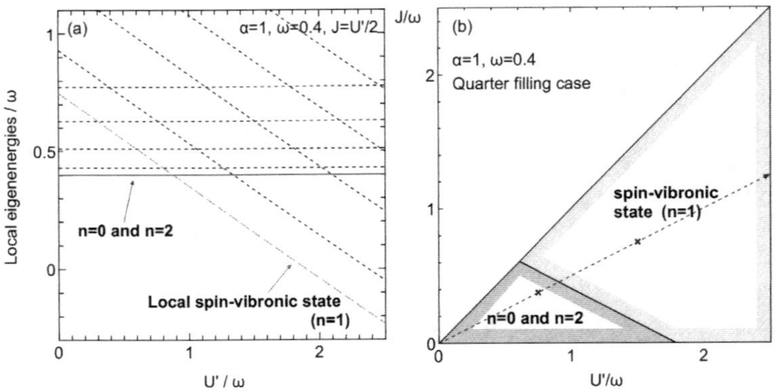

Figure 1. (a) Eigenenergies of $H_{\rm loc}$ vs. U' for $J = U'/2$ and $\alpha = 1$. The solid line indicates the lowest eigenenergy of the local degenerate ground states of $n = 0$ and $n = 2$, while the dashed line denotes the lowest eigenenergy of the spin-vibronic state of $n = 1$. The dotted lines indicate the eigenenergies of several low-lying states. (b) The local ground-state phase diagram on the plane of U' and J. The dashed line denotes the path of the parameter change to depict Fig. 1(a). The crosses indicate the typical parameters of the two types of the local ground states, which will be discussed in Figs. 2 and 3. Note that ω is set as the energy unit.

and

$$\begin{aligned} H_{\rm loc} = (U'+J)\sum_{\gamma} n_{\gamma\uparrow}n_{\gamma\downarrow} + (U'+J)n_+n_- + 2J\sum_{\sigma,\sigma'} d^\dagger_{+\sigma}d^\dagger_{-\sigma'}d_{+\sigma'}d_{-\sigma} \\ + E_d n + \sqrt{2}\alpha\omega[(a_+ + a^\dagger_-)\tau_+ + (a_- + a^\dagger_+)\tau_-] + \omega(a^\dagger_+ a_+ + a^\dagger_- a_- + 1), \end{aligned} \tag{5}$$

where $n_{\gamma\sigma} = d^\dagger_{\gamma\sigma}d_{\gamma\sigma}$, $n_\gamma = \sum_\sigma n_{\gamma\sigma}$, and $\tau_\pm = \sum_\sigma d^\dagger_{\pm\sigma}d_{\mp\sigma}$. Note that n is invariant under this transformation and we use the relations of $U = U' + J + J'$ and $J = J'$. Form the comparison of eqs. (2) and (5), we remark that the pair-hopping term disappears after the transformation. In order to obtain this form, it is important to include Coulomb interaction terms correctly in the original Hamiltonian from the viewpoint of the symmetry.

When we use the transformed Hamiltonian, there exists an advantageous point that we can adopt the total angular momentum J_z, defined by $J_z = T_z + L_z$, as the conserved quantity, where $T_z = \sum_\sigma (n_{+\sigma} - n_{-\sigma})/2$ and $L_z = a^\dagger_+ a_+ - a^\dagger_- a_-$. Namely, J_z is composed of the electron orbital moment and the rotational moment of Jahn-Teller phonons. Note that this conserved quantity is not easily found in the form of the original Hamiltonian, eqs. (1) and (2). From the viewpoint of the numerical calculations, it is useful to introduce the conserved quantity J_z in addition to total electron number and total spin moment, since we can save the computer resources.

2.2. Local electron states

Before proceeding to the numerical results of the Anderson model, let us discuss the local electron state. Since we are interested in the dynamical Jahn-Teller effect, we consider the quarter-filling case with one electron on the impurity site, by setting E_d so as to satisfy $\langle n \rangle = 1$, where $\langle \cdots \rangle$ denotes the operation to take thermal average. In order to treat the phonon part, we truncate the phonon basis for each Jahn-Teller mode at a finite number $N_{\rm ph}$, which is set as $N_{\rm ph} = 50$.

In Fig. 1(a), we show the eigenenergies of $H_{\rm loc}$ vs. U' for $J = U'/2$ and $\alpha = 1$. Note that we set ω as the energy unit for the local problem. For small U', the ground states with double degeneracy are composed of the states of $n = 0$ and $n = 2$ with the same weights, leading to

$\langle n\rangle = 1$. Note that the state of $n = 2$ is characterized by the linear combination of two double-occupied orbital states, $d^\dagger_{a\uparrow}d^\dagger_{a\downarrow}|0\rangle$ and $d^\dagger_{b\uparrow}d^\dagger_{b\downarrow}|0\rangle$, where $|0\rangle$ denotes the vacuum state. On the basis of eq. (5), it is possible to derive the effective interactions mediated by dynamical Jahn-Teller phonons and the pair-hopping attraction is enhanced among several kinds of interactions [24]. Thus, in order to gain the attractive interaction, double-occupied orbital states are stabilized for small U'.

However, when we increase the Coulomb repulsion, such double-occupied orbital states become unstable. Instead of them, there appears a single-occupied orbital state with active spin degree of freedom. Thus, we expect the quartet ground state, when we increase the magnitude of U', as observed in Fig. 1(a). It is called the spin-vibronic state from the following reason: Since this state includes one electron in the impurity site, the dynamical Jahn-Teller effect becomes active. The k-th eigenstate of $H_{\rm loc}$ labeled by σ and J_z is given by

$$|\sigma, J_z; k\rangle = \sum_{j=0}^{\infty}\left[p^{(k,j)}_{J_z} d^\dagger_{+\sigma}|J_z - 1/2; j\rangle + q^{(k,j)}_{J_z} d^\dagger_{-\sigma}|J_z + 1/2; j\rangle\right]. \tag{6}$$

Here $|L;j\rangle = |L + j, j\rangle$ for $L \geq 0$ and $|j, j + |L|\rangle$ for $L < 0$, where $|n_+, n_-\rangle = (1/\sqrt{n_+!n_-!})(a^\dagger_+)^{n_+}(a^\dagger_-)^{n_-}|0\rangle$. Note that J_z takes half-odd-integer values as $J_z = \pm 1/2$, $\pm 3/2$, $\pm 5/2$, $\cdots$ and the coefficients satisfy the relations of $p^{(k,j)}_{-J_z} = q^{(k,j)}_{J_z}$ and $q^{(k,j)}_{-J_z} = p^{(k,j)}_{J_z}$. For a fixed spin σ, we find the double degeneracy in each eigen-state originates from the rotational mode of Jahn-Teller phonons. The ground state is specified by $k = 1$ and $J_z = \pm 1/2$ for each $\sigma =\uparrow$ and $\downarrow$, suggesting that the ground state is quartet. We understand that the phonon state with $L_z = J_z + 1/2$ ($J_z - 1/2$) is coupled with the electron state with pseudo-orbital $T_z = -1/2$ ($+1/2$), leading to the vibronic state specified by total angular momentum J_z.

In Fig. 1(b), we show the local ground-state phase diagram on the plane of U' and J. Note that we draw the figure in the region for physically realistic parameters of $U' \geq J$. As easily understood from Fig. 1(a), we find the degenerate states of $n = 0$ and $n = 2$ in the region of small U'. On the other hand, we observe the spin-vibronic state, when we increase the value of U'. These two states are converted in a manner of the level crossing.

3. Calculated Results

Now we include the effect of hybridization in the model. In order to reveal electronic and phononic properties of H at low temperatures, it is necessary to evaluate corresponding susceptibilities. The susceptibility of an arbitrary operator A is given by

$$\chi_A = \frac{1}{Z}\sum_{k,j}\frac{e^{-E_k/T} - e^{-E_j/T}}{E_j - E_k}|\langle j|(A - \langle A\rangle)|k\rangle|^2, \tag{7}$$

where E_k is the eigenenergy of the k-th eigenstate $|k\rangle$ of H, Z is the partition function given by $Z = \sum_k e^{-E_k/T}$, and $\langle A\rangle = \sum_k e^{-E_k/T}\langle k|A|k\rangle/Z$. In this paper, we calculate the susceptibilities of charge, spin, and total angular momentum, which are defined as χ_c, χ_s, and χ_J, respectively.

For the evaluation of susceptibilities, in this paper, we employ a numerical renormalization group (NRG) method [21, 22]. In this technique, it is possible to include efficiently the conduction electron states in the vicinity of the Fermi energy by discretizing momentum space logarithmically. In actual calculations, we introduce a cut-off Λ for the logarithmic discretization of the conduction band. Due to the limitation of computer resources, we remain m low-energy states in each renormalization step. In this paper, we set $\Lambda = 5$ and $m = 5,000$.

In the NRG calculations, the energy unit is D, which is a half of the conduction electron bandwidth W. Hereafter D is set as unity. In this energy unit, a temperature T is defined

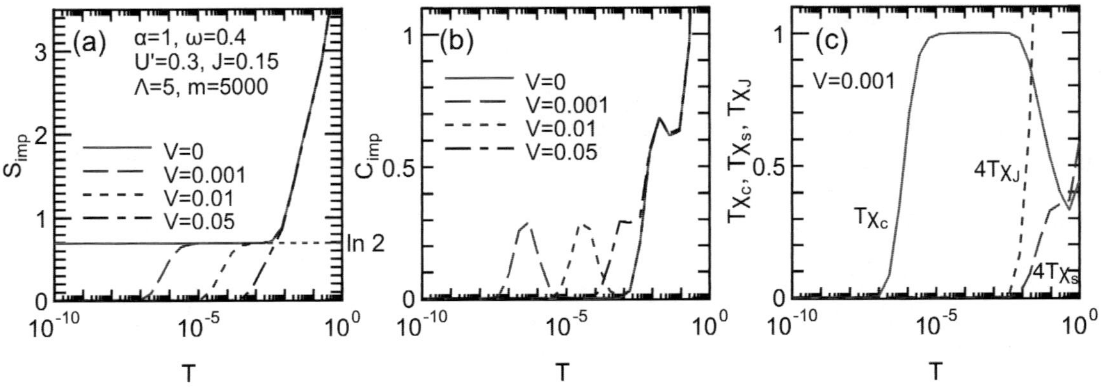

Figure 2. The temperature T dependences of (a) the entropy $S_{\rm imp}$ and (b) specific heat $C_{\rm imp}$ for several V for $U' = 0.3$ and $J = 0.15$. (c) The susceptibilities for charge $T\chi_c$, spin $T\chi_s$, and total angular momentum $T\chi_J$ vs. T for $U' = 0.3$, $J = 0.15$, and $V = 0.001$.

as $T_i = \Lambda^{-(i-1)/2}$ in the NRG calculation, where i is the number of the renormalization step. By using the NRG method, we also evaluate entropy $S_{\rm imp}$ and specific heat $C_{\rm imp}$ of localized electron. For the purpose, we calculate it by the numerical derivative of the entropy $S_{\rm imp}$ through the relation of $C_{\rm imp} = \partial S_{\rm imp}/\partial \log T = (S_{{\rm imp},i-1} - S_{{\rm imp},i})/\log(\Lambda^{1/2})$, where $S_{{\rm imp},i}$ is the entropy at the step i.

First we consider the case of the local degenerate ground states of $n = 0$ and $n = 2$. In Figs. 2(a) and 2(b), we depict the results of entropy $S_{\rm imp}$ and specific heat $C_{\rm imp}$ for various values of V for $U' = 0.3$ and $J = 0.15$. At $V = 0$, we find the double degeneracy originating from the local ground states of $n = 0$ and $n = 2$. For high temperatures as $T \gtrsim 10^{-2}$, irrespective of the value of V, $S_{\rm imp}$ shows large value due to the phonon excited states. With decreasing temperature, $S_{\rm imp}$ rapidly decreases and indicates a plateau of $\ln 2$ for $V < 0.05$, while $S_{\rm imp}$ directly goes to zero for $V > 0.05$ at relatively high temperatures.

For $V < 0.05$, in the plateau region, $C_{\rm imp}$ becomes almost zero. When we further decrease temperature, we observe the release of entropy $\ln 2$ and $C_{\rm imp}$ forms a clear peak. Here we define the Kondo temperature $T_{\rm K}$ as the temperature at which $C_{\rm imp}$ shows the peak. Finally, for $T < T_{\rm K}$, both $S_{\rm imp}$ and $C_{\rm imp}$ eventually vanish. For $V > 0.05$, without showing the peak in $C_{\rm imp}$ at $T = T_{\rm K}$, $C_{\rm imp}$ becomes zero. At temperatures lower than $T_{\rm K}$, we expect that the system is in the local Fermi-liquid state.

In Fig. 2(c), we depict the susceptibilities for charge, spin, and total angular momentum for $V = 0.001$, in order to clarify what is screened in the present Kondo-like phenomenon. We observe that $T\chi_s$ and $T\chi_J$ are rapidly suppressed at the temperature higher than $T_{\rm K}$, while $T\chi_c$ becomes zero around at $T_{\rm K}$. The behavior of χ_c is quite similar to that of the Anderson-Holstein model with attractive interaction larger than Coulomb repulsion. Namely, we confirm that the charge degree of freedom is relevant to the Kondo-like phenomenon due to the entropy release of $\ln 2$, as naively expected from the local degenerate ground states of $n = 0$ and $n = 2$.

Next we show the results for the local spin-vibronic ground state. In Figs. 3(a) and 3(b), we show the temperature dependence of $S_{\rm imp}$ and $C_{\rm imp}$ for several values of V. At $V = 0$, we find the four-fold degeneracy due to the local spin-vibronic ground state, leading to $\ln 4$ as residual entropy. In the high-temperature region such as $T \gtrsim 10^{-2}$, we also observe large $S_{\rm imp}$ due to phonon degrees of freedom. With decreasing temperature, $S_{\rm imp}$ rapidly decreases. For $V > 0.35$, we do not find a plateau and $S_{\rm imp}$ smoothly goes to zero. On the other hand, for $V < 0.35$, a plateau of $\ln 4$ is clearly observed and in this region, $C_{\rm imp}$ becomes zero. When the entropy $\ln 4$ is released, we find a peak in $C_{\rm imp}$, which also defines the Kondo temperature $T_{\rm K}$. For $T < T_{\rm K}$,

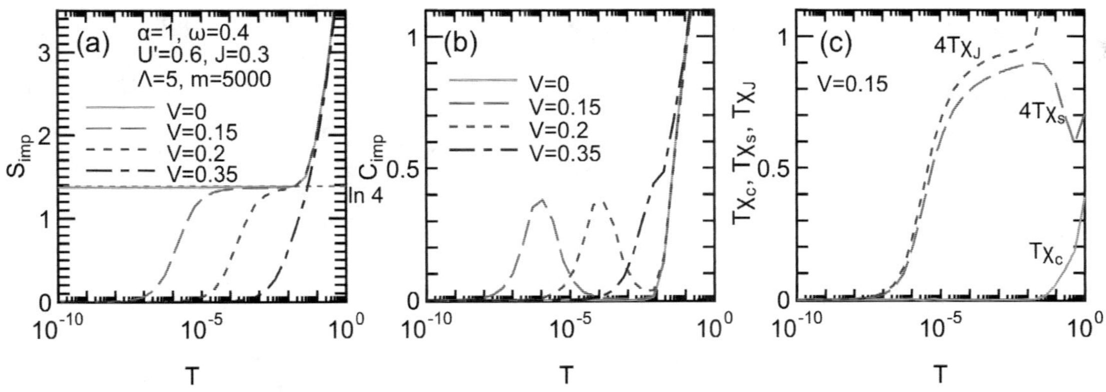

Figure 3. The temperature T dependences of (a) the entropy $S_{\rm imp}$ and (b) specific heat $C_{\rm imp}$ for several V for $U' = 0.6$ and $J = 0.3$. (c) The susceptibilities for charge $T\chi_c$, spin $T\chi_s$, and total angular momentum $T\chi_J$ vs. T for $U' = 0.6$, $J = 0.3$, and $V = 0.15$.

the system is also expected to be the local Fermi-liquid state.

In Fig. 3(c), we depict the susceptibilities of charge, spin, and total angular momentum. In sharp contrast to Fig. 2(c), we find that $T\chi_c$ is rapidly suppressed for $T \gg T_{\rm K}$, while both $T\chi_s$ and $T\chi_J$ go to zero around at $T_{\rm K}$. Namely, spin and total angular momenta are simultaneously screened by conduction electrons, leading to the release of $\ln 4$.

4. Discussion and Summary

We have discussed the Kondo effect of the two-orbital Anderson model which is dynamically coupled with local Jahn-Teller phonons. It has been found that spin and total angular momenta are simultaneously screened by conduction electrons, leading to the release of $\ln 4$, for the case with the local spin-vibronic ground state.

Here we provide a brief comment on the heavy-electron state, where the electronic specific heat coefficient γ becomes $10 \sim 10^3$ times larger than that of the normal metal. From the relation of $C_{\rm imp} = \gamma T$, we consider that the large peak of $C_{\rm imp}$ implies the enhancement of γ. On the basis of the local Fermi-liquid theory, we obtain the scaling relation of $\gamma \sim 1/T_{\rm K}$. For the parameter region of the local spin-vibronic ground state, we find that $T_{\rm K}$ is lower than that for the case of the local degenerate ground states of $n = 0$ and $n = 2$, when we use the same value of V. Namely, we expect that γ becomes large in the local spin-vibronic ground state. In our preliminary numerical results on γ (not shown here), we observe that γ is rapidly increased, when we enter the region of the local spin-vibronic state with the increase of U' for a fixed value of J. The results seem to indicate a possibility of heavy-electron state due to the spin-vibronic Kondo effect. It is one of our future problems to confirm this issue.

Note that in the spin-vibronic Kondo effect, we have observed the release of the entropy $\ln 4$, but it is not obvious that the release of entropy $\ln 4$ always occurs irrespective of the parameters. In particular, the adiabaticity seems to be a key issue. In this paper, we set $\omega = 0.4$ in the unit of $D = 1$, but it indicates $\omega/W = 0.2$, where W denotes the electron bandwidth. This value is not large enough to violate immediately the adiabatic condition, but in actual materials, ω seems to be smaller than the present values. In the adiabatic region, one may consider a possibility of the separation of the screening of spin and total angular momenta. On the other hand, in a purely electronic two-orbital Anderson model, we expect the occurrence of local spin-orbital quartet ground state, leading to the simultaneous release of the entropy $\ln 4$ of spin and orbital degrees of freedom. If the spin-vibronic ground state is continuously connected to the spin-orbital state of the purely electronic model, there may always occur the simultaneous release of the entropy ln 4. In any case, it is another future problem to clarify precisely how the entropy release of ln 4 occurs in the present model.

In summary, we have analyzed the two-orbital Anderson model coupled with local Jahn-Teller phonons. Depending on the parameters, we have found a couple of local ground states. Among them, we have emphasized the entropy release of ln 4, originating from the local spin-vibronic quartet ground state. After the release of ln 4, we expect the appearance of heavy-electron state due to the spin-vibronic Kondo effect. The condition of the occurrence of entropy release of ln 4 will be discussed elsewhere in future.

Acknowledgments

This work has been supported by a Grant-in-Aid for Scientific Research on Innovative Areas "Heavy Electrons" (No. 20102008) for the Ministry of Education, Culture, Sports, Science, and Technology, Japan and a Grant-in-Aid for Scientific Research (C) (No. 24540379) from Japan Society for the Promotion of Science. The computation in this work has been partly done using the facilities of the Supercomputer Center of Institute for Solid State Physics, University of Tokyo.

References

[1] See, for instance, *Proceedings of International Conference on New Quantum Phenomena in Skutterudite and Related Systems* (Skutterudite 2007), J. Phys. Soc. Jpn. **77** (2008) Suppl. A.
[2] S. Sanada, Y. Aoki, H. Aoki. A. Tsuchiya, D. Kikuchi, H. Sugawara, and H. Sato: J. Phys. Soc. Jpn. **74** (2005) 246.
[3] K. Hattori, Y. Hirayama, and K. Miyake: J. Phys. Soc. Jpn. **74** (2005) 3306.
[4] K. Hattori, Y. Hirayama and K. Miyake: J. Phys. Soc. Jpn. **75** Suppl. (2006) 238.
[5] K. Mitsumoto and Y. Ōno: Physica C **426-431** (2005) 330.
[6] K. Mitsumoto and Y. Ōno: Physica B **378-380** (2006) 265.
[7] T. Hotta: J. Phys. Soc. Jpn. **76** (2007) 084702.
[8] T. Hotta: J. Phys. Soc. Jpn. **77** (2008) 103711.
[9] T. Hotta: J. Phys. Soc. Jpn. **78** (2009) 073707.
[10] K. Oshiba and T. Hotta: J. Phys. Soc. Jpn. **80** (2011) SA134.
[11] K. Oshiba and T. Hotta: J. Phys. Soc. Jpn. **80** (2011) 094712.
[12] T. Fuse and Y. Ōno: J. Phys. Soc. Jpn. **79** (2010) 093702.
[13] T. Fuse and Y. Ōno: J. Phys. Soc. Jpn. **80** (2011) SA136.
[14] T. Fuse, Y. Ōno, and T. Hotta: J. Phys. Soc. Jpn. **81** (2012) 044701.
[15] T. Fuse and T. Hotta: arXiv:1207.2599.
[16] T. Hotta: Phys. Rev. Lett. **96** (2006) 197201.
[17] T. Hotta: Physica B **403** (2006) 1371.
[18] T. Hotta: J. Phys. Soc. Jpn. **76** (2007) 023705.
[19] T. Hotta: J. Phys. Soc. Jpn. **76** (2007) 034713.
[20] K. Mitsumoto and Y. Ōno: J. Phys. Soc. Jpn **79** (2010) 054707.
[21] K. G. Wilson: Rev. Mod. Phys. **47** (1975) 773.
[22] H. R. Krishna-murthy, J. W. Wilkins, and K. G. Wilson, Phys. Rev. B **21** (1980) 1003.
[23] Y. Takada: Phys. Rev. B **61** (2000) 8631.
[24] T. Hotta: J. Phys. Soc. Jpn. **79** (2010) 023709.

Journal of Physics: Conference Series **428** (2013) 012014 doi:10.1088/1742-6596/428/1/012014

Effective Coulomb interaction in multiorbital system

Izumi Hase and Takashi Yanagisawa

Electronics and Photonics Institute, AIST, Central-2, Umezono 1-1-4, Tsukuba, Ibaraki, Japan

E-mail: i.hase@aist.go.jp

Abstract. Transition metal atom generally takes various valences, and sometimes there are some 'missing valences', for example Fe usually takes 2+,3+ and 5+ , but does not take other valences so often. We have calculated the atomic multiplet energies for the high-spin and low-spin configurations within the ligand-field theory and the Hartree-Fock approximation, and found that the Coulomb interaction energy ($U_{\rm eff}$) becomes small when the valence is 'missing'. In case $U_{\rm eff} < 0$, this valence is not stable and segregates to other valences, and if the system is metallic, this cation site behaves as an attractive center between two conduction electrons. This result might explain one of the mystery of the iron-pnictide superconductor, e.g. the LDA+U approach can reproduce the experimental magnetic moment $\sim 0.3\mu_B$/Fe only when $U < 0$ is adopted. We have also investigated the effect of the distortion with symmetry lower than the regular octahedron. We have found that the value of $U_{\rm eff}$ increased in most cases, but in some special cases $U_{\rm eff}$ decreases and falls below the value $U - 3J$, which is the least value of the undistorted system.

1. Introduction

Every atom in an ionic compound takes a valence, for example, Fe in FeO takes 2+ valence. Generally speaking, transition metal atom takes many valences, for example, Fe takes +2,+3,+5 valences. However, Fe scarcely takes +4 valence. In fact, $SrFeO_3$ takes Fe^{4+} valence, but isovalent $CaFeO_3$ and $Sr_3Fe_2O_7$ takes mixed valence[1, 2, 3], i.e. $Fe^{3+} : Fe^{5+} = 1 : 1$. This clearly shows that Fe^{4+} valence is not stable for some reason. This kind of 'missing valence' is also found in IIIB- or VB-group elements. For example, Bi in $BaBiO_3$ does not takes Bi^{4+} but segregates into Bi^{3+} and Bi^{5+} valences. In this Bi site the electrons are 'attractive', because when the outermost valence ($6s$) electron is singly occupied (Bi^{4+}) it is unstable, and favors the $(6s)^0$ (i.e. Bi^{3+}) and $(6s)^2$ (i.e. Bi^{5+}) configurations. This attractive potential can induce the high-T_c superconductivity in (Ba,K)BiO_3[4, 5, 6]. However, why the missing valence takes place in the d-electron system is still not clear, because the Coulomb repulsion between the d-electrons is usually very large.

In this paper we consider a transition metal ion M surrounded by the A_6 octahedron, where A denotes an anion. We investigate that how the effective Coulomb interaction changes with varying n, which is the number of d-electrons on the M site. This paper is organized as follows: In section 2, we calculate the atomic multiplet energies and the effective Coulomb interaction energies $U_{\rm eff}$ for the high-spin and low-spin configurations for the MA_6 octahedron using the Hartree-Fock approximation. General trend of $U_{\rm eff}$ for the number of d-electrons is discussed. In section 3, we take the distortion of the MA_6 octahedron into account and show that in some special cases, $U_{\rm eff}$ decreases and falls below the value $U - 3J$, which is the least value of the undistorted system. Summary is given in section 4.

2. Octahedral case

Our treatment is based on the ligand-field theory. Since we are interested in the on-site effective Coulomb interaction of a MA_6 octahedron, we consider the following MA_6 cluster multiorbital d-p Hamiltonian:

$$H = H_t + H_U, \quad (1)$$
$$H_t = \sum_{i\sigma} \epsilon_d d^\dagger_{i\sigma} d_{i\sigma} + \sum_{j\sigma} \epsilon_p p^\dagger_{j\sigma} p_{j\sigma} + \sum_{ij\sigma} t_{ij}(d^\dagger_{i\sigma} p_{j\sigma} + h.c.) \quad (2)$$
$$H_U = U \sum_i n_{i\uparrow} n_{i\downarrow} + U' \sum_{i \neq j} n_{i\uparrow} n_{j\downarrow} + J \sum_{ij} (d^\dagger_{i\uparrow} d_{j\uparrow} d^\dagger_{j\downarrow} d_{i\downarrow} + h.c.) \quad (3)$$

Here, ϵ_d (ϵ_p) denotes the one-electron energy of d- (p-) orbitals on the cation (anion). t_{ij} denotes the transfer energy between the i'th d-orbital of cation and the j'th p-orbital of anion. The suffix j also denotes the six anion sites. σ is the spin index. For simplicity, we use the Kanamori parameter U, U' and J for the Coulomb and exchange interaction terms[7]. U and U' are the intra-orbital and inter-orbital Coulomb interaction energy. J is the exchange integral which leads to the Hund's coupling. These parameters have a relation $U' = U - 2J$.

First we consider the one-electron term H_t. If the cation M is surrounded by A_6 octahedron, the degeneracy of the 5-fold d-orbitals is partially lifted by the hybridization with anion p-orbitals. The one-electron energies of the d-orbitals are

$$\epsilon(d\gamma) = \frac{1}{2}\{(\epsilon_d + \epsilon_p) + \sqrt{(\epsilon_d - \epsilon_p)^2 + 8(pd\sigma)^2}\}, \quad (4)$$
$$\epsilon(d\epsilon) = \frac{1}{2}\{(\epsilon_d + \epsilon_p) + \sqrt{(\epsilon_d - \epsilon_p)^2 + 8(pd\pi)^2}\} \quad (5)$$

Here $(pd\sigma)$ and $(pd\pi)$ are the standard matrix elements between the neighboring d- and p-orbitals. Since $(pd\sigma) > (pd\pi)$, $\epsilon(d\gamma) > \epsilon(d\epsilon)$ if $\epsilon_d > \epsilon_p$. This splitting $\Delta_c \equiv \epsilon(d\gamma) - \epsilon(d\epsilon)$ leads to the octahedral crystal field $10Dq$ in the crystal field theory.

Next we consider the Coulomb interaction term H_U. We adopt the Hartree-Fock approximation and the atomic energy for every d^{n+m} electron configuration $(t_{2g})^{n_\uparrow}(t_{2g})^{n_\downarrow}(e_g)^{m_\uparrow}(e_g)^{m_\downarrow}$ can be obtained. The result is

$$E((t_{2g})^{n_\uparrow}(t_{2g})^{n_\downarrow}(e_g)^{m_\uparrow}(e_g)^{m_\downarrow}) = U * (\#\text{ of pairs which occupy a single orbital}) \quad (6)$$
$$+ \; U' * (\#\text{ of pairs other than } U\text{-pairs}) \quad (7)$$
$$- \; J * (\#\text{ of pairs which have the same spins}) \quad (8)$$

We can calculate the Hartree-Fock energy for each spin configuration in this way. For a given n (number of d electrons), many spin and orbital configurations are possible, but as is well known, if the crystal field is octahedral, then the spin configuration of the ground state is either the high-spin (HS) state or the low-spin (LS) state. Which spin state is most stable depends on the magnitude of Δ_c. Table 1 shows the ground state energy for each d^n configuration. Here we set the origin of energy is at $\epsilon(d\epsilon)$, and added the crystal-field splitting Δ_c to the energy of e_g orbital. We also included the effect of the distortion of the octahedron, as described in section 3. For undistorted case, Δ_t=$\Delta_t^{'}$=0. In this case, the HS state becomes the ground state if $\Delta_c < 5J$ for n=4,5 and $\Delta_c < 4J$ for n=6,7, respectively. Therefore, if $\Delta_c < 4J$ then the ground state is always the HS state, and if $\Delta_c > 5J$ then the ground state is always the LS state.

Table 1. The ground state energy for each d^n configuration. The intermediate spin configuration is $(t_{2g})^4(e_g)^1$ and $(t_{2g})^4(e_g)^2$ for n=5 and 6, respectively.

n	High spin	Low spin	Intermediate spin
1	$-\Delta_t'$		
2	$U-3J-\Delta_t'$		
3	$3U-9J$		
4	$6U-18J+\Delta_c-\Delta_t$	$6U-13J-\Delta_t'$	
5	$10U-30J+2\Delta_c$	$10U-20J-\Delta_t'$	$10U-24J+\Delta_c-\Delta_t-\Delta_t'$
6	$15U-38J+2\Delta_c-\Delta_t'$	$15U-30J$	$15U-33J+\Delta_c-\Delta_t-\Delta_t'$
7	$21U-49J+2\Delta_c-\Delta_t'$	$21U-45J+\Delta_c-\Delta_t$	
8	$28U-63J+2\Delta_c$		
9	$36U-80J+3\Delta_c-\Delta_t$		

The effective Coulomb interaction U_{eff} is written by

$$U_{\text{eff}}(n) = E(d^{n+1}) + E(d^{n-1}) - 2E(d^n) \tag{9}$$

where $E(d^n)$ denotes the ground state energy within the d^n configurations, given in Table 1.

Figure 1 shows the curves that effective Coulomb interaction $U_{\text{eff}}(n)$=0. Therefore, if U/J is below these lines then $U_{\text{eff}}(n) < 0$, namely, this valence is unstable. In other words, this site behaves as an attractive potential for electron pairs. The region $U_{\text{eff}}(n) < 0$ is the shadow area in Fig.1. The upper panel is for the LS state. For the LS state, this $U_{\text{eff}}(n) < 0$ area is wide when n=1,2,4,5,7,9. Or we can say that if n=3,6,8 then the valence is more stable. This is natural because in this case t_{2g} and e_g orbitals are 'closed'. The lower panel is for the HS state. Similarly, the stable states are n=3,5,8, in which the t_{2g} and e_g orbital are 'closed'. If we set Δ_c=0, $U_{\text{eff}}(n) = 0$ curve has a large dip at n=5. In this case the 5-fold degeneracy of the d electrons are not lifted, and the only one stable state is n=5. These results agree with the previous report[8] for the HS case.

Here we comment for the tetrahedral field. The crystal field induced by the tetrahedral configuration of anions has a duality with the octahedral one, i.e. 5-fold d-orbitals split into two-fold e_g and three-fold t_{2g} orbitals, and the energy of e_g orbital is lower than that of t_{2g} . Therefore, we can use electron-hole symmetry for the tetrahedral field. For example, the Fe atom in LaOFeAs is surrounded by four As atoms, forming $FeAs_4$ tetrahedron. Its valence state is Fe^{2+}, i.e. the d^6 configuration. Using the elentron-hole symmetry, it corresponds to the d^4 configuration in the octahedral ligand field. This d^4 configuration is the missing valence, as seen in $CaFeO_3$ and $Sr_3Fe_2O_7$[1, 2, 3]. Katayama-Yoshida *et al.* also claims that the d^4 state can be superconducting[6]. This concept of 'missing valence' can explain one of the mystery of iron-pnictide superconductor, which is that the LDA+U approach can reproduce the experimental magnetic moment $\sim 0.3\mu_B$/Fe only when $U < 0$ is adopted[9].

3. Distorted case

Next we consider a small D_{4h} distortion of the MA_6 octahedron. In the crystal field theory, due to the anisotropic crystal field, the degeneracy of the t_{2g} and e_g orbitals are lifted. In view of the ligand-field theory, this energy splitting is ascribed as the anisotropic transfer matrix $(pd\sigma)$ and $(pd\pi)$ in eqn.(4) and (5). If the MA_6 octahedron is elongated along z-direction, then $(pd\sigma)$ and $(pd\pi)$ along z-direction decreases, the energy of the d_{z^2}, d_{zx} and d_{yz} orbitals decreases, while

Journal of Physics: Conference Series **428** (2013) 012014 doi:10.1088/1742-6596/428/1/012014

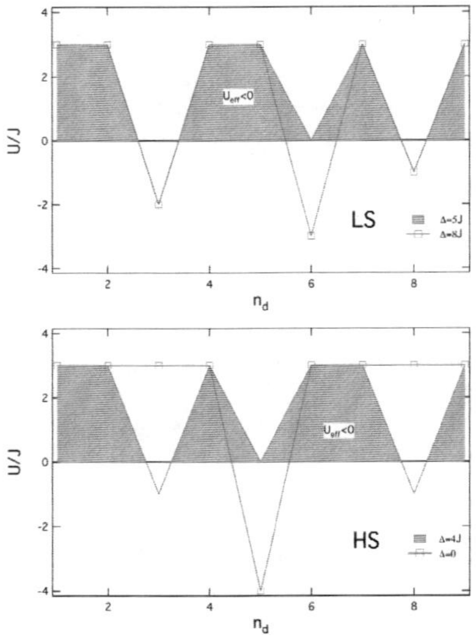

Figure 1. A phase diagram for U vs n. Shaded region means $U_{\text{eff}}(n) < 0$, for $\Delta_c = 5J$(HS) and $\Delta_c = 4J$(LS). Only $U > 0$ has physical meaning.

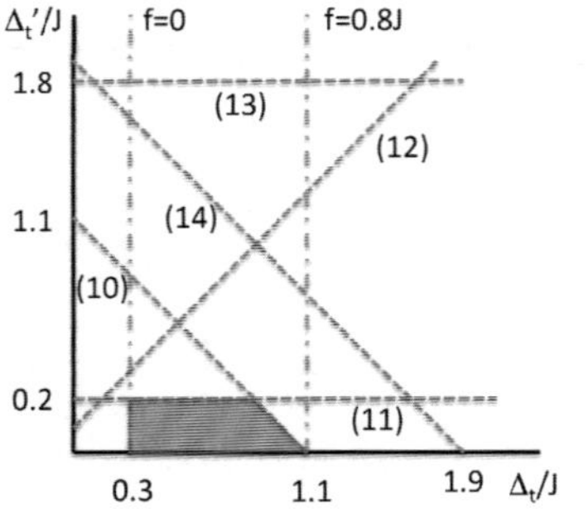

Figure 2. The region which satisfies eqn.(10-14) in the $(\Delta_t, \Delta_t^{'})$ plane. For example, (13) shows the line which satisfies eqn.(13). In the shaded area $f = 3\Delta_c + \Delta_t - 15J$ is positive, and $U_{\text{eff}}(5)$ is lower than $U - 3J$. In this figure we set $\Delta_c = 4.9J$.

$d_{x^2-y^2}$ and d_{xy} do not change the energy. Anyway, we write this energy splitting as $\tilde{\Delta_t}$ for e_g orbitals, and $\tilde{\Delta_t^{'}}$ for t_{2g} orbitals, respectively. If we fix the volume of the MA_6 octahedron, the center of gravity of the orbital energy is unchanged. And if the t_{2g} and e_g orbitals are partially occupied, then the system can gain energy by the distortion. As for the e^1 and e^3 configurations, the energy gain is $\Delta_t \equiv \tilde{\Delta_t}/2$. As for the t_2^1 and t_2^4 (t_2^2 and t_2^5) configurations, the collapse (elongation) of the MA_6 octahedron lowers the electronic energy, and the energy gain is $\Delta_t^{'} \equiv \frac{2}{3}\tilde{\Delta_t^{'}}$. This is the Jahn-Teller effect[10]. We included this energy for $E(d^n)$, and the result is shown in Table 1. In case of $\Delta_t \neq 0$ and $\Delta_t^{'} \neq 0$, the intermediate spin (IS) state can become the ground state for n=5,6. The energy $E(d^n)$ of IS states are also shown in Table 1.

Next we calculate $U_{\text{eff}}(n)$. In case of the undistorted octahedron, the least value of $U_{\text{eff}}(n)$ is $U - 3J$. However, in case of the distorted octahedron, $U_{\text{eff}}(n)$ can become further smaller. We consider the case that the ground states of d^4 and d^5 are the HS state, and the ground state of d^6 is the LS state. From $E(d^n)$ given in Table 1, these conditions are described as follows:

$$\Delta_t + \Delta_t^{'} < 6J - \Delta_c \tag{10}$$
$$\Delta_t^{'} < 10J - 2\Delta_c \tag{11}$$
$$-\Delta_t + \Delta_t^{'} < 5J - \Delta_c \tag{12}$$
$$\Delta_t^{'} < -8J + 2\Delta_c \tag{13}$$
$$\Delta_t + \Delta_t^{'} < -3J + \Delta_c \tag{14}$$

If we put a value on Δ_c, these inequalities give a region in the $(\Delta_t, \Delta_t^{'})$ plane. In this case $U_{\text{eff}}(5) = U + 12J - 3\Delta_c - \Delta_t$, and if $f = 3\Delta_c + \Delta_t - 15J > 0$, $U_{\text{eff}}(5)$ becomes lower than $U - 3J$. Figure 2 shows this region for $\Delta_c = 4.9J$. In this case we obtain $f = 0.8J$ at $(\Delta_t, \Delta_t^{'}) = (1.1J, 0)$.

However, since Δ_t and $\Delta_t^{'}$ both come from the same distortion of the octahedron, Δ_t and $\Delta_t^{'}$ are not independent each other. The relation between Δ_t and $\Delta_t^{'}$ changes when we change ϵ_d, ϵ_p, and the wave functions of anions. Thus we cannot determine the minimum value of $U_{\text{eff}}(5)$, but it can still be smaller than $U - 3J$, which is the least value of *any* $U_{\text{eff}}(n)$ in the undistorted case.

Similarly, in the case that the ground states of d^5 is the HS state, and the ground state of d^6 and d^7 are the LS state, $U_{\text{eff}}(6)$ becomes below $U - 3J$ in some region in the $(\Delta_t, \Delta_t^{'})$ plane. Interestingly, if we set $\Delta_c = 4.1J$, this region becomes identical with the region of Fig.2. More generally speaking, the $(\Delta_t, \Delta_t^{'})$ region which makes $U_{\text{eff}}(6) < U - 3J$ corresponds to the region which makes $U_{\text{eff}}(5) < U - 3J$, if we perform the transformation $\Delta_c \to 9J - \Delta_c$.

4. Summary

Using the Hartree-Fock approach, we have calculated the effective Coulomb interaction energy $U_{\text{eff}}(n)$ in the octahedral and distorted-octahedral ligand field. In the octahedral ligand field, the ground state is the either high-spin state or the low-spin state, depending on the strength of the octahedral crystal field. In the distorted-octahedral ligand field, intermediate spin can also become the ground state. Moreover, as for the d^5 and d^6 case, $U_{\text{eff}}(n)$ can become lower than $U - 3J$, which is the least value of $U_{\text{eff}}(n)$ for the octahedral ligand field.

References

[1] Takano M, Nakanishi N, Takeda Y, Naka S, Takada S 1977 *Mat. Res. Bull.* **12** 923
[2] Takano M, Nasu S, Abe T, Yamamoto K, Endo S, Takeda Y, Goodenough JB 1991 *Phys. Rev. Lett.* **67** 3267
[3] Kuzushita K, Morimoto S, Nasu S, Nakamura S 2000 *J. Phys. Soc. Jpn.* **69** 2767
[4] Varma CM 1988 *Phys. Rev. Lett.* **61** 2713
[5] Hase I, Yanagisawa T 2007 *Phys. Rev.* B **76** 174103
[6] Katayama-Yoshida H, Kusakabe K, Kizaki H, Nakanishi A 2008 *Appl. Phys. Exp.* B **1** 081703
[7] Brandow B H 1977 *Adv. Phys.* **26** 651
[8] Bouquet A E, Saitoh T, Mizokawa T, Fujimori A 1992 *Solid State Commun.* **83** 11
[9] Nakamura H 2009 *Physica* C **469** 908
[10] Here we did not take the elastic energy into account. If we also consider the elastic energy loss due to the lattice distortion, the total energy gain is $\Delta_t = \tilde{\Delta}_t/4$ and $\Delta_t^{'} = \frac{1}{3}\tilde{\Delta_t^{'}}$, respectively. However, the discussion in this section still works.

Journal of Physics: Conference Series **428** (2013) 012015 doi:10.1088/1742-6596/428/1/012015

Jahn-Teller theory beyond the standard model

S Bhattacharyya[1], D Opalka[1,3], L V Poluyanov[2] and W Domcke[1]

[1] Department of Chemistry, Technische Universität München, D-85747 Garching, Germany

[2] Institute of Chemical Physics, Academy of Sciences, Chernogolovka, Moscow 142432, Russia

E-mail: swarnendu.bhattacharyya@ch.tum.de, domcke@ch.tum.de

Abstract. In the standard model of Jahn-Teller (JT) theory, the spin-free electronic Hamiltonian is expanded up to second order in normal-mode displacements, while the spin-orbit (SO) coupling operator is approximated in zeroth order. In this article, the systematic extension of JT theory beyond the standard model is outlined. For the classic $E \times e$ and $T_2 \times t_2$ JT problems in C_{3v} and T_d symmetry, respectively, it is shown how the conventional Taylor expansion of the spin-free Hamiltonian can be replaced by an expansion in invariant polynomials up to arbitrarily high orders. The theory of SO coupling in JT systems is extended beyond the standard model by an expansion of the Breit-Pauli SO-coupling operator up to first order in normal-mode displacements. The spectroscopic relevance of the higher-order electrostatic and the linear relativistic JT coupling terms is illustrated for the vibronic structures of the photoelectron spectra of tetrahedral P_4 and Sb_4.

1. Introduction

The theory of vibronic coupling, which includes Jahn-Teller (JT) theory [1] and Renner theory [2] as special cases, is based on the following concepts :

(i) Representation of the electronic Hamiltonian in a basis of diabatic [3] electronic states.

(ii) Expansion of the electronic Hamiltonian in a Taylor series at the reference geometry of high symmetry.

(iii) Use of symmetry selection rules to determine the non-vanishing matrix elements.

In the "standard model" of JT theory, which is described in numerous reviews, monographs and edited volumes on the JT effect [4–11], the electrostatic (spin-free) Hamiltonian is expanded up to second order in normal-mode displacements. The spin-orbit (SO) coupling operator, which becomes relevant for systems containing heavier elements (e.g. transition metals, lanthanides or actinides) is approximated at zeroth order, that is, at the high-symmetry reference geometry [4–11].

Extensions of the electrostatic JT Hamiltonian beyond the standard model were occasionally considered for specific systems. For example, the third-order and fourth-order terms in the $E \times e$ Hamiltonian [12] for trigonal systems have been included in some investigations of static and dynamic JT effects in clusters and solids [13–16]. The systematic expansion of the electrostatic $E \times e$ JT Hamiltonian for trigonal systems up to sixth order has first been given by Viel and Eisfeld [17]. Applications of such high-order $E \times e$ Hamiltonians have been reported, for

[3] Present address: Department of Chemistry, University of Cambridge, Cambridge CB2 1EW, UK

example, for the methoxy and NO_3 radicals [18, 19], the ammonia cation [20] and transition-metal trifluorides [21]. It was discovered by *ab initio* calculations that "intramolecular collisions" of the ligand atoms at large amplitudes of the JT-active bending mode result in a pronounced positive anharmonicity of the *ab initio* bending potentials, which requires a JT expansion up to at least sixth-order in the bending mode [21, 22].

A general symmetry-adapted polynomial expansion of electrostatic $T \times t$ and $T \times e$ JT Hamiltonians in tetrahedral systems has been developed by Opalka and Domcke [23, 24]. This approach combines JT theory with the theory of invariant polynomials [25]. Symmetry-adapted polynomials up to high orders were explicitly given and a combinatorial scheme was developed to express terms of arbitrary order as products of a small number of invariant polynomials. The method was applied to the methane cation in its triply degenerate ($^2T_2$) ground state. Terms up to 12th order in the t_2 bending mode, up to 8th order in t_2 stretching mode and up to 10th order in the bending mode of e symmetry were included. The parameters were determined by a nonlinear least-squares fitting of a large set of *ab initio* data obtained at the full-valence CASSCF/MRCI/cc-pVTZ level [23, 24].

In the vast literature on the JT effect in solids, the SO coupling operator has been approximated by an effective atom-like SO operator, which, as such, is independent of the nuclear coordinates [4–7]. In treatments of joint JT and SO couplings in molecules, the SO splitting likewise was taken as independent of the nuclear coordinates and approximated by its value at the high-symmetry reference geometry [8–10]. A theoretical description of the SO coupling beyond the standard model has been developed only recently in the work by Poluyanov and Domcke for the trigonal, tetragonal, tetrahedral and cubic point groups [26–30]. In this approach, the SO interaction is considered as a perturbation of the electrostatic electronic Hamiltonian $\mathcal{H}_{el}$,

$$\mathcal{H}_{el} = \mathcal{H}_{es} + \mathcal{H}_{SO} \tag{1}$$

where $\mathcal{H}_{SO}$ is the Breit-Pauli SO operator [31]. The latter can be derived from the Dirac-Coulomb-Breit operator either by the elimination of the small components or by a perturbation expansion in c^{-1} [32]. A nonrelativistic spin-orbital electronic basis is employed and the Breit-Pauli operator is treated exactly as in the nonrelativistic vibronic coupling theory, that is,

(i) Representation of the Breit-Pauli operator in a basis of (nonrelativistic) diabatic [3] electronic states.

(ii) Expansion of the Breit-Pauli operator in powers of normal-mode displacements at the reference geometry.

(iii) Use of symmetry selection rules to determine the nonvanishing matrix elements.

Assuming that the SO coupling is a relatively weak perturbation of the electrostatic Hamiltonian, the Taylor expansion of the SO operator was terminated at first order. This way, the relativistic JT forces (arising from the SO operator) are included in the JT Hamiltonian. The existence of relativistic JT forces was revealed for the tetragonal, tetrahedral and cubic point groups [27–30]. In trigonal systems, on the other hand, the linear relativistic JT coupling constants are zero by symmetry [26].

In this article, we give a synopsis of recent theoretical developments which extend the JT theory systematically beyond the standard model. The motivation for these developments arises primarily from modern quantum chemistry. In order to harness the phenomenal progress in *ab initio* electronic-structure theory, including relativistic quantum chemistry, it is a necessity to develop JT theory substantially beyond the standard model. The latter was conceived during the 1940s and 1950s, at a time when the JT and SO coupling parameters had to be determined via the fitting of at best a few low-resolution spectral data, which required that the number of adjustable parameters had to be kept to a minimum. Nowadays, an essentially unlimited amount of data can be generated by accurate *ab initio* electronic-structure calculations, which eliminates the restriction of JT and SO coupling Hamiltonians to low-order expansion terms with just few coupling parameters.

2. Expansion of the spin-free Jahn-Teller Hamiltonian beyond second order

2.1. Polynomial invariant theory and Jahn-Teller Hamiltonian

The potential-energy (PE) surfaces of polyatomic molecules containing N identical nuclei are subject to permutational invariance of like nuclei. The permutations of the identical nuclei in a molecule form a group, the so-called Complete Nuclear Permutation (CNP) group which is nothing but the symmetry group of N identical nuclei (S_N) [33, 34]. Being a proper subgroup of the full symmetry group of the molecular system, the elements of the CNP group commute with the Hamiltonian. The molecular point group, on the other hand, is defined (locally) if a well-defined equilibrium geometry exists in a certain region of the PE surface. In recent years, the exploitation of permutation symmetry has been of increasing interest in the construction of analytic representations of global PE surfaces. Polynomials, invariant in the CNP group, have been employed to form a set of invariant functions which provide an approximation space for analytic PE surfaces [35].

The established description of the JT effect relies on the symmetry of the irreducible representations of the molecular point group defined at the configuration of highest symmetry. It will be shown below that in many cases of interest the irreducible representations of the molecular point group (matrix groups) are isomorphic to those of the permutation symmetry group. Therefore, the methods of invariant theory [25] can be employed for the efficient derivation of JT Hamiltonians.

The invariance of a polynomial under a linear algebraic group G can be defined as follows

$$\sigma \circ p(\mathbf{v}) = p(\sigma^{-1}\mathbf{v}) = p(\mathbf{v}) \tag{2}$$

where σ is a matrix representation of all the elements in G, p is an element of the ring of invariant polynomials under the group G ($\mathbb{K}[V]^G$) and $\mathbf{v}$ is an element of the underlying vector space V. The evaluation of the group action on the polynomials requires the matrix representations of the elements of the group. The generating set of invariant polynomials under the action of a finite linear group G can be obtained by the successive application of the Reynolds operator (which is a G-invariant projection) to all terms of a general polynomial expansion

$$\mathcal{R}(p) = \frac{1}{|G|}\sum_{\sigma \in G} \sigma \circ p. \tag{3}$$

Hilbert has given the proof (Hilbert's finiteness theorem) [36] that the ring of invariant polynomials $\mathbb{K}[V]^G$ under the group G is finitely generated by the generating set of homogeneous invariant polynomials $(p_1, \ldots, p_r)$ which is a subset of $\mathbb{K}[V]^G$, i.e.

$$\mathbb{K}[V]^G = \mathbb{K}[p_1, \cdots, p_r]. \tag{4}$$

The generating set of invariant polynomials has an *a priori* upper bound of their degree which is given by the order of the group $|G|$ (Noether degree bound) [37].

The linearly independent elements of $\mathbb{K}[p_1, \cdots, p_r]$ form a vector space and the adiabatic PE surface can be approximated by restricting the expansion up to a certain order. In the case of a single PE surface, defined by $\langle\varphi|\mathcal{H}_{el}|\varphi\rangle$, where $|\varphi\rangle$ is an adiabatic electronic state, the Hamiltonian $\mathcal{H}_{el}$ is a totally symmetric function of nuclear coordinates and the electronic wave function $|\varphi\rangle$ is invariant under symmetry operations. Multi-sheeted intersecting adiabatic PE surfaces, on the other hand, cannot be approximated by polynomial expansions due to the presence of cusps at conical intersections. Therefore, a representation of the multi-sheeted PE surface in a diabatic basis must be found to express the elements of the diabatic PE matrix as smooth functions of the nuclear coordinates.

The adiabatic electronic wave functions of a manifold of m intersecting electronic PE surfaces can be represented as superpositions in an m-dimensional diabatic basis, which forms an electronic sub-Hilbert space. Assuming that all other electronic states are sufficiently far apart in energy, the wave function is written as

$$|\Psi\rangle = \sum_{i=1}^{n} c_i |\varphi_i^d\rangle \tag{5}$$

where the $|\varphi_i^d\rangle$ are diabatic [3, 38–40] electronic basis states. The expectation value of the electronic Hamiltonian can be expressed as a function of nuclear coordinates and electronic coefficients

$$\begin{aligned}\langle \mathcal{H}_{el} \rangle &= \langle \Psi | \mathcal{H}_{el} | \Psi \rangle \\ &= \sum_{i,j} c_i \langle \varphi_i^d | \mathcal{H}_{el} | \varphi_j^d \rangle c_j \\ &= \sum_{i,j} c_i (\mathcal{H}_{el})_{ij} c_j.\end{aligned} \tag{6}$$

The energy expectation value (6) is quadratic in the electronic coefficients and $\mathcal{H}_{el}$ is expanded in terms of polynomials in symmetry-adapted nuclear coordinates. Therefore, we have to find the second-order invariants in the vector space V^{el} and the invariant polynomials in the vector space V^{nu} of the symmetry-adapted nuclear coordinates.

The most transparent case arises when the nuclear and the electronic basis functions transform according to the same irreducible representation, such as in the $E \times e$ and $T_2 \times t_2$ JT effects. In this case, the nuclear coordinates and electronic states form identical vector spaces and their matrix group representations are also identical in both coordinate spaces ($\Gamma^{el} = \Gamma^{nu}$). Weyl's polarization method can then be used to obtain the generating polynomials from the generators of the invariant ring of a single copy of these [41, 42]. The JT vibronic matrix can then very efficiently be constructed as the Hessian (with respect to the electronic degrees of freedom) of the invariant polynomials [23, 24].

2.2. The $E\times e$ Jahn-Teller Hamiltonian

Molecules of C_{3v} symmetry possess degenerate electronic states of E symmetry and degenerate vibrational modes of e symmetry [12], which transform like x, y in the C_{3v} point group. The polynomial invariants of the E representation in C_{3v} form a ring and are finitely generated by two polynomials of degree 2 and 3 respectively, which form the so-called generating set of the invariant ring. These are [42]

$$\begin{aligned} f_1 &= x^2 + y^2 \\ f_2 &= x^3 - 3xy^2. \end{aligned} \tag{7}$$

All the elements of the ring $\mathbb{K}[x, y]^{C_{3v}}$ can be constructed in terms of these generators

$$\mathbb{K}[x, y]^{C_{3v}} = \mathbb{K}[f_1, f_2]. \tag{8}$$

The ring of the C_{3v}-invariant polynomials in the direct sum of two vector spaces of E symmetry (electrons) and e symmetry (nuclei), $\mathbb{K}[V^E \oplus V^e]^{C_{3v}}$ can be computed from

$\mathbb{K}[V^e]^{C_{3v}} = \mathbb{K}[f_1, f_2]$ by Weyl's polarization method [41]. As Weyl's polarization operator is a linear differential operator, the JT matrix expansion of any order n can be found by taking the Hessian of all possible products of the primary invariants of order $(n+2)$. After eliminating the redundant terms, it is straightforward to represent the $E \times e$ JT expansion of any order as a sum of the trace and a traceless matrix [17]

$$\mathcal{H}_{es}^{(n)}[E \times e] = \begin{pmatrix} \mathcal{V}^{(n)} & 0 \\ 0 & \mathcal{V}^{(n)} \end{pmatrix} + \begin{pmatrix} \mathcal{W}^{(n)} & \mathcal{Z}^{(n)} \\ \mathcal{Z}^{(n)} & -\mathcal{W}^{(n)} \end{pmatrix}. \tag{9}$$

The well-known first-order and second-order JT Hamiltonians are the Hessians of f_2 and f_1^2 respectively.

The expansion of the $E \times e$ JT Hamiltonian up to sixth order is given by

$$\begin{aligned}
\mathcal{V}^{(1)} &= 0 \\
\mathcal{V}^{(2)} &= a_1^{(2)}(x^2 + y^2) \\
\mathcal{V}^{(3)} &= a_1^{(3)}(x^3 - 3xy^2) \\
\mathcal{V}^{(4)} &= a_1^{(4)}(x^4 + 2x^2y^2 + y^4) \\
\mathcal{V}^{(5)} &= a_1^{(5)}(x^5 - 2x^3y^2 - 3xy^4) \\
\mathcal{V}^{(6)} &= a_1^{(6)}(x^6 + 3x^4y^2 + 3x^2y^4 + y^6) + a_2^{(6)}(x^6 - 6x^4y^2 + 9x^2y^4)
\end{aligned} \tag{10}$$

$$\begin{aligned}
\mathcal{W}^{(1)} &= \lambda_1^{(1)} x \\
\mathcal{W}^{(2)} &= \lambda_1^{(2)}(x^2 - y^2) \\
\mathcal{W}^{(3)} &= \lambda_1^{(3)}(x^3 + xy^2) \\
\mathcal{W}^{(4)} &= \lambda_1^{(4)}(x^4 - y^4) + \lambda_2^{(4)}(x^4 - 6x^2y^2 + y^4) \\
\mathcal{W}^{(5)} &= \lambda_1^{(5)}(x^5 + 2x^3y^2 + xy^4) + \lambda_2^{(5)}(x^5 - 4x^3y^2 + 3xy^4) \\
\mathcal{W}^{(6)} &= \lambda_1^{(6)}(x^6 + x^4y^2 - x^2y^4 - y^6) + \lambda_2^{(6)}(x^6 - 5x^4y^2 - 5x^2y^4 + y^6)
\end{aligned} \tag{11}$$

$$\begin{aligned}
\mathcal{Z}^{(1)} &= \lambda_1^{(1)} y \\
\mathcal{Z}^{(2)} &= \lambda_1^{(2)}(-2xy) \\
\mathcal{Z}^{(3)} &= \lambda_1^{(3)}(x^2y + y^3) \\
\mathcal{Z}^{(4)} &= \lambda_1^{(4)}(-2x^3y - 2xy^3) + \lambda_2^{(4)}(4x^3y - 4xy^3) \\
\mathcal{Z}^{(5)} &= \lambda_1^{(5)}(x^4y + 2x^2y^3 + y^5) + \lambda_2^{(5)}(-2x^4y + 6x^2y^3) \\
\mathcal{Z}^{(6)} &= \lambda_1^{(6)}(-2x^5y - 4x^3y^3 - 2xy^5) + \lambda_2^{(6)}(4x^5y - 4xy^5)
\end{aligned} \tag{12}$$

where $\mathcal{V}^{(n)}$ is the nth order of the trace and $\mathcal{W}^{(n)}$ and $\mathcal{Z}^{(n)}$ are the nth order diagonal and off-diagonal elements, respectively, of the traceless JT Hamiltonian. The polynomial expansions presented here differ, at the first look, from those of Viel and Eisfeld [17]. However, they are interconvertable by taking linear combinations. As an illustration, the number of free parameters to be optimized for an 8th order expansion is given in tabular form in Table 1. Note that the number of independent optimization parameters grows very slowly with the order of the expansion, which reflects the high inherent symmetry of the $E \times e$ JT Hamiltonian.

Table 1. Number of parameters in the trace and the diagonal and off-diagonal terms in the $E \times e$ JT expansion in each order.

order	1	2	3	4	5	6	7	8
parameters $\mathcal{W}, \mathcal{Z}$	1	1	1	2	2	2	3	3
parameters $\mathcal{V}$	0	1	1	1	1	2	1	2
total	1	2	2	3	3	4	4	5

As an example, the sixth order $E \times e$ JT Hamiltonian matrix is given explicitly in Eq. (13). It can be seen that the sixth-order JT Hamiltonian contains only 4 independent optimization parameters.

$$\mathcal{H}^{(6)} = [a_1^{(6)}(x^6 + 3x^4y^2 + 3x^2y^4 + y^6) + a_2^{(6)}(x^6 - 6x^4y^2 + 9x^2y^4)]\,\mathbf{1}$$

$$+ \begin{pmatrix} \lambda_1^{(6)}(x^6 + x^4y^2 - x^2y^4 - y^6) & \lambda_1^{(6)}(-2x^5y - 4x^3y^3 - 2xy^5) \\ +\lambda_2^{(6)}(x^6 - 5x^4y^2 - 5x^2y^4 + y^6) & +\lambda_2^{(6)}(4x^5y - 4xy^5) \\ & \\ \lambda_1^{(6)}(-2x^5y - 4x^3y^3 - 2xy^5) & -\lambda_1^{(6)}(x^6 + x^4y^2 - x^2y^4 - y^6) \\ +\lambda_2^{(6)}(4x^5y - 4xy^5) & -\lambda_2^{(6)}(x^6 - 5x^4y^2 - 5x^2y^4 + y^6) \end{pmatrix}. \tag{13}$$

Using Weyl's polarization method [41], it is straightforward to include an arbitrary number of e modes to obtain the $E \times (e + e + \cdots)$ JT Hamiltonian up to arbitrary order.

2.3. The $T_2 \times t_2$ Jahn-Teller Hamiltonian

The three-sheeted $T_2 \times t_2$ PE surface is represented by three diabatic electronic states of T_2 symmetry, denoted conveniently as x, y, z. The nuclear coordinates of t_2 symmetry are also denoted by x, y, z to reveal the high inherent symmetry of the $T_2 \times t_2$ JT Hamiltonian. The ring of invariant polynomials of the t_2 representation in T_d is finitely generated by a set of three polynomials of degree 2, 3 and 4 in the coordinates x, y, z [23, 42]

$$\begin{aligned} f_1 &= x^2 + y^2 + z^2 \\ f_2 &= xyz \\ f_3 &= x^4 + y^4 + z^4. \end{aligned} \tag{14}$$

Any member of the ring $\mathbb{K}[x, y, z]^{T_d}$ can be represented in terms of these generating polynomials, that is

$$\mathbb{K}[x, y, z]^{T_d} = \mathbb{K}[f_1, f_2, f_3]. \tag{15}$$

The JT vibronic matrix is given by the doubly polarized invariant polynomials in the combined vector spaces of the electronic coefficients and the nuclear coordinates, which transform identically under the group $T_d \simeq S_4$. Any term of the Hamiltonian matrix expansion is just the Hessian of an invariant polynomial of the ring $\mathbb{K}[x, y, z]^{T_d}$ up to multiplication with a constant factor.

An elegant combinatorial scheme has been developed to represent the $T_2 \times t_2$ JT PE matrix in terms of the generating polynomials in Ref. [23]. The JT vibronic matrix has the highly symmetric structure

$$\mathcal{H}_{el} = \begin{pmatrix} \mathcal{W}(x,y,z) & \mathcal{Z}(z,x,y) & \mathcal{Z}(y,x,z) \\ \mathcal{Z}(z,x,y) & \mathcal{W}(y,x,z) & \mathcal{Z}(x,y,z) \\ \mathcal{Z}(y,x,z) & \mathcal{Z}(x,y,z) & \mathcal{W}(z,x,y) \end{pmatrix}. \tag{16}$$

Where $\mathcal{W}$ and $\mathcal{Z}$ are the diagonal and off-diagonal elements of the JT matrix, respectively. Their expansion up to sixth order reads

$$\begin{aligned}
\mathcal{W}^{(1)}(x,y,z) &= 0 \\
\mathcal{W}^{(2)}(x,y,z) &= a_1^{(2)}x^2 + a_2^{(2)}(y^2+z^2) \\
\mathcal{W}^{(3)}(x,y,z) &= a_1^{(3)}xyz \\
\mathcal{W}^{(4)}(x,y,z) &= a_1^{(4)}x^4 + a_2^{(4)}(y^4+z^4) \\
&\quad + a_3^{(4)}(x^2y^2+x^2z^2+y^2z^2) \\
\mathcal{W}^{(5)}(x,y,z) &= a_1^{(5)}x^3yz + a_2^{(5)}(xy^3z+xyz^3) \\
\mathcal{W}^{(6)}(x,y,z) &= a_1^{(6)}(y^6+z^6) + a_2^{(6)}x^6 + a_3^{(6)}(x^4y^2+x^4z^2) \\
&\quad + a_4^{(6)}(x^2y^4+x^2z^4) + a_5^{(6)}(y^4z^2+y^2z^4) + a_6^{(6)}x^2y^2z^2 \\
\mathcal{Z}^{(1)}(x,y,z) &= b_1^{(1)}x \\
\mathcal{Z}^{(2)}(x,y,z) &= b_1^{(2)}yz \\
\mathcal{Z}^{(3)}(x,y,z) &= b_1^{(3)}x^3 + b_2^{(3)}(xy^2+xz^2) \\
\mathcal{Z}^{(4)}(x,y,z) &= b_1^{(4)}x^2yz + b_2^{(4)}(y^3z+yz^3) \\
\mathcal{Z}^{(5)}(x,y,z) &= b_1^{(5)}x^5 + b_2^{(5)}(x^3y^2+x^3z^2) \\
&\quad + b_3^{(5)}(xy^4+xz^4) + b_4^{(5)}(xy^2z^2) \\
\mathcal{Z}^{(6)}(x,y,z) &= b_1^{(6)}y^3z^3 + b_2^{(6)}(y^4z^2+y^2z^4) \\
&\quad + b_3^{(6)}x^4yz + b_4^{(6)}(x^2y^3z+x^2yz^3).
\end{aligned} \tag{17}$$

The expansion terms up to 12th order can be found in Ref. [23]. The beauty of this representation is that the elements of the 3×3 JT matrix consist of just two functions, whose position in the matrix is determined by the first argument. There are only two kinds of parameters, $a_i^{(n)}$ and $b_i^{(n)}$ which are to be determined by a least-squares fitting of *ab initio* data. The number of parameters to be optimized for an 8th-order expansion of the $T_2\times t_2$ PE matrix is summarized in Table 2.

Table 2. Number of parameters in the diagonal and off-diagonal terms in the $T_2\times t_2$ JT matrix in each order.

order	1	2	3	4	5	6	7	8
parameters $\mathcal{W}$	0	2	1	3	2	6	4	9
parameters $\mathcal{Z}$	1	1	2	2	4	4	6	6
total	1	3	3	5	6	10	10	15

The 8th order PE matrix, for example, contains 53 parameters.

Journal of Physics: Conference Series **428** (2013) 012015 doi:10.1088/1742-6596/428/1/012015

It is straightforward to include an arbitrary number of t_2 modes in this approach. Weyl's polarization method provides a convenient scheme to obtain the generators for several t_2 coordinates. There are 12 generators in total for the $T_2 \times (t_2 + t_2)$ matrix expansion, for example [23].

2.4. Application : The $E\times e$ Jahn-Teller effect in P_4^+

P_4^+ is a system exhibiting an extremely strong JT effect in its electronic ground state of ^2E symmetry involving the doubly degenerate (e) bending mode. The $^2E \times e$ JT effect in P_4^+ with a dimensionless linear JT coupling parameter of about 5.0 [43, 57] may be the strongest $E \times e$ JT effect known in nature. The first *ab initio* investigation of the JT effect in the ^2E state of P_4^+ was carried out by Meiswinkel and Köppel [43] in the early 1990s. A more recent investigation was performed by Opalka et al.

The *ab initio* PE data were calculated at the CASSCF/cc-pVDZ level of theory using the MOLPRO quantum chemistry package [44]. The parameters of the polynomial expansion have been obtained by fitting the eigenvalues of the PE matrix to the *ab initio* data. *Ab initio* energies up to 2 eV above the energy at the reference geometry (the equilibrium geometry of P_4) have been calculated. A total of 1100 energy points have been included in the least-squares fitting. The difference between a fit including terms up to second order and a fit including terms up to sixth order is illustrated in Fig. 1. The figure confirms the necessity to extend the JT expansion beyond the standard model in this system. For the second-order fit, the *ab initio* data in the range $-0.35 \leq s_{2y} \leq 0.35$ (in Å) have been included. For the 6th-order fit, the *ab initio* data in the range $-0.8 \leq s_{2y} \leq 0.8$ (in Å) have been taken into account. Fig. (1) clearly shows that the second-order approximation is insufficient for an accurate representation of the *ab initio* PE surface of the ^2E ground state of P_4^+. Due to intra-molecular collisions which occur at large displacements from the equilibrium geometry in the e modes, the $E \times e$ JT PE surface exhibits a strong positive anharmonicity. An rms residual error of 0.02 eV has been estimated for the 6th-order surface for energies below 2.0 eV.

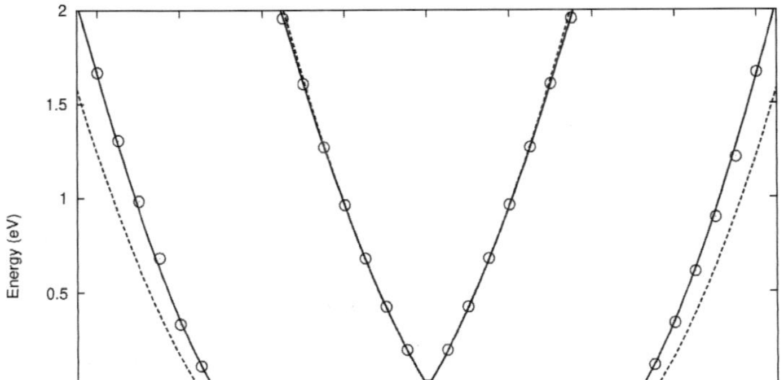

Figure 1. Cut of the 2nd order (dashed line) and the 6th order (full line) PE surface of the electronic ground state of P_4^+ along the symmetry coordinate s_{2y} compared with the *ab initio* data (circles).

The importance of JT coupling terms up to 6th order for an accurate representation of $E \times e$ JT PE surfaces has also been demonstrated for NH_3^+ [20], NO_3 [19] and various transition metal trifluorides of D_{3h} symmetry [21, 22].

The experimental photoelectron spectra of P_4, As_4 and Sb_4 were reported by several groups more than 20 years ago [53–56]. It was found that these cluster cations exhibit very strong JT effects, involving the e mode in the ^2E ground state and the t_2 mode in the $^2T_2$ excited state. In addition, the effects of (zeroth-order) SO splitting were identified in the JT spectra of As_4^+ and Sb_4^+ [53–56].

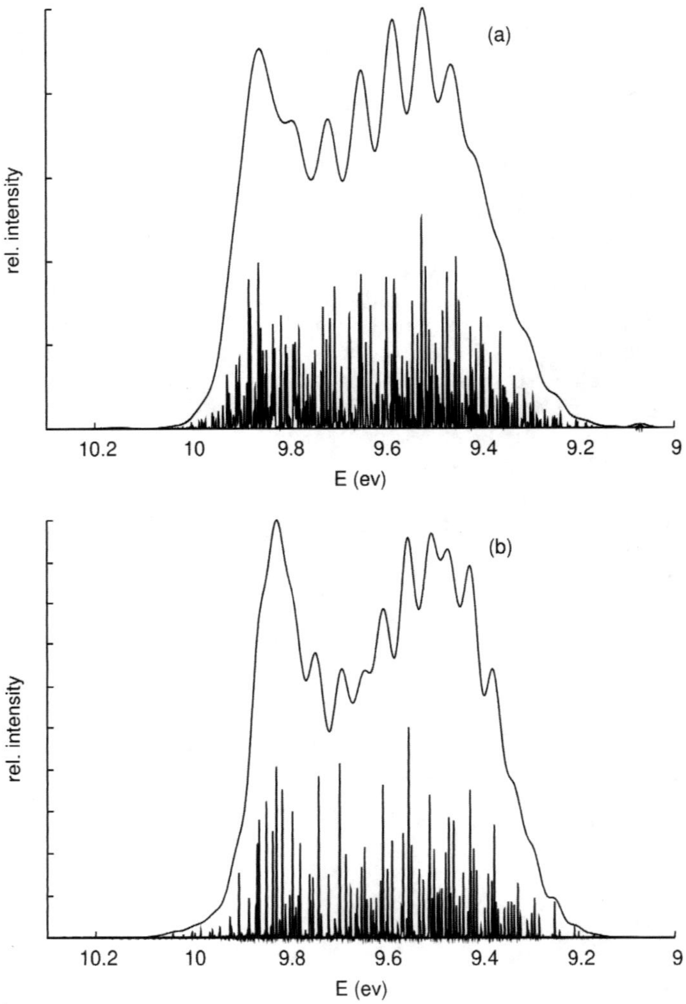

Figure 2. $^2E \times e$ JT spectrum of P_4^+ obtained with the quadratic (a) and the 6th-order (b) JT Hamiltonian

The photoelectron spectrum of the ^2E state of P_4^+ was calculated via the time-dependent approach to molecular electronic spectra [59, 60] as the Fourier transform of the autocorrelation function [58]. The propagation of the wave packet on the coupled diabatic PE surfaces was performed with the Chebysheff method [61] (see Ref. [58] for details). The so-called "low-resolution spectrum" is obtained by convolution with a Gaussian of 35 meV full width at half maximum. To obtain a "high-resolution spectrum" the full width at half maximum has been reduced by a factor of 1/30. The ^2E photoelectron spectrum of P_4^+ including up to quadratic JT coupling is shown in Fig. 2a. The clear double-hump structure of the envelope is due to a very strong linear JT coupling. The high line density and irregular structure of the high-resolution spectrum arise from quadratic JT coupling. Fig. 2b shows the photoelectron spectrum calculated with the inclusion of JT coupling terms up to 6th order. While the low-resolution envelopes of the two spectra are similar, the line density is reduced in the 6th-order spectrum as a result of the significant positive anharmonocity of the potential.

3. Expansion of the spin-orbit coupling Hamiltonian beyond zeroth order

3.1. Relativistically generalized Jahn-Teller selection rules

The symmetry group of the Hamiltonian including electron spin and SO coupling is the so-called spin double group of the respective point group [45]. The elements of the spin double group are of the form

$$Z_n = C_n U_n^\dagger \tag{18}$$

where the C_n are the symmetry operators of the point group and the U_n are unitary 2×2 matrices acting on the two-component spinors (or the spin functions α, β in the nonrelativistic limit). To close the group, the operator $-Z_n$ has to be included for every Z_n of Eq. (18), which doubles the order of the group [45]. The character tables of the common spin double groups can be found, for example, in Ref. [46].

Consider the group C_{3v} as one of the simplest non-Abelian point groups. The spin-free JT selection rule in the C_{3v} point group is [1]

$$[E^2] = A + E \tag{19}$$

where $[\Gamma^2]$ denotes the symmetrized square of the irreducible representation Γ [45]. According to Eq. (19), vibrational modes of e symmetry are JT active in first order. The relativistically generalized selection rules in the spin double group C'_{3v} are (for a single unpaired electron)

$$^2E = E \times E_{1/2} = E_{1/2} + E_{3/2} \tag{20a}$$

$$E_{1/2} \times E_{3/2} = 2E \tag{20b}$$

where $E_{1/2}$ and $E_{3/2}$ are two-dimensional two-valued irreducible representations of C'_{3v}. The spin function of a single electron transforms as $E_{1/2}$ in C'_{3v}. According to Eq. (20a), the four-fold degenerate ^2E state splits into two-fold degenerate $E_{1/2}$ and $E_{3/2}$ states by SO coupling. According to Eq. (20b), vibrational modes of e symmetry can couple the $E_{1/2}$ and $E_{3/2}$ SO-split electronic states. Since the e modes are already JT active in the electrostatic Hamiltonian, no new JT couplings arise from the SO operator. This has been confirmed by the explicit construction of the $^2E \times e$ JT Hamiltonian with inclusion of SO coupling [26]. In C_{3v} systems, the SO operator thus lifts the degeneracy of the nonrelativistic ^2E state, while the JT forces are of electrostatic origin [47].

Interestingly, a different situation is encountered in the tetragonal groups (D_{2d}, C_{4v}, D_{4h}). In the D_{2d} group, the spin-free selection rule is

$$[E^2] = A_1 + B_1 + B_2. \tag{21}$$

As is well known, the nondegenerate b_1 and b_2 normal modes are JT active (in first order) in tetragonal systems [4–6, 10]. The relativistically generalized selection rules in the D'_{2d} spin double group are

$$^2E = E \times E_{1/2} = E_{1/2} + E_{3/2} \tag{22a}$$

$$E_{1/2} \times E_{3/2} = B_1 + B_2 + E. \tag{22b}$$

According to Eq. (22a), the SO interaction lifts the four-fold degeneracy of the ^2E state. According to Eq. (22b), the $E_{1/2}$ and $E_{3/2}$ components of the SO-split ^2E state are coupled in first order by b_1, b_2 and e modes. Since the e modes are not JT active at the electrostatic level (see Eq. (21)), the JT activity of the e modes must arise from the SO operator. The same selection rules apply in C'_{4v} and D'_{4h}. The relativistic JT forces are thus complementary to the electrostatic JT forces in tetragonal systems.

In tetrahedral systems (point group T_d), the spin-free selection rules are [4–7, 10]

$$[E^2] = A_1 + E \tag{23a}$$

$$[T_{1,2}^2] = A_1 + E + T_2. \tag{23b}$$

The vibrational modes transform as a_1, e or t_2 in T_d. In electronic states of E symmetry, only the modes of e symmetry are JT active in first order (Eq. (23a)). In electronic states of T_1 or T_2 symmetry, the t_2 modes as well as the e modes are JT active (Eq. (23b)). $^2T_1$, $^2T_2$ and ^2E electronic states transform as follows in T'_d:

$$^2T_1 = T_1 \times E_{1/2} = G_{3/2} + E_{1/2} \tag{24a}$$

$$^2T_2 = T_2 \times E_{1/2} = G_{3/2} + E_{5/2} \tag{24b}$$

$$^2E = E \times E_{1/2} = G_{3/2} \tag{24c}$$

where $E_{1/2}$, $E_{5/2}$ are two-dimensional double-valued irreducible representations, while $G_{3/2}$ is a four-dimensional double-valued irreducible representation of T'_d [48]. According to Eqs. (24a, 24b), $^2T_1$ ($^2T_2$) electronic states split into $G_{3/2}$ and $E_{1/2}$ ($E_{5/2}$) irreducible components by SO coupling. The four-fold degeneracy of a ^2E electronic state, on the other hand, is not lifted by SO coupling at the tetrahedral reference geometry (Eq. 24c). While the two-fold degeneracy of the $E_{1/2}$ and $E_{5/2}$ levels cannot be lifted by any intramolecular interaction (Kramers degeneracy [49]), the energy levels of $G_{3/2}$ can split into two two-fold degenerate levels.

The JT selection rules in the T'_d spin double group are

$$E_{1/2} \times G_{3/2} = E + T_1 + T_2 \tag{25a}$$

$$E_{5/2} \times G_{3/2} = E + T_1 + T_2 \tag{25b}$$

$$\left\{G_{3/2}^2\right\} = A_1 + E + T_2. \tag{25c}$$

Here $\{\Gamma^2\}$ denotes the antisymmetrized product of Γ [45, 50]. Eq. (25c) reveals that the degeneracy of a $G_{3/2}$ electronic state is unstable with respect to deformations via e and t_2 modes. For a ^2E state, which transforms as $G_{3/2}$ in T'_d (Eq. (24c)), the t_2 mode is not JT active at the electrostatic level, see Eq. (23a). It follows that the JT activity of the t_2 modes in ^2E states must arise form the SO operator. A more detailed analysis (see below) reveals that in $^2T_{1,2}$ states the t_2 modes and the e modes are JT active (in first order) via electrostatic as well as relativistic forces. In ^2E states, on the other hand, the JT activity of the e modes is of electrostatic origin, while the JT activity of the t_2 modes is of relativistic origin [27, 28]. The JT selection rules in cubic symmetry (O_h, O'_h) are the same as in Eq. (23-25), apart from the additional inversion symmetry.

The relativistically generalized selection rules for electronic states of higher spin multiplicity in odd-electron systems can be obtained by a straightforward extension of the above analysis, since the representation of quartet, sextet, etc. spin states can be decomposed into the two-valued irreducible representations of the spin double group. In even-electron systems, the electronic states transform according to single-valued irreducible representations of the spin double group.

3.2. Relativistic $^2E \times e$ Jahn-Teller effect in D_{2d} systems

Let us consider, as the simplest model of a D_{2d}-symmetric object, a covalently bonded five-atomic YX_4 system, where Y is the central atom and the YX_1X_2 and YX_3X_4 atoms are in planes which are oriented perpendicular to each other. Aluminium tetroxide (AlO_4) is an example.

For the purpose of symmetry analysis, the electrostatic Hamiltonian is written as (in atomic units)

$$\mathcal{H}_{es} = -\frac{1}{2}\nabla^2 - \Phi(\mathbf{r}) \tag{26a}$$

$$\Phi(\mathbf{r}) = \frac{Q_0}{r} + \sum_{k=1}^{4} \frac{Q_1}{r_k} \tag{26b}$$

where Q_0 and Q_1 are effective atomic charges, $\mathbf{r}$ is the radius vector of the electron, $r = |\mathbf{r}|$ and $r_k = |\mathbf{r} - \mathbf{R}_k|$. The $\mathbf{R}_k$, $k = 1 \ldots 4$, are the radius vectors to the atoms $X_1 \ldots X_4$. The Y atom is at the origin of the coordinate system.

The Breit-Pauli operator for this system reads [31]

$$\mathcal{H}_{SO} = -ig_e\beta_e^2\mathbf{S}\left[\frac{Q_0}{r^3}(\mathbf{r}\times\boldsymbol{\nabla}) + Q_1\sum_{k=1}^{4}\frac{1}{r_k^3}(\mathbf{r}_k\times\boldsymbol{\nabla})\right] \tag{27a}$$

where

$$\mathbf{S} = \frac{1}{2}(\mathbf{i}\sigma_x + \mathbf{j}\sigma_y + \mathbf{k}\sigma_z), \tag{27b}$$

σ_x, σ_y, σ_z are the Pauli spin matrices, β_e is the Bohr magneton, g_e is the g-factor of the electron, and $\mathbf{i}$, $\mathbf{j}$, $\mathbf{k}$ are the Cartesian unit vectors.

We consider an isolated electronic E orbital with diabatic wave functions ψ_x, ψ_y which transform as x, y in the D_{2d} point group. The four-dimensional Hilbert space of a single electron in the E orbital is spanned by the spin-orbital basis functions

$$\{\psi_x\alpha, \psi_y\alpha, \psi_y\beta, \psi_x\beta\}\,. \tag{28}$$

The nine vibrational coordinates of the YX_4 molecule transform as

$$\Gamma_v = 2a_1 + b_1 + 2b_2 + 2e \tag{29}$$

in the D_{2d} point group. According to the selection rules of Eqs. (21,22), the b_1, b_2 and e modes are JT active in first order. In the following we consider, for brevity and clarity, only one representative of the b_1, b_2 and e modes, respectively. We denote the nondegenerate normal coordinates of b_1, b_2 symmetry as q_1, q_2, and the normal coordinates of e symmetry as q_x, q_y.

The spin-free Hamiltonian $\mathcal{H}_{es}$ is expanded, as usual, up to second order in normal coordinates of b_1, b_2 and e symmetry (see Ref. [30] for details). The Breit-Pauli operator is expanded up to first order in the normal modes. This expansion can be written as

$$\mathcal{H}_{SO} = \mathcal{H}_{SO}^{(0)} + \mathcal{H}_{SO}^{(1)} \tag{30a}$$

$$\mathcal{H}_{SO}^{(0)} = [h^y(e)\sigma_y + h^x(e)\sigma_x] + h^z(a_2)\sigma_z \tag{30b}$$

$$\begin{aligned}\mathcal{H}_{SO}^{(1)} =& [h_1^x(e)q_1\sigma_x + h_1^y(e)q_1\sigma_y] + [h_2^y(e)q_2\sigma_y - h_2^x(e)q_2\sigma_x] \\ &+ [h_y^z(e)q_y\sigma_z - h_x^z(e)q_x\sigma_z] + h_1^z(b_2)q_1\sigma_z + h_2^z(b_1)q_2\sigma_z \\ &+ h_+(b_1)(q_x\sigma_x + q_y\sigma_y) + h_-(a_2)(q_x\sigma_x - q_y\sigma_y) \\ &+ h^+(a_1)(q_x\sigma_y + q_y\sigma_x) + h^-(b_2)(q_x\sigma_y - q_y\sigma_x).\end{aligned} \tag{30c}$$

Journal of Physics: Conference Series **428** (2013) 012015 doi:10.1088/1742-6596/428/1/012015

The expansion coefficients in Eqs. (30b, 30c) are operators in electronic coordinate space. Their transformation properties are indicated in parentheses. The square brackets indicate symmetry-adapted linear combinations.

The $^2E \times (b_1 + b_2 + e)$ SO Hamiltonian matrix is obtained by calculating the matrix elements of $\mathcal{H}_{SO}$ with the electronic basis functions (28). The result is [30]

$$\mathcal{H}_{SO}[^2E \times e] = \begin{pmatrix} 0 & i\Delta & i\gamma q_+ & 0 \\ -i\Delta & 0 & 0 & -i\gamma q_+ \\ -i\gamma q_- & 0 & 0 & i\Delta \\ 0 & i\gamma q_- & -i\Delta & 0 \end{pmatrix} \tag{31a}$$

where

$$q_\pm = q_x \pm i q_y. \tag{31b}$$

Here Δ is the zeroth-order SO coupling which gives rise to the SO splitting of the ^2E state at the reference geometry. The real parameter γ represents the linear JT coupling arising from the SO operator. In the real-valued spin-orbital basis (28) the diagonal elements of the Breit-Pauli operator are necessarily zero, since the latter is purely imaginary (see Eq. (27a)). However, at the reference geometry ($q_x = q_y = 0$) the Hamiltonian matrix can be diagonalized, yielding the eigenvalues $-\Delta$, $-\Delta$, Δ, Δ. Due to the Kramers degeneracy of the eigenvalues, the unitary transformation is not unique. A suitable choice is [30]

$$U = \frac{1}{\sqrt{2}} \begin{pmatrix} 1 & 1 & 0 & 0 \\ -i & i & 0 & 0 \\ 0 & 0 & 1 & 1 \\ 0 & 0 & -i & i \end{pmatrix}. \tag{32}$$

This transformation defines a SO-adapted diabatic electronic basis.

Including the electrostatic JT coupling terms up to first order, the linear $^2E \times (b_1 + b_2 + e)$ JT coupling matrix in the SO-adapted basis reads

$$\widetilde{\mathcal{H}}^{(1)}[^2E \times (b_1 + b_2 + e)] = E_E \mathbf{1}_4 + \begin{pmatrix} \Delta & bq_2 + iaq_1 & 0 & i\gamma q_+ \\ bq_2 - iaq_1 & -\Delta & i\gamma q_+ & 0 \\ 0 & -i\gamma q_- & \Delta & -bq_2 + iaq_1 \\ -i\gamma q_- & 0 & -bq_2 - iaq_1 & -\Delta \end{pmatrix}. \tag{33}$$

Here $\mathbf{1}_4$ denotes the 4×4 unit matrix and a, b are the linear JT coupling constants of the b_1, b_2 modes. When the electrostatic JT couplings are neglected ($a = b = 0$), the Hamiltonian matrix (37) becomes isomorphic with the linear $^2E \times e$ JT+SO coupling Hamiltonian of trigonal systems [47].

The vibronic matrix (33) can easily be diagonalized, since the quartic secular polynomial factorizes into two quadratic polynomials due to time-reversal symmetry. The eigenvalues are given by

$$V_1 = V_2 = E_E - \sqrt{\Delta^2 + a^2 q_1^2 + b^2 q_2^2 + \gamma^2 (q_x^2 + q_y^2)} \tag{34a}$$

$$V_3 = V_4 = E_E + \sqrt{\Delta^2 + a^2 q_1^2 + b^2 q_2^2 + \gamma^2 (q_x^2 + q_y^2)}. \tag{34b}$$

They represent five-dimensional elliptic hyperboloids (the energies as a function of four nuclear coordinates). The Hamiltonian matrix and adiabatic potentials with inclusion of the quadratic electrostatic expansion terms can be found in Ref. [30].

Although the D_{2d} group is one of the less common point groups, it certainly is of relevance for the theoretical description of photochemical reactions. The D_{2d} group, being a subgroup of the tetrahedral and cubic groups, is a frequently occurring epikernel group in tetrahedral and cubic systems, that is, it is associated with local minima of the multi-dimensional JT PE surfaces [51]. The determination of JT and SO couplings at stationary points of D_{2d} symmetry is thus of widespread relevance in the theoretical analysis of transition-metal complexes, organometallic compounds and crystals.

3.3. Relativistic $^2T_2 \times t_2$ and $^2E \times t_2$ Jahn-Teller effect in tetrahedral and octahedral systems
Let us consider a four-atomic tetrahedral system with a single unpaired electron. Examples are the cluster cations P_4^+, As_4^+, Sb_4^+, Bi_4^+. The point group of the electrostatic Hamiltonian is T_d; the point group of the SO operator is T_d'. For the purpose of symmetry analysis, the electrostatic Hamiltonian of the unpaired electron is approximated as

$$\mathcal{H}_{es} = -\frac{1}{2}\nabla^2 - \Phi(\mathbf{r}) \tag{35a}$$

$$\Phi(\mathbf{r}) = Q\sum_{k=1}^{4}\frac{1}{r_k} \tag{35b}$$

where Q is the effective nuclear charge of the four equivalent atoms. The r_k are defined analogously to Eq. (26).

The Breit-Pauli operator for this system reads [31]

$$\mathcal{H}_{SO} = -ig_e\beta_e^2 Q\mathbf{S}\sum_{k=1}^{4}\frac{1}{r_k^3}(\mathbf{r}_k \times \boldsymbol{\nabla}) \tag{36}$$

where $\mathbf{S}$ is defined in Eq. (27b).

Molecular orbitals transforming according to the T_2 representation can be constructed as linear combinations of atomic p orbitals $\varphi_x^{(k)}$, $\varphi_y^{(k)}$, $\varphi_z^{(k)}$ on the four atoms

$$\psi_j = \frac{1}{2}\sum_{k=1}^{4}\varphi_j^{(k)}, \qquad j = x, y, z. \tag{37}$$

The six-dimensional Hilbert space of a single electron in the T_2 orbital is spanned by the spin-orbital basis functions

$$\{\psi_x\alpha, \psi_y\alpha, \psi_z\alpha, \psi_z\beta, \psi_y\beta, \psi_x\beta\}\,. \tag{38}$$

The six vibrational coordinates of the X_4 system transform as

$$\Gamma_v = a_1 + t_2 + e \tag{39}$$

in the T_d point group. According to the selection rule (23b), the t_2 and e modes are JT active in first order. We consider only the normal modes of t_2 symmetry in what follows and denote the normal coordinates of t_2 symmetry as Q_x, Q_y, Q_z.

The spin-free Hamiltonian $\mathcal{H}_{es}$ is expanded, as usual, up to second order in the normal coordinates of t_2 symmetry. The Breit-Pauli operator is expanded up to first order

$$\mathcal{H}_{SO} = \mathcal{H}_{SO}^{(0)} + \mathcal{H}_{SO}^{(1)} \tag{40a}$$

$$\mathcal{H}_{SO}^{(0)} = h^x\sigma_x + h^y\sigma_y + h^z\sigma_z \tag{40b}$$

$$\begin{aligned}\mathcal{H}_{SO}^{(1)} =& h(a_2)P_{a_2} + h(e_a)P_{e_a} + h(e_b)P_{e_b}\\ &+ h(t_{1x})P_{t_{1x}} + h(t_{1y})P_{t_{1y}} + h(t_{1z})P_{t_{1z}}\\ &+ h(t_{2x})P_{t_{2x}} + h(t_{2y})P_{t_{2y}} + h(t_{2z})P_{t_{2z}}\end{aligned} \tag{40c}$$

where the P_j are first-order symmetry adapted polynomials in $\mathbf{Q}$, defined as

$$P_{a_2}(\mathbf{Q},\boldsymbol{\sigma}) = \frac{1}{\sqrt{3}}(Q_x\sigma_x + Q_y\sigma_y + Q_z\sigma_z) \tag{41a}$$

$$P_{e_a}(\mathbf{Q},\boldsymbol{\sigma}) = \frac{1}{\sqrt{6}}(2Q_x\sigma_x - Q_y\sigma_y - Q_z\sigma_z) \tag{41b}$$

$$P_{e_b}(\mathbf{Q},\boldsymbol{\sigma}) = \frac{1}{\sqrt{2}}(Q_y\sigma_y - Q_z\sigma_z) \tag{41c}$$

$$P_{t_{1x}}(\mathbf{Q},\boldsymbol{\sigma}) = \frac{1}{\sqrt{2}}(Q_z\sigma_y - Q_y\sigma_z) \tag{41d}$$

$$P_{t_{2x}}(\mathbf{Q},\boldsymbol{\sigma}) = \frac{1}{\sqrt{2}}(Q_z\sigma_y - Q_y\sigma_z). \tag{41e}$$

The $P_{t_{1y}}$, $P_{t_{1z}}$ and $P_{t_{2y}}$, $P_{t_{2z}}$ follow by cyclic permutation.

The ${}^2T_2 \times t_2$ SO Hamiltonian matrix is obtained by calculating the matrix elements of $\mathcal{H}_{SO}$ with the electronic basis functions (38). The result is [28]

$$\mathcal{H}^{(0)}_{SO}[T_2] = \Delta \begin{pmatrix} 0 & i & 0 & -1 & 0 & 0 \\ -i & 0 & 0 & i & 0 & 0 \\ 0 & 0 & 0 & 0 & -i & 1 \\ -1 & -i & 0 & 0 & 0 & 0 \\ 0 & 0 & i & 0 & 0 & i \\ 0 & 0 & 1 & 0 & -i & 0 \end{pmatrix} \tag{42}$$

$$\mathcal{H}^{(1)}_{SO}[T_2 \times t_2] = \alpha \begin{pmatrix} 0 & 0 & -iQ_x & -iQ_z & Q_+ & 0 \\ 0 & 0 & iQ_y & Q_z & 0 & -Q_+ \\ iQ_x & -iQ_y & 0 & 0 & -Q_z & iQ_z \\ iQ_z & Q_z & 0 & 0 & iQ_y & -iQ_x \\ Q_- & 0 & -Q_z & -iQ_y & 0 & 0 \\ 0 & -Q_- & -iQ_z & iQ_x & 0 & 0 \end{pmatrix} \tag{43}$$

where

$$Q_\pm = Q_x \pm iQ_y. \tag{44}$$

Here Δ is the zeroth-order SO splitting of the ${}^2T_2$ state and α is the linear relativistic ${}^2T_2 \times t_2$ JT coupling constant.

The Hermitian matrix $\mathcal{H}^{(0)}_{SO}$ of Eq. (42) can be diagonalized by a unitary transformation U. Since the transformed matrix $\widetilde{\mathcal{H}}^{(0)}_{SO}$ has degenerate eigenvalues, the unitary matrix U is not unique. A suitable choice is given in Appendix B of Ref. [28] and defines a SO-adapted electronic basis. The zeroth-order SO vibronic matrix takes the form

$$\mathcal{H}^{(0)}_{SO}[T_2] = \mathrm{diag}(-\Delta, -\Delta, -\Delta, -\Delta, 2\Delta, 2\Delta). \tag{45}$$

In agreement with the group-theoretical result (24b), the zeroth-order SO coupling splits the six-fold degenerate ${}^2T_2$ manifold into a quadruply degenerate manifold ($G_{3/2}$) and a doubly degenerate manifold ($E_{5/2}$) [48].

Including the electrostatic JT coupling terms up to first order, the linear ${}^2T_2 \times t_2$ JT coupling

Hamiltonian in the SO adapted basis reads [28]

$$\widetilde{\mathcal{H}}^{(1)}[^2T_2 \times t_2] = E_{T_2}\mathbf{1}_6 + \begin{pmatrix} -\Delta & -i\tilde{a}_1Q_+ & 0 & i\tilde{a}_1Q_z & 0 & i\tilde{a}_2Q_- \\ i\tilde{a}_1Q_- & -\Delta & -i\tilde{a}_1Q_z & 0 & -\frac{i}{\sqrt{3}}\tilde{a}_2Q_- & \frac{2i}{\sqrt{3}}\tilde{a}_2Q_z \\ 0 & i\tilde{a}_1Q_z & -\Delta & i\tilde{a}_1Q_- & -i\tilde{a}_2Q_+ & 0 \\ -i\tilde{a}_1Q_z & 0 & -i\tilde{a}_1Q_+ & -\Delta & -\frac{2i}{\sqrt{3}}\tilde{a}_2Q_z & -\frac{i}{\sqrt{3}}\tilde{a}_2Q_+ \\ 0 & \frac{i}{\sqrt{3}}\tilde{a}_2Q_+ & i\tilde{a}_2Q_- & \frac{2i}{\sqrt{3}}\tilde{a}_2Q_z & 2\Delta & 0 \\ -i\tilde{a}_2Q_+ & -\frac{2i}{\sqrt{3}}\tilde{a}_2Q_z & 0 & \frac{i}{\sqrt{3}}\tilde{a}_2Q_- & 0 & 2\Delta \end{pmatrix} \tag{46a}$$

where

$$\tilde{a}_1 = \frac{1}{\sqrt{3}}(a + 2\alpha) \tag{46b}$$

$$\tilde{a}_2 = \frac{1}{\sqrt{2}}(a - \alpha) \tag{46c}$$

and a is the linear electrostatic $T_2 \times t_2$ JT coupling constant. It is seen that the effective linear JT coupling parameters in the SO-adapted basis have an electrostatic (a) and relativistic (α) contribution. The parameter $\tilde{a}_1$ is the effective linear JT coupling constant of the $G_{3/2}$ manifold. The parameter $\tilde{a}_2$ describes the pseudo-JT (PJT) coupling of the $G_{3/2}$ and $E_{5/2}$ manifolds. Depending on the signs of a and α, there is constructive or destructive interference of the electrostatic and relativistic JT couplings.

For sufficiently large values of the SO splitting 3Δ, the $G_{3/2}$ manifold (with eigenvalues $E_{T_2} - \Delta$) can be considered to be approximately decoupled from the $E_{5/2}$ manifold (with eigenvalue $E_{T_2} + 2\Delta$) [52]. In this approximation, the $^2T_2 \times t_2$ JT Hamiltonian is reduced to a 4×4 matrix

$$\widetilde{\mathcal{H}}^{(1)}[G_{3/2} \times t_2] = (E_{T_2} - \Delta)\mathbf{1}_4 + i\tilde{a}_1 \begin{pmatrix} 0 & -Q_+ & 0 & Q_z \\ Q_- & 0 & -Q_z & 0 \\ 0 & Q_z & 0 & Q_- \\ -Q_z & 0 & -Q_+ & 0 \end{pmatrix}. \tag{47}$$

The eigenvalues of this vibronic matrix are

$$V_1 = V_2 = E_{T_2} - \Delta - |\tilde{a}_1|R \tag{48a}$$

$$V_3 = V_4 = E_{T_2} - \Delta + |\tilde{a}_1|R \tag{48b}$$

where

$$R = (Q_x^2 + Q_y^2 + Q_z^2)^{\frac{1}{2}}. \tag{48c}$$

The adiabatic potentials (48) are doubly degenerate (Kramers degeneracy) and represent a double cone in four-dimensional space (the energies as a function of three nuclear coordinates).

The analysis for a ^2E state, which transforms as $G_{3/2}$ in T_d', is analogous and has been discussed in Ref. [27]. For brevity, we consider here only the t_2 vibrational mode. The Hamiltonian matrix in the linear approximation reads [27]

$$\mathcal{H}^{(1)}[^2E \times t_2] = E_E\mathbf{1}_4 + i\gamma \begin{pmatrix} 0 & Q_z & Q_- & 0 \\ -Q_z & 0 & 0 & -Q_- \\ -Q_+ & 0 & 0 & Q_z \\ 0 & Q_+ & -Q_z & 0 \end{pmatrix} \tag{49}$$

As discussed in Section 3.1, the zeroth-order SO splitting vanishes for a ^2E state in tetrahedral symmetry. Moreover, the linear electrostatic $^2E \times t_2$ JT coupling constant is zero by symmetry (see Eq. (23a)). The linear JT coupling constant γ in Eq. (49) is thus of purely relativistic origin. The eigenvalues of the vibronic matrix (49) are given by Eq. (48), with $|\tilde{a}_1|$ replaced by $|\gamma|$.

The $^2T_2 \times e$ and $^2E \times e$ JT Hamiltonians of tetrahedral systems are not discussed in detail in this synopsis for brevity. The results can be found in Refs. [27, 28]. The $^2E \times e$ case in tetrahedral symmetry is remarkable. As is well known, the zeroth-order SO splitting is zero for a ^2E state in tetrahedral symmetry [4–6, 10]. A detailed analysis reveals that the SO coupling within a ^2E state vanishes in any order of the normal mode of e symmetry. The $^2E \times e$ JT effect in tetrahedral systems is thus strictly unaffected by SO coupling.

As is well known, the electrostatic $^2T_2 \times (t_2 + e)$ and $^2E \times (t_2 + e)$ JT Hamiltonians up to second order are the same in tetrahedral and cubic systems, apart from the obvious additional restrictions due to inversion symmetry [4–11]. The relativistic JT forces, on the other hand, are not identical in tetrahedral and cubic systems. A detailed analysis reveals that the $^2T_2 \times e$ JT coupling has electrostatic as well as relativistic contributions in tetrahedral systems [28]. In octahedral and cubic systems, on the other hand, the linear relativistic $^2T_{2g} \times e_g$ coupling parameter is zero by symmetry [29]. This fact reveals that the symmetry selection rules for the SO operator differ from those for the electrostatic Hamiltonian in a subtle manner.

3.4. Application: The $^2T_2 \times t_2$ Jahn-Teller effect in Sb_4^+

Recent *ab initio* calculations of the matrix elements of the Breit-Pauli operator with nonrelativistic CASSCF wave functions and diagonalization of the SO-CI matrix have established the zeroth-order SO splittings of the T_2 state and have confirmed, moreover, the existence of substantial linear relativistic JT couplings for the heavier group-V tetramers [57].

Herein, we restrict our attention to JT and SO coupling in Sb_4^+, referring to Ref. [58] for a comparative discussion of the photoelectron spectra of P_4, As_4, Sb_4 and Bi_4. The electrostatic as well as relativistic JT couplings of the e mode in the $^2T_2$ state are rather weak and have little impact on the JT spectra [58]. For clarity, we therefore discuss the pure $T_2 \times t_2$ effect in Sb_4^+.

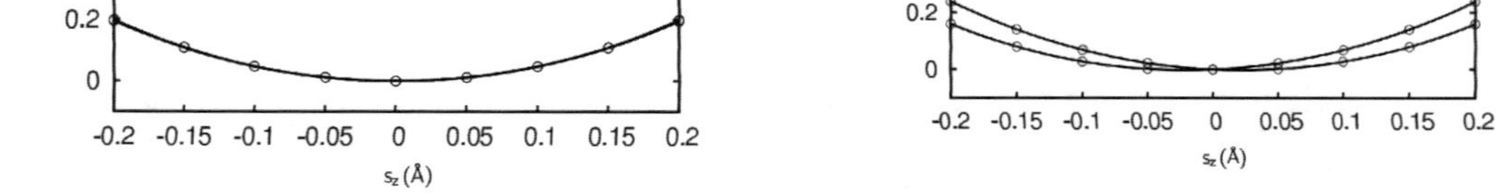

Figure 3. Electrostatic (a) and relativistic (b) PE surfaces of the electronic ^2E ground state and the $^2T_2$ first excited state for Sb_4^+ (circles: *ab initio* data; solid lines: fit)

.

Cuts of the PE surfaces of the ^2E and $^2T_2$ states of Sb_4^+ along the t_2 symmetry coordinate s_z, obtained without (a) and with SO coupling (b), are shown in Fig. 3. Fig. 3a reveals the very strong electrostatic $T_2 \times t_2$ JT splitting in the $^2T_2$ state upon distortion in the t_2 mode. Fig. 3b shows the PE functions obtained with inclusion of SO coupling. It illustrates the zeroth-order SO splitting of the $^2T_2$ state into $G_{3/2}$ and $E_{5/2}$ states. Moreover, it can be seen that the JT splitting of the $G_{3/2}$ state in (b) is significantly weaker than the JT splitting of the nonrelativistic $^2T_2$ state in (a). This feature reflects the destructive interference between a and α in Eq. (46b) ($a/\omega_{t_2} = 3.2, \alpha/\omega_{t_2} = -0.58$). Moreover, it is seen that the PE functions of the ^2E state in Fig. 3b exhibit a splitting as a function of s_z, which is the signature of the relativistic $^2E \times t_2$ JT coupling ($\gamma/\omega_e = 0.02$).

The photoelectron spectrum of the $^2T_2$ state of Sb_4^+, calculated in the nonrelativistic approximation, is displayed in Fig. 4. The sharp lines in Fig. 4 correspond to a resolution of 0.5 meV (which is determined by the total propagation time of the calculation). The envelope in Fig. 4 represents a "low-resolution spectrum", which was obtained from the high-resolution spectrum by convolution with a Gaussian of 15 meV full width at half maximum [58]. The photoelectron band of the $^2T_2$ state of Sb_4^+ is characteristic for a very strong $^2T_2 \times t_2$ JT effect ($a/\omega_{t_2} = 3.2$). The energy levels as well as the intensities are highly irregular, reflecting the pronounced anharmonicity of the adiabatic PE surfaces and strong nonadiabatic couplings. The low-resolution envelope, on the other hand, exhibits a quasi-regular progression, see Fig. 4.

The $^2T_2$ photoelectron spectrum obtained with inclusion of zeroth-order SO splitting ($\Delta = -71.45$ meV) is shown in Fig. 5a. The photoelectron band splits into two sub-bands with a separation of approximately 3Δ (0.21 eV). In addition, we observe a significant reduction of the extension of the quasi-progression in the t_2 mode, which reflects the quenching of the $T_2 \times t_2$ JT coupling by the zeroth-order SO splitting.

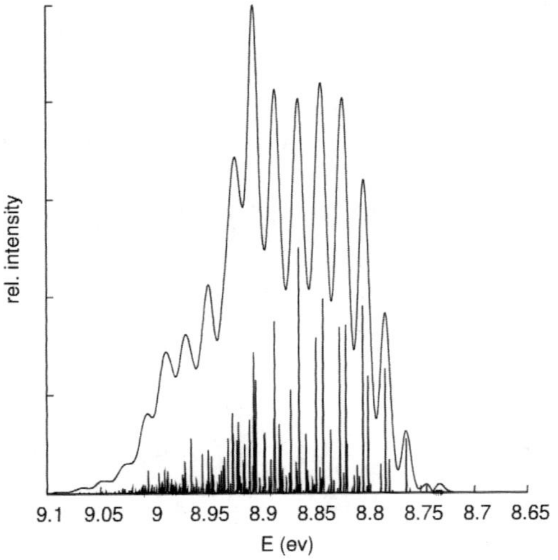

Figure 4. $^2T_2 \times t_2$ JT spectrum of Sb_4^+

Journal of Physics: Conference Series **428** (2013) 012015 doi:10.1088/1742-6596/428/1/012015

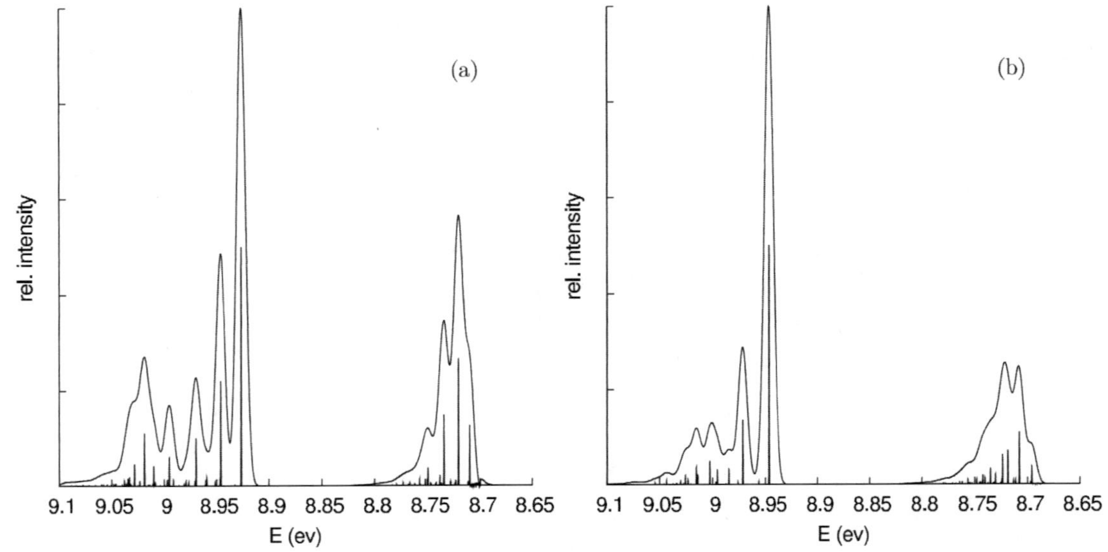

Figure 5. ${}^2T_2 \times t_2$ JT spectrum of Sb_4^+ including zeroth-order (a) and zeroth plus first-order (b) SO coupling

.

When the linear relativistic SO coupling constant α is additionally included, the ${}^2T_2$ photoelectron band displayed in Fig. 5b is obtained. It can be seen that the quasi-progression in the upper ($G_{3/2}$) state is shorter than in Fig. 5a, while the quasi-progression in the lower ($E_{5/2}$) state is more extended than in Fig. 5a. These features reflect the destructive and constructive interference in $\tilde{a}_1$ and $\tilde{a}_2$, respectively (see Eq. (46b, c)). The inclusion of the linear relativistic JT couplings thus is necessary for an accurate prediction of the vibronic structure of the photoelectron spectrum of Sb_4. As expected, these effects are even larger in the photoelectron spectrum of Bi_4 [58].

The available experimental photoelectron spectra of P_4, As_4 and Sb_4 are strongly broadened due to the high temperature of the clusters [53–56]. As a result, the existing photoelectron spectrum of Sb_4 is structureless [55, 56]. A high-resolution photoelectron spectrum of cold tetrahedral Sb_4 clusters would be of great interest for the future development of JT theory.

4. Summary and outlook

In this synopsis, we sketched the concepts of a systematic extension of JT theory beyond the standard model. The latter has been the paradigm for the analysis of static and dynamic JT effects in Physics and Chemistry since many decades. The motivation for a substantial extension of JT theory arises from modern computational electronic-structure theory, which can provide an essentially unlimited amount of PE data for JT systems, including an accurate description of SO coupling effects. The traditional expansions of the electrostatic PEs up to second order and the SO coupling energies up to zeroth order are insufficient for an accurate modeling of the *ab initio* data.

For the electrostatic (spin-less) Hamiltonian, we have summarized herein the high-order expansions of the two most generic and common JT Hamiltonians: the $E \times e$ Hamiltonian in C_{3v} symmetry and the $T_2 \times t_2$ Hamiltonian in T_d symmetry. Since C_{3v} is a subgroup of C_{6h}, T_d, and O_h, the $E \times e$ master JT Hamiltonian is also valid for these groups. Likewise, T_d is a subgroup of O_h. Therefore, the $T_2 \times t_2$ master JT Hamiltonian applies also in cubic symmetry. The conventional Taylor expansion of the electronic Hamiltonian has been replaced by an expansion in invariant polynomials, employing the powerful tools of invariant theory [25, 41]. Invariant theory allows the construction of JT expansions up to arbitrary order. Since the $E \times e$ and $T_2 \times t_2$ JT Hamiltonians are symmetric in the electronic and nuclear vector spaces, Weyl's polarization method can be employed and results in an elegant construction method of high-order JT Hamiltonians.

It has recently been shown by *ab initio* electronic-structure calculations for representative JT systems (e.g. NH_3^+ [20], NO_3 [19], CH_4^+ [23, 24] and transition-metal trifluorides [21, 22]) that for strong JT effects involving bending modes the expansion of the electrostatic potential up to second order is generally insufficient. Intramolecular collisions of the ligand atoms lead to pronounced positive anharmonicities. As a result, expansions of the PEs of the bending modes at least up to sixth order are a necessity.

The second restriction of the standard model of JT theory is the approximation of the SO coupling operator in zeroth-order of the normal-mode Taylor expansion. Recently, the theory of SO coupling in JT systems was extended beyond the standard model by the Taylor expansion of the Breit-Pauli SO operator up to first order in normal-mode displacements for the common spin double groups (trigonal, tetragonal, tetrahedral and cubic systems) [26–30]. If considered necessary, the expansion of the Breit-Pauli operator could be extended to include second-order terms. Since a nonrelativistic spin-orbital electronic basis was employed, the SO coupling appears as a perturbation of the nonrelativistic electronic Hamiltonian. The contributions of the electrostatic (Coulombic) interactions and the electrodynamic (current-induced) contributions are thus explicitly exhibited for each JT coupling constant. The different scaling of the electrostatic and relativistic interactions with the nuclear charge Z of heavy elements is thus transparent. Moreover, the SO-induced JT couplings are *a priori* obtained in a diabatic form. The diabatic basis of the nonrelativistic Hamiltonian is sufficient; no additional diabatization procedure is necessary. From a practical point of view, it is helpful that the matrix elements of the full (two-electron) Breit-Pauli operator with nonrelativistic CASSCF wave functions are efficiently implemented in several of the widely available electronic-structure codes (e.g. GAMESS, MOLPRO, MOLCAS).

For molecules, molecular complexes and crystals containing very heavy elements, the so-called two-step approach to relativistic electronic-structure (treating electron correlation first at the electrostatic level, followed by inclusion of SO-coupling effects by the diagonalization of the Breit-Pauli operator) may become slowly convergent and therefore inefficient [62]. In this case, it may be preferable to formulate the JT theory by employing a two-component electronic spinor basis (provided, e.g., by the ZORA, Douglas-Kroll-Hess or exact two-component (X2C) methods [32, 63]). A quasi-diabatic electronic basis has then to be defined at the relativistic level. The methods of invariant theory outlined in Section 2 may then be employed in the framework of the spin double groups.

References

[1] Jahn H A and Teller E 1937 *Proc. Roy. Soc. (London)* A **161** 220
[2] Renner E 1934 *Z. Physik* **92** 172
[3] Longuet-Higgins H C 1961 *Advances in Spectroscopy* vol 2, ed H W Thompson (New York: Interscience) p 429
[4] Sturge M D 1967 *Solid State Phys.* **20** 91
[5] Englman R 1972 *The Jahn-Teller Effect* (New York: Wiley)
[6] Bersuker I B and Polinger V Z 1989 *Vibronic Interactions in Molecules and Crystals* (Heidelberg: Springer)
[7] O'Brien M C M and Chancey C C 1993 *Am. J. Phys.* **61** 688
[8] Bersuker I B 2001 *Chem. Rev.* **101** 1067
[9] Applegate B E, Barckholtz and Miller T A 2003 *Chem. Soc. Rev.* **32** 38
[10] Bersuker I B 2006 *The Jahn-Teller Effect* (Cambridge University Press)
[11] Köppel H, Yarkony D R and Barentzen H (eds.) 2009 *The Jahn-Teller Effect* (Heidelberg: Springer)

[12] We follow the convention that uses the capital letters for the symmetry labels of electronic states and lower-case letters for the symmetry labels of vibrational modes.
[13] García-Fernández P, Bersuker I B, Aramburu J A, Barriuso M T and Moreno M 2005 *Phys. Rev.* B **71** 184117
[14] Marenich A V and Boggs J E 2005 *Chem. Phys. Lett.* **404** 351
[15] Hauser A W, Callegari C, Soldán P and Ernst W E 2010 *Chem. Phys.* **375** 73
[16] Hauser A W, Auböck G, Callegari C and Ernst W E 2010 *J. Chem. Phys.* **132** 164310
[17] Viel A and Eisfeld W 2004 *J. Chem. Phys.* **120** 4603
[18] Marenich A V and Boggs J E 2005 *J. Chem. Phys.* **122** 024308
[19] Faraji S, Köppel H, Eisfeld W and Mahapatra S 2008 *Chem. Phys.* **347** 110
[20] Viel A, Eisfeld W, Neumann S, Domcke W and Manthe U 2006 *J. Chem. Phys.* **124** 214306
[21] Mondal P, Opalka D, Poluyanov L V and Domcke W 2011 *Chem. Phys.* **387** 56
[22] Mondal P, Opalka D, Poluyanov L V and Domcke W 2012 *J. Chem. Phys.* **136** 084308
[23] Opalka D and Domcke W 2010 *J. Chem. Phys.* **132** 154108
[24] Opalka D and Domcke W 2010 *Chem. Phys. Lett.* **494** 134
[25] Derksen H and Kemper G 2002 *Computational Invariant Theory* (New York: Springer)
[26] Domcke W, Mishra S and Poluyanov L V 2006 *Chem. Phys.* **322** 405
[27] Poluyanov L V and Domcke W 2008 *J. Chem. Phys.* **129** 224102
[28] Poluyanov L V and Domcke W 2010 *Chem. Phys.* **374** 86
[29] Poluyanov L V and Domcke W 2012 *J. Chem. Phys.* **137** 114101
[30] Poluyanov L V and Domcke W 2012 *Chem. Phys.* **407** 1
[31] Bethe H A and Salpeter E E 1957 *Quantum Mechanics for One and Two Electron Atoms* (Berlin: Springer)
[32] Dyall K G and Faegri K 2007 *Relativistic Quantum Chemistry* (Oxford University Press)
[33] Bunker P R and Jensen P 1998 *Molecular Symmetry and Spectroscopy (second edition)* (Ottawa: NRC Research Press)
[34] Collins M A and Parsons D F 1993 *J. Chem. Phys.* **99** 6756
[35] Braams B J and Bowman J M 2009 *Int. Rev. Phys. Chem.* **28** 577
[36] Hilbert D 1890 *Math. Ann.* **36** 473
[37] Noether E 1915 *Math. Ann.* **77** 89
[38] Lichten W 1967 *Phys. Rev.* **164** 131
[39] Smith F T 1969 *Phys. Rev.* **179** 111
[40] Pacher T, Cederbaum L S and Köppel H 1993 *Adv. Chem. Phys.* **84** 293
[41] Weyl H 1946 *The Classical Groups. Their Invariants and Representations* (Princeton: Princeton University Press)
[42] Ascher E and Gay D 1985 *J. Phys.* A: Math. Gen. **18** 397
[43] Meiswinkel R and Köppel H 1993 *Chem. Phys. Lett.* **201** 449
[44] Werner H J, Knowles P J, et al., MOLPRO, Version 2006.1 and 2008, see http://www.molpro.net
[45] Hamermesh M 1962 *Group Theory and its Application to Physical Problems* (Reading: Addison-Wesley)
[46] Balasubramanian K 1997 *Relativistic Effects in Chemistry,* Part A (New York: Wiley)
[47] Chau F T and Karlsson L 1977 *Phys. Scripta* **16** 258
[48] We use symbols $E_{1/2}$, $E_{5/2}$, $G_{3/2}$ of Hamermesh for the irreducible representations of T_d' and O_h'. In the solid-state literature, these representations are denoted as Γ_6, Γ_7, Γ_8.
[49] Wigner E 1959 *Group Theory* (New York: Academic Press)
[50] Jahn H A 1938 *Proc. Roy. Soc. (London)* A **164** 117
[51] Ceulemans A 1987 *J. Chem. Phys.* **87** 5374
[52] Moffitt W and Thorson W 1957 *Phys. Rev.* **108** 1251
[53] Elbel S, Kudnig J, Grodzicki M and Lempka H J 1984 *Chem. Phys. Lett.* **109** 312
[54] Dyke J M, Morris A and Stevens J C H 1986 *Chem. Phys.* **102** 29
[55] Wang L-S, Niu B, Lee Y T, Shirley D A, Ghelichkhani E and Grant E R 1990 *J. Chem. Phys.* **93** 6318
[56] Wang L-S, Niu B, Lee Y T, Shirley D A, Ghelichkhani E and Grant E R 1990 *J. Chem. Phys.* **93** 6327
[57] Opalka D, Segado M, Poluyanov L V and Domcke W 2010 *Phys. Rev.* A **81** 042501
[58] Opalka D, Poluyanov L V and Domcke W 2011 *J. Chem. Phys.* **135** 104108
[59] Heller E J 1981 *Acc. Chem. Res.* **14** 368
[60] Schinke R 1993 *Photodissociation Dynamics* (Cambridge, UK: Cambridge University Press)
[61] Tal-Ezer H and Kosloff R 1984 *J. Chem. Phys.* **81** 3967
[62] Fleig T 2012 *Chem. Phys.* **395** 2
[63] Liu W 2010 *Mol. Phys.* **108** 1679

XXIst International Symposium on the Jahn–Teller Effect 2012 IOP Publishing
Journal of Physics: Conference Series **428** (2013) 012016 doi:10.1088/1742-6596/428/1/012016

Resonance and decay phenomena lead to quantum mechanical time asymmetry

A Bohm and H V Bui

Department of Physics, University of Texas at Austin, Austin TX 78712, USA

E-mail: bohm@physics.utexas.edu, viethai@physics.utexas.edu

Abstract. The states (Schrödinger picture) and observables (Heisenberg picture) in the standard quantum theory evolve symmetrically in time, given by the unitary group with time extending over $-\infty < t < +\infty$. This time evolution is a mathematical consequence of the Hilbert space boundary condition for the dynamical differential equations. However, this unitary group evolution violates causality. Moreover, it does not solve an old puzzle of Wigner: How does one describe excited states of atoms which decay exponentially, and how is their lifetime τ related to the Lorentzian width Γ? These question can be answered if one replaces the Hilbert space boundary condition by new, Hardy space boundary conditions. These Hardy space boundary conditions allow for a distinction between states (prepared by a preparation apparatus) and observables (detected by a registration apparatus). The new Hardy space quantum theory is time asymmetric, i.e, the time evolution is given by the semigroup with $t_0 \leq t < +\infty$, which predicts a finite "beginning of time" t_0, where t_0 is the ensemble of time at which each individual system has been prepared. The Hardy space axiom also leads to the new prediction: the width Γ and the lifetime τ are exactly related by $\tau = \hbar/\Gamma$.

1. Introduction

Most physicists believe that quantum theory, non-relativistic and relativistic, is time-symmetric which means that time t extends over $-\infty < t < +\infty$. But asymmetric time evolution with $t_0 < t < +\infty$, t_0 finite, have been discussed in a variety of forms, e.g, [1] [2] and references thereof. Here we want to discuss the quantum mechanical time asymmetry due to the boundary condition for the the dynamical equations of quantum mechanics.

The main reason for the belief in time symmetric quantum theory with $-\infty < t < +\infty$ is probably of mathematical origin: To solve the dynamical differential equations (Schrödinger equation for states, or Heisenberg equation for observables in the Heisenberg picture) one has to use boundary conditions. Von Neumann's choice for these boundary conditions was the Hilbert space (i.e. Lebesgue-square integrable wave functions). From a mathematical theorem (Stone-von Neumann theorem [3]) then follows the (unitary) group evolution with $-\infty < t < +\infty$, for the states (in the Schrödinger picture) or for the observables (in the Heisenberg picture). A similar theorem, not as well known holds if one uses the Schwartz space boundary condition of the Dirac ket formalism [4].

But Hilbert space and Schwartz space are not the only choice for boundary conditions of the Schrödinger and Heisenberg equation. In Lax-Phillips scattering theory [5], Hardy spaces were used in applications to scattering of classical waves. Hardy spaces have also been suggested more recently in the quantum theory of scattering. They are needed to obtain a quantum theory of resonance phenomena and decaying states [6, 7, 8, 9, 10, 11, 12, 13].

That scattering resonances and decaying states may be connected with each other has long been suspected, but an exact theory that unifies these phenomena was not possible within the conventional mathematical frame of (non-relativistic and relativistic) quantum theory provided by the Hilbert space, or by the Schwartz space triplet for Dirac kets. Hardy spaces are perhaps the only mathematical tools that allow an unification of resonance scattering and of decay phenomena.

2. Decaying States and Resonances

Many people have been thinking about the nature of the relationship between resonance and decay phenomena. According to an anecdote we heard from Larry Schulman, Wigner was going around Princeton in the 1960s asking people: "Tell me, what is the state vector of the $2\,P^{1/2}$ of the H-atom?" (The $2\,P^{1/2}$ state is the first excited energy level which decays into the ground state $1\,S^{1/2}$). The ground state vector $|n=1\rangle$ is an eigenstate of the Hamiltonian, $H\,|E_{n=1}\rangle = E_1\,|E_{n=1}\rangle$. For the excited and decaying states, like the $2\,P^{1/2}$ $|n=2\rangle$, Gamow [14] had postulated (1928) a vector with complex eigenvalues: $|n=2\rangle = \phi^G = |E_2 - i\hbar/2\tau\rangle$. This was unacceptable at that time since H was to be selfadjoint and $|n=2\rangle$ was to be a vector of the Hilbert space.

Wigner knew from his works on the Weisskopf-Wigner approximation [15] and on Breit-Wigner (or Lorentzian) lineshape [16] that in quantum mechanics there could *not* be a state vector like the Gamow vector with exponential time evolution:

$$\phi^G(t) = e^{-\frac{iHt}{\hbar}}\phi^G = e^{-\frac{i(E_R - i\hbar/2\tau)t}{\hbar}}\phi^G = e^{-\frac{iE_Rt}{\hbar}}e^{-\frac{t}{2\tau}}\phi^G\,. \tag{1}$$

For Gamow vectors like (1), the probability to detect an observable $A = |\psi^-\rangle\langle\psi^-|$ in such Gamow state $\phi^G(t)$, would be exponentially decreasing:

$$\mathcal{P}_{\phi^G(t)}(\psi^-) = \mathrm{Tr}(|\psi^-\rangle\langle\psi^-|\phi^G(t)\rangle\langle\phi^G(t)|) = |\langle\psi^-|\phi^G(t)\rangle|^2 = e^{-\frac{t}{\tau}}\,|\langle\psi^-|\phi^G(0)\rangle|^2\,. \tag{2}$$

But using the Weisskopf-Wigner approximation methods, the probability to detect $|\psi^-\rangle\langle\psi^-|$ in a prepared resonance state $\phi^{BW}(t)$ with Breit-Wigner energy distribution of width Γ,

$$\phi^{BW}(E) = \frac{c}{E - (E_R - i\Gamma/2)}\,, \tag{3}$$

had been calculated [17] with the result that

$$\mathcal{P}_{\phi^{BW}(t)}(\psi^-) = \mathcal{P}_{\phi^{BW}}(\psi^-(t)) \sim e^{-\frac{\Gamma t}{\hbar}} + (\Gamma \times (\textit{additional term}))\,. \tag{4}$$

Thus, in addition to the exponential with lifetime $\tau = \frac{\hbar}{\Gamma}$, there is the *additional* term proportional to Γ in (4). As a consequence of this term in the Weisskopf-Wigner approximation (4), the state vector ϕ^{BW} with Breit-Wigner energy wave function $\phi^{BW}(E)$ given in (3), *cannot* have purely exponential time evolution.

Furthermore, a mathematical theorem [18] asserted that in the Hilbert space $\mathcal{H}$ there is no vector ϕ^G with exponential time evolution.

The conclusion from these and other results is:

1. Using strictly Hilbert space methods one *cannot* have a state vector which has exact exponential decay.

2. Starting with Breit-Wigner energy wave function (3), there will be deviations from exponential decay law (4). A state, with the Breit-Wigner energy wave function restricted to the positive real axis $0 \leq E \leq \infty$ as in (3), cannot have purely exponential time evolution.

In spite of these mathematical results, many people think that a Breit-Wigner resonance state with something like the energy distribution (3), and a decaying Gamow state, with the time evolution (1) or (2), are the same physical entities:

A Breit-Wigner resonance is an exponentially decaying state.

This is in particular the prevailing opinion in non-relativistic quantum mechanics, and for relativistic resonances one still speaks of energy dependent width. Therefore, one may ask the questions: Can one construct a *mathematical* theory in which a resonance of width Γ and an exponentially decaying state with exponential lifetime τ,

$$\tau = \frac{\hbar}{\Gamma}, \tag{5}$$

are just two different observations of the same quasi-stable physical state? Can one find a generalized state vector, which is described by something like a Dirac ket, or by something which is a little more "generalized" than the Dirac ket?

In standard Hilbert space quantum mechanics, there is no vector like a Dirac ket or something a little "more" generalized "than a Dirac ket" and "... there does not exit (in 1959 [19])... a rigorous theory to which these various [Weisskopf-Wigner] methods can be considered as approximations".

The task, therefore is: to find a mathematical theory which contains new generalized vectors, or kets, and other mathematical entities which

1. describe Breit-Wigner lineshape, i.e, something like (3), and
2. also describe exact exponential time evolution, like the Gamow vector (1), and for which in addition the relation (5) holds.

If such a theory can be found, does such a theory have some surprising properties?

3. Unifying Exponential Time Evolution and Breit-Wigner Energy Lineshape

The mathematical ingredients of quantum theory are: linear superposition, scalar products and operators in linear spaces, dynamical equations (Schrödinger for states and Heisenberg for observables), (and symmetry transformations of non-relativistic (Galileo group) and relativistic space-time (Poincare group) which will not be discussed here.)

Dynamical differential equations need to be solved under boundary conditions. The standard boundary condition for the dynamical differential equations in quantum mechanics is the Hilbert space boundary condition. Does one really have to use the Hilbert space axiom as the boundary conditions for the dynamical (Schrödinger or Heisenberg) equations?

The Hilbert space of von Neumann requires Lebesgue integrals in order that every Cauchy sequence has a limit element in the Hilbert space $\mathcal{H}$ (completeness). But physicists do not use Lebesgue integrals; the energy wave functions are not given by a *class* of Lebesgue square integrable function $f(E)$ ($0 \leq E < \infty$), but by one Schwartz space function $\varphi(E') = \langle E' | \varphi \rangle$ which is smooth, infinitely differentiable, rapidly decreasing function. The Dirac kets $|E\rangle$ are continuous (in Schwartz space convergence) antilinear functionals $F_E(E')$:

$$F_E(E') = \langle E' | E \rangle = \delta(E' - E) \text{ for } 0 \leq E < \infty \text{ (or also for } -\infty < E < \infty), \tag{6}$$

not vectors.

For the Schwartz space Φ, one uses Riemann integrals, rather than Lebesgue integrals; however Dirac kets $|E\rangle$ are in the space of antilinear continuous functionals, $F_E = |E\rangle$, $|E\rangle \in \Phi^\times$, and one has a triplet of spaces:

$$\Phi \subset \mathcal{H} = \mathcal{H}^\times \subset \Phi^\times . \tag{7}$$

The Schwartz space and the Dirac kets $|E\rangle \in \Phi^\times$, have been well accepted by the majority of physicists. But even the Dirac kets in $\Phi^\times$ are still insufficient to describe exponentially decaying states with complex energy $E_R - i\hbar/2\tau$, like the Gamow vector $|E - i\hbar/2\tau^-\rangle$. There is no such vector in the dual space $\Phi^\times$ of the Schwartz space. But one can modify the mathematical axioms of quantum physics further, such that the Gamow state vectors ϕ^G with the property (1), and with the Breit-Wigner energy distribution like (3), are mathematically well defined, and that in addition $\tau = \frac{\hbar}{\Gamma}$ holds as exact relation. In this theory, a resonance is an exponentially decaying state. Furthermore, both have a common mathematical origin, the pole of the S-matrix.

Physicists use only Riemann integrals. But they also use Dirac kets $|E\rangle$. Furthermore, they associate resonances with poles of the S-matrix at $z_R = E_R - i\Gamma/2$. For Gamow vectors, one can use mathematically defined kets with complex eigenvalues $|E - i\Gamma/2^-\rangle$. The Lippmann-Schwinger kets [23] also suggest analytic continuation into the complex energy plane. If all of these properties are to be put together into a mathematical theory, one needs energy wave functions, which are not only smooth and rapidly decreasing, like the $\langle E\,|\phi^+\rangle = \phi^+(E) \in \mathcal{S}$ (Schwartz functions), but which can also be analytically continued into the complex energy plane.

Before looking for a theory with complex energy variable that unifies the Breit-Wigner resonances and exponentially decaying states, one needs to look at the experimental situation:

In contrast to the discussions of the Weisskopf-Wigner methods and the consequences of the mathematical theorems for Hilbert space vectors, the analysis of the experimental data is usually done in terms of exponential time evolution and Breit- Wigner energy distribution.

A compilation of the experimental data, for a large number of spontaneously decaying quantum systems in different areas of physics, shows good agreement with the exponential law $\mathcal{P}(t) \sim e^{-t/\tau}$ [20]. In addition, the relation $\tau = \hbar/\Gamma$ has been tested more recently for the 3 $^2P_{3/2}$ of Na to very high accuracy. Both line-width Γ [21] and lifetime τ [22] have been measured with sufficiently high accuracy:

The Breit-Wigner line width measurement gives:

$$\Gamma = (40.537 \pm 0.087)\, 10^{-9}\ \text{eV}\ ,\ \tfrac{\hbar}{\Gamma} = (16.237 \pm 0.035)\ \text{ns}$$

and the exponential lifetime measurement gives $\quad \tau = (16.254 \pm 0.022)$ ns.

On the basis of these and other experimental data, it is not unjustified to make the hypothesis that:

> A scattering resonance $(E_R,\ \Gamma)$ (but without the non-resonant background $B_j(E)$ which is also present in the data of scattering experiments), and an exponentially decaying state $(E_D,\ \tau)$ are the same physical entities. The center of the Breit-Wigner resonance energy E_R
> usually
> incorporate corresponds to the energy of the decaying state E_D and the width Γ of the Breit-Wigner lineshape corresponds to the inverse lifetime τ of the decaying state:

$$(E_R,\ \Gamma) = \left(E_D, \frac{\hbar}{\tau}\right) . \tag{8}$$

The standard mathematical theory with Hilbert space axiom as well as the Schwartz space theory with Dirac kets do not contain the mathematical means to incorporate a state with the property (8). To incorporate these new "state" vectors $|E_R - i\Gamma/2^-\rangle$, one needs to solve the dynamical equations of quantum mechanics, (9) and (10) below, under new boundary conditions.

4. Causality and Various Solutions of the Dynamical Differential Equations of Quantum Mechanics

The dynamical evolution of quantum systems can be described as the evolution of the observable, in which the states do not change in time (Heisenberg picture), or as the evolution of the state with the observable kept fixed (Schrödinger picture).

The Heisenberg equation for the observable $A(t)$ or in the special case $A(t) = |\psi^-(t)\rangle\langle\psi^-(t)|$ for the observable vector $\psi^-(t)$ are:

$$i\hbar\frac{\partial A(t)}{\partial t} = -[H, A(t)] \tag{9a}$$

$$i\hbar\frac{\partial}{\partial t}\psi^-(t) = -H\psi^-(t)\,. \tag{9b}$$

The Schrödinger equation for the state operator $\rho(t)$ ("density matrix") or for the state vector $\phi^+(t)$ are:

$$i\hbar\frac{\partial\rho(t)}{\partial t} = [H, \rho(t)] \tag{10a}$$

$$i\hbar\frac{\partial}{\partial t}\phi^+(t) = H\phi^+(t)\,. \tag{10b}$$

We use the label $-$ for the observables and $+$ for the state in order to emphasize that H is the exact Hamiltonian which includes the interaction and shall not use the interaction-free state vectors usually denoted $\phi = \phi^{in}$ and correspondingly $\psi = \psi^{out}$ which evolve with a hypothetical free Hamiltonian called $K = H_0 = H - V$, where V is an interaction Hamiltonian. The label $+$ and $-$ at the in-state ϕ^+ and the out-observable ψ^- have been taken to conform with the Lippmann-Schwinger equation (22) below.

The predictions of the theory that are compared with the experimental data (measured as ratios of detector count $N(t)/N$) are the Born probabilities:

$$\mathrm{Tr}\,(A(t)\rho) = \mathrm{Tr}\,(A\rho(t)) \sim \frac{N(t)}{N}\,. \tag{11}$$

For the special case of the observable $|\psi^-(t)\rangle\langle\psi^-(t)|$ in the state ϕ^+, these probabilities are calculated as:

$$\begin{aligned} \mathcal{P}_{\phi^+}(|\psi^-(t)\rangle\langle\psi^-(t)|) &= \mathrm{Tr}\,(|\psi^-(t)\rangle\langle\psi^-(t)|\phi^+\rangle\langle\phi^+|) \\ &= |\langle\psi^-(t)|\phi^+\rangle|^2 \qquad \text{(in Heisenberg picture)} \\ &= |\langle\psi^-|\phi^+(t)\rangle|^2 \qquad \text{(in Schrödinger picture)}\,. \end{aligned} \tag{12}$$

In order to calculate these probabilities, one has to solve the dynamical equation (9) or (10). To solve differential equations requires the choice of a *boundary condition*, i.e a statement what kind of the solutions should be admitted as possible solutions of (9a) (9b) or of (11) (12). In standard quantum mechanics, *boundary conditions* of the dynamical equations are given by the Hilbert space axiom:

$$\begin{aligned} &\text{for the admitted set of state vectors } \{\phi^+\} \\ &\text{choose the Hilbert space } \mathcal{H} \text{ (Schrödinger picture)}, \\ &\textit{or} \\ &\text{for the set of vectors representing observables } \{\psi^-\} \\ &\text{choose the Hilbert space } \mathcal{H} \text{ (Heisenberg picture)}. \end{aligned} \tag{13}$$

In place of the Hilbert space axiom, one can also choose other possible boundary conditions. For instance, one can choose the Schwartz space boundary condition (i.e. using the mathematical version of Dirac's ket formalism):

$$\begin{aligned}&\text{for the set of admitted state vectors } \{\phi^+\} \text{ choose the Schwartz space } \Phi\,,\\ &\textit{or}\\ &\text{for set of observable vectors } \{\psi^-\} \text{ choose the Schwartz space } \Phi.\end{aligned} \tag{14}$$

The solutions of the Heisenberg equation (9b) for $\psi(t)$ as well as the solutions of the Schrödinger equation (12) for $\phi^+(t)$ are according to a mathematical theorem, (the Stone-von Neumann theorem [3] for the Hilbert space $\mathcal{H}$ (i.e., for Lebesgue integrable $|\phi^+(E)|^2$)) given by the unitary group evolution. Thus, as a consequence of this Stone-von Neumann theorem, one obtains from the Hilbert space axiom of mathematics:

$$\psi^-(t) = U(t)\,\psi^- = \mathrm{e}^{iHt/\hbar}\,\psi^-\,, \quad -\infty < t < +\infty \qquad \textit{in the Heisenberg picture}, \tag{15}$$

or

$$\phi^+(t) = U^\dagger(t)\,\phi^+(t) = \mathrm{e}^{-iHt/\hbar}\,\phi^+\,, \quad -\infty < t < +\infty \qquad \textit{in the Schrödinger picture}. \tag{16}$$

Consequently: Using the Hilbert space boundary conditions (13), the probability for the observable $|\psi^-(t)\rangle\langle\psi^-(t)|$ in the state ϕ^+ is predicted for "all time" $-\infty \leq t \leq +\infty$:

$$\mathcal{P}_{\phi^+}(|\psi^-(t)\rangle\langle\psi^-(t)|) = |\langle\psi^-(t)\,|\phi^+\rangle|^2 \qquad \text{for } \psi^-,\ \phi^+ \in \mathcal{H} \text{ and for } -\infty \leq t \leq +\infty\,. \tag{17}$$

By a similar theorem [4], one obtains under the Schwartz space boundary condition (14) the same prediction:

$$\mathcal{P}_{\phi^+}(|\psi^-(t)\rangle\langle\psi^-(t)|) = |\langle\psi^-(t)\,|\phi^+\rangle|^2 \qquad \text{for } \psi^-,\ \phi^+ \in \Phi \text{ and for all } -\infty \leq t \leq +\infty\,. \tag{18}$$

Summarizing, if one solves the dynamical equations of quantum mechanics (9b) or (12) under the Hilbert space axiom or under the Schwartz space boundary condition, one predicts from (15) the probability $|\langle\psi^-(t)\,|\phi^+\rangle|^2$ of the observable $|\psi^-(t)\rangle\langle\psi^-(t)|$ (or $A(t)$) in the state ϕ^+ for the time: $-\infty \leq t \leq +\infty$.

What is the physical evidence that we should have a time evolution *group* (15) or (16) i.e that t in $U(t) = \mathrm{e}^{iHt/\hbar}$ extends over $-\infty < t < +\infty$?

The prediction (15) or (16) would mean that the probability $|\langle\psi^-(t)\,|\phi^+\rangle|^2$ for the observable $|\psi^-(t)\rangle\langle\psi^-(t)|$ in the state ϕ^+ is predicted for all time $-\infty \leq t \leq +\infty$, even for $t \to -\infty$. But it is obvious that an observable $|\psi^-(t)\rangle\langle\psi^-(t)|$ can not be measured in a state ϕ^+ at the time $t \leq t_0$, where t_0 is the finite time at which the state ϕ^+ has been prepared, since a state ϕ^+ must be prepared first before an observable $|\psi^-(t)\rangle\langle\psi^-(t)|$ can be measured in this state (causality). E.g. a detector for $|\psi^-(t)\rangle\langle\psi^-(t)|$ *cannot* count the decay products of a decaying state $\phi^G(t_0)$ before the decaying state $\phi^G(t_0)$ has been prepared at a time $t_0 < t$, but only for $t > t_0$. Experiments indicate that there is such a finite time t_0 for a decaying quantum system, which of course will be measured as an ensemble of finite times $t_0^{(i)}$[23]. The onset times of the dark periods in quantum jump experiments might provide an experimental conformation of this ensemble of times [24].

This means the Born probability to measure the observable $|\psi^-(t)\rangle\langle\psi^-(t)|$ or $A(t)$ in the state ϕ^+:

$$\begin{aligned}\mathcal{P}_{\phi^+}(A(t)) &= \mathrm{Tr}(A(t)\,|\phi^+(t_0)\rangle\langle\phi^+(t_0)\,|)\\ \mathcal{P}_{\phi^+}(\psi^-(t)) &= |\langle\psi^-(t)\,|\phi^+\rangle|^2 = |\langle e^{iHt}\psi^-\,|\phi^+\rangle|^2 = |\langle\psi^-\,|e^{-iHt}\phi^+\rangle|^2 = |\langle\psi^-\,|\phi^+(t)\rangle|^2\\ &\quad \text{exists } (\textit{experimentally}) \text{ only for } t \geq t_0 (= 0)\,.\end{aligned} \tag{19}$$

Here t_0 is the time at which the state ϕ^+ has been prepared, thus the detection time of the decay products t must be later than t_0. Causality provides the intuitive justification that $t \geq t_0$: the evolution in the real world is *time-asymmetric*.

This causality should be reflected in the theory. Therefore, it would be good to find a new theory (precisely *new boundary conditions for the dynamical equations* (9) or (10)), replacing the Hilbert space axiom (13); and similarly, the Schwartz space axiom (14).

The new boundary conditions for the dynamical equation need to be chosen such, that the solutions of the Schrödinger equation (12) for the state $\phi^+(t)$ will be given by the *semigroup*:

$$\mathcal{U}^\times(t) = \mathrm{e}^{-iHt/\hbar} \qquad 0 \leq t < \infty \qquad \text{for state } \phi^+ . \tag{20}$$

Similarly using the Heisenberg picture, the solutions of the Heisenberg equation (9b) for the observable, $\psi^-(t)$, must be given by the *semigroup*:

$$\mathcal{U}(t) = \mathrm{e}^{+iHt/\hbar} \qquad 0 \leq t < \infty \qquad \text{for observable } \psi^- . \tag{21}$$

The dynamical equations (9a) (9b), or (11) (12) remain the same, but the boundary conditions (13) and (14) need to be revised.

5. What the $i\epsilon$ in the Lippmann-Schwinger Equation Suggests

To conjecture such a new "causal" theory in which the time evolution will be given by (20) or (21), we start with the phenomenological theory of scattering and decay.

One uses in- and out- plane wave "state" $|E^+\rangle$ and $|E^-\rangle$ which fulfill the Lippmann-Schwinger equation [25]:

$$|E^\pm\rangle = |E \pm i\epsilon\rangle = |E\rangle + \frac{1}{E - H \pm i\epsilon} V|E\rangle = \Omega^\pm|E\rangle, \quad \epsilon \to +0 \tag{22}$$

One speaks of complex energies (as dictated by the $\pm i\epsilon$ in (22):

for the analytic S matrix [26]	$S_j(E) \to S_j(z)$
for the Gamow states [14] ϕ^G	$z_R = E_R - i\Gamma/2$
or uses the $\pm i\epsilon$ in the propagator of field theory [27]	$z = E \pm i\epsilon$, ϵ infinitesimal

The $|E^+\rangle$ of (22) are taken as basis system for the Dirac basis vector expansion of in-*state* vectors:

$$\phi^+ = \sum_{j,j_3,\eta} \int_0^\infty dE|E, j, j_3, \eta^+\rangle\langle^+E, j, j_3, \eta|\phi^+\rangle = \int_0^\infty dE|E^+\rangle\langle^+E|\phi^+\rangle . \tag{23}$$

The $|E^-\rangle$ of (22) are taken as basis system of vectors ψ^-

$$\psi^- = \sum_{j,j_3,\eta} \int_0^\infty dE|E, j, j_3, \eta^-\rangle\langle^-E, j, j_3, \eta|\psi^-\rangle = \int_0^\infty dE|E^-\rangle\langle^-E|\psi^-\rangle , \tag{24}$$

which are now postulated to represent *observables* and therefore fulfill the Heisenberg equation.

The measured quantities are the Born probabilities

$$|(\psi^- , \phi^+)|^2 = \mathrm{Tr}(|\psi^-\rangle\langle\psi^- |\phi^+\rangle\langle\phi^+ |) , \tag{25}$$

for the observable $|\psi^-\rangle\langle\psi^-|$ in the state ϕ^+. These vectors ψ^- and the observable A in general represent the registration apparatus (e.g. a detector); note that $|\psi^-\rangle$ of (24), will not be

Journal of Physics: Conference Series **428** (2013) 012016 doi:10.1088/1742-6596/428/1/012016

interpreted as an "out-state" (which makes no physical sense), but as an observable representing the detector.

Causality requires that the observable $|\psi^-\rangle\langle\psi^-|$ cannot be measured in the state ϕ^+ before the state ϕ^+ has been prepared at t_0 (which in (20) and (21) has been chosen as $t_0 = 0$, but which could be any *finite* time). ($t_0 = 0$ in (20) and (21) represents an ensemble of times $t_0^{(1)}, t_0^{(2)}, \ldots$ for an ensemble of quantum particles, cf. [23, 24].)

The Dirac basis vector expansions (23), (24) use two different kind of kets:

$|Ejj_3\eta^{\mp}\rangle \in \Phi_{\pm}^{\times}$ suggested by the Lippmann-Schwinger "out-plane waves" $|E^-\rangle$ and in-plane waves $|E^+\rangle$, respectively.

Because of the $+i\epsilon$ in the Lippmann-Schwinger equation (22), the energy wave function of the prepared in-state ϕ^+,

$$\phi^+(E) = \langle^+Ejj_3\eta|\phi^+\rangle = \overline{\langle\phi^+|Ejj_3\eta^+\rangle}\,, \tag{26}$$

is the boundary value of an **analytic function in the lower complex energy semi-plane** (for complex energy $z = \overline{E + i\epsilon} = E - i\epsilon$ immediately below the cut along the real axis on the second sheet of the S-matrix $S_j(z)$, cf. Fig 1 below.)

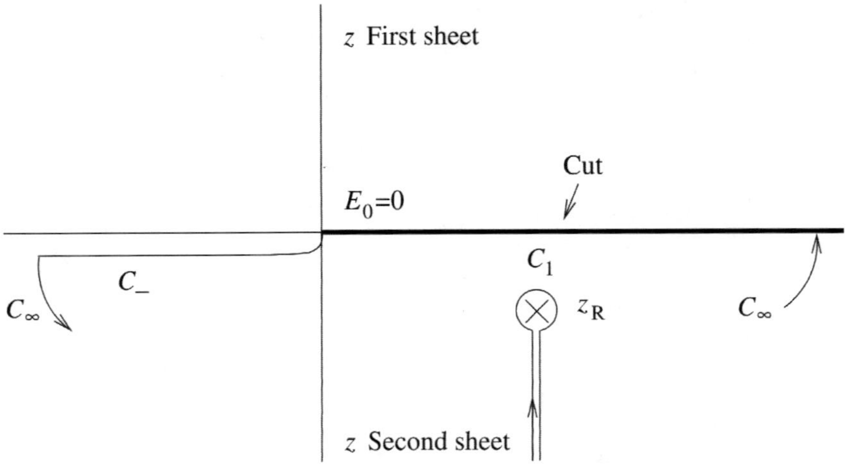

Figure 1. Deformation of the contour of integration from the cut along the positive real axis into complex plane second sheet of the S-matrix, where the resonance pole is located.

Similarly, the energy wave function of the observable $|\psi^-\rangle\langle\psi^-|$,

$$\psi^-(E) \equiv \langle^-Ejj_3\eta|\psi^-\rangle\,, \tag{27a}$$

needs to be a function that can be extended to an **analytic function in the upper complex energy semi-plane** so that its complex conjugate

$$\overline{\psi^-(E)} = \langle\psi^-|E\,j\,j_3\,\eta^-\rangle\,, \tag{27b}$$

can be extended to an analytic function in the lower complex semi-plane.

6. Calculating the Born Probabilities

In order to calculate the Born probability amplitude $(\psi^-|\phi^+)$ for the *observable* ψ^- in the *state* ϕ^+, one uses basis vector expansions (23), (24). In this amplitude $(\psi^-|\phi^+)$, the S-matrix element $\langle^-E', j', j'_3, \eta'|E, j, j_3, \eta^+\rangle$ will appear, which is the bra-ket of the basis system $|E, j, j_3, \eta^+\rangle$ of the interacting in-state vector ϕ^+ in (23) and the basis system $|E, j, j_3, \eta^-\rangle$ of the interacting out-observable vector ψ^- in (24). The modules squared: $|(\psi^-, \phi^+|^2$, represents the Born probabilities (11), (12), and (25) for the observable ψ^- in the state ϕ^+.

This matrix element can be simplified using energy- and angular momentum conservation:

$$\langle^-E', j', j'_3, \eta'|E, j, j_3, \eta^+\rangle = \delta(E'-E)\delta_{j'_3 j_3}\delta_{j'j} S_j^{\eta'\eta}(E)\,. \tag{28}$$

Thus, the matrix element has been expressed in terms of a "reduced" S-matrix elements $S_j^{\eta'\eta}(E)$, where η', η are the channel quantum numbers or particle species quantum numbers.

Consider the simplest case that there is just one first order pole of $S_j(E)$ at $z = z_R$ and there are no other singularities. The reduced S-matrix $S_j^{\eta'\eta}(E)$ encapsulates the properties of the specific interaction between target and projectile, which show up as such specific features like Breit-Wigner bumps (3) in the cross section, and are due to a pole of the S-matrix at $z_R = E_R - i\Gamma/2$:

$$S_j^{\eta'\eta}(E) = 2ia_j^{\eta'\eta}(E) + \delta^{\eta'\eta} = 2i\frac{R^{\eta'\eta}}{E - z_R} + \delta^{\eta'\eta} + B^{\eta'\eta}(E)\,. \tag{29}$$

$B^{\eta'\eta}(E)$ denotes some slowly varying background.

From the analyticity property of the energy wave functions (26) (27b) follows that $\langle\psi^-|E^-\rangle\langle E^+|\phi^+\rangle S_j(E)$ can be analytically continued into the lower complex plane (second sheet), except for singularities of $S_j(E) = S_j^{\eta'\eta}(E)$. A singularity can be a first order pole at $z = z_R = E_R - i\Gamma/2$ indicating that there is a resonance in the S-matrix element $S_j(E) = 2\,i\,a_j(E) + 1$, at the position z_R as shown in Fig 1.

The contour of integration for the S-matrix element can be deformed from the cut along the positive real axis as shown in Fig.1. Using (23), (24) and (28), the Born probability amplitude (ψ^-, ϕ^+) for the observable ψ^- in the state ϕ^+ is then written as

$$(\psi^{\rm out}, S\phi^{\rm in}) \equiv (\psi^-, \phi^+) = \sum_j \int_{E_0=0}^{\infty} dE \sum_{j_3}\sum_{\eta,\eta'} \langle\psi^-|E, j, j_3, \eta'^-\rangle S_j^{\eta'\eta}(E)\langle^+E, j, j_3, \eta|\phi^+\rangle \tag{30}$$

For the present purpose, $(\psi^{\rm out}, S\phi^{\rm in})$ is just another way of writing (ψ^-, ϕ^+). In simplified notation, this equation (30) is also written as:

$$(\psi^-, \phi^+) = \int_{E_0=0}^{\infty} dE\langle\psi^-|E^-\rangle S_j(E)\langle^+E|\phi^+\rangle = \int_{\rm Cut} \overline{\psi^-(E)}\; S_j(E)\,\phi^+(E)\,. \tag{31}$$

With the right properties of the energy wave functions $\langle\psi^-|E^-\rangle\langle^+E|\phi^+\rangle$, the contour of integration in (31) can be deformed from the Cut along the positive real axis first sheet, (which is also the positive real axis second sheet) into the contour in the lower complex energy plane on second sheet given by C_-, C_∞ and along the circle C_1 around the pole at z_R on the second sheet of the Riemann surface, (where the resonance pole of the S-matrix element or of the scattering amplitude $a_j(E)$) is located).

Here we want to consider the integral along the contour C_1 around the pole term $\frac{R^{\eta'\eta}}{z - z_R}$ of the S-matrix (29):

$$\oint_{C_1} dz\;\langle\psi^-|z^-\rangle\;\frac{R^{\eta'\eta}}{z - z_R}\;\langle^+z|\phi^+\rangle\,. \tag{32a}$$

The contour of this integral can be deformed into the integral:

$$\int_{-\infty_{II}}^{+\infty_I} dE\, \langle\psi^-|E^-\rangle\, \frac{R^{\eta'\eta}}{E-z_R}\, \langle^+E|\phi^+\rangle\,. \tag{32b}$$

The program is now to determine the property of the spaces Φ_- (of $\{\phi^+\}$ representing states) and Φ_+ (of $\{\psi^-\}$ representing observables), that means to determine the mathematical properties of the energy wave functions $\psi^-(E) \equiv \langle^-E|\psi^-\rangle$ and $\phi^+(E) = \langle^+E|\phi^+\rangle$ of (23) and (24).

We start from the integral along C_1 in (32b). The mathematical properties of $\langle\psi^-|E^-\rangle$ and $\langle^+E|\phi^+\rangle$ are then determined, i.e, conjectured, in such a way that a scattering resonance with Breit-Wigner lineshape at $z = z_R$ in (29) and a decaying state $|\,z_R^-\rangle$ with exponential time evolution,

$$e^{-iHt/\hbar}\,|z_R^-\rangle = e^{-iE_Rt/\hbar}e^{-(\tau/2)t}\,|z_R^-\rangle\,, \tag{33}$$

are both derived from the S-matrix pole at z_R. If that is possible, it would mean that the S-matrix pole represents the exponential decaying state $|z_R^-\rangle$, as well as the Breit-Wigner resonance with scattering amplitude $a_j^{BW} = \frac{R}{E-z_R}$.

7. Conjecturing the Hardy space axiom from the Pole Term of the S-Matrix

We focus our attention on the S-matrix pole term and the contour of integration C_1 around the pole at z_R, and take the S-matrix pole at $z_R = E_R - i\Gamma/2$ as the definition for a resonance. The scattering amplitude of a resonance at z_{R_i} is then $a_j^{BW} = \frac{1}{2i} S_j^{BW}(E) = \frac{R_i}{E-z_{R_i}}$. The integral along the infinite semi-circle C_∞ on the second sheet of the S-matrix needs to vanish (this is part of the conditions which the energy wave function, whose properties we wish to conjecture by this process, needs to fulfill in order to obtain the new axiom -new boundary conditions for solutions of the dynamical equations, which will replace the Hilbert space boundary condition (13) and Schwartz space boundary condition (14)). The integral around the first order pole z_R of the S-matrix is deformed into an integral along $-\infty_{II}$ to 0 and an integral along the cut from 0 to $+\infty$, i.e, into an integral along the negative real axis $-\infty_{II} < E < 0$ and along the cut $0 < E < +\infty$ as indicated in (32b).

The Breit-Wigner resonance amplitude is thus related to the integral from $-\infty_{II}$ to 0 on the second sheet and on to $+\infty_I$ along the cut: $-\infty_{II} < E < +\infty_I$.

In this way, at the S-matrix pole $z_R = E_R - i\Gamma/2$:

The Breit-Wigner resonance amplitude:	will be related to	a Gamow vector $\|z_R, j, j_3, \eta^-\rangle$ with Breit-Wigner energy wave function:

$$\langle\psi^-|\phi_j^G\rangle = \langle\psi^-|z_R, j, j_3, \eta^-\rangle\sqrt{2\pi\Gamma} = \\ = \int_{-\infty_{II}}^{+\infty} dE\,\langle\psi^-|E, j, j_3, \eta^-\rangle \frac{i\sqrt{\frac{\Gamma}{2\pi}}}{E-z_R} \tag{34a}$$

Thus the Breit-Winger:	will be related to	the ket, i.e the functional $\|z_R, j, j_3, \eta^-\rangle$ on the space of observables $\{\psi^-\} \equiv \Phi_+$

$$a_j^{BW}(E) = \frac{R}{E-z_R} \quad \Longleftrightarrow \quad \phi_j^G = |z_R, j, j_3, \eta^-\rangle\sqrt{2\pi\Gamma} = \\ = \int_{-\infty_{II}}^{+\infty} dE\,|E, j, j_3, \eta^-\rangle \frac{i\sqrt{\frac{\Gamma}{2\pi}}}{E-z_R} \tag{34b}$$

This vector $|E_R - i\Gamma/2,\, j,\, j_3,\, \eta^-\rangle = |z_R,\, j,\, j_3,\, \eta^-\rangle = |z_R^-\rangle = |E_R - i\Gamma/2^-\rangle$ can then be shown to:

1. be an eigenket of H with the discrete complex eigenvalue $(E_R - i\Gamma/2)$ (as Gamow had wanted):

$$\langle H\psi^- |z_R^-\rangle = \langle \psi^- |\, H^\times\, |z_R^-\rangle \;= z_R\, \langle \psi^- |z_R^-\rangle \quad \text{for all } \psi^- \in \Phi_+\,, \tag{35}$$

or in ket notation

$$H^\times |E_R - i\Gamma/2^-\rangle \;= (E_R - i\Gamma/2)\, |E_R - i\Gamma/2^-\rangle\,.$$

2. have the property:

$$\begin{aligned}\langle e^{iHt/\hbar}\,\psi^- |E_R - i\Gamma/2^-\rangle &\equiv \langle \psi^- | e^{-iH^\times t/\hbar} |E_R - i\Gamma/2^-\rangle \\ &= e^{-iE_R t/\hbar} e^{-(\Gamma/2)t/\hbar} \langle \psi^- |E_R - i\Gamma/2^-\rangle\,.\end{aligned} \tag{36}$$

This means that the generalized eigenvector $|\, z_R, j, j_3, \eta^-\rangle$ on the space of $\{\psi^-\} \equiv \Phi_+$ represents a scattering resonance with resonance energy E_R and Breit-Wigner width Γ, as shown in (34b) and it is also an exponentially decaying state of lifetime τ as shown in (36). Thus

$$\text{a resonance of width } \Gamma \text{ in (34b)} \equiv \text{a decaying state with lifetime } \tau = \frac{\hbar}{\Gamma} \text{ in (36)}. \tag{37}$$

Therewith we would have a theory that unifies the Breit-Wigner resonances with the exponentially decaying Gamow vectors.

In order to be able to make the contour deformations described above, and to derive the results (34b), (35), and (36), *new mathematical properties* have to be assumed for the energy functions $\langle^+ E\,|\phi^+\rangle$ and $\langle^- E\,|\psi^-\rangle$. These new mathematical properties turn out to agree exactly with mathematical conditions that define the smooth Hardy functions. This is an example of what Wigner calls the "unreasonable effectiveness of mathematics in physical sciences".

Summarizing: In order that a Breit-Wigner resonance in (29) can be associated to a Gamow vector as done in (34b), a new hypothesis has to be made, which replaces the Hilbert space axiom (13) of von Neumann, and the Schwartz space axiom (14) of Dirac's formalism, by a new axiom: The Hardy space axiom for the dynamical equations of quantum mechanics:

The energy wave functions of a state (obeying the Schrödinger equation) are smooth Hardy functions analytic on the lower complex energy plane $\mathbb{C}_-$:

$$\phi^+(E) = \langle^+ E\,|\phi^+\rangle \in (\mathcal{H}^2_- \cap \mathcal{S})_{\mathbb{R}_+} = \text{“smooth Hardy functions” of } \mathbb{C}_- \tag{38}$$

The energy wave functions of an observable vector (obeying the Heisenberg equation) are smooth Hardy functions analytic on the upper complex energy plane $\mathbb{C}_+$:

$$\psi^-(E) = \langle^- E\,|\psi^-\rangle \in (\mathcal{H}^2_+ \cap \mathcal{S})_{\mathbb{R}_+} \text{“smooth Hardy functions” of } \mathbb{C}_+ \tag{39}$$

In (38) (39), the $\mathcal{H}^2_\pm$ denote the Hardy *class* spaces and $\mathcal{S}$ denotes the space of Schwartz functions.

It is not clear that the smooth Hardy functions (38) (39) are the only functions that lead to the relation (34b). If one insists on Lebesgue integrals, one can probable obtain something similar using Hardy classes $\mathcal{H}^2_\pm$ in place of smooth Hardy functions $(\mathcal{H}^2_\mp \cap \mathcal{S})$. But for quantum physics, the beautiful Dirac kets of Schwartz space $\mathcal{S}$ should be retained. Therefore, (38) (39) is probably the best choice for the energy wave functions. This is expressed by the Hardy space axiom (40), (41) for the space of in-states and out-observables.

Therewith we have arrived at a new **axiom**, the Hardy space axiom:

The set of prepared (in-) states $\{\phi^+\}$ defined by the preparation apparatus (e.g. accelerator etc.) is:

$$\{\phi^+\} = \Phi_- \subset \mathcal{H} \subset \Phi_-^\times \,, \qquad |E, j, j_3, \eta^+\rangle \in \Phi_-^\times \,. \tag{40}$$

The set of (out-) observables $\{\psi^-\}$ defined by the registration apparatus (e.g. detector) is

$$\{\psi^-\} = \Phi_+ \subset \mathcal{H} \subset \Phi_+^\times \,, \qquad |E, j, j_3, \eta^-\rangle \in \Phi_+^\times \,. \tag{41}$$

Here $\Phi_\mp$ denotes the Hardy spaces of the complex semiplanes $\mathbb{C}_\mp$ whose energy wave functions are given by (38) (39), and $\Phi_\mp^\times$ are their duals (spaces of continuous antilinear functionals).

The discrepancy in the label $^\pm$ for the kets $|E^\pm\rangle$ in (22) and the label $^\mp$ for the Hardy spaces $\Phi_\mp$ in (40) (41) and the Hardy classes $\mathcal{H}^2_\mp$ in (38) (39) has its origin in the different conventions in physics for the Lippmann-Schwinger equation of (22) and the convention in mathematics for the Hardy spaces, $\Phi_\mp$ and $\mathcal{H}^2_\mp$. The kets $|z^\pm\rangle$, $|E^\pm\rangle$, $|E_R - i\Gamma/2^-\rangle$ are now defined as continuous antilinear functionals on the Hardy spaces $\Phi_\mp$: $|E^\pm\rangle \in \Phi_\mp^\times$. One has finally a consistent mathematical theory which unifies resonance and decay phenomena (with $\Gamma = \frac{\hbar}{\tau}$ as an exact equality).

With the mathematical theory defined by the Hardy space axiom (40) (41), one can use mathematical theorems: (similar to the Stone-von Neumann theorem which applied to the Hilbert space boundary condition (13) of the dynamical equation, and which led to the unitary group evolution $-\infty < t < +\infty$.) Solving the same dynamical equations (9b) and (12) (Heisenberg or Schrödinger) now under the Hardy space boundary conditions (40) (41) leads, (by the Paley-Wiener theorem [28]), to a *semigroup evolution*:

For the observable ψ^- obeying the Heisenberg equation (9b), one obtains:

$$\psi^-(t) = e^{iH(t-t_0)}\psi^-(t_0) \qquad t_0 \le t < \infty \,. \tag{42}$$

For the prepared state obeying the Schrödinger equation (12), one obtains:

$$\phi^+(t) = e^{-iH(t-t_0)}\phi^+(t_0) \qquad t_0 \le t < \infty. \tag{43}$$

This means that the Hardy space boundary condition, which was needed in order to establish $\tau = \hbar/\Gamma$ as an exact relation, predicts a time *asymmetric* dynamical evolution (42) or (43) and therewith a beginning of time $t_0 =$ *finite*, for quantum systems.

A straightforward consequence of the semigroup solutions (42) (43) is that it predicts the Born probabilities

$$\mathcal{P}_{\phi^+}(\psi^-(t)) = |\langle\psi^-(t)\,|\phi^+(t_0)\rangle|^2 \quad \text{only for} \quad t \ge t_0 \,(=0)\,. \tag{44}$$

This solves the causality problem (it also avoids the exponential catastrophe for Gamow states): The observable $|\psi^-(t)\rangle\langle\psi^-(t)|$ or any other observable $A(t)$ (solution of the Heisenberg equation) is predicted *only* for time t after the preparation time t_0 of the state ϕ^+.

The finite quantum mechanical beginning of time t_0 is probably observed as an ensemble of finite times [23, 24]. Some people may want to consider the big bang time as an example of a quantum mechanical beginning of time, then one can speculate about many more bangs. A more important consequence of the beginning of time is probably that it provides a mathematical theory which unifies the resonance and decay phenomena of quantum physics into an exact mathematical theory, with lifetime $\tau =$ inverse width $\hbar/\Gamma$.

Acknowledgments

The kind hospitality at Tsukuba University is gratefully acknowledged.

References

[1] Halliwell J J , Pérez-Mercader J and Zurek W H 1994 *Physical Origins of Time Asymmetry* (Cambridge University Press)
[2] Bohm A, Doebner H D and Kielanowski P 1998 *Irreversibility and Causality* Lecture Notes in Physics Vol 504 (Berlin: Springer-Verlag)
[3] Stone M H 1932 *Ann. Math.* **33** No. 3 643; Neumann V J 1932 *Ann. Math.* **33** No. 3 567
[4] Bohm A and Gadella M 1989 *Dirac Kets, Gamow Vectors and Gelfand Triplet* Lecture Note in Physics Vol 348 (New York and Berlin: Springer-Verlag) Proposition II p 82
[5] Lax P D and Phillips R S 1967 *scattering Theory* 1st Edition (New York: Academic Press)
[6] Bohm A 1978 *Proc. 7th Int. Group Theory Coll.* Lecture Note in Physics Vol 94, ed Bohm A *et al* p 245
[7] Bohm A 1979 *Quantum Mechanics: Foundations and Applications* 1st Edition (New York: Springer)
[8] Strauss Y, Horwitz L P and Eisenberg E 2000 *J. Math. Phys.* **41** No. 12 8050
[9] Strauss Y 2003 *Int. J. Theor. Phys.* **42** 2285
[10] Strauss Y, Horwitz L P and Volovick A 2006 *J. Math. Phys.* **47**, 123505
[11] Baumgärtel H 2007 *Int. J. Theor. Phys.* **46** 1959
[12] Baumgärtel H 2010 *J. Math. Phys.* **51** 113508
[13] Baumgärtel H 2006 *Rev. Math. Phys.* **18** 61
[14] Gamow G 1928 *Z. Phys.* **51** 204-12
[15] Weisskopf V and Wigner E P 1930 *Z. Phys.* **65** 18
[16] Breit G and Wigner E P 1936 *Phys. Rev.* **49** 519
[17] Goldberger M and Watson K 1964 *Collision theory* (New York: Wiley) chapter 8
[18] Khalfin L A 1958 *Sov. Phys. JETP* **6** 1053-63
[19] Levy M 1959 *Nuovo Cimento* **13** 115
[20] Norman E B 1988 *Phys. Rev. Lett.* **60** 2246
[21] Oates C W, Vogel K R and Hall J L 1996 *Phys. Rev. Lett.* **76** 2866
[22] Volz U, Majerus M, Liebel H, Schmitt A and Schmoranzer A 1996 *Phys. Rev. Lett.* **76** 2862
[23] Bohm A 1999 *Phys. Rev.* A **60** 861
[24] Bohm A, Kielanowski P and Wickramasekara S 2006 *Ann. Phys.* **321** 2299
[25] Lippmann B A and Schwinger J 1950 *Phys. Rev.* **79** 469480; Gell-Mann M and Goldberger M L 1953 *Phys. Rev.* **91** 398
[26] Eden R J, Landshoff P V, Olive V P and Polkinghorne J C 1962 *The Analytic S-Matrix* (New York: Cambridge University Press); Taylor J R 1972 *Scattering Theory* (John Wiley & Sons, Inc)
[27] Weinberg S 1995 *The Quantum Theory of Field* Vol 1 (Cambridge University Press)
[28] Paley R and Wiener N 1934 *Fourier Transform in the Complex Domain* (New York: American Mathematical Society)

Process induced pre-existing defects in non-crystalline SiO_2 and GeO_2 at vacated O-atom bonding sites and comparisons with Ge-S(Se) alloy bonding sites

Gerald Lucovsky*, Daniel Zeller, JinWoo Kim, and Kun Wu
North Carolina State University, Raleigh NC, USA

***e-mail: lucovsky@ncsu.edu**

Abstract: Three related topics are addressed in this article: (i) X-ray spectroscopy (XAS) studies of remote plasma deposited (RPD) nc-SiO_2 and nc-GeO_2 emphasizing (a) band-edge states and (b) pre-existing bonding defects; (ii) interpretation of X-ray absorption and photoemission spectra based on many electron theory, and in particluar charge transfer multiplets (CTs); and (iii) band-edge electronic structure and intrinsic defects in nc-SiO_2 and nc-GeO_2 thin films and their interfaces with Si and Ge substrates. The most significant result is the identification of local atomic structure in medium range order (MRO) clusters in which the pre-existing defects are embedded. These defects are vacated O-atom sites in which O-atoms have never been resident. They are confined to 1 nm scale chemically-ordered clusters distributed non-periodically with *quartz-structured 4-fold coordinated Si(Ge)* and *2-fold coordinated O bonding in* 12-atom regular rings. Vacated O-atom sites defects are formed during processing and annealing, and reducing macroscopic as well as local bond-strain strain. They are qualitatively different, and readily distinguished from defects introduced by electrical, and by X-ray, γ-ray or high energy electron stressing of nc-SiO_2 and nc-GeO_2.

1. Introduction

It is significant to note that ESR-active pre-existing defects detected in nc-SiO_2 and nc-GeO_2, and in Se-rich $Ge_{1-x}Se_x$ alloys are qualitatively different. This is due quantitative differences in medium range order (MRO) between oxides and chalcogenides. MRO is order is by definition order beyond the regimes of (i) bond-lengths or 2-atom atom distances, and (iii) 3-atom correlated bond-angles. S-atoms and Se-atoms form S-S and Se-Se bonds, where O-atoms do not except in peroxides which have not been detected in irradiated SiO_2 bulk glasses or films. MRO is correlated with d-state ligand field splittings, ΔLF. These are non-zero when molecular orbital theory descriptions of valence band states in which case the number of valence band features is either 3 or more in compound and alloy oxides and chalcogenides [1]. The ΔLF are energy differences in 5-fold degenerate atomic core level d-states induced by non-spherical local symmetry in the valence band structure, or in virtual bound states in the conduction band regime as well. Of particular importance are the values of ΔLF associated with O-vacancy defect sites. These are split into 2-fold degenerate eg states and 3-fold degenerate t2g spectral regimes. This splitting is a direct consequence the non-spherical local symmetry and the medium range order (MRO) at the vacated O-atom sites [1].

Values of ΔLF, in electron volts (eV), are obtained by X-ray absorption (XAS) and electron energy loss spectroscopy (EELS). The same values can also be extracted from dielectric constants obtained from a dispersion analysis of (i) reflectivity (R), and (ii) spectroscopic ellipsometry (SE) spectral data. ΔLF is local symmetry, atomic coordination, and atom specific. It is equally important to distinguish the spin-orbit splitting, ΔSO of the centro-symmetric core states from the final state values of ΔSO' in O K edge spectra.

Additionally, and of particular importance for nc-SiO_2, GeO_2, and $Ge(S,SE)_2$, a non-zero value of ΔLF requires a scale of *coherent-order* extending up to, and generally beyond, 3rd next-nearest-neighbors, NN. Medium range order (MRO) regime includes discrete values of dihedral angles, and correlated/symmetric dihedral angle arrangements as well. Therefore, a non-vanishing ΔLF is a spectroscopic indicator of MRO, complementing X-ray and neutron diffraction determinations of the position, and width of the first sharp diffraction peak (FSDP) in reciprocal space [2]. This approach to MRO provides a significant and direct connection with the existence of pre-existing ESR active defects in nc-SiO_2, and nc-GeO_2 as well.

2. Medium Range Order in Non-crystalline Oxides

MRO regime bonding correlations are obtained from the position FSDP in the structure factor from X-ray or neutron diffraction [2,3]. MRO correlation lengths, λcorr, are 0.4 nm in SiO_2 and GeO_2 [2,3] and extend to ~0.65 nm in $GeSe_2$ [2,3,4]. These are in the nm-regimes of 3rd and 4th NN distances. The coherence length, λcoh, for molecular cluster formation is calculated from the width of FSDP and is beyond the MRO regime. For example, λcoh, is ~1 nm in Si(Ge)O_2, [3] and increases to ~2.9 nm in $GeSe_2$ [4]. These clusters are *non-periodically-distributed* in both oxides and chalcogenides, contributing to macroscopic strain relief. These clusters are more constrained locally than the non-crystalline networks in which they are immersed. It is therefore not surprising that the nm-scale regions in which pre-existing, locally strain-reducing vacated O-bonding site defects are created.

Ab-initio calculations on terminated small O-Si-O-Si-O clusters with 0.5 nm dimensions demonstrate, that MPO is derived from back-donation of electrons from occupied O 2pπ non-bonding orbitals into empty Si atom 3d orbitals [5]. The back-donation also reduced bond-ionicity and is common in when bond-ionicities exceed 50%, The back-donation is facilitated by the large Si-O-Si bond angle of ~150 degrees [5]. The local bonding is qualitatively different in chalcogenides. Bond angles centered on S or Se sites are between ~95 and ~105 degrees, and significant back-donation as described above is not present.

However, it is mandatory to describe the atomic states of atoms in both nc-oxides and nc-chalcogenides in the same way using representations that are non-spherical. Proceeding in this way the 2pπ states of Si in nc-SiO_2 have a T1u character, and the Si 3d states on Si, a T2g character [5].

The continuous random network (CRN) description for glasses and nc-thin films was first proposed in a paper published by Zacharaisen in 1932 [6]. Since that time it has generally been accepted as being a good model for non-crystalline glasses and then films, being extended to chalcogenides as well. Zacharaisen's description of a CRN includes only 1st and 2nd NN correlations, or equivalently, bond-lengths and bond-angles, respectively [7, 8, 9]. This definition was initially supported by X-ray studies of a group in the UK, who demonstrated that nc-SiO_2 glass was not microcrystalline [10]. However, Zacharaisen *did not reference* a second paper by the same authors in which they observed what is now designated as the FSDP [11]. They assigned this X-ray diffraction feature in nc-SiO_2 to an inter-atomic spacing of 0.43 nm, essentially the same as that more recently reported in nc-SiO_2 by similar X-ray diffraction, including the Fourier transformed structures factor [2,3]. The detection of MRO in nc-Si(Ge)O_2, and nc-Ge(S,S) and As(S,Se) compound phases and alloys, combined with local-strain reduction in clustered regions is what reduces total strain, both local- and macroscopic in the inhomogeneous networks in which they are immersed. This non-periodic nano-scale *hard-soft* nixtures is the mechanism for total strain-relief rather than an average bond-coordination number matched to the degrees of vibrational disorder [7].

Journal of Physics: Conference Series **428** (2013) 012017 doi:10.1088/1742-6596/428/1/012017

3. Experimental Procedures

The nc-SiO_2 films of this chapter have been were prepared by remote plasma enhanced chemical-vapor-deposition (RPECVD) or simply RPD, onto HF-last rinsed Si substrates. The depositions were made with a substrate temperature of 300°C. Thin films were then annealed in Ar to temperatures up to 950°C. [12]. These films and nc-GeO_2 films will henceforth be designated as remote plasma deposited nc-Si(GeO_2). Nc-SiO_2 films formed by oxidation of Si in dry-O_2 ambients at temperatures to 900°C and subsequently annealed at higher temperatures cannot be distinguished by spectroscopic methods from 300°C RPD nc-SiO_2 films annealed to the same temperatures.

RPD nc-GeO_2 films were deposited onto Ge substrates at 300°C, after a rinse. in NH_4OH. RPECVD depositions used down-stream injected 2% GeH_4 in He. Deposition was activated by O-species from an up-stream localized O_2–He plasma [12]. Reactive deposition was on either:(i) chemically cleaned Ge surfaces, and (ii) Ge surfaces subjected to either: (a) a remote plasma-assisted oxidation, RPO, (b) a remote plasma-assisted nitridation, RPN, or (c) an RPO followed by RPN [12].

Si and Ge $L_{2,3}$ soft X-ray XAS spectra were obtained using total electron yield (TEY) data acquisition on Beam-Line 8-1 at the Stanford Synchrotron Research Lightsource (SSRL). In the TEY approach, the photoelectron yield vs. X-ray photon energy is X-ray irradiation determined by measuring the current necessary to return the sample to neutral. The current is proportional to the linear absorption coefficient with a precision of a few percent [13]. Finally, in these XAS studies, and also the XAS studies performed on Beam-Line 10, the nc-SiO_2/GeO_2 thin films are deposited on Si or Ge substrates, and these are mounted on an Al metal support with stretched Ta wire in contact with the top surface of the sample.

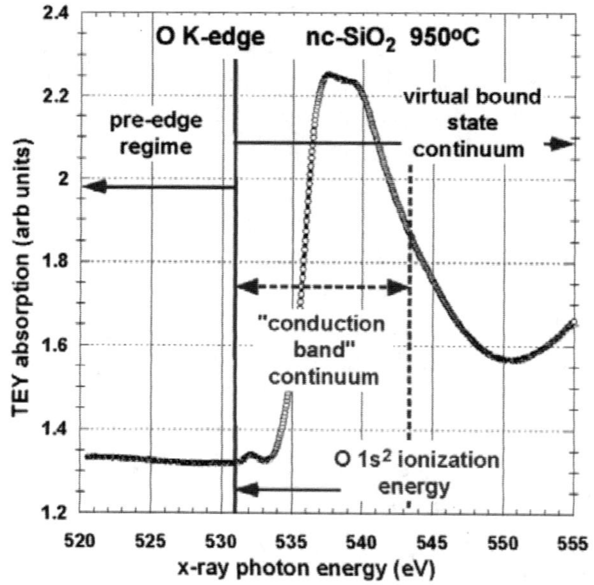

Fig. 1. OK pre-edge, conduction band and O K post edge spectral regimes identified for nc-SiO^2 formed by RPD and annealed at 950°C for 1 minute in Ar.

4. Experimental Results and Analysis of Spectral Data

Figure 1 is XAS O K edge spectrum for an RPD nc-SiO_2 annealed at 950°C in an Ar ambient. TEY absorption is plotted for X-ray photon energies between 520 eV and 570 eV. This spectrum is marked to indicate 3 spectral domains of spectroscopic significance: (a) 525 to 532 eV, O K pre-edge that includes mid-band-gap pre-existing. shake-up defect state features; (b) 532 to ~550 eV, this so-called O K edge includes intrinsic virtual bound states above the O-atom ionization (IP) energy and generally referred to as conduction band final-states; and (c) 550 to 570-80 eV, the O K post-edge, or *ZANES* region that includes shake-up features associated with localized O and Si core states with symmetric closed shell bonding arrangements [14]. These are local resonances in the a continuum regime above the ionization limit and require a three-electron state charge transfer multiplet (CTM) description.

Spectra in the pre-edge region combined with ab initio theory calculations indicate that (i) pre-existing defects in thin film SiO_2 are vacated O-atom sites, rather than *bona-fide* O-vacancies in which only one of the SiO-atom bonds has been broken with an Si-O terminal group attached to the network structure. As such *O-vacancy defects* contain one Si-atom dangling bond and one non-bonding O-hole center (NBOHC) with a terminal Si-O neutral bond. In contrast, the *O-vacated site defects* of this chapter contain two Si-atom strongly-correlated dangling bonds. This pair of state with overlapping radial wave-function requires X-ray activation for detection in electron spin resonance (ESR) measurements [14].

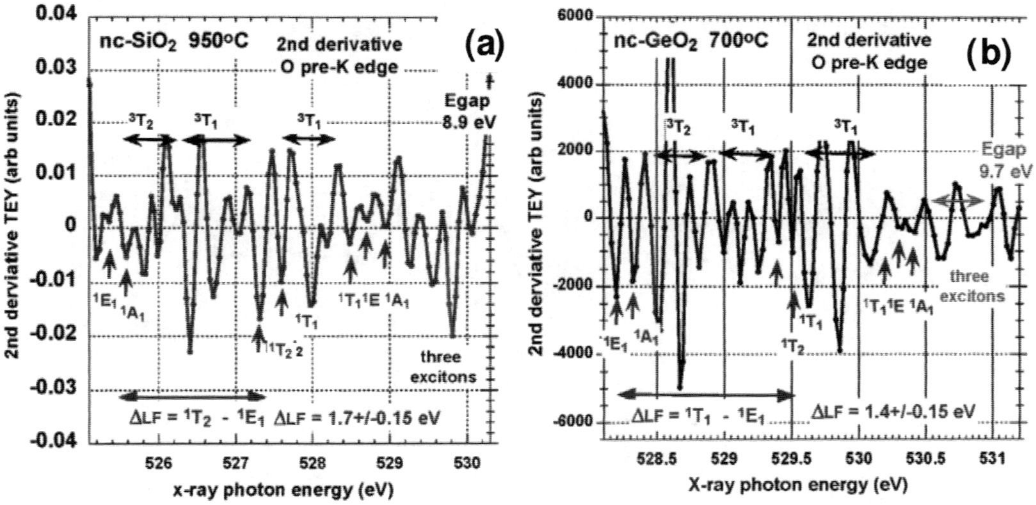

Fig. 2. O K pre-edge 2nd derivative spectrum for remote plasma deposited (a) SiO_2 annealed at 900°C for 1 minute in Ar, and (b) GeO_2 annealed at 700°C for 1 minute in Ar

A relationship between pre-existing defects in RPD nc-SiO_2 and MRO results in discrete dihedral angles extending to 3rd and generally 4th nearest-neighbors [9]. This scale of MRO is established by position of the FSDP in the structure factor from X-ray or neutron diffraction [2,3]. Analysis of the FSDP yields a λcorr in SiO_2 of ~0.4 to 0.5 nm corresponding to discrete dihedral angles bridging Si-O-Si bonds, and a larger coherence length, λcoh identifies non-periodically organized clusters with dimensions ~1 nm. As such, nc-SiO_2 is not a CRN according to Zacharaisen which is *by definition* has no MRO [6].

Figure 2(a) is a high resolution 2nd derivative O K pre-edge X-ray absorption spectra for an SiO_2 film annealed in Ar at 950°C: (i) the band-gap, (ii) 3 band edge excitons, (iii) singlet states *inter-mixed* and separated from triplet terms, and (iv) singlet and triplet ground states. Fig. 2(b) is a similar plot for REPCVD GeO_2 with the post-deposition, pre-metallization anneal at a lower temperature of 700°C.

Returning to Fig. 2(a), the effective resolution in this spectral regime of the excited defect states, is ~0.1 eV; however, the data are obtained continuously with sampling rate spectral widths of 0.025 to 0.30 eV. This allows 2- and 3-point smoothing of the data prior to differentiation, as well as 2- and 3-point smoothing of differentiated TEY spectra. Fermion wave functions for electrons in the Si-atom dangling bonds are anti-symmetric and include 4-vector spin states (see Figs. 2(a) and 2(b)). The symmetry labeling of features in Fig. 2 is from the d^8 Tanabe-Sugano (T-S) diagram for tetrahedral-bonded Si at intermediate values of ΔLF [1]. The intermediate value is an indicator of partially covalent bonding, as contrasted with the definition of crystal field splitting which assumed 100% ionic bonding. The T-S diagrams were developed for occupied d-states on single transition metal (TM)- and rare-earth (RE) atom, however, the same group theoretically-allowed transitions also apply to 2-occupied d-states on different atoms, e.g., n = 3, in close proximity, e.g., up to 0.3 to 0.5 nm. In nc-SiO_2 ,this is a regime in which the respective radial Si-atom wave-functions overlap forming relatively weak Si-Si bond insuring strong-correlation. This observation is equivalent to an extension of the Pauli Exclusion principle from single atoms to pairs of inter-acting atoms with strongly-correlated Fermion states. More importantly, the relative strengths of defect terms in SiO_2 films display approximately the same temperature dependence for films annealed at ~875° to 900ºC and 950°C as the D2 Raman peak recorded by Galeener [15,16].

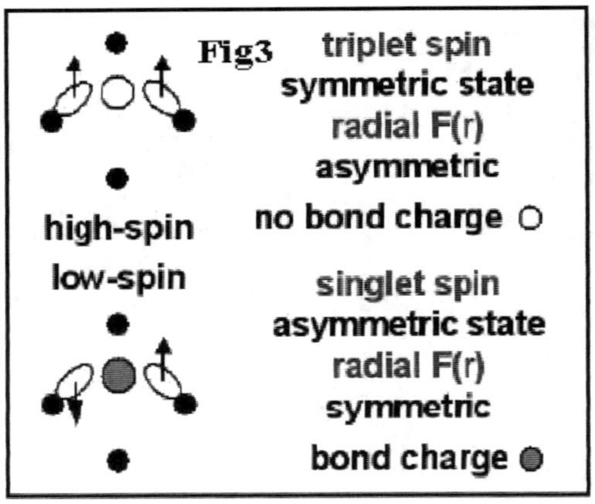

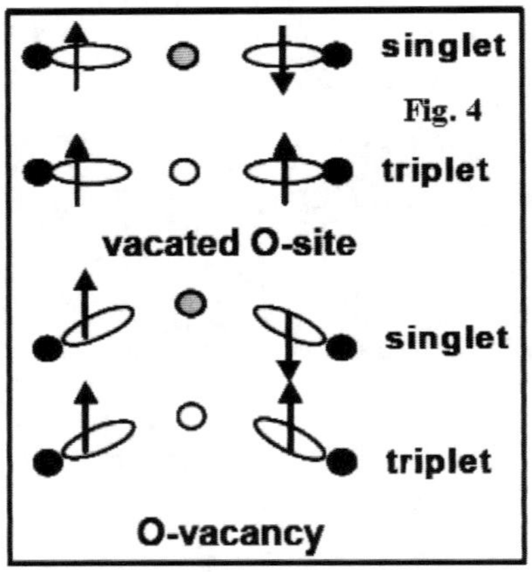

Fig. 3. Schematic representation of radial spatial and spin contributions to singlet and triplet two-electron Fermions.

Fig. 4. Schematic representation of orbitals in relaxed O-bond sites and in un-relaxed O-sites.

Figures 2(a) and 2(b) includes the symmetries of the radial spatial and spin contributions to the singlet and triplet two electron Fermion wave functions. The important differences in orbital geometries between process-induced vacated O-bonding sites and O-vacancies created by stress are indicated in Fig. 3. The orbital geometry for the O-vacancy has the same Si-O-Si bond-angle as in SiO_2, 148 degrees and the Si orbitals are not co-linear as they shown symbolically for the vacated O-site [5]. When the distance between two Si atoms bordering the vacancy is increased, the 3-fold symmetry is maintained for Si-atoms back-bonded to the network.

Figures 2(a) and 2(b) indicate O pre-edge spectra respectively for nc-SiO_2 and GeO_2. Values of ΔLF have been extracted from the analysis of these spectra, and are respectively 1.7 eV and 1.3 eV, each with an experimental uncertainty of 0.15 eV. The respective values are derived from the 4th NN local molecular orbital bonding at the defect sites. This is sensitive to the formal ionic charge of the atomic dangling bonds extending into the vacated sites, and the bonding coordination of the dangling bond atom. These charges are 4+ for Si and Ge in the respective oxides. Each of these dangling bond atoms is 4-fold coordinated with 3 equivalent O-atoms back-bonded the dangling bonds. The respective 4+ ionic radii are 0.40 nm for Si^{4+} and 0.53 nm for Ge^{4+}. Based on electrostatics, the ratio of values (SiO_2/GeO_2) = 1.31, and is expected to be equal to the inverse ratio of the respective +4 ionic radii (Ge/Si) =1.32. This demonstrates that values for the ΔLF extracted from O pre-K edge spectra are localized at the defect site, and include: (i) the two dangling bonds projected in the site, (ii) the vacated O-atom site, and finally (iii) the three equivalent O-atoms back-bonded to the dangling bonds, thereby making a total of 5-atoms. This local group has MRO, and therefore a localized molecular orbital. The same values of the ΔLF are extracted from Si and Ge, 2p and 3d core state excitations and the O K-edge, as well. This leads to an interesting and significant connection between the values of ΔLF and Madelung constant truncated at 5 atom nm-scale clusters. This truncation is the same as in a primitive unit cell in the crystalline forms of SiO_2 and GeO_2. This equivalence of local ΔLF values at defects and in nm-scale clusters, with Madelung constants at third and fourth NN's has important implications for the validity of one-electron band structures. It effectively invalidates the almost universally-accepted concept of determining an empty structure, filling it available electrons, and using this as a basis to identify un-occupied conduction band states. In the ab-initio calculations the conduction band state are virtual, and qualitatively the same as virtual molecular orbital final states for optical and thermal excitations. Further discussions of his flawed concept are clearly beyond the scope of this chapter. As such, the text below continues a critique of the assignments made by the Galeener group for the pre-existing defect state geometries.

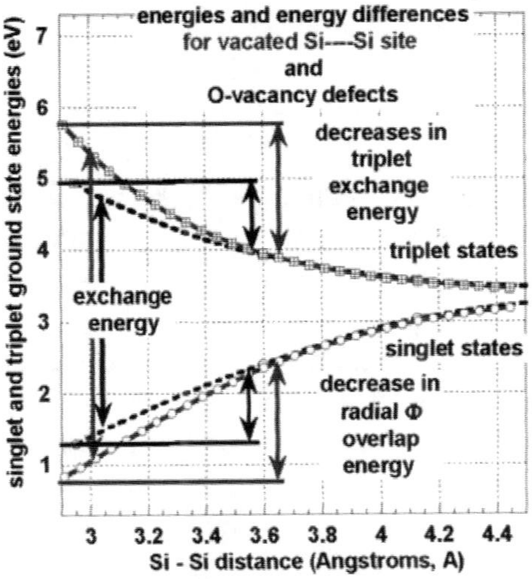

Fig. 5. Variation of single and triplet ground state energies as function of the distance between Si atoms for a vacated O-atom site, and an O-vacancy created by X-ray, γ-ray or high energy electron irradiation.

The Raman and IR frequencies for 4-symmetric Si-O bond-stretching vibrations are 800 cm^{-1}. Scaled to 3-symmetric Si-O bonds, and localized at the Si-dangling bond O-vacated defects yields a frequency of ~600 cm^{-1}, thereby providing a new assignment for the Raman D2 feature [16]. The results of ab-initio calculations based on two electron wave functions for the two geometries in Fig. 4 are included in Fig. 5, which displays the variation of the single and triplet ground state energies as function of the distance between the Si atoms. As expected, both the exchange energy and the radial wave function, F(r) overlap are larger for the vacated site for the smallest spacing. However, for Si-Si distances >3.6 Å and extending to 4.5 to 5.0 Å, the ground state energies are essentially the same for singlet and triplet state spin occupancy. The density functional theory (DFT) studies of the Robertson and Slugger groups based on 1-electron states indicate that TM atoms bordering a vacancy show relatively small relaxations of order 0.1-0.2 Å, and as such the only possible ground state and excited states in their respective DFT studies must have a singlet character [17,18].

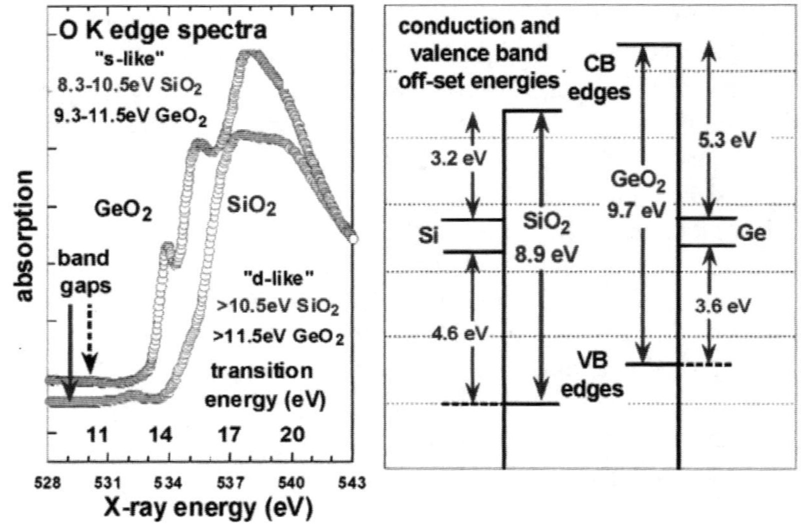

Fig. 6. (a) O K-edge spectra of ~5 nm t films of nc-SiO_2 and nc-GeO_2 remote plasma deposited at 300°C onto SiON surfaces of Si. The nc-SiO_2 film was annealed in Ar at 900°C, and the nc-GeO_2 was not annealed' after deposition. (b) Band offset energies for SiO_2 on Si and GeO_2 on Ge.

The results in Fig. 5 explain the fundamental differences and similarities between ab initio calculations and the DFT results. First, the only possible ground states in DFT calculations are singlet states, so that they cannot explain all of the spectroscopic study results of the Lucovsky group which yield both singlet and triplet features in O pre-K-edge XAS shake-up spectra for nc-SiO_2 and nc-GeO_2, and additionally for TM elemental and composite or mixed oxides as well [19]. The ground state energies of the singlet ground states calculated by the Robertson group are >3 eV above the valence band edge and have only a singlet character. Their estimate is in agreement with other one-electron calculations in Ref. 20. However, it is not consistent with electrical measurements, which indicate the singly-occupied ESR active states of states prior to post-metal annealing are significantly closer to the valence band edge, < 0.3 eV.

n marked contrast, ab initio quantum chemistry calculations yield both singlet and triplet features for pre-existing defects in TM oxides and SiO_2 consistent with Fig. 4 and T-S

Journal of Physics: Conference Series **428** (2013) 012017 doi:10.1088/1742-6596/428/1/012017

diagrams [1,19,20]. In summary above, XAS studies on SiO_2 films as-deposited and annealed at different temperatures establish (i) as-deposited films display both singlet and triplet features, and (ii) singlet features increase in strength as annealing temperatures are increased consistent with the new proposed interpretation for D2. Figure 6(b) conduction and valence band offset energies for $Si\text{-}SiO_2$ and $Ge\text{-}GeO_2$ interfaces.

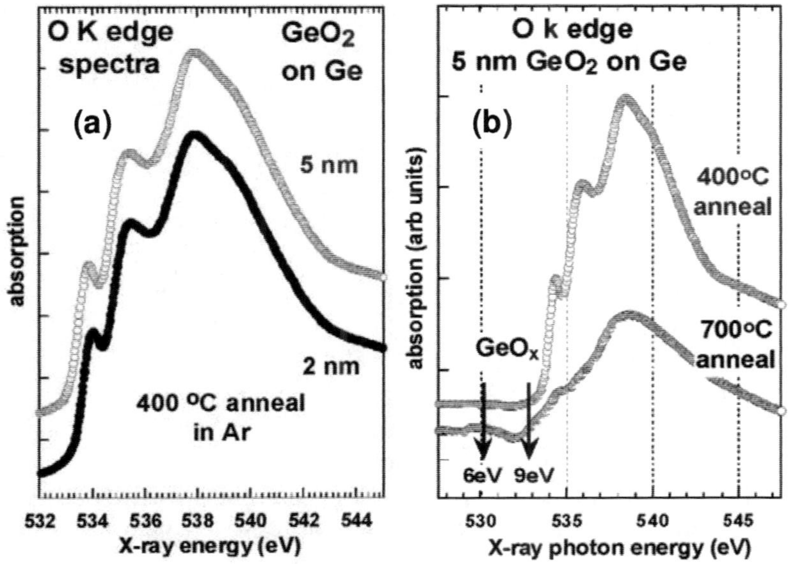

Fig, 7(a). O K edge spectra for two GeO_2 films of different thickness, 2 nm and 5 nmon chemically-cleaned Ge, and annealed at 400 °C. Fig. 7(b). Changes in spectra between 5 nm film of Fig. 7(a) annealed at 400°C and 700°C.

Figure 7(a) is a comparison between O K edge spectra of 5 nm thick films of nc-GeO_2 annealed at 400°C. The spectral features at the band edge and above the respective threshold energies for the transition from "s-like" to "d-like" features are the same. This is consistent with the nm-dimensions of the embedded clusters.

Figure 7(b) presents a comparison between n O K pre-edge spectra for the 5 nm thick GeO_2 film deposited on chemically-cleaned Ge at 300°C, and annealed for one minute in Ar at 400°C and 700°C. Decomposition reactions between nc-GeO_2 and c-Ge substrates at c-Ge/nc-GeO_2 interfaces at temperatures of 700°C release gaseous GeO, and create Ge suboxide bonding throughout the Ge film. This reaction leads to the differences in the spectra in this figure. Combined with other experiments the maximum processing temperatures for nc-GeO_2 (i) at nitrided c-Ge surfaces is between 550° and 600°C, and (ii) increases to about 700°C an c-Si surfaces interfacial layers formed by sequential remote plasma oxidations and nitridations with a total interfacial layer thickness no more than 0.7 to 0.8 nm.

5. Discussion

ESR features in SiO_2 films exposed to X-rays and γ-rays. and high energy electrons have been interpreted with an O-vacancy mechanism in which one of two Si-O bands at an Si-O-Si bonding site has been broken [19] The ESR E' center is a singly occupied dangling bond attached to three-fold Si, requiring activation by high-energy electrons, X-rays or γ-rays. The

Journal of Physics: Conference Series **428** (2013) 012017 doi:10.1088/1742-6596/428/1/012017

second ESR signal is a non-bonding O-associated hole center (NBOHC) defect. Therefore in irradiated films or glasses a pair of neutral defects comprised of an E' center and a NBOHC is created [16,19,20].

ESR studies by Galeener and co-workers on bulk-quenched silica (nc-SiO_2) have distinguished between pre-existing, and X-ray and γ-ray radiation induced defects [16] Pre-existing defect densities increase exponentially with increasing quenching and annealing temperatures and in "dry silicas" these are assigned to E' centers which are singly occupied Si-atom dangling bonds. NBOHC defects are detected in "dry silicas", but only after X-ray or γ-ray irradiation, and also reported in "wet silicas" with detectable Si-OH bonding [13]. Pre-existing defects have also been detected by 2nd derivative O K pre-edge X-ray absorption spectroscopy in remote plasma deposited and thermally grown nc-SiO_2 and nc-GeO_2 thin films [19,20].

It is important to re-emphasize the differences between (i) pre-existing and (ii) radiation, electron and neutron induced defects in "wet" and "dry" silica, [16,20] since this differentiation is important for understanding differences between ESR defects in nc-$GeS(Se)_2$ [20,21,22]. The "dry-silica" results also apply to pre-existing defects in plasma-deposited and annealed thin films of nc-GeO_2 and nc-SiO_2, and nc-GeS_2 bulk and thin film samples as well. In contrast, ESR defects in S-rich nc-$Ge_{1-x}S_x$ alloys are similar to "wet-silica" results. The pre-existing "wet silica" defects are NBOHC's and have qualitatively different properties when compared with pre-existing defects in "dry silicas" [16]. These defects always require activation with optical or UV radiation for ESR detection. They are present in the S-rich alloys because the S-S bond energy is less than the Ge-S bond energy. This means that it costs more energy to break a Ge-S-Ge bonding arrangement that a Ge-S-S-Ge arrangement. This bond cleave gives a pair of non-bridging S-hole centers (NBSHC's), the S (and also Se) analog of NBOHC's.

The results reported this section, as well as the comments made in Section 4 relative to a new interpretation of the origin of the D2 feature are consistent with the results presented for nc-GeS_2 and S-rich nc-$Ge_{1-x}S_x$ alloys. The conclusion as that vacated sites are responsible for pre-existing defects in nc-GeO_2 and nc-SiO_2, and nc-GeS_2 as well, and not planar small ring structure.

As already noted above, and similar to plasma-deposited SiO_2 and GeO_2, the stoichiometric GeS_2 compositions contain small concentrations of detectable pre-existing ESR defects [20,21,22]. However, for Ge concentrations between about 20 at. % and about 31.5 at. %, two defect signatures are evident, one characteristic of an E' center on the Ge atoms, and the second is an NBSHC local terminal group on one of the Ge-atoms. These conjugate defects result from local strain induced breaking a Ge-S bond, into a Ge dangling bond back-bonded to the three S-atoms of the network, and a terminal and neutral singly occupied S dangling bond. Paired pre-existing defects of this sort have not been reported in non-crystalline S-rich concentrations $As_{1-x}(S,Se)_x$ alloys [23,24]. Local strain in SiO_2 and GeO_2 is relieved by vacating bridging O-atoms, whereas the lower average coordination in chalcogenide rich As-S and As-Se alloys is insufficient to force defect formation by vacating or bond-breaking mechanisms, and as such there are no reports of pre-existing defects in these alloys, or in stoichiometric compounds as well.

Complementary studies of amorphous As and Se, and As_2Se_3 were performed by the NRL group [23,24]. These studies performed at low temperatures, demonstrated an optically induced ESR signal in As_2Se_3 by excitation in the Urbach tail, absorption constant ~100 cm^{-1}, and accompanied by mid-band gap absorption. The NRL group has attributed the ESR response in As_2S_3 and As_2Se_3, respectively, to a single electron occupancy in non-bonding S 2pπ and Se 3pπ lone pair orbitals, that is to NBS(Se)HCs [23,24]. Finally, the Xerox PARC group of Biegelsen and Street demonstrated qualitatively different ESR signals for optically induced ESR in As_2Se_3 in the Urbach tail, and at higher photon energy as which photo-darkening occurred, [25] confirming significant SRO and MRO scale bonding changes in the photo-darkened state. These local changes in SRO and MRO have identified in EXAFS studies, [20] and ab-initio calculations [26].

6. **Summary and Conclusions**

(i) Quantification of MRO has been achieved by analysis of the FSDP in the structure factor obtained from analysis of X-ray and neutron [2,3]. This extends to 3rd and 4th NN's and as such includes discrete values of dihedral angles, and in many instances it also includes constant and coupled phase relationships between the discrete dihedral angles [9].

(ii) MRO has also been detected by spectroscopic determinations of d-state ligand field splittings, ΔLF's, that derive from the valence band states comprised of symmetry-adapted linear combinations (SALC's) of atomic states [1,19]. These splittings are detected in (i) Si L2,3 and Ge M3,4 spectra, respectively, for nc-SiO_2 and nc-GeO_2, (ii) UPS and XPS valence band spectra, and in (iii) *conduction band* virtual bound state spectra obtained by spectroscopic ellipsometry and reflectivity, as well as the analysis of O K edge XAS spectra.
(iii) ΔLF's, and evidence for MRO are also obtained from analysis of O K pre-edge defect spectra. This approach treats the 2-electrons localized at an O-vacancy, or in a vacated O-atom site as a strongly-correlated pair on two different atoms, and then applies the appropriate T-S diagrams for assigning symmetries and ordering of the excited states, thereby extracting values of ΔLF's [1,19].
(iv) Proceeding in this the vacated O-atom site model has been verified by a combination of ab-initio quantum chemistry calculations and the O K pre-edge spectra for nc-SiO_2, and nc-GeO_2.
(v) The vacated O-atom site model has been extended to a vacated S (Se)-atom model for nc-$GeS(Se)_2$. Due to catenation of S- and Se-atoms in S(Se)-rich nc-$Ge_{1-x}S(Se)_x$ and also in nc-$As_{1-x}S(Se)_x$ alloys, the vacated S(Se)-atom site models do not apply.
Finally, (vi) it is significant to note that pre-existing defects in nc-SiO_2 and nc-GeO_2 are thermally-activated by removal of O-atoms from Si-O-Si and Ge-O-Ge bonding sites within chemically self-organized cluster geometries obtained from the analysis of the width of FSDP in reciprocal space. In a similar way, the limiting defect densities in a-Si(H) with 10 at.% H are associated with thermal activation of Si-H vibrations at the perimeter of the self-organized clusters in a-Si(H) alloys [27].

Acknowledgements
The authors acknowledge support from the National Science Foundation, NSF, and from the Defense Threat Reduction Agency, DTRA. They also acknowledge the contribution of Cheng Cheng and Brian Papas at NC State University. The authors acknowledge helpful discussions with Professors David Aspens Jerry Whitten at NC State University. Lastly the authors acknowledge the use of the Beam-lines 10-1 and 8-1 at the Stanford Synchrotron Research Lightsource, SSRL, in Menlo Park CA, and assistance of Dr. Dennis Norland, a beam-line scientist.

[1] F.A. Cotton, Chemical Applications of Group Theory, 2nd Ed. (Wiley Nescience, New York, 1963) Chapter 8.
[2] J. Du, R.L Corrales, Phys. Rev. B 72, 092201 (2005).
[3] S.C. Moss and D.L. Price: in Physics of Disordered Materials, edited b. D. Adler and H. Fritsch and S.R. Vishinski (Plenum Press, New York, 1985), p. 77; D.L. Price et al., J. Phys. Condense. Matter 1, 1005 (1989).
[4] N.R. Rae, et al., J. Non-Cryst. Solids 240, 221 (1998).
[5] J. L. Whitten, et al., J. Vac. Sci..Technol. B 20 (2002) 1710.
[6] W.H. Zachariasen, J. Am. Chem. Soc. 54, 3341 (1932).
[7] G. Lucovsky and J.C. Phillips, J. Phys.: Condens. Mater. 19 455218 (2007), and references therein.
[8] B. Kramer, K. Maschke and P. Thomas, phys. stat. sol. (b) 49, 525 (1972).
[9] G. Lucovsky and F.L. Galeener, J. of Non-Cryst. Solids 35 & 36 (1980) 1209.
[10] J.T. Randall, H.P. Rooksby, B.S. Cooper, Z. Krist, 75, 196 (1930).
[11] J.T. Randall, H.P. Rooksby, B.S. Cooper, Nature (supplement), 458 (1930).
[12] G. Lucovsky and D. Zeller, J. Nanoscience and Nanotechnology, 11, 7974-7981((2011).
[13] Owens, S.C. Bayliss, G.W. Fraser, S.J. Gurman, Nucl. Instrum. Methods Phys. 280 Res. Sect. A 385, 556 (1997).
[14]. 25. F. deGroot and A. Kotani, Core Level Spectroscopy of Solids (CRC Press, Boca Raton, 2008).
[15] F.L. Galeener and G. Lucovsky, Phys. Rev. Lett. 37, 1474 (1976).
[16] F.L. Galeener, J. Non-Cryst. Solids 71, 373 (1985); Phys. Rev. B 47, 7760 (1993).
[17] K. Xiong, et al., Appl. Phys. Lett. 87, 183505 (2005).
[18] L. Gavartin, et al., Appl. Phys. Lett. 89, 082908 (2006).
[19] G. Lucovsky, Japan J. Appl. Phys. 50, 04DC09 (2011).
[20] D.L. Griscom, Ceramic Soc. Jap. 90, 823 (1981), and Refs. therein.
[21] R.A. Street and D.J. Biegelsen, J. of Non-Cryst. Solids 32 (1979) 339.
[22] K. Arai and H Namikawa, Solid State Commun. 13, 1167 (1973)/
[23] S.G. Bishop et al., Phys. Rev. Lett. 21, 1346 (1975)
[24] S.G. Bishop et al., Phys. Rev. B 15, 2294 (1977).
[25] J..M. Lee, M.A. Paesler, D.E. Sayers, A. Fontaine, J. of Non-Cryst. Solids 123, 295 (1990).
[26] R. Mowrer, G. Lucovsky, L.S. Sremaniak, J.L. Whitten, J. of Non-Cryst. Solids 338-340, 543 (1994), and Refs. therein.
[27] G. Lucovsky and H. Yang, Jpn. J. Appl. Phys. 37, 1082 (1998).

XXIst International Symposium on the Jahn–Teller Effect 2012 IOP Publishing
Journal of Physics: Conference Series **428** (2013) 012018 doi:10.1088/1742-6596/428/1/012018

Band Jahn-Teller effects and Peierls Instability in $IrTe_2$

D. Ootsuki, S. Pyon[1], K. Kudo[1], M. Nohara[1], M. Horio[2], T. Yoshida[2], A. Fujimori[2], M. Arita[3], H. Anzai[3], H. Namatame[3], M. Taniguchi[3,4], N. L. Saini[5,1], T. Mizokawa

Department of Physics and Department of Complexity Science and Engineering, University of Tokyo, 5-1-5 Kashiwanoha, Kashiwa, Chiba 277-0882, Japan
[1]Department of Physics, Okayama University, Kita-ku, Okayama 700-8530, Japan
[2]Department of Physics, University of Tokyo, 7-3-1 Hongo, Tokyo 113-0033, Japan
[3]Hiroshima Synchrotron Radiation Center, Hiroshima University, Higashi-hiroshima 739-0046, Japan
[4]Graduate School of Science, Hiroshima University, Higashi-hiroshima 739-8526, Japan
[5]Department of Physics, University of Roma "La Sapienza" Piazzale Aldo Moro 2, 00185 Roma, Italy

E-mail: `ootsuki@sces.k.u-tokyo.ac.jp`

Abstract. We report a photoemission spectroscopy study on $IrTe_2$ which exhibits an interesting lattice distortion below 270 K. Ir $4f$ core-level photoemission spectrum is broadened in going from 300K to 40K. This result indicates the charge moderation of Ir $5d$ electrons. Angle-resolved photoemission spectroscopy(ARPES) result at 300K shows multi-band Fermi surfaces with six-fold symmetry which are basically consistent with band structure calculations. At the distorted phase, the inner Fermi surfaces become quasi one-dimensional Fermi surface due to the lattice distortion.

1. Introduction

Recently, Pyon *et al.*[1] and Yang *et al.* [2] have discovered interesting interplay between lattice distortion and superconductivity in triangular lattice $Ir_{1\ x}Pt_xTe_2$ in which multi-band Fermi surfaces are expected to play signi cant roles. The parent material $IrTe_2$ exhibits a structural phase transition at 270 K from the trigonal (P3m-1) to the monoclinic (C2/m) structure accompanied by anomalies of electrical resistivity and magnetic susceptibility [3]. When the lattice distortion is suppressed by chemical substitution of Pt or Pd for Ir or intercalation of Pd, $IrTe_2$ becomes superconductors [1][2]. An electron di raction result by Yang *et al.* shows that the structural transition is accompanied by superstructure with wave vector of q = (1/5, 0, -1/5). Such superstructure is often observed in charge density wave (CDW) materials [2]. In multi-band Fermi surfaces with Ir $5d$ and Te $5p$ orbital degeneracy, the nesting character can be enhanced by orbitally-induced Peierls mechanism [4]. On the other hand, a recent optical study by Fang *et al.* on single crystal samples shows that there is no gap opening expected for CDW and, instead, band structure is reconstructed over a broad energy scale up to 2 eV [5]. Fang *et al.* conclude that the structural transition of $IrTe_2$ is not of CDW type but of a novel type driven by Te $5p$ holes [5].

Journal of Physics: Conference Series **428** (2013) 012018 doi:10.1088/1742-6596/428/1/012018

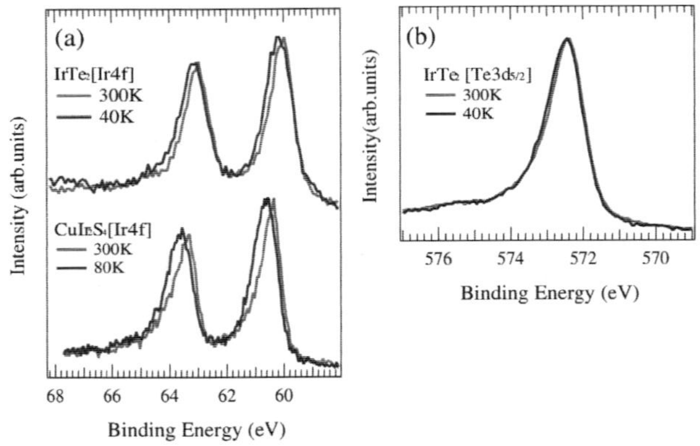

Figure 1. (color online) (a) Ir $4f$ core-level photoemission spectra of $IrTe_2$ compared to that of $CuIr_2S_4$.[6] (b) Te $3d_{5/2}$ core-level photoemission spectra of $IrTe_2$.

2. Experimental Results and Disscussion

2.1. X-ray Photoemission Spectroscopy

The polycrystalline sample of $IrTe_2$ was prepared as reported in ref. 1.[1] $IrTe_2$ is nonsuperconducting. The x-ray photoemission spectroscopy (XPS) was carried out at 300 K and 40 K using JEOL JPS9200 analyzer. Monochromatic Al K (1486.6 eV) was used as x-ray source. The total energy resolution was about 0.6 eV. The base pressure of the chamber was in the 10 ^{7} Pa range. The binding energy was calibrated using the Au $4f$ core level of the gold reference sample. We fractured the polycrystalline samples of $IrTe_2$ at 300 K for the XPS measurements.

Ir $4f$ and Te $3d$ core-level photoemission spectra of $IrTe_2$ are displayed in Fig. 1. The Ir $4f$ peak width of $IrTe_2$ slightly increases in going from 300 K to 40 K while the Te $3d$ peaks do not show any such changes with temperature. The increase of peak width indicates that the density of Ir $5d$ t_{2g} electrons is modulated in the low temperature phase of $IrTe_2$. Here, it should be noted that the Ir $4f$ peak width increase of $IrTe_2$ is comparable to that of $CuIr_2S_4$ in which the octamer Ir^{3+}/Ir^{4+} charge ordering was established [6] and the charge di erence between the Ir^{3+} site and the Ir^{4+} site was observed in the Ir $4f$ XPS [6]. Interestingly, the Ir $4f$ binding energy of $IrTe_2$ is smaller than that of $CuIr_2S_4$ as shown in Fig. 1(a), suggesting that the actual number of Ir $5d$ electrons of $IrTe_2$ (formally Ir^{4+}) is larger than that of $CuIr_2S_4$ (formally $Ir^{3.5+}$).

2.2. Angle-resolved Photoemission Spectroscopy

The single crystal samples of $IrTe_2$ were prepared using a self-flux method [1, 5]. The photoemission measurements were performed at beamline 9A, Hiroshima Synchrotron Radiation Center using a SCIENTA R4000 analyzer with circularly polarized light. The total energy resolution was set to 18 meV for excitation energy of h = 23 eV. The angular resolution was set to 0.3° that gives the momentum resolution of 0.015 Å ^{1} for h =23 eV. The incident beam is 50° o the sample surface. The base pressure of the spectrometer was in the 10 ^{9} Pa range. The samples were oriented by *ex situ* Laue measurements. The spectra were acquired within 8 hours after the cleavage. The binding energy was calibrated using the Fermi edge of gold reference samples. We cleaved the single crystalline samples of $IrTe_2$ at 300 K for the ARPES measurements and then cooled the samples.

Figure 2(a) shows the Fermi surface mapping of $IrTe_2$ at 300 K above the structural transition temperature. The ARPES data are taken at h = 23 eV and, for the Fermi surface mapping, the ARPES intensity is integrated within an energy window of 5 meV at the Fermi level (E_F).

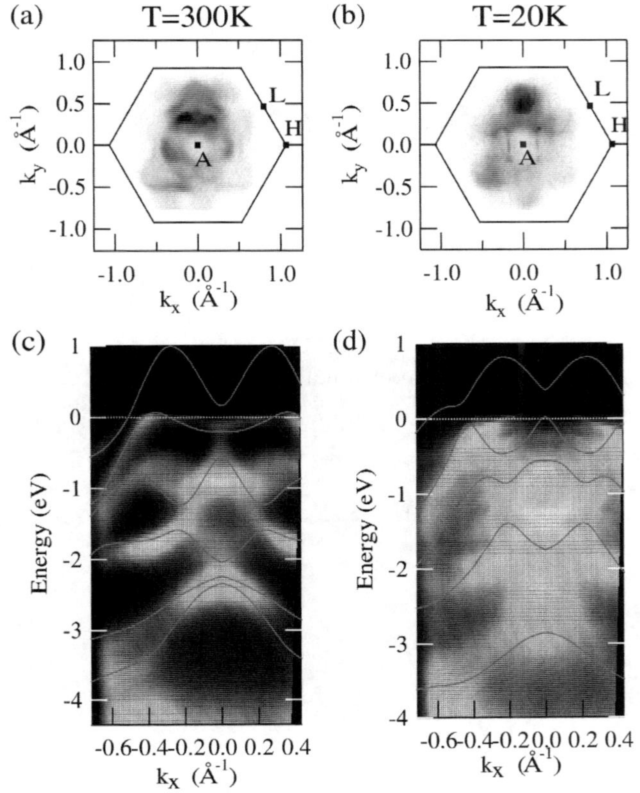

Figure 2. (color online) Fermi surface mapping of $IrTe_2$ for h = 23 eV taken at 300 K (a) and at 20 K (b). The hexagonal Brillouin zone boundary is shown by solid lines. The center of the hexagon roughly corresponds to the A point for h = 23 eV. Comparison of the ARPES data and the band calculation results [5] along the A-H direction for h = 23 eV taken at 300 K (c) and at 20 K (d).

At h = 23 eV, the momentum perpendicular to the Ir plane approximately corresponds to $/c$, where c is the out-of-plane lattice constant. Therefore, the center of the hexagonal Brillouin zone is the A point. The direction from the A point to the L (H) point corresponds to the direction of Ir-Ir (Ir-Te) bond. In agreement with the band structure calculations [2, 5], the flower-shaped outer Fermi surface with six-fold symmetry is observed as shown in Fig. 2(a). In addition, the inner Fermi surfaces like six connected beads can be identi ed although e ect of thermal excitations at 300 K tends to obscure the relatively small Fermi pockets. The inner Fermi surfaces observed around the A point at 300 K are roughly consistent with the prediction of the band-structure calculations [2, 5].

The Fermi surface mapping at 20 K well below the transition temperature is displayed in Fig. 2(b). Across the structural transition, the flower shape of the outer Fermi surface does not change appreciably. However, in the region where the outer Fermi surface is close to the inner Fermi surfaces, while the outer Fermi surface is separated from the inner Fermi surfaces at 300 K, the outer Fermi surface overlaps with the inner Fermi surfaces at 20 K. In contrast to the small e ect on the outer Fermi surface, the inner Fermi surfaces dramatically change their shapes by the structural transition. At 20 K, two straight portions of Fermi surfaces are observed at 20 K. Such geometry of inner Fermi surfaces at 20 K deviates from the prediction of the calculation. The straight portions are perpendicular to the A-H direction or the direction of

Ir-Te bond. Therefore, both of the Te $5p$ and Ir $5d$ orbitals would be involved in the structural transition if the straight Fermi surfaces are driven by orbitally-induced Peierls mechanism [4]. Figures 2(c) and (d) show the wide-range band dispersions along the A-H direction at 300 K and 20 K, respectively. Whereas the band dispersions at 300 K agree with the predictions of the band-structure calculations, those at 20 K deviate from the predictions [2, 5].

3. Conclusion

In conclusion, we have measured the XPS and ARPES spectra of triangular lattice chalcogenide $IrTe_2$. Ir 4f core-level photoemission spectra show Ir $5d$ t_{2g} charge modulation established in the low-temperature phase of $IrTe_2$. ARPES result at 300K shows multi-band Fermi surfaces with six-fold symmetry which are basically consistent with the band structure calculations. At the distorted phase, the inner Fermi surfaces become quasi one-dimensional by the lattice distortion suggesting that nesting character is enhanced.

Acknowledgments

We would like to thank valuable discussions with Prof. D. I. Khomskii.

References

[1] S. Pyon, K. Kudo, and M. Nohara, J. Phys. Soc. Jpn. **81**, 053701 (2012).
[2] J. J. Yang, Y. J. Choi, Y. S. Oh, A. Hogan, Y. Horibe, K. Kim, B. I. Min, and S-W. Cheong, Phys. Rev. Lett. **108**, 116402 (2012).
[3] N. Matsumoto, K. Taniguchi, R. Endoh, H. Takano, and S. Nagata, J. Low Temp. Phys. **117**, 1129 (1999).
[4] D. I. Khomskii and T. Mizokawa, Phys. Rev. Lett. **94**, 156402 (2005).
[5] A. F. Fang, G. Xu, T. Dong, P. Zheng, and N. L. Wang, arXiv:1203.4061.
[6] K. Takubo, S. Hirata, J. Y. Son, J. W. Quilty, T. Mizokawa, N. Matsumoto and S. Nagata, Phys. Rev. Lett. **95**, 246401 (2005).

Lithium intercalation properties in manganese-iron Prussian blue analogues

T Matsuda[1], Y Kurihara[1] and Y Moritomo[1,2,3]

[1] Graduated School of Pure and Applied Sciences, University of Tsukuba, Tsukuba 305-8571, Japan
[2] Tsukuba Research Center for Interdisciplinary Materials Science (TIMS), University of Tsukuba, Tsukuba 305-8571, Japan

E-mail: moritomo@sakura.cc.tsukuba.ac.jp

Abstract. We investigated the electronic states and structural properties of $Li_xMn[Fe(CN)_6]_{0.83}{\cdot}3.5H_2O$, and $Li_xMn[Fe(CN)_6]_{0.87}{\cdot}2.6H_2O$, which have different amount of $[Fe(CN)_6]$ vacancies. X-ray absorption spectra near the Fe and Mn K-edges revealed that the Li intercalation/deintercalation process is two-electron reaction, i.e. Mn^{II}—NC—Fe^{II}, Mn^{II}—NC—Fe^{III}, and Mn^{III}—NC—Fe^{III}. The crystal structure of $Li_xMn[Fe(CN)_6]_{0.83}{\cdot}3.5H_2O$ remains cubic and single phase throughout the Li intercalation/deintercalation process. The crystal structure of $Li_xMn[Fe(CN)_6]_{0.87}{\cdot}2.6H_2O$ is also cubic, while phase separation of Li^+-rich phase and Li^+-poor phase was observed during the Mn^{II}/Mn^{III} reduction/oxidation process.

1. Introduction

Nanoporous materials are attracting the increasing interests from materials scientists, since such materials can be utilized in various devices, *e.g.*, Li ion secondary batteries, electrochromic devices, hydrogen storage devices, molecular sensors, and molecular filters. Among nanoporous materials, Prussian blue analogues [1-16], $A_xM_A[M_B(CN)_6]_y{\cdot}zH_2O$ (M_A and M_B are transition metal ions, and A is an alkali metal ion), have three-dimensional nanoporous frameworks of cyano-bridged transition metal ions. In the interstitial site of the nanoporous framework, the alkali cations and the a part of the water molecules (zeolitic waters) are located, and the alkali cation concentration (x) can be electrochemically controlled via the electronic states of the transition metal ions to maintain charge neutrality[3,4]. The remaining water molecules (ligand waters) coordinate to the M_A ion at the vacancy sites of $[M_B(CN)_6]$. As for the crystal structure, the typical compounds show face-centered cubic structures ($Fm{-}3m$) in $M_A[M_B(CN)_6]_{2/3}{\cdot}zH_2O$. In a stoichiometric compound, $AM_A[M_B(CN)_6]$, where the alkali cations occupy just one half of the interstitial sites, the compounds show several space groups by the ordering of the alkali cations [5]. Moreover, by the utilization of the cooperative Jahn-Teller instability inherent to Mn^{III} and Cu^{II}, the compounds exhibit the tetragonal structures ($I{-}4m2$) in $RbMn[Fe(CN)_6]$, $CsCu[Co(CN)_6]$, and $CsCu[Cr(CN)_6]$ [5-7].

Recently, Okubo *et al.* reported the reversible Li ion intercalation properties in $K_{0.14}Mn_{1.43}[Fe(CN)_6]{\cdot}6H_2O$ and $Rb_{0.7}Mn_{1.15}[Fe(CN)_6]{\cdot}2.5H_2O$ [14]. We have also demonstrated that a thin film of $Li_{1.32}Mn[Fe(CN)_6]_{0.83}{\cdot}3.5H_2O$ shows a large charge capacity (= 128 mA h g^{-1}) as well as excellent cyclability, which indicates that the Prussian blue analogues are one of the promising

[3] To whom any correspondence should be addressed.

cathode materials for the Li ion secondary batteries [15,16]. In this Li ion intercalation process, Mn^{II}/Mn^{III} reduction/oxidation occurs and the cooperative Jahn-Teller instability can be expected. In this work, we prepared two $Li_xMn[Fe(CN)_6]_y{\cdot}zH_2O$, which have different amount of $[Fe(CN)_6]$ vacancy sites and investigated the electronic states and structural properties.

2. Experimental

2.1. Preparation of $Li_xMn[Fe(CN)_6]_y{\cdot}zH_2O$

Thin films of $Na_{1.32}Mn[Fe(CN)_6]_{0.83}{\cdot}3.5H_2O$ and $Na_{1.48}Mn[Fe(CN)_6]_{0.87}{\cdot}2.6H_2O$ were electrochemically synthesized on an indium tin oxide (ITO) transparent electrode under potentiostatic conditions at −0.50 V vs a standard Ag/AgCl electrode in an aqueous solution containing $K_3[Fe(CN)_6]$ (1.0 mmol dm^{-3}), $MnCl_2{\cdot}6H_2O$ (1.5 mmol dm^{-3}), and NaCl (1 and 3 mol dm^{-3}). Before the film growth, the surface of the ITO electrode was purified by electrolysis of water for 3–5 min. The obtained film was transparent with a thickness of around 1 μm. Chemical compositions of the films were determined by the inductively coupled plasma (ICP) method and CHN organic elementary analysis (Perkin-Elmer 2400 CHN Elemental Analyzer). The Na ions were electrochemically substituted by Li ions, using lithium metal as the reference and counter electrode, and ethylene carbonate (EC) / diethyl carbonate (DEC) solution containing $LiClO_4$ (1 mol dm^{-1}) as an electrolyte. We electrochemically contained the Li concentration (x), and thus obtained thin films of $Li_xMn[Fe(CN)_6]_{0.83}{\cdot}3.5H_2O$ (L_xMF83), and $Li_xMn[Fe(CN)_6]_{0.87}{\cdot}2.6H_2O$ (L_xMF87).

2.2. X-ray absorption measurements

The electronic states of the Mn and Fe sites of L_xMF83 against x were determined by the *ex-situ* X-ray absorption spectra (XAS) around the Mn and Fe K-edge. The XAS measurements were conducted at the beamline 7C of the Photon Factory, KEK. The XAS spectra were recorded by a Lytle detector in a fluorescent yield mode with a Si(111) double-crystal monochromator at 300 K. The background subtraction and normalization were conducted by ATHENA program [17].

2.3. Powder X-ray diffraction measurements

The *ex-situ* powder X-ray diffraction (XRD) measurements of L_xMF83, and L_xMF87 against x were performed at the beamline 8A of KEK-PF, equipped with an imaging plate detector. The samples were washed with DEC, and carefully removed from the ITO glasses. The obtained powders were sealed in 300 μm glass capillaries. XRD patterns were measured at 300 K and the exposure time was 5 min. Wavelength of the X-ray were 0.77516 Å for L_xMF 83, and 0.77476 Å for L_xMF 87, respectively. The lattice constants of each compounds were refined by the RIETAN-FP program [18].

3. Results

3.1. Voltage dependence and electronic states on Li ion intercalation processes

Figure 1 shows the voltage against Li metal dependence on x of $Li_xMn[Fe(CN)_6]_{0.83}{\cdot}3.5H_2O$ (L_xMF83) and $Li_xMn[Fe(CN)_6]_{0.87}{\cdot}2.6H_2O$ (L_xMF87) during the Li ion intercalation process. Here, the possible x regions are $0 \leq x \leq 1.32$ for L_xMF83 and $0 \leq x \leq 1.48$ for L_xMF87 due to the charge neutrality. On both films, the electrochemical reactions were observed at around 3.8, and 3.4 V.

The electronic states for each electrochemical reactions are studied by the Mn and Fe K-edge XAS spectra of L_xMF83 against x. Figure 2 shows the Mn K-edge XAS spectra of L_xMF83 against x [16]. The peak around 6549 eV, which is ascribed to Mn^{II} [10-12], was observed in the whole-x region. On the contrary, the shoulder peak around 6552 eV, which is ascribed to Mn^{III} [10-12], was observed in the small-x region. These spectra indicates the reduction/oxidization of Mn (high-spin Mn^{II}, and high-spin Mn^{III}) occurs in the small-x region. Figure 3 shows the Fe K-edge XAS spectra of L_xMF83 against x [16]. The absorption peak energies (E) in the Fe K-edge spectra show clear blue shift with decrease in x, and saturate below $x = 0.41$. These data indicates blue shift is ascribed to the oxidization of Fe (low-spin Fe^{II} → low-spin Fe^{III}) [10-13], and indicates the reduction/oxidization of Fe takes place in the large-x region. Considering the formula and the range of x, the reduction/oxidization of Fe occurs in $0.49 \le x \le 1.32$ region, while the reduction/oxidation of Mn occurs in $0 \le x \le 0.49$ region in L_xMF83. The same scenario can also be applied to L_xMF87, and the reduction/oxidization of Fe occurs in $0.61 \le x \le 1.48$ region, while the reduction/oxidation of Mn occurs in $0 \le x \le 0.61$ region. Hence, the first plateau in Figure 1 corresponds to the Mn^{3+}/Mn^{2+} reduction process, and the second plateau in Figure 1 corresponds to $[Fe^{III}(CN)_6]^{3-}/[Fe^{II}(CN)_6]^{4-}$ reduction process. We can control x at will by using these data.

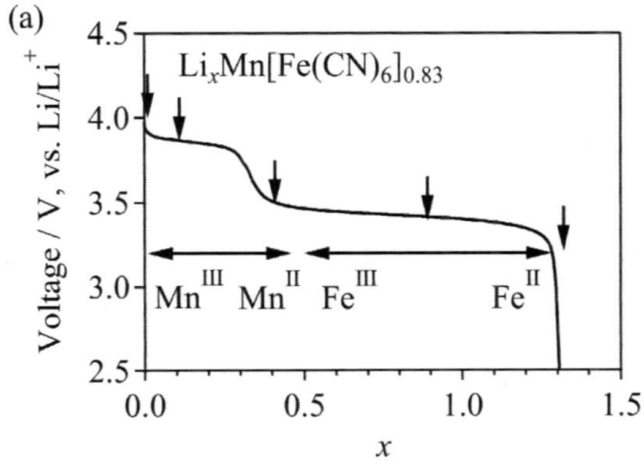

Figure 1. Voltage dependence on Li ion concentration (x) of $Li_xMn[Fe(CN)_6]_{0.83}$ (L_xMF83) (a), and $Li_xMn[Fe(CN)_6]_{0.87}$ (L_xMF87) (b). The arrows indicate x for the XAS measurements.

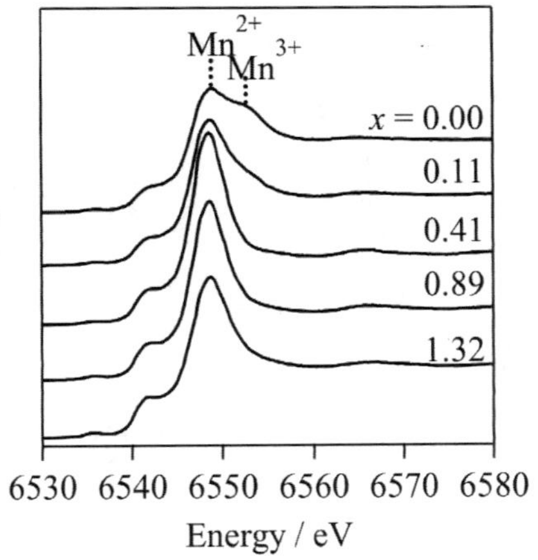

Figure 2. Mn K-edge XAS spectra of $Li_xMn[Fe(CN)_6]_{0.83}\cdot 3.5H_2O$ (L_xMF83).

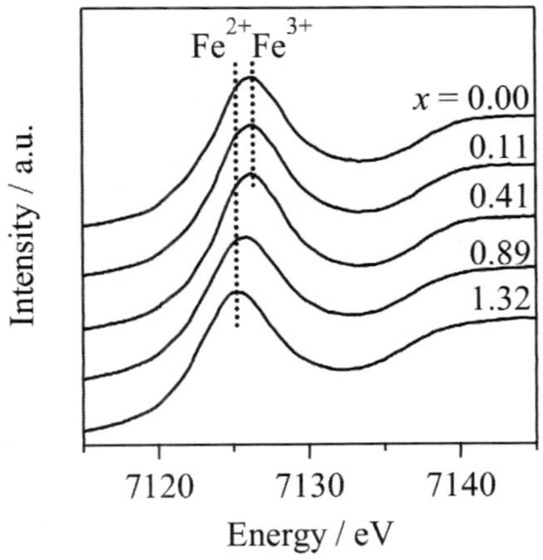

Figure 3. Fe K-edge XAS spectra of $Li_xMn[Fe(CN)_6]_{0.83}\cdot 3.5H_2O$ (L_xMF83).

3.2. Structural properties.

Figure 4a shows the XRD patterns of L_xMF83 [16]. The peak patterns are all similar, which indicates that the lattice structure remains the face-centered cubic (*Fm*–3*m*; $Z = 4$) throughout whole x region. Figure 4b shows the magnified 200 peak, and the reflection shifts to the lower-angle side above x = 0.41 while shifts to the lower-angle side below $x = 0.41$. The value of $x = 0.41$ is close to the value of x = 0.49 where reduction/oxidation switch between Mn and Fe occurs.

Figure 5a shows the XRD patterns of L_xMF87. The lattice structures were the face-centered cubic (*Fm*–3*m*; $Z = 4$) above $x = 0.61$. In $x = 0.15$, each reflection peaks were split into two, which indicates the occurrence of the phase separation, and again, the face-centered cubic structure (*Fm*–3*m*; $Z = 4$)

was observed in x = 0.00. Figure 5b shows the magnified 200 peak, and the reflection shifts to the lower-angle side above x = 0.61. The new peak appeared at the higher angle side in x = 0.15, and the peak at the lower angle side disappeared in x = 0.00. Overall feature of the XRD patterns of x = 0.15 can be reproduced by two cubic ($Fm\text{–}3m$; Z = 4) phases model with different lattice constants. The value of x = 0.61 is where reduction/oxidation switch between Mn and Fe occurs, and the phase separation occurs in the reduction/oxidation of Mn region.

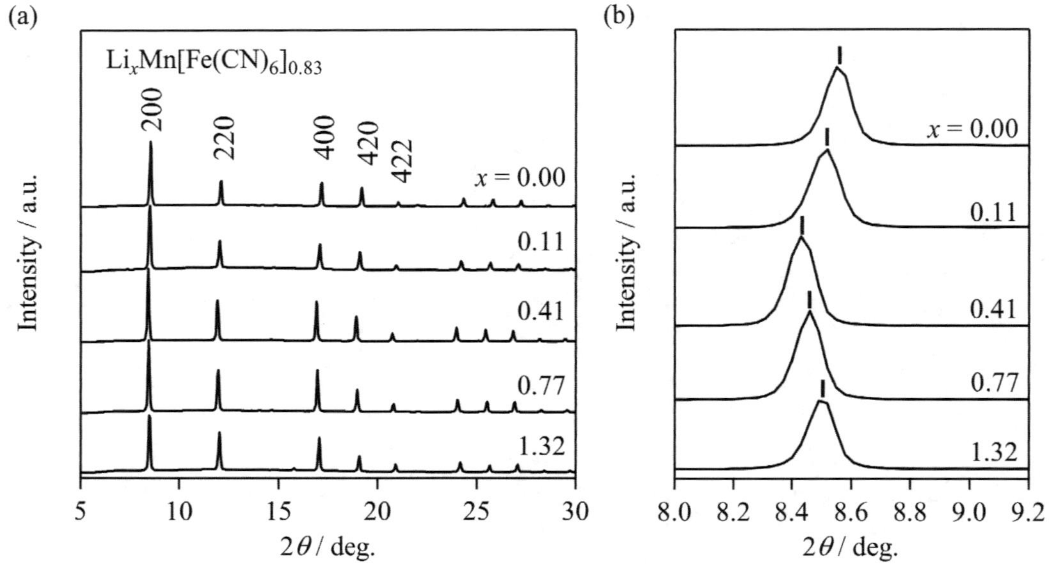

Figure 4. (a) The powder XRD patterns for $Li_xMn[Fe(CN)_6]_{0.83}\cdot 3.5H_2O$ (L_xMF83). (b) Magnified 200 peaks. The ticks indicate the reflection peak positions. The wavelength of the X-ray is 0.77516 Å.

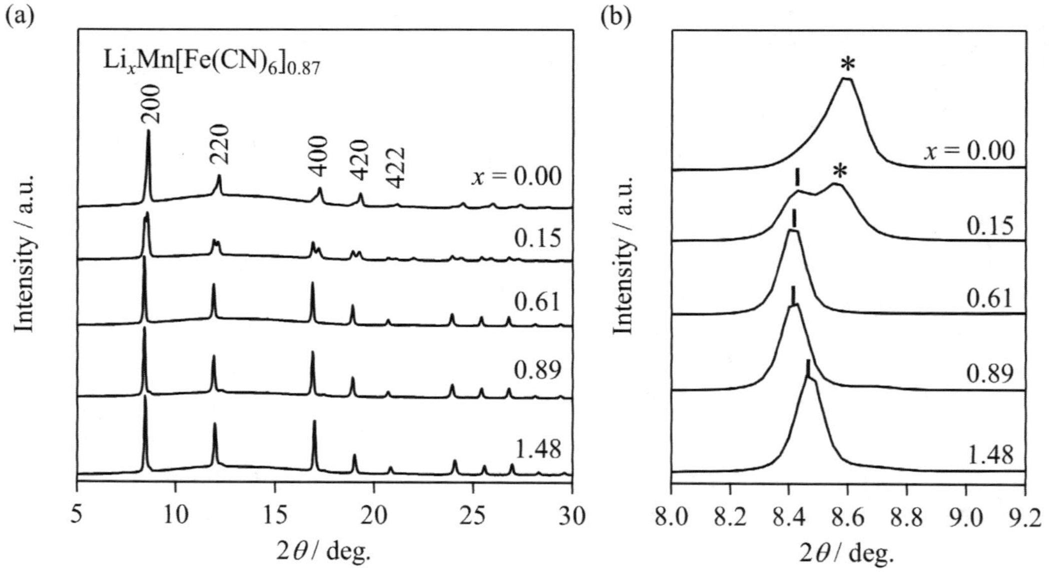

Figure 5. (a) The XRD patterns for $Li_xMn[Fe(CN)_6]_{0.87}\cdot 2.6H_2O$ (L_xMF87). (b) Magnified 200 peaks. The ticks indicate the reflection peaks of Mn^{II} phase and the stars indicate the reflection peaks of Mn^{III} phase. The wavelength of the X-ray is 0.77476 Å.

Figure 6a shows the lattice constants against x of L$_x$MF83 refined by the Rietveld analyses. In the reduction/oxidation of Fe region ($0.49 \leq x \leq 1.32$), the lattice constants decreases with x. The decrease is ascribed to the smaller size of $[Fe^{II}(CN)_6]^{4-}$ as compared with that of $[Fe^{III}(CN)_6]^{3-}$. In the reduction/oxidation of Mn region ($0 \leq x \leq 0.49$), the lattice constants increase with x. The increase is ascribed to the larger ionic radius of high-spin Mn^{II} (0.83 Å) than high-spin Mn^{III} (0.645 Å).

Figure 6b shows the lattice constants against x of L$_x$MF87 refined by the Rietveld analyses. In the reduction/oxidation of Fe region ($0.61 \leq x \leq 1.48$), the lattice constants decreases with x as observed in L$_x$MF83. On the contrary, in the reduction/oxidation of Mn region ($0 \leq x \leq 0.61$), the phase separation occurred and two cubic phase appeared. The lattice constants of larger volume phase is almost same as $x = 0.61$, while the lattice constants of the smaller volume phase is similar as $x = 0$ of L$_x$MF83, i.e., $Mn^{II}_{0.57}Mn^{III}_{0.43}[Fe^{III}(CN)_6]_{0.83}\cdot 3.5H_2O$. Hence, the larger volume phase can be ascribed to the Mn^{II}-rich, or Li^+-rich phase, and the smaller volume phase can be ascribed to the Mn^{III}-rich, or Li^+-poor phase. This phase separation was caused by the volume change for the difference in the ionic radii of Mn^{II} and Mn^{III}.

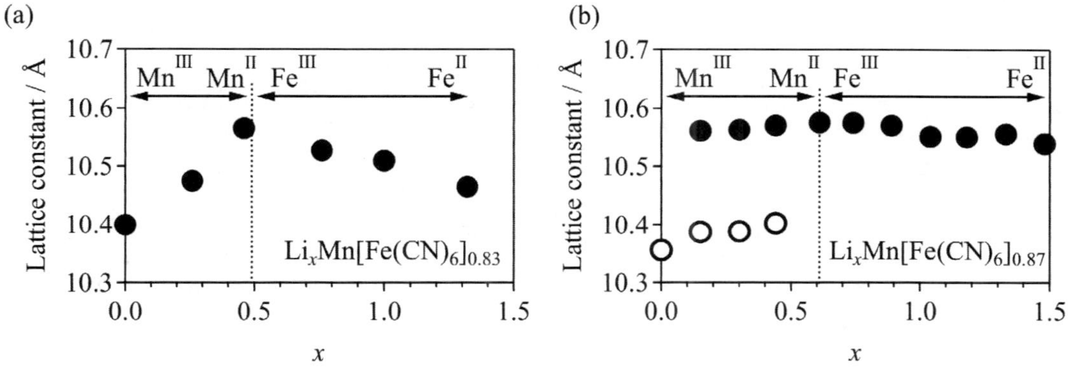

Figure 6. Li ion concentration (x) dependence on the lattice constants of $Li_xMn[Fe(CN)_6]_{0.83}$ (L$_x$MF83) (a), and $Li_xMn[Fe(CN)_6]_{0.87}$ (L$_x$MF87) (b).

4. Discussion

Here, we discuss the crystal structures of Prussian blue analogues with Jahn-Teller active sites.. The crystals structures of $RbMn[Fe(CN)_6]$, $CsCu[Co(CN)_6]$, and $CsCu[Cr(CN)_6]$ are tetragonal due to the cooperative Jahn-Teller instability inherent to Mn^{III} and Cu^{II} in [4-6]. The crystal structure dependences against the amounts of $[Fe(CN)_6]$ sites (y) and Mn^{III} sites are inreported in $Rb_{3y-2}Mn[Fe(CN)_6]_y\cdot zH_2O$, and $Cs_xMn[Fe(CN)_6]\cdot zH_2O$ systems [6-9]. In $Rb_{3y-2}Mn[Fe(CN)_6]_y\cdot zH_2O$, the tetragonal crystal structure ($I\bar{4}m2$) can be observed in $y > 0.95$, and the driving force of the structural distortion is ascribed to the cooperative Jahn-Teller instability of the Mn^{III} sites. On the contrary, the cubic structure ($Fm\bar{3}m$) appeared in $y = 0.91$, because the Jahn-Teller elongation of Mn^{III} cannot occur in only one direction of the lattice due to the low $Mn^{III}/(Mn^{II} + Mn^{III})$ ratio of 0.65 and much $[Fe(CN)_6]$ vacancies of 9%. The same cubic structures were observed in $Cs_xMn[Fe(CN)_6]\cdot zH_2O$ due to the low $Mn^{III}/(Mn^{II} + Mn^{III})$ ratio.

In L$_x$MF87, the maximum of $Mn^{III}/(Mn^{II} + Mn^{III})$ ratio is 0.61 in $x = 0$ and $[Fe(CN)_6]$ vacancy sites of 13% exists. These values are close to that of $y = 0.91$ compound in $Rb_{3y-2}Mn[Fe(CN)_6]_y\cdot zH_2O$. Hence, Mn^{III}-rich phase in L$_x$MF87 is cubic structure. Moreover, L$_x$MF83 which have lower $Mn^{III}/(Mn^{II} + Mn^{III})$ ratio of 0.49 and more $[Fe(CN)_6]$ vacancy sites of 17% is also cubic, and the suppression of the phase separation may be due to much $[Fe(CN)_6]$ vacancy sites.

5. Conclusions

In summary, we investigated the Li ion intercalation properties of $Li_xMn[Fe(CN)_6]_{0.83}\cdot 3.5H_2O$, and $Li_xMn[Fe(CN)_6]_{0.87}\cdot 2.6H_2O$. The XAS spectra clearly indicates the two-electron reduction/oxidation of

Mn and Fe. The crystal structure of $Li_xMn[Fe(CN)_6]_{0.83}\cdot 3.5H_2O$ remains cubic and single phase throughout the Li intercalation/deintercalation process. The crystal structure of $Li_xMn[Fe(CN)_6]_{0.87}\cdot 2.6H_2O$ was also cubic, while the phase separation of Mn^{II}-rich, or Li^+-rich phase and Mn^{III}-rich, or Li^+-poor phase was observed during the Mn^{II}/Mn^{III} reduction/oxidization process. The phase separation observed in $Li_xMn[Fe(CN)_6]_{0.87}\cdot 2.6H_2O$ is ascribed to the difference in the ionic radii of Mn^{II} and Mn^{III}. The cubic structure observed in Mn^{III}-rich, or Li^+-poor phase is ascribed to the Jahn-Teller elongation is randomly oriented by the $[Fe(CN)_6]$ vacancy sites.

Acknowledgements
This work was supported by a Grant-in-Aid (21244052) for Scientific Research from the Ministry of Education, Culture, Sports, Science and Technology. Elementary analysis was performed at Chemical Analysis Division, Research Facility Center for Science and Engineering, University of Tsukuba. The X-ray powder diffraction and X-ray absorption experiments were performed under the approval of the Photon Factory Program Advisory Committee (Proposal No. 2010G502 and 2011G501).

References
[1] Verdaguer M, Bleuzen A, Marvaud V, Vaissermann J, Seuleiman M, Desplanches C, Scuiller A, Train C, Garde R, Gelly G, Lomenech C, Rosenman I, Veillet P, Cartier C and Villain F 1999 *Coord. Chem. Rev.* **190–192** 1023
[2] Ludi A, and Güdel H U 1973 *Struct.Bonding* **14** 1
[3] Itaya K, Ataka T and Toshima S 1982 *J. Am. Chem. Soc.* **104** 4767
[4] Moritomo Y, Igarashi K, Matsuda T and Kim J E 2009 *J. Phys. Soc. Jpn.* **78** 074602
[5] Matsuda T, Kim J E and Moritomo Y 2012 *Dalton Trans.* **41** 7620
[6] Ohkoshi S, Tokoro H, Utsunomiya M, Mizuno M, Abe M and Hashimoto K. 2002 *J. Phys. Chem. B* **106** 2423
[7] Kato K, Moritomo Y, Takata M, Sakata M, Umekawa M, Hamada N, Ohkoshi S, Tokoro H and Hashimoto K 2003 *Phys. Rev. Lett.* **91** 255502
[8] Ohkoshi S, Matsuda T, Tokoro H and Hashimoto K 2005 *Chem. Mater.* **17** 81
[9] Matsuda T, Tokoro H, Hashimoto K and Ohkoshi S 2006 *Dalton Trans.* 5046
[10] Yokoyama T, Tokoro H, Ohkoshi S, Hashimoto K, Okamoto K and Ohta T 2002 *Phys. Rev. B* **66** 184111
[11] Osawa H, Iwazumi T, Tokoro H, Ohkoshi S, Hashimoto K, Shoji H, Hirai E, Nakamura T, Nanao S and Isozumi Y 2003 *Sol. St. Commun.* **125** 237
[12] Ishiji K, Deguchi M, Kawakami K, Nakajima N, Matsuda T, Tokoro H, Ohkoshi S and Iwazumi T 2010 *J. Phys. Soc. Jpn.* **79** 074801
[13] Omura A, Kurihara Y, Matsuda T, Tanida H, Uruga T and Moritomo Y 2011 *J. J. Appl. Phys.* **50** 125802
[14] Okubo M, Asakura D, Mizuno Y, J-D Kim, T Mizokawa, T Kudo and I Honma 2010 *J. Phys. Chem. Lett.* **1** 2063
[15] Matsuda T and Moritomo Y 2011 *Appl. Phys. Exp.* **4** 047101
[16] Matsuda T and Moritomo Y 2012 *J. Nanotech.* **2012** 568147
[17] Ravel B and Newville M 2005 *J. Synch. Rad.* **12** 537
[18] Izumi F and Momma K 2007 *Sol. St. Phenom.* **130** 15

Jahn-Teller effects in transition-metal compounds with small charge-transfer energy

Takashi Mizokawa

The University of Tokyo, Department of Complexity Science and Engineering, 5-1-5 Kashiwanoha, Kashiwa, Chiba 277-0882, Japan

E-mail: mizokawa@k.u-tokyo.ac.jp

Abstract. We have studied Jahn-Teller effects in $Cs_2Au_2Br_6$, $ACu_3Co_4O_{12}$(A=Ca or Y), and $IrTe_2$ in which the ligand p-to-transition-metal d charge-transfer energy is small or negative. The Au^+/Au^{3+} charge disproportionation of $Cs_2Au_2Br_6$ manifests in Au $4f$ photoemission spectra. In $Cs_2Au_2Br_6$ with negative Δ and intermediate U, the charge disproportionation can be described using effective d orbitals constructed from the Au $5d$ and Br $4p$ orbitals and is stabilized by the Jahn-Teller distortion of the Au^{3+} site with low-spin d^8 configuration. In $ACu_3Co_4O_{12}$, Δs for Cu^{3+} and Co^{4+} are negative and Us are very large. The Zhang-Rice picture is valid to describe the electronic state, and the valence change from Cu^{2+}/Co^{4+} to Cu^{3+}/Co^{3+} can be viewed as the O $2p$ hole transfer from Co to Cu or $d^9 + d^6\underline{L} \rightarrow d^9\underline{L} + d^6$. In $IrTe_2$, both Δ and U are small and the Ir $5d$ and Te $5p$ electrons are itinerant to form the multi-band Fermi surfaces. The ideas of band Jahn-Teller transition and Peierls transition are useful to describe the structural instabilities.

1. Introduction

Insulating transition-metal compounds are classified into (i) Mott-Hubbard type insulators where Mott gap E_G is mainly determined by Coulomb interaction U between the transition-metal d electrons and (ii) charge-transfer type insulators where E_G is determined by charge-transfer energy Δ from the ligand p state to the transition-metal d state (Zaanen-Sawatzky-Allen scheme) [1, 2]. In transition-metal compounds with relatively large Δ, orbital orderings can be described by the Jahn-Teller mechanism, in which lattice distortion removes the orbital degeneracy, and by the Kugel-Khomskii mechanism, in which the superexchange interaction $J \sim -2t^2/E_G$ (t: transfer integral, E_G: Mott gap) governs the spin-orbital orderings.

For example, in $LaMnO_3$ with $\Delta \sim 4$ eV, the Mn^{3+} ion takes high-spin d^n ($n = 4$) configuration and the MnO_6 octahedron is accompanied by Jahn-Teller distortion. The A-type antiferromagnetic state of $LaMnO_3$ can be described by the superexchange interaction between the high-spin d^4 states. On the other hand, there are some transition-metal compounds known to have negative Δ. In particular, high valence transition-metal oxides such as $NaCuO_2$ and $CaFeO_3$ tend to have negative Δ [3, 4]. In $CaFeO_3$, Fe valence is formally 4+ and the Fe^{4+} ion is expected to take high-spin d^n ($n = 4$) configuration. However, $CaFeO_3$ with negative Δ takes $d^{n+1}\underline{L}$ ($\underline{L}$ represents a ligand hole) configuration instead of d^n [5]. In such case, the Jahn-Teller instability expected for high-spin d^4 state can be avoided. Actually, $CaFeO_3$ does not exhibit Jahn-Teller distortion of FeO_6 octahedron and shows charge disproportionation with breathing distortion of FeO_6 octahedron [6].

In this article, we would like to discuss possible Jahn-Teller effects in negative-Δ or small-Δ transition-metal compounds including $Cs_2Au_2Br_6$, (Ca,Y)$Cu_3Co_4O_{12}$ and $IrTe_2$. As for $IrTe_2$, the relationship between structural distortions and orbital orderings is analyzed on the basis of itinerant electron models considering band Jahn-Teller effect and Peierls instability. In perovskite-type $Cs_2Au_2Br_6$ and $CaCu_3Co_4O_{12}$, the Jahn-Teller effects tend to stabilize the Au^+/Au^{3+} and Cu^{3+}/Co^{3+} valence states respectively, whereas the smallness of Δ introduces additional charge and/or orbital instabilities.

2. $Cs_2Au_2Br_6$

Perovskite-type $Cs_2Au_2Br_6$ exhibits the charge disproportionation $Au^{2+} + Au^{2+} \rightarrow Au^+ + Au^{3+}$ [7] as illustrated in Fig. 1(a). The Au^+ and Au^{3+} sites take $d^{10}(t_{2g}^6 e_g^4)$ and low-spin $d^8(t_{2g}^6 e_g^2)$ configurations respectively. The $AuBr_6$ octahedron with low-spin d^8 configuration is expected to have Jahn-Teller instability in which the doubly degenerate e_g levels are split by the compression of the $AuBr_6$ octahedron and the stabilized $x^2 - y^2$ orbitals are occupied by the pair of electrons [Fig. 1 (b)].

As shown in Fig. 1(c), the Au $4f$ core level peak of $Cs_2Au_2Br_6$ is split into the Au^+ and Au^{3+} components [8]. The energy difference between the Au^+ and Au^{3+} components is ~ 2 eV, and this situation of $Cs_2Au_2Br_6$ is in sharp contrast to $CaFeO_3$ in which the Fe $2p$ core level is hardly split in spite of the charge disproportionation of Fe^{3+} and Fe^{5+}. In case of $CaFeO_3$, the small (negative) Δ and the large U provide the $d^5\underline{L}$ ground state and, consequently, the charge disproportionation of $Fe^{4+} + Fe^{4+} \rightarrow Fe^{3+} + Fe^{5+}$ corresponds to the configuration change of $d^5\underline{L} + d^5\underline{L} \rightarrow d^5 + d^5\underline{L}^2$. Therefore, the number of Fe $3d$ electrons as well as the energy of Fe $2p$ core level can be more or less same between the Fe^{3+} and Fe^{5+} sites. Namely, when Δ is small and U is large, Zhang-Rice description of $d^{n+1}\underline{L}$ configuration is useful to describe the electronic state.

As for $Cs_2Au_2Br_6$, a recent theoretical study shows that the Au $5d$ and Br $4p$ states are almost degenerate and the charge-transfer energy Δ is negative [10]. If the d-d Coulomb interaction energy U for Au $5d$ is large, the charge disproportionation of $Au^{2+} + Au^{2+} \rightarrow Au^+ + Au^{3+}$ may correspond to the configuration change of $d^{10}\underline{L} + d^{10}\underline{L} \rightarrow d^{10} + d^{10}\underline{L}^2$. In such case, the Jahn-Teller instability at the Au^{3+} site is expected to be suppressed. However, the d-d Coulomb interaction energy U of $Cs_2Au_2Br_6$ is not large enough and the hybridization between the d^n and $d^{n+1}\underline{L}$ states is substantial. In such case, effective d orbitals can be constructed from the Au $5d$ and Br $4p$ orbitals and the effective d orbitals accommodate the valence electrons. Therefore, the Au^{3+} site with low-spin d^8 configuration should have strong Jahn-Teller instability. The charge disproportionation of $Cs_2Au_2Br_6$ is stabilized by the Jahn-Teller effect of the Au^{3+} site. Here it should be noted that the Au^+/Au^{3+} state can be destroyed by moderate light illumination [8]. The light illumination probably disturbs the Jahn-Teller distortion and destabilizes the Au^+/Au^{3+} state.

3. (Ca,Y)$Cu_3Co_4O_{12}$

In A-site ordered perovskite AA'$_3B_4O_{12}$, the three quarters of A sites are occupied by Jahn-Teller active Cu^{2+} or Mn^{3+} ions and the BO_6 octahedra are tilted to stabilize the Cu^{2+} or Mn^{3+} ions with the pseudo-square planar coordination as shown in Fig. 2(a). Recently, $ACu_3Co_4O_{12}$ (A=Ca or Y) has been synthesized using high pressure synthesis and the transport study has revealed that $CaCu_3Co_4O_{12}$ is metallic and $YCu_3Co_4O_{12}$ is insulating [11]. However, if Cu valence is +2, then Co valence is expected to be +4 in $CaCu_3Co_4O_{12}$ and +3.75 in $YCu_3Co_4O_{12}$. Since the Cu and Co ions have integer number of $3d$ electrons with relatively large U, $CaCu_3Co_4O_{12}$ is expected to be a Mott insulator. On the other hand, if Cu is formally +3, then Co valence is +3.25 in $CaCu_3Co_4O_{12}$ and +3 in $YCu_3Co_4O_{12}$, which is consistent with the experimental result. Since Δs of Cu^{3+} oxides and Co^{4+} oxides tend to be negative, ligand

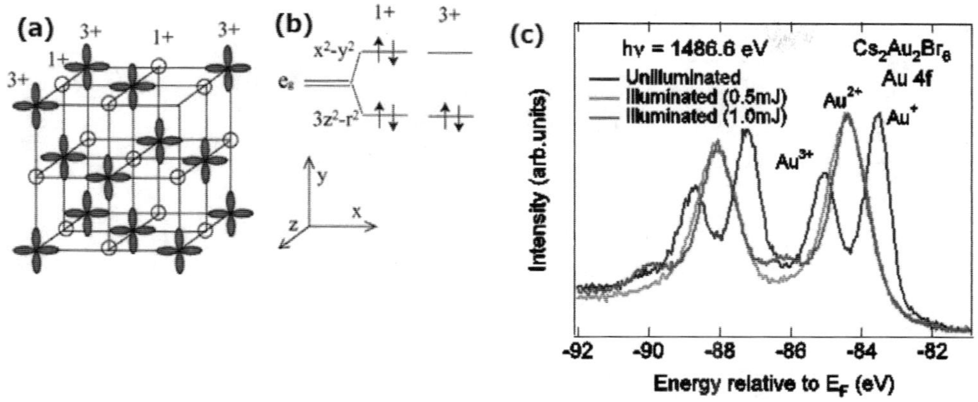

Figure 1. (a) Au^+/Au^{3+} charge disproportionation of $Cs_2Au_2Br_6$. (b) Electronic configurations of the Au^+ and Au^{3+} sites. (c) Au $4f$ photoemission spectrum of $Cs_2Au_2Br_6$.

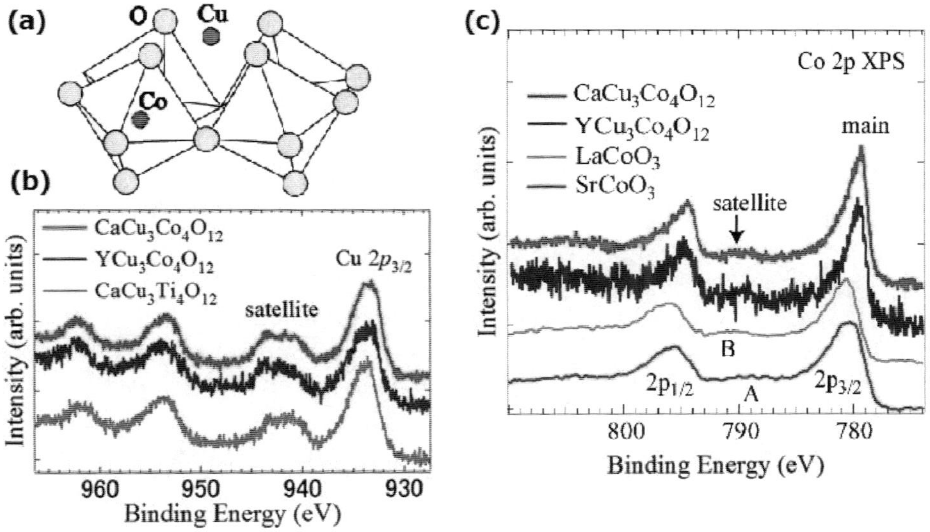

Figure 2. (a) Crystal structure of $ACu_3Co_4O_{12}$. (b) Cu $2p$ photoemission spectra for A = Ca and Y compared with that of $CaCu_3Ti_4O_{12}$. (c) Co $2p$ photoemission spectra for A = Ca and Y compared with those of $LaCoO_3$ and $SrCoO_3$ [13].

(O $2p$) hole should be considered explicitly in order to understand the electronic structure of $ACu_3Co_4O_{12}$.

Figure 2(b) shows the Cu $2p$ photoemission spectra of $CaCu_3Co_4O_{12}$ and $YCu_3Co_4O_{12}$ which are compared with that of $CaCu_3Ti_4O_{12}$. The Cu $2p$ binding energy of $CaCu_3Co_4O_{12}$ and $YCu_3Co_4O_{12}$ is similar to that of Cu^{2+} (d^9) state in $CaCu_3^{2+}Ti_4O_{12}$. The Cu $2p$ binding energy re ects the number of Cu $3d$ electrons. Therefore, if Cu is formally 3+, the Cu ion should take d^9L configuration instead of d^8. In Fig. 2(c), the Co $2p$ photoemission spectra of $CaCu_3Co_4O_{12}$ and $YCu_3Co_4O_{12}$ are compared with those of $LaCo^{3+}O_3$ and $SrCo^{4+}O_3$ [13]. The position and the shape of the charge-transfer satellite of $CaCu_3Co_4O_{12}$ and $YCu_3Co_4O_{12}$ are very similar

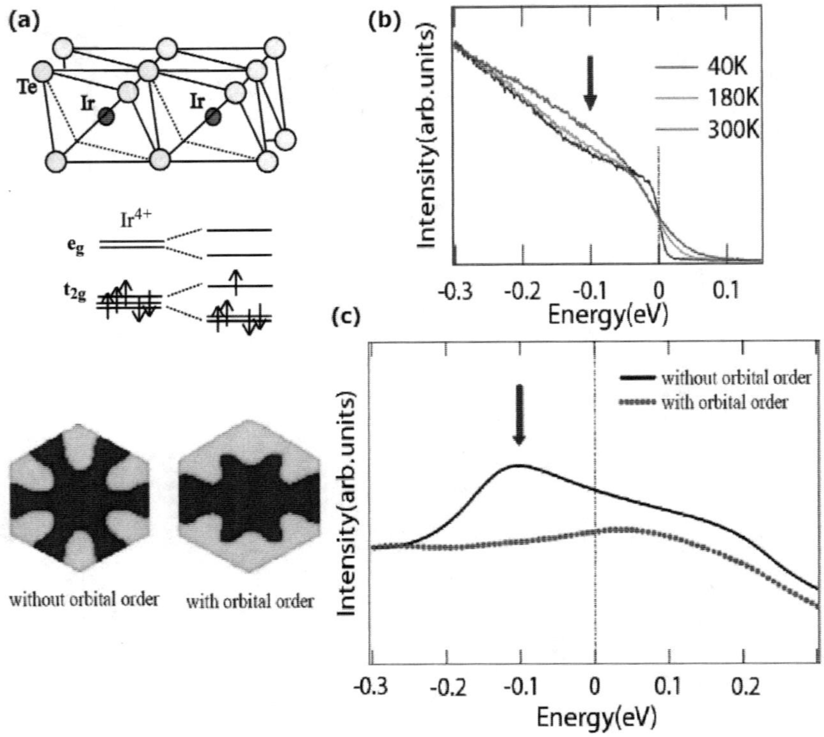

Figure 3. (a) Crystal structure of $IrTe_2$, electronic configuration and expected Jahn-Teller splitting for Ir^{4+}, and expected Fermi surface change due to band Jahn-Teller effect. (b) Valence-band photoemission spectra above and below the structural transition. (c) Calculated density of states with and without orbital order due to band Jahn-Teller effect.

to $LaCoO_3$ indicating that Co valence is close to +3 for A=Ca and Y. The combination of Cu 2*p* and Co 2*p* spectra shows that the Cu valence is formally +3 and the Co valence is +3 in $YCu_3Co_4O_{12}$ consistent with the fact that $YCu_3Co_4O_{12}$ is insulating and $CaCu_3Co_4O_{12}$ is metallic. The valence change from Cu^{2+}/Co^{4+} to Cu^{3+}/Co^{3+} can be viewed as the O 2*p* hole transfer from Co to Cu or $d^9 + d^6L \rightarrow d^9L + d^6$ [12]. Since Δ is small and U is large in the late 3*d* transition-metal oxides, the Zhang-Rice description or ligand hole description becomes valid.

4. $IrTe2_2$

Both Δ and U tend to decrease in going from oxides to chalcogenides [4]. Some transition-metal chalcogenides such as $CuIr_2S_4$ exhibit exotic phase transitions originating from the orbital degeneracy of transition-metal *d* orbitals [14, 15]. Among them, $IrTe_2$ shows a structural phase transition at $\sim$ 270 K which is probably due to Ir 5*d* orbital order [16]. Recently, it has been found that Pt doping suppresses the static orbital order, and the superconductivity is induced around the quantum critical point for the orbital or bond order [17]. Valence-band photoemission spectrum of $IrTe_2$ is shown in Fig. 2(a). Across the orbital or bond order temperature, the spectral weight at -0.1 eV is suppressed instead of that at the Fermi level. The spectral weight suppression is reproduced by the calculation with $\Delta \sim 1$ eV which is relatively small among various transition-metal chalcogenides such as NiS and FeS.

The Ir $4f$ photoemission spectra reported by Ootsuki *et al.* indicates that the low temperature phase of $IrTe_2$ is accompanied by weak modulation of Ir $5d$ electron density [18]. The charge modulation or charge density wave can be induced by Fermi surface nesting due to the orbital order (orbitally-induced Peierls effect) [19]. Actually, the calculated Fermi surfaces show this tendency as shown in Fig. 3(a). Without the orbital order, the Fermi surfaces of the Ir $5d$ bands are made up from the Ir $5d$ yz, zx, and xy orbitals and have the six-fold symmetry as expected from the trigonal structure. Under the compression along b-axis and the orbital order, the Fermi surface is calculated to be quasi-one-dimensional with Ir $5d$ yz character. Such quasi-one-dimensional Fermi surface is expected to have instability to charge or spin density wave.

5. Concluding remarks

We have discussed Jahn-Teller effects in $Cs_2Au_2Br_6$, $ACu_3Co_4O_{12}$(A=Ca or Y), and $IrTe_2$ in which the ligand p-to-transition-metal d charge-transfer energy is small or negative. In IrTe2, both Δ and U are small and the Ir $5d$ and Te $5p$ electrons are itinerant to form the multi-band Fermi surfaces. In such case, the ideas of band Jahn-Teller transition and Peierls transtion are valid to describe the structural instabilities. In $Cs_2Au_2Br_6$, Δ is negative and U is intermediate. In such case, the Au^+/Au^{3+} charge disproportionation can be described using effective "d orbitals" constructed from the Au $5d$ and Br $4p$ orbitals. The Au^{3+} site with low-spin d^8 configuration undergoes the Jahn-Teller distortion. As for $ACu_3Co_4O_{12}$, Δs for Cu^{3+} and Co^{4+} are negative and the Zhang-Rice picture is valid to describe the electronic state. The valence change from Cu^{2+}/Co^{4+} to Cu^{3+}/Co^{3+} can be viewed as the O $2p$ hole transfer from Co to Cu or $d^9 + d^6L \rightarrow d^9L + d^6$.

Acknowledgments

We would like to thank valuable collaborations and discussions with D. Ootsuki, J.-Y. Son, K. Takubo, Y. Morita, Y. Wakisaka, T. Sudayama, M. Arita, H. Anzai, H. Namatame, M. Taniguchi, I. Yamada, M. Azuma, Y. Shimakawa, M. Takano, S. Pyon, K. Kudo. M. Nohara, K. Ikeda, N. Kojima, A. Fujimori, and D. I. Khomskii.

References

[1] Fujimori A and Minami F 1984 Phys. Rev. B **30** 957
[2] Zaanen J, Sawatzky G A, Allen J W 1985 Phys. Rev. Lett. **55** 418
[3] Mizokawa T, Namatame H, Fujimori A, Akeyama K, Kondoh H, Kuroda H, and Kosugi N 1991 Phys. Rev. Lett. **67** 1683
[4] Bocquet A E, Fujimori A, Mizokawa T, Saitoh T, Namatame H, Suga S, Kimizuka N, Takeda Y, and Takano M 1992 Phys. Rev. B **45** 1561
[5] Takano M, Nakanishi N, Takeda Y, Naka S and Takada T 1977 Mater. Res. Bull. **12** 923
[6] Mizokawa T, Khomskii D I, Sawatzky G A 2000 Phys. Rev. B **61** 11263
[7] Liu X J, Moritomo Y, Ichida M, Nakamura A, and Kojima N 2000 Phys. Rev. B **61** 20
[8] Son J Y, Mizokawa T, Quilty J W, Takubo K, Ikeda K, and Kojima N 2005 Phys. Rev. B **72** 235105
[9] Matsuno J, Mizokawa T, Fujimori A, Takeda Y, Kawasaki S, and Takano M 2002 Phys. Rev. B **66** 193013
[10] Ushakov A V, Streltsov S V, and Khomskii D I 2011 J. Phys.: Condens. Matter **23** 445601
[11] Yamada I, Ishiwata S, Terasaki I, Azuma M, Shimakawa Y, and Takano M 2010 Chem. Mater. **22** 5328
[12] Mizokawa T, Morita Y, Sudayama T, Takubo K, Yamada I, Azuma M, Takano M, and Shimakawa Y 2009 Phys. Rev. B **80** 125105
[13] Saitoh T, Mizokawa T, Fujimori A, Abbate M, Takeda Y, and Takano M 1997 Phys. Rev. B **56** 1290
[14] Nagata S, Hagino T, Seki Y, and Bitoh T 1994 Physica B **194 – 196**, 1077
[15] Radaelli P G, Horibe Y, Gutmann M J, Ishibashi H, Chen C H, Ibberson R M, Koyama Y, Hor Y S, Kirykhin V, and Cheong S W 2002 Nature **416** 155
[16] Matsumoto N, Taniguchi K, Endoh R, Takano H, and Nagata S 1999 J. Low Temp. Phys. **117** 1129
[17] Pyon S, Kudo K, and Nohara M 2012 J. Phys. Soc. Jpn. **81** 053701

Journal of Physics: Conference Series **428** (2013) 012020 doi:10.1088/1742-6596/428/1/012020

[18] Ootsuki D, Wakisaka Y, Pyon S, Kudo K, Nohara M, Arita M, Anzai H, Namatame H, Taniguchi M, Saini N L, Mizokawa T 2012 Phys. Rev. B **86** 014519
[19] Khomskii D I and Mizokawa T 2005 Phys. Rev. L ett. **94** 156402

Journal of Physics: Conference Series **428** (2013) 012021 doi:10.1088/1742-6596/428/1/012021

Electrochemical lithium intercalation into $Bi_2Sr_2CaCu_2O_{8+\delta}$

Takahiro Shimono[1], Wataru Kobayashi[1,2], Hiroaki Nitani[3], Reiji Kumai[4], Yutaka Moritomo[1,2]

[1] Graduate School of Pure and Applied Sciences, University of Tsukuba, Ibaraki 305-8571, Japan
[2] Tsukuba Research Center for Interdisciplinary Materials Science (TIMS), University of Tsukuba, Ibaraki 305-8571, Japan
[3] Institute of Materials Structure Science, High Energy Accelerator Research Organization, Ibaraki 305-0801, Japan
[4] Condensed Matter Research Center and Photon Factory, Institute of Materials Structure Science, High Energy Accelerator Research Organization, Tsukuba 305-0801, Japan

E-mail: `kobayashi.wataru.gf@u.tsukuba.ac.jp`

Abstract. We have prepared Li-intercalated $Li_xBi_2Sr_2CaCu_2O_{8+\delta}$ (x =0-2.0) samples by using electrochemical method, and performed synchrotron x-ray di raction, Cu K-edge x-ray absorption ne structure (XAFS), and magnetic susceptibility measurements. With increasing x, a- and c-lattice parameters monotonically increase, which shows lithium intercalation into $Bi_2Sr_2CaCu_2O_{8+\delta}$. Accompanied by the lithium insertion, the valence of Cu ion changes from Cu^{2+}/Cu^{3+} to Cu^{1+}/Cu^{2+} to realize charge neutrality. This change of the valence was detected by Cu K-edge XAFS measurement. A clear increase of spectral weight that corresponds to 1s→ 4p ($3d^{10}\underline{L}$) was observed at around 8982 eV with x. The superconducting (SC) transition temperature T_C signi cantly changes from 74 K for $x = 0$ to 90 K for $x = 0.8$, which is attributed to modi ed density of states by the decrease of hole concentration. A volume fraction of the superconducting phase was 1-2 % for x 0.6 implying phase separation where Li-rich non SC phase and Li-poor SC phase coexist. Such a phase separation is universally seen in electrode active materials.

1. Introduction

Transition-metal oxides have attracted much interest because of fascinating phenomena such as high-temperature superconductivity in cuprates and oxipnictides [1], giant magnetoresistance in perovskite Mn oxide [2], large thermoelectric properties in rock-salt-type layered cobalt oxide [3], and multiferroic e ect in hexagonal Mn oxide [4]. In these materials, carrier concentration is one of the most important parameter for controlling these properties, which is mainly tuned by chemical doping and/or oxygen nonstoichiometry.

Recently, another doping, namely electrostatic doping has been demonstrated using eld-e ect transistors (FET) that uses a liquid electrolyte as a gate insulator. When electric eld is applied for the FET, a Helmholtz electric double layer occurs at the surface. The accumulated charge carrier can exhibit superconductivity even in insulating material $KTaO_3$ [5]. In addition to this study, this FET structure is also used to control superconducting transition temperature T_C of $La_{1\ x}Sr_xCuO_4$ system [6]. In spite of the clear control of the superconductivity, there are two disadvantages shown below; (1) atomically flat surface is essential; (2) in principle, bulk properties such as magnetization, and thermodynamic properties can not be detected.

$Bi_2Sr_2CaCu_2O_{8+\delta}$(Bi2212) is one of the most interesting high T_C superconductors with critical temperatures in the range of 70-100 K. Bi2212 has a layered structure consisting of double CuO_2 planes sandwiched by BiO-SrO rocksalt-type layers [7], in which carrier concentration is modi ed by substitutions of Pb ions into Bi sites or oxygen nonstoichiometry [8]. In addition to the chemical doping, iodine and lithium ions have been found to be intercalated owing to the layered structure weakly connected by Van der Waals interaction [9, 10, 11]. Fujiwara *et al.* have performed lithium intercalation into $Bi_2Sr_2CaCu_2O_{8+\delta}$ by using electrochemical method and found that T_C changes with increasing lithium content [10] implying that lithium intercalation changes the electronic structure of Bi2212. Thus, bulk properties of superconductors can be tuned by the electrochemical intercalation, which will be another way to control superconductivity. We have prepared Li-intercalated $Li_xBi_2Sr_2CaCu_2O_{8+\delta}$ (x =0-2.0) samples by using electrochemical method, and performed synchrotron x-ray di raction, Cu K-edge x-ray absorption ne structure (XAFS), and magnetic susceptibility measurements. We have succeeded in detecting the modi cation of the valence of Cu ion using XAFS, which is consistent with structural and magnetic properties. The superconducting (SC) transition temperature T_C signi cantly changes from 74 K for x = 0 to 90 K for x = 0.8, which is attributed to modi ed density of states yielded by the lithium intercalation.

2. Experiments

Polycrystalline sample of $Bi_2Sr_2CaCu_2O_{8+\delta}$ was prepared by solid state reaction. Stoichiometric amounts of Bi_2O_3, $SrCO_3$, $CaCO_3$ and CuO were mixed, and the mixture was sintered at 780 °C for 12 h. Then, the product was nely ground, and sintered at 820 °C for 24 h. The nal product was again nely ground, and mixed with a binder polymer [polyvinylidene difluoride (PVDF)] to be Bi2212:PVDF=95:5 (weight ratio). Then, the mixture as an active material for the positive electrode was coated using a spin coater on an indium tin oxide (ITO) transparent electrode where the surface was beforehand cleansed by UV. Finally, the positive electrode was dried at 120 °C for 2h in vacuum.

Electrochemical reaction was performed in a galvanic cell with the electrolyte made of ethylenecarbonate(EC)/diethylcarbonate(DEC) solution containing 1 mol/l $LiClO_4$. As a negative electrode, stainless steel with lithium metal was used. The theoretical capacity of $Li_xBi_2Sr_2CaCu_2O_{8+\delta}$ was calculated to be $33x$ mAh/g so that the charge/discharge current density of $6.6x$ mA/g was applied for 5h to obtain lithium intercalated $Li_xBi_2Sr_2CaCu_2O_{8+\delta}$ ($x = 0.2, 0.4, 0.6, 0.8, 1.0$, and 2.0). The cut-o voltage was from 1.5 to 3.7 V. Lithium content x was evaluated using the current density.

Synchrotron powder x-ray di raction measurements with wavelength of 0.7749 Å were carried out at the BL-8A at the Photon Factory, KEK, and the structural re nements were performed by Rietveld analysis (RIETAN-FP). X-ray absorption ne structure(XAFS) measurement was performed using the BL-9C at the photon Factory, KEK. The CuK-edge was recorded by a Lytle detector in fluorescent yield mode with a Si(111) double-crystal monochromator and was normalized by the intensity at 9300 eV. The magnetization was measured from 5 to 120 K by a commercial superconducting quantum interference device (SQUID, Quantum Design MPMS). The lithium intercalated sample was treated in a glove box lled with Ar gas, and was sealed by adhesive tape for these measurements. We would like to note that the same sample was used for the three di erent measurements.

3. Results and Discussion

Figure 1(a) shows the synchrotron powder x-ray di raction pattern of as-grown Bi2212 sample and Rietveld tting (green line). Almost all the peaks were well tted by tetragonal symmetry

Journal of Physics: Conference Series **428** (2013) 012021 doi:10.1088/1742-6596/428/1/012021

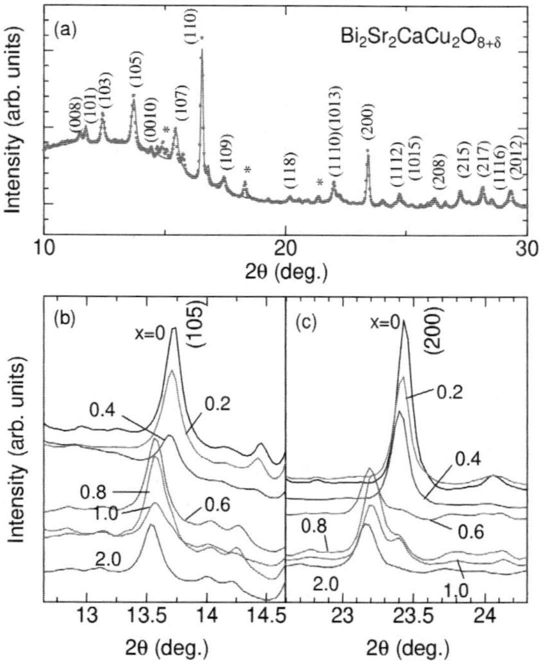

Figure 1. (Color online) (a) Synchrotron powder x-ray di raction pattern of as-grown Bi2212 sample, the magni ed di raction patterns in the range of (b) 12.6-14.7° and (c) 22.6-26.3° for $x = 0$ 2.0.

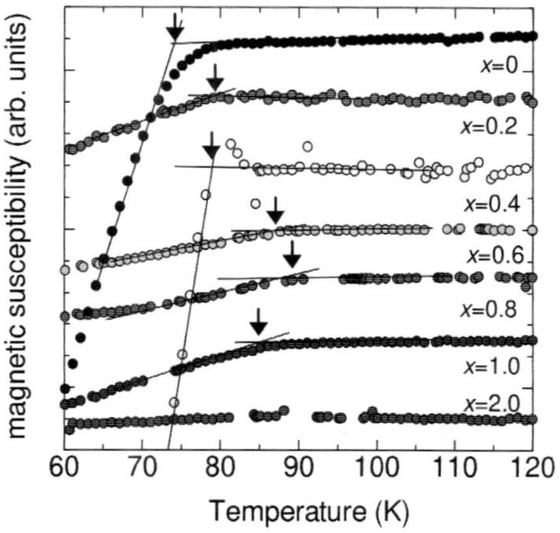

Figure 2. (Color online) Temperature dependence of magnetization for $Li_xBi_2Sr_2CaCu_2O_{8+\delta}$ ($x = 0$ 2.0).

(I4/mmm) with lattice parameters of a =3.8134(2) Å and c = 30.817(6) Å, respectively. A tiny amount of CuO and $Bi_{0.5}(Sr, Ca)_{0.5}O_z$ was observed as impurity phases [12]. A broad peak-like background below 2 =18° is attributed to Kapton tape which wrapped the sample to prevent its exposure to the air. Figures 1(b) and (c) show the magni ed x-ray di raction patterns in the range of 12.6-14.7° and 22.6-26.3°. With increasing x, both the peaks clearly shift to the lower angles implying that the lattice parameter increases. This can be attributed to lithium intercalation into Bi2212 whose layers are weakly connected by Van der Waals interaction.

Figure 2 shows the temperature dependence of the magnetization for $Li_xBi_2Sr_2CaCu_2O_{8+\delta}$ ($x = 0$ 2.0) in the temperature range of 60-120 K. With increasing x, the superconducting transition temperature T_C (onset) increases, shows the maximum of 90 K at $x = 1$, and nally vanishes at $x = 2$. This tendency is qualitatively explained by hole concentration per Cu site. According to P. H. Duvigneaud *et al.*, c-axis lattice parameter has a correlation with both the oxygen content and T_C [13]. Since the c-axis parameter of the as-grown sample is $c = 30.817(6)$ Å, the oxygen content is evaluated to be 0.17 implying that the present sample is in the overdoped region. With increasing x, the hole concentration would decrease due to the charge neutrality. Thus, the lithium intercalated sample experiences the under-doped region through the optimum-doped region with the maximum T_C of 90 K.

Figure 3 shows Cu K-edge spectra for $x = 0$ 2.0 samples. The inset shows the raw spectral data in the range between 8960 and 9020 eV. Characteristic shoulder and peak structures are seen at around 8982 and 8993 eV, respectively, which are consistent with previous works [11, 14, 15]. (we indicate these features as A and B, respectively.) According to the polarized Cu K-edge x-ray-absorption-near-edge structure for the Bi2212 single crystal performed by Bianconi *et al.*

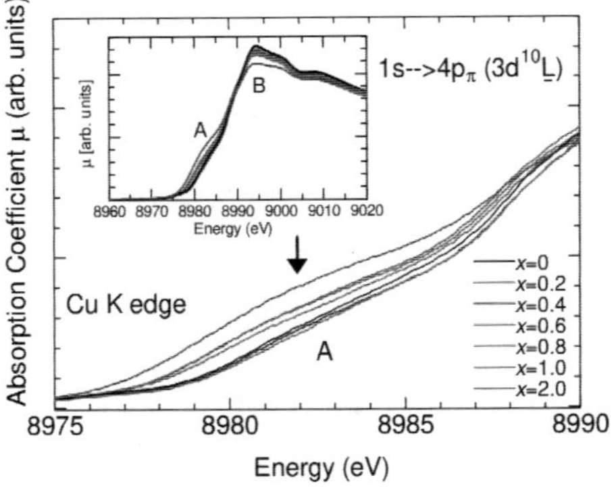

Figure 3. (Color online) Cu K-edge spectra for $x = 0$ 2.0 samples. The inset shows the raw spectral data in the range between 8960 and 9020 eV.

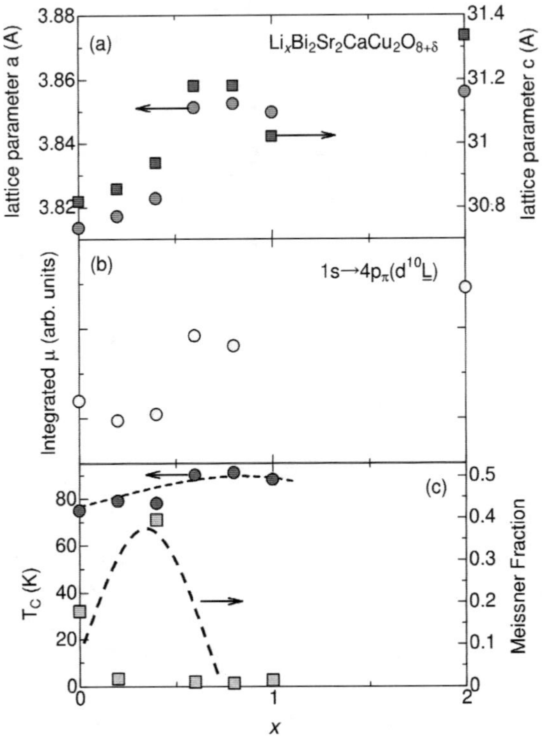

Figure 4. (Color online) (a) a- and c-axis lattice parameters, (b) integrated spectral weight at around 8982 eV, (c) T_c and Meissner fraction as a function of x.

[14], the shoulder A is assigned to the 1s to out-of-plane 4p state (3d^{10}L), while the peak B is assigned to the 1s to in-plane 4p state (3d^{10}L), where underline and L denote the hole state and ligand, respectively. Combined with the calculation using the mu n-tin potential with di erent cluster size, these features are well tted and interpreted [14] as follows; the shoulder A feature is determined by an unoccupied high-energy conduction band, where the nal-state electron is delocalized over a large cluster involving also the positions of the Sr and Ca ions. On the other hand, the peak B feature is given by the multiple scattering of the photoelectron by O and Cu ions in the four neighbor Cu sites in the CuO_2 plane. Thus, looking at the spectral data shown in Fig. 3, the spectral weight of the shoulder A at 8982 eV increases and that of the peak B decreases with increasing x, which implies decrease of the multiple scattering of photoelectron in CuO_2 plane and modi cation of positions of the Sr and Ca ions by lithium intercalation. The decrease of the multiple scattering may be due to modi ed CuO_2 plane by the lithium intercalation and/or decrease of ligand hole concentration. Note that Cu_2O with monovalent Cu ion has a peak at 8980 eV that is close to 8982 eV of the peak A [11]. Thus, the increased spectral weight of the shoulder A may also be interpreted by increased proportion of Cu^+. Furthermore, the peak B position for $x = 1$ shifted to lower energies by 0.5 eV compared with that for $x = 0$, which shows the valence of Cu ion decreases. A similar behavior is observed in insulating $Bi_2Sr_2YCu_2O_{8+\delta}$ [14].

Figure 4 summarizes (a)lattice parameters, (b)integrated spectral weight at 8982 eV, (c)T_C and volume fraction of superconductivity as a function of x. As shown in Fig. 4(a), a- and c-axis lattice parameters increase with increasing x, exhibit jumps at around $x = 0.6$, and again slightly increase up to $x = 2$. a- and c-axis lattice parameters for $x = 2$ were larger by 1.5 and 2 % than those for $x = 0$, respectively. As shown in Figs. 1(b) and (c), small shoulder-like peaks are observed at around 13.7 and 23.4°, respectively, which indicates phase separation. Such a phase separation is generally seen in electrode materials where several phases can exist in a narrow composition range. The phase separation is related to the discontinuous jump in the lattice parameters. As shown in Fig. 4(b), x dependence of the integrated spectral weight at 8982 eV is similar to that of c-axis lattice parameter, which indicates a strong correlation between c-axis lattice parameter and the shoulder A feature. This is consistent with the discussion stated in the previous paragraph. With increasing x, T_C increases and shows the maximum of 90 K at $x = 1$, and then nally vanishes at $x = 2$. On the other hand, volume fraction of superconductivity of the intercalated sample was low except for the $x = 0.4$ sample, which again indicates phase separation where lithium-rich and -poor phases coexist. These results clearly shows that the electrochemical intercalation will give an original way to control superconductivity.

4. Summary

We have investigated the electrochemical lithium intercalation into $Bi_2Sr_2CaCu_2O_{8+\delta}$, and found that lattice parameters, integrated spectral weight of x-ray absorption ne structure, and magnetization systematically change with the intercalated lithium content x. The increased lattice parameters and the enhanced spectral weight at 8982 eV show indirect evidences of the lithium intercalation into $Bi_2Sr_2CaCu_2O_{8+\delta}$. We have also succeeded to control the superconducting transition temperature T_c and the volume fraction of superconductivity by introducing lithium ions. T_C signi cantly changes from 74 K for $x = 0$ to 90 K for $x = 0.8$, which can be attributed to the decrease of the hole concentration/Cu ion. A volume fraction of the superconducting phase was 1-2 % for x 0.6 implying phase separation where Li-rich non SC phase and Li-poor SC phase coexist.

Acknowledgments

This work was supported by a Grant-in-Aid for Scienti c Research (No.23684022) from the Ministry of Education, Culture, Sports, Science and Technology, Japan. The X-ray powder di raction and X-ray absorption experiments were performed under the approval of the Photon Factory Program Advisory Committee (Proposal Nos. 2010G502 and 2011G501).

[1] Kamihara Y, Watanabe T, Hirano M, and Hosono H 2008 *J. Am. Chem. Soc.* **130** 3296
[2] Imada M, Fujimori A, and Tokura Y 1998 *Rev. Mod. Phys.* **70** 1039
[3] Terasaki I, Sasago Y, and Uchinokura K 1997 *Phys. Rev.* B **56** R12685
[4] Kimura T, Goto T, Shintani H, Ishizaka K, Arima T and Tokura Y 2003 *Nature* **426** 55
[5] Ueno M, Nakamura S, Shimotani H, Yuan HT, Kimura N, Nojima T, Aoki H, Iwasa Y, and Kawasaki M 2011 *Nature Nanotechnology* **6** 408
[6] Bollinger AT, Dubuis G, Yoon J, Pavuna D, Misewich J, and Bozovic I 2011 *Nature* **472** 458
[7] Tarascon JM, Le Page Y, Barboux P, Bagley BG, Greene LH, McKinnon WR, Hull GW, Giroud M, and Hwang DM 1988 *Phys. Rev.* B **37** 9382
[8] Munakata F, Kawano T, Yamauchi H, and Inoue Y 1992 *Physica* C **190** 471
[9] Fleischer NA, Manassen J, Coppens P, Lee P, Gao Y, and Greenbaum SG 1992 *Physica* C **190** 367
[10] Fujiwara M, Nakanishi M, Kusano Y, Fujii T, Takada J, Takeda Y, and Ikeda Y 1997 *Physica* C **279** 219
[11] Liang G, Sahiner A, Croft M, Xu W, Xiang X.-D, Badresingh D, Li W, Chen J, Peng J, Zettl A and Lu F 1993 *Phys. Rev.* B **47** 1029
[12] Fujiwara M, Nagae M, Kusano Y, Fujii T, and Takada J 1997 *Physica* C **274** 317
[13] Duvigneaud P H, De Boeck C, and Guo Y F 1998 *Supercond. Sci. Technol.* **11** 116
[14] Bianconi A, Li C, Campanella F, Della Longa S, Pettiti I, Pompa M, Turtu S, and Udron D 1991 *Phys. Rev.* B **44** 4560
[15] Saini NL, Lanzara A, Bianconi A, and Oyanagi H 1998 *Phys. Rev.* B **58** 11768

Ligand field splittings in core level transitions for transition metal (TM) oxides: Tanabe-Sugano diagrams and (TM) dangling bonds in vacated O-atom defects

Gerry Lucovsky, Kun Wu, Brian Pappas, and Jerry Whitten
North Carolina State University, Raleigh NC, USA

e-mail: lucovsky@ncsu.edu

Abstract: Defect states in the forbidden band-gap below the conduction band edge are active as electron traps in nano-grain high-) transition metal (TM) oxides with thickness >0.3 nm, e.g., ZrO_2 and HfO_2. These oxides have received considerable attention as gate-dielectrics in complementary metal oxide semiconductor (CMOS) devices, and more recently are emerging as candidates for charge storage and memory devices. To provide a theoretical basis for device functionality, ab-initio many-electron theory is combined with X-ray absorption spectroscopy (XAS) to study O K edge and TM core level transitions. These studies identify ligand field splittings (ΔLF) for defect state features,. When compared with those obtained from O-atom and TM-atom core spectroscopic transitions, this provides direct information about defect state sun-nm bonding arrangements. comparisons are made for (i) elemental TiO_2 and Ti_2O_3 with different formal ionic charges, Ti^{4+} and Ti^{3+}, and for (ii) Magneli Phase alloys, Ti_nO_{2n-1}, n is an integer $9 \geq n > 3$, and $(TiO_2)_x(HfO_2)_{1-x}$ alloys. The alloys display multi-valent behavior from (i) different ionic-charge states, (ii} local bond-strain, and (iii) metallic hopping transport. The intrinsic bonding defects in TM oxides are identified as pairs of singly occupied dangling bonds. For 6-fold coordinated Ti-oxides defect excited states in 2nd derivative O K pre-edge spectra are essentially the same as single Ti-atom d^2 transitions in Tanabe-Sugano (T-S) diagrams. O-vacated site defects in 8-fold coordinated ZrO_2 and HfO_2 are described by d^8 T-S diagrams. T-S defect state ordering and splittings are functions of the coordination and symmetry of vacated site bordering TM atoms. ΔLF values from the analysis of T-S diagrams indicate medium range order (MRO) extending to 3rd and 4th nearest-neighbor (NN) TM-atoms. Values are different for 6-fold Ti, and 8-fold ZrO_2 and HfO_2, and scale inversely with differences in respective formal ionic radii. O-vacated site bonding defects in TM nano-grain oxides are qualitatively similar to vacant-site defects in non-crystalline SiO_2 and GeO_2 for ulta-thin films, < 0.2 nm thick, and yield similar performance in MOSCAPs on Ge substrates heralding applications in aggressively-scale CMOS devices.

1. Introduction and Chapter-Overview

There is considerable interest in indentifying the local bonding of intrinsic chemical defects in ultra-thin nano-grain (ng-) high-κ TM oxides, ng-HfO_2 and ng-ZrO_2 [1-5]. These have been *targeted* as gate dielectrics for the ultimate limits of integration of Si devices. One proposal is for Ge channel regions grown in wells on a Si wafer where higher hole mobilities reduce the *foot-print* CMOS unit cells. The Robertson and Shlugger groups have employed DFT methods to study high-k vacancy defects [1,2]. These results are applicable only to singlet TM-atom defect states with a maximum occupancy of two [6]. In contrast, the Lucovsky group, has proposed a qualitatively different approach initially designated as an *effective d^2-model* [3-5]. The alternative model assigns two-electron wave functions to TM-atom excited X-ray excited states. This includes singlet and triplet spin excited states including ground-state holes. This approach takes specific account of the hole generated in an X-ray, visible or vacuum ultra-violet excitation process. As such, the d^2-model is based on a CTM formalism using symmetry adapted combinations (SALC) of atomic states) [7,8,9,10]. This approach has been applied to nc-SiO_2 and ng-HfO_2, as examples (nc- and ng- not be used from this point forwar). This chapter discusses in detail the use X-ray absorption spectroscopy (XAS), applied first for the assignment of pre-existing, process-induced defect for SiO_2 and GeO_2 and used here as for the TM-atom studies as well [10]

A combination of XAS O pre-K edge spectra and ab-initio theory, coupled with previously published electron spin resonance (ESR) studies has established that the pre-existing defects of Galeener in SiO_2 [10], are vacated O-atom Si-O-Si bonding groups that *have never contained* an O-atom [10,11]. The defects are a pair of nearest neighbor (NN) Si- or Ge-atom neutral dangling bonds projecting in the vacated site with sufficient radial wave-function overlap to satisfy conditions for Fermion wave functions, and hence Pauli exclusion to apply [7]. The appropriated CTMs are associated with two-electron Fermion wave-functions with 3-vector spatial, and 4-vector spin components [7-9,13-15]. The NN dangling bonds are then functionally equivalent atomic states of TM occupied states. For TM defect excite states the d^2 diagram is used for 2-electron ground-states, and the d^8 diagram for 2-hole ground states. These are (i) d^2 diagrams for Ti oxides with octahedral or distorted octaheral primitive unit cells, and (ii) d8 diagrams for Zr and Hf oxides with 2-hole ground states, d^8, in tetrahedral, cubic, tetragonal or monoclinic arrangements in ZrO_2 and HfO_2.

Tunabe-Sugano (T-S) diagrams are used to characterize the dangling-bond pair excited states singlet and triplet states, in particular their ordering in energy and two-elecron state symmetries [14]. This chapter establishes that ligand field splittings of the defects, ΔLF, are determined by the energy separation between eg and t2g excited states are local properties of the dangling bond site, including up to 3rd and 4th NNs. The values of ΔLF are readily extracted from XAS O pre-K edge defect spectra, are the same to within the uncertainty of these measurements, ~±0.2 V, as the values of ΔLF, respectively obtained TM core level atomic spectra. For Ti-oxides, elemental oxides and composite or complex mixed oxides the ΔLF values from Ti L2,3 spectra that are approximately the same with small differenes associated for different formal ionic charges, +4 in TiO_2, and +3 in Ti_2O_3. Local bond-strain induced multi-valency accounts in for most of the multiplity of Ti L2,3 spectral features in Magneli-phase and TiO_2-HfO^2 composie oxides. There is also additional multi-valency explained by a hopping assisted transport mechanism for metallic conductivity present in Ti_2O_3, and the Ti-oxide based alloys.

Background studies in Ref. 11 for SiO_2 and GeO_2 indicate that there is intrinsic medium range order (MRO) at the vacated O-atom defect sites. The MRO extends to 3rd NN Si-O distances, and 4th NN O-O distances as well. One important manifestation of MRO is that the basis set of vectors that apply are comprised of SALC directed along site-determined cartesian-coordinate axes [16]. This representation includes the local site symmetry of the Si or Ge atom, and their respective back-bonded O-atom neighbors. The MRO includes the singly occupied Si or Ge dangling bonds projected into the vacated O-atom site. The vacated O-atom must be *counted* defining a *1st nearest neighbor* to each of the 2dangling bonds. This means the value of ΔLF obtained from an analysis of defect state symmetries and degeneracies includes the energy difference between a pair of eg and t2g singlet states locallized at the defect site.. This is approach is exact for T-S diagrams that apply to a pair of dangling bonds with some degree of radial wave function overlap. A test case for this model is demonstrated in Section 4 where it is shown that defect state symmetries and ordering are the same for TiO_2 with a formal valence of +4 for Ti, and in Ti_2O_3 where the formal valence for Ti is +3. The values of ΔLF differer by ~12% due differences in formal ionic radii, ~0.40 nm for the +4 valence and 0.52 nm for the +3 valences, When these are taken into accoount the results are by the experimentally-determined uncertainty, and the resolution of the X-ray monochromater, at most of order ±0.15 eV.

2. Experimental methods: deposition and characterization of electronic structure of TM oxide

2.1. Thin film deposition. Nano-grain thin films of TiO_2, and Ti_nO_{2n-1}alloys, ZrO_2 and HfO_2 were deposited at 300°C by remote plasma-enhanced chemical vapor deposition (RPECVD or simply RPD onto (i) SiON passivated crystalline Si and (ii) remote plasma nitrided crystalline Ge [17]. Prior to deposition onto Ge, native oxides were removed by a wet-chemical etching/rising process [18]. The films were subsequently annealed at temperatures between 400°C and 950°C in Ar. Thin nano-grain films of TiO_2-HfO_2 alloys and ZrO_2 were also deposited by reactive evaporation on oxidized Si substrates, and annealed at temperatures between 700°C and 900°C in Ar as well [19]. The TM core-level spectra of the HfO_2 and ZrO_2 were qualitatively and quantitatively the same as those obtained from RPECVD (RPD) films.

2.2. XAS spectra at the Stanford synchrotron research lightsource.

The O K edge, and pre-O K edge XAS spectra were obtained on Beam-line 10-1 at the Stanford Synchrotron Research Lightsource (SSRL) in Menlo Park, Ca. Energies of conduction band state, the pre-conduction band edge defect states electronic transitions have been obtained by X-ray absorption spectroscopy (XAS) using the total electron yield (TEY) detection method [20]. This is based on a measurement of current required to neutralize positive charge created by relaxation of XAS final virtua bound excited states by photoemission events. This approach provides an excellent measure of the linear absorption coefficient with a precision of a few percent. The signal to noise ratio (>3000:1) and the spectral resolution (~ 0.1 eV) for these measurements are sufficient for performing a 2nd derivative analysis of the O K edge and O K pre-edge regimes, and core level 2p to 3d core level L2,3 transitions for Ti oxides, as well p-to-d transitions applicable to other TM oxides with 3p (ZrO_2 and Y_2O_3) and 5p (HfO_2) core-level states. The Si L2,3 spectra, and Ge M3,4 spectra were also obtained by the same TEY method, but on BL 8-1 at SSRL. BL 10-1 covers a spectral range of X-ray energies between about 200 and 1200 eV, and BL 8-1 covers a spectral range from about 50 eV to 200 eV. Equally important O K edge, and pre-K-edge data were taken on BL 10-1 with continuous scanning, and with sampling intervals ranging from 30 to 40 meV, well inside a resolution limit of ~0.1 eV as determined by the X-ray monochrometer aperture and grating. Finally, O-pre-K edge absorption for TiO_2 and c-Ti_2O_3 done at different times with different interval sampling, 30 meV and 50 meV, yield the same number and ordering of pre-edge defect states. This establishes that background noise is not an issue dn this experimental X-ray spectroscopy results presented below, particularly the defect relared 2nd derivative O K pre-edge spectra.

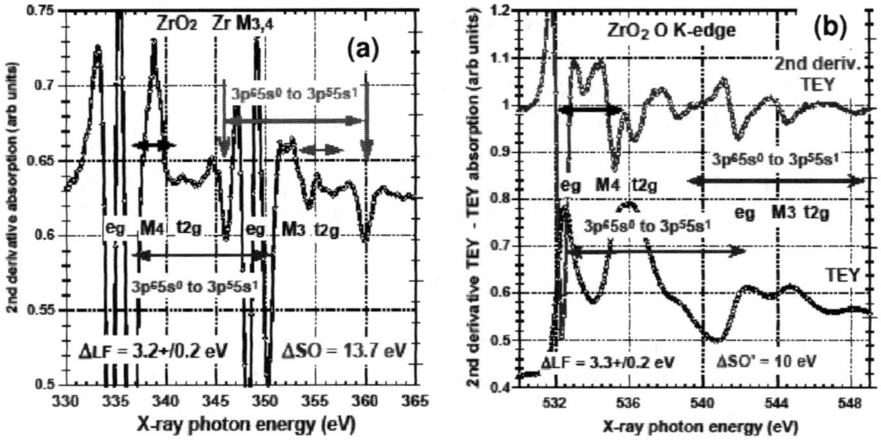

Fig. 1. (a). The second derivative Zr M3,4 spectrum for ZrO_2;
(b). The O K edge spectrum, and the O K edge 2nd derivative spectrum.

3. Ligand Field Determinations from ZrO_2: Zr M3,4 and ZrO_2 O K-edge spectra

Zr M3,4 and ZrO_2 O K-edge spectra are presented for ZrO_2 to identify the spectral change resulting from differences between ΔSO in the Zr M3,4 spectrum, and an effective ΔSO' in the ZrO_2 O K-edge replicated spectral features. ΔSO' is a splitting between 4pσ and 4pπ states in the valence band molecular orbital states of ZrO_2.

The three-electron process that is relevant also increases the spectral width of features in the M3 and M4 features in the O K-edge spectrum relative to the Zr M3,4 spectrum. Figure 1(a) is the second derivative Zr M3,4 spectrum for ZrO_2. Respective values of ΔLF and ΔSO' are 3.2±0.2 eV, and 13.7±0.2 eV. ΔLF is the energy difference between the eg and t2g features in the M3 and M4 energy regimes, and ΔSO' is the energy difference between the corresponding eg features in the M3 and M4 regimes. ΔSO is the spin-orbit splitting of the Zr 3p core level. A comparison between the $3p^6 4d^0$ to $3p^5 4d^1$ transitions, and the $3p^6 5d^0$ to $3p^5 5s^1$ transitions gives an energy difference of ~8 eV between the lower lying Zr 3d states and the higher lying 5s states.

Fig. 1(b) includes the O K edge spectrum, and the 2nd derivative of that spectrum as well. Of particular importance are the photon energies for the d-state derived features, and the s-state derived features. The respective values of ΔLF and ΔSO' are 3.3±0.2 eV, and 10±0.2 eV. As aniticipated the extracted ΔLF values are the same to within the experimental uncertainties. On the other hand, the values of ΔSO' and ΔSO are significantly different. This difference in ΔSO' is related to the nature of the excitation process the results in M3 and M4 features in O K edge conduction band regime. This difference is the same as the difference in energy between $3p^6 4d^0$ to $3p^5 4d^1$ transitions, and the $3p^6 5d^0$ to $3p^5 5s^1$ transitions. This is reduced by 1.5 eV from 8.0 eV in the Zr M3,4 spectrum to 6.5 eV in the ZrO_2, at the same time the ligand field splitting remains effectively the same in each spectrum.

The O K-edge spectrum plot extends from 530 eV to 549 eV, with an extrapolated X-ray photon energy of 531.5 eV. Each of the d-state dervied features in the Zr M3,4 spectrum has the same relative energy in each spectrum, but with markedly different matrix elements. Equally important and as noted above, the effective spin-orbit splitting, defined above as ΔSO' is significantly smaller. This is explained in the context of *three-step coherent process* that gives rise to M3 and M4 features. The O K-edge spectral features aare the result of *three-concurrent and choherent* CTM transitions [25]. The CTM transitons are initiated by excitations from occupied $1s^2$ O-core level to two virtual-bound states above the level that defines the Ionizaation Potential (IP) of an O $1s^2$ core level that takes into account a partial covalent charcter, and is shifted up in energy by more than10 eV. This initiating excitation is concurrent and coherent with two other transitions. The first of these is: (i) the filling of the O 1s empty core levels by a transition in which electrons from occupied Zr 4p-derived molecular orbital states in the Zr valence band fill the O 1s core holes, and (ii) a shake-up energy conserving transition to two virtual bound resonance states in the conduction band continuum [25]. The filling of the two $1s^2$ core-level holes is accompanied by emission of X-rays. The *shake-up process* is effectively a *p-d transition* with a large transition energy matrix element. This process explains: (i) the difference between ΔSO and ΔSO′, (ii) the effective shift of the relative final d-state and s-state transitions, and (iii) the different spectral line shapes between the Zr M3,4 and ZrO_2 O K-edge spectra. Similar differences occur for the HfO_2 O K-edge spectrum, and these are displayed in Fig. 2.

Figure 2 compares OK edge spectra of two HfO_2 films marked to indicate features in Hf core level O5,6 spectral features. Two nm thick films annealed at 700°C have a tetragonal (T) HfO_2 structure, whereas thicker 5 nm HfO_2 films, also annealed at 700°C, display a mixture of tetragonal (T) and monoclinic (M) nano-grains [11]. These films are prepared on nitrided Ge substrates. When annealed at a lower temperature of 500°C after remote plasma deposition at 300°C, the features in both 2 nm and 5 nm thick films are significantly broader as indicated in Fig. 2(b). These differences result from an in-plane Ge surface templating effect at the higher temperature that is absent at the lower temperature. Based on N K-edge spectra, the Ge-N disordered interface remains largely unchanged from its initial spectral character after 500°C annealing, but is removed completely except for on narrow feature after the 700°C anneal. Templating dominates in the thinner 2 nm film, and is

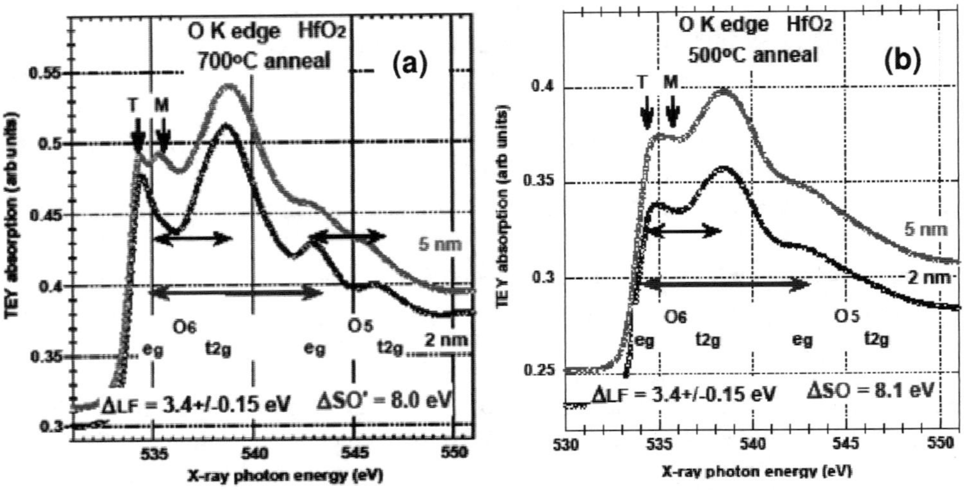

Fig. 2 OK edge spectra of 2 nm and 5 mn HfO2 films annealed
(a) at 700°C; (b) at 500°C.

diminshed in the thicker 5 nm yielding changes from T to a combined T-M films texture after the 700°C anneal. The features in thee 500°C films have often been mistaken as being an indicator of the film being amorphous. This is not the case. These films like the higher temperature 700°C annealed films display the same values of ΔLF and ΔSO'. Equally important for technology applications, e.g., CMOS gate dielectrics, the low temperature films do not show strong bias-dependent increases in current for increases in temperature from 25°C to 125°C substrate heating. The resulting J-V characteristics are similar to SiO_2, where electron injection is reduced significantly by band-edge excitons situated between the photo-conducting band edge and defect states with the SiO_2 layer. Additionally and consistent with XRD results for these films which indicate a significant grain-size dependent broadening, the grain-size is limited by film thickness, and defect sites resent in individual grains requiring inter-grain hopping. This is in contrast with the thicker films where defects reside within grains and on grain-boundaries as well, resulting in hopping-conductivity that is strongly temperature dependent.

Returning to the HfO_2 results in Figure 2, the values of ΔLF and ΔSO', respectively in the O K-edge spectra are 3.4±0.15 eV and 8.0±0.15 eV. The 3.4 eV value is the same as in the O5,6 core level transition, but the ΔSO′ value is smaller than 8.6 eV the O5,6 core level spin-orbit splitting. Fig. 2(a) that compares OK edge spectra of these films is marked to high-light the features in Hf O5,6 spectral features. The relative $5p^6 6s^0$ to $5p^5 6s^1$ position is therefore changed because of the 0.6 eV reduction in the O5 and 06 average energy difference. The changes and relative energy differences parallel the similar changes in ZrO_2 films, and also results from the three step coherent process for soft X-ray electron excitations out of the occupied O $1s^2$ state to the virtual bound regime.

4. Experimental Results: Ti elemental-oxides and alloys

4.1 Ti elemental oxides.

There is a corelation between ΔLF extracted from the analysis of Ti L2,3 spectra, and the spectral widths (ΔSW)of vacated O-atom site singlet and triplet defect spectral features in the O pre K-edge regime [8,9,14]. An approximate empirical relationship is obtained from the T-S diagrams: ΔSW

Journal of Physics: Conference Series **428** (2013) 012022 doi:10.1088/1742-6596/428/1/012022

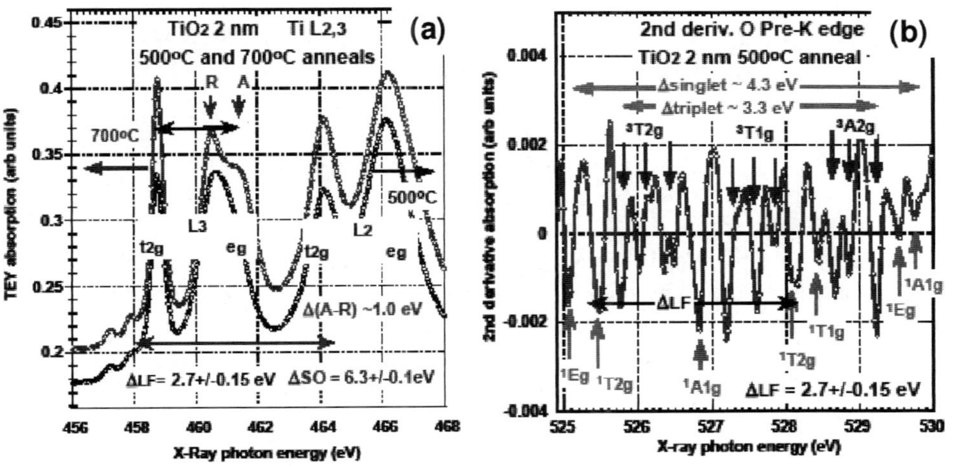

Fig. 3. XAS spectra and Second derivative XAS spectra: for TiO_2: (a) Ti L2,3 edge; and (b) O K pre-edge vacated O- site excited defect spectrum.

≈1.8±0.15xΔLF. This relationship holds for ZrO_2 and HfO_2 and applies to Ti-oxides as well. This scaling relationship indicates the importance of ΔLF for both (i) intrinsic final TM d-state dervied features, as well as (ii) final states of strongly correlated pairs of TM-atom dangling bond defect transitions. The determination of ΔLF is based on the spitting of eg and t2g features in the T-S diagrams that apply; e.g. d^2 for TiO_2, and d^8 for tetragonal HfO_2 [15,16]. These are then compared with values of ΔLF extracted directly from the TEY 2nd derivative core-level absorption spectra transitions, O K edge and Ti L2,3 spectra as well. These transistions are interpreted using symmetry adapted combinations of atomic states (SALC) [17] appropriate to the local symmetry of Ti in the nano-grains of different Ti oxides. X-ray photoemission studies identify the ground state energies of the occupied Ti dangling bond sites that project in the vacancy. This theory bassed approximation for ΔSW provides a determination the vacated O-site singlet and triplet state energies relative to the conduction and valence edges that define the forbidden energy gap, ΔEg. This is the first time this approach has been applied to the Ti_nTi_{2n-1} and TiO_2-HfO_2 alloys.

Figure 3 includes XAS spectra and Second derivative XAS spectra for TiO_2 films in which rutile edge-connected octaheral bonding dominates after a 500°C anneal, and anatase bonding dominates after a 700°C anneal. Fig. 3(a) presents the Ti L2,3 plots for the 2 annealing conditions. The removal of the eg into 1 eV split into A_{1g} and B_{1g} components is indicated. The A_{1g} terms are different for anatatse (A) and rutile (R) interconnections of the primitive octahedral unit cell. The B_{1g} terms are not evident in the Ti L2,3 spectra. Edge sharing dominates in the (R) structure. This increases the c-axis length and reduces the energy of the A_{1g} term.

Spectra for anatase (A) dominant grains are present in TiO_2 films annealed at higher temperatures (not shown), and these indicate a decrease the energy of the A_{1g} split component with a shift to higher X-ray photon energies. The following energy differences have been obtained for this film: (i) ΔLF = 2.7 eV and (ii) ΔA-R, the energy separation between the A and R features is ~ 1eV. Finally, ΔSO = 6.2 eV and has been estimated from the energy difference between the t2g features in the L3 and L2 spectral regimes. The inter-connection of octahedra in the anatase (A) structure is via corners, and this leads to contractions in the c-axis and a shift ot the A_{1g} feature to a higher energy than in the rutile (R) structured nano-regions.

The Ti L2,3 spectra are dipole allowed X-ray transitions from occupied, spin-orbit split Ti $2p_{3/2}$ and $2p_{1/2}$ states at ~460 eV and 454 eV, respectively to empty Ti 3d states. There also dipole allowed transitions from Ti $2p_{3/2}$ and $2p_{1/2}$ states to empty 4s states [24,25]. Holes in the Ti 2p core level states play a significant role in the symmetries of the final sates for these excitations. The initial and final state symmetries are described in the context charge transfer multiplet (CTM) theory [24,25]. These transitions are represented symbolically in Eqns. 1(a) and 1(b):

$$2p^6 3d^0 \rightarrow 2p^5 3d^1 \quad 1(a) \text{ and}$$

$$2p^6 4s^0 \rightarrow 2p^5 4s^1 \quad 1(b).$$

The final states of the 1(b) transitions are higher in photon energy than those of the 1(a) transitions, and in a different and separated spectral regime centered about 8 eV higher in photon energy. A complete analysis of the final state properties for the transitions in Eqn. 1(b) is beyond the scope of this article, and detailed discussions are included in Ref. 25.

More important, for the L2,3 transitions in the Ti-elemental and complex oxides of this chapter, the ΔLF values are essentially the same as the value found for these Ti oxides. The value of ΔSO is also approximately the same in all of the oxides studied, 6.3±0.2 eV. The spread of values includes the nominal value of the 2p core state spin-orbit splitting: ΔSO = 6.4±0.1 eV. More important, the relative values of ΔLF and ΔSO are sufficiently different and give rise to separate and distinct L_2 and L_3 energy regimes in the $L_{2,3}$ spectrum. Stated in more quantitative way, as long as ΔSO > 2xΔLF, the L2 and L3 regions are separate and distinct. This is also the case for the other group IV oxides, ZrO_2 and HfO_2, addressed above. Moving on the O-K pre-edge spectrum in Fig. 3(c), and has 16 features. This number is the same as of number features in the Ti L2,3 spectra in Fig. 3(a). However in this case the number is determined by the symmetry and coordination of the Ti-atom dangling bonds projected into the vacated O-atom site. It is a property of the excited states, and consistent with the number of detectable features in the d^2 T-S diagram that applies. The sixteen (16) triplet and singlet features as a function of increasing energy corresponds to the T-S d^2 plot for cubic octahedral local symmetry. This includes the 3-fold degeneracy of ^{3}T2g and ^{3}T1g states removed by final state effects associated with the spin-orbit splitting of the final virtual bound states in these transitions [24]. There accounts for 9 triplet features with triplet symmetries, e.g., ^{3}T1g. The remaining 7 spectral features are assigned to non-degenerate singlet ^{1}Ag, ^{1}Eg, ^{1}T1g and ^{1}T2g features. The ordering of the singlets and triplets follows the T-S diagram for octahedral symmetry with 2 occupied strongly-correlated d-states. The spectrum is for an intermediate strength effective ligand field splitting defined by the ^{1}Eg to ^{1}T2g energy difference of 2.7±0.15eV and essentially the same as values in the Ti L2,3 spectrum.

Before presenting and analyzing the XAS spectra for the Magneli Phase Ti_nO_{2n-1}, and TiO_2-HfO_2 alloys, the 2nd derivative Ti L2,3 for Ti_2O_3 is presented in Figs. 4(a) and the second derivative Ti_2O_3 O pre K-edge spectra is displayed in Fig. 4(b). These spectra are obtained form a single crystal with the corrundum structure. As such, they include contributions for X-ray photons with components of the E-field aligned with both the c-axis in the z-direction, and two a-axis directions in the x-y base plane. Equally important, the formal valence state of Ti in Ti_2O_3 is +3 rather than +4 in TiO_2. The ionic radius is increased from $R(Ti^{4+})$ = 0.0745 nm to $R(Ti^{3+})$ = 0.081 nm, equivalently the ratio of the +4 charge state to +3 charge state radii is 0.92.

The 2nd derivative spectra Ti L2,3 spectrum is in Fig. 4(a) and the 2nd derivative O K pre-edge spectrum in Fig. 4(b) spectra have been obtained with the X-ray beam at approximately normal incidence on the the base plane, and as such the electric field vector E is perpendicular to the c-axis. Qualitatively and quantitatively similar Ti L2,3 spectra have been obtained with the X-ray incident on a vertical face, with the E-vector perpendicular to both the c- and a-axis direction: these spectra are not shown. Values of ΔLF = 2.5±0.15 eV and ΔSO = 6.2±0.15 eV extracted from the Ti L2,3 vertical face spectrum of Ti_2O_3.are the same within the experimental uncertainty as those for of extraced from the Ti L2,3 base face spectrum of TiO_2.

Mre significant, the value of ΔLF = 2.35±0.15 eV extracted from the Ti_2O_3 O pre-K-edge base face spectrum is different than the value of ΔLF = 2.7±0.15 eV obtained from the TiO_2 O pre-K-edge

spectrum. The ratio of the these values is 0.87±0.1, and close to the ratio of the ionic radii of Ti^{4+} to Ti^{3+}, 0.92. Since R (Ti^{3+}) > R(Ti^{4}) ΔLF is smaller;
Finally, the spectral widths for the singlet and triplet transitions, respectively in the O K pre-edge spectrum, 4.0±0.1 eV and 3.0±0.1 eV, are the same as those determined by ab-initio calculations and are consistent with spectral widths from T-S diagrams [14].

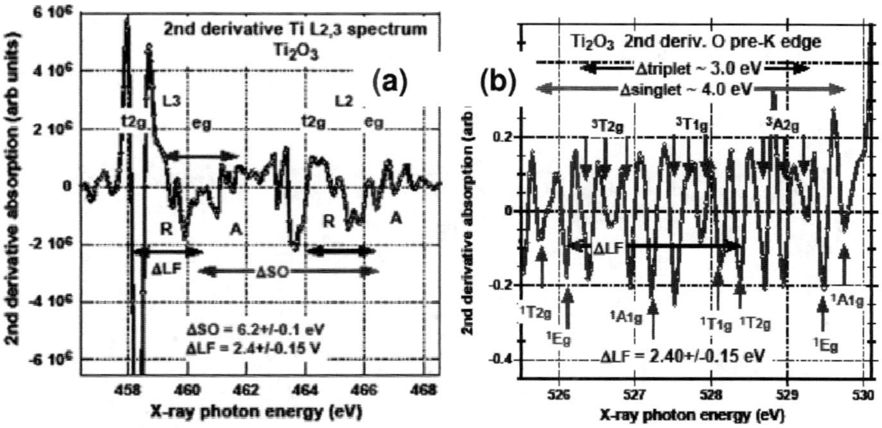

Fig. 4. XAS spectra for single crystal corundum Ti2O3:
(a) 2nd derivative Ti L2,3; (b) 2nd derivative O pre-K edge.

4.2. XAS spectra for Magneli Phase Ti_nO_{2n} and $(TiO_2)_x(HfO_2)_{1-x}$ alloys.

XAS spectra are presented analyzed for: (i) Magneli-phase alloys, Ti_nO_{2n-1}, n is an integer 0>n≥ 3, and (ii) TiO_2-HfO_2 mixed oxide alloys for TiO_2-rich (67% TiO_2), and HfO_2-rich (67% HfO_2) compositions. Emphasis is on respective Ti L2,3 spectra, and the extraction of ΔLF, and comparisons with ΔLF extracted from O K pre-edge spectra. The TiO_2 and Ti_2O_3 results are important for the interpretation of these comparisons since (i) the Magneli-phase alloys are mixtures Ti^{4+} TiO_2 and Ti^{3+} Ti_2O_3, (ii) the TiO_2-HfO_2 alloys display metallic hopping conductivity, similar to Ti_2O_3.
Figure 5(a) is the second derivative Ti L2,3, spectrum for a Magneli-phase alloy writen as TiO_{2-x}, x~0.15. This spectrum, including includes (A) and (R) connections. The stronger low photon energy component indicatesg mostly (R) edge-corner sharing. The following values have been extracted from the analysis: ΔLF = 2.40±0.15 eV, and ΔSO = 6.3±0.1 eV. ΔSO is essentially the same as the value for

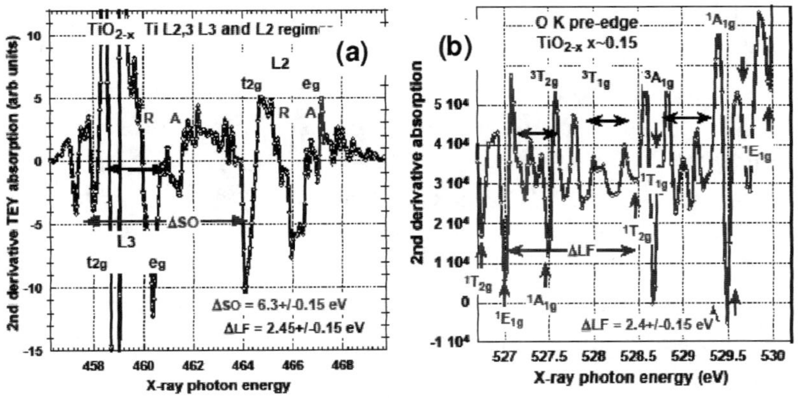

Fig. 5. XAS spectra for Magneli-phase Ti_nO_{2n-1} alloy second derivative: (a) Ti L2,3; and (b) O K pre-edge.

Journal of Physics: Conference Series **428** (2013) 012022 doi:10.1088/1742-6596/428/1/012022

TiO_2, and other Ti-oxides reported in this article.
The value of $\Delta LF = 2.40\pm0.15$ eV is smaller compared to $\Delta LF = 2.70\pm0.1$ eV for TiO_2, but not as small as the value for Ti_2O_3, 2.4 ± 0.15 eV. As discussed above, the values of ΔLF from Ti L2,3 spectra reflect the average properties of the nano-grains, and distribution of Ti^{3+} and Ti^{4+} atomic radii. The ΔLF from O K pre-edge spectra are determined by a similar average over the valence of the Ti-atoms that are directed into the vacated O-atom bonding site.
Figure 5(b) is the second derivative TiO_{2-x} O K pre-edge spectrum. The number and ordering of symmetry designated singlet and triplet terms is the same is essentially the same as those in TiO_2 and Ti_2O_3. More significant are the spectral widths of singlet and triplet states: 4.0 eV and 3,0 eV, respectively and the same as those for Ti_2O_3 and TiO_2.

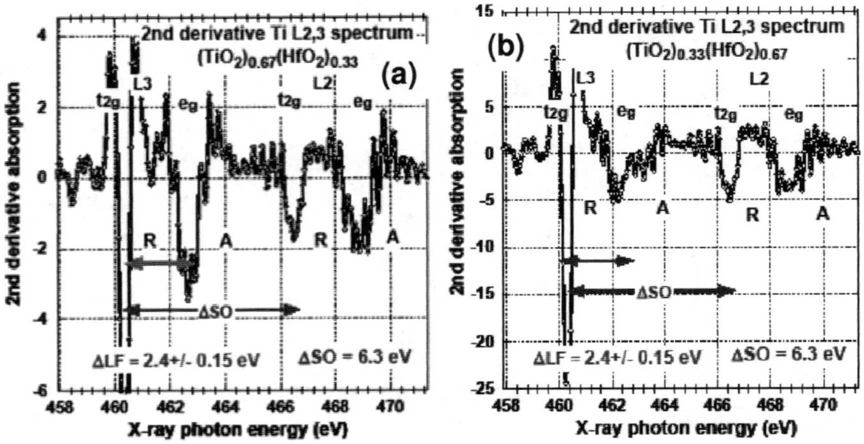

Fig. 6. Second derivative TI L2,3 $(TiO_2)_x(HfO_2)_{1-x}$: (a) x = 0.67; and (b) x = 0.33

.Moving on, results are presented for TiO_2-HfO_2, $(TiO_2)_x(HfO_2)_{1-x}$ alloys. This are of interest because differences in ionic radii and local bonding coordination are too large for mixed Ti-O and Hf-O bonding in the same nano-grain within a thin film alloy. The spectral features are addressed for two alloys with different Ti to H ratios, H is 2 for x = 0.67, and 0.5 for x = 0.33. Figures 6(a) and 6(b) are respectively the 2nd derivative Ti L2,3 plots in (a) for an alloy with 67% TiO_2 and 33% HfO_2, and in (b) for a *mirror-image* alloy with 33% TiO_2 and 67% HfO_2. These alloys display metallic hopping conductivity, so that hopping induced multi-valency is a contributor to the 16 features in the respective L3 and L2 regimes.Since there is a relatively strong dependence on alloy concentration, a second contributing mechanism also involve grain size Ti multi-valency is strain-induced by the large radius difference between Hf^{4+} and Ti^{4+} contributing to Ti^{3+} bonding as well as Ti^{2+} bonding. This explanation requires a larger size for rutile (R) nano-grains. This is correlated with the higher fraction of TiO_2, and consistent with the XAS spectra.The extracted values from Ti L2,3 spectra, $\Delta SO = 6.4$ eV, and $\Delta LF = 2.4$ eV are effectively the samethose of the Magneli-phase TiO_{2-x} and Ti_2O_3 films. The values of $\Delta SO' = 6.0$ eV are also effectively the same as those for the TiO_2, and the TiO_{2-x} and Ti_2O_3 films, when experimental uncertainties are included. In contrast the $\Delta LF = 2.4$ eV, including experimental uncertainties is close to those of the TiO_{2-x} and Ti_2O_3, and smaller than $\Delta LF = 2.70$ for TiO_2.

Journal of Physics: Conference Series **428** (2013) 012022 doi:10.1088/1742-6596/428/1/012022

Independent of the detailed models that apply to the specific alloys, the metallic conduction in each of the alloys, the Magneli-phase TiO_{2-x} alloy, and the TiO_2-HfO_2 alloys must both be considered as candidate thin film materials for memory, switching and other emerging device technologies.

5. Theory of Defect Spectra - Vacated O-atom Site Transitions

Two theory-based models have been proposed for the electronic structure of O-vacancy states. The Robertson and Shlugger groups have applied density functional theory (DFT) based models primarily to HfO_2 and ZrO_2, in many instances using cubic phases with a CaF_2 lattice structure. These models have identified singlet O-vacancy (VO), and interstitial O-atoms (IO) states in the band gap [9,11]. Determined in this way the band gap is significantly smaller than that reported in experimental studies of X-ray photoemission spectroscopy (XPS) for ground state energies, and O pre-edge XAS [11]. There are modifications to ΔFT theory in which photons, corresponding to band gaps, are added to the ground states. These approaches do not take into account the holes created by optical or thermal excitation. A detailed critique of these additions to DFT theory is beyond the scope of this article. In contrast, Lucovsky and co-workers have applied two-electron T-S diagrams in this article and other cited references [12] to: (i) two Si(Ge) dangling bond states with d-symmetries projecting into the vacated O-atom site in SiO_2 and GeO_2, respectively, and (ii) two dangling bond states in TM oxides taking note of different local site symmetries for vacancy sites and the TM sites as well [5-7]. This model for SiO_2 and GeO_2 predicts singlet states and triplet terms after a significant relaxation that reduces the repulsion between NN dangling bond defects [12,14].

This approach and ΔFT yield different ground state energies in the lower half of the band-gap, as well as different energies of singlet state trapping states in the upper half of the band gap. After a structural relaxation in the tetrahedrally-bonded vacancy site in the DFT calculations, two electrons are distributed equally over four TM dangling bonds.

In contrast in the quantum chemistry ab-initio calculations and supported by measurements, there are displacive motions in vacated O-atom sites yielding different ground states for singlet and triplet spin states. The displacive motion leads a small separateion betweeen the singlet and triplet ground states and explains the observable singlet and triplet features in the excited 2nd derivative O K pre-edge spectra.

In SiO_2 and GeO_2, the pair of dangling bonds in located within a rigid and ordered cluster comprised of 6-member rings with 1 nm dimensions [14]. The vacated site reduces local strain with in the cluster. The clusters are imbedded in compliant, and more disordered matrix of 5-and 7-member rings. The vacated O-atom sites in nano-grain TM oxides are qualitatively different, and have been addressed in two recently published chapters of a monograph dealing with the Jahn-Teller effect [6,7]. Based on results in those papers, and the more recent understanding of the properties of dangling on the 2 Si- and Ge-atoms bordering a vacated, two fold coordinated bonding O-bonding site, the electronic structure for vacancies in (a) TiO_2 and (b) HfO_2 with tetragonal, and mixed tetragonal-monoclinic textures are addressed. The focus is on the local symmetry and coordination of the TM atoms, and the coordination(s) of the O-atoms which assumed to be within the nano-grains with 2 nm to 5 nm dimensions. Ab-initio calculations are currently being performed on model bonding geometries for the O-atom vacated sites: e.g., 4-fold coordinated sites in tetragonal HfO_2, as well as 3-fold coordinated O-atom sites in monoclinic HfO_2 and ZrO_2, and in rutile and anatase TiO_2. Two theory-based models have been proposed for the electronic structure of O-vacancy states. The Robertson and Shlugger groups have applied density functional theory (DFT) based models primarily to HfO_2 and ZrO_2, in many instances using cubic phases with a CaF2 lattice structure. These models have, identified singlet O-vacancy (VO), and interstitial O-atoms (IO) states in the band gap [9,11]. Determined in this way the band gap is significantly smaller that what has been reported in experimental studies of X-ray photoemission spectroscopy (XPS) for ground state energies, and O pre-edge XAS [11]. There are some modifications to DFT theory in which photons, corresponding to band gaps, are added to the ground states. These approaches however do not take into account the holes created by optical or thermal excitation. A detailed critique of these additions to DFT theory is beyond the scope of this article.

In contrast, Lucovsky and co-workers have applied two-electron T-S diagrams in this article and other cited references [12] to: (i) two Si(Ge) dangling bond states with d-symmetries projecting into the vacated O-atom site in SiO_2 and GeO_2, respectively, and (ii) two dangling bond states in TM oxides taking note of different local site symmetries for vacancy sites and the TM sites as well [5-7]. This model for SiO_2 and GeO_2 predicts singlet states and triplet terms but only if there is a significant relaxation that reduces the repulsion between NN dangling bond defects [12,14].

This approach and the the DFT yield different ground state energies in the lower half of the band-gap, as well as different energies of singlet state trapping states in the upper half of the band gap. In addition after a structural relaxation in the tetrahedrally-bonded vacancy site, the two electrons are distributed equal over the four TM dangling bond.

In constrast and supported by measurements, there are displacive motions in vacate O-atom sites in the quantum chemistry ab-initio calculations which yield two different electron ground states for singlet and triplet spin states. These displacive motions lead to at most a small separateion betweeen the singlet and triplet ground states, thereby accounting for the singlet and triplet features in the excited state spectra as determined from the 2nd derivative of the O K pre-edge spectra.

Proceeding in this way and using SiO_2 as an example, as the distance between the bordering atoms is increased, the exchange energy for triplet terms is decreased, and the overlap energy, associated with two electron symmetric radial wave function for singlet sates is decreased as well, This has a profound effect on the spectrum; e.g. it promotes increased mixing of singlet and triplet excited states, and eventually makes it possible to observed singlet and triplet states in the same O K pre-edge spectrum. If Si and/or Ge atoms bordering the vacated O-site atoms relax completely to relieve local-strain then singlet and triplet features, as indicated in T-S diagrams, occur with approximately equal weighting factors in the defect spectrum. For SiO_2 and GeO_2, the respective Si and Ge neutral dangling bonds are located in small, ordered molecular locally-rigid clusters identified from their respective FSDPs in X-ray diffraction measurements [32]. The degree of relaxation which makes the triplet and singlet ground state energies effectively equal comes from the compliant covalent random bonding present in the non-crystalline matrix in which encapsulates the clusters [12,14]. A more limited relaxation dangling bond-states with partial overlap of nearest-neighbor dangling bonds applies the nanograin TM oxides of this final contract report [8,9], limits the observed spectral to the singlet transitions identified in the appropriate, local coordination and symmetry T-S diagrams [15,16].

The O-vacancy sites in nanograin TM oxides are qualitatively different, and have been addressed in two recently published chapters of a monograph dealing with the Jahn-Teller effect [6,7]. Based on results in those papers, and the more recent understanding of the properties of dangling on the 2 Si- and Ge-atoms bordering a vacated, two fold coordinated bonding O-bonding site, the electronic structure for vacancies in (a) TiO_2 and (b) HfO_2 with tetragonal, and mixed tetragonal-monoclinic textures are addressed in this final chapter. For TM oxides, the focus is on the local symmetry and coordination of the TM atoms, and the coordination(s) of the O-atoms. Ab initio calculations are currently being performed on model bonding geometries for the O-atom vacated sites: e.g., 4-fold coordinated sites in tetragonal HfO_2, as well as 3-fold coordinated O-atom sites in monoclinic HfO_2 and ZrO_2, and in rutile and anatase TiO_2. The same model holds for the defect states since they are atom-dependent; i.e. associated with neutral dangling bonds projecting in to the O-atom vacated with of the respective defect centers.

6. Discussion: (i) Scales of order of Ligand Field Spittings, and (ii) Vacated O-atom states in TM oxides and Application in Device Technology.

Several different splittings of final states in XAS spectra have been addressed for ZrO_2 and HfO_2, which provide model systems for comparing a larger number of Ti-elemental and complex/mixed oxides which include multivalency in formal Ti-atoms charges as well. These splittings include transitions from core level occupied p-states to virtual bound final states derived from symmetry adapted combinations of d-states in one energy regime, and s-states in a second. This core level

Journal of Physics: Conference Series **428** (2013) 012022 doi:10.1088/1742-6596/428/1/012022

splitting accounts for the separation of the final states into two energy regimes for the d-state features. In TiO_2 are two energy regimes: a lower energy L3 regime, and a higer energy L2 regime. There is also a separate and distinct regime at higher energies above the L2 regime for the transitions to evirtual bound s-dervied states.

The average L_3-L_2 differenes in Ti-oxide Ti L2,3 spectra has been determined for TiO_2, Ti_2O_3,

Magneli-phase alloys, Ti_nO_{2n-1}, and HfO_2-TiO_2 alloys all values so obtained are in excellent agreement with the spin-orbit splitting to the limit of expermental uncertainty, the 2p core-level splitting, $\Delta SO = 6.4\pm0.15$ eV.

The group theoretical formalism for determining the number and symmetry of the final states is based on CTM approaches in Refs. 24 and 25. Since defect states in vacated O-sites leave the site neutral, the sites have two singly occupied dangling bond states projecting into the vacancy site. This means the multiplets have two-electron ground states, as well as singlet and triplet excited state two electron terms.

The ground state splittings for these transitions are replicated with a different splitting, in this article ΔSO'. in virtual bound that requires a three-electron state for transitions form the O $1s_2$ core-level. The three electron transition is coherent with three transitions occurring simultaneously. These are: (i) a first transition from the O $1s_2$ state to a resonant virtual bound state above the ionization threshold for the O-core state, and generally reduced in energy by an bond-ionicty determined positive shift of the ground state energy, this is followed by: (ii) a radiative decay from two p-derived states in the valence band separated by an energy defined here as ΔSO' and replicated in the O K-edge sepectra, generally smaller than the core level ΔSO (in ZrO_2 and HfO_2), and at most equal to ΔSO (inTi-oxides); and finally (iii) the energy difference between the X-ray photon, and the virtual bound state is in the energy regime above the O-atom ionization energy. This differences from the commonly-accepted and incorrect characterization of empty conduction band states which are simply-filled by electrons. In constrast the virtual bound states, or virtual molecular orbital states require a compensated hole for each electron-occupied virtual bound state. These holes are in the core level state, the valence band, or an electrode that is either tunneling or direct contact with the this energy regime.

The more interesting energy difference in the ligand field splitting, ΔLF, defined by energy separation between d-state derived e_g and t_{2g} states in final state spectral features. The focus on this paper is the relationship between and feature in the final states of excitations with the vacated atom sites. For example, to a very good approximation limited by experimental uncertainy, the values of ΔLF in Ti-oxides, and both elemental and alloy, is the same in three spectra: (i) the Ti L2,3 spectrum, (ii) the O K-edge spectrum, and (iii) the O pre-K-edge spectrum as well.

6.1 Scales of Order for Ligand Field Splittings.

The observation of the same Ti and Hf ligand field splittings, ΔLF, in the respective Ti L2,3 and Hf O5,6 spectra, in the O K edge spectra, and the O pre-K edge defect state spectra has been found within experimental uncertainties. This observation, and the application of the Tanabe-Sugano diagrams is in consistent with with the historical evolution from crystal field splitting (ΔCF) to the ligand field splittings. Following the discussion in Ref. 16: (i) Hans Bethe developed a crystal field theory in which the bonding was 100% ionic and defined a crystal field splitting, ΔCF, accordingly. Van Fleck pointed out that "even in the case of highly covalent bonding, which is best treated by molecular orbital theory, the symmetry properties and requirements remain exactly the same for the crystal field model and the ligand field model". Finally Cotton, showed that whenever molecular orbital theory applies, and when there was sufficient chemical ordering to extend beyond SRO, bond-lengths and bond-angles, extend into the MRO domain. This regime includes dihedral angles, and correlations among them. In this case, ΔLF is non-vanishing, but ia atom specific, and coordination and formal charge specific to the same degree as was noted and discussed with respect to Ti in Section 4.

Using Zr-oxides and Hf-oxides as
prototypical examples, the extraction of ligand field splitting from spectra is different for (i) the Ti

L2,3 core level transition, and the O pre-K-edge spectra, and (ii) the replication of the Ti L2,3 features in the so-called conduction band regime. The observablity of a TM L2,3 spectrum, and the extraction of a value for ΔLF, simply requires a range of order in the material extending into the MRO regime of at least 3-atom bond correlations as in dihedral angles, and generally to 4-atom bond correlations. This includes materials such as SiO_2 and GeO_2 nano-grain dimensions of order 1nm in chemically-ordered clusters imbedded in a more compliant completely disordered network without MRO. The same description also applies to hydrogenated amorphous Si with about 5 to 10 atomic percent H in mono-hydride (Si-H) arrangments. The detection of in nano-grain materials is dependent on grain-size for L2,3, M3,4 and O5,6 replication in O K-egde spectra, and indepenent of grain-size for the O pre-K-edge spectra for vacated O-atom defect sites. For example, ΔLF can be extracted from HfO_2 films 2 nm thick, if depositied on SiON passivated Si surfaces that have broad features, as well as in films 3.5 to 4 nm and thicker on the same passivated substrates which display sharper features in the "conduction band" regime of virtual bound states. This is includes Jahn-Teller effect (JTE) degeneracy remove of "conducton band edge" degeneracies. In marked constrast, the detection of ΔLF in the excited state vacated O-vacancy sides depends only MRO at the defect site extending to at least 3rd NN, and sometimes 4th NN atom correlations, and is not correlated with the existence of larger grain sizes required for JTE band edge degeneracy removal.

6.2. Vacated O-atom states in TM oxides and Technology Applications

The O-vacancy sites in nano-grain TM oxides are qualitatively different than those of SiO_2, GeO_2 and GeS_2. The TM defects have been addressed in several previous publications in two recently published chapters of a monograph dealing with the JTE, but based on new insights required in unconstrained relaxation in nc-oxides and chalcogenides in a 0.1 0.2 nm regime. The TM oxides require additional considerations addressed below [5,6].

The TiO_2 O K-pre-edge spectra in Fig 3(d) is qualitatively similar to the pre-existing defect spectra in nc-SiO_2 and GeO_2, All three of these spectra have been analyzed in the context of the appropriate T-S diagrams. This is consistent with the partial overlap of the radial wave functions and the; their Fermion wave functions. These obey Pauli exclusion and explain the observed singlet and triplet features.

The sub-oxide Magneli phase alloys, e.g. Ti_4O_7 and the $TiO_{1.85}$ alloy of this report have similar band edge defects. Their position relative to the conduction and valence band edges makes ideal candidates for memory and thin film transistor applications.

The observation in the last paragraph holds for TiO_2-HfO_2 alloys. The Ti defects are lower in the band-gap determined by the TiO_2 fraction, and as such these alloys are also candidate materials for memory and thin film transistor applications.

The single states are important for applications based on electron transport, since they (a) can accept electrons with either spin-up or spin-down, and (b) provide near neighbor trapping for variable range hopping or Poole-Frenkel transport paths [30].

Multi-valency is also responsible the metallic conductivity of Ti_2O_3 above the insulator to metal transition temperature as discussed in Ref. 10. The significant aspect of alloy-controlled effective multivalency is that it provides a way of changing and/or controlling the density of O-vacancy defects in the two alloys of this study. Electrons can be injected into the singlet states near the band edge from negatively biased Si, Ge, and other semiconductor substrates, and metals, each with different offset energies [4-7]. Injection can occur over a considerable portion of the band-gap as well. The singlet states closest the conduction band edge serve as relatively shallow traps for injection into oxide layers. The deepest traps capture of injected charge provide an electrical basis for distinguishing between occupied and empty deep traps [30]. Fast erasing is initiated by injection of holes in the occupied singlet and triplet ground states. The transition form the negatively charge deep traps to these positively charged deep traps is radiative, effectively a “p to s” transition with a short life-time for recombination.

Two terminal devices with asymmetric current-voltage characteristics provide options for memory

devices [4-6]. The results of this paper are qualitatively different than mechanisms suggested elsewhere. These alternative mechanisms attribute the switching and memory effects to movement of O-atom vacancies of Magneli phases, rather than electron, and possibly hole-transport through electronic states associated with vacancies [4-7]. The proposed electronic transport mechanism is currently being tested in two terminal device structures with metal or doped semiconductor contacts optimized for injection of electron and holes.
Finally, it is significant to recognize the marked qualitative and quantitative differences between tTi L2,3 spectra of the (i) Ti elemental oxides, TiO_2 and Ti_2O_3, and (ii) Magneli Phase Ti_nO_{2n-1} and the $(TiO_2)_x(HfO_2)_{1-x}$ alloys. TiO_2 is and insulator with a band-gap of ~ 3.1 eV, and Ti_2O_3 is a narrow band-gap semiconductor with an insulator metal transition that is sensitive to the stoichiometry. In contrast, both the Phase Ti_nO_{2n-1} and the $(TiO_2)_x(HfO_2)_{1-x}$ alloys exhibit metallic hopping conductivity transport which results in increases the number of Ti formal charge states from Ti^{4+} to both Ti^{3+} and Ti^{2+}. This in turm creates difference in ground state, and singlet state energies that can be used to provide electron and hole injection from two different electrodes provide a configuration that will result in controlled electron trapping to on-state, and controlled injection of holes to reduce trapping and provide the an off-state.

Acknowledgements
The PIs acknowledge support from the National Science Foundation (NSF), and the Defense Threat Reduction Agency (DTRA). They also acknowledge the contributions to the support for graduate students in the Lucovsky group at NC State University, including Dan Zeller, Kun Wu, and Cheng Cheng who have benefited as well from research support funding. The authors acknowledge helpful discussions with Professor Jerry Whitten and David Aspnes at NC State University. Lastly the authors acknowledge the use of the Beam-lines 10-1 and 8-1 at the Stanford Synchrotron Research Lightsource, SSRL, in Menlo Park CA, and in Particular Jan Luning and Dennis Nordlund who are members of the SSRL technical staff.

1. K. Xiong, J. Robertson and S.J. Clark: Appl. Phys. Lett. 87, 183505 (2006).
2. J. L. Gavartin, D. Muñoz Ramo, and A. L. Shluger, Appl. Phys. Lett. 89 082908 (2006).
3. J. L. Whitten, et al., J. Vac. Sci..Technol. B 20 (2002) 1710.
4. R. Waser, R. Dittmann, G. Staikov, and K. Szot: Adv. Mater. 21, 2632 (2209).
5. F. Zhang, et al., J. Appl. Phys, 109, 104504 (2011
6. J. P. Strachan, et. al, Adv. Mater. 22, 3573 (2010).
7. G. Lucovsky, L. Miotti and K.B, Pastos, Jpn. J. Appl. Phys. 50 10PF04 (2011).
8. G. Lucovsky, L. Miotti and K. PasBastos, Progress in Theoretical Chemistry and Physics 23, 193-211 (2012).
9. G. Lucovsky, L. Miotti and K. PasBastos, Progress in Theoretical Chemistry and Physics 23, 361-376 (2012).
10. P.A. Cox, *Transition Metal Oxides*, (Oxford Univ. Press, Oxford, 2010)
11. G. Lucovsky and D. Zeller, J. of Nanoscience and Technology, 11, 7974 (2011).
12. G. Lucovsky, J. Kim, K. Wu et al., J. of Optoelectronics and Advanced Materials 13, 1359-1363 (2011).
13. G. Lucovsky, H. Seo, S. Lee, L.B. Fleming, M.D. Ulrich, J. Lüning, P. Lysaght, and G. Bersuker, Japan J. Appl. Phys. 46 1899 (2007).
14. K. Wu, B. Papas, G. Lucovsky, and J.L. Whitten, unpublished.
15. Y. Tanabe and S. Sugano: J. Phy. Soc. Japan 11, 864 (1956).
16. F.A. Cotton, *Chemical Applications of Group Theory, 2nd Ed.* (Wiley Interscience, New York, 1963) Chapter 8.
17. H. Niimi and G. Lucovsky, J. Vac. Sci. Technol A 17, 3185 (1999); B 17, 2610 (1999).
18. H. Seo, K.B. Chung, J.P. Long, and G. Lucovsky., J. Electrochem. Soc. 156, H813 (2009).
19. S-G. Lim., S. Kriventsov, T.N. Jackson, J.H. Haeni, D.G. Schlom, A.M. Balbashov, R. Uecker, P. Reiche, J.L. Freeouf, and G. Lucovsky: J. Appl. Phys. 91, 4500 (2002).
20. A. Owens, S.C. Bayliss, G.W. Fraser, S.J. Gurman, Nucl. Instrum. Methods Phys. 280 Res. Sect. A 385, 556 (1997).
21. Y. Hwu, Y.D. Yao, N.F. Cheng, et al., NanoStructure Materials 9, 335 (1997).
22. D.L. Griscom, Ceramic Soc. Jap. 90, 823 (1981), and Refs. therein
23. ruby@colorado.edu/~smyth/tio2.html
24. F.M.F. de Groot, J.C. Fluggle, et al., Phys. Rev. 41, 928 (1990).
25. F. deGroot and A. Kotani, *Core Level Spectroscopy of Solids* (CRC Press, Boca Raton, 2008).
26. H. Takeuchi, D. Ha, T.J. King, J. Vac. Sci. Technol. A, 22, 1337 (2004).
27. G. Lucovsky, L. Miotti and K. PasBastos, Jap. J. Appl. Phys. 50, 10PF04 (2011).
28. J.E. Huheey, *Inorganic Chemistry, 2nd Edition* (Harper and Roe, New York, 1978), Chap. 3.
29. T*he Jahn-Teller Effect and Beyond*, Ed. by J.E. Boggs and Z. Polliner, (The Academy of Sciences, Moldavia, 2008), Chaps. 3 and 4.
30. J-L. Autran, D. Munteanu and M. Houssa, in *High-K Gate Dielectrics*, M. Houssa, Editor, (Inst. of Physics, Bristol, 2004) Chap. 3.4.

XXIst International Symposium on the Jahn–Teller Effect 2012 IOP Publishing
Journal of Physics: Conference Series **428** (2013) 012023 doi:10.1088/1742-6596/428/1/012023

Ligand Field Theory: An ever-modern theory

Claude A. Daul

Department of Chemistry, University of Fribourg / Switzerland

E-mail: claude.daul@unifr.ch

Abstract: The Ligand Field (LF) model in molecular science or the Crystal Field model in condensed matter science has been introduced more than eighty years ago. Since then, this theory plays a central role each time that molecules containing d- and/or f-elements with open shells are adressed. No doubt, this fact is related to the dominant localisation of the frontier orbitals within the metal-sphere. This common feature enables us to describe approximately the electronic structure of these molecules using orbitals that are centred on a single atom and to treat their interaction with the chemical environment essentially as a perturbation. Another reason for the success of this simple theory is the fact that the more accurate molecular orbital theory does generally over-estimate covalence of transition metal atoms and thus yields wave functions that are too delocalized.
We give here a survey of the development of LF theory since its introduction simultaneously by Hans Bethe and John Hasbrouck van Vleck more than eighty years ago. Over the years, LF theory was a semi-empirical model with adjustable parameter until the end of last century when we introduced non-empirical LF theory that is based on DFT calculations. The results of this first-principle prediction are in very good agreement with the experimental observations. Sample calculations of tetrahedral and octahedral Cr-complexes and hexa-acquo Ni(II)-complexes are used to validate the new model and to analyse the calculated parameters

Introduction

Hans Bethe and John Hasbrouck van Vleck have introduced the Ligand Field (LF) model more than eighty years ago simultaneously. It is a semi-empirical model with adjustable parameter. Twenty years later Christian Klixbüll Jorgensen and Klaus Erik Schäffer proposed the Angular Overlap Model (AOM) that is a revisited version of LF theory with more chemical insight, still using adjustable parameters, but also the angular geometry of the metal complex. About twenty years ago we proposed a new, non-empirical, Density Functional Theory (DFT) based Ligand Field model [1] based on a multi-determinantal description of the multiplet fine structure of the whole ligandfield manifold. We propose to call this model LFDFT. The key feature of this approach is the explicit treatment of near degeneracy correlation using ad hoc Configuration Interaction (CI) within the active space of Kohn-Sham (KS) orbitals with dominant d- or f-character. The calculation of the CI-matrices is based on a symmetry decomposition and/or on a ligandfield analysis of the energies of all single determinants (micro-states) calculated according to Density Functional Theory (DFT) for frozen KS-orbitals corresponding to the averaged configuration, eventually with fractional occupations, of the d- or f-orbitals. This procedure yields multiplet energies with an accuracy of a few hundred wave numbers and fine structure splitting accurate to less than a tenth of this amount. With this procedure we have been able to calculate all customary molecular properties e.g. Zero Field Splitting (ZFS) [2], Zeeman interaction [3], Hyper-Fine Splitting (HFS) [3], magnetic exchange coupling [4], shielding constants [5], etc. …
An LFDFT calculation involves two steps: (i) an Average Of Configuration (AOC), with equal occupation of the d- or f-orbitals is carried out, (ii) with these orbitals kept frozen, the energies of all Single Determinants (SD) within the whole LF-manifold is performed. These energies are then used to estimate all the Racah- and LF-parameters needed in a conventional LF-calculation. The results of this first-principle prediction are in very good agreement with the experimental values.
We will describe, next, briefly the LF and the AOM model followed by LFDFT. The final part of

this survey will be devoted to four typical applications.

Angular Overlap Model (AOM) and Ligand Field (LF) model

a) Ligand field model

We want to represent the action of the chemical environment through an electrostatic perturbation

$$\mathbf{h} = \mathbf{h_0} + \mathbf{v_{LF}}$$

where $\mathbf{h_0}$ is the hamiltonien of the free ion and $\mathbf{v_{LF}}$ the electrostatic potential of ligand charge density at the metallic centre. If $\rho(\mathbf{r})$ the charge density due to the ligands one obtains for the electrostatic potential

$$v_{LF}(\mathbf{r}) = -e\int \frac{\rho(\mathbf{R})}{|\mathbf{R}\text{-}\mathbf{r}|} d\mathbf{R}$$

The symbols used are defined in the fig. below and represents the charge of the electron.

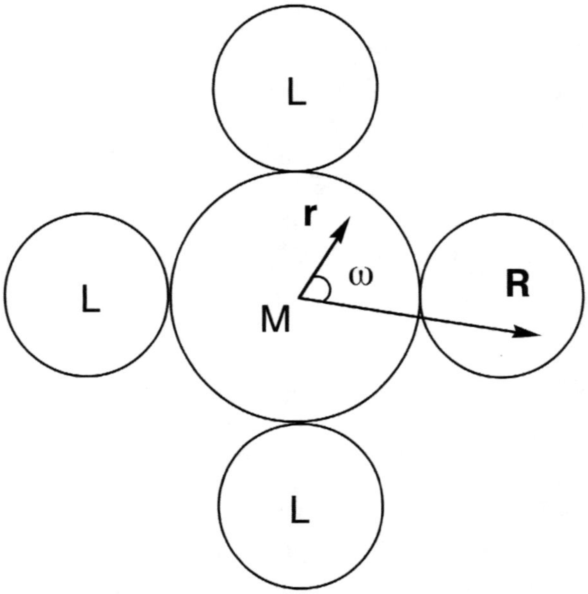

Expand 1/|**r**-**R**| to obtain

$$\frac{1}{|\mathbf{R}\text{-}\mathbf{r}|} = \frac{1}{r_>}\sum_{k=0}^{\infty}\left(\frac{r_<}{r_>}\right)^k P_k(\cos\omega)$$

where $r_< = \min\{r,R\}$, $r_> = \max\{r,R\}$ and $P_k(x)$ is a Legendre polynomial of order k.

We suppose (van Vleck and Bethe, 1930) that : r < R. With this hypothesis and writing: $r_< = r$ and $r_> = R$ we obtain

$$\frac{1}{|\mathbf{R}\text{-}\mathbf{r}|} = \frac{1}{R}\sum_{k=0}^{\infty}\left(\frac{r}{R}\right)^k P_k(\cos\omega) = \frac{1}{R}\sum_{k=0}^{\infty}\frac{4\pi}{2k+1}\left(\frac{r}{R}\right)^k \sum_{m=-k}^{+k} Y_{km}(\theta,\phi)Y_{km}(\Theta,\Phi)$$

That yields v_{LF} as

Journal of Physics: Conference Series **428** (2013) 012023 doi:10.1088/1742-6596/428/1/012023

$$v_{LF}(r,\theta,\phi) = \sum_{k=0}^{2l} \sum_{q=-k}^{+k} h_{kq}(r) Y_{kq}(\theta,\phi)$$

It is sometimes usefull to represent v_{LF} in the basis of d- or f-orbitals :

$$\langle lm | v_{LF} | lm' \rangle = \sum_{k=0}^{2l} \sum_{q=-k}^{+k} h_{kq} \langle lm | Y_{kq} | lm' \rangle$$

where the parameters h_{kq} are adjustable and describe the ligand field. The coeffients <lm|kq|lm'> are simple coupling coefficients : $\langle lm | kq | lm' \rangle = \int d\Omega Y_{lm}(\Omega) \cdot Y_{kq}(\Omega) \cdot Y_{lm'}(\Omega)$ that can easily be calculated by numerical quadrature (vide infra). This theory is on the basis of the celebrated Tanabe-Sugano diagrams (cf. Fig. 1). An equivalent result can be obtained with the AOM discussed in the following section.

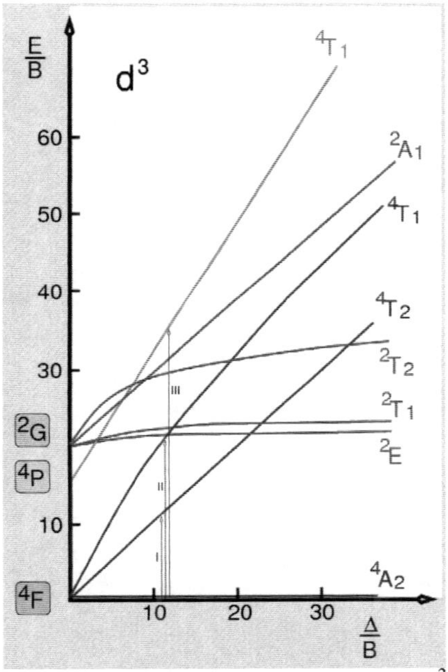

Fig.1 Tanabe-Sugano diagram for d^3

b) Angular Overlap Model (AOM)

The AOM is based on the following four postulates:

- The LF acting on the open-shell electrons of a paramagnetic ion is the resultant of a sum of contributions coming from individual ligands in the molecule or crystal.
- Only contributions from the neighbouring *ligands* are taken into account.
- Each single ligand contribution in the sum is axially symmetric about the line joining its centre to that of the metal ion.
- Single-ligand contributions are dependent only on the nature of the ligand and its distance from the paramagnetic ion, and do not depend on other properties of the chemical environment or host crystal.

Consider the most common M-L interactions:

no nodal plane:	σ
one nodal plane:	π
two nodal planes:	δ (only observed when the ligand is a metal complex)

Journal of Physics: Conference Series **428** (2013) 012023 doi:10.1088/1742-6596/428/1/012023

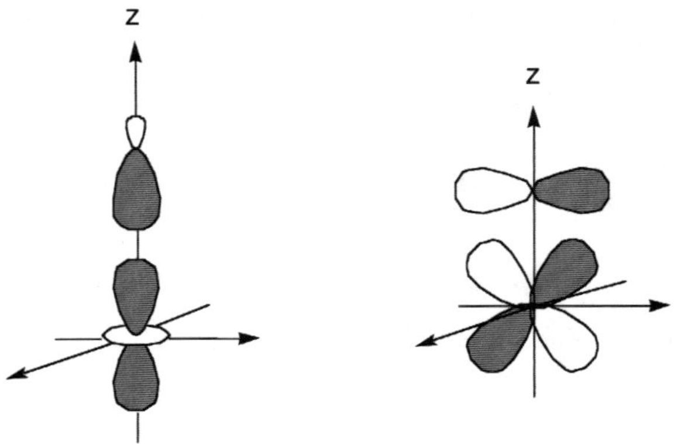

Définition de e_σ Définition de e_π

E.g. the splitting of five d-orbitals due to a local M-L interaction is:

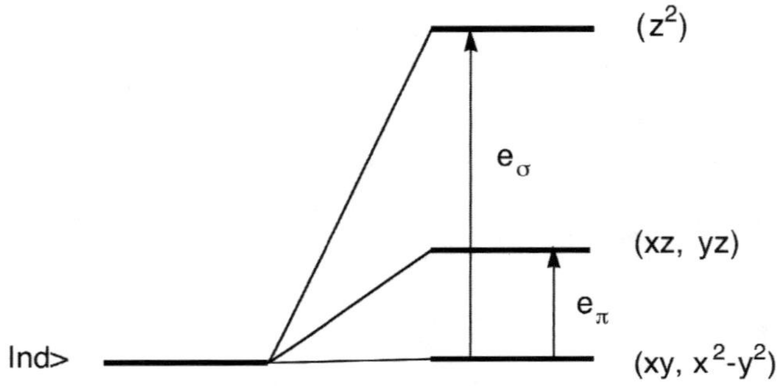

A semi-quantitative physical interpretation of the parameters e_σ et e_π can be estimated as follows:

$$e_k \approx \frac{\langle\psi_M|\mathbf{h}|\psi_{L_k}\rangle^2}{\langle\psi_M|\mathbf{h}|\psi_M\rangle - \langle\psi_{L_k}|\mathbf{h}|\psi_{L_k}\rangle} \approx \text{const.}\langle\psi_M|\psi_L\rangle^2$$

where **h** is an effective hamiltonian for the M-L entity and ψ represents respectively the orbitals of the metal M and of the ligand L, i.e. σ or π.
If we suppose additivity and transferability of the AOM parameters we obtain for the ligand field

$$\mathbf{v} = \sum_L \mathbf{v}_L$$

The calculation of the elements $\langle d_i|\mathbf{v}|d_j\rangle$ in the basis of d-orbitals is illustrated below.
Let us first consider the interaction of a ligand L in axial z position with the d-orbitals , i.e. $\langle d_i|\mathbf{v}^z{}_L|d_j\rangle = v^z{}_{ij,L}$

$v^z{}_L$	z^2	yz	xz	xy	x^2-y^2
z^2	σ_L				
yz		π_L			
xz			π_L		
xy				0	
x^2-y^2					0

Journal of Physics: Conference Series **428** (2013) 012023 doi:10.1088/1742-6596/428/1/012023

Next consider the interaction of ligand L on an arbitrary position on an (θ, ϕ) direction :

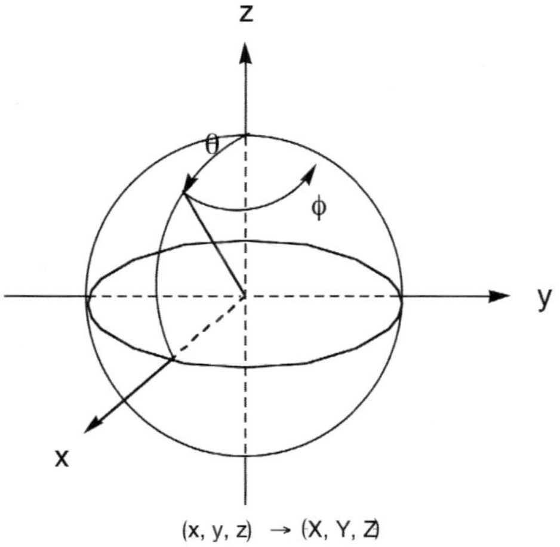

This involves simply a rotation in 3D using Eulerian angles {α,β,γ). If one applies this transformation to the whole set of d-orbitals we obtain

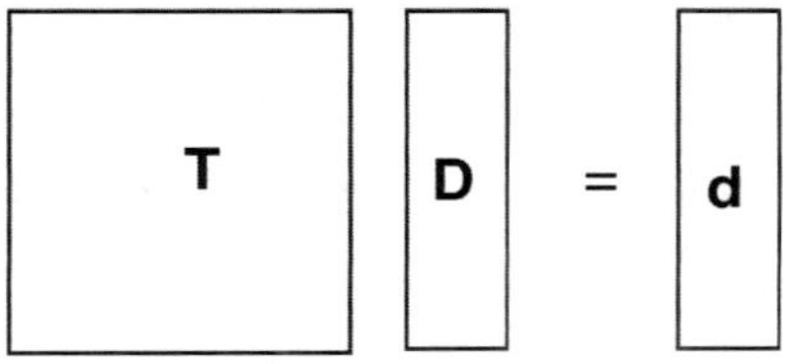

where the elements of **T** are nothing but the well-known Wigner Dastellungsmatrizen:

$$Y_{lm}(\theta,\phi)=\sum_{\sigma=-l}^{l} D^{l}_{m\sigma}(\alpha,\beta,\gamma)Y_{l\sigma}(\Theta,\Phi)$$

where

$$D^{l}_{m\sigma}(\alpha,\beta,\gamma)=\int Y_{lm}(\theta,\phi)Y_{lm}(\Theta,\Phi)d\Omega$$

Wigner's matrices are most conviniently obtained using Lebedev's quadrature. According to this quadrature rule, the surface integral of a function over the unit sphere:

$$I(f)=\frac{1}{4\pi}\int d\Omega f(\Omega)=\frac{1}{4\pi}\int_0^{\pi}\sin(\theta)d\theta\int_0^{2\pi}d\varphi f(\theta,\varphi)$$

is approximated in the Lebedev scheme as :

$$\tilde{I}[f]=\sum_{i=1}^{N}w_i f(\theta_i,\varphi_i)$$

This formula is exact for linear combinations of spherical harmonics up to degree 2N-1 ! The nodes and weights in the formula above can be obtained from internet: (http://www.ccl.net/cca/software/SOURCES/FORTRAN/Lebedev-Laikov-Grids/)

That is, in practice an AOM calculation is as follows :

(i) Calculate $\mathbf{R}_L = \begin{bmatrix} \cos\gamma\cos\beta\cos\alpha - \sin\gamma\sin\alpha & \cos\gamma\cos\beta\sin\alpha + \sin\gamma\sin\alpha & -\cos\gamma\sin\beta \\ -\sin\gamma\cos\beta\cos\alpha - \cos\gamma\sin\alpha & -\sin\gamma\cos\beta\sin\alpha + \cos\gamma\cos\alpha & \sin\gamma\sin\beta \\ \sin\beta\cos\alpha & \sin\beta\sin\alpha & \cos\beta \end{bmatrix}$

(ii) Calculate $\mathbf{D}^l_{m\mu}(\alpha_L,\beta_L,\gamma_L) = \int d\Omega Y_{l\mu}(\mathbf{R}_L\Omega) Y_{lm}(\Omega)$

(iii) Calculate $\mathrm{v}^{(\alpha,\beta,\gamma)}_{mm',L} = \sum_{k=\sigma,\pi_x,\pi_y} D^l_{mk}(\alpha_L,\beta_L,\gamma_L) e_k D^l_{m'k}(\alpha_L,\beta_L,\gamma_L)$

(iv) Sum up $\langle lm|v_{LF}|lm'\rangle = \sum_L \mathrm{v}^{(\alpha,\beta,\gamma)}_{mm',L}$

(v) Use this matrix to obtain the LF splitting of the full open shell manifold

Density Functional Theory calculation of the Ligand-Field manifold (LFDFT)

Ligand field theory has been used with success to describe ground and excited electronic states originating from d^n or f^n elements in their complexes. Both crystal/ligand field theory (LFT) and developments - the angular overlap model (AOM) parameterize the Hamiltonian in terms of one-electron (ligand field) parameters and two-electrons repulsion integrals within the manifold of d-electrons. The latter ones are treated as atomic-like, thus preserving spherical symmetry, while the former take full account for the lowering of symmetry when a spherical metal atom or ion is introduced in a complex. Thus, for a d^n manifold in cubic symmetry, only one energy parameter 10Dq – the energy difference between the σ and π, e_g and t_{2g} for octahedral field (or between the $\sigma+\pi$ and π, t_2 and e orbitals for tetrahedral field) is introduced in addition to Racah's inter-electronic repulsion parameters B and C. These three parameters are usually determined from a fit to electronic absorption spectra in high resolution. From these data, two general observations can be made:
i) Ligands order in a sequence of increasing values of 10Dq, called spectrochemical series, where negatively charged ligands such as I < Br < Cl < F possess smaller 10Dq values, than neutral molecules H_2O < NH_3 < pyridine, CN^- and CO being the strongest ligand due to their ability for back bonding.
ii) B and C parameters in complexes are smaller than those for the free ions - the phenomenon being discovered by Schäffer and Jørgensen and rationalized in terms of the electronic cloud expansion of the d-orbitals when going from free TM ions to complexes (nephelauxetic effect). The more reducing and softer ligands show a stronger reduction than the more oxidizing and harder ones.

Description of the method

The LFDFT method, which has been recently proposed by Atanasov and Daul is briefly summarised here. Let us assume that we know the complex geometry, either from a first principle geometry optimisation or from X-ray data. The first step consists in a spin-restricted SCF DFT calculation of the average of the l^n configuration (AOC), providing an equal occupation n/(2l+1) on each MO dominated by the l-orbitals. The Kohn-Sham orbitals, which are constructed using this

AOC are best suited for a treatment in which, interelectronic repulsion is - as is done in LF theory, approximated by atomic-like Racah parameters B and C. The next step consists in a spin-unrestricted calculation of the manifold of all Slater determinants (SD) originating from e.g. the d^n shell, i.e. 45, 120, 210 and 252 SD for $d^{2,8}$, $d^{3,7}$, $d^{4,6}$ and d^5 metal ions, respectively. These SD energies are used in the third step to extract the parameters of the one-electron 5x5 LF matrix h_{ab} as well as Racah's parameters B and C in a procedure, which we describe below. Finally, we introduce these parameters as input for a LF program allowing to calculate all the multiplets using CI of the full LF manifold utilizing the symmetry as much as possible. We should note that in classical LF theory, it is only the LF h_{ab} matrix, which carries information about the symmetry and the actual bonding in the complex, thus providing useful chemical information.

E.g. for octahedral complexes of a d^n TM the LF matrix reduces to one parameter – 10Dq. We, thus have, for each SD energy the simple linear expression in terms of B, C and 10Dq:

$$E\left(SD_\mu^d\right) = E\left(\det\left|d_{i(\mu,1)}\sigma_{i(\mu,1)}d_{i(\mu,2)}\sigma_{i(\mu,2)}\cdots d_{i(\mu,n)}\sigma_{i(\mu,n)}\right|\right) = \sum_i \langle d_i | h_{LF} | d_i \rangle + \sum_{i<j}\left(J_{ij} - K_{ij}\delta_{\sigma_i\sigma_j}\right)$$
$$= \frac{3m_\mu - 2n_\mu}{5} 10D_q + \beta_\mu B + \gamma_\mu C + E_0$$

The single determinants SD_μ^d are labelled with the subscript $\mu = 1, \ldots, \binom{10}{n}$ and with the superscript d to refer to pure d-spinorbitals. The values of m_μ and n_μ specify the electronic configuration $t_{2g}^{n_\mu} e_g^{m_\mu}$, while the β_μ and γ_μ are coefficients obtained after substituting standard expressions for the Coulomb J_{ij} and exchange K_{ij} integrals in terms of d-only orbitals d_i and spin functions σ_i. E_0 represents the gauge origin of energy.

Having obtained energy expressions for each SD_μ^d: 10Dq, B, C and E_0 are estimated using a least-squares procedure. Using matrix notation, we thus obtain an *overdetermined* system of linear equations with the unknown parameters stored in $\vec{\mathbf{X}}$ and given below.

$$\vec{\mathbf{E}} = \mathbf{A}\vec{\mathbf{X}}, \qquad \text{i.e.} \qquad \vec{\mathbf{X}} = (\mathbf{A}^T\mathbf{A})^{-1}\mathbf{A}^T\vec{\mathbf{E}}$$

Comparing SD energies from DFT with those calculated using the LF parameter values, we can state for all considered cases, that the LF parameterization scheme is remarkably compatible with SD energies from DFT; standard deviations between DFT-SD energies and their LFDFT values are generally between 0.01 and 0.05 eV as shown in Fig. 2 below:

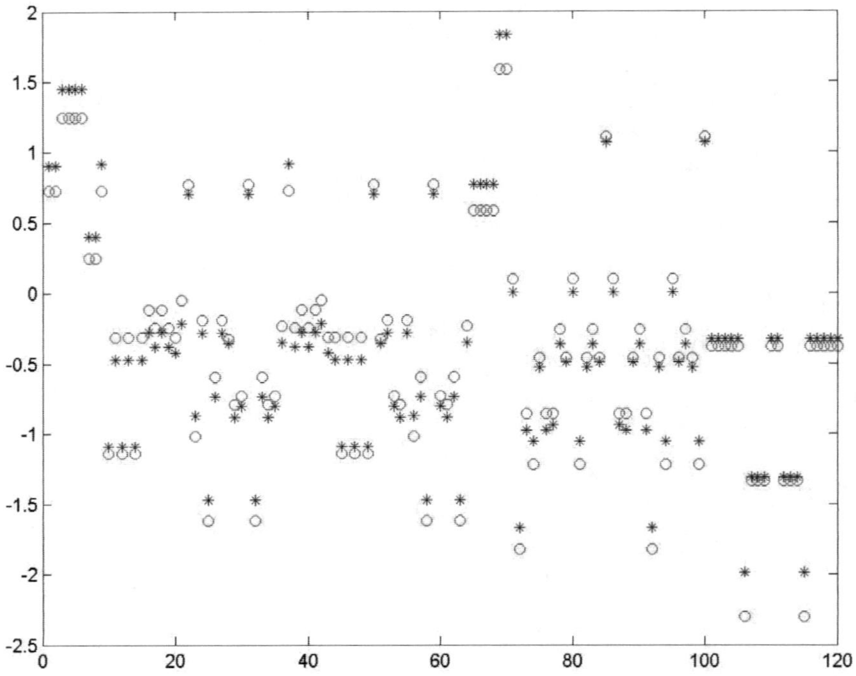

Fig. 2: Predicted LF energies (o) and calculated DFT energies (*) of the 120 micro-states of a d^3 manifold.

This model can obviously be generalized allowing to treat systems with symmetry lower than cubic or even without any symmetry (C_1). Here we make use of the general observation that the KS orbitals and the set of SD considered in eq.(1) convey all the information needed to setup the LF matrix.

Following the effective Hamiltonian approach, let us consider the KS orbitals dominated by d-functions which result from an AOC d^n DFT-SCF calculation. From the components of the eigenvector matrix built up from such MOs one takes only the components corresponding to the d functions. Let us denote the square matrix composed of the new column vectors by **U** and introduce the overlap matrix **S**:

$$\mathbf{S} = \mathbf{U}\mathbf{U}^{\mathrm{T}}$$

Since **U** is in general not orthogonal, we use Löwdin's symmetric orthogonalisation procedure to obtain an equivalent set of orthogonal eigenvectors (**C**):

$$\mathbf{C} = \mathbf{S}^{-\frac{1}{2}}\mathbf{U}$$

We identify now these vectors as the eigenfunctions of the effective LF Hamiltonian $h_{\mathrm{LF}}^{\mathrm{eff}}$ sought, as

$$\varphi_i = \sum_{\mu=1}^{5} c_{\mu i} d_{\mu}$$

Journal of Physics: Conference Series **428** (2013) 012023 doi:10.1088/1742-6596/428/1/012023

Thus, the fitting procedure described in the previous section will enable us to estimate $h_{ii} = \langle \varphi_i | h_{LF}^{eff} | \varphi_i \rangle$ and hence the full representation matrix of h_{LF}^{eff} as

$$h_{\mu\nu} = \langle d_\mu | h_{LF}^{eff} | d_\nu \rangle = \sum_{i=1}^{5} c_{\mu i} h_{ii} c_{\nu i}$$

The next step is now to generalize the fitting procedure for the case of no or low symmetry. The energy of a single determinant becomes thus:

$$E\left(SD_k^\varphi\right) = E\left(\det\left|\varphi_{i(k,1)}\sigma_{i(k,1)}\varphi_{i(k,2)}\sigma_{i(k,2)}\cdots\varphi_{i(k,n)}\sigma_{i(k,n)}\right|\right) = \sum_i \langle \varphi_i | h_{LF} | \varphi_i \rangle + \sum_{i<j} \left(J_{ij} - K_{ij}\delta_{\sigma_i \sigma_j}\right)$$

Where SD_k^φ is composed of the spinorbitals mentioned earlier.

In order to calculate the electrostatic contribution, it is useful to consider the transformation from the basis of SD_k^φ to the one of SD_μ^d. Using basic linear algebra, we get:

$$\left|SD_k^\varphi\right\rangle = \sum_\mu T_{k\mu} \left|SD_\mu^d\right\rangle$$

Where $T_{k\mu} = \det\left|c_{i(k,1:n),j(\mu,1:n)}\right|$ i.e. the determinant of a $n \times n$ sub-matrix of $\mathbf{C} \otimes \sigma$

$$\begin{vmatrix} c_{i(k,1),j(\mu,1)} & c_{i(k,1),j(\mu,2)} & \cdots & c_{i(k,1:n),j(\mu,n)} \\ c_{i(k,2),j(\mu,1)} & c_{i(k,2),j(\mu,2)} & \cdots & c_{i(k,2),j(\mu,n)} \\ \cdots & \cdots & \cdots & \cdots \\ c_{i(k,n),j(\mu,1)} & c_{i(k,n),j(\mu,2)} & \cdots & c_{i(k,n),j(\mu,n)} \end{vmatrix}$$

With the indices of the spinorbitals $\varphi_{i(k,1)}\sigma_{i(k,1)}, \varphi_{i(k,2)}\sigma_{i(k,2)}, \ldots, \varphi_{i(k,n)}\sigma_{i(k,n)}$ and $d_{j(\mu,1)}\sigma_{j(\mu,1)}, d_{j(\mu,2)}\sigma_{j(\mu,2)}, \ldots, d_{j(\mu,n)}\sigma_{j(\mu,n)}$ respectively. Note that these indices are in fact a two-dimensional array of (number of SD) x (number of electrons or holes) integers. Finally the energy of a SD can be rewritten as

$$E_k = E\left(SD_k^\varphi\right) = \sum_i \langle \varphi_i | h_{LF} | \varphi_i \rangle + \sum_{\mu,\nu} T_{k\mu} T_{k\nu} \left\langle SD_\mu^d \right| G \left| SD_\nu^d \right\rangle$$

Where $G = 1/r_{12}$ i.e. the electrostatic repulsion of all electron pairs in the LF manifold.

The matrix elements are readily obtained using Slater's rules and the resulting electrostatic two-electron integrals <ab|cd> in terms of Racah or Slater-Codon parameters. Thus the final equation to estimate h_{ii}, ($i = 1,\ldots, 5$), B and C from the DFT energies E_k of all the SD within the LF manifold will be of the same shape as eqs.(4) and (5) where $\vec{X} = (h_{11},\ldots,h_{55},B,C)$ and **A** is calculated as indicated above

Applications

Computational details

All DFT calculations have been performed using the ADF program package [6]. The approximate SCF KS one-electron equations are solved by employing an expansion of the molecular orbitals in a basis set of Slater-type orbitals (STO). All atoms were described through triple-ς STO basis sets given in the program database (basis set IV) and the core-orbitals up to 3p for the TM and up to 1s (for O, N, C), 2p (Cl) and 3d (Br) were kept frozen. We used the local density approximation (LDA), where exchange-correlation potential and energies have been computed according to the Vosko, Wilk and Nusair's (VWN) [7] parameterization of the electron gas data.

a) Calculated and experimental photophysical properties of Cs_2NaYCl_6:Cr^{3+} and Cs_2NaYBr_6:Cr^{3+}
The ground and excited state properties of the Cr^{3+} ion doped into the cubic host lattices Cs_2NaYCl_6 and Cs_2NaYBr_6 is studied using LF-DFT (vide supra). We calculate the energies of the multiplets $^4A_{2g}$, $^4T_{2g}$, ^2E_g , and $^4T_{1g}$. The adiabatic energy surface of both ground state and excited states of the CrX_6^{3-} cluster along the symmetry-adapted displacements (cf. fig. 3) of the six nearest halogen neighbours of the chromium impurity was calculated. The experimental Cr–X distance in the ground state can be reproduced to within 0.01 Å for both chloride and bromide systems. The calculated d – d excitation energies agree within 2'000 cm^{-1}. An energy lowering is obtained in the first $^4T_{2g}$ excited state when the octahedral symmetry of CrX_6^{3-} is relaxed along the e_g Jahn–Teller coordinate. The geometry corresponding to the energy minimum is in excellent agreement with the $^4T_{2g}$ geometry derived from high-resolution optical spectroscopy of Cs_2NaYCl_6:Cr^{3+}. It corresponds to an axially compressed and equatorially elongated CrX_6^{3-} octahedron. An exhaustive analysis of the problem is given in ref. [8].

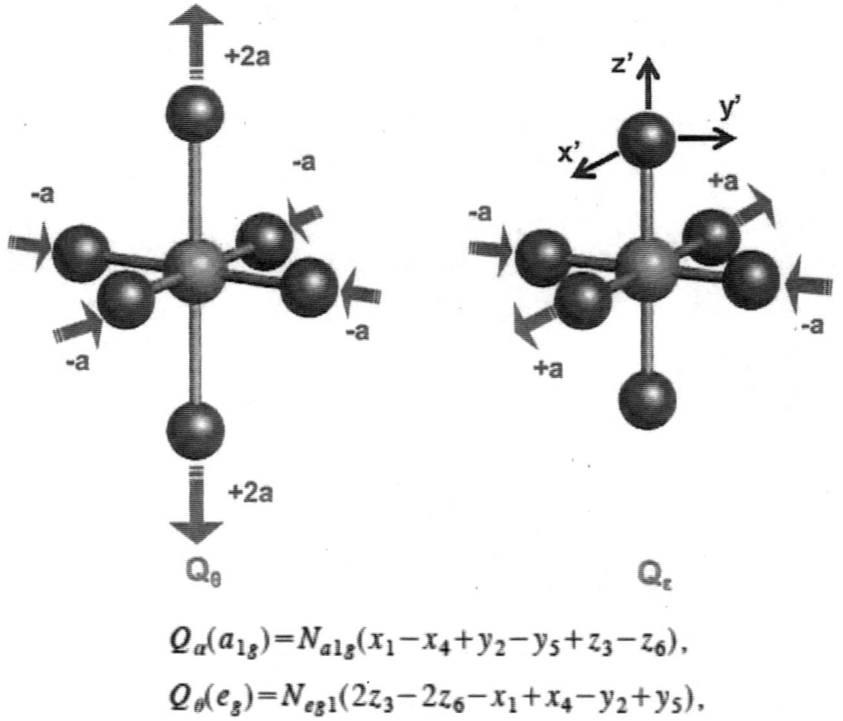

$$Q_a(a_{1g})=N_{a1g}(x_1-x_4+y_2-y_5+z_3-z_6),$$
$$Q_\theta(e_g)=N_{eg1}(2z_3-2z_6-x_1+x_4-y_2+y_5),$$
$$Q_\epsilon(e_g)=N_{eg2}(x_1-x_4-y_2+y_5),$$

Fig. 3: Symmetry-adapted displacements of the six nearest halogen neighbours of the chromium impurity

Fig. 4 below displays the contour plot of the $^4A_2$ potential energy surface as a function of both Q_θ and Q_ε.

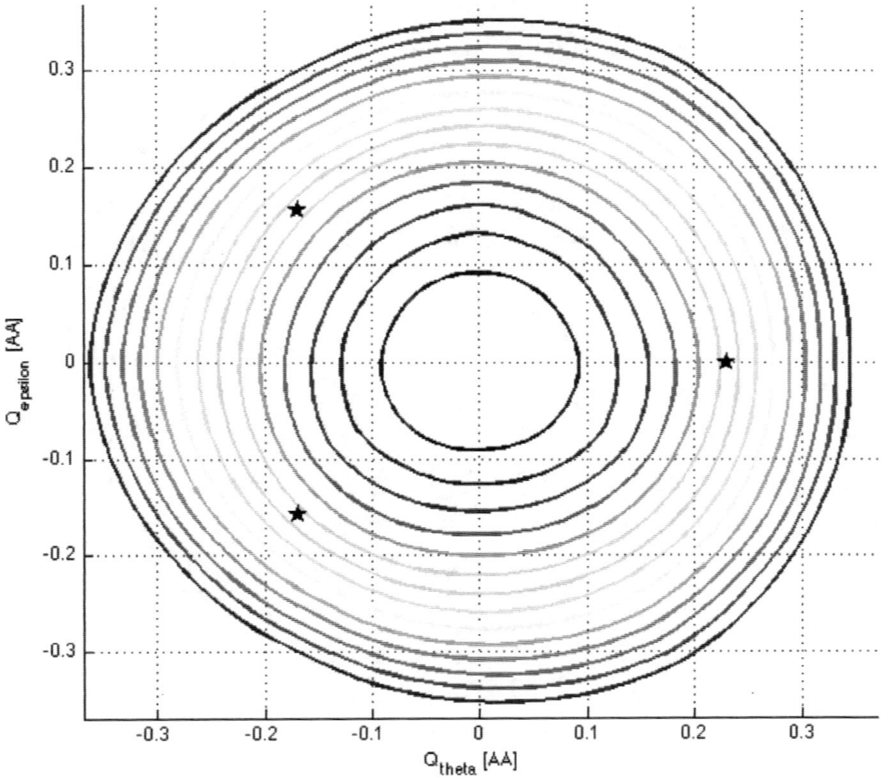

Fig. 4: Contour plot of the $^4A_2$ potential energy surface as a function of both Q_θ and Q_ε. The asterisks indicate the position of the three equivalent excited states $^4T_2$ distorted geometries

Fig. 5 below shows the contour plot of the $^4T_2$ potential energy surface as a function of both Q_q and Q_e

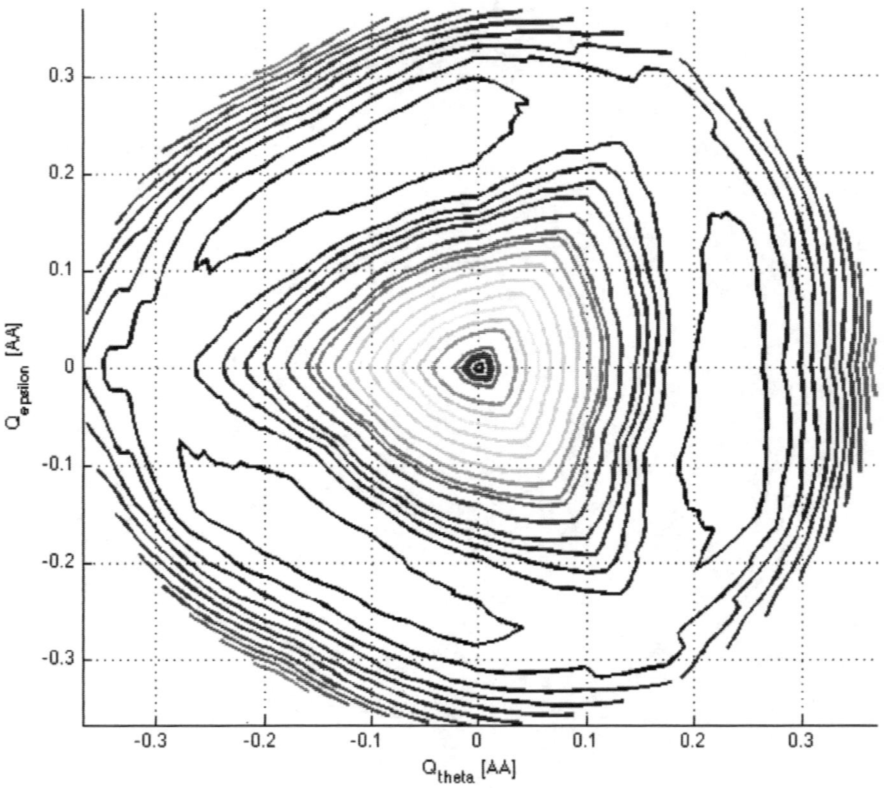

Fig. 5: Contour plot of the $^4T_2$ potential energy surface as a function of both Q_θ and Q_ε

Fig. 6 below represents the cross section through the $^4T_2$ potential energy surface along the Jahn-Teller active distortion Q_θ

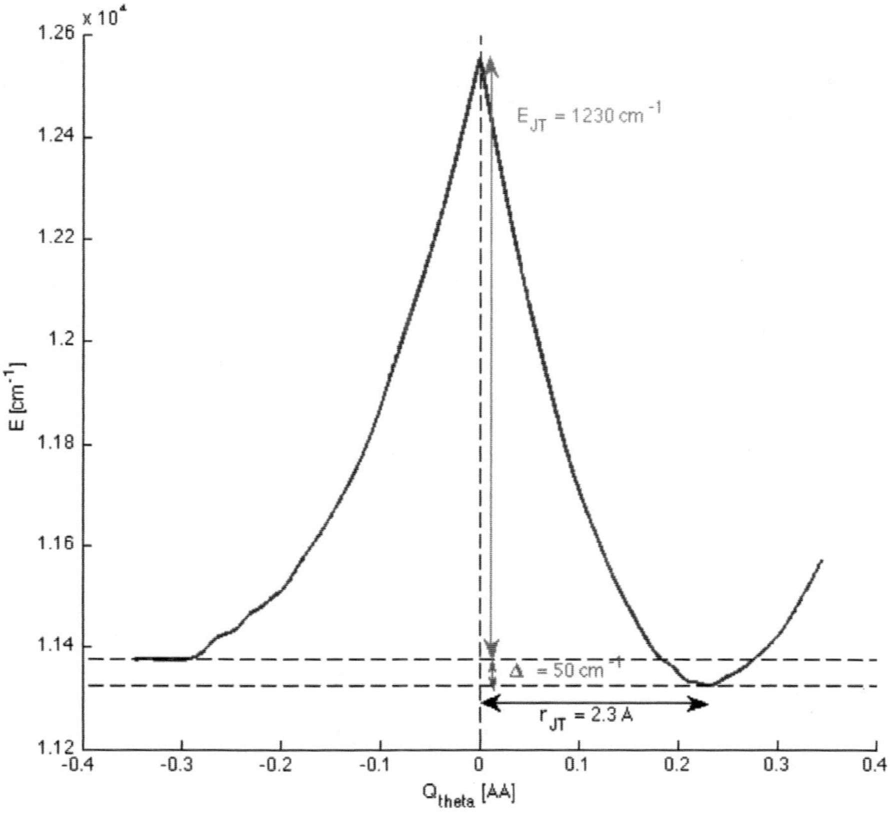

Fig. 6: Cross section through the $^4T_2$ potential energy surface along the Jahn-Teller active distortion Q_θ

Fig. 7 below shows the cross section through the $^4T_2$ and $^4A_2$ potential energy surfaces along the Jahn-Teller active distortion Q_θ:

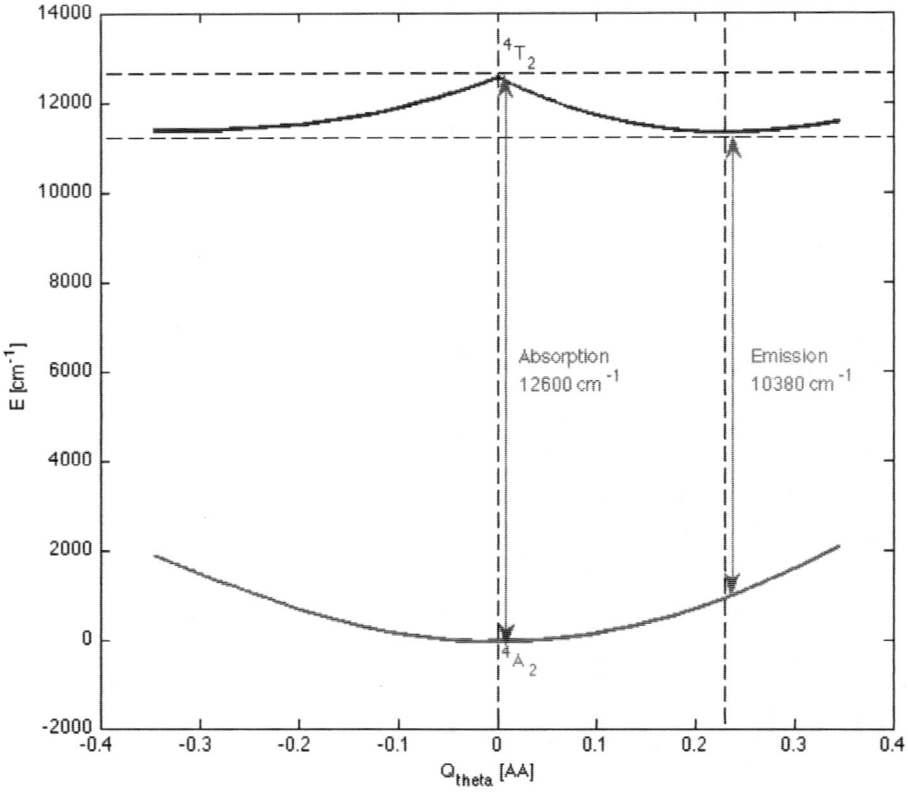

Fig. 7: Cross section through the $^4T_2$ and $^4A_2$ potential energy surfaces along the Jahn-Teller active distortion Q_θ

High-resolution absorption, MCD and luminescence spectra have been reported for $Cs_2NaYCl_6:Cr^{3+}$. Accurate values of the excited state energies are therefore known and included in Table 1 and 2. In particular, the region of $^4A_{2g} \rightarrow {}^4T_{2g}$ electronic origins was investigated in detail [9]. As it can be seen from inspection of these tables, the agreement between observed and predicted data, is excellent.

Journal of Physics: Conference Series **428** (2013) 012023 doi:10.1088/1742-6596/428/1/012023

Table 1. Calculated and experimental spectroscopic and structural properties for $Cs_2NaYCl_6:Cr^{3+}$. Units are Å and cm^{-1}.

		VWN	**BP**	**PW91**	**Exp.**
Geometry $^4A_2$	d(Cr-Cl) [Å]	2.382	2.462	2.401	2.44
Geometry $^4T_2$	d(Cr-Cl)(x,y) [Å]	2.492	2.592	2.512	2.54
	d(Cr-Cl)(z) [Å]	2.312	2.382	2.356	2.41
Absorption Max. [cm^{-1}]	$^4A_2 \rightarrow {}^4T_2$	12550	9760	12600	12800
Emission Max . [cm^{-1}]	$^4T_2 \rightarrow {}^4A_2$	11030	8382	10380	11000
Stokes Shifft [cm^{-1}]	$^4A_2 \rightarrow {}^4T_2$	1520	1378	2220	1800
Jahn-Teller Energy [cm^{-1}]	$^4T_2$	900	662	1230	
Absorption Max. [cm^{-1}]	$^4A_2 \rightarrow {}^4T_1$	15890	14650	17455	18200
Absorption Max. [cm^{-1}]	$^4A_2 \rightarrow {}^2E$	12320	11509	11850	14430

Table 2. Calculated and experimental spectroscopic and structural properties for $Cs_2NaYBr_6:Cr^{3+}$. Units are Å and cm^{-1}.

		VWN	**BP**	**PW91**	**Exp.**
Geometry $^4A_2$	d(Cr-Br) [Å]	2.525	2.615	2.579	2.60
Geometry $^4T_2$	d(Cr-Br)(x,y) [Å]	2.635	2.755	2.694	2.75
	d(Cr-Br)(z) [Å]	2.495	2.550	2.530	2.55
Absorption Max. [cm^{-1}]	$^4A_2 \rightarrow {}^4T_2$	11590	9500	11800	12400
Emission Max. [cm^{-1}]	$^4T_2 \rightarrow {}^4A_2$	10525	8055	9845	10200
Stokes Shifft [cm^{-1}]	$^4A_2 \rightarrow {}^4T_2$	1365	1445	1955	2200
Jahn-Teller Energy [cm^{-1}]	$^4T_2$	469	680	1000	
Absorption Max. [cm^{-1}]	$^4A_2 \rightarrow {}^4T_1$	14930	14550	15990	17700
Absorption Max. [cm^{-1}]	$^4A_2 \rightarrow {}^2E$	11390	11100	12100	13900

b) Tetrahedral d^2 $CrX_4(g)$ where $X = Cl^-, Br^-$

Tetrahedral d^2 complexes possess a $^3A_2(e^2)$ ground state as well as $^3A_2 \rightarrow {}^3T_2$ and $^3A_2 \rightarrow {}^3T_1$, e → t_2 singly excited states. They give rise to broad d-d transitions in the optical spectra. In addition, spin-flip transitions within the e^2 configuration lead to sharp line excitations. Multiplet energies from LDA agree within a few hundred cm^{-1} with experimental data. In particular the $^3A_2 \rightarrow {}^3T_2$ transition energy and thus 10Dq nicely agrees with experiment as is seen from inspection of table 3. Experimental transition energies for $CrCl_4$ and $CrBr_4$ as well as values of B, C and 10Dq deduced from a fit to experiment for $CrCl_4$ are also listed.

Table 3. Electronic transition energies of CrX_4, X=Cl and Br, with geometries optimized using LDA functional and calculated using values of B, C and 10Dq from least square fit to DFT energies of the Slater determinants according to the method described earlier.

	$CrCl_4$			$CrBr_4$	
Term	This work	LF-fit	Exp.[a]	This work	Exp.[a]
$^3A_2(e^2)$	0	0	0	0	0
$^1E(e^2)$	6542	6089	-	6373	6666
$^1A_1(e^2)$	11114	10586	-	10698	10869
$^3T_2(e^1t_2^1)$	7008	7010	7250	6163	-
$^3T_1(e^1t_2^1)$	10316	10440	10000	9269	-
$^1T_2(e^1t_2^1)$	13454	12991	12000	12434	-
$^1T_1(e^1t_2^1)$	15074	14718	-	14037	-
$^1A_1(t_2^2)$	32099	30599	-	30120	-
$^1E(t_2^2)$	21121	20716	-	19271	-
$^3T_1(t_2^2)$	16033	16229	16666	14424	13258
$^1T_2(t_2^2)$	21217	20822	-	19373	-
R(M-X)	2.104	-	-	2.264	-
B	355	376	-	347	-
C	1903	1579	-	1855	-
10Dq	7008	7250	-	6162	-
Std Dev.	0.030	-	-	0.030	-

Optimized geometries correspond to stable minima in the ground state potential surface with harmonic frequencies of the α_1, ε, $\tau_2(1)$ and $\tau_2(2)$ vibrational modes of 359, 113, 126 and 464 cm^{-1} (for $CrCl_4$); 207, 63, 82 and 350 cm^{-1} (for $CrBr_4$); respectively; [a] Studer P, Thesis, University of Fribourg, 1975.

c) Octahedral Cr^{III} d^3 complexes.

In Table 4 we list the predicted (this work), adjusted (LF fit to exp.) and observed (Exp.) multiplet energies for CrX_6^{3-} ($X = F^-, Cl^-, Br^-, CN^-$) complex ions. We used a LDA functional to calculate the Cr^{III}-X bond lengths and we compare these results with energies from a LF calculation utilizing values of B, C and 10Dq obtained from a best fit to the spectra. Bond lengths are too long while values of 10Dq are too small compared to experiment. The situation improves if instead of optimized, experimental bond lengths are taken for the calculation. Even in this case, spin-forbidden transitions come out by 3000-4000 cm^{-1} too low in energy compared to experiment. Clearly, in this example of highly charges species, our prediction is much less accurate. In order to

Table 4. Electronic transition energies of CrX_6^{3-}, X = F^-, Cl^-, Br^- and CN^- molecules with geometries optimized using LDA functionals calculated using values of B, C and 10Dq from least square fit to DFT energies of the Slater determinants and to experiment. The values of $(10Dq)_{orb}$ as deduced from the e_g-t_{2g} KS-orbital energy difference taken from the $\ldots t_{2g}^{1.8} eg^{1.2}$ SCF KS-energies are also listed. Experimental transition energies are also listed.

Term	CrF_6^{3-}			$CrCl_6^{3-}$			$CrBr_6^{3-}$			$Cr(CN)_6^{3-}$		
	This work	LFT fit to exp.	Exp.	This work	LFT fit to exp.	Exp.	This work	LFT fit to exp.	Exp.	This work	LFT fit to exp.	Exp.
$^4A_{2g}(t_{2g}^3)$	0	0	0	0	0	0	0	0	0	0	0	0
$^2E_g(t_{2g}^3)$	12497	15802	16300[a]	10756	14426	14430[b]	10333	13900	13900[b]	9413	12034	12460[f]
$^2T_{1g}(t_{2g}^3)$	13044	16461	16300[a]	11180	14873	-	10694	14348	-	9682	12425	13070[f]
$^2T_{2g}(t_{2g}^3)$	18628	23260	23000[a]	15918	21037	-	15185	20281	-	15234	19084	18370[f]
$^4T_{2g}(t_{2g}^1e_g^1)$	13569	15298	15200[a]	10911	12800	12800[b]	9816	12400	12400[b]	30760	26595	26700[f]
$^4T_{1g}(t_{2g}^1e_g^1)$	19443	22262	21800[a]	15618	18198	18200[b]	13992	17700	17700[b]	35910	32743	32680[f]
$^2A_{1g}(t_{2g}^1e_g^1)$	24071	28709	-	20056	25351	-	18709	24459	-	38324	36488	-
$^2T_{1g}(t_{2g}^1e_g^1)$	26348	31473	-	21878	27421	-	20316	26503	-	40022	38569	-
$^2T_{2g}(t_{2g}^1e_g^1)$	25959	30970	-	21568	27079	-	20047	26159	-	39983	38438	-
$^2E_g(t_{2g}^1e_g^1)$	27819	33341	-	23147	29098	-	21530	28126	-	41084	39952	-
$^4T_{1g}(t_{2g}^1e_g^1)$	30339	34636	35000[a]	24375	28455	-	21861	27643	-	63147	55352	-
R(M-X)	1.957	-	1.933[c]	2.419	-	2.335[d]	2.588	-	2.47[e]	2.071	-	2.077[g]
B	605	734	-	484	550	-	427	543	-	452	554	-
C	2694	3492	-	2403	3450	-	2395	3296	-	1919	2559	-
10Dq	13598	15297	-	10911	12800	-	9816	12400	-	30760	26595	-
SD	0.113	-	-	0.105	-	-	0.113	-	-	0.105	-	-
$(10Dq)_{orb}$[h]	13928	-	-	10775	-	-	9622	-	-	30953	-	-

[a] K_3CrF_6: Allen GC, El-Sharkawy AM, Warren KD (1971) Inorg.Chem. 10:2538
[b] $Cs_2NaYCl[Br]_6$: Cr^{3+}, Schwartz RW (1976) Inorg.Chem. 15:2817
[c] Knox K, Mitchell DW (1961) J. Inorg. Nucl. Chem. 21:253
[d, e] estimated for $Cs_2NaCrCl_6$ and $Cs_2NaCrBr_6$, Gilardoni F., Weber J., Bellafrouh K., Daul C., Guedel H.-U., J. Chem. Phys. 104 (1996) 7624
[f] Witzke H, Theoret. Chim.Acta 20:171(1974)
[g] Jagner S, Ljungstroem E, Vannerberg N-G (1974) Acta Chem. Scand. A28:623
[h] This entry lists the KS-eigenvalue differences $\varepsilon(e_g)$- ε (t_{2g}) for the sake of comparison

improve the quality of the prediction we obviously need to consider the environment of the CrX_6^{3-} chromophore by adding an appropriate embedding potential to the KS-hamiltonian. Already the use of experimental bond lengths does significantly improve the precision of our calculation as mentioned before. A full analysis of this problem is given in ref. [10]. For the $Cr(CN)_6^{3-}$ ion we note a value of 10Dq which is by 4000 cm^{-1} higher than the experimental one. This is unusual and not expected: calculated and experimental Cr-C bond distances are indeed very close. We can conclude that for the latter system, existing DFT functionals do not perform properly. In addition the standard deviations are too high. More work is needed to better understand the phenomenon and to improve this.

d) Octahedral Hexaacquo Ni(ii) d^8 complexes.

In Table 5 we list the multiplet energies for the $Ni(H_2O)_6^{2+}$ obtained according to the method outlined in section 2. The experimental electronic spectrum of Ni(ii) ion in aqueous solution exhibit three absorption peaks centered at 8'800cm^{-1} ($^3A_2 \rightarrow {}^3T_2$), at 15'000$cm^{-1}$ ($^3A_2 \rightarrow {}^3T_1$) and at 25'500$cm^{-1}$ ($^3A_2 \rightarrow {}^3T_1$). The agreement with the predicted multiplets is perfect in this case.

Table 5. Predicted multiplet energies of $Ni(H_2O)_6^{++}$ in D_{2h} symmetry based on the five KS-orbitals with dominant metal-3d character. The fine structure (spin-orbit coupling) is also included

Degeneracy (Multiplicity)	Multiplet Energy [10^3cm^{-1}]	S(S+1)	Spin-Orbit splitting [10^3cm^{-1}]		
3	0.0000000	2.000	0.0000000	0.0001074	0.0001645
3	8.8252743	2.000	8.5714426	8.6014099	8.7127700
3	8.8414083	2.000	8.7298366	8.7386983	9.0699081
3	8.8737000	2.000	9.0847504	9.0917825	9.2288776
3	14.5809908	2.000	13.9177850	14.1466476	14.1493461
3	14.5921568	2.000	14.3609969	14.3613509	14.3698568
3	14.5934211	2.000	15.0584802	15.0595351	15.0618965
1	14.6214912	0.000	15.7200626		
1	14.6221451	0.000	15.7224490		
1	23.0538531	0.000	22.8226585		
1	23.0674913	0.000	22.8343287		
1	23.0966616	0.000	22.8599397		
1	23.5866071	0.000	23.7076105		
3	25.2933052	2.000	25.2187733	25.2277340	25.5319615
3	25.3050996	2.000	25.5347498	25.5451571	25.5868215
3	25.3063639	2.000	25.5873367	25.5920497	25.7705637
1	27.5214320	0.000	27.6420000		
1	27.5214702	0.000	27.6420403		
1	27.5376041	0.000	27.6580624		
1	34.1878765	0.000	34.3128908		
1	34.2156132	0.000	34.3405845		
1	34.5899024	0.000	34.7083317		
1	34.5948335	0.000	34.7133008		
1	34.6084717	0.000	34.7269772		
1	60.3671270	0.000	60.5117478		

From the spin-orbit splitting of the $^3A_2$ ground state it is possible to obtain the ZFS D-tensor using a conventional spin hamiltonian approach

$$H_{ZFS} = \vec{S} \cdot \mathbf{D} \cdot \vec{S} = D\left(\hat{S}_z^2 - \frac{2}{3}\right) + E\left(\hat{S}_x^2 - \hat{S}_y^2\right)$$

and equating the energies of the spin-orbit components of the $^3A_2$ to the eigenvalues of this spin hamiltonian. The result is given below :

$$\begin{bmatrix} 0.0000 \\ 0.1074 \\ 0.1645 \end{bmatrix} = \begin{bmatrix} -2/3 & 0 & 1 \\ 1/3 & -1 & 1 \\ 1/3 & 1 & 1 \end{bmatrix} \begin{bmatrix} D \\ E \\ C \end{bmatrix}$$

The solution of this set of three linear equations is: D = 0.1360 cm^{-1} and E = 0.0285 cm^{-1}. The third parameter C describes the arbitrary energy origin. There are experimental measurements of the ZFS for $Ni(H_2O)_6^{++}$ in $NiSiF_6.6H_2O$ available. At 90K the ZFS is 0.17 cm^{-1}. However, D is very sensitive on alteration of the temperature and pressure. Thus, an exhaustive description of the observed ZFS needs a dynamical study of the process which goes far beyond the goal of this work.

Aknowledgement: This work was supported by the Swiss National sciences Foundation.

References

1 M. Atanasov, C. A. Daul, C. Rauzy, Chem. Phys. Lett. 367 (2003) 737-746
M. Atanasov, C. A. Daul, C. Rauzy, Structure & Bonding 106 (2004) 97-125
2 A. Borel, C. A. Daul, L. Helm, Chem. Phys. Lett., 383 (2004) 584-591
3 Atanasov, M.; Baerends, E. J.; Baettig, P.; Bruyndonckx, R.; Daul, C.; Rauzy, C.; Zbiri, M. Chem. Phys. Letters (2004), 399(4-6), 433-439.
4 M. Atanasov, C. A. Daul, Chem. Phys. Lett. 379 (2003) 209
M. Atanasov, C. A. Daul, Chem. Phys. Lett. 381 (2003) 584-591
5 Senn, F.; Daul, C. A.; J. Molec. Struct.: THEOCHEM (2010), 954(1-3), 105-108.
6 E.J.Baerends, D.E.Ellis, P. Ros, Chem.Phys. 2 (1993) 42; P.M.Boerrigter, G. te Velde, E.J.Baerends, Int. J. Quantum Chem. 33 (1988) 87; G. teVelde, E.J.Baerends (1992) Comput.Phys. 99(1992)84.
7 S.H.Vosko, L.Wilk, M. Nusair, Canadian J. Phys. 58 (1980) 1200.
8 K. Bellafrouh, C. A. Daul, H. U. Güdel, F. Gilardoni, J. Weber: J. Chem. Phys. 104, 7624 (1996)
9 R. Knochenmuss, C. Reber, M. V. Rajasekharan, and H. U. Güdel, J. Chem. Phys. 85, 4280 (1986).
10 K. Bellafrouh, C. A. Daul, H. U. Güdel, F. Gilardoni, J. Weber, Theor. Chim. Acta 91 (1995) 215-224

Historical remarks on the birth of the ligand field diagrams and the advent of laser

Satoru Sugano and Yukito Tanabe, adapted by H. Koizumi

Abstract. Tanabe and Sugano's personal accounts are given for the birth of the Tanabe-Sugano diagrams and the advent of laser.

1. Introduction

The first laser was operated on May 16, 1960, by T. H. Maiman, using ruby (Al_2O_3 crystal Cr^{3+} impurities doped) R-line [1]. The R-line corresponds to the transitions between $^4A_2$ and ^2E states in Fig. 1. For this achievement, energy diagrams of d electrons in the ligand field calculated by Y. Tanabe and S. Sugano (Tanabe-Sugano diagrams [2]) played a pivotal role. Their calculation was not meant to be utilized for the development of lasers; Tanabe did it for pure scientific interest since he was fascinated by the beautiful Racah's method; Sugano was also fascinated by the Racah's method, however he seemed to have an inspiration that some practical application might arise from their calculation. In this paper, a brief account is given how the calculation of the Tanabe-Sugano diagram was started, and how it contributed to the advent of laser.

2. The Confirmation of Validity

The validity of the ligand field theory was confirmed by studying optical transitions between the terms with the same orbital configurations. In the diagram, where the ground level is taken as the horizontal coordinate axis, the excited terms with the same orbital configuration as that of the ground term are almost parallel to the horizontal axis. This means that the transition energies are independent of the magnitude of a cubic crystal-field. Tanabe and Sugano noticed that optical transitions of these excited terms should be observed experimentally as a group of relatively sharp spectral lines. Actually, coexistence of such sharp spectral lines without orbital excitations and broad absorption bands with orbital excitation is the important result derived from the energy levels diagram in the ligand-field theory (or the Tanabe-Sugano diagarm) [2].

In ruby sharp spectral lines are observed as optical transitions between terms with orbital configuration t_2^3. These transitions of ruby are shown as R, R', and B in Fig. 1, where lift of Kramers degeneracy at right numerical figures are completely neglected [5].

2.1. Optical Studies of Transitions

It was in the nineteen-fifties that the crystal field theory of H. Bethe proposed in 1929 was applied, for the first time, to highly excited states such as those optically excited states of d-

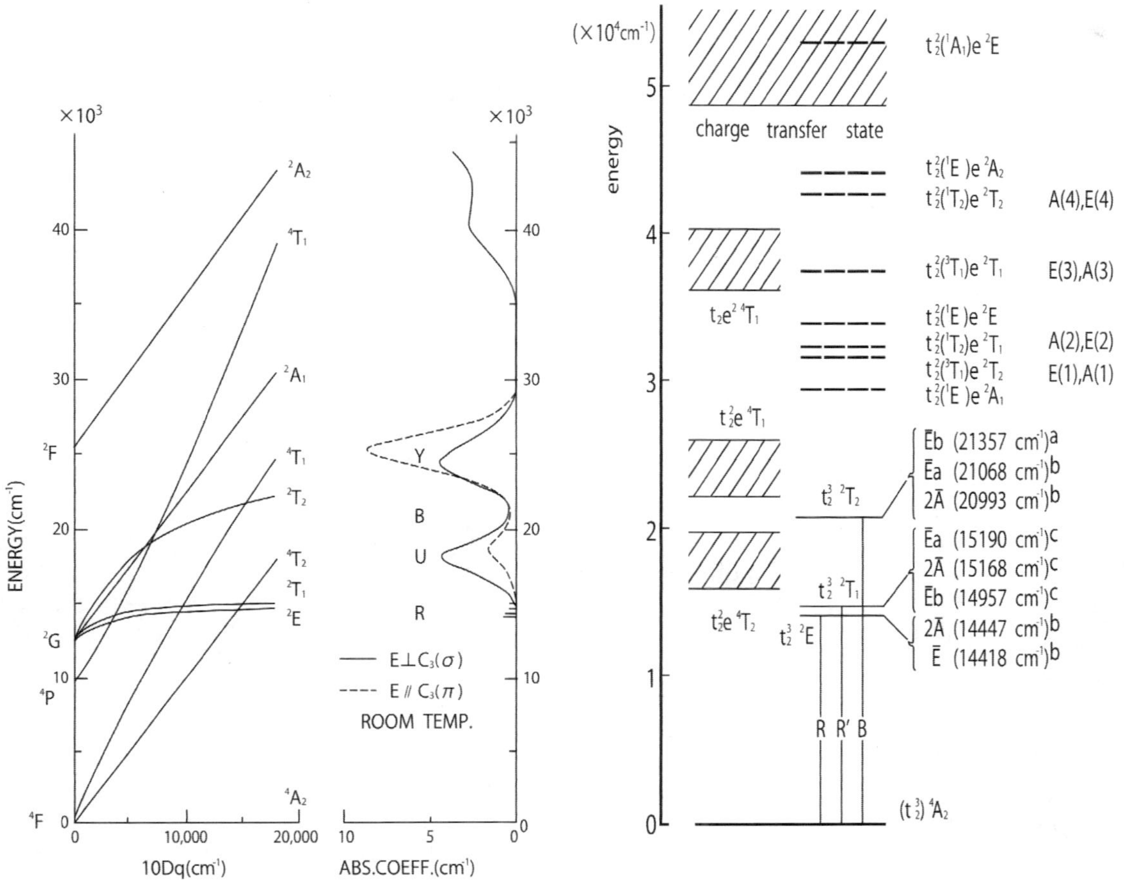

Figure 1. Left: The Tanabe-Sugano diagram for the d^3 system [2]. Middle: The absorption spectrum of ruby [3]. Right: The energy level diagram of ruby [4]. Note that in the Figs. 5.1-5.7 of Ref. [2], $d\epsilon$ and $d\gamma$ should be read as t_2 and e, respectively, in the present article.

electron compounds, though the theory has been applied to low excited states where magnetic and thermal properties are discussed.

We now show that the extended theory of Bethe, called ligand-field theory may be applied to highly excited states up to ~ 5 eV. In Fig. 2, the observed absorption spectrum of optically pumped ruby in the t_2^3 state by T. Kushida is compared with calculated one [6].

2.2. Fine Structure Spectral Line

In order to make detailed assignment of a group of sharp spectral lines coming from the transitions without orbital excitation, we have to study fine-structure of the terms caused by crystal fields of low symmetries, caused by application of external electric (Stark effect) and external magnetic fields (Zeeman effect). The best example of these transitions are found in ruby. In Fig. 2, the observed Zeeman pattern of the R1 and R2 absorption lines of ruby is compared with the theory [7, 8]; Pseudo-Stark splittingΔ γ of the R-lines of ruby are also compared [9].

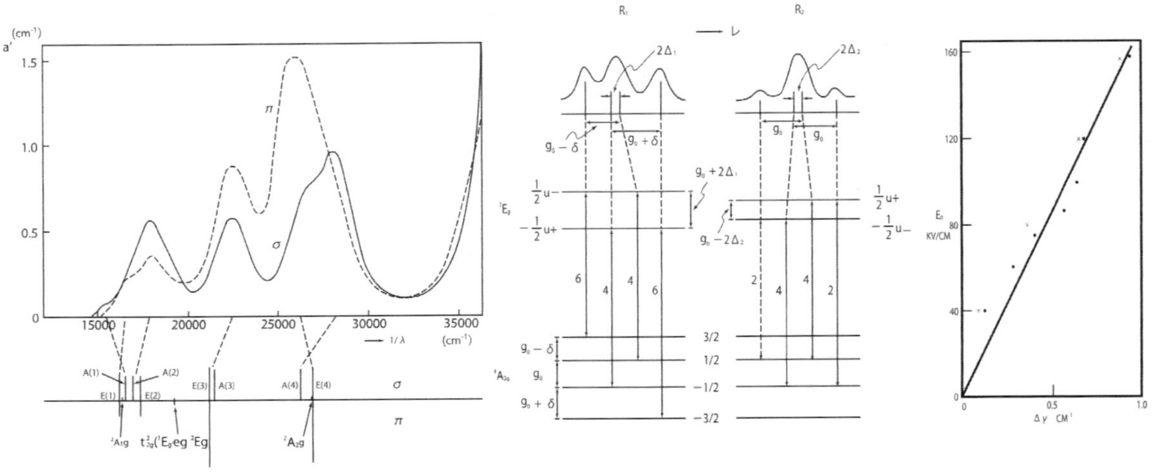

Figure 2. Left: The observed absorption spectrum of optically pumped ruby in the t_2^3 state by T. Kushida [6]. Middle: The observed Zeeman pattern of the R1 and R2 absorption lines of ruby at 20 K for a magnetic field parallel to the trigonal axis and the polarizatio perpendicular to it [7, 8]. Right: Pseudo-Stark splitting$\Delta\ \gamma$ of the R-lines of ruby vs applied electric field E_0; Data are obtained from the R1(•) and R2(×) lines [9].

3. At Bell Labs

Sugano stayed in Bell Labs from September 1959 to August 1961, and associated with A. Schawlow, D. Wood and W. Kaiser. His host was A. Liehr.

In September 1959, at the 1st quantum electronic conference, Schawlow talked about ruby as a possible laser medium; however, he dismissed pink ruby because its three-level system would require drastically depleting the ground-state population and also because *its fluorescence efficiency was then believed to be low* [10].

Sugano knew that the pink ruby' s fluorescence efficiency is high through the communication with G. Kuwabara and A. Misu. He told about this to Lier, then, his association with Schawlow, Wood, and Kaiser started. But somehow the high fluorescence efficiency of the pink ruby did not reach Schawlow.

But Schawlow and coworkers F. Varsanyi and Wood found in December 1959 that pink ruby actually has a high fluorescence efficiency. In June 1960 a paper by Maiman also reported that he had seen partial ground state depletion under flash-lamp excitation in pink ruby [1].

4. Epilog

Tanabe recalled that he never thought that laser was possible in the ruby which he have been so familiar and is an expert on it; however, retrospectively, it is totally incomprehensible why such a possibility did not come to his mind.

Sugano was impressed by the paper by Schawlow and Townes on the possibility of the optical maser [11]; however, he felt that the ruby could be used for the laser emission from association with experimentalists dealing with ruby. As seen in Fig. 1 of Ref. [5], the longest wavelength photoemission from ruby is very strong.

Anyway, things like this (even you are next to a big discovery you do not realize it and miss it) happen very often in reality; but we can say that what you are studying is more fundamental, it will more likely lead to discoveries that are of very good use [12].

Journal of Physics: Conference Series **428** (2013) 012024 doi:10.1088/1742-6596/428/1/012024

References

[1] Maiman T H 1960 *Phys. Rev. Lett.* **4** 564
[2] Tanabe Y and Sugano S 1954 *J. Phys. Soc. Japan* **9** 766
[3] Sugano S, Tanabe Y and Kamimura H 1970 *Multiplets of transition-metal ions in crystals* (New York: Academic Press)
[4] Kushida T 1966 *J. Phys. Soc. Japan* **21** 1331
[5] Sugano S 1987 *A Tribute to Arthur L. Schawlow* ed Yen W M and Levenson M D (Springer-Verlag) p 224
[6] Shinada M, Sugano S and Kushida T 1966 *J. Phys. Soc. Japan* **21** 1342
[7] Sugano S and Tanabe Y 1958 *J. Phys. Soc. Jpn.* **13** 880
[8] Sugano S and Tsujikawa I 1958 *J. Phys. Soc. Japan* **13** 899
[9] Kaiser W, Sugano S and Wood D L 1961 *Phys. Rev. Lett.* **6** 605
[10] Nelson D F, Collins R J and Kaiser W 2010 *Phys. Today* **Jan** 40
[11] Schawlow A L and Townes C H 1958 *Phys. Rev.* **112** 1940
[12] Tanabe Y 1988 *Busturi (Nihon Butsuri Gakkai Shi)* **43** 907

Journal of Physics: Conference Series **428** (2013) 012025 doi:10.1088/1742-6596/428/1/012025

PROTON-LATTICE COUPLING AND VIBRONIC EFFECTS IN KDP-FAMILY THERMODYNAMICS

A.A.Levin, S.P.Dolin, T.Yu.Mikhailova, M.V.Solin
N.S.Kurnakov Institute of General and Inorganic Chemistry RAS
Leninskii pr. 31 Moscow 119991 Russian Federation

Mail to: levin@igic.ras.ru

Based on the vibronic theory of heteroligand complexes a new model Hamiltonian is proposed for the ferroelectrics of KDP (KH_2PO_4) family. In local representation three items of this Hamiltonian describe the H-bond protons, the potential energy of lattice's oscillators and the coupling of these two subsystems respectively. In Ising form, convenient for numerical calculations, the Hamiltonian offered includes explicitly the characteristics of the electronic structure as well as the orbital vibronic constants of the AO_4 tetrahedra. The Ising version of theory is then applied to numerical studying of some questions for KDP thermodynamics.

Introduction

Recently [1,2 and references therein] we demonstrated that vibronic theory of heteroligand molecular systems (VTHS) may be applied in the microscopic theory of H-bonded ferroelectrics like KDP and squaric acid $H_2C_4O_4$. In present work we use this approach to handle some thermodynamic problems for the KDP family materials. With the account for proton-lattice coupling (PLC) on the base of four-particle Bethe cluster approximation we consider the PLC contribution to the critical temperature T_c value for KDP, this family forefather.

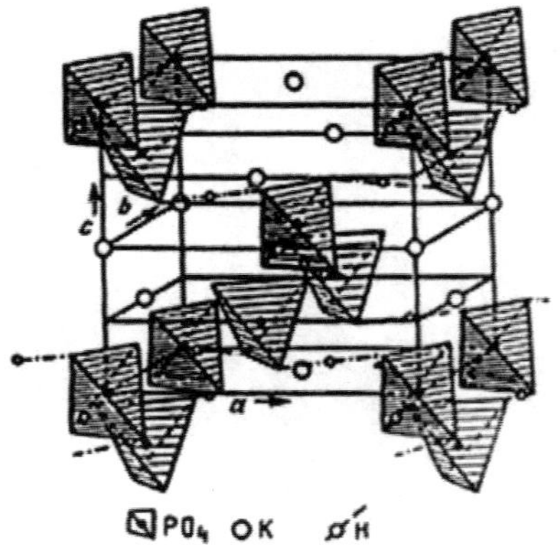

Fig.1. KDP crystal structure

Two subsystems in KDP-like materials.

The space structure of the KDP-like ferroelectrics (FE) MH_2AO_4 (M=K,Rb, A=P,As) (Fig.1) clearly includes two main subsystems – a "non-hydrogen" frame that is formed by mutually bounded tetrahedral anionic complexes AO_4 and a set of hydrogen atoms on the symmetric double-well H-bonds O–H…O ↔ O…H–O, that connect the AO_4 tetrahedra. Two equilibrium positions of H-atom at each H-bond are close to the minima of the proton's adiabatic potential (here and below for brevity's sake we say 'hydrogen' and 'proton' meaning hydrogen as well as deuterium or its nuclei). We don't include the M atoms in the non-hydrogen subsystem although it is possible, see [3,4]. From structural chemistry viewpoint these atoms may play most probably the role of electron donors for the AO_4-framework rather than of complex-forming agents like P and As, thus for materials at hand we handle the PLC only with due account taken of protons and AO_4-tetrahedra.

Protons' mobility for the FE considered (according to conventional theory [5,6]) is related to their tunneling between the minima of the potential curves. Thus several (not more than four) H atoms may be situated at the nearest sites around each AO_4-tetrahedron. Hence each oxygen atom in the crystal may be ascribed to one of two kinds: O (-) or else O(…) in dependence on the character of its bonding with the nearest hydrogen – covalent (O-H) or hydrogen (O…H). Then any AO_4-tetrahedron may be considered as a heteroligand complex $AO(-)_kO(\ldots)_{4-k}$ (k=0,1…4) and be treated by means of the VTHS approaches.

VTHS basics.

The VTHS [2] treats the heteroligand complex $AL_{N-k}XY\ldots Z$ by means of the second-order perturbation theory as the derivative of the homoligand AL_N and for our purposes one may restrict itself with the case of closed electron shells for both complexes. Then the lowest energy surface $E(Q)$ of the adiabatic potential (APES) for the heteroligand complex has the form

$$E(Q) = S_0 + \frac{1}{2}\sum_v k_v Q_v^2 - 2\sum_{n,m}\frac{S_{nm}^2}{\omega_{nm}} - 4\sum_v\sum_{n,m}\frac{S_{nm}A_{nm}^v}{\omega_{nm}}Q_v \,. \qquad (1)$$

Here $Q = \{Q_v\}$ is the set of normal (or symmetry) coordinates for the AL_N system, k_v and A_{nm}^v are its force constants and orbital vibronic constants, written in the basis of occupied (n) and vacant (m) molecular orbitals (MO) of the non-perturbed complex AL_N and ω_{nm} -

energy gaps between these MO. Notations S_{nm} in (1) respond to matrix elements in the MO basis for the "substitution operator" H_S [2], that defines the influence of ligand substitution and S_0 is the first-order perturbation term. One can easily find from (1) the main parameters of the APES considered – its minimum coordinates Q_ν^f and the depth $E(Q^f)$ of this minimum - with regard to the analogous parameters of unperturbed homoligand system:

$$Q_\nu^f = \frac{4}{k_\nu}\sum_{n,m}\frac{S_{nm}A_{nm}^\nu}{\omega_{nm}}. \tag{2}$$

$$E(Q^f) = S_0 - 2\sum_{n,m}\frac{S_{nm}^2}{\omega_{nm}} - 8\sum_\nu \frac{1}{k_\nu}\left(\sum_{n,m}\frac{S_{nm}A_{nm}^\nu}{\omega_{nm}}\right)^2, \tag{3}$$

where relation (2) describes the geometry change of homoligand complex upon complete or partial ligand replacement and (3) – the corresponding change of its potential energy. To use (2), (3) we need to determine S_{nm}. Let's adopt [2] that the replacement of an atom in the complex is equivalent to the replacement of the AOs' energies of the eliminated atom by the AO's energies of the atom-substituent. Then one has for S_{nm}

$$S_{nm} = \sum_i c_{in}c_{im}\Delta\alpha_i. \tag{4}$$

where c_{in}, c_{im}- are the coefficients at the *i*-th AO in *n*-th and *m*-th substituted atom MO and $\Delta\alpha_i$ - the change of AO's orbital energy (one-center Coulomb integral) upon this replacement. Let's mention in this connection that comparison with experiments confirms the use of (2) and (4) to describe quite well the observed geometry changes upon ligands' substitution [2].

The use of VTHS for KDP

In KDP-like FE any proton distribution determines a definite distribution of two kinds of oxygens over the lattice i.e. over the vertices of every AO_4 tetrahedron. The «inner» potential energies of such complexes are determined by (3) if compared with standard unperturbed tetrahedral complex in the hypothetic "protophase", where all protons are situated at the centres of H-bonds. Then it seems reasonable to consider the potential energy *H* of the macroscopic FE pattern to be equal to the sum of the "inner" energies of the tetrahedral complexes [1,7]. In so doing one should keep in mind that two oxygens of different kinds

always participate in each H-bond. This reason restricts the composition of neighbouring complexes and removes the first-order terms as S_0 in (3) and (1) from the Hamiltonian.
Another important property of H deals with S_{nm} matrix elements in relations (2), (3). It seems clear that proton transfer from the H-bond centre to the position correspondent to O(-) and O(…) formation at the ends of this bond influence oppositely the levels of the AOs for these two oxygens. Thus the valence levels of oxygens through the whole crystal depend on proton distribution. In particular, it concerns the σ-orbitals of O atoms that take part in the bonds A-O(-) and A-O(…). For these AO's the changes of the Coulomb integrals may be written as:

$$\Delta\alpha_i = |\Delta\alpha|\sigma_i^z \qquad (5),$$

where $\sigma_i^z = \pm 1$ are the discrete dichotomic variables (pseudospins) [1].

Local representation of the Hamiltonian.

The relations above yield in two different forms of H. The first one may be obtain by summation of (3) for all AO_4 tetrahedra (t) of the KDP pattern over the points of the APES'es minima. Taking into account the H-bonds connections between the tetrahedra and using the formulae (3), (4) and (5) then gives [7]

$$H = -\frac{1}{2}\sum_{t,\alpha\neq\beta} J^{el}_{t\alpha\beta}\sigma^z_{t\alpha}\sigma^z_{t\beta} + \frac{1}{2}\sum_{t,\nu} k_{t\nu}Q^2_{t\nu} - \sum_{t,\nu,\alpha} k_{t\alpha\nu}Q_{t\nu}\sigma^z_{t\alpha}. \qquad (6)$$

It is seen from (6), that the obtained Hamiltonian includes three terms. The first one deal with the proton subsystem of the material, the second one characterises the harmonic potential energy of the 'lattice' described. Finally the third term contains the products of pseudospins and oscillator variables and responds to PLC in its local form [3,4]. The obtained form of H clearly demonstrates that the use of VTHS leads 'automatically' to the account of PLC with resulting Hamiltonian (6) describing coupled proton-lattice system. Here let's mention that relation (6) resembles that suggested for KDP in the frames of local single-mode model [3,4]. The difference is in the fact that the model considered here is not supposed to be single-mode one and admits the presence of several distortional modes describing the A(=P or As) atoms displacements not only along z axis but maybe along all three axes of the crystal lattice. Besides the relations obtained do not include indefinite constants thus all terms in the rhs of (6) may be calculated by quantum chemistry methods.

The Ising model for Hamiltonian.

An Ising-type representation [5,6] may be constructed for H by the following way. Let's write the (potential) energy of the macroscopic KDP-type crystal as the sum of inner potential energies E_t of the AO_4 tetrahedra:

$$H = \sum_t E_t \qquad (7),$$

and after it let's use for each E_i relations (3), (4) and (5).

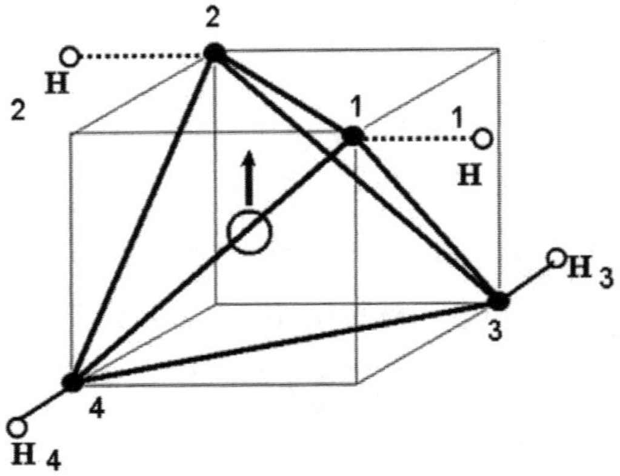

Fig.2. Numeration of O and H atoms in the PO_4 tetrahedron

Then an expression for H arises:

$$H = -\tfrac{1}{2} \Sigma_{(i \neq j)} J_{ij} \sigma^z_i \sigma^z_j , \qquad (8),$$

where J_{ij} are the Ising model interaction parameters. In the theory of KDP-like FE one usually takes into account only two of them, the largest by absolute value, denoted as U and V [5,6]. Here U describes the effective pairwise interaction between two 'upper' or two 'lower' protons connected with a chosen tetrahedron, and V – the interaction between one 'upper' and one 'lower' proton (see Fig.2), while the explicit relations for U and V are the following [7]:

$$U = c_a^2 \Delta\alpha^2 \left(\frac{l_b^2}{4\omega_\perp} - \frac{l_e^2}{2\omega_\parallel} + \frac{l_b^2 A_\perp^2}{\omega_\perp^2 k_\perp} - \frac{2 l_e^2 A_\parallel^2}{\omega_\parallel^2 k_\parallel} \right) \qquad (9)$$

$$V= c_a^2 \Delta\alpha^2 \left(\frac{l_b^2}{4\omega_\perp} + \frac{l_b^2 A_\perp^2}{\omega_\perp^2 k_\perp} \right) \tag{10}$$

Symbols $\omega_\|$ and $\omega_\perp$ denote the energy gaps e-a* and b-a* between the e, b MO's (HOMO) and a* MO (LUMO) for the AO_4 -tetrahedron with S_4 symmetry in protophase, see Fig.3. The electronic structure of a tetrahedron is also characterised in (9) and (10) by the $|\Delta\alpha|$ value and the coefficients l_e, l_b, c_a of the symmetrized oxygen AOs in the MOs. Here the values $k_\perp$, $k_\|$ and $A_\perp$, $A_\|$ denote force constants and orbital vibronic constants [8] of the tetrahedron responding to its central atom displacement perpendicular to xy plane ($\perp$) or within this plane ($\|$). Then in accordance with (9), (10) each Ising parameter contains two components:

$$U= U_{el} + U_{vib} \text{ , } V= V_{el}+ V_{vib} \text{ ,} \tag{11}$$

one of them depends only on the MOs of the tetrahedron in protophase while the other however depends on force and vibronic constants as well due to the PLC account.

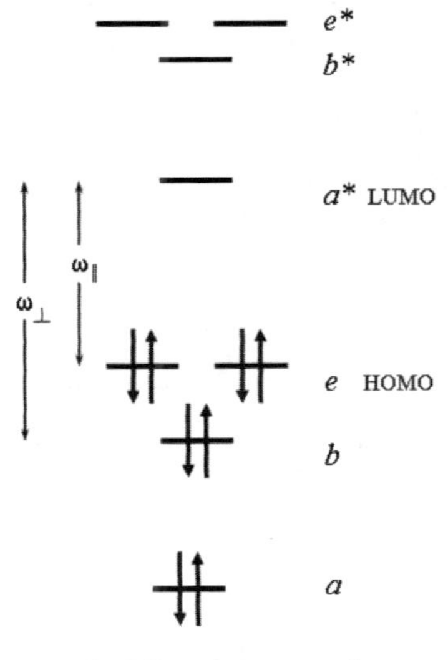

Fig.3. The MO diagram for PO_4 unit in protophase; σ – MO approximation

Numerical data for Ising parameters

To use the relations (9)-(10) for the analysis of the thermodynamics of KDP-like materials we need the values of U and V together with its electronic (el) and vibronic (vib) components. In fact we've already calculated earlier [9] these these U and V by means of *ab initio* methods

Journal of Physics: Conference Series **428** (2013) 012025 doi:10.1088/1742-6596/428/1/012025

using different types of clusters, modelling the space structure of KDP/DKDP in protophase. The details of the geometry for the clusters were borrowed from neutron diffraction data [10] for an averaged paraelectric (disordered) phase, its structure is believed to reproduce all possible interactions in crystals, PLC including. The results of the most reliable calculations are collected in Table 1.

Table 1. The Ising parameters (all in K) for KDP and DKDP evaluated by cluster MP2 calculations *.

Basis \ Parameter	-U	V	-U	V
	KDP		**DKDP**	
cluster $\{P[O\text{-}H...OPH_3]_4\}^+$				
6-311+G(d,p)	311	353	444	504
6-311+G(2d,2p)	302	349	422	488
6-311++G(2d,2p)	301	348	422	488
cluster $\{P[O\text{-}H...OP(OH)_3]_4\}^+$				
6-311+G(d,p)	301	345	430	491
6-311+G(2d,2p)	296	343	412	477
6-311++G(2d,2p)	295	342	412	477

* results for three largest basis sets are presented

The el- and vib- components values for U and V may be obtained in principle directly from (9), (10) though it requires additional calculations of MOs and of force and orbital vibronic constants. However the useful estimates for these values may be found much easier. As shown in [7] the V_{vib} value may be estimated from the data [10] on relative displacement of phosphorous atom along z axis upon the temperature variation at the vicinity of the point of FE transition. In particular, for KDP the correspondent estimate of V_{vib} gives the value 120K and, if compared with Table 1, yields in the ratio $V_{vib}/V \approx 0.35$.

Although there is no similar way to estimate U_{vib} (as the FE transition causes the P displacement namely along z and not along the orthogonal axes), it seems conceivable that

$$U_{vib}/U \approx V_{vib}/V \qquad (12)$$

This approximated equality is bolstered by the following reasoning. In the protophase considered any AO_4 tetrahedron (Fig.2) with actual S_4 symmetry is in fact close to an ideal form with T_d symmetry, its three local axes ***x, y, z*** being almost equivalent. As a result, one-electron b,e levels arise due to only slight splitting of triple-degenerated level t_2 into double-degenerated e and non-degenerated b, that leads to relation (12) (see also [11]).

Proton tunneling and Tc calculation.

The Ising parameters determination are necessary but not enough to study the role of PLC in thermodynamics of H-bonded FE, as one should also take into account the tunneling of protons along H-bonds. In the FE microscopic theory [5, 6] it is done by the replacement of classical Hamiltonian (8) by its quantum analogue:

$$H = -\Omega\sum_i \sigma_i^x - \frac{1}{2}\sum_{i\neq j} J_{ij}\sigma_i^z\sigma_j^z \ . \qquad (13).$$

Here Ω is the tunneling parameter describing the frequency of proton quantum motion along the H-bond and σ_i^x, σ_i^z are treated as Pauli operators for the spin ½ that act on the wavefunctions of protons, localized at the minima of its adiabatic potential along *i*-th H-bond. Then for T_c calculation the numerical procedure may be applied [12] that uses Bethe cluster method with tunneling and allows to obtain T_c value as a function on Ω and Ising parameters utilized.

PLC influence on critical temperature

The results of such calculation for KDP are shown in Fig.4 for three cases including the simplest one that takes into account only the nearest-neighbours interactions of pseudospins, i.e. at γ=0. For Ising parameters the values U= -300K, V=345K were used, calculated *ab initio* by MP2 procedure (see Table 1) and for tunneling parameter the value Ω=370K [13] was adopted.

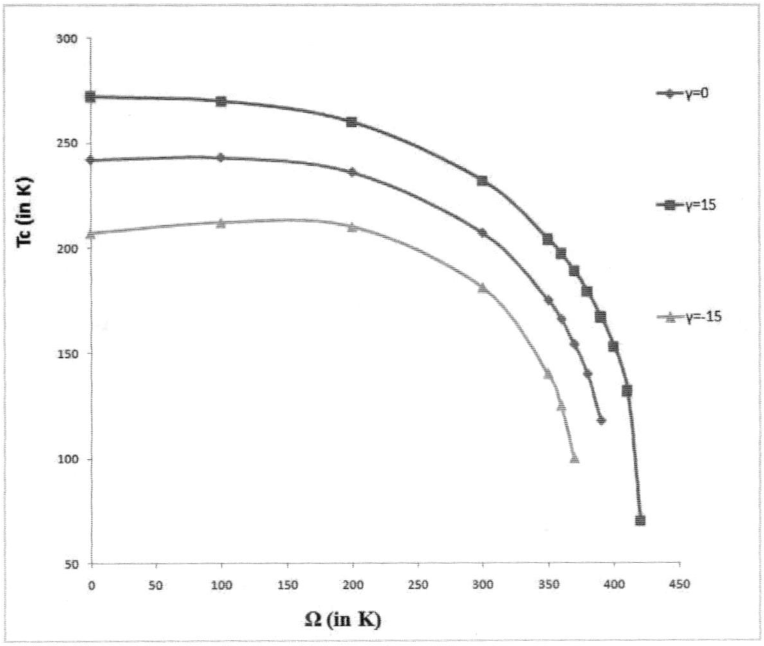

Fig.4. T_c(KDP) *versus* Ω using U = -300K, V = 345K as the Ising parameters.

Journal of Physics: Conference Series **428** (2013) 012025 doi:10.1088/1742-6596/428/1/012025

As we see from Fig.4 the corresponding T_c is about 150K that is in reasonable agreement with the experimental data (123K) thus corroborating the use of the averaged diffraction data of the paraelectric phase to obtain the parameters of (13).

To estimate in theory the PLC influence over T_c value an analogous calculation (γ=0) was performed with the U, V values substituted by the values which it would have had without PLC, i.e. by the pure electronic components U_{el} = U-U_{vib}, V_{el} = V-V_{vib}. The results are presented at Fig.5 and suggest the absence of the structural phase transition into an ordered phase due to its complete suppression by tunneling. Hence Figs.4 and 5, taken together, evidence that both the presence of the ferroelectric phase transition in KDP and the value of its critical temperature are determined mainly by PLC.

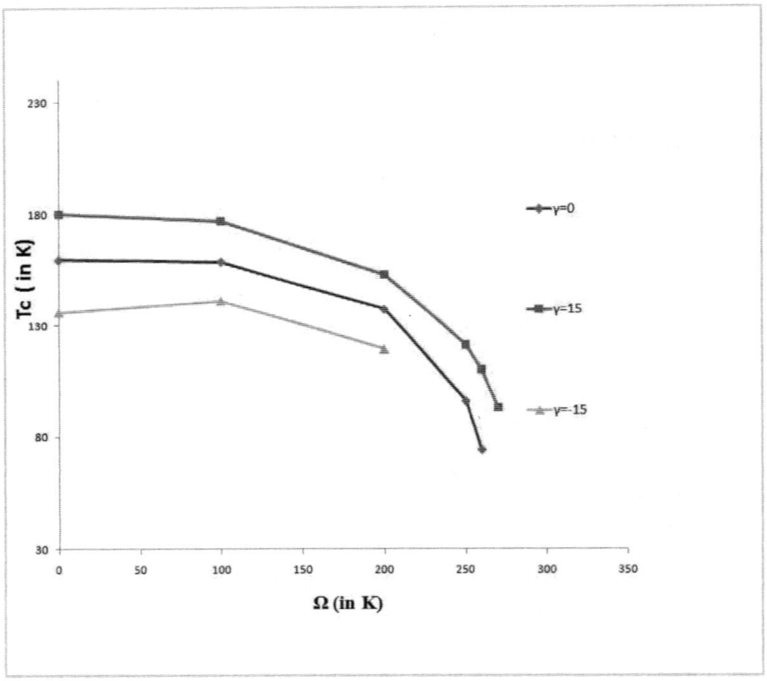

Fig.5. T_c(KDP) *versus* Ω using U_{el}, V_{el} components as the Ising parameters.

Up to now we proceeded from assumption that the (averaged) geometry of the paraelectic phase reflects completely all the interactions in the materials considered, including PLC. Let's discuss here an opposite situation – what if the paraelectric phase geometry doesn't reproduce the PLC effect so the paper [9] and Table 1 present in fact namely electronic components of the Ising parameters. If so, in accordance with (11), complete U, V values are to be obtained by adding U_{vib}, V_{vib} to the data from Table 1. The corresponding calculations for KDP (Fig.6, γ=0) give however for T_c the value about 300K that overestimates at least twice both the experimental data and our result obtained above. Similar but less drastic variations of U, V values and T_c estimate are to be observed also if the geometry of paraelectric phase reflects PLC partially. Thus one may conclude that initially adopted assumption about “complete reflection” of PLC by the structure of paraelectric phase of KDP yields (within the frames of model used) in the best approximation of the theoretical estimate for T_c to the experimental value.

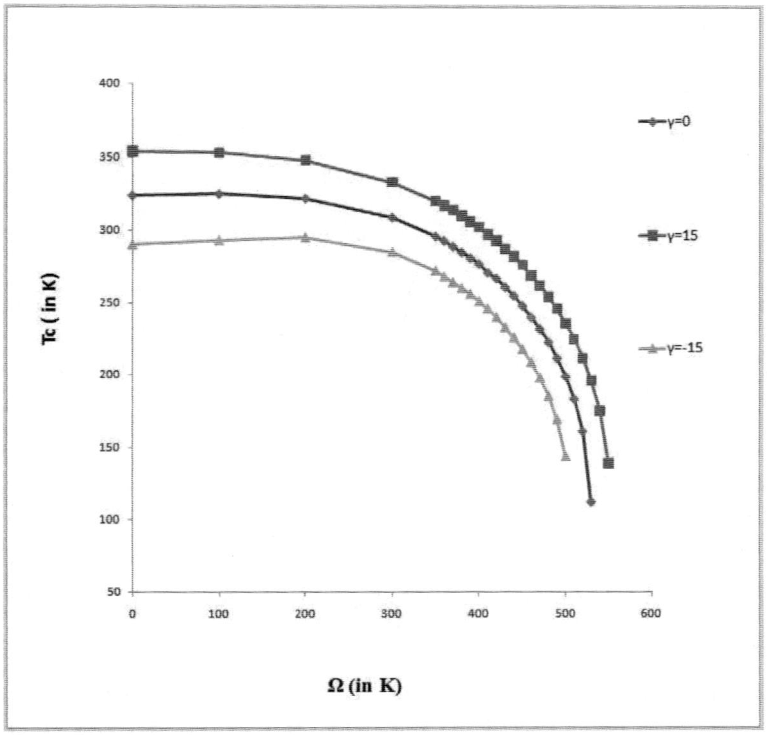

Fig.6. T_c(KDP) *versus* Ω treating U = -300K, V = 345K as electronic components of the Ising parameters.

Conclusions above are made by the examination of the curves from Figs.4-6 with $\gamma = 0$. Meanwhile it is clear that these conclusions remain valid also for $\gamma = \pm 15$K i.e. despite some variations of the Ising parameters set due to approximate modeling of non-neighboring pseudospin couplings, tangible influencing the Hamiltonian (13). It is seen from Fig. 4 that $\gamma = 15$K leads to the worsening of agreement between the calculated $T_c \approx 150$K and the experiment. At the same time $\gamma = -15$K slightly improves this agreement that maybe indicates some preference of ‘negative contribution’ in long-range coupling of pseudospins for KDP alike mentioned earlier in [14].

This work is supported by RFBR, project 11-03-00349.

Journal of Physics: Conference Series **428** (2013) 012025 doi:10.1088/1742-6596/428/1/012025

References

[1] Levin A A and Dolin S P 1996 *J. Phys.Chem.* **100** 6258

[2] Levin A A and D'yachkov P N 2002 *Heteroligand molecular systems, Bonding, shapes and isomer stabilities*, (Taylor & Francis, London/New York)

[3] Blinc R 1982 *Croatica Chem. Acta* **55** 7

[4] Blinc R and Žekš B 1982 *J.Phys.C: Solid State Phys.* **15** 4661

[5] Blinc R and Žekš B 1974 *Soft Modes in Ferroelectrics and Antiferroelectrics* (North Holland, Amsterdam, Oxford, Elsevier, New York)

[6] Strukov B A and Levanyuk A P 1998 *Ferroelectric Phenomena in Crystals* (Springer, Berlin)

[7] Dolin S P *et al* 2012 *Vibronic Interactions and Jahn-Teller Effect. Theory and Applications, Progress in Theoretical Chemistry and Physics* ed. M Atanasov (Springer) **6** 419

[8] Bersuker I B 1984 *The Jahn-Teller Effect and Vibronic Interactions in Modern Chemistry* (Plenum Press, New York)

[9] Dolin S P *et al* 2010 *Int. J. Quant. Chem.* **110** 77

[10] Nelmes R J et al 1987 *Ferroelectrics* **71** 125

[11] Dolin S P *et al* 2012 *Int. J. Quant. Chem.* **112** 2976

[12] Blinc R and Svetina S 1966 *Phys. Rev.* **147** 430

[13] Dolin S P *et al* 2010 *J. Mol. Struct.* **972** 115

[14] Abalmasov V A and Yurkov A S 2012 *Solid State Physics* **54** 924 (in Russian)

Ferroelectric phase transitions in cubic perovskites

Victor Polinger

Department of Physics, University of Washington, Seattle, WA 98195-1560, USA

E-mail: polinv@uw.edu

Abstract. In low-temperature phase, due to the pseudo Jahn-Teller effect in unit cells, atoms are coherently localized in one of eight trigonal wells of the adiabatic potential energy surface. With temperature, nuclear motion delocalizes over symmetry-equivalent wells. Depending on the size of potential barriers, the two limiting cases are considered: (*a*) shallow wells, and (*b*) deep wells. They correspond to two different mechanisms of delocalization. When potential barriers are shallow compared to kT_C, the delocalization is due to over-the-barrier activation of Arrhenius type. Alternatively, when the over-the-barrier activation is thermally inaccessible, the delocalization is due to tunnelling. Temperature dependence of spontaneous polarization is treated in terms of the mean-field approximation. Using experimental data for $BaTiO_3$, numerical values of the corresponding parameters are estimated. The qualitative description is followed by a numeric evaluation of the temperature sequence of ferroelectric phase transitions for the two cases of shallow and deep wells. The theory agrees with the experimental data.

1. Introduction

Discovered in 1920, ferroelectricity is over ninety years old already. Yet some fundamental questions about its microscopic nature remain unanswered. Here, in this brief introduction, we cannot give a comprehensive review of this vast trend in physics of solid state. The references listed below are to just provide a general perspective of some ideas and basic concepts in the modern theory of ferroelectricity. More elaborate introduction to the problem is given by Bersuker [1].

The traditional understanding of ferroelectricity is based on the phenomenological approach of Ginsburg [2], Devonshire [3], Cochran [4], Anderson [5], and Landau [6]. It explains ferroelectric phase transitions as resulting from competition of the short-range inter-ionic repulsion with the long-range dipole-dipole attraction via the limiting-phonon TO mode. This led to the concept of soft mode that goes to zero at the point of a phase transition. It was experimentally observed in many ferroelectric crystals.

In its present form, the phenomenological theory can be found in different texts on solid-state physics and, also, in specialized monographs and review papers on ferroelectricity, particularly, in Lines and Glass [7], Rabe et al. [8], Smolenskii [9], and Dawber et al. [10]. The phenomenological theory was not intended to and does not provide any systematic way to deduce how chemical composition of a particular crystal determines its ferroelectric properties.

The early attempts to answer this question went in two different directions. One is the so-called electronic theory. Launched by Wigner [11] and Jaynes [12, 13], it relates the bulk polarization of a ferroelectric crystal solely to the electron polarizability of the composition ions. The polarizability implies electron hybridization of the ground-state electron wave function with some excited (unoccupied) opposite-parity electron states. Providing flexibility to the ground-state wave function, electron hybridization allows it induce a local dipole moment.

For cubic perovskites ABO_3, the band gap to the polar states is of the order of 3 eV. With this gap, the ionic polarization is an order of magnitude less than the one experimentally observed. From another hand, to obtain the measured value of polarization, one has to assume an energy gap as small as 0.13 eV [10] which is 20 times smaller than the observed gap. Due to this inconsistency the electronic theory was rejected.

The other way was assuming a significant anharmonicity of the crystal lattice. The overall polarization of the crystal was presumed to be purely due to displacements of crystal-lattice atoms into low-symmetry off-centre minimum positions on the adiabatic potential energy surface (APES). In a central-symmetrical crystal structure, polar displacements of charged ions create local dipole moments. Originating from Mason and Matthias [14] this idea was developed by Devonshire [3, 15] and Slater [16]. Atomic displacements were measured quite accurately in X-ray diffraction experiments. To explain the observed polarization, unusual electric charges were assumed (the so-called Born charges) of the displaced atoms. For example, in $BaTiO_3$, the Born charge of Ti was found to be close to +7 [8-10]. Evidently, sharing and donating its four valence electrons, Ti cannot have an electric charge greater than +4.

It is important to note that both approaches turned out to be quantitatively inconsistent and failed when applied to cubic perovskites. The corresponding discussion of their deficiency can be found elsewhere [7 - 10]. Also, today there are quite a number of papers that combine both approaches. They start with a model Hamiltonian with its parameters taken from a first-principles electronic structure calculation of the lowest APES; e.g. see [17, 18] and references therein.

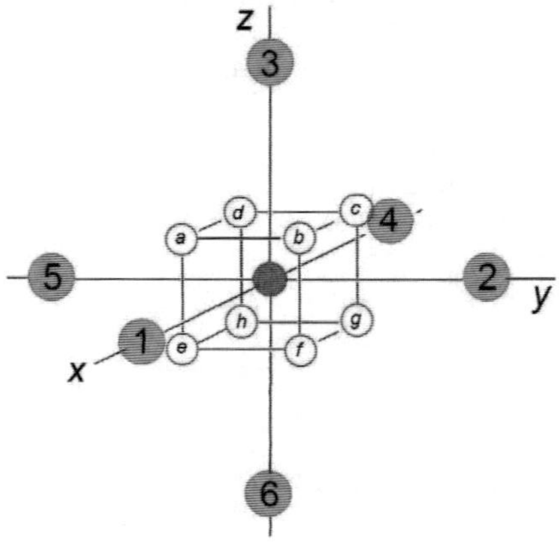

Figure 1. The octahedral fragment of the cubic perovskite crystal structure ABO_3, with the transition metal atom B at the centre (solid red) and six oxygen atoms (blue) at the apexes of the octahedron. The characters *a, b, c*, etc. denote the eight off-centre positions (not to scale) of the atom B induced by the pseudo JTE. The counter-phase displacement of the oxygen atoms is not shown.

In this paper we follow the vibronic approach first introduced by Bersuker in 1966 [19]. Its more elaborate version was published in 1967 [20]. The basic idea is local instability of octahedral fragments [BO_6] in a cubic perovskite ABO_3 (Fig.1) due to the pseudo Jahn-Teller effect (JTE). The physical nature of the pseudo JTE is the electron-vibrational (vibronic) hybridization of valence (occupied) electron orbitals with some excited (unoccupied) electron states involved in this coupling. The general idea of the pseudo JTE was introduced in 1957 by Öpik and Pryce [21] for the simple example of linear triatomic molecules, *ABA*. Somewhat different version of the vibronic approach in terms of electron and phonon bands (sometimes called "the two-band theory" or "the interband model") was introduced by Kristoffel and Konsin in 1967 [22].

A general review of the pseudo JTE can be found in the books [23, 24]. The vibronic coupling is known to result in a significant anharmonicity of the lowest APES. The high-symmetry point can be a minimum or a maximum depending on the relative value of the Jahn-Teller stabilization energy, E_{JT}, and the energy gap, Δ, separating the valence (occupied) electron shell from the excited (unoccupied) orbitals involved in the vibronic coupling (Sect.2).

In cubic perovskites, the vibronic coupling involves three polar modes, *X*, *Y*, and *Z*, of a unit cell. They correspond to an off-center displacement of the cation *B* and a counter-direction displacement of the oxygen octahedron. In barium titanate, the band gap is of the order of 3 eV. For the Jahn-Teller stabilization energy, E_{JT}, a rough estimate gives $E_{JT} \approx 4$ eV, greater than the energy gap (Sect. 2); so the vibronic coupling is strong enough and the criterion of the pseudo Jahn-Teller instability holds.

This paper is organized in the following way. In Section 2, we start with the ground-state APES for the octahedral fragment $[BO_6]$ in a perovskite structure ABO_3 (Fig.1) in the form that results from the vibronic hybridization (the pseudo JTE) of the highest-occupied molecular orbitals (HOMO) with the lowest unoccupied molecular orbitals (LUMO). Particular attention is paid to the case of barium titanate. The corresponding parameters of the theory are estimated from measured data. In Section 3.1, the mean-field approximation is introduced. It reduces the issue of structural instability of the ferroelectric crystal to the problem of instability of one unit cell in the average field of all other unit cells of the crystal. In Section 3.2, this problem is solved in a self-consistent way and the corresponding phase diagrams are discussed. We will consider the two limiting cases: (*a*) shallow wells, and (*b*) deep wells. As will be shown, they correspond to two different mechanisms of delocalization. When potential barriers are shallow compared to kT_C, the delocalization is due to over-the-barrier activation of Arrhenius type. Alternatively, when the over-the-barrier activation is thermally inaccessible, the delocalization is due to tunneling of the system between symmetry-equivalent potential wells. The respective phase transitions are treated as order-disorder ones. In Section 4 we discuss different pro- and contra- arguments for the order-disorder character of ferroelectric phase transitions in cubic perovskites. From entropy arguments it follows that the actual crystal structures include ordered clusters of a finite size. From this point of view, discussed in this paper order-disorder phase transition should be understood as a rather simplified description of the real-life case.

2. Zero temperature: the ground-state APES in barium titanate

Following Bersuker [1], consider the octahedral fragment of the unit cell centered at the transition metal atom *B* (solid red circle in Fig. 1). It is surrounded by six oxygen atoms (blue circles in Fig. 1). In the MO LCAO presentation, the HOMO are formed by nine $2p_\pi$ orbitals of the six oxygen atoms. They transform as the irreducible representations t_{1u}, t_{2u}, and t_{1g} of the octahedral symmetry group. The LUMO are $t_{2g}(3d)$ orbitals of the transition metal *B*, namely, $|yz(B)\rangle$, $|xz(B)\rangle$, and $|xy(B)\rangle$. As one can see in Fig.2, in the high-symmetry configuration, when the atom *B* is in the center of oxygen octahedron, the HOMO-LUMO overlap is zero. Therefore, when the transition atom *B* is at the center of the oxygen octahedron, these orbitals are non-bonding; they do not contribute to the chemical bonding of the transition metal with the oxygen atoms.

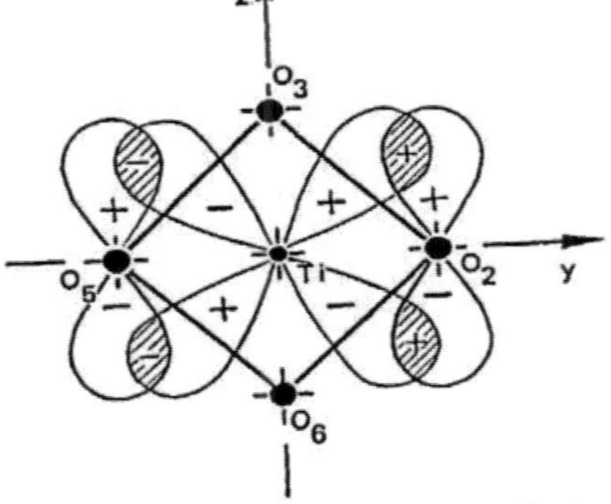

Figure 2 Zero overlap (shaded) of the oxygen HOMO with the metal LUMO in the high-symmetry cubic configuration. Shown is the particular case of the oxygen symmetry-adapted orbital $|T_{1u}x\rangle$ and the $|T_{2g}yz\rangle$ orbital of Ti^{4+}. An off-centre displacement of Ti^{4+} creates a non-zero overlap resulting in an additional covalent chemical bond of oxygen atoms with Ti^{4+} [1, 25].

An off-center displacement of the central ion, $\mathbf{Q} = \langle X, Y, Z \rangle$, changes the HOMO-LUMO overlap to a non-zero value forming an additional covalent bond [19, 20 24-27]. Correspondingly, with the additional bond, the total energy of the unit cell lowers. This energy benefit is the origin and physical nature of the pseudo JTE in cubic perovskites. If strong enough, it may trigger the dipole instability in every unit cell resulting in ferroelectric phase transitions.

For the octahedral fragment $[BO_6]$, the one-electron Hamiltonian can be expanded in a power series with respect to **Q**. With up to linear terms included, we come to the Jahn-Teller Hamiltonian, $H_{\text{JT}} = \Delta\hat{\mathbf{A}} + F_0\mathbf{Q}\cdot\hat{\mathbf{d}}$. It includes the zero-order term of the expansion (the energy gap term), $\Delta\hat{\mathbf{A}}$, and the linear term of the Jahn-Teller coupling, $F_0\mathbf{Q}\cdot\hat{\mathbf{d}}$. Here $\hat{\mathbf{A}}$ is the diagonal matrix 9×9 acting within the manifold of the nine electron orbitals, namely, the six HOMO and the three LUMO, so the energy gap between HOMO and LUMO is 2Δ. F_0 is the constant of linear pseudo Jahn-Teller coupling, the so-called orbital vibronic coupling constant,

$$F_0 = \left\langle 2p_z(\text{O}) \left| \left(\frac{\partial H}{\partial Q_x} \right)_0 \right| 3d_{xy}(\text{Ti}) \right\rangle. \tag{1}$$

Also, we introduced three matrices 9×9, $\hat{\mathbf{d}} = \langle \hat{d}_x, \hat{d}_y, \hat{d}_z \rangle$ of the electric dipole moment of electrons, **d**. Its reduced matrix element is incorporated in F_0.

The Jahn-Teller Hamiltonian, H_{JT}, is a matrix 9×9 acting in the basis set of the six HOMO and the three LUMO. Its nine eigenvalues correspond to one-electron energy levels that change adiabatically with the off-center position of the transition-metal B. In the case of barium titanate, the octahedron is $[TiO_6]$ and there are twelve valence electrons populating the hybridized HOMO. Benefiting in energy from the new covalency created by the polar distortions, Q_x, Q_y, and Q_z, these twelve electrons contribute to the following potential energy of the lowest APES [1, 19, 21, 24-26],

$$U(\mathbf{Q}) = \tfrac{1}{2}K_0 Q^2 - 2\left[\sqrt{\Delta^2 + 2F_0^2(Q^2 - X^2)} + \sqrt{\Delta^2 + 2F_0^2(Q^2 - Y^2)} + \sqrt{\Delta^2 + 2F_0^2(Q^2 - Z^2)}\right]. \tag{2}$$

Here, in addition to the eigenvalues of the Jahn-Teller Hamiltonian, H_{JT}, we introduced the elastic energy, $\tfrac{1}{2}K_0(X^2 + Y^2 + Z^2) = \tfrac{1}{2}K_0 Q^2$, associated with the off-center displacement of the central ion. K_0 is the respective primary force constant. Expressed by K_0, the corresponding elasticity is due to σ orbitals and ionic interactions. The expression (2) for the ground-state APES follows from first principles applied to the pseudo JTE in cubic perovskites. Its elaborate explanation, detailed derivation, and analysis was given by Bersuker [19, 20, 24-27]; also, see [1].

Introducing dimensionless parameters u, δ, x, y, and z, such that

$$U = u\frac{F_0^2}{K_0}, \quad \Delta = \delta\frac{F_0^2}{K_0}, \quad X = x\frac{F_0}{K_0}, \quad Y = y\frac{F_0}{K_0}, \quad Z = z\frac{F_0}{K_0}, \tag{3}$$

the ground-state APES can be presented in the dimensionless form,

$$u(x,y,z) = \tfrac{1}{2}r^2 - 2\left[\sqrt{\delta^2 + 2(r^2 - x^2)} + \sqrt{\delta^2 + 2(r^2 - y^2)} + \sqrt{\delta^2 + 2(r^2 - z^2)}\right], \tag{4}$$

with $r^2 = x^2 + y^2 + z^2$.

In a close vicinity of the high-symmetry point, $x = y = z = 0$, the ground-state APES is a spherical paraboloid,

$$u \approx \frac{1}{2\delta}(\delta - 8)r^2 - 6\delta. \tag{5}$$

The respective (dimensionless) curvature is $1 - 8/\delta$. The parabola (5) is open down when $1 - 8/\delta < 0$ and is open up in the opposite case. Correspondingly, on the ground-state APES expressed by the Eq.1, the high-symmetry point, $X = Y = Z = 0$, is a maximum when $0 < \delta < 8$ (equivalently, $\Delta <$

Journal of Physics: Conference Series **428** (2013) 012026 doi:10.1088/1742-6596/428/1/012026

$8(F_0)^2/K_0$) and is a minimum in the opposite case, when $\delta > 8$ (equivalently, when $\Delta > 8(F_0)^2/K_0$). In the former event, the octahedron [TiO_6] is unstable with respect to the polar distortions, X, Y, and Z. As mentioned above, this instability is due to the pseudo JTE [1, 19]. The inequality $0 \le \delta \le 8$ [or, equivalently, $\Delta < 8(F_0)^2/K_0$] is *the condition of instability* in this particular case. There is a rigorous proof [27] that the pseudo JTE is the *only possible* source of this kind of instability in cubic perovskites.

In the case of a relatively strong vibronic coupling, when $\delta < 8$, in addition to the high-symmetry maximum at $x = y = z = 0$, the potential surface (4) has three kinds of extreme points. For the dimensionless APES expressed by Eq.4, the equipotential cross-sections are mapped onto the surface of the unit sphere shown in Fig. 3. There are eight trigonal wells at $(\pm r_0, \pm r_0, \pm r_0)$ with $r_0 = ½(64 - \delta^2)^{1/2}$. The corresponding stabilization energy, $u(0, 0, 0) - u(r_0, r_0, r_0)$, is $E_{JT}(111) = ⅜(8 - \delta)^2$. The trigonal wells are separated by twelve orthorhombic potential barriers at $(\pm r_1, \pm r_1, 0)$, $(\pm r_1, 0, \pm r_1)$, and $(0, \pm r_1, \pm r_1)$ with $r_1 \approx 1.2 r_0 = 0.6(64 - \delta^2)^{1/2}$. The stabilization energy, $u(0, 0, 0) - u(r_1, r_1, 0)$, is

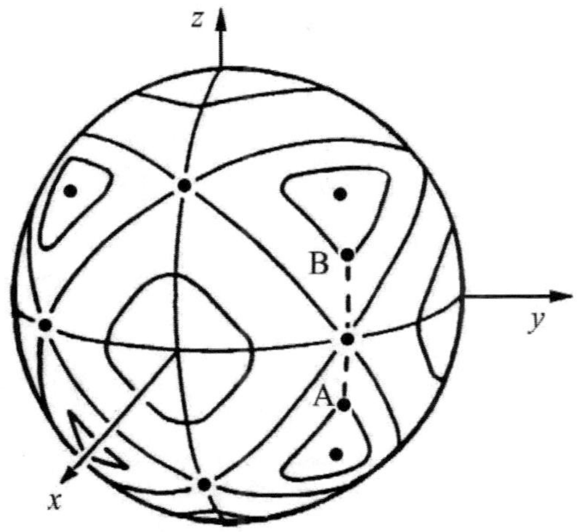

Figure 3. Equipotential cross-sections of the APES (4) mapped onto the unit sphere around the high-symmetry position at the centre of the oxygen octahedron. The trigonal minimum points correspond to the eight off-centre positions of the atom *B* denoted as *a, b, c,* etc. in Fig.1. The broken curve is path of steepest descent from the orthorhombic saddle point to two near-neighbour wells A and B.

$E_{JT}(110) \approx 0.35(8 - \delta)^2$. The height of the orthorhombic potential barriers is $\Delta u = E_{JT}(110) - E_{JT}(111) \approx 0.028(8 - \delta)^2$. Finally, there are six tetragonal saddle points at $(\pm r_2, 0, 0)$, $(0, \pm r_2, 0)$, $(0, 0, \pm r_2)$ with $r_1 = r_0\sqrt{2} = ½[2(64 - \delta^2)]^{1/2}$. The corresponding stabilization energy is $u(0, 0, 0) - u(r_2, 0, 0) = E_{JT}(100) = ¼(8 - \delta)^2$.

The curvature of the APES at the trigonal minimum points is obtained from the eigenvalue problem for the matrix of second derivatives of $u(x, y, z)$ in (4). At each trigonal well, the three normal frequencies split into a non-degenerate and a two-fold degenerate one. The non-degenerate mode corresponds to a longitudinal vibration along [111]. The two-fold degenerate E mode corresponds to transverse vibrations perpendicular to the direction [111]. The respective curvatures are $K_A = 1 - (\delta/8)^2$ and $K_E = ¼K_A$.

2.1. Estimating parameters of the APES for barium titanate.

For $BaTiO_3$, available experimental data and, also, the results of *ab initio* calculations allow us to estimate the parameters of the APES. Of the three independent parameters that determine the polar instability, Δ, F, and K_0, the energy gap, Δ, is relatively well known, $2\Delta \approx 3.25$ eV. The remaining two parameters can be obtained from the experimental value of the off-centre displacement of the ion Ti^{4+} in the trigonal well, $Q_0 \approx 0.11$ Å, and from the Jahn-Teller stabilization energy, $E_{JT}(111)$, in the trigonal well. From XAFS, this energy is about 0.2 eV [28] while from *ab initio* calculations, $E_{JT}(111) \approx 0.3$ eV [29], slightly different but of the same order of magnitude. The difference is quite understandable. The latter value is calculated for the rhombohedral phase where the distortion is enhanced by the mean field of the crystal, whereas the XAFS experiments are performed in the paraelectric phase, and hence reflect just the local distortion. For our local model the former value, $E_{JT}(111) \approx 0.2$ eV, is more appropriate.

Another option to involve experimental data is to exploit vibrational frequencies. In particular, in the trigonal well, the transversal vibration, $\hbar\omega_E$ (the one perpendicular to the axis [111]), may be assumed to approximately represent the soft mode at low temperatures. The usual estimate is $\hbar\omega_E \approx 100$ cm^{-1} [30]. The vibrational frequencies and, respectively, the vibrational quanta, $\hbar\omega_0$, $\hbar\omega_A$, and $\hbar\omega_E$, depend on the reduced mass, $M = m_1m_2/(m_1 + m_2)$, where m_1 is mass of the transition-metal atom and m_2 is cumulative mass of oxygen atoms participating in the polar displacements. In cubic perovskites, ABO_3, there are three oxygen atoms per each atom B. Then $m_2 = 3(16\ \text{u}) = 48$ u. In the case of barium titanate, $m_1 = 48$ u and, therefore, $M = 24$ u. The two versions of the numerical estimate of all the relevant parameters are given in the Table 1.

As one can see, in spite of essentially different experimental data used in the two estimates, the numeric values are quite reasonable and rather consistent (within the expected orders of magnitude). Although the electron energy gap, $2\Delta = 3.25$eV, is relatively large, the pseudo JTE is even stronger providing the polar instability in both versions of the Table 1. The corresponding stabilization energy, $16F^2/K_0$, equal to 3.98 eV and 3.75 eV in the versions I and II respectively, is greater than the gap, $2\Delta = 3.25$eV. On the other hand, the condition of instability may be very restricting, which is why it is not the case in other similar systems.

Table 1. Numerical values of the parameters of the pseudo JTE, introduced in Eqs.2 and 3, obtained from the comparison of experimental data with the calculated APES. Two versions based on the two sets of data are given.

Main parameters	**Version I** [$E_{JT}(111) = 0.2$ eV]	**Version II** ($\hbar\omega_E = 100$ cm^{-1})
K_0	13.3 eV/Å^2	11.0 eV/Å^2
$\hbar\omega_0$	388 cm^{-1}	354 cm^{-1}
F_0	1.82 eV/Å	1.65 eV/Å
$(F_0)^2/K_0$	0.249 eV	0.246 eV
F_0/K_0	0.137 Å	0.149 Å
δ	6.54	6.60
$E_{JT}(111)$	0.20 eV	0.181 eV
$Q_0 = (F/K_0)r_0$	0.11 Å	0.11 Å
$\hbar\omega_A$	224 cm^{-1}	200 cm^{-1}
$\hbar\omega_E$	112 cm^{-1}	100 cm^{-1}
$E_{JT}(110)$	0.1867 eV	0.169 eV
$\Delta E_{JT} = E_{JT}(111) - E_{JT}(110)$	108 cm^{-1}	98 cm^{-1}

In the ground-state APES (1) and its dimensionless equivalent (4), the significant anharmonicity is obtained without involving the anharmonic terms in the Hamiltonian; only the linear vibronic coupling is included. On the other hand, without the pseudo JTE included in the Hamiltonian, just the singlet anharmonicity with even up to the forth-order terms does not create the eight-minimum APES [15]. Thus, with its eight trigonal wells, twelve orthorhombic saddle points, and six tetragonal saddle points, the APES (1) is a sole result of the pseudo JTE [1,19,20]. By the vibronic hybridization of the ground and excited electronic states, it generates the so-called "vibronic anharmonicity" [24] which is much more ample than the singlet anharmonicity introduced directly in the Hamiltonian.

3. Rising the temperature. The mean-field approximation.

In ferroelectrics the temperature dependence of spontaneous polarization is due to the special anharmonic shape of the APES (Section 2). On the APES of cubic perovskites, the only kind of minimum points are the trigonal wells. Therefore, at low temperatures, in each unit cell the vibrational motion of atoms is localized at the bottom of a trigonal well. It corresponds to an off-center displacement of the transition metal ion along one of the eight equivalent trigonal directions. The inter-cell interaction is minimized when the off-center displacements in different unit cells and the corresponding dipole moments are ordered forming a low-symmetry phase.

When the local dipole moments are ordered, they determine the overall polarization of the crystal as a whole. Respectively, the spontaneous polarization of the unit cells induces a non-zero electric field, the mean field of all unit cells. Depending on the character and sign of the inter-cell coupling, the ordered phase can be ferroelectric, antiferroelectric, or of a more complex kind. For barium titanate, the vibronic approach predicts the low-temperature ferroelectric phase with all Ti atoms in a correlated off-center position along the same trigonal direction [1, 19, 20].

The mean field and the ordered off-center position of the Ti atoms are interdependent in a self-consistent way. With temperature, Boltzmann population of excited energy levels grows. When the temperature excitation in trigonal wells approaches the height of next-in-energy orthorhombic potential barriers, the system averages out over two nearest-neighbor trigonal wells along, say, $\left[111\right]$ and $\left[11\bar{1}\right]$, by hoping or tunneling over the orthorhombic barrier. The crystal becomes disordered along one of its symmetry axes, the z axis in this example, remaining ordered along the other two. At average, the crystal is polarized along [110]. Macroscopically, this type of ordering is equivalent to the orthorhombic phase. Correspondingly, the orientation of the mean field changes to [110]. Still, lowered by the mean field, this pair of potential wells is separated from the other equivalent ones by tetragonal potential barriers that are higher in energy.

By a farther increase in temperature, higher-energy states become temperature populated, their energy reaching the higher potential barrier with the tetragonal saddle point, say, [100] on its top. The crystal structure transforms into the tetragonal phase with four directions of disorder. The average polarization and the induced electric field change to the one along [100]. This is the so-called tetragonal phase (Table 2). Finally, when the temperature is high enough, the maximum of the adiabatic potential at $X = Y = Z = 0$ is overcome. The local dipole moments become averaged over all eight trigonal wells, and the crystal reaches the completely disordered paraelectric phase. Worth to mention, deduced from the vibronic approach, this qualitative description of ferroelectric phase transitions in $BaTiO_3$ was first predicted in 1966 [19]. Today it is a common knowledge supported by multiple experiments [7].

3.1. The mean-field theory.

To quantify this qualitative description, consider the Hamiltonian of a cubic perovskite crystal ABO_3 as the following sum over different unit cells:

$$H = \sum_{\mathbf{m}} H_{\mathbf{m}} + \sum_{\mathbf{m}\neq\mathbf{n}} H_{\mathbf{mn}}. \tag{6}$$

Unit cells are labeled with vector indexes $\mathbf{m} = \langle m_x, m_y, m_z\rangle$ and $\mathbf{n} = \langle n_x, n_y, n_z\rangle$. The one-cell Hamiltonian, $H_{\mathbf{m}} = H_0(\mathbf{m}) + H_{JT}(\mathbf{m})$ is the same as the one introduced above in Section 2 with the only difference, this time it has the index $\mathbf{m}$ of the unit cell.

The off-center displacement of the atom B creates a nuclear dipole moment which, in its turn, induces a local dipole moment of valence electrons. The respective mechanism involves a non-zero HOMO-LUMO overlap discussed in Section 2 (Fig.2). Employing the operator of the electric dipole moment, $\mathbf{d_m}$, of electrons associated with **m**th unit cell, the Hamiltonian of inter-cell coupling can be presented as follows:

$$H_{\mathbf{mn}} = \frac{1}{2} f_{\mathbf{mn}} \hat{\mathbf{d}}_{\mathbf{m}} \cdot \hat{\mathbf{d}}_{\mathbf{n}} + F_{\mathbf{mn}} \mathbf{Q}_{\mathbf{m}} \cdot \hat{\mathbf{d}}_{\mathbf{n}} + \frac{1}{2} K_{\mathbf{mn}} \mathbf{Q}_{\mathbf{m}} \cdot \mathbf{Q}_{\mathbf{n}}. \tag{7}$$

The reduced matrix element of the operator $\mathbf{d_m}$ is incorporated in the coupling parameters $f_{\mathbf{mn}}$ and $F_{\mathbf{mn}}$. They decrease with the inter-cell distance as $|\mathbf{m} - \mathbf{n}|^{-3}$.

The inter-cell coupling (7) can be derived directly from the first principles. In what follows, the coupling parameters, $f_{\mathbf{mn}}$ and $F_{\mathbf{mn}}$, will combine into the inter-cell correlation constant. It will be determined from fitting to experimental data. Therefore, the direct relation of the Hamiltonian (7) to the first-principle parameters of the theory is unnecessary here.

To decouple the Hamiltonian (7), we apply the *mean-field approximation*. It is introduced in the traditional way, similar to how it is done in the theory of magnetic phase transitions. In a ferroelectric phase, the temperature averages $\langle \hat{\mathbf{d}}_{\mathbf{m}} \rangle$ and $\langle \mathbf{Q_m} \rangle$ do not depend on the cell index **m**. This means we can assume $\langle \hat{\mathbf{d}}_{\mathbf{m}} \rangle = \langle \mathbf{d} \rangle$ and $\langle \mathbf{Q_m} \rangle = \langle \mathbf{Q} \rangle$. Also, compared to the average values, $\langle \mathbf{d} \rangle$ and $\langle \mathbf{Q} \rangle$, the fluctuations, $\Delta\hat{\mathbf{d}}_{\mathbf{m}} = \hat{\mathbf{d}}_{\mathbf{m}} - \langle \mathbf{d} \rangle$ and $\Delta\mathbf{Q_m} = \mathbf{Q_m} - \langle \mathbf{Q} \rangle$, are assumed to be small. Plugging $\hat{\mathbf{d}}_{\mathbf{m}} = \langle \mathbf{d} \rangle + \Delta\hat{\mathbf{d}}_{\mathbf{m}}$ and $\mathbf{Q_m} = \langle \mathbf{Q} \rangle + \Delta\mathbf{Q_m}$ into the Hamiltonian (7), it is transformed into a quadratic form of the fluctuations $\Delta\hat{\mathbf{d}}_{\mathbf{m}}$ and $\Delta\mathbf{Q}_{\mathbf{m}}$. As the fluctuations are assumed to be small, the terms quadratic in $\Delta\hat{\mathbf{d}}_{\mathbf{m}}$ and $\Delta\mathbf{Q}_{\mathbf{m}}$ are neglected.

In the remaining linear terms, the back-substitution of $\Delta\hat{\mathbf{d}}_{\mathbf{m}} = \hat{\mathbf{d}}_{\mathbf{m}} - \langle \mathbf{d} \rangle$ and $\Delta\mathbf{Q_m} = \mathbf{Q_m} - \langle \mathbf{Q} \rangle$ returns back to the original operators, $\hat{\mathbf{d}}_{\mathbf{m}}$ and $\mathbf{Q_m}$. This gives the energy dependence in the **n**th unit cell due to the average dipole moment induced by the off-centre distortion in the **m**th unit cell. Omitting constant terms and plugging the resultant approximated expression of $H_{\mathbf{mn}}$ into (6) gives the decoupled Hamiltonian of the whole crystal. It is a sum of independent (mean-field) Hamiltonians, $H_{\mathrm{MF}}(\mathbf{m})$, of decoupled unit cells, $H = \Sigma_{\mathbf{m}} H_{\mathrm{MF}}(\mathbf{m})$, additive with respect to different unit cells. Due to the translational symmetry, the one-cell Hamiltonians are all the same. They do not depend on the cell index **m**,

$$H_{\mathrm{MF}} = \frac{1}{2M} P^2 + \frac{1}{2} K_0 Q^2 + \Delta\hat{\mathbf{A}} + F_0 (\mathbf{E} + \mathbf{Q}) \cdot \hat{\mathbf{d}} + K_0 \mathbf{D} \cdot \mathbf{Q}. \tag{8}$$

Here M is the mass of the atom B, the vector $\mathbf{P} = \langle P_x, P_y, P_z \rangle$ is its linear momentum conjugated to the off-centre displacement **Q**. Vectors **E** and **D**,

$$\mathbf{E} = \frac{f}{F_0} \langle \mathbf{d} \rangle + \left(\frac{F}{F_0} - 1 \right) \langle \mathbf{Q} \rangle, \qquad \mathbf{D} = \frac{F - F_0}{K_0} \langle \mathbf{d} \rangle + \left(\frac{K}{K_0} - 1 \right) \langle \mathbf{Q} \rangle, \tag{9}$$

describe the induced mean field. They act upon the valence electrons and nuclei respectively. In (9), we introduced the inter-cell correlation parameters, $f = \Sigma_{\mathbf{n}} f_{\mathbf{n}0}$, $F = \Sigma_{\mathbf{n}} F_{\mathbf{n}0}$, and $K = \Sigma_{\mathbf{n}} K_{\mathbf{n}0}$. Also, for $\mathbf{n} = 0$, we assume that $f_{\mathbf{n}0} = 0$, $F_{\mathbf{n}0} = F_0$ and $K_{\mathbf{n}0} = K_0$. One of the mean-field terms, $F_0 \mathbf{E} \cdot \hat{\mathbf{d}}$, can be eliminated by the shift transformation, $\mathbf{Q} \to \mathbf{Q} - \mathbf{E}$. It changes the Hamiltonian (8) into $H_{\mathrm{MF}} = H_0 + H_{\mathrm{JT}} + K_0(\mathbf{D} - \mathbf{E}) \cdot \mathbf{Q}$ with $H_0 = ½\,(P^2/M + K_0 Q^2)$. As above, the constant terms, the ones that do not include the dynamic operators, **Q**, **P**, and $\hat{\mathbf{d}}$, are omitted.

In the octahedron $[BO_6]$, due to the vibronic coupling, an off-centre displacement $\langle \mathbf{Q} \rangle$ of the atom B induces a low-symmetry redistribution of the electron charge on the six oxygen atoms creating a non-zero electron dipole moment, $\langle \mathbf{d} \rangle$. Vice versa, any low-symmetry distribution of the electric charge induces an off-centre displacement of the atom B. Therefore, expressed by (9), the two contributions to the mean electric field are interdependent and proportional to one another. Consider the commutator of the Hamiltonian, H_{MF}, with the linear momentum $\mathbf{P}$ of the atom B, $[H_{MF}, \mathbf{P}]$. It brings to the equation $[H_{MF}, \mathbf{P}] = K_0(\mathbf{Q}+\mathbf{D}) + F_0\hat{\mathbf{d}}$. As the average commutator is zero, by averaging this equation, we find the corresponding interdependence, $F_0\langle\mathbf{d}\rangle + K_0(\mathbf{D} + \langle\mathbf{Q}\rangle) = 0$. Combining it with the equations (9) we come to

$$F\langle\mathbf{d}\rangle = -K\langle\mathbf{Q}\rangle. \tag{10}$$

Plugging the equation (10) into the expressions (9), we find both contribution in the mean field, $\mathbf{D}$ and $\mathbf{E}$, proportional to the average off-centre displacement $\langle\mathbf{Q}\rangle$. Correspondingly, the one-cell Hamiltonian (8) takes the form,

$$H = H_0 + H_{JT} + LK_0\langle\mathbf{Q}\rangle\cdot\mathbf{Q}. \tag{11}$$

with the inter-cell correlation parameter,

$$L = \frac{F}{F_0} - \frac{Kf}{FF_0} - \frac{F_0K}{FK_0}. \tag{12}$$

As in the high-symmetry case discussed in Section 2, the one-electron Hamiltonian (11) is matrix 9 × 9 acting in the basis set of the six HOMO and the three LUMO. Its nine eigenvalues correspond to one-electron energy levels that change adiabatically with the off-centre displacements of the transition-metal B. Compared to the one-cell Hamiltonian introduced in Section 2, the only difference of the Hamiltonian (11) is the mean-field term, $LK_0\langle\mathbf{Q}\rangle\cdot\mathbf{Q}$.

In the case of barium titanate, there are twelve valence electrons populating the hybrid HOMO. The resultant APES is similar to the one introduced in (2) with the addition of the low-symmetry term, $\lambda K_0\langle\mathbf{Q}\rangle\cdot\mathbf{Q}$.

$$H = \frac{P^2}{2M} + U(\mathbf{Q}) + \lambda K_0\langle\mathbf{Q}\rangle\cdot\mathbf{Q}. \tag{13}$$

Here $U(\mathbf{Q})$ is the same as in Eq.2 and $\lambda = 12L$ with L from Eq.12.

3.2. *Self-consistent evaluation of order parameters.*

At $T = 0$K, the system is locked at the bottom of one of the eight wells, the one deepened by the mean field. The latter is determined by the average value of the off-centre displacement, $\langle\mathbf{Q}\rangle$. It has to be evaluated in a self-consistent way, similar to how it is done in the theory of magnetic phase transitions.

At higher temperatures the nuclear motion delocalizes over different wells. In what follows, we consider the two limiting cases: (*a*) very deep trigonal wells and (*b*) shallow trigonal minima. The two cases correspond to two significantly different mechanisms of delocalization. In the case (a), when over-the-barrier activation is thermally inaccessible, the delocalization is of a tunnelling nature. In the case (b), when the potential barriers are not very high, it is over-the-barrier activation of Arrhenius type (similar to the Orbach-type relaxation applied to pseudo-spin).

In the case of barium titanate, according to the estimates in Section 2.2, potential barriers between trigonal wells, $\Delta E_{JT} = E_{JT}(110) - E_{JT}(111)$, are of the order of 100 cm^{-1}. For the transversal vibration, the one perpendicular to the trigonal axis [111], the energy quantum, $\hbar\omega_E$, is of the same order of magnitude, 100 cm^{-1}. Therefore, in $BaTiO_3$ we have the case of a weak-to-moderate coupling, closer to the case (*b*). In general, in cubic perovskites, we find a wide range of coupling parameters. To some cubic crystals ABO_3, one of the two limiting cases applies quite accurately.

3.2.1. The tunnelling model for deep trigonal wells. General theory of tunnelling splitting in Jahn-Teller systems was developed in [31] (see also [24]). The particular case of eight trigonal wells, (the so-called 8-site model), was considered by Gomez et al. [32]. In the paraelectric phase, when the mean field is zero, $\langle \mathbf{Q} \rangle = 0$, the one-cell Hamiltonian (13) has cubic symmetry. Its potential energy, the APES, has eight trigonal wells separated by orthorhombic and tetragonal potential barriers discussed in Section 2. If the potential barriers are high enough, $\Delta E_{JT} \gtrsim \hbar\omega_E$, vibrational motion of the system is locked in one well. Ground state of the unit cell corresponds to eight Born-Oppenheimer states in the wells, $|a\rangle$, $|b\rangle$, $|c\rangle$, …, $|h\rangle$, as labelled in Fig.1.

When the temperature is low enough, $kT \lesssim \Delta E_{JT}$, the potential barriers are significantly higher than the energy gaps separating temperature populated states. In this case the set of vibronic states participating in the temperature averaging can be truncated including just the eight vibronic tunnelling states, the ones with the lowest energy. Neglecting tunnelling, if the wells are symmetry equivalent and the ground state is eight-fold degenerate. Tunnelling lifts the eight-fold degeneracy. The ground multiplet splits into two triplets and two singlets, $A_{1g} + T_{1u} + A_{2u} + T_{2g}$.

The tunnelling Hamiltonian (14) was diagonalized and the tunnelling energy levels were found by Gomez et al. [32]. Energy gaps between the tunnelling energy levels depend on the three overlap integrals, $\langle a|b\rangle$, $\langle a|c\rangle$, and $\langle a|g\rangle$. Evidently, for neighbouring wells, the overlap integral $\langle a|b\rangle$ has dominating value. In a good approximation, the other two, $\langle a|c\rangle$ and $\langle a|g\rangle$, can be neglected. In this case the tunnelling multiplet is equidistant. Measured from the ground-state energy level in the infinite-deep wells, the tunnelling energy levels are equidistant,

$$E\left(A_{1g}\right) \approx 3\Gamma, \quad E\left(T_{1u}\right) \approx \Gamma, \quad E\left(T_{2g}\right) = -\Gamma, \quad E\left(A_{2u}\right) = -3\Gamma. \tag{14}$$

Here Γ is the tunneling parameter,

$$\Gamma \approx \frac{\hbar\omega_E}{2\pi} \exp\left(-\frac{1}{\hbar} \int_A^B \sqrt{2m\left[U(s) - E_0\right]} ds \right), \tag{15}$$

$\hbar\omega_E$ being the energy quantum of the transversal vibration in a trigonal well, the one perpendicular to the axis [111]. The integral is taken over the arc length s along the path of steepest descent from the orthorhombic saddle point to the closest trigonal minimum of the APES. In Fig. 4 it is the broken curve. $U(s)$ is the potential energy (the APES) introduced in Eq.2, and A and B are classical turning points where the system hits the barrier wall.

Using eight tunnelling states as the basis set, the Hamiltonian (13) is represented by the following matrix [32],

$$\begin{matrix} & |A_{2u}\rangle & |T_{2g}\xi\rangle & |T_{2g}\eta\rangle & |T_{2g}\zeta\rangle & |T_{1u}x\rangle & |T_{1u}y\rangle & |T_{1u}z\rangle & |A_{1g}\rangle \end{matrix}$$

$$H(\boldsymbol{\alpha}) = \Gamma \begin{pmatrix} 3 & -\alpha_x & -\alpha_y & -\alpha_z & 0 & 0 & 0 & 0 \\ -\alpha_x & 1 & 0 & 0 & 0 & -\alpha_z & -\alpha_y & 0 \\ -\alpha_y & 0 & 1 & 0 & -\alpha_z & 0 & -\alpha_x & 0 \\ -\alpha_z & 0 & 0 & 1 & -\alpha_y & -\alpha_x & 0 & 0 \\ 0 & 0 & -\alpha_z & -\alpha_y & -1 & 0 & 0 & -\alpha_x \\ 0 & -\alpha_z & 0 & -\alpha_x & 0 & -1 & 0 & -\alpha_y \\ 0 & -\alpha_y & -\alpha_x & 0 & 0 & 0 & -1 & -\alpha_z \\ 0 & 0 & 0 & 0 & -\alpha_x & -\alpha_y & -\alpha_z & -3 \end{pmatrix}. \tag{16}$$

Here the vector $\boldsymbol{\alpha} = j\langle \mathbf{d} \rangle$ is introduced where $j = \lambda K_0 Q_0^2/\Gamma$ is the dimensionless correlation constant in terms of the tunnelling parameter Γ and $\langle \mathbf{d} \rangle$ is the (dimensionless) average dipole moment of electrons in one unit cell. The components of $\boldsymbol{\alpha}$, namely, α_x, α_y, and α_z, are order parameters of the corresponding ferroelectric phase. They are solutions to the system of the following transcendental equations:

$$\alpha_i = \frac{j}{\Gamma}\left\langle \hat{d}_i \right\rangle_\alpha, \ \left\langle \hat{d}_i \right\rangle_\alpha = \frac{1}{Z_\alpha}\mathrm{Tr}\left(e^{-H(\alpha)/kT}\hat{d}_i\right), \ Z_\alpha = \mathrm{Tr}\left(e^{-H(\alpha)/kT}\right), \ i = x, y, z. \tag{17}$$

By substitution, the system of equations (17) can be reduced to three coupled equations,

$$\Gamma\alpha_i = jkT\frac{\partial}{\partial\alpha_i}\left\{\ln \mathrm{Tr}\left[e^{-H(\alpha)/kT}\right]\right\}, \quad i = x, y, z. \tag{18}$$

In the lowest-temperature *rhombohedral phase*, all atoms B are shifted into one of the trigonal wells, say, in the direction [111]. In this case, $\alpha_x = \alpha_y = \alpha_z = \alpha$ and we have just one transcendental equation to solve. The respective minimum is deepened by the mean field. If it is strong enough compared to Γ, tunnelling is quenched, and the system is locked in the respective well. Rising temperature weakens the mean field and its locking effect is reduced. The corresponding temperature dependence of the order parameter, α, for different values of correlation parameter $j = \lambda K_0 Q_0^2/\Gamma$ is shown in Fig.4. Plugging the resultant value of α into the expression for the Helmholtz free energy, $\Phi = -\partial(\ln Z_\alpha)/\partial\beta$ with $\beta = 1/kT$, we find the free energy of the rhombohedral phase, Φ(111).

Similarly, in the *orthorhombic phase*, all atoms B are "displaced" along the two-fold symmetry axis [110]. As there is no minimum on the APES in this direction, this "displacement" should be interpreted as an averaged motion over two near-neighbour wells divided by the orthorhombic potential barrier. In this case, $\alpha_x = \alpha_y = \alpha$ and $\alpha_z = 0$ and again, we have just one equation to solve. Plugging the evaluated value of $\alpha_x = \alpha_y = \alpha$ and $\alpha_z = 0$ into the Helmholtz free energy, we find its value

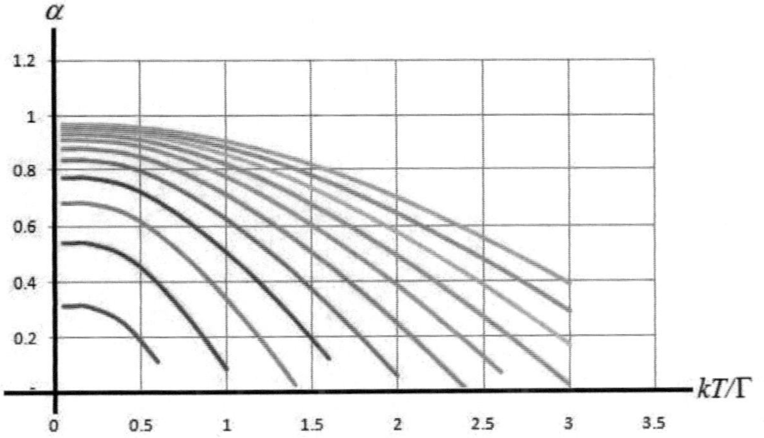

Figure 4. Temperature dependence of the order parameter, α, in the rhombohedral phase when the tunnelling mechanism of disorder dominates. The correlation parameter, $j = \lambda K_0 Q_0^2/\Gamma$, varies from $j = 1$ for the left-bottom graph to $j = 3$ for the graph at the right top with the step, $\Delta j = 0.2$.

in the orthorhombic phase, Φ(110). Solving the equation Φ(111) = Φ(110) for temperature, we can find the T_C value for the *rhombohedral-to-orthorhombic* phase transition.

For different ferroelectric phases, the temperature dependence of the Helmholtz free energy is shown in Fig. 5 for the case when $j = 5$ chosen as explained below. As above, j is the dimensionless coupling constant, $j = \lambda K_0 Q_0^2/\Gamma$, and $t = kT/\Gamma$ is the dimensionless temperature in terms of the tunnelling parameter Γ. At about $t = 3.8$ ($kT = 3.8\Gamma$) the two graphs intersect. Above 3.8Γ, the free energy of the orthorhombic phase becomes lower than that of the rhombohedral phase causing the rhombohedral to orthorhombic phase transition. The respective Curie temperature changes with the inter-cell coupling parameter $j = J/\Gamma$ are almost linear (Fig.6).

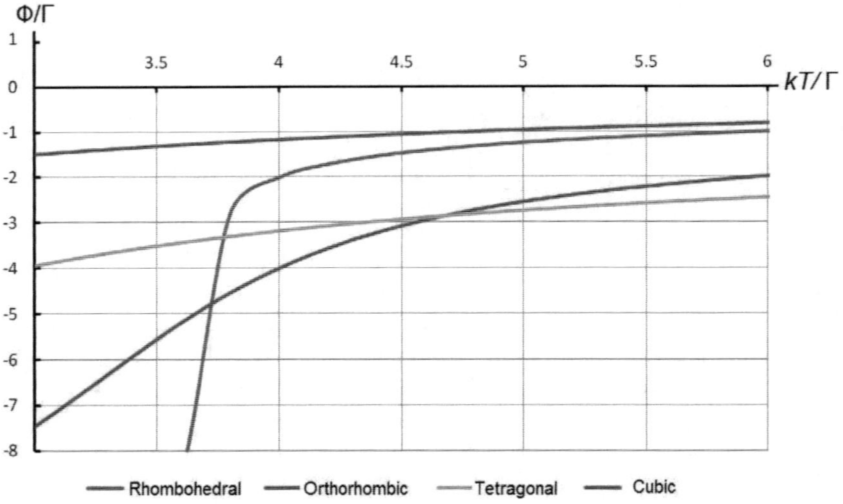

Figure 5. Temperature dependence of the Helmholtz free energy Φ (in units of Γ) for the four phases, rhombohedral, orthorhombic, tetragonal, and the paraelectric, at $j = 5$. At $kT < 3.8\Gamma$, the free energy of the rhombohedral phase is the lowest. With temperature, at $kT = 3.8\Gamma$, the free energy of the orthorhombic phase becomes the lowest causing the rhombohedral to orthorhombic phase transition. Above $kT \approx 4.6\Gamma$, the tetragonal phase becomes most energy-advantageous.

In the tetragonal phase the atom B is in tetragonal position, "displaced" along one of the four-fold symmetry axes. Similar to the orthorhombic phase, this "displacement" results from averaging over four close-neighbour trigonal wells by tunnelling through four potential barriers separating them. Therefore, the respective mean-field order parameter can be set as $\alpha_x = \alpha$ and $\alpha_y = \alpha_z = 0$. Plugging it

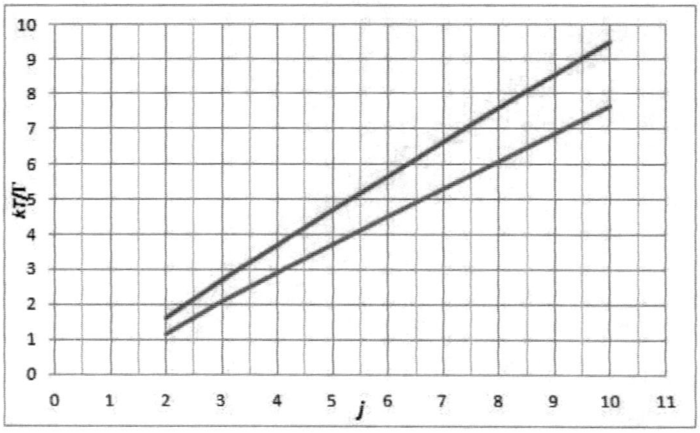

Figure 6. Temperatures of the two phase transitions, rhombohedral-to-orthorhombic (lower graph, blue) and orthorhombic-to-tetragonal (upper graph, red) versus the dimensionless coupling parameter j.

into the Helmholtz free energy, $\Phi(100)$, and solving the equation $\Phi(110) = \Phi(100)$, we evaluated the temperature of the *orthorhombic-to-tetragonal* phase transition shown in Fig.6. For the same $j = 5$ the phase transition takes place at $t \approx 4.67$ ($kT \approx 4.67\Gamma$ Similar to previous case, the Curie temperature of

Journal of Physics: Conference Series **428** (2013) 012026 doi:10.1088/1742-6596/428/1/012026

the orthorhombic-to-tetragonal phase transition changes with j almost linearly. With the estimated value of the tunnelling parameter, $\Gamma \approx 38$ cm^{-1}, a reasonable fit with the observed temperatures of ferroelectric phase transitions, as shown in Table 2, is achieved at $j \approx 5$.

Table 2. Calculated values of the two low-temperature phase transitions, rhombohedral-to-orthorhombic T_C(I) and orthorhombic-to-tetragonal T_C(II), compared to the experimental values T_C(I) = 178K and T_C(II) = 278K. The percentage errors are evaluated as $[(T_C - T_{exp})/T_{exp}] \times 100\%$.

$j = \lambda K_0 Q_0^2/\Gamma$	4.6	4.7	4,8	4.9	5.0	5.1	5.2	5.3
$t_1 = kT_C(\mathrm{I})/\Gamma$	3.41	3.49	3.57	3.65	3.73	3.81	3.89	3.97
T_C(I) (in K)	186	191	195	200	204	208	213	217
Percentage error for T_C(I)	4	7	9.5	12	15	17	20	22
$t_2 = kT_C(\mathrm{II})/\Gamma$	4.28	4.38	4.48	4.58	4.67	4.77	4.87	4.97
T_C(II) (in K)	234	239	245	250	256	261	266	271
Percentage error for T_C(II)	16	14	12	10	8	6	4	2.5

As one can see from the Table 2, the lower-temperature phase transition is better described with lower values of the mean field parameter λ than the higher-temperature one (with the ratio of about 4.6/5.3). This is quite understandable in view of the larger amplitudes of vibration in the minima at higher temperatures and hence larger intercell interaction of the distortions.

An important side of the tunnelling model is its first-order nature. The above mentioned equations (14) through (18) follow from the first-order perturbation approach to the eight-fold degenerate ground state. In this theory, the small parameter is the tunnelling overlap Γ. In the case of barium titanate, as one can see from Fig.5, the tunnelling model fails to provide a reasonable description of the high-temperature phase transition, tetragonal-to-cubic. It could happen by averaging over all eight trigonal wells. This implies a coherent tunnelling through all twelve orthorhombic barriers. Obviously, due to a relatively high entropy factor, the respective probability is too low. An alternative tunnelling path is through the high-symmetry point, at $X = Y = Z = 0$. In the oxygen octahedron, it corresponds to the on-centre position of the atom of transition metal. This potential barrier is of the order of 0.2 eV = 1600 cm^{-1}. In the case of Ti, the atom mass is 48 u and the 0.2-eV barrier is too high for a reasonable magnitude of the tunnelling energy gap.

3.2.2. The case of relatively shallow trigonal wells. On the APES, there may be a two-dimensional trough with a relatively weak warping. The latter produces moderately low potential barriers between trigonal minimum points. At the same time, as in the above-mentioned case of barium titanate, the high-symmetry point, $X = Y = Z = 0$, may be high in energy, so the first-order tunnelling is impossible. In this case the disorder is achieved by second-order processes through a temperature population of the excited states close in energy to the top of the corresponding potential barrier(s). This is over-the-barrier activation similar to the Arrhenius-type activation of chemical reactions and, also, resembling the Orbach relaxation in magnetic resonance.

The temperature of the ferroelectric phase transition, T_C, is assumed to be of the order of or higher in magnitude than the energy height of the potential barriers, $kT_C \gtrsim \Delta E_{JT}$. Then the hindered rotation of the off-centre atom B about its on-centre position can be treated in terms of classical mechanics and classical statistical physics. Using the dimensionless parameters introduced in Eq.3, for the average displacement, $\langle \mathbf{Q} \rangle$, we introduce the dimensionless vector $\bar{\mathbf{r}} = \langle \mathbf{Q} \rangle K_0/F_0$ with components $\bar{x}$, $\bar{y}$, and

$\bar{z}$. The temperature averaging is over the states of the Hamiltonian (13). Since H includes the vector $\langle \mathbf{Q} \rangle$ as a parameter, it brings to the coupled system of three transcendental equations,

$$\bar{x} = -\frac{t}{\lambda}\frac{\partial \ln Z}{\partial \bar{x}}, \quad \bar{y} = -\frac{t}{\lambda}\frac{\partial \ln Z}{\partial \bar{y}}, \quad \bar{z} = -\frac{t}{\lambda}\frac{\partial \ln Z}{\partial \bar{z}}. \tag{19}$$

Here, as above in Eq.17, Z, is the partition function, $Z = \mathrm{Tr}(e^{-H/kT})$.

For relatively *high temperatures*, classical statistics applies, so the partition function can be expressed as the following integral [6]:

$$Z = \int e^{-H(p,q)/kT} dq dp = c t^{3/2} \int e^{-[u(x,y,z)+\lambda \bar{\mathbf{r}}\cdot\mathbf{r}]/t} dx dy dz. \tag{20}$$

Here $c = \left[\left(E_{\mathrm{JT}} / \omega_0 \right) \sqrt{2\pi} \right]^3$ is a constant factor with $\omega_0 = \sqrt{(K_0/M)}$ and $E_{\mathrm{JT}} = F_0^2/K_0$, $t = kT/E_{\mathrm{JT}} = kTK_0/F_0^2$ is the dimensionless temperature, and $u(x, y, z)$ is the potential energy from Eq.3. In the tetragonal phase the atom B is displaced in one of the three tetragonal directions, say, along the x axis. In this case, $\bar{x} \neq 0$ and $\bar{y} = \bar{z} = 0$. Out of the three equations (19), we are left with just one where the partition sum, Z, is from (20) with the displacement vector $\bar{\mathbf{r}} = \langle \bar{x},\ 0,\ 0 \rangle$.

For $\lambda = 0.15$, the same value as used above in the tunnelling model, the numerical solution of this equation is given in Fig.7. With temperature, the dimensionless displacement, $\bar{x}$, decreases smoothly

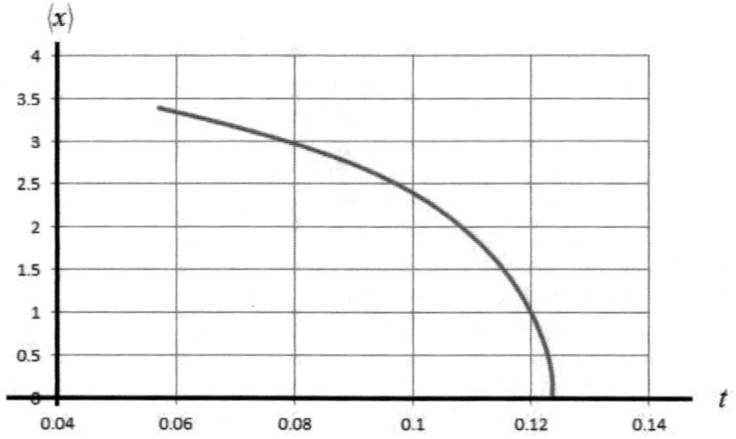

Figure 7. Temperature dependence of the order parameter, $\bar{x}$, in the tetragonal phase. At t = 0.1236, the off-centre displacement of the atom B smoothly drops to zero manifesting the second-order phase transition.

to zero at $t \approx 0.1236$. In the case of barium titanate with $F_0^2/K_0 \approx 0.25$ eV (see the parameter estimate in the Table 1 in Section 2.1), this corresponds to $kT_C \approx 250$ cm^{-1} = 359 K, reasonably close to the experimental value of 373 K. As distinguished from the low-temperature phase transitions, trigonal-to-rhombohedral and rhombohedral-to-tetragonal, the transformation into the cubic phase is of a second-order.

4. Discussion.

Compared to other theories of ferroelectricity, the vibronic approach has two attractive advantages. One is its direct relation to first principles. For a given crystal, it answers the important question about the chemical and structural origin of ferroelectricity. All parameters of the pseudo Jahn-Teller Hamiltonian, the ones introduced in the Eqs.1, 2, and 3, are rigorously defined and can be numerically evaluated from the corresponding electronic structure. Another important advantage of the vibronic theory is the direct and qualitatively full explanation of the origin of all the experimentally observed properties of the ferroelectric crystals.

The only existent first-principle alternative is a direct calculation of the lowest APES point-by-point as, for example, is described in [17, 18]. However, in this approach the chemical nature of the anharmonicity of the APES, most important for ferroelectricity, is left in the shade of the extensive computer work. The pseudo Jahn-Teller theory reveals the vibronic nature of the anharmonicity and, accompanied with the corresponding group-theoretical analysis, indicates the molecular orbitals responsible for ferroelectricity. In this paper, the main characteristics of cubic perovskites, namely, the local anharmonicity and inter-cell correlation parameter determining the long-range inter-cell coupling were derived from the cooperative pseudo Jahn-Teller effect. For $BaTiO_3$ the parameters of the theory are in reasonable agreement with the experimental data.

4.1 Order-disorder versus displacive character of ferroelectric phase transitions.

In this paper, following Bersuker [19], we assumed the order-disorder character of ferroelectric phase transitions in cubic perovskites. It results from the local setup of the pseudo Jahn-Teller effect in each unit cell. Supporting this assumption, the most striking example is the off-centre displacement of the atom of Ti in barium titanate observed in the paraelectric phase way above the Curie temperature of the tetragonal-to-cubic phase transition [28]. As accurate measurements indicate, Ti remains displaced closely along [111] directions throughout all of the $BaTiO_3$ phases. The magnitude of the off-center displacement decreases monotonically by only13% from 35 to 590 K, showing no steps at the phase transitions [29]. Confirmed by a wide variety of experimental data, including EXAFS [29], optical measurements [34-37], X-ray absorption [38, 39], EPR [40], NMR [41], neutron scattering [28], etc., this observation is in strong disagreement with the theory of displacive phase transitions [4].

At the same time, for phase transitions in barium titanate, a number of experiments seem to support their displacive nature. In particular, the low-frequency (soft) mode was observed in neutron scattering [43], IR absorption [44], hyper-Raman scattering [45], etc. This apparent contradiction has two sides, theoretical and experimental.

4.1.1. The theoretical aspects. Usually, critics of the vibronic approach include its failure to predict the soft mode near the Curie temperature. This argument is wrong. The existence of the soft mode is not rejected by the vibronic theory and follows directly from the temperature dependence of the Helmholtz' free energy, Φ, with respect to the limiting-phonon TO mode, $[\omega_{\text{eff}}(T)]^2 = (\partial^2\Phi/\partial q^2)_0$. In this paper, the consideration is limited with the simple version of the mean-field approximation. It does not provide any details about phonon dispersion. Its analysis goes beyond the purpose of the present paper. This study was done in the two-band approach [22, 46-48]. The vibronic hybridization was found responsible for the temperature dependence of phonon frequencies including the temperature drop of the frequency of the limiting-phonon TO mode close to Curie temperature. Its mechanism is in growing population of the low-energy portion of the electron energy band with lowering temperature. This strengthens the pseudo Jahn-Teller effect and reduces the respective curvature, $(\partial^2\Phi/\partial q^2)_0$.

Extended in this paper, the vibronic approach of Bersuker [19] and the two-band theory of Konsin and Kristoffel [22, 46-48] explore the same physical mechanism of ferroelectricity, namely, the pseudo Jahn-Teller hybridization of electron states. The major difference is in the way the vibronic Hamiltonian is treated. In this paper, the mean-field approximation is used *before* any Fourier transformation to crystal waves is applied. It addresses the orientation degrees of freedom of the pseudo spin at each unit cell. Therefore, obviously, it provides a better description of the order-disorder phase transitions. In the two-band theory, the mean-field approximation is applied *after* the Jahn-Teller Hamiltonian is presented in crystal plane-wave form. It operates with crystal lattice modes that are uniform over the whole crystal. Evidently, it is adjusted to describe the displacive transitions.

This raises the question of which way is closer to the real-life situation in cubic perovskites. Introduced above (Section 3.2.1) the tunnelling mechanism of disorder operates with the vibronic pseudo-spin at each metal site. Mathematically, this model is analogous to the one used in magnetic phase transitions in a crystal with non-zero electron spin at each site. No one argues against the order-disorder character of magnetic phase transitions; there is nothing to "displace". Similarly, as long as the vibronic pseudo spin does not change its magnitude and just changes its orientation when the atom *B* tunnels from one trigonal well into another, the tunnelling mechanism is of the order-disorder type.

In the rhombohedral phase, all metal atoms are shifted in the same direction. Is it possible that all metal atoms of the crystal hope coherently in a different well creating the displaced orthorhombic phase? The phase transition takes place when the Helmholtz free energy of the two phases becomes equal, $\Phi(\mathrm{I}) = \Phi(\mathrm{II})$. As $\Phi = U - TS$, the condition of phase equilibrium can be expressed as $\Delta U = -T\Delta S$. When all metal atoms of the crystal coherently hope into another trigonal well, the number of states changes by an order of 10^{23}. Correspondingly, ΔS is of the order of $\ln(10^{23}) = 23\ln(10) = 53$. For barium titanate, the lowest Curie temperature, $T = 187$ K. This gives the right side of the equation, $T\Delta S \approx 9430$ K $= 6550$ cm^{-1}. From the estimates of the Table 1 (Section 2.1), the orthorhombic potential barrier is of the order of $\Delta U \approx 100$ cm^{-1}, two orders of magnitudes smaller. This means the entropy barrier is too big to be compensated by the left-hand side, ΔU. Therefore, no coherent transition to another trigonal well can occur.

Interestingly enough, often the entropy argument is used against the vibronic approach in favor of the displacive phase transitions. For tetragonal-to cubic transition in $BaTiO_3$, from the estimated value of ΔU of about 0.133 eV ≈ 1000 cm^{-1} and $T = 373$ K $= 260$ cm^{-1}, one can evaluate the *experimental* value, $\Delta S \approx (1000\ \mathrm{cm}^{-1})/(260\ \mathrm{cm}^{-1}) \approx 4.13$. In a one-cell disorder, the number of states changes from 1 to 6. This gives $\Delta S \approx 1.79$, which is too small compared to the *experimental* value. The equilibrium can be achieved if we accept the compromising assumption: the number of states changes from 1 to about 62. This means the intercell correlation creates ordered clusters of the size of up to about 10 unit cells [49, 50]. Thus the truth is somewhere in between the two limiting cases of an order-disorder or displacive phase transitions. From this point of view, discussed in the present paper order-disorder phase transition should be understood as rather a simplified description of the real-life case.

4.1.2. The experimental aspects. For the ferroelectric phase transitions in cubic perovskites, the apparent contradiction in experimental manifestations can be explained by their relativity to the means of experimental technique used. Observed with different methods, the same phenomenon can create different impressions. In every unit cell, the vibronic motion has its characteristic time, τ, and dimension, l. Regarding time, it is the time, τ, it takes the atom *B* to penetrate (by tunneling or hopping) from one trigonal well to another. It is of the order of $\tau \approx \hbar/\Gamma$. Here, as in Eq.15, Γ is the tunneling parameter. In cubic perovskites, Γ ranges from 0.01 cm^{-1} to 50 cm^{-1}. Respectively, the characteristic time is from $\tau \approx 10^{-9}$ s = 1 ns to $\tau \approx 10^{-13}$ s = 0.1 ps. Regarding dimension, l is the size of one elementary cell where the averaging occurs. For barium titanate, it is of the order of $l \approx 4$ Å $= 4 \times 10^{-10}$ m = 0.4 nm.

The interpretation of the experimental data has to be related to these two important parameters.

- The observed effect should be attributed to the time scale of the method used. It may be essentially different in different experimental technique. The "time of measurement", τ_{exp}, has to be compared to the time τ of the orientation hoping of the metal ion between trigonal wells of the APES [51]. Compared to the characteristic time, the experiment can be short-lasting or long-lasting (Table 3). Discussed by Stern [51], NMR technique can be used as an example. It measures the nuclear quadruple transitions and motionally averages dynamic displacements that vary in times with characteristic time less than 10^{-8} s.

- Also, the observed effect has to be attributed to the characteristic size of the method used, the "experimental length", l_{exp}. Depending on the experimental technique, the method can be short-range and long-range (Table 3). For example, all optical methods, including IR absorption, hyper-Raman, or Brillouin scattering of light, are long-range methods. Their experimental length, l_{exp}, is of the order of wavelength of the incident or scattered light, $l_{exp} \sim 10$ nm ÷ 10 μm. This is by several orders in magnitude longer than the crystal characteristic size, $l \approx 4$ Å = 0.4 nm. Therefore, the long-range methods provide the information averaged over many unit cells. The soft-mode data are correct, however, they do not provide enough reasons to decide (reject or accept) about the order-disorder or displacive nature of the ferroelectric phase transition.

As an alternative technique that can reveal the local disorder, one can use XAFS which is a much faster measurement with a short averaging time, $\tau_{exp} < 10^{-15}$ s and short range, $l_{exp} < 0.01$ nm.

Table 3. Some examples of short-lasting/long-lasting and short-range/long-range experimental techniques.

Experimental technique used	Short lasting ($\tau_{exp} \lesssim 10^{-14}$ s)	Long lasting ($\tau_{exp} \gtrsim 10^{-12}$ s)
Short-range ($l_{exp} \lesssim 0.4$ nm)	XAFS	X-ray diffraction Neutron scattering
Long-range ($l_{exp} \gtrsim 10$ nm)	Optical pulse spectroscopy	IR reflectivity NMR Brillouin Scattering Hyper-Raman Scattering

Acknowledgements

The author thanks Professors I B Bersuker and M D Kaplan for many helpful discussions of the fundamentals of ferroelectric phase transitions in cubic perovskites.

References

[1] Bersuker I B 2012 *This issue*
[2] Ginzburg V L 1945 *J. Exp. Theor. Phys.* **15** 739
Ginzburg V L 1949 *Uspekhi Fizicheskikh Nauk* **38** 490
[3] Devonshire A F 1954 *Theory of Ferroelectrics* (Adv. Phys. vol 3) (Taylor & Francis) p 85
Devonshire A F 1964 *Repts Progr. Phys.* **27** 1
[4] Cochran W 1959 *Phys. Rev. Lett.* **3** 412
Cochran W 1960 *Crystal Stability and the Theory of Ferroelectricity* (Adv. Phys. vol 9) (Taylor & Francis) p 387
[5] Anderson P W 1959 *Fizika Dielectrikov* (AN SSSR) p 290
[6] Landau L D and Lifshitz E M 2000 *Statistical Physics, Part I* (Butterworth-Heinemann)
[7] Lines M E and Glass A M 1977 *Principles and Applications of Ferroelectric and Related Materials* (Clarendon)
[8] Rabe K M, Ahn C H, and Triscone J-M 2007 *Physics of Ferroelectrics: A Modern Perspective* (Springer Series *Topics in Applied Physics* vol 105) (Springer)
[9] Smolenskii G 1984 *Ferroelectrics and Related Materials* (*Ferroelectricity and Related Phenomena*) (Gordon and Breach)
[10] Dawber M, Rabe K M, and Scott J F, 2005 *Rev. Mod. Phys.* **77** 1083
[11] Wigner E P 1949 *unpublished*
[12] Jaynes E T 1950 *PhD Thesis* (Princeton)
Jaynes E T 1950 *Phys. Rev.* **79** 1008
[13] Jaynes E T 1953 *Ferroelectricity* (Princeton Univ. Press)
[14] Mason W P and B. T. Matthias B T 1948 *Phys. Rev.* **74** 1622
Mason W P 1950 *Piezoelectric Crystals and Their Applications to Ultrasonics* (Van Nostrand)
[15] Devonshire A F 1949 *Phil. Mag.* **40** 1040
[16] Slater J C 1950 *Phys. Rev.* **78** 748

[17] Maksimov E G 2009 *Phys.-Usp.* **52** 603
[18] King-Smith R D and Vanderbilt D 1994 *Phys. Rev.* B **49** 5828
Vanderbilt D 2004 *Ferroelectrics* **301** 9
[19] Bersuker I B 1966 *Phys. Lett.* **20** 589
[20] Bersuker I B and Vekhter B G 1967 *Sov. Phys. – Solid State* **9** 2452
[21] Öpik U and Pryce M H L 1957 *Proc. Roy. Soc. London* A**238** 425
[22] Kristoffel N N and Konsin P I 1967 *Phys. Stat. Sol.* **21** K39
[23] Englman R 1972 The Jahn-Teller Effect in Molecules and Crystals (Wiley)
[24] Bersuker I B and V. Z. Polinger 1989 *Vibronic Interactions in Molecules and Crystals* (Springer)
Bersuker I B 2006 *The Jahn-Teller Effect* (Cambridge University Press)
[25] Bersuker I B 2010 *Electronic Structure and Properties of Transition Metal Compounds. Introduction to the Theory* (Wiley) pp 371-373
[26] Bersuker I B 1995 *Ferroelectrics* **164,** 75
[27] Bersuker I B and Polinger V 2012 *Phys. Rev. B* submitted for publication
[28] Jeong I-K, Lee Seunghun, Jeong Se-Young, Won C J, Hur N, and A. Llobet A 2011 *Phys. Rev.* B**84**, 064125
[29] Hanske-Petitpierre O, Yacoby Y, Mustre de Leon J, Stern E A, and Rehr J J 1992 *Phys. Rev.* B**44**, 6700
Sicron N, Ravel B, Yacoby Y, Stern E A, Dogan F, and Rehr J J 1994 *Phys. Rev.* B**50**, 13168
Ravel B, Stern E A, Vedrinskii R I, and Kraisman V 1998 *Ferroelelctrics* **206-207** 407
[30] Garcia-Fernandez P 2012 *unpublished*
[31] Ghosez P 2002 *Microscopic Properties of Ferroelectric Oxides from First-Principles : Selected Topics* (Troisieme Cycle de la Physique en Suisse Romande, Lausanne).
[32] Bersuker I B 1963 *Sov. Phys. - JETP* **16** 933
[33] Gomez M, Bowen S P, and Krumhansl J A 1967 *Phys. Rev.* **153** 1009
[34] Quitet A M, Lambert M, and Guinier A, 1973 *Solid State Commun.* **12** 1053
Comes R, Currat R, Denoyer F, Lambert M, and Quittet M 1976 *Ferroelectrics* **12** 3
[35] Burns G and Dacol F 1981 *Ferroelectrics* **37** 661
[36] Gervais F 1984 *Ferroelectrics* **53** 91
[37] Dougherty T P, Wiederrecht G P, Nelson K A, Garrett M H, Jensen H P, and Warde C 1992 *Science* **258** 770
[38] Ehses K H, Bock H, and Fischer K 1981 *Ferroelectrics* **37** 507
[39] Itoh K, Zeng L Z, Nakamura E, and Mishima N 1985 *Ferroelectrics* **63** 29
[40] Müller K A and Berlinger W 1986 *Phys. Rev.* B **34** 6130
Müller K A and Berlinger W 1986 *Phys. Rev.* B **34** 6130
Müller K A, Berlinger W, Blazey K W, and Albers J 1987 *Solid State Commun.* **61** 21
Volkel G and Muller K A 2007 *Phys. Rev.* B **76** 094105
[41] Zalar B, Laguta V V, and Blinc R 2003 *Phys. Rev. Lett.* **90** 037601
Zalar B, Labar A, Seliger J, and Blinc R 2005 *Phys. Rev.* B **71** 064107
Blinc R, Zalar B, Laguta V V, and Itoh M 2005 *Phys. Rev. Lett.* **94** 147601
[42] Jeong I-K, Lee Seunghun, Jeong Se-Young, Won C J, Hur N, and A. Llobet A 2011 *Phys. Rev.* B **84**, 064125
[43] Yamada Y, Shirane G, Linz A 1969 *Phys. Rev.* **177** 848
Harada J, Axe J D, Shirane G 1971 *Phys. Rev.* B **4** 155
[44] Luspin Y, Servoin J L, Gervais F 1980 *J. Phys.* C **13** 3761
[45] Vogt H, Sanjurjo J A, Rossbroich G 1982 *Phys. Rev.* B **26** 5904
[46] Girshberg Ya G and Tamarchenko V I 1976 *Fiz. Tverd. Tela* **18** 1066
Girshberg Ya G and Tamarchenko V I 1976 *Fiz. Tverd. Tela* **18** 3340
[47] Plakida N M and Mailyan G L 1977 *Fiz. Tverd. Tela* **19** 121
[48] Mailyan G L and Plakida N M 1977 *Phys. Status Solidi* B **80** 543
[49] Comez R, Lambert M, and Guinier A, 1968 *Solid State Commun.* **6** 715
[50] Rebane T K 1984 *Teor. i Eksp. Khim.* **20** 532
[51] Stern E A 2004 *Phys. Rev.Lett.* **93**, 037601

Journal of Physics: Conference Series **428** (2013) 012027 doi:10.1088/1742-6596/428/1/012027

Dynamic Polar Nanoregions and Broken Local Symmetry in Relaxor Ferroelectrics Probed by Inelastic Light Scattering

Seiji Kojima*, Ryu Ohta, Takuma Ariizumi, and Junta Zushi
Graduate School of Pure and Applied Sciences, University of Tsukuba, Tsukuba, Ibaraki 305-8573, Japan
**kojima@bk.tsukuba.ac.jp*

Abstract

The precursor dynamics of relaxor ferroelectrics are studied by Brillouin and Raman scattering. Above T_C, Raman scattering spectra reflect the existence of polar nanoregions (PNRs) by broken local symmetry in a cubic matrix. The polarization fluctuations in dynamic PNRs induce elastic anomaly and an intense broad central peak below Burns temperature T_B. The average size of dynamic PNRs was estimated as a function of temperature assuming the local piezoelectricity in PNRs owing to the lack of the center of symmetry. The marked growth of the size towards T_C was found in a cubic paraelectric phase. In normal ferroelctric KTN, $BaTiO_3$ also show the precursor dynamics related to dynamic PNRs.

KEYWORDS: relaxor, ferroelectric, polar nanoregion, broken symmetry, elastic anomaly, central peak, Brillouin scattering

1. Introduction

Ferroelectric oxides with perovskite structure have been mostly studied by the structural simplicity and rich of application as functional materials. The origin of ferroelectricity in $BaTiO_3$ has been discussed extensively by the displacive model in which the condensation of anharmonic soft optic mode causes spontaneous polarization.[1] However, the order-disorder nature was also discussed on the basis of the pseudo Jahn-Teller effect (PJTE)[2] and the diffraction study.[3] Recently it was reported that the coexistence of displacive and order-disorder processes still remain even in the lowest temperature rhombohedral phase.[4] Such a coexistence in perovskite ferroelctrics was also discussed by the polarizability model and it was predicted that both dynamics appear in different time scales.[5] The recent Brillouin scattering study of $BaTiO_3$ clarified the order-disorder behavior in a gigahertz range,[6] and SHG signals were observed above T_C owing to the broken local symmetry.[7]

In contrast to normal ferroelectric $BaTiO_3$, relaxor ferroelectrics (RFs) such as $Pb(Mg_{1/3}Nb_{2/3})O_3$ (PMN), $Pb(Sc_{1/2}Nb_{1/2})O_3$ (PSN) with B-site disorder undergoes diffusive phase transitions and remarkable dielectric dispersion near Curie temperature T_C.[8] Another interest on RFs is the colossal piezoelectric response observed in PMN-$PbTiO_3$ solid solutions. The origin of RF behaviors has been attributed to the nanoscale heterogeneity, namely, chemical ordered regions (CORs) and the polar nanoregions (PNRs), which are related to random fields. However, the recent study shows that CORs only contribute diffuseness partly and PNRs plays the dominant role in RFs.[9] The current interests of PNRs are the temperature evolution of PNRs[10] and the roles of static/dynamic PNRs.[11] The schematic temperature evolution of PNRs is shown in Fig. 1.[12] At Burns temperature T_B which is typically a few hundreds degrees above T_C, the nucleation of PNRs starts and their size/volume gradually increase on cooling.[13] T_B was defined by the change of refractive index owing to the second order electro-optic (Kerr) effect induced by polarization islands in a nanometer scale. It was suggested that T_B can be associated with the formation of short lived correlations between the off-centered ions without local strain fields. For further cooling from T_B, the existence of the intermediate temperature T^* was detected by acoustic emissions (AEs) in several kinds of relaxor ferroelectrics.[14] It is speculated that at T^* PNRs start to correlate to each other, transforming from dynamic to static. The precursor dynamics related to the evolution of PNRs is not yet understood well and it becomes one of the most important topics in condensed matter physics.

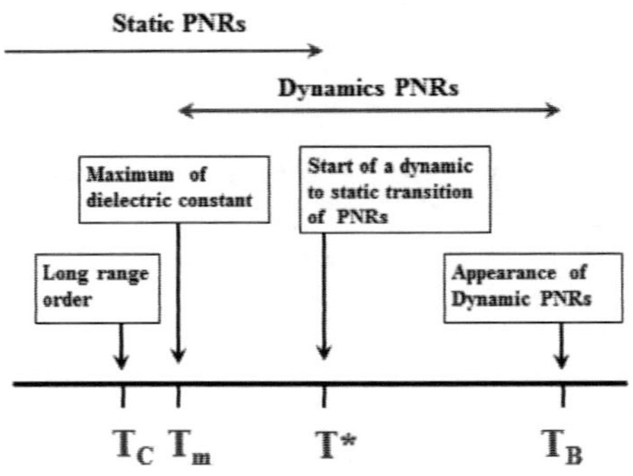

Fig. 1 Evolution of dynamic and static polar nanoregions in relaxor ferroelectrics.,where T_B, T^*, T_m, and T_C denote Burns temperature, intermediate temperature, dielectric maximum temperature, and Curie temperature, respectively.

In the high-symmetry paraelectric phase, local symmetry breaking occurs in PNRs. Both static and dynamic PNRs induce the Raman scattering of forbidden modes in a high-symmetry paraelectric matrix in the terahertz range. Brillouin scattering in the gigahertz range shows elastic anomaly through the local piezoelectric coupling in dynamic PNRs and broad central peak owing to the local polarization fluctuations in dynamic PNRs. This paper reports the recent progress to understand dynamic PNRs in RFs. The precursor dynamics and the temperature evolution of dynamic PNRs are investigated using the broadband inelastic light scattering spectroscopy. The precursor dynamics of normal ferroelectric phase transitions is also discussed.

2. Broken local symmetry in a paraelectric cubic phase of relaxor ferroelectrics

In the most of perovskite oxide ferroelectrics, the high-temperature prototypic paraelectric phase is cubic system with the space group of $Pm\bar{3}m$ (Z=1). In the first-order Raman selection rule, optical modes have Raman inactive T_{1u} and T_{2u} symmetry. It is known that Raman scattering is very sensitive to the local symmetry breaking by the observation of the forbidden optical bands arising from local symmetry breaking. As a typical case, the appearance of the first order Raman scattering in a cubic paraelectric phase is shown in Fig. 2. It shows Raman scattering spectra of relaxor ferroelectric $0.85Pb(Zn_{1/3}Nb_{2/3})O_3$-$0.15PbTiO_3$ (PZN-15PT) above and below T_C= 481 K observed at the backward scattering geometry $a(cc)\bar{a}$.[15] Main Raman peaks are the Pb-localized mode near 55 cm^{-1}, the B-cation localized mode near 255 cm^{-1}, and the polar oxygen octahedral mode near 815 cm^{-1}. The intense first order Raman scattering still remain above T_C, it indicates the existence of local lattice distortion which destroys the center of symmetry in cubic matrix. As the origin of the broken cubic symmetry, the fluctuating local polarization along [111] direction predicted by PJTE[2] may appears in a dynamic PNR below T_B. Such local polarizations break the center of symmetry through the local rhombohedral distortions induced by the local piezoelectric coupling in PNRs.

Figure 2 shows the another important feature of relaxor ferroelectrics. In displacive-type ferroelectrics, a soft optic mode is usually observed below 100 cm^{-1}. However, the Pb-localized mode near 55 cm^{-1} does not show such a softening. In general, Pb-based perovskite relaxor ferroelectrics do not show any soft mode, while show a temperature dependent broad central peak related to polarization fluctuations of dynamic PNRs as discussed in 3.2.

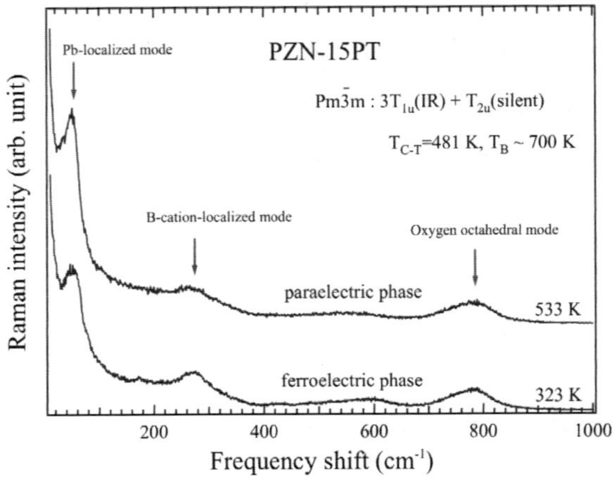

Fig. 2 Raman scattering spectra of PZN-15PT above and below T_C= 481 K observed at the backward scattering geometry $a(cc)\bar{a}$. The first order Raman scattering still remain above T_C.

3. Precursor dynamics of relaxor ferroelectrics

In this section, dynamics related to the temperature evolution of PNRs in a paraelectric phase are discussed in relation with elastic anomaly, central peak, and size of dynamic PNRs.

3.1 *Elastic anomaly*

When the LA mode propagates in RFs, it interacts with local polarization fluctuations of dynamic PNRs through electrostrictive coupling in cubic matrix and local piezoelectric coupling in PNRs.[10] The dominant change of phase velocity of LA mode, V_L, occurs according to the following equation as similar to the Kerr effect which is proportional to $<P_iP_j>$,

$$V_L = V_L^0 + V_L^1 <P_iP_j> \qquad (1)$$

where V_L^0 and V_L^1 are constants, < > means ensemble average.

In a paraelectric phase above T_C, there is no first order term of polarization according to Neumann's principle of symmetry. At very high temperatures, it holds that

$$<P_iP_j> = 0 \qquad \text{for } T > T_B . \qquad (2)$$

However, below T_B dynamic PNRs with local polarization appears. Then,

$$<P_iP_j> \neq 0 \qquad \text{for } T < T_B . \qquad (3)$$

Since the V_L^0 shows the linear temperature dependence, V_L shows the deviation below T_B as shown in Fig. 3 for the case of B-site disordered PSN.[16] Below T_B, the second term of

$V_L^2\langle P_iP_j\rangle$ increases gradually owing to the growth of PNRs toward T_C. Such an anomaly is generally observed in relaxor ferroelectrics.[10,17] The growth process of PNRs is important and the temperature variation of the size of dynamic PNRs is discussed in 3.1.

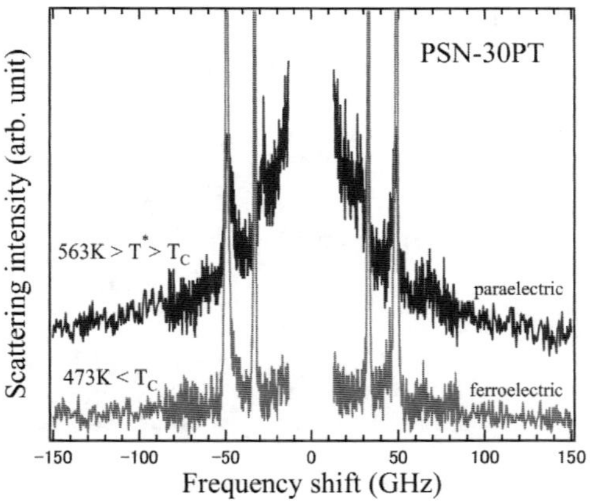

Fig. 3 Broadband Brillouin scattering spectram of PZN-30PT above and below T_C= 500 K observed at the backward scattering geometry $a(cc)\bar{a}$. The sharp peaks at 32 and 50 GHz are TA and LA peaks, respectively. The broad central peak of qusasi-elastic scattering from relaxation mode of local polarization fluctuations becomes very week above T*~560 K.

3.2 Central peak

The quasi-elastic scattering due to relaxation process appears as broad Rayleigh wings in an inelastic light scattering spectrum. It is called a central peak (CP), which has been observed not only in ferroelectrics but also in supercooled liquids. The width of CP is approximately proportional the reciprocal of relaxation time. The critical slowing down of relaxation time is observed in the order-disorder type phase transition. As a general aspect in Brillouin scattering spectra of relaxor ferroelectrics an intense broad CP appears below T_B and its intensity increases toward T_C, and becomes weak below T_C as shown in Fig. 4 for the case of 0.70PSN-0.30PT (PSN-30PT). Simultaneously the CP width shows remarkable narrowing from T_B down to T_C. Under the assumption of a single Debye relaxation process, the relaxation time τ_{CP} is determined by the relation $\pi \times$ (CP width) $\times\ \tau_{CP} = 1$. The temperature dependence of the reciprocal of τ_{CP} is plotted in Fig. 5. It shows the critical slowing down of dynamic PNRs toward T_C = 500 K. Such a critical slowing down has been generally observed in relaxor ferroelectrics not only perovskite but also tungsten bronze structures.[10,17,18,19]

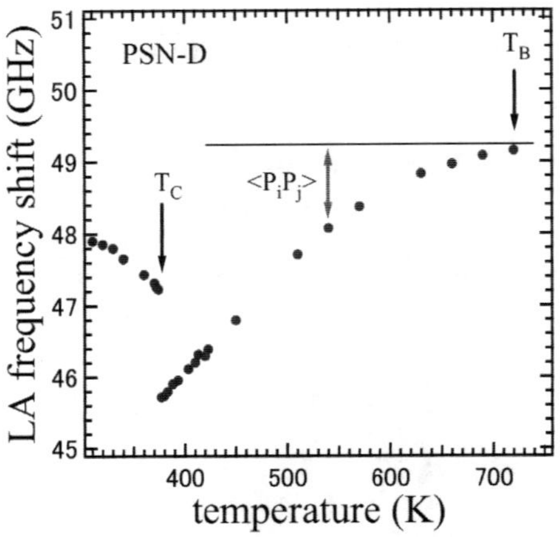

Fig. 4 Temperature dependence of LA mode frequency of PZN-30PT related to elastic anomaly above and below T_C= 500 K. The deviation from the linear part appears below T_B.

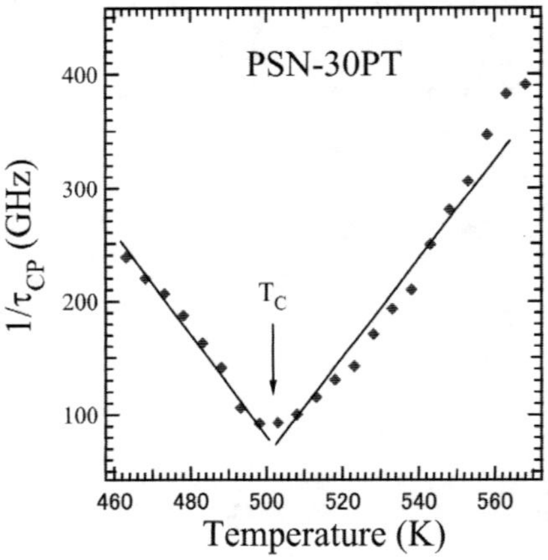

Fig. 5 Plots of reciprocal relaxation time which shows critical slowing down in PZN-30PT which obeys Curie-Weiss law.

3.3 Size of dynamic polar nanoregions

The growing process of PNRs is one of the most important topics in ferroelectrics. It is possible to observe the images of static PNRs by TEM or piezoforce microscopy.[20] However, it is difficult to observe fluctuating dynamic PNRs by such conventional experimental methods. Therefore, the average size of dynamic PNRs was estimated by Brillouin scattering measurement.[12,21]

The thermally fluctuating of the local spontaneous polarization in a particular PNR induces flipping of local strain by the local piezoelectric coupling due to the lack of the center of symmetry in PNRs. By the bilinear coupling between polarization and strain, the speed of the propagation of local polarizations in a PNR can be equal to the speed of the propagation of local strain flipping motion in a PNR. Therefore, the size of a flipping PNR can be approximately equal to the propagation length of the LA mode during the flipping period. The length, $l_{PNR}= V_{LA} \times \tau_{CP}$, may give the rough estimation of the average size of dynamic PNRs. The temperature dependence of the average size of PNRs of PSN-30PT is shown in Fig. 6.[12] Below the intermediate temperature T^*, at which the rapid growth of PNRs starts, the remarkable increase of the size occurs towards T_C. The CP is very sensitive for the fluctuations of dynamic PNRs, while is not sensitive to static local polarizations in static PNRs. Therefore, it is expected that the size of static PNRs become much larger than that of dynamic one just above T_C.

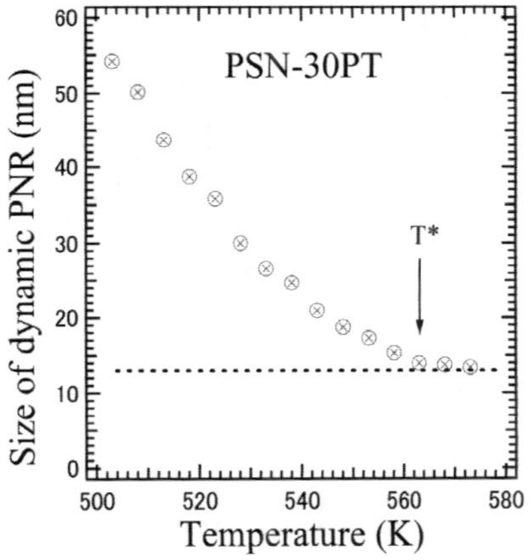

Fig. 6 Temperature dependences of size of dynamic polar nanoregions of PSN-30PT with T_C= 500 K.

4. Precursor dynamics of normal ferroelectrics

Nowadays the existence of PNRs and the role in relaxor nature have been widely accepted. In this section, precursor dynamics of normal ferroelectric are discussed. The dielectric anomaly of normal ferroelectrics obeys Curie-Weiss law, in contrast with diffuse and dispersive dielectric anomalies in RFs.

In $BaTiO_3$, Burns reported the deviation of refractive index from the linear temperature dependence about 120 K above T_C as similar to that of RFs.[22] The existence of the intense first order Raman scattering was also observed above T_C. These results suggested broken local symmetry in a cubic paraelectric phase with space group $Pm\bar{3}m$ (Z=1). To clarify the precursor dynamics intensively, the temperature dependence of SHG was measured accurately.[7] In a $BaTiO_3$ crystal, the SHG intensity appears clearly above T_C. The power law of the SHG integrated intensity showed the crossover, near T_B~580 K, from the exponent γ=1, reflecting the scattering by polarization fluctuations, into the larger exponent γ=2, reflecting the scattering by PNRs having strongly correlated polar displacements. According to PJTE, the displacement of a Ti ion along a [111] direction from the center of an oxygen octahedron was predicted in agreement with the x-ray diffraction study.[2,3] The local polarizations coupled to the local rhombohedral distortions may enhance the first order Raman scattering in a cubic paraelectric phase. Recently the critical slowing down of order-disorder nature has been observed in the vicinity of T_C by Brillouin scattering study.[6]

$KTa_{1-x}Nb_xO_3$ (KTN) undergoes a ferroelectric phase transition similar to $BaTiO_3$. The intense first order Raman scattering appears also in a paraelectric phase with space group $Pm\bar{3}m$. KTN32 (x=0.32) crystals with $T_{C\text{-}T}$=258 K were studied by broadband Brillouin scattering.[23] The elastic anomaly and the broad central peak owing to local polarization fluctuations of PNRs were observed above T_C. The acoustic emission count rate was also studied as a function of temperature, and T_B=620 K and T^*=310 K were found.[24] The average size of a dynamic PNR was estimated as a function of temperature, and the marked growth of the size starts at T^* predicted by the recent AE study[24] as shown in Fig. 7. The size increases gradually in relaxor PSN-30PT, while the size of KTN32 shows the rapid increase just above T_C. Such a difference has the correlation with the very sharp dielectric anomaly of KTN32 in the vicinity of T_C.

According to these experimental results it is concluded that, at least for perovskite oxide ferroelectrics, the precursor dynamics of PNRs can be common nature in a cubic paraelectric phase not only RF but also normal ferroelectrics. The difference of dynamics between normal and RFs can be ascribed to static PNRs and CORs related to the random fields of RFs, which suppress the growth of dynamic PNRs and enhance the dynamic to static transition of PNRs below T^*. Brillouin scattering is also a powerful tool to study diffusive nature caused by static PNRs and CORs through the observations of the diffuse elastic anomaly and the extended slowing down of the relaxation time.[21,25]

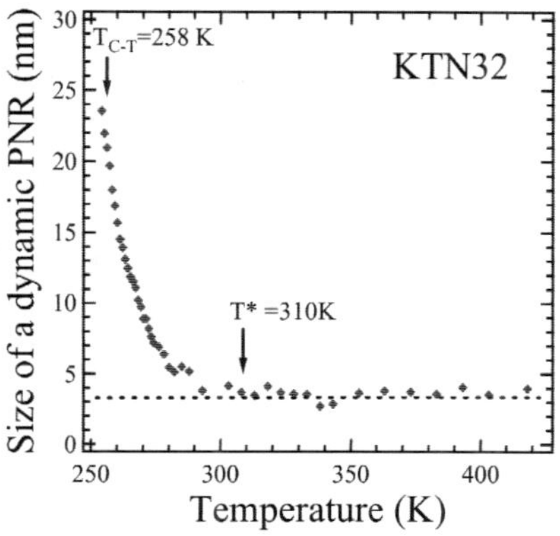

Fig. 7 Temperature dependences of size of dynamic polar nanoregions of KTN32 with T_C= 258 K.

5. Conclusion

The dynamical precursor dynamics of relaxor ferroelectrics are studied by Brillouin and Raman scattering. Above T_C, Raman scattering spectra reflect the existence of dynamic PNRs by broken local symmetry in a cubic matrix. The polarization fluctuations in dynamic PNRs induces elastic anomaly related to the second order effect of polarization and the intense broad central peak by the relaxation process, which shows the critical slowing down of relaxation time τ_{cp} above T_C. The average size of a dynamic PNR was estimated as a function of temperature, and the marked growth of the size toward T_C was found in a cubic paraelectric phase. In normal ferroelctric KTN, $BaTiO_3$ also show the precursor dynamics related to dynamic PNRs. The difference between relaxor and normal ferroelectrics is attributed to the static PNRs and CORs related to random fields of relaxor ferroelectrics.

Acknowledgements

The authors are thankful for the collaboration to Prof. Z.-G. Ye, Prof. K.-H. Ko, Prof. S. Tsukada, Dr. M. Ahart, Dr. G. Shabbir, and Dr. V. Sibasbramanian. The author (S. K.) is grateful for the discussion with Prof. A. Bussmann-Holder, Prof. I. B. Bersuker, Prof. M. Kaplan, and Prof. H. Koizumi during the 21st International Symposium of the Jahn-Teller Effect , 26-31 August, 2012, Tsukuba.

References

1. W. Cochran: Adv. Phys. **9** (1960) 387.
2. I. B. Bersuker: Phys. Lett. **20** (1966) 589, Phys. Rev. Lett. **108** (2012) 137202.
3. R. Comes, M. Lambert, and A. Ginner, Solid State Commun. **6**, 715 (1968).
4. G. Völkel and K. A. Müller, Phys. Rev. B **76** (2007) 094105.

5. Annette Bussmann-Holder, J. Phys.: Condens. Matter **24** (2012) 273202.
6. J.-H. Ko, T.H. Kim, K. Roleder, D. Rytz, and S. Kojima: Phys. Rev. B **84** (2011) 094123.
7. A. M. Pugachev, V. I. KOvalevskii, N. V. Surovtsev, S. Kojima, S. A. Prosandeev, I. P. Raevski, and S. I. Raevskaya, Phys. Rev. Lett. **108**, 247601 (2012).
8. G.A. Smolenskii, J. Phys. Soc. Jpn. **28** (1970) 26.
9. X. Zhao, W. Qu, X. Tan, A.A. Bokov, and Z.-G. Ye, Phys. Rev. B **79** (2009) 144101.
10. S. Kojima. M. Ahart, V. Sibasbrumanian, A.A. Bokov, and Z.-G. Ye, J. Adv. Diele. **2** (2012) 1241004.
11. Y. P. Shi and A. K. Soh: Appl. Phys. Lett. **99** (2012) 092908.
12. S. Kojima, S. Tsukada, A.A. Bokov, and Z.-G. Ye, J. Appl. Phys. **109**, 084114 (2011).
13. G. Burns and F. H. Dacol, Phys. Rev. B **28**, 2527 (1983).
14. M. Roth, E. Mojaev, E. Dul'kin, P. Gemeiner, and B. Dkhil, Phys. Rev. Lett. **98**, 265701 (2007).
15. Md. S. Islam, S. Tsukada, W. Chen[4], Z.-G. Ye, and S. Kojima, (to be submitted).
16. M. Ahart, A. Hushur, Y. Bing, Z. Ye, R. J. Hemley, and S. Kojima: Appl. Phys. Lett. **94** (2009) 142906.
17. S. Kojima and J.-H. Ko: Curr. Appl. Phys. **11** (2011) S22.
18. V. Sivasburamanian and S. Kojima, Phys. Rev. B **85** (2012) 054104.
19. J.-H. Ko and S. Kojima: Appl. Phys. Lett. **91** (2007) 082903.
20. V. V. Shvartman and A. L. Kholkin: J. Adv. Diele. **2** (2012)1241003.
21. S. Kojima and S. Tsukada, Ferroelectrics **405** (2010) 32.
22. G. Burns and F. H. Dacol, Ferroelectrics **104**, 25 (1990).
23. R. Ohta, J. Zushi, T. Ariizumi, and S. Kojima, Appl. Phys. Lett. **98**, 092909 (2011).
24. E. Dul'kin, S. Kojima, and M. Roth, Euro. Phys. Lett., **97**, 57004 (2012).
25. S. Tsukada and S. Kojima: Phys. Rev. B **78** (2008) 144106.

Journal of Physics: Conference Series **428** (2013) 012028 doi:10.1088/1742-6596/428/1/012028

A Local Approach to Solid State Problems: Pseudo Jahn-Teller origin of Ferroelectricity and Multiferroicity

I B Bersuker, Institute for Theoretical Chemistry, The University of Texas at Austin, Austin, Texas 78712, USA, bersuker@cm.utexas.edu

Abstract

This is a partially review paper in which, in continuation of previous work, it is shown that in perovskite crystals of ABO_3 type the spontaneous polarization is triggered by local vibronic interactions, the pseudo Jahn-Teller effect (PJTE). The driving force of the latter is added covalency by distortion, which is essentially of local (chemical) origin. The local origin of polar instability in crystals is confirmed by the theorem of structural instability proved earlier. For crystals of $BaTiO_3$ type local PJT interactions of the metal ion with the oxygen environment results in a peculiar adiabatic potential energy surface (APES) which has eight trigonal [111] type minima, twelve [110] type saddle points between them, six higher in energy [100] type saddle points at the top of the barrier connecting four minima, and a maximum at the cubic symmetry. The temperature dependence of the free energy with this potential explains the origin of all the four phases in such crystals; only the lowest rhombohedral phase is fully ordered, the other two ones at higher temperatures are partially disordered, and the paraelectric phase is fully (three-dimensionally) disordered. For $BaTiO_3$ this picture is confirmed by numerical estimates and ab initio DFT calculations.

An important further development of this theory was reached recently by showing that not only B ions with electronic d^0 configurations are subject to the PJTE instability to produce ferroelectricity in perovskite crystals ("the d^0 mystery"), but some other specific d^n configurations with unpaired electrons may be dipolar active too, the crystal being thus both magnetic and ferroelectric (multiferroic), and the necessary conditions for such multiferroicity are formulated for the whole d^0-d^n variety of perovskites. Moreover, the condition of multiferroicity was shown to depend also on the spin state, high-spin or low-spin, which in conditions of spin crossover leads to a magnetic-ferroelectric crossover that can be manipulated by external perturbations.

1. Introduction. The pseudo Jahn-Teller effects in crystals is of local origin

In any approximate treatment of multi-particle problems a hierarchy of interactions should be taken into account, the strong interactions to be considered first followed by weaker interactions. In ionic dielectric crystals some local interaction may be stronger than long-range forces. In particular, this is true for vibronic coupling of electronic states with local low-symmetry distortions in the Jahn-Teller effect (JTE) and pseudo JTE (PJTE). When such local interactions are stronger than the cooperative forces, many solid state properties are triggered first of all by the local effects (that are independent of the translational symmetry). In this sense the approach to such solid state problem should be local: first the local vibronic coupling problem should be solved followed by the cooperative interaction of such local formations. This leads to cooperative JTE (CJTE) and cooperative PJTE (CPJTE) (see, e.g., in [1-3]).

With regard to the JTE, there may be still room for non-local, direct cooperative vibronic coupling called "band JTE" which may take place when the degenerate electronic states are rather delocalized [2, 3]. This is hardly possible for the PJTE; the latter is in essence of local origin. Indeed, consider first the two effects in the simplest case when there are two electronic states, |1> and |2>, interacting with one low-symmetry mode Q, a "fingerprint" of the JTE and PJTE illustrated in Fig 1. There are only two parameters to describe the JT instability, the vibronic coupling constant F

$$F_{12} = \langle 1|(\partial H/\partial Q)_0|2\rangle \tag{1}$$

and the primary force constant,

$$K_0 = \langle 1|(\partial^2 H/\partial Q^2)_0|1\rangle \tag{2}$$

($Q_0 = \dfrac{F}{K_0}$ and $E_{JT} = \dfrac{F^2}{2K_0}$ in Fig. 1*a*), and only one additional parameter, the energy gap Δ, is required to describe the instability condition in the PJTE (Fig. 1*b*),

$$\Delta < (F^2/K_0) \tag{3}$$

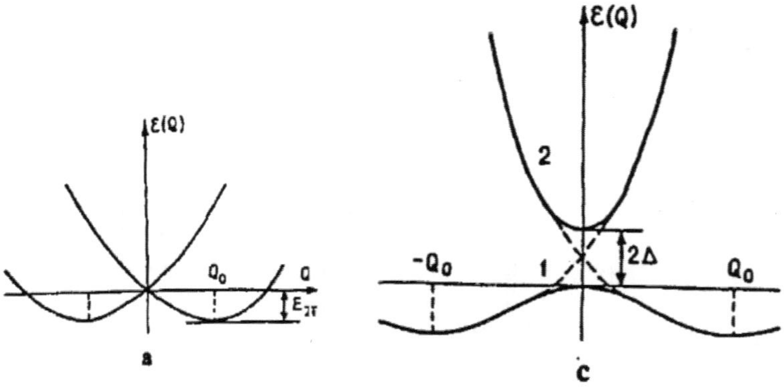

Journal of Physics: Conference Series **428** (2013) 012028 doi:10.1088/1742-6596/428/1/012028

Fig. 1. Energy profiles in a two-level pseudo Jahn-Teller effect (b) versus the Jahn-Teller effect (a) with a single distortion mode. The pseudo crossing in the PJTE case is shown by dashed lines.

At first sight the two pictures are similar indeed, but this similarity does not go very far; it comes out that the differences are more important [4]. First we refer to often encountered in literature two misunderstandings (misrepresentations) of the PJTE that (1) the energy gap 2Δ between the interacting states should be small, and (2) the two mixing states belong to the same irreducible representation. They stem from the initial paper of Opik and Pryce [5] who introduced the PJTE to show that a small splitting of the degenerate state, akin to pseudodegeneracy, does not destroy completely the JT instability, provided the inequality (3) is valid. For a long time, the paradigm that 2Δ should be small served as a serious obstacle to the development of this trend. In fact, however, the condition of PJTE instability (3) contains three parameters, so there are no limitations on the values of any one of them, taken separately. With the progress of computational methods it was shown that small 2Δ value is not a necessary condition of the PJTE instability, as it may be compensated in Eq. (3) by larger F values and smaller K_0 values. For instance, the conversion of the unstable planar configuration of NH_3 to the pyramidal one takes place due to the PJTE coupling to the excited state at Δ~14 eV [6], the PJTE bending of the linear molecules L-M-L with M=Ca, Sr, Ba, and L=Cl, Br, I, takes place at Δ~10 eV [7], and the PJTE ferroelectric distortion in $BaTiO_3$ is produced at the energy gap Δ~3.2 eV (see below in Section 3).

The fact that the PJTE may take place at any reasonable 2Δ values is one of the main points in modern understanding of the PJTE. It extends the number of possible systems subjected to the PJTE to be practically unlimited, including also the systems with the JTE that may, too, be subject to the PJTE. Additionally, the two coupled states in the PJTE may be of any symmetry and, accordingly, the distortion Q can take any direction controlled by the product of symmetries of the two mixing states in the expression (1) of F, whereas in the JTE the two degenerate states belong to the same irreducible representation of the symmetry group of the system that limits the possible JTE distortions Q.

Most important for the goal of this paper is the *difference in the driving force* that causes the symmetry breakings in these two effects. As seen from Fig. 1*a*, the derivative of the potential near the point of degeneracy in the JTE is nonzero, meaning a nonzero force causing the distortion in the Q direction (the force originates from the lower symmetry of the charge distribution in any one of the states of the degenerate term [2]). Contrary to this, in the PJTE there is no distorting force: the first derivative of the potential at the point of highest symmetry is zero (Fig. 1*b*). So what drives the distortion in the PJTE? The analysis shows [2, 4] that the driving force of symmetry breaking in the PJTE is the *added covalence bonding* between the near-neighbor atoms by distortion (see also Section 3). *Covalency is a net local (chemical) effect* dependent on the electronic states of the near-neighbor atoms and facilitated by the PJTE vibronic coupling. In this sense *the approach to the solid state problems with PJTE centers should be essentially local*, meaning revealing the local PJTE on the centers first and their influence of the crystal properties.

This important statement follows more rigorously from the theorem of lattice instability which proves that the PJTE is the only source of symmetry breaking in polyatomic systems in non-degenerate states [6, 8, 9], and the spontaneous polarization in ferroelectric perovskites is triggered by local polar distortions [10, 11]. Indeed, consider the APES, $E(Q)$, of an ionic cubic crystal ABO_3 with perovskite structure near its high-symmetry (cubic) configuration as a function of low-symmetry displacements Q that shift the sublattice of positive (metal B) ions with respect to the negative ones (oxygen atoms) - a limit optical phonon displacement; the displacement of the A atom is ignored. The crystal becomes unstable with respect to polarizing nuclear displacements if the curvature K of the APES in the Q direction at the point of cubic symmetry is negative,

$$K = \left(\frac{\partial^2 E}{\partial Q_\alpha^2}\right)_0 < 0, \qquad \alpha = x, y, z, \tag{4}$$

Putting $E(Q) = \langle \psi_0 | H | \psi_0 \rangle$, where H is the Hamiltonian and ψ_0 is the ground-state adiabatic wave function, we get

$$K = \left\langle \psi_0 \left| \left(\frac{\partial^2 H}{\partial Q_\alpha^2}\right)_0 \right| \psi_0 \right\rangle + 2\left\langle \psi_0 \left| \left(\frac{\partial H}{\partial Q_\alpha}\right)_0 \right| \frac{\partial \psi_0}{\partial Q_\alpha} \right\rangle \tag{5}$$

The first term in (5) is the introduced above K_0 magnitude in Eq. (2). Denoting the second one as K_v, $K = K_0 + K_v$, we notice that $K_v < 0$. Indeed, presenting ψ_0' in terms of the second-order perturbation theory, we get directly

$$K_v = -2\sum_n \frac{|<\psi_0|(\partial H/\partial Q_\alpha)_0|\psi_n>|^2}{E_n - E_0} \tag{6}$$

where the matrix elements in the numerator are the vibronic coupling constants F_{on} introduced in Eq. (1) which together with the energy gap $2\Delta_{0n} = E_n - E_0$ between the mixing states are the main parameters of the PJTE. Therefore K_v is termed *vibronic contribution*. Thus, being always negative, the K_v term contributes to the instability of the system. If $K_0 > 0$, then the negative term K_v is the only source of instability.

In a series of papers [6, 8, 9] (see also in [2, 3]) the following statement was proved. *For any polyatomic system, in the high-symmetry nuclear configuration*, where $(\partial E(Q)/\partial Q)_0 = 0$ (the force-equilibrated configuration in the direction Q), the non-vibronic contribution is always positive,

$$K_0 > 0, \tag{7}$$

and, therefore, *the vibronic contribution is the only source of instability of any polyatomic system in a non-degenerate state*. In case of degenerate states with the JTE, the PJTE produces additional instability, or instability in other (than the JTE) directions. This is the *theorem of instability*.

For crystal structures the importance of this result is also in the different nature of the two contributions K_0 and K_v. According to Eq. (2), the primary force constant K_0 is a diagonal matrix element of the second derivative with respect to Q which includes a combination of odd-type local distortions of all unit cells over the crystal; reduced to local in-cell displacements $Q_\alpha^{(m)}$, K_0 can be presented as a sum of two contributions [12],

$$K_0^{(\alpha)} = \frac{1}{N}\sum_{\mathrm{m}}\langle 0|\left(\frac{\partial^2 H}{\partial Q_\alpha^{(\mathrm{m})2}}\right)_0|0\rangle + \frac{1}{N}\sum_{\mathrm{m}}\sum_{\mathrm{n}\neq\mathrm{m}}\langle 0|\left(\frac{\partial^2 H}{\partial Q_\alpha^{(\mathrm{m})}\partial Q_\alpha^{(\mathrm{n})}}\right)_0|0\rangle. \tag{8}$$

where the first term is the sum of in-sell contributions, and the second term stand for the inter-sell (akin to elastic) interactions. Due to the overwhelming local nature of the chemical bonding in dielectrics, it decreases with the distance, but this decrease is relatively slow, and the respective radius of elastic correlation covers macroscopically large domains in the crystal. K_0 is thus a whole-crystal long-range interaction magnitude.

On the contrary, in K_v in Eq. (6) there are only off-diagonal matrix elements of first derivatives of the Hamiltonian, and therefore, translated to local distortions, it does not contain

Journal of Physics: Conference Series **428** (2013) 012028 doi:10.1088/1742-6596/428/1/012028

inter-cell interaction terms. Keeping overlap of atomic wavefunctions of only near-neighbor atoms |0> and |n>, we get approximately that the total PJTE vibronic coupling is a sum of local center coupling :

$$K_v^{(\alpha)} = \sum_n \frac{\left| \langle 0 | \left(\frac{\partial H}{\partial Q_\alpha} \right)_0 | n \rangle \right|^2}{E_0 - E_n(0)} \qquad (9)$$

The matrix elements in Eq. (9) are nonzero when the overlap between the corresponding wave function of the near-neighbor atoms of the two sublattices, the ground state of one of them with excited state of the other one (ψ_0 and ψ_n in Eq. (6)), increases by the Q displacements.

The local nature of the negative K_v contribution to the curvature means that the instability is essentially of local origin. The long-range interactions presented by K_0 are still most important as they play a key role in the realization of the condition of instability $K < 0$ that takes place at $|K_v| > K_0$ (and results in Eq. (3) in a two-level approximation). *The trigger mechanism of spontaneous polarization of the crystal is* thus *of local PJTE origin* directly related to its atomic (electronic) structure, but the magnitude of local distortion and the crystal polarization depends also on the long-range interactions. In other words, the long-range interactions K_0 by themselves, without the negative local contribution K_v (without the added covalency[2]) cannot produce the spontaneous polarization of the crystal, whereas $|K_v|$ may be too small yielding just the weak PJTE without instability or the local instability by itself may be not strong enough to overcome the rigid lattice. K_0 and K_v depend also on the third ion A in the perovskite (Ba in $BaTiO_3$) as it defines the size of the unit cell which influences all the Ti-O interactions (see below).

2. PJTE origin of ferroelectricity in perovskite crystals

The theoretical investigation of the origin of ferroelectricity and ferroelectric phase transitions has already a long history (see [13] and references therein). Below we describe the modern aspects of our vibronic (PJTE) theory of this phenomenon. As it follows from the theorem of instability, the ferroelectric displacements in a cubic perovskite crystal structure ABO_3 are triggered by the local PJTE coupling in the transition-metal center. Therefore, we start with the Hamiltonian of one unit cell in the high-symmetry cubic configuration. Consider the octahedral fragment of the unit cell centered at the transition metal atom B (red circle in Fig. 2). It is surrounded by six oxygen atoms (blue circles in Fig. 2). Cation atoms A are not shown.

For the octahedral fragment $[BO_6]^{8-}$ shown in Fig. 2, in the MO LCAO presentation (e.g., see [14]) the highest occupied MO (HOMO) are formed by the atomic $2p_\pi$ orbitals of the six oxygen atoms. They belong to the representations T_{1u}, T_{2u}, and T_{1g} of the octahedral symmetry group and are denoted in Fig. 3 by t_{1u}, t_{2u}, and t_{1g}, respectively. The lowest unoccupied MO (LUMO) are $t_{2g}(3d)$ orbitals of the transition metal: $|yz(B)>$, $|xz(B)>$, and $|xy(B)>$ (Fig. 3). In the regular cubic configuration, these HOMO-LUMO orbitals are orthogonal and do not participate in the Ti—O bonding (the bonding is realized via the inner σ orbitals and ionic interactions). Obviously, any off-center displacement of the atom B lowers the cubic symmetry and makes their overlap nonzero. Fig. 4 illustrates this effect for the orbital $|yz(Ti)>$. This orbital overlap means additional covalence in the Ti—O bonding, thus explaining the origin of the PJTE as due to added covalence. It is the driving force in the vibronic coupling (PJTE) origin of the local dipolar instability that produces the ferroelectric polarization of the crystal.

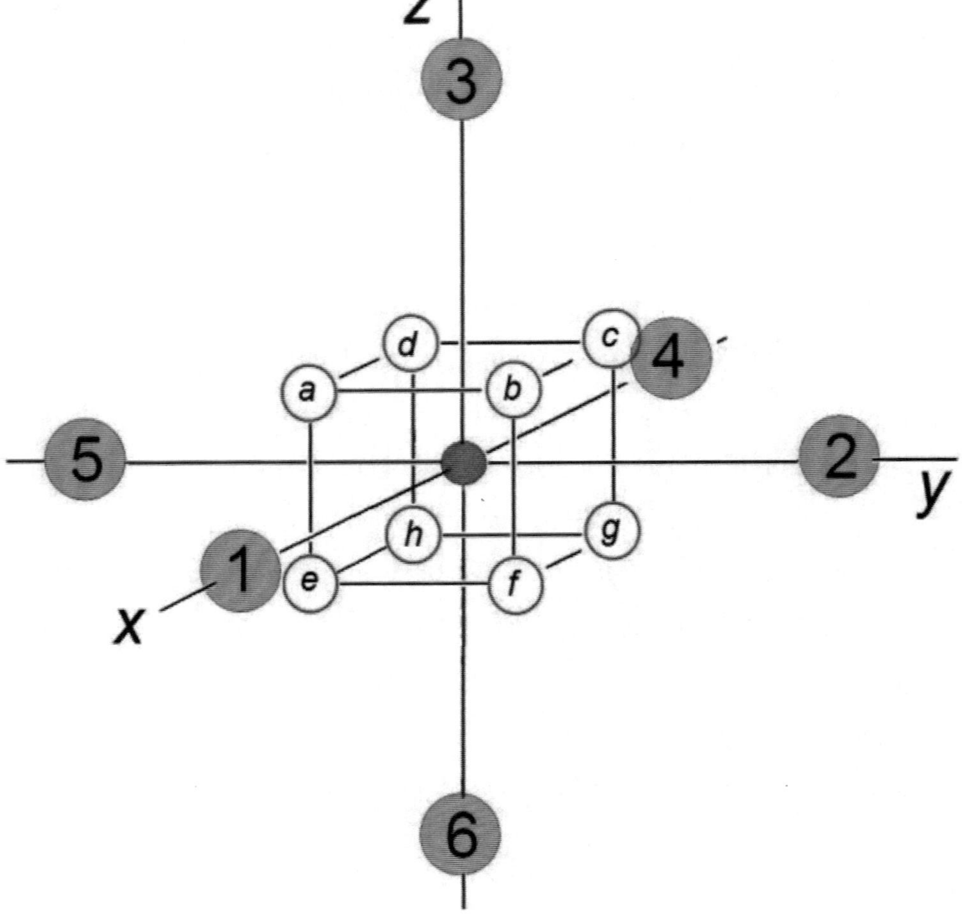

Fig. 2. (Color online) The octahedral fragment of the perovskite crystal structure ABO_3 with the transition metal atom B at the center (red) and six oxygen atoms (blue) at the apexes of the octahedron. The letters a, b, c, …denote the induced by the PJTE off-center positions of the atom B in the eight wells of the APES.

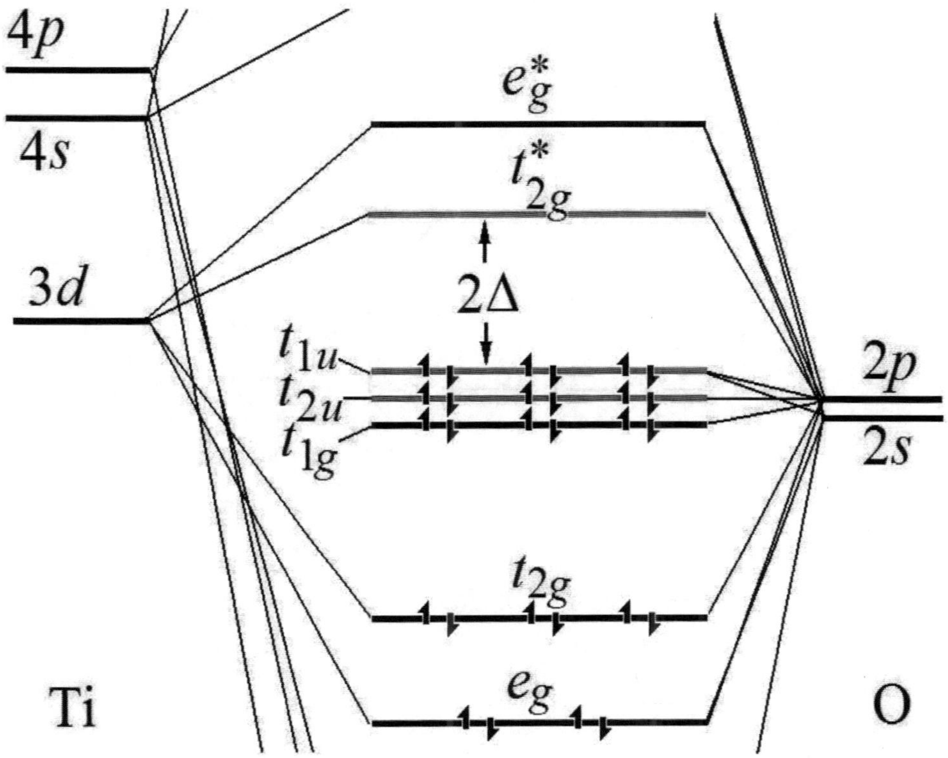

Fig. 3. (Color online). The energy level diagram for the octahedral cluster TiO_6^{8-}. The HOMO t_{1u} and t_{2u} (mostly oxygen) and LUMO t_{2g} (mostly titanium) levels that are vibronic coupled in the PJTE are shown in red.

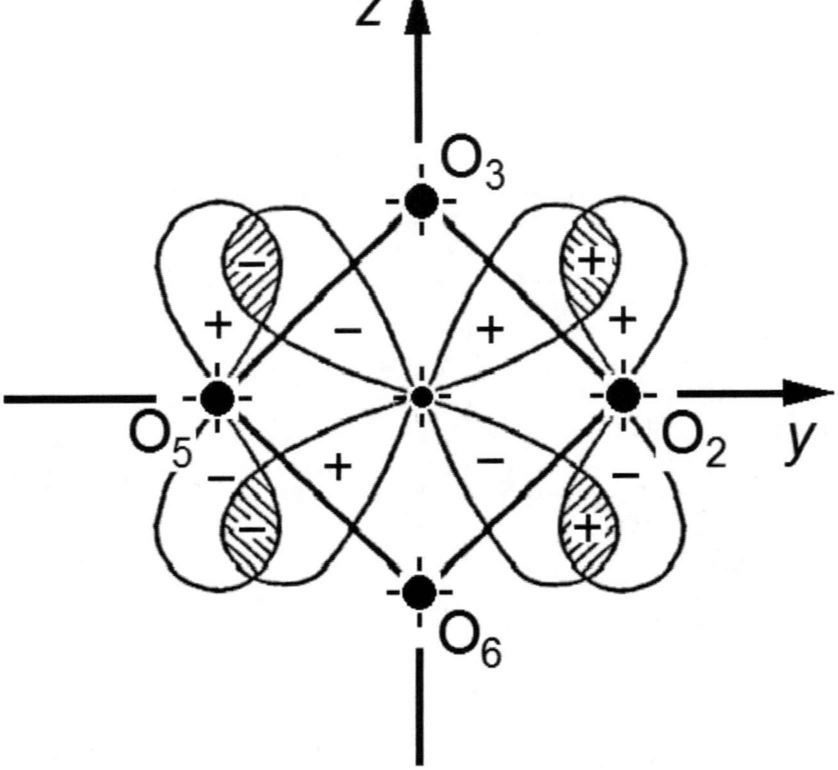

Fig. 4. Illustration to the added covalence in the PJTE: the net overlap of the 3*d* orbital $|T_{2u}(\mathrm{Ti})yz\rangle$ with π orbitals of the surrounding oxygen atoms (shaded) is zero when the metal ion is at the center and becomes non-zero in its y-direction off-center displacement inducing additional covalence bonding Ti—O.

With the fully occupied HOMO, the ground state is A_{1g}. Since the three Cartesian coordinates of the central atom, Q_x, Q_x, and Q_x, transform as T_{1u}, the general PJT vibronic coupling problem is $(A_{1g} + T_{1u}) \otimes t_{1u}$, where T_{1u} is the excited state formed by the one-electron excitation HOMO → LUMO ($t_{1u} \rightarrow t_{2g}$). There may be many such excitations, so the problem is better formulated as the vibronic coupling of the six HOMOs t_{1u} and t_{2u} to the three LUMOs t_{2g} via the dipole-active displacements of the central atom, Q_x, Q_y, and Q_z. With the linear vibronic coupling included, the Hamiltonian of the PJTE becomes as follows [2, 3, 12]:

Journal of Physics: Conference Series **428** (2013) 012028 doi:10.1088/1742-6596/428/1/012028

$$
H_{\mathrm{PJT}} = \begin{pmatrix}
\Delta & 0 & 0 & 0 & FQ_z & FQ_y & 0 & FQ_z & -FQ_y \\
0 & \Delta & 0 & FQ_z & 0 & FQ_x & -FQ_z & 0 & FQ_x \\
0 & 0 & \Delta & FQ_y & FQ_x & 0 & FQ_y & -FQ_x & 0 \\
0 & FQ_z & FQ_y & -\Delta & 0 & 0 & 0 & 0 & 0 \\
FQ_z & 0 & FQ_x & 0 & -\Delta & 0 & 0 & 0 & 0 \\
FQ_y & FQ_x & 0 & 0 & 0 & -\Delta & 0 & 0 & 0 \\
0 & -FQ_z & FQ_y & 0 & 0 & 0 & -\Delta & 0 & 0 \\
FQ_z & 0 & -FQ_x & 0 & 0 & 0 & 0 & -\Delta & 0 \\
-FQ_y & FQ_x & 0 & 0 & 0 & 0 & 0 & 0 & -\Delta
\end{pmatrix} \tag{10}
$$

Here the HOMO-LUMO energy gap is denoted by 2Δ (the energy levels of the HOMO t_{1u} and t_{2u} are approximately degenerate) and the vibronic coupling constant is

$$
F = \left\langle 2p_z(\mathrm{O}) \middle| \left(\frac{\partial H}{\partial Q_x}\right)_0 \middle| 3d_{xy}(\mathrm{Ti}) \right\rangle. \tag{11}
$$

The particular form of the matrix Hamiltonian (14) follows from symmetry considerations. It has the following nine eigenvalues:

$$
\epsilon_1 = \sigma_x,\ \epsilon_2 = -\sigma_x,\ \epsilon_3 = \sigma_y,\ \epsilon_4 = -\sigma_y,\ \epsilon_5 = \sigma_z,\ \epsilon_6 = -\sigma_z,\ \epsilon_{7,8,9} = -\Delta. \tag{12}
$$

Here $\sigma_i = \sqrt{\Delta^2 + 2F^2\left(Q^2 - Q_i^2\right)}$ and $Q^2 = Q_x^2 + Q_y^2 + Q_z^2$. For eigenvalues ϵ_1, ϵ_2, and ϵ_7, the corresponding eigenvectors are:

$$
\begin{aligned}
\psi_1^\dagger &= N_1\left(\Delta + \sigma_x,\ 0,\ 0,\ 0,\ FQ_z,\ FQ_y,\ 0,\ FQ_z,\ -FQ_y\right) \\
\psi_2^\dagger &= N_2\left(\Delta - \sigma_x,\ 0,\ 0,\ 0,\ FQ_z,\ FQ_y,\ 0,\ FQ_z,\ -FQ_y\right), \\
\psi_7^\dagger &= N_7\left(0,\ 0,\ 0,\ 0,\ Q_y,\ -Q_z,\ 0,\ Q_y,\ Q_z\right),
\end{aligned} \tag{13}
$$

with the normalization factors $N_1 = [2\sigma_x(\sigma_x + \Delta)]^{-1/2}$, $N_2 = [2\sigma_x(\sigma_x - \Delta)]^{-1/2}$, and $N_7 = [2(Q^2 - Q_x^2)]^{-1/2}$. The remaining six eigenvectors can be obtained from Eq. (17) by symmetry operations of the cubic group.

Populated with 12 electrons, the lowest one-electron orbital energy levels (16) generate the ground-state adiabatic potential energy surface (APES) reported in the first paper on the vibronic theory in 1966 [10]:

$$
U(Q) = \tfrac{1}{2}K_0 Q^2 - 2\left[\sqrt{\Delta^2 + 2F^2\left(Q^2 - Q_x^2\right)} + \sqrt{\Delta^2 + 2F^2\left(Q^2 - Q_y^2\right)} + \sqrt{\Delta^2 + 2F^2\left(Q^2 - Q_z^2\right)}\right]. \tag{14}
$$

Here we added the elastic energy of the off-center distortion ½ K_0Q^2, where K_0, as above, is the respective primary force constant. It is due to the σ bonds and other (electrostatic) interactions between the oxygen and titanium atoms. The APES (14) includes three parameters, K_0, Δ, and F, and three nuclear coordinates, Q_x, Q_y, and Q_z. By introducing dimensionless u, δ, x, y, and z,

$$U = u\frac{F^2}{K_0},\quad \Delta = \delta\frac{F^2}{K_0},\quad Q_x = x\frac{F}{K_0},\quad Q_y = y\frac{F}{K_0},\quad Q_z = z\frac{F}{K_0}, \tag{15}$$

we reduce the number of relevant parameters to just one, δ. The dimensionless APES is ($r^2 = x^2 + y^2 + z^2$):

$$u(x,y,z) = \tfrac{1}{2}r^2 - 2\left[\sqrt{\delta^2 + 2(y^2+z^2)} + \sqrt{\delta^2 + 2(x^2+z^2)} + \sqrt{\delta^2 + 2(x^2+y^2)}\right]. \tag{16}$$

Under condition $0 \le \delta \le 8$, or

$$\Delta < \frac{8F^2}{K_0}, \tag{17}$$

the APES (14), [or its dimensionless version (16)], has eight trigonal minimum points, $r_0(1, -1, 1)$, $r_0(1, 1, 1)$, $r_0(-1, 1, 1)$, $r_0(-1, -1, 1)$, $r_0(1, -1, -1)$, $r_0(1, 1, -1)$, $r_0(-1, 1, -1)$, and $r_0(-1, -1, -1)$. In Fig.1, they are labeled *a*, *b*, *c*, *d*, *e*, *f*, *g*, and *h*, respectively. At these points we have $r_0 = \frac{1}{2}(64 - \delta^2)^{1/2}$ and the stabilization energy $E_{JT}(111) = \frac{3}{8}(8 - \delta)^2$. Returning back to the variables (19), we have

$$Q_0 = \frac{F}{2K_0}\sqrt{64-\delta^2},\quad E_{JT}(111) = \frac{3F^2}{8K_0}(8-\delta)^2. \tag{18}$$

The inequality $0 \le \delta \le 8$ or (17) is *the condition of instability*. When the energy gap is larger than $8F^2/K_0$, or $\delta > 8$, the vibronic coupling is not strong enough to provide the dipole instability. In this case, there is just one stable configuration, the on-center position of the atom B, and ferroelectricity is impossible.

The curvature of the APES at the minimum points follows from the eigenvalue problem for the matrix of second derivatives of $u(x, y, z)$ in (16). At each trigonal well, the three normal frequencies split into a non-degenerate and a two-fold degenerate one. The non-degenerate mode corresponds to a longitudinal vibration along [111]. The respective curvature is $k_A = \frac{1}{64}(64-\delta^2)$. The two-fold mode corresponds to transverse vibrations perpendicular to the direction [111]. Its curvature is $k_E = \frac{1}{4}k_A$. In the variables (15) the respective expressions are

$$K_A = \tfrac{1}{64}K_0(64-\delta^2),\quad K_E = \tfrac{1}{4}K_A \tag{19}$$

In addition to the minima the APES (14) has two types of saddle points: orthorhombic and tetragonal. The twelve orthorhombic saddle points, $Q_1(1, 1, 0)$, $Q_1(1, -1, 0)$, $Q_1(-1, 1, 0)$, $Q_1(-1, -1, 0)$, $Q_1(1, 0, 1)$, $Q_1(1, 0, -1)$, $Q_1(-1, 0, 1)$, $Q_1(-1, 0, -1)$, $Q_1(0, 1, 1)$, $Q_1(0, 1, -1)$, $Q_1(0, -1, 1)$, and $Q_1(0, -1, -1)$ with

$$Q_1 \approx \frac{F}{3K_0}\sqrt{64-\delta^2} = \tfrac{2}{3}Q_0, \quad E_{JT}(110) \approx 0.34\frac{F^2}{K_0}(8-\delta)^2 \approx 0.91 E_{JT}(111). \tag{20}$$

The orthorhombic saddle points separate the eight trigonal wells from one another with potential barriers

$$\Delta E_{JT} = E_{JT}(111) - E_{JT}(110) \approx 0.09\frac{F^2}{K_0}(8-\delta)^2. \tag{21}$$

The six tetragonal saddle points are at $Q_2(1, 0, 0)$, $Q_2(-1, 0, 0)$, $Q_2(0, 1, 0)$, $Q_2(0, -1, 0)$, $Q_2(0, 0, 1)$, and $Q_2(0, 0, -1)$ with

$$Q_2 = \frac{F}{2K_0}\sqrt{2\left(64-\delta^2\right)} = Q_0\sqrt{2}, \quad E_{JT}(100) = \frac{F^2}{4K_0}(8-\delta)^2 = \frac{2}{3}E_{JT}(111). \tag{22}$$

Available experimental and *ab initio* calculations data allow us to estimate the APES parameters for $BaTiO_3$, at least by order of magnitude. From the three independent parameters that control the local PJTE displacements, the energy gap 2Δ to the PJT-active excited state is relatively well known, $2\Delta = 3.25$ eV. Another parameter can be obtained from the experimental value of the off-center displacement of the Ti ion in the trigonal minimum of the APES, $Q_0 \approx 0.11$ Å. It is a function of the vibronic coupling constant F and the primary force constant K_0. The third independent parameter can be obtained from the value of the PJT stabilization energy in the APES minimum $E_{JT}(111)$. From the *ab initio* calculations (see, e.g., Section 4) we get $\Delta E_{JT}(\text{calc}) \approx 0.3$ eV while from XAFS experiments $\Delta E_{JT}(\text{exp}) \approx 0.2$ eV [15]. Notably, the two values do not differ essentially, and the origin of their difference is quite understandable: the calculated value is determined from the crystal ferroelectric phase where the distortion is enhanced by the cooperative interactions, whereas the XAFS experiments are performed on the paraelectric phase, and hence reflect just the local distortion. For our local model the latter value is more appropriate.

Another option to involve experimental data is to exploit vibrational frequencies. In particular, in the APES minimum the transversal vibration $\hbar\omega_E$ (the one perpendicular to the axis [111]) may be assumed to represent approximately the soft mode behavior at low temperatures.

The usual estimate[13]$\hbar\omega_E$ ~100 cm^{-1}. From this value, applying Eqs.(14) and (18), we find the other parameters. The two versions of the numerical estimate of all the relevant parameters are given in the Table 1.

Table 1. Numerical values of the PJTE parameters obtained from the comparison of experimental data with the calculated APES, given in two versions based on two sets of data [12].

Main parameters	Version I [$E_{JT}(111) = 0.2$ eV]	Version II ($\hbar\omega_E = 100$ cm^{-1})
K_0	163 eV/Å^2	114 eV/Å^2
$\hbar\omega_0$	963 cm^{-1}	805 cm^{-1}
F	6.37 eV/Å	5.17 eV/Å
δ	6.54	6.94
$E_{JT}(111)$	0.2 eV	0.098 eV
$\hbar\omega_E$	277 cm^{-1}	100 cm^{-1}
$E_{JT}(110)$	0.1804 eV	0.0876 eV
$\Delta E_{JT} = E_{JT}(110) - E_{JT}(111)$	150 cm^{-1}	78 cm^{-1}

We see that, in spite of essentially different experimental data used in the two estimates, the numerical values are quite reasonable and rather consistent (within the expected orders of magnitude). With these parameters we can verify the PJTE criterion of instability $\Delta < 8F^2/K_0$ in Eq. (17). In both versions of Table 1 the electron energy gap $2\Delta = 3.25$eV is less than the PJT stabilization energy $16F^2/K_0$ equal to 3.38 eV and 3.74 eV in the versions I and II, respectively. This result shows that the PJTE is not limited to narrow energy gaps between interacting states, and it is feasible for perovskite systems. On the other hand, the condition of instability may be very restricting, which is why it is not realized in other similar systems (see Section 4). From these estimates we see that the performed above approximate evaluation of the PJTE yields a reasonable physical picture of the phenomenon of ferroelectricity compatible with the experimental observations.

Note that the essentially anharmonic APES (14) is obtained without involving the anharmonic terms in the Hamiltonian; only linear terms of vibronic coupling are included in Eq. (10). On the other hand, it was shown (Section 4) that just anharmonicity included in the Hamiltonian up to forth-order terms, does not produce the eight-minimum potential surfaces. Thus, with its eight trigonal wells, twelve orthorhombic saddle points, and six tetragonal saddle points, the APES (14) is a sole result of the PJTE. By the PJT hybridization of the ground and

Journal of Physics: Conference Series **428** (2013) 012028 doi:10.1088/1742-6596/428/1/012028

excited electronic states it generates the so-called "vibronic anharmonicity"[2, 3] which is much more ample than the singlet anharmonicity introduced directly in the Hamiltonian.

3. Origin of multiferroicity and magnetic-ferroelectric crossover

Coexistence of ferroelectric and magnetic properties (multiferroicity) implies that the ferroelectric crystal, which is a dielectric, has also a nonzero magnetic moment, meaning unpaired electrons. In the ferroelectric $BaTiO_3$ with perovskite ABO_3 structure the d^0 configuration of the Ti^{4+} ion has no unpaired electrons, and attempts to obtain ferroelectricity in perovskites with d^n, n>0, transition metals B ions were unsuccessful for a long time. This prompted some authors to term the situation as a "*d^0 mystery*" for ferroelectrics. However, more recently quite a number of ferroelectrics-multiferroics, mostly perovskites with configurations d^3-d^7, were obtained and studied [16, 17]. The origin of these special properties of perovskite ferroelectrics with d^n configurations does not follow directly from displacive theories, and so far had no general explanation. In this paper we show that the vibronic (PJTE) theory of ferroelectricity, outlined above, explains also directly the origin of perovskite multiferroics with proper ferroelectricity and formulates the necessary conditions that ABO_3 perovskites with a magnetic d^n configuration of the B ion may be ferroelectric. Moreover, we show also that the multiferroics conditions that emerge from the PJTE for d^n ions with n=3, 4, 5, 6, and 7, are directly influenced by the well known transition metal high-spin/low-spin crossover, resulting in the coexistence of three phenomena: ferroelectricity (FE), magnetism (M), and spin crossover (SCO). This, in turn, leads to a quite novel phenomenon: *magnetic-ferroelectric (multiferroics) crossover* (MFCO), creating a rich variety of possible magnetoelectric and related effects [18].

The typical molecular-orbital (MO) energy scheme for an octahedral cluster BO_6^{8-} in the approximation of linear combinations of atomic orbitals is shown above in Fig. 2 with the MO electron population of the d^0 configuration, e.g., when B=Ti. The highest occupied MO (HOMO) in this case is t_{1u} which is a threefold degenerate odd-parity (ungerade) linear combination of mostly oxygen p_π orbitals, while the lowest unoccupied MO (LUMO) is t_{2g}, mostly atomic three d_π orbitals of the transition metal ion B, and the next excited MO is the double degenerate one e_g (the non-bonding oxygen b_{1g} MO is not shown as irrelevant). For the d^0 case the HOMO configuration is thus $(t_{1u})^6=(t_{1u}\downarrow)^3(t_{1u}\uparrow)^3$, where the arrows up and down indicate the two spin states; the energy term of this configuration is $^1A_{1g}$. The excited state with opposite parity is formed by the one-electron $(t_{1u}\uparrow)\rightarrow(t_{2g}\uparrow)$ or $(t_{1u}\uparrow)\rightarrow(e_g\uparrow)$ excitation resulting in the lowest excited ungerade term $^1T_{1u}$ at the energy gap 2Δ. The PJTE vibronic coupling constant under the polar displacements of the B atom along [111] is given above in Eq. (11), while the condition of instability $\Delta<8F^2/K_0$ follows from Eqs. (17).

For the d^1 configuration of the B ion instead of the d^0 the HOMO becomes $(t_{1u}\downarrow)^3(t_{1u}\uparrow)^3(t_{2g}\uparrow)^1$ with the term $^2T_{2g}$, and LUMO (taken into account Hund's rule) is $(t_{1u}\downarrow)^2(t_{1u}\uparrow)^3(t_{2g}\uparrow)^2$ with the lowest excited ungerade term $^4T_{1u}$. Hence the two closest terms of different parity (that otherwise could mix under nuclear displacements to produce the PJT dipolar distortion), *possess different spin multiplicity*, and hence *they do not mix by the vibronic coupling*; the latter does not contain spin operators [1-3]. In principle, there may be higher in energy electronic configurations of opposite parity with the same spin as the ground state one, but they are at much larger energy gaps Δ, and therefore hardly satisfying the condition of instability (17) (numerical estimate show that the condition (17) is very limiting [41]). A similar picture emerges for d^2 configurations for which the two lowest terms of opposite parity are $^2T_{2g}$ and $^5T_{1u}$ (see Table 2).

The situation changes for d^3. Indeed in this case the HOMO is $(t_{1u}\downarrow)^3(t_{1u}\uparrow)^3(t_{2g}\uparrow)^3$ with the ground state term $^4A_{1g}$, and in the low-spin conditions of the strong ligand fields (large t_{2g}-e_g separation in Fig. 3, see, e.g., [14]) the LUMO is $(t_{1u}\downarrow)^2(t_{1u}\uparrow)^3(t_{2g}\uparrow)^3(t_{2g}\downarrow)^1$ with the lowest ungerade term $^4T_{1u}$. Therefore for d^3 configurations (e.g., Mn^{4+}) in sufficiently strong ligand fields the situation becomes again favorable for the PJTE and polar distortions, but in this case, distinguished from the d^0 case, the system possess also a magnetic moment created by three unpaired electrons. However, if the ligand field is weak and the separation t_{2g}-e_g is small, the high-spin arrangement of the excited electronic configuration takes place, and the excitation electron occupies the $e_g\uparrow$ orbital instead of $t_{2g}\downarrow$; the LUMO configuration under Hund's rule becomes $(t_{1u}\downarrow)^2(t_{1u}\uparrow)^3(t_{2g}\uparrow)^3(e_g\uparrow)^1$ with the lowest ungerade state $^6T_{1u}$. Here again there is no PJTE on dipolar distortions and no ferroelectric instability. In this way we considered all the d^n configurations with n=0, 1, 2, …10, shown in Table 2.

It follows from this table that according to the PJTE only B ions with configurations d^3-low-spin, d^4-low-spin, d^5-low-spin and high-spin, d^6-high-spin and intermediate-spin, d^7-high-spin, d^8, and d^9 can, in principle, produce multiferroics, provided the criterion of instability (8) is

Journal of Physics: Conference Series **428** (2013) 012028 doi:10.1088/1742-6596/428/1/012028

fulfilled. Transition metal ions B with configurations d^1, d^2, d^3-high-spin d^4-high-spin, d^6-low-spin, d^7-low-spin, and d^{10} are not expected to produce multiferroics under this mechanism. Experimentally obtained perovskite multiferroics with such B ions, for example, $Mn^{4+}(d^3)$, $Cr^{3+}(d^3)$, $Mn^{3+}(d^4)$, $Fe^{3+}(d^5)$, $Fe^{2+}(d^6)$, $Co^{2+}(d^7)$, etc. [16, 17], fit well with the conclusions of Table 2; there are no multiferroics with d^0, d^1, d^2, and d^{10} configurations.

Table 2. Necessary conditions that ABO_3 perovskites with the electronic d^n configuration of the B ion possess both ferroelectric and magnetic properties simultaneously [18]; EC=electronic configuration, GS = ground state, LUES = lowest ungerade excited state, FE = ferroelectric, MM = magnetic, MF = multiferroic, LS = low-spin, HS = high-spin; IS = Intermediate spin; LS* and HS* are two spin configurations of d^3 in the one-electron excitation. $(t_{1u})^6 = (t_{1u}\downarrow)^3(t_{1u}\uparrow)^3$; $(t_{1u})^5 = (t_{1u}\downarrow)^2(t_{1u}\uparrow)^3$; $(t_{2g})^6 = (t_{2g}\uparrow)^3(t_{2g}\uparrow)^3$; $(e_g)^4 = (e_g\uparrow)^2(e_g\downarrow)^2$.

d^n	*Example*	*HOMO EC and GS term*	*LUMO EC and LUES term*	FE	MM	MF
d^0	Ti^{4+}	$(t_{1u})^6$, ${}^1A_{1g}$	$(t_{1u})^5(t_{2g}\uparrow)^1$, ${}^1T_{1u}$	yes	no	no
d^1	Ti^{3+}, V^{4+}	$(t_{1u})^6(t_{2g}\uparrow)^1$, ${}^2T_{2g}$	$(t_{1u})^5(t_{2g}\uparrow)^2$, ${}^4T_{1u}$	no	yes	no
d^2	V^{3+}, Cr^{4+}	$(t_{1u})^6(t_{2g}\uparrow)^2$, ${}^3T_{2g}$	$(t_{1u})^5(t_{2g}\uparrow)^3$, ${}^5T_{1u}$	no	yes	no
d^3, LS	Cr^{3+}, Mn^{4+}	$(t_{1u})^6(t_{2g}\uparrow)^3$, ${}^4A_{2g}$	$(t_{1u})^5(t_{2g}\uparrow)^3(t_{2g}\downarrow)^1$, ${}^4T_{1u}$	yes	yes	yes
d^3, HS		$(t_{1u})^6(t_{2g}\uparrow)^3$, ${}^4A_{2g}$	$(t_{1u})^5(t_{2g}\uparrow)^3(e_g\uparrow)^1$, ${}^6T_{1u}$	no	yes	no
d^4, LS	Mn^{3+}, Fe^{4+}	$(t_{1u})^6(t_{2g}\uparrow)^3(t_{2g}\downarrow)^1$, ${}^3T_{2g}$	$(t_{1u})^5(t_{2g}\uparrow)^3(t_{2g}\downarrow)^2$, ${}^3T_{1u}$	yes	yes	yes
d^4, HS		$(t_{1u})^6(t_{2g}\uparrow)^3(e_g\uparrow)^1$, ${}^5T_{2g}$	$(t_{1u})^5(t_{2g}\uparrow)^3(e_g\uparrow)^2$, ${}^7T_{1u}$	no	yes	no
d^5, LS	Mn^{2+}, Fe^{3+}	$(t_{1u})^6(t_{2g}\uparrow)^3(t_{2g}\downarrow)^2$, ${}^2T_{2g}$	$(t_{1u})^5(t_{2g})^6$, ${}^2T_{1u}$	yes	yes	yes
d^5, HS		$(t_{1u})^6(t_{2g}\uparrow)^3(e_g\uparrow)^2$, ${}^6A_{1g}$	$(t_{1u})^5(t_{2g})^4(e_g\uparrow)^2$, ${}^6T_{1u}$	yes	yes	yes
d^6, LS	Fe^{2+}, Co^{3+}	$(t_{1u})^6(t_{2g})^6$, ${}^1A_{1g}$	$(t_{1u})^5(t_{2g})^6(e_g\uparrow)^1$, ${}^3T_{1u}$	no	no	no
d^6, IS_1		$(t_{1u})^6(t_{2g})^5(e_g\uparrow)^1$, ${}^3T_{1g}$	$(t_{1u})^5(t_{2g})^6(e_g\uparrow)^1$, ${}^3T_{1u}$	yes	yes	yes
d^6, IS_2		$(t_{1u})^6(t_{2g})^5(e_g\uparrow)^1$, ${}^3T_{1g}$	$(t_{1u})^5(t_{2g})^5(e_g\uparrow)^2$, ${}^5T_{1u}$	no	yes	no
d^6, HS		$(t_{1u})^6(t_{2g})^4(e_g\uparrow)^2$, ${}^5T_{2g}$	$(t_{1u})^5(t_{2g})^5(e_g\uparrow)^2$, ${}^5T_{1u}$	yes	yes	yes
d^7, LS	Co^{2+}, Ni^{3+}	$(t_{1u})^6(t_{2g})^6(e_g\uparrow)^1$, ${}^2E_g$	$(t_{1u})^5(t_{2g})^6(e_g\uparrow)^2$, ${}^4T_{1u}$	no	yes	no

d^7, HS		$(t_{1u})^6(t_{2g})^5(e_g\uparrow)^2$, $^4T_{2g}$	$(t_{1u})^5(t_{2g})^6(e_g\uparrow)^2$, $^4T_{1u}$	yes	yes	yes
d^8	Ni^{2+}, Cu^{3+}	$(t_{1u})^6(t_{2g})^6(e_g\uparrow)^2$, $^3A_{1g}$	$(t_{1u})^5(t_{2g})^6(e_g)^3$, $^3T_{1u}$	yes	yes	yes
d^9	Cu^{2+}	$(t_{1u})^6(t_{2g})^6(e_g)^3$, ^2E_g	$(t_{1u})^5(t_{2g})^6(e_g)^4$, $^2T_{1u}$	yes	yes	yes
d^{10}	Zn^{2+}	$(t_{1u})^6(t_{2g})^6(e_g)^4$, ^1A_g	$(t_{1u})^5(t_{2g})^6(e_g)^4(ns\uparrow)^1$, $^3T_{1u}$	no	no	no

Of special importance is also the fact that some d^n ions with n=3, 4, 5, 6, and 7, dependent on the ligands of the octahedral environment, may produce two types of magnetic centers, high-spin (HS) and low-spin (LS), and in d^6 case there maybe also intermediate spin (IS) states (d^3 has two spin configurations in the one-electron excitation). According to our analysis, only d^5 systems follow the necessary condition of potential multiferroics in both spin states, but the PJTE condition of instability (8) and the magnetic moments are different in these two cases. For d^3, d^4, d^6, and d^7 ions only one of the two spin states may serve as a candidate of potential multiferroics. On the other hand, in many cases the two spin states are close in energy producing a well known phenomenon of transition metal *spin crossover* (SCO), in which case the system can be relatively easily transferred from one spin state to another by external perturbations like heat, light, and magnetic fields (for reviews see [19]). Since, as shown above, the change of the spin state changes also the ferroelectric state, *the SCO is simultaneously a magnetic-ferroelectric (multiferroic) crossover* (MFCO). This coexisting magnetic, ferroelectric, and SCO phenomena opens a variety of new possibilities to manipulate the properties of the system with exciting novel functionalities to electronics and spintronics. Here are several examples:

1. For d^3 and d^4 (Cr^{3+}, Mn^{4+}, Mn^{3+}, Fe^{4+}, etc.) ferroelectrics in the LS state in conditions of MFCO, magnetic fields facilitate the LS→HS transition that destroys the ferroelectricity (and multiferroicity), while an electric field in the HS non-ferroelectric state may transfer the system to the ferroelectric (multiferroic) LS state, thus realizing *electric demagnetiz*ation;

2. For d^5 ferroelectrics in conditions of MFCO, if the ferroelectricity is (most probable) different in the two spin states, an electric field may change the spin state (*electric magnetization or demagnetization*);

3. For d^6 and d^7 (Fe^{2+}, Co^{3+}, Co^{2+}, Ni^{3+}, etc) in the non-ferroelectric LS state under conditions of MFCO, magnetic fields facilitates the LS→HS transition that induces ferroelectricity and hence multiferroicity in a *strong magnetoelectric effect* (the d^6 LS state is nonmagnetic); in the non-ferroelectric LS state in MFCO conditions, an electric field may transfer the system to the multiferroic state (*electric magnetization*);

4. The SCO phenomenon is well known to be influenced also by stress, heat, light, and cooperative effects in crystals [19], hence these perturbations can be used to manipulate the MFCO and all the consequent properties including those mentioned above. The dependence of the MFCO on pressure adds a ferroelastic order to the magnetic and ferroelectric ones;

5. There is already a long history of attempts to use transition metal SCO systems as units of magnetic bistability; the difficulty is in the fast relaxation (short lifetime) of the higher in energy spin state [20]. By choosing a system in the MFCO condition, one can increase the lifetime of the excited dipolar (multiferroic) state by applying an external electric field.

6. An important feature of the revealed MFCO is that it is of local origin and hence it does not necessarily require strong cooperative interactions, meaning that, in principle, it may take place as a magnetic-dipolar effect in separate molecular systems, clusters, thin films, etc., provided the condition of instability (17) takes place.

4. Further comparison with experimental data

One of the main features of the vibronic theory of ferroelectricity under consideration as compared with other approaches is that it yields a direct and qualitatively full explanation of the origin of all the experimentally observed properties of the ferroelectric crystals [21-36]. Below we list several basic properties that so far can be explained exclusively by the vibronic theory.

1.*Off-center displacement of the B (metal) atom in the paraelectric phase.* This property is confirmed by a variety of experimental data mentioned above, noticeable EXAFS [29-31], optical measurements [22-24, 32], X-ray absorption [25,26], EPR experiments [27, 28], NMR [33, 34], neutron scattering [35], etc., in disagreement with all displacement theories in which the metal off-center displacement occurs as a result of the phase transition. Not only is the Ti ion displaced along the [111] type direction in the paraelectric phase where the averaged symmetry is cubic, but it is displaced in this direction in the tetragonal phase where the crystal symmetry and the macroscopic polarization are tetragonal [35]. In the vibronic theory the off-center displacement of the B atom along [111] is due to the local PJTE that is present in all the phases. In $BaTiO_3$, except the lowest rhombohedral phase, the dipole moment formed by the B atom resonates between equivalent minima, and its possible observation depends on the "frequency scale" of the experimental method (see below).

2.The d^0- d^{10} problem and origin of multiferroicity is one of the up-to-date issues widely discussed in the literature. The first question is why, in the variety of transition metal perovskites ABO_3, for a long time only those with the transition metal B atom that has the d^0 configuration were considered as possible ferroelectrics (the "d^0 mystery" [16*b*]), and only recently a limited number of some other systems with n>0 were shown to exhibit ferroelectric properties. For instance, why is $BaTiO_3$ ferroelectric, whereas $BaVO_3$ with the nearest neighbor transition metal V that has the d^1 configuration is not ferroelectric [37]. This feature is important in search for ferroelectric magnets (multiferroics) because only d^n perovskites with n>0 may have unpaired electrons with nonzero magnetic moments. The vibronic theory elucidates directly this problem as shown in Section 3. Table 2 summarizes these conditions; it explains the origin of the observed ferroelectricity in a variety of perovskites [16, 17], for instance in $CaMnO_3$ [38] and $LaCrO_3$ [39]. The vibronic theory directly explains also why the ferroelectricity in $CaMnO_3$ increases with negative pressure and by substitution of Ca with Ba: it increases the volume of the unit cell and the Mn—O distance decreasing the rigidity K_0 in the condition of instability $\Delta<4F^2/K_0$ (see below). In addition to formulating the necessary conditions of multiferroicity the PJTE predicts the phenomenon of magnetic-ferroelectric crossover with a variety of magnetoelectric effects.

3. *Direct Dependence on chemical composition.* As in the examples above, the PJTE is very sensitive to the chemical composition of the system. In addition to the considered in more detail dependence of the ferroelectric and multiferroics properties on the d^n configuration of the B ion in the ABO_3 perovskites, the dependence of these properties on the kind of transition metal with this configuration, namely $3d^n$, $4d^n$, $5d^n$,..., as well as on the A ion can also be elucidated qualitatively in the vibronic theory. First, similar to the negative pressure mentioned above, the size of the A atom in ABO_3 determines the B—O distance (Fig. 2) and the strength of its inner σ bonds that contribute to the K_0 value: with increase of the B—O distance the K_0 value decreases

(and the derivative in the expression for F in Eq. (11) increases) facilitating the condition of instability $\Delta<4F^2/K_0$. This explains why, in the series of $3d^0$ titanates $CaTiO_3$, $SrTiO_3$, $BaTiO_3$ with increasing size of the A atom from left to right, only $BaTiO_3$ is ferroelectric, $SrTiO_3$ being a virtual ferroelectric, and $CaTiO_3$ is non-ferroelectric. Approximate numerical calculations confirm this conclusion [40]. This effect is also observed in $CaMnO_3$ crystal [38] where the substitution of Ca with Ba increases significantly its ferroelectric properties. For the same reason the crystal $Sr_{1-x}Ba_xMnO_3$ becomes ferroelectric only at $x \geq 0.45$ [41]. For a similar reason only $BaTiO_3$ is ferroelectric in the series $BaTiO_3$, $BaZrO_3$, and $BaHfO_3$; the B—O bonding distance in this series decreases (and hence K_0 increases) from left to right because of the increase of the size of the B atom in the approximately fixed positions of the oxygens that is controlled by the size of the A atom (Ba). As mentioned above, we assume here that the A ion is "inactive" in the sense that it does not produce a strong PJTE with distortions, but there may be cases when this ion is the center of the trigger mechanism of ferroelectric distortions instead of (or in combination with) the B ion. Note also that the multiferroic properties depend also on B ion environment that produces either low-spin or high-spin configurations [18].

4. *Disorder and dynamics in the higher-temperature phases and the order-disorder nature of the phase transitions*. The experiments on X-ray diffuse scattering[8] show that only the lowest-temperature phase in $BaTiO_3$ and $KNbO_3$ are fully ordered, whereas the higher-temperature phases are partially disordered and the paraelectric phase is fully disordered. This picture was predicted qualitatively in the first paper on the vibronic theory in 1966 [10]. Further confirmation of this phenomenon was obtained in the EPR experiments with probing ions [27, 28], in NMR measurements [33, 34], etc. In the EPR experiments the Mn^{4+} ions that substitute Ti^{4+} ions was shown to follow the Ti^{4+} positions (according to Table 2 this ion in perovskites is "ferroelectric"); in the rhombohedral phase the Mn ion is shifted off-center to the trigonal [111] minimum of the APES, as expected, but at the phase transition to the orthorhombic phase the EPR signal broadens and disappears, while relaxation measurements show that in this phase the ion performs hoping between two near-neighbor equivalent but differently oriented minima with a characteristic correlation time of the reorientational hoping $\tau\sim 10^{-9}$-10^{-10} s. No displacive theories can explain these experimental observations that follow directly from the vibronic theory.

5. *Relativity to the means of observation due to the dynamics of the ferroelectric ion.* This is yet another feature of the ferroelectrics under consideration that has nothing to do with displacive theories and get full explanation in the vibronic theory. Because of the dynamics of the ferroelectric ions there may be differences in experimental observations of the same phenomena with different methods. This effect can be attributed to the essential different time-scales of these methods, different "time of measurement" τ' in comparison with the time τ of the reorientational hoping of the metal ion between the minima of the APES, as well as to different space-scale measurements as compared with the unit cell dimensions [28, 31, 33].

With regard to the time-scale, the conclusions from the experimental data may be different for experiments with $\tau'>\tau$ and $\tau'<\tau$ because of the full or partial averages over the reorientation positions (APES minima) observed in the former case. In particular, the origin of apparent differences in the interpretation of the experiments on NMR [33] and XAFS [31] at the tetragonal--paraelectric phase transition in $BaTiO_3$ were explained as due to the essential difference in the time-scales of the two methods of measurement [31]: in NMR $\tau' \sim 10^{-8}$ s which is larger than $\tau \sim 10^{-9}$-10^{-10} s,[14, 15] whereas in XAFS $\tau' \sim 10^{-15}$ s, which is smaller than τ.

As for the partial displacive contribution to the phase transition which is seen in some of the experiments, it can be attributed to the space averages. Indeed, for instance, in the tetragonal phase, according to the vibronic theory described above, the system resonates between four minima situated along one of the tetragonal axes, so as in average the unit cell is tetragonally distorted. Since in the tetragonal phase the lattice is ordered in this tetragonal direction, the crystal is tetragonally distorted too. This is seen as a displacive component in the paraelectric-tetragonal phase transition in experiments that perceive the averaged picture [28, 36]. For transitions between the tetragonal to orthorhombic and the latter to rhombohedral phases such effects are expected to be less pronounced because they are accompanied by changes in dynamics of only two minima. Note that the wavelength in the usually employed optical measurements of soft modes is much larger than the unit cell where the dynamics of the ions at the order-disorder phase transition takes place; the observed picture in such measurements is averaged over many unit cells, thus seen as a displacive component.

6. *Anharmonicity, orbital hybridization, and unusual atomic charges.* In some approaches to the theory of ferroelectricity it is assumed that the ferroelectric properties occur due to significant lattice anharmonicity; the eight-minimum APES that emerges from PJTE theory is also a kind of

anharmonicity. In particular, in the theory of ferroelectricity (see, e.g., [42]) the main interaction potential between the ions in taken expanded in series including anharmonic terms up to fourth order. Meanwhile, there is a rigorously proven theorem showing that potentials expanded in series up to the fourth order do not produce the eight-minimum surface [43]. This means that the proper (lattice) anharmonicity that emerges from the higher order terms in the Hamiltonian may be insufficient to reproduce the phenomenon. In addition, the lattice anharmonicity is expected to be approximately the same in related perovskites, and it is not clear why this anharmonicity is so specific and so strongly dependent on the chemical composition and electronic structure in ferroelectrics.

This situation is fully clarified in the vibronic theory. Indeed, it is known that the PJTE, by mixing electronic states under nuclear displacements produces a special additional anharmonicity which may be much stronger than the lattice one (*vibronic anharmonicity* [2, 3]). The vibronic anharmonicity is strongly related to the atomic and electronic structure of the crystal and may produce rather sophisticated APES (see Eq. (14)) thus explaining directly the dependence of the ferroelectricity on the composition of the crystal. *The PJTE is not equivalent to proper (lattice) anharmonicity*, it takes place without the anharmonicity terms in the Hamiltonian, and this is one of the important features of the vibronic theory.

Another important feature of ferroelectricity that emerges from some theoretical calculations is the hybridization of the B metal *d* orbitals with the 2*p* orbitals of the oxygen atoms due to the ferroelectric distortion. This true observation follows directly from the PJTE for which the orbital mixing producing added covalence is the driving force. With regard to unusual atomic charges that some authors assume to explain ferroelectric properties (see in [13]), there is neither (based on first principles) theoretical, nor experimental evidence of their feasibility. On the contrary, such charges contradict to the existing experience of electronic structure calculations, and they are hardly possible from a thermodynamic point of view.

5. Conclusions

One of the main goals of this paper is to affirm that as a PJTE ferroelectric symmetry breaking in perovskite lattices is of local origin, and to reverse the long-standing understanding of the origin of ferroelectricity in ionic crystals as due to the compensation of the local repulsion between the ions by the long-range dipole-dipole interactions, by showing that under certain conditions the local dipolar distortions occur due to the PJTE independent of the long-range interaction, and the ferroelectric phase transitions in these cases are not displacive, but of order-disorder type.

This basic idea was first formulated in 1966, and to present day got full experimental confirmation. It emerges directly from the *theorem of instability* which proves that the ferroelectric sublattice displacements are triggered by the local dipolar instability induced by the PJTE; dependent on the parameter values, the latter can be either enhanced or damped by the long-range interactions. DFT calculations for the $BaTiO_3$ crystal as a whole and its several clusters confirm that the local off-center displacements of the Ti ion are not induced by long-range interactions. The vibronic PJTE coupling between the ground state (formed mainly by oxygen orbitals) and excited states (mainly Ti ion orbitals) produces a local eight-minimum APES with two kinds of saddle points, confirmed independently by the DFT calculations. The theory solves also the d^0-d^{10} problem, showing directly how and why proper ferroelectric and multiferroic properties may depend on the number of *d* electrons and their spin arrangement. In combination with the spin-crossover phenomenon a novel effect of *magnetic-ferroelectric crossover* is predicted. For $BaTiO_3$ the parameters of the theory are in reasonable agreement with the experimental data.

Most important are also the general qualitative aspects of the theory which explain the major properties of perovskite ABO_3 ferroelectrics and multiferroics that cannot be understood in the usual displacive theories. These include the trigonal off-center displacement of the B atom in the paraelectric and ferroelectric tetragonal phases, the characteristic dependence of the properties on the chemical composition (A and B ions) and the d^n electronic configuration of the B atom, the origin of the observed B atom dynamics in the higher-temperature phases and order-disorder phase transition, and the role of anharmonicity and orbital hybridization. Of special interest is also the predicted by the theory dependence of the experimental observations on the method of measurement which is due to the local (tunneling or hoping) dynamics of the ferroelectric ion B, and the relation of the time scale of the method of measurement to this dynamics. For this reason, for instance, EPR, NMR, and XAFS measurements show different results. Because of the local character of the ion dynamics, the experimental measurements depend also on the wavelength of the employed irradiation; as this wavelength usually covers many unit cells, the outcome of the experiment is seen as an averaged "displacive" component (which is in fact not there).

References

1. Englman R 1972 *The Jahn-Teller Effect in Molecules and Crystals* (New York: Wiley)
2. Bersuker I B 2006 *The Jahn-Teller Effect* (Cambridge, UK:Cambridge University Press)
3. Bersuker I B and Polinger V Z 1989 *Vibronic Interactions in Molecules and Crystals* (Berlin: Springer)
4. Bersuker I B 2013 *Chem. Rev.*, DOI: 10.1021/cr300279n
5. Öpik U and Pryce M H L 1957 *Proc. R. Soc. London* A **238** 425.
6. Bersuker I B, Gorinchoi N N and Polinger V Z 1984 *Teor. Chim. Acta* **66** 161
7. Garcia-Fernandez P, Bersuker I B and J. E. Boggs 2007 *J. Phys. Chem.* A **111** 10409
8. Bersuker I B 1988 *Fiz. Tverd. Tela (Sov. Phys. Solid State)* **30** 1738; Bersuker I B 1980 *Teor. Eksp. Khim.* **16** 291; Bersuker I B 1980 *Nouv. J. Chim.* **4** 139
9. Bersuker I B, Polinger V Z and Gorinchoi N N 1992 *J. Mol. Struct. (Theochem)* **270** 369
10. Bersuker I B 1966 *Phys. Lett.* **20** 589
11. Bersuker I B and Vekhter B G 1967 Fizika *Tverdogo Tela (Sov. Phys. – Solid State)* **9** 2452; Bersuker I B and Vekhter B G 1978 *Ferroelectrics* **19** 137; Bersuker I B 1995 *Ferroelectrics*, **164** 75
12. Bersuker I B and Polinger V Z, *to be published*
13. Cochran W 1959 *Phys. Rev. Lett.* **3** 412; Cochran W 1961 *Adv. Phys.* **10** 40; Anderson P. W. 1959 In "*Fizika Dielectrikov*" (Moscow: AN SSSR) p 290; Ginzburg V. L. 1945 *J. Exp. Theor. Phys.* **15** 739; Ginzburg V. L. 1960 *Fizika Tverdogo Tela* **2** 2031; Landau L and Lifshitz E. M. 1994 *Statistical Physics* 3rd ed. *Part I* (Oxford: Butterworth-Heinemann) p 446; Lines M E and Glass A M 1977 *Principles and Applications of Ferroelectric and Related Materials* (Oxford: Clarendon); Smolenskii G A, Bokov V A, Isupov V A, Krainik N N, Pasinkov P E and Shur M S 1971 *Ferroelectrics and Antiferroelectrics* (Leningrad: Nauka); Ghosez Ph 2002 *Microscopic Properties of Ferroelectric Oxides from First-Principles : Selected Topics*, Part 1, (Lausanne :Troisieme Cycle de la Physique en Suisse Romande)
14. Bersuker I B 2010 *Electronic Structure and Properties of Transition Metal Compounds. Introduction to the Theory*, Second Edition (New York: Wiley)

15. Ravel B, Stern E A, Vedrinskii R I and Kraisman V 1998 *Ferroelectrics* **206-207** 407

16. (a) Cheong S-W and Mostovoy M 2007 *Nature Mater.* **6** 13; Ramesh R and Spaldin N 2007 *Nature Mater.* **6**, 21; Picozzi S and Ederer C 2009 *J. Phys. Condens. Matter* **21** 303201; Spaldin N A and Fiebig M 2005 *Science* **309** 391; (b) Hill N A 2000 *J. Phys. Chem.* **104** 6694

17. Wang K F, Liu J-M and Ren Z. F. 2009 *Adv. Phys.* **58**, 321; Bibes M, Villegas J E and Barthelemy A 2011 *Adv. Phys.* **60** 5; Heron J T, Trassin M, Ashraf K., Gajek M, He Q, Yang S Y, Nikonov D E, Chu Y-H, Salahuddin S and Ramesh R 2011 *Phys. Rev. Lett.* **107** 217202; Stroppa A and Picozzi S 2012 *Eur. Phys. J.* B **85** 240

18. Bersuker I B 2012 *Phys. Rev. Lett.* **108** 137202.

19. Gutlich P and Goodwin H A Eds. 2004 *Spin Crossover in Transition Metal Compounds,* vol. I- III (Heidelberg: Springer)

20. Hauser A, Enachescu C, Daku M L, Vargas A and Amstutz N 2006 *Coord. Chem. Rev.* **250** 1642

21. Comes R, Lambert M and Guinner A 1968 *Solid State Commun.* **6** 715

22. Quitet A M, Lambert M and Guinner A 1973 *Solid State Commun.* **12** 1053; Comes R, Currat R, Denoyer F, Lambert M and M. Quittet, Ferroelectrics **12**, 3 (1976).

23. G. Burns and F. Dacol, Ferroelectrics **37**, 661, (1981).

24. F. Gervais, Ferroelectrics **53,** 91 (1984).

25. K. H. Ehses, H. Bock, and K. Fischer, Ferroelectrics **37**, 507 (1981).

26. K. Itoh, L. Z. Zeng, E. Nakamura, and N. Mishima, Ferroelectrics, **63**, 29 (1985).

27. K. A. Muller and W. Berlinger, Phys. Rev. B **34**, 6130 (1986); K. A. Muller , W. Berlinger, K. W. Blazey, and J. Albers, Solid State Commun. **61**, 21 (1987).

28. G, Volkel and K. A. Muller, Phys. Rev. B **76**, 094105 (2007).

29. O. Hanske-Petitpierre, Y. Yacoby, J. Mustre de Leon, E. A. Stern, and J. J. Rehr, Phys. Rev. B **44**, 6700 (1992); N. Sicron, B. Ravel, Y. Yacoby, E. A. Stern, F. Dogan, and J. J. Rehr, Phys. Rev. B **50**, 13168 (1994-II).

30. B. Ravel, E. A. Stern, R. I. Vedrinskii, and V. Kraisman, Ferroelectrics **206-207**, 407 (1998).

31. E. Stern, Phys. Rev. Lett. **93**, 037601 (2004).

32. T.P. Dougherty, G.P. Wiederrecht, K.A. Nelson, M.H. Garrett, H.P. Jensen, and C. Warde, Science **258**, 770 (1992).

33. B. Zalar, V. V. Laguta, and R. Blinc, Phys. Rev. Lett. **90**, 037601 (2003); B. Zalar, A. Labar, J. Seliger, and R. Blinc, Phys. Rev. B **71**, 064107 (2005).

34. B. Zalar, V. V. Laguta, and R. Blinc, Phys. Rev. Lett. **90**, 037601 (2003); R. Blinc, B. Zalar, V. V. Laguta, and M. Itoh, Phys. Rev. Lett. **94**, 147601 (2005) .

35. I.-K. Jeong, Seunghun Lee, Se-Young Jeong, C. J. Won, N. Hur, and A. Llobet, Phys. Rev. B **84**, 064125 (2011)

36. A. Bussman-Holder, H. Beige, and G. Volkel, Phys. Rev. B **79,** 184111 (2009).

37. S.-Y. Yan, Y. Xie, T. Liu, and H.-T. Yu, J. Phys. : Condens. Matter **22**, 125501 (2010).

38. S. Bhattarcharjee, E. Bousquet, and Ph. Ghosez, Phys. Rev. Lett. **102**, 117602 (2009).

39. C. Ederer, T. Harris, and R. Kovacik, Phys. Rev. B **83**, 054110 (2011).

40. I. B. Bersuker, N. N. Gorinchoy and T. A. Fedorco**,** *Ferroelectrics,* **153**, 1 (1994).

41. H. Sakai, J. Fujioka, T. Fukuda, D. Okuyama, D. Hashizume, F. Kagawa, H. Nakao, Y. Murakami, T. Arima, A. Q. R. Baron, Y. Taguchi, and Y. Tokura, Phys. Rev. Lett. **107**, 137601 (2011).

42. W. Zhong, D. Vanderbilt , and K. M. Rabe, Phys. Rev. B **52**, 6301 (1995-I)

43. A. F. Devonshire, Phil. Mag. 40, 1040 (1949).

Polar and Magneto-Electric Properties of Anti-Ferrodistortive Ordered Jahn-Teller Distortions in a multiferroic metal-organic framework

Alessandro Stroppa
CNR-SPIN, Via Vetoio, 67100, L'Aquila-Italy
E-mail: alessandro.stroppa@spin.cnr.it

Abstract. There is great interest in hybrid organic-inorganic materials such as metal-organic frameworks (MOFs). Here, we focus on $[C(NH_2)_3]Cu(HCOO)_3$, a MOF with perovskite topology which crystallizes in polar space group $Pna2_1$. In inorganic compounds, octahedral tilting and Jahn-Teller structural distortions are usually non-polar distortions. However, in this MOF cooperative interactions between the antiferro-distortive distortions of the framework and the $C(NH2)_3$ organic cation via hydrogen bonding breaks the inversion symmetry and induces a ferroelectric polarization. Our ab-initio study supports the picture of an orbital-order-induced ferroelectricity, a rare example of dipolar ordering caused by electronic degrees of freedom. Moreover, we show that the switching of polarization direction implies the reversal of the weak ferromagnetic component. These results therefore offer an important starting point for tailoring multiferroic properties in this emerging class of materials for various technological applications.

1. Introduction

Metal-organic frameworks (MOFs) are a new class of hybrid crystalline compounds comprised of extended networks formed by organic linkers and metal cations, forming porous materials at the interface between molecular coordination chemistry and materials science. They show unique properties arising from the organic-inorganic duality[1]. There have been a huge interest in these compounds not only because they show a variety of physical properties and applications, such as gas storage, gas exchange or separation, catalysis, drug delivery, optics, magnetism [2] but also for the infinite possibilities of new MOFs that can be created by varying the inorganic/organic components[3-6]. Very recently, more dense hybrid frameworks with perovskite ABX_3 topology have attracted much attention, for their interesting magnetic, optical, electronic and dielectric properties along with improved thermal stability[3,7,8]. These compounds provide access to more oxide-like properties and the flourishing diversity of structures and dimensionality seen in these materials has led to their growth into a major field of research[3]. The ABX_3 topology is ubiquitous in inorganic material science. We may think of an organic-inorganic hybrid perovskite by ideally replacing the inorganic A component with an organic cation and the inorganic X anion with an organic ligand. The ABX_3 structure is the basis of many transition metal oxides with important properties such as high T_c superconductivity, colossal magnetoresistance,

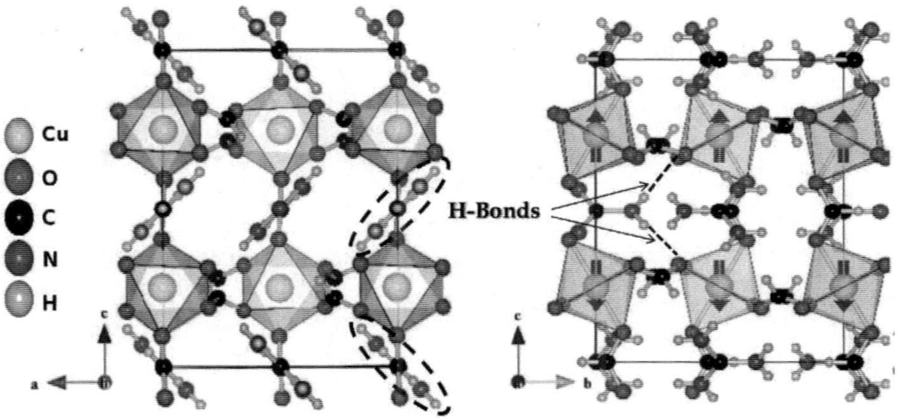

Figure 1. Perspectives views of the crystalline unit cell. The molecular plane of the guanidinium and the stacking along the c axis is highlighted in the left part; the hydrogen bonds between the guanidinium and the BX_3 framework is highlighted in the right part.

ferroelectricity (FE), charge ordering, nonlinear optical behaviour, and so on[5]. Last but not least, multiferroicity[9].

Multiferroics (MFs) present at least two coexisting orders among the electric, magnetic and/or elastic ones. The study of multiferroic compounds has become a hot field in materials science[10-13]. The most appealing combination involves electric and magnetic orders, opening new routes to spintronic devices[11]. MFs can be divided in two main groups: those having independent electric and magnetic orders and those exhibiting a magnetoelectric (ME) coupling. The former allows, for example, the conception of a four-state memory[11-15]. Probably ME compounds are more interesting, since they may offer the possibility of controlling the magnetic ordering via an electric field and viceversa[16]. MFs are highly desirable [9] although, the coexistence of electric and magnetic orders is rare[13]. When one combines the Renaissaissance of Magnetoelectric MFs [16] and the Explosion of papers in MOFs[17] it is clear that MF MOFs is a newborn field of research where only a few papers have been published so far.

The very first $ab-initio$ study has been performed by A. Stroppa et al.[18] We have studied the compound $[C(NH_2)_3]Cu[(HCOO)_3]$, synthesized in 2009[19]. We call it Cu-MOF hearafter. We found that the Cu-MOF i) is multiferroic with a ferroelectric (FE) polarization estimated of 0.37 $\mu C/cm^2$ and it shows a weak- ferromagnetic component (weak-FM); ii) it is magnetoelectric, *i.e.* the weak-FM is coupled to the FE polarization, thus allowing the electric control of magnetic properties; iii) the FE polarization is due to a peculiar interplay between Jahn-Teller (JT) distortions and organic A-group through hydrogen bondings[18]. To the best of our knowledge, this is a new and very peculiar mechanism, which demonstrates the rich field for new effects in this class of materials. We want to remark that while the ab-initio design and characterization of multiferroic inorganic materials can be considered nowadays at a mature stage, this is certainly not true for MF MOFs.

2. Structural properties and antiferro-distortive order

The compound crystallizes in the $Pna2_1$ space group. The structure is shown in Figs 1,2. An important feature is the presence of JT active Cu^{+2} ions (d^9 electronic configuration). The two-fold degeneracy of this configuration is removed by an octahedral structural distortion, which, in turn, introduces long and short Cu-O bonds in the ab plan (Q_2 JT mode).

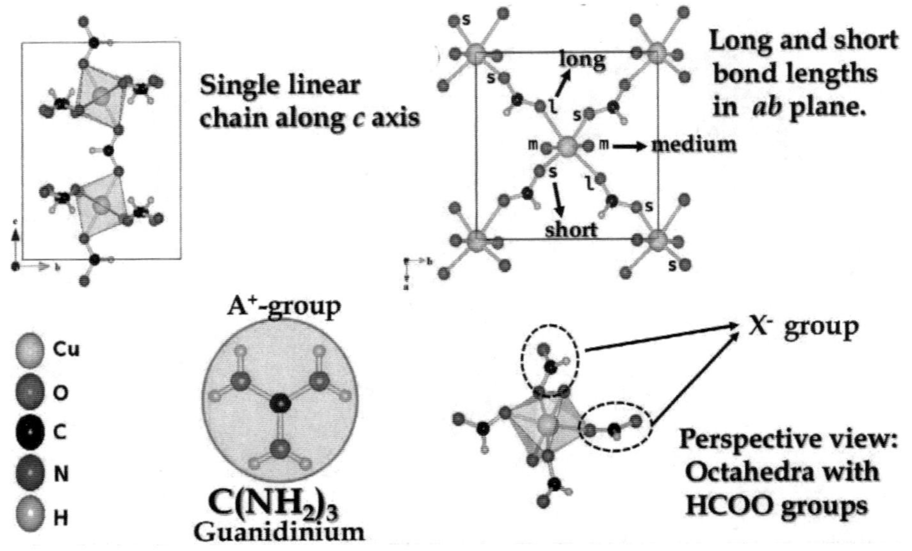

Figure 2. Structural details of crystalline unit cell. Top part: Single linear chain along the c axis showing the tilting of the octahedra; projected view of the the octahedra into the ab plane showing the long, medium and short (l, m, s) bond distances. Bottom part: details of the guanidinium cations and perspective view of the octahedra and the basal ligands.

The different orientations of the long Cu-O bonds on different Cu sites give rise to a characteristic Anti-Ferro-Distortive (AFD) pattern[20]: the long Cu-O bond on a Cu site is almost perpendicular to the long Cu-O bond on the neighboring Cu site in the ab plane. It is possible to introduce a function of the JT modes through the JT phase ϕ which quantifies the degree of anti-ferro distortion (AFD):[15] when ϕ is $\sim 4\pi/3$ or $2\pi/3$, one has a maximum distortion in the crystal, while for $\phi \sim \pi$ one has that Cu-O long and short bonds become equal, and the AFD disappears. For details, see Ref. [20]. It must be noted that AFD modes are non-polar in standard inorganic ABX_3 compound[20]. The ground-state is antiferromagnetic, with Cu spins ferromagnetically coupled within ab plane while antiferromagnetically coupled between parallel ab planes. Note that, by introducing the spin- orbit coupling in the calculations, a weak- ferromagnetic component appears, in agreement with experiments[19]. This has important consequences for the magnetoelectric behavior[18].

3. Ferroelectric order

In order to study the ferroelectric order we consider a paraelectric (centric) structure[13] in terms of which one can present the polar crystal structure as due to symmetry-lowering structural distortion from the high symmetry centric group. It is useful to introduce a parameter λ which is the normalized amplitude of the distortion connecting the centric to the polar structure: λ=1 represents the compound in the $Pna2_1$ space group, λ=0 represents the Pnna centric space group. For each λ between 0 and 1, one can linearly interpolate the atomic positions between the two end-points, defining a continuous path as a function of λ, the so-called switching path. We found that the system has a FE polarization of P=$\pm$ 0.37 $\mu C/cm^2$ along the polar c axis. Obviously, P=0 for λ=0 and P=P(λ) is an odd function of the polar distortion. The calculated energy behaviour reproduces the expected double-well profile characteristic of a switchable ferroelectric system[20]. Surprisingly, by monitoring the changes of ϕ along the path, we have found a clear correlation between P=P(λ) and $\phi(\lambda)$. In principle, one has $\phi_{\alpha,\beta}(\lambda)$, where α,β denotes the two

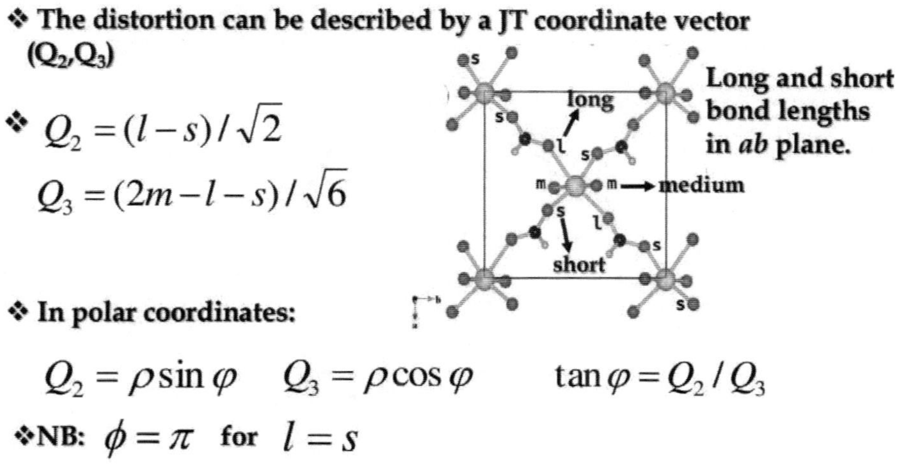

Figure 3. Jahn-Teller distortions and the description thorugh the Q_2, Q_3 variables and Jahn-Teller phase.

Cu-sublattices in the *ab* plane [see Fig. 3 and 4]. Each sublattice has the same orientation of the long Cu-O bonds, but the orientation "rotates" by $\pi/2$ in the neighboring Cu site (the so called JT pseudo-rotation)[20]. One can see that P=$P(\lambda)$ and $\phi(\lambda)$ have the same behavior, thus suggesting that AFD distortions can be the source of the ferroelectric polarization. This is counterintuitive since AFD distortions are non-polar[20].

In Fig. 4 (a),(b),(c) we show the energy E, the polarization P, and the "amplitude" of the AFD distortions along the switching path. The microscopic explanation of the ferroelectric polarization is as follows: the $[C(NH_2)_3]^+$ molecular group interacts with the BX_3 framework through hydrogen bonding N-HO, where the H atoms belong to NH_2 of the $[C(NH_2)_3]^+$ group, and the O atom is bond to Cu. Both the H atoms of NH_2 group interact via H bond, but while at λ=0 state, the H bond strengths are the same, for λ larger than 0, *i.e.* towards the ferroelectric state, the H bonding strengths become asymmetric due to the JT mode[13]:[since the AFD mode controls the oxygen positions, the change of the AFD mode amplitude (changing ϕ along λ), induces asymmetric distortions in the NH_2 part of the A-group, *i.e.* there is an induced dipole moment which then sum up to give a finite polarization[18].

4. Conclusions

We have shown that the Cu-MOF, belonging to the metal-organic framework family with ABX_3 topology, shows a rich and fascinating physics. Our study supports the following picture: i) It is ferroelectric with an estimated polarization of 0.37 $\mu C/cm^2$. ii) The origin of the ferroelectricity is very intriguing: non-polar structural distortions (*i.e.* antiferro- distortions) give rise to a loss of inversion symmetry, paving the way for the appearance of polarization; note that this mechanism is not found in classical ABX_3 inorganic compounds. iii) Hydrogen bonding plays a major role in coupling the nonpolar distortions acting in the BX_3 framework with the organic cation. iv) A weak ferromagnetic component is allowed and it is coupled to the ferroelectricity: by using this very rare property, it would be possible to control the polarization by an applied magnetic field, and vice versa. This is called magnetoelectricity. This feature has not been discussed here, but we refer to Ref. [20] for more details.

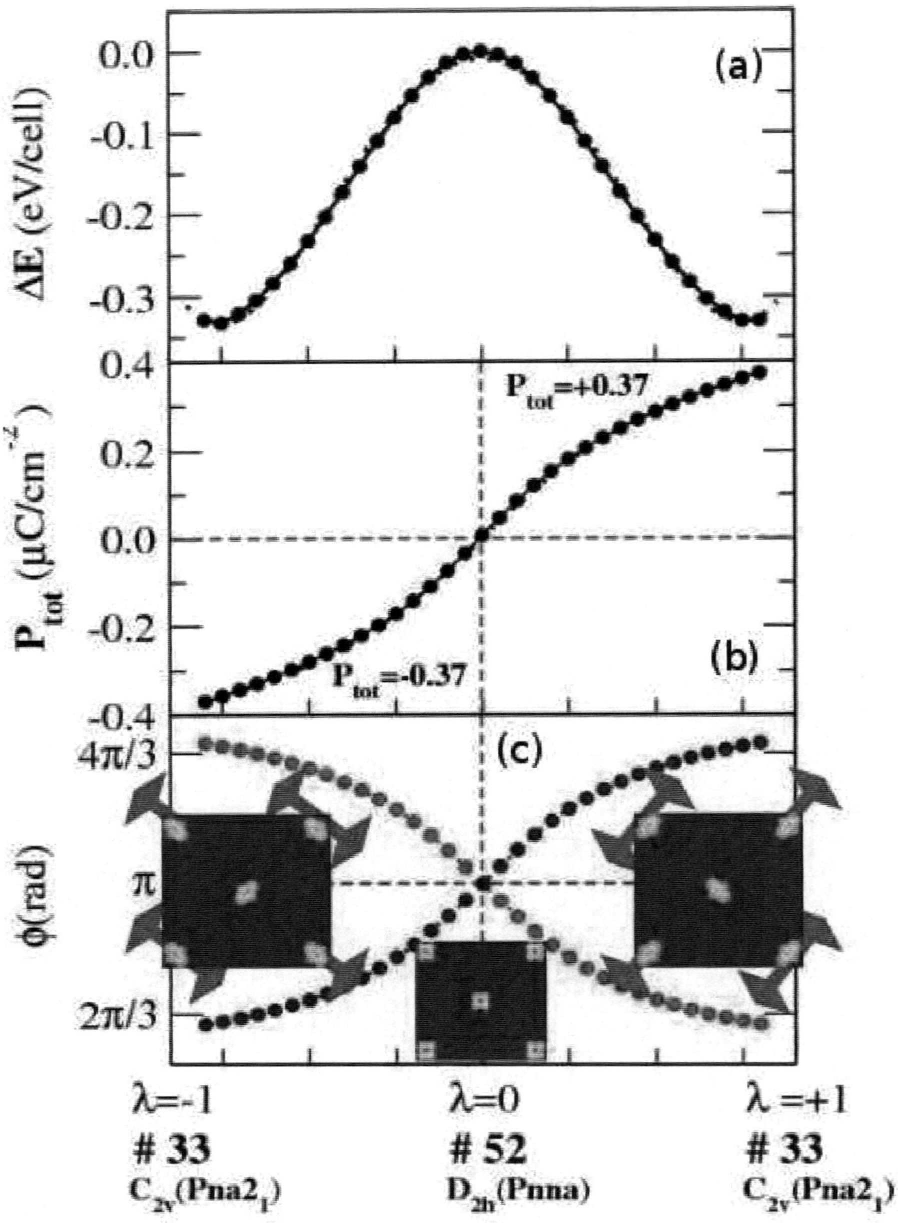

Figure 4. (a) Energy profile, (b) ferroelectric polarization and (c) Jahn-Teller ϕ phases along the swithing path. In (c) the orbital ordering underlying the antiferro-distortive ordering is also shown.

4.1. Acknowledgments
This research was funded by the European Research Council under the European Community, 7th Framework Programme (20072013)/ERC Grant Agreement no. 203523 won by Dr. S. Picozzi. Computational support by CASPUR Supercomputing Center in Rome through Standard HPC Grant 2011 is gratefully acknowledged. A.S. wishes to express his special thanks to Prof. I. Bersuker for interesting discussions about the Jahn-Teller physics and for the invitation to the XXIst International Symposium on the Jahn-Teller Effect. Interesting discussions with Dr. P. Barone and Prof. J. M. Perez-Mato are also acknowledged. All the calculations have been done by using the VASP package[21]. Figures were plotted using the VESTA package.[22] For the symmetry analysis we used the tools of teh Bilbao Crystallographic server[23], in particular the PSEUDOSYMMETRY[24] and AMPLIMODES crystallographic softwares[25].

5. References

[1] Yaghi O M, O'Keeffe M, Ockwig N W, Chae H K, Eddaoudi M and Kim J 2003 *Nature* **423** 705
[2] Férey G and Cheetham A K 1999 *Science* **283** 1125
[3] C N R Rao C N R, Cheetham A K and Thirumurugan A 2008 *J. Phys.: Condens. Matter* **20** 083202
[4] Eddaoudi M, Kim J, Rosi N, Vodak D, Wachter J, O'Keeffe M and Yaghi O M 2002 *Science* **295** 469
[5] Green M A 2010 *Nat. Mat.* **9** 539
[6] 2009 *Chem. Soc. Rev.* **38** 1201
[7] Jain P, Dalal N S, Toby B H, Kroto H W and Cheetham A K 2008 *J. Am. Chem. Soc.* **130** 10450
[8] Clausen H F, Poulsen R D, Bond A D, Chevallier M A S and Iversen B B 2005 *J. Solid State Chem.* **178** 3342
[9] Tokura Y and Nagaosa N 2000 *Science* **288** 389
[10] Ramesh R and Spaldin N A 2007 *Nat. Mater.* **6** 21
[11] Chu Y H, Martin L W, Holcomb M B, Gajek M, Han S J, He Q, Balke N, Yang C H, Lee D, Hu W, Zhan Q, Yang P L, Fraile-Rodgriguez A, Scholl A, Wang S X and Ramesh R 2008 *Nat. Mater.* **7** 478
[12] Cheong S W and Mostovoy M 2007 *Nat. Mater.* **6** 13
[13] Hill N A 2000 *J. Phys. Chem. B* **104** 6694
[14] Khomskii D I 2001 *Bull. Am. Phys. Soc. C* 21.002
[15] Bersuker I B 2012 *Phys. Rev. Lett.* **108** 137202
[16] Spaldin N A and Fiebig M 2005 *Science* **309** 391
[17] G. Férey 2008 *Chem. Soc. Rev.* **37** 191
[18] Stroppa A, Jain P, Barone P, Marsman M, Perez-Mato J M, Cheetham A K, Kroto H W, Picozzi S 2011 *Angew. Chemie Int. Ed.* **50** 5847
[19] Hu K L, Kurmoo M, Wang Z M and Gao S 2009 *Chem. Eur. J.* **15** 12050
[20] Bersuker I B 2006 *The Jahn-Teller Effect* (Cambridge University Press, Cambridge)
[21] Kresse G and Furthmüller J 1996 Phys. Rev. B **54** 11169
[22] Momma K and Izumi F 2008 J. Appl. Crystallogr. **41** 653
[23] http://www.cryst.ehu.es/
[24] Kroumova E, Aroyo M I, Perez-Mato J M, Ivantchev S, Igartua J M, Wondratschek H J. 2001 Appl. Crystallogr. **34** 783
[25] Orobengoa D, Capillas C, Aroyo M I and Perez-Mato J M 2009 J. Appl. Crystallogr. **42** 820

Journal of Physics: Conference Series **428** (2013) 012030 doi:10.1088/1742-6596/428/1/012030

Electronic structure of Cu in ferromagnetic $CaCu_3Sn_4O_{12}$

Masaichiro Mizumaki

Japan Synchrotron Radiation Research Institute, SPring-8, 1-1-1 Kouto, Sayo-cho, Sayo-gun, Hyogo 679-5198

Japan Science and Technology Agency, CREST, 5 Sanbancho, Chiyoda-ku, Tokyo 102-0075, Japan

E-mail: mizumaki@spring8.or.jp

Takashi Saito

Japan Science and Technology Agency, CREST, 5 Sanbancho, Chiyoda-ku, Tokyo 102-0075, Japan

Institute for Chemical Research, Kyoto University, Gokasho, Uji, Kyoto 611-0011, Japan

Takayuki Uozumi

Osaka Prefecture University, 1-1 Gakuen-cho, Sakai, Osaka 599-8531, Japan

Yuichi Shimakawa

Japan Science and Technology Agency, CREST, 5 Sanbancho, Chiyoda-ku, Tokyo 102-0075, Japan

Institute for Chemical Research, Kyoto University, Gokasho, Uji, Kyoto 611-0011, Japan

Abstract. We have investigated the electronic structure of Cu ions in ferromagnetic compound $CaCu_3Sn_4O_{12}$ by using X-ray absorption and magnetic circular dichroism (MCD) measurements around the Cu $L_{3,2}$- and O K-edges. In both edges, the signal of MCD were clearly observed. By using sum rules, we could obtained M_{spin}= 0.87 ± 0.01 μ_B, M_{orb}= 0.07 ± 0.01 μ_B, and M_{total}= 0.93 ± 0.02 μ_B, per Cu ion. The orbital magnetic moments of Cu ions are significant and are originated from the spin-orbit interactions.

1. Introduction

A-site-ordered perovskites $AA'_3B_4O_{12}$ show diverse and intriguing physical properties. For example, $CaCu_3Ti_4O_{12}$, exhibits a colossal dielectric constant over a wide temperature range [1]. $CaCu_3Mn_4O_{12}$ is a ferrimagnetic semiconductor with a Curie temperature above room temperature [2]. $CaCu_3Ru_4O_{12}$ is a Pauli-paramagnetic metal and its magneto-transport properties are reported to show a heavy-fermion like behavior [3, 4].

It is very difficult to understand the origin of these compounds magnetic properties due to the balance among the strength of interaction between A'-A', A'-B, and B-B sites. Especially, it was widely known that magnetic Jahn-Teller ions of Cu at the square-planar coordinated A'-site play a crucial role in all properties of these perovskites. Therefore , we have tried to extract A'-A' interaction by introduced non-magnetic ions to the B-site. $CaCu_3B_4O_{12}$ (B = Ge, Sn, and Ti) and their solid solutions show unusual transitions from ferromagnetic $CaCu_3Ge_4O_{12}$ to antiferromagnetic $CaCu_3Ti_4O_{12}$ to ferromagnetic $CaCu_3Sn_4O_{12}$. The ferro- to antiferromagnetic transition mechanism in these materials is not based on the superexchange interaction [5]. While the Sn-ions in the B-sites of the $CaCu_3Sn_4O_{12}$ (CCSO) are non-magnetic, the materials show the ferromagnetic ordering below 10 K [6]. There is no study of electronic and magnetic states of Cu ions from the microscopic point of view. Therefore, we carried out the X-ray absorption (XAS) and magnetic circular dichroism (MCD) measurements in order to elucidate the electronic and magnetic states of Cu ions and O ions in CCSO.

2. Experimental

The XAS and MCD for Cu $L_{3,2}$ and O K-edges for the disproportionated CCSO were measured at 9 K by a total electron yield method at BL25SU of SPring-8 in Japan. The energy resolution $E/\Delta E$ at the Cu $L_{3,2}$- and O K-edges was greater than 5000. The incident photon energy was calibrated by measuring the energies of the Ti $L_{3,2}$-edges of TiO_2 and the Ni $L_{3,2}$-edges of NiO. Powder sample was pasted uniformly on a sample holder by using carbon tape. The spectra were obtained using parallel (I_+) and antiparallel (I_-) photon spins along the magnetization direction of the sample, to which a static magnetic field of 1.9 Tesla was applied. The MCD intensity was defined by the difference between the two absorption spectra ($I_{MCD} = I_- - I_+$).

3. Results and Discussion

Figure 1 shows the I_+ and I_- XAS spectra at the Cu $L_{3,2}$-edges. This spectral shape is similar to that of $BiCu_3Mn_4O_{12}$ containing typical divalent Cu ions. [7, 8] The MCD intensities at the Cu L_3-edge (~931.5 eV) are negative while those at the Cu L_2-edge (~951.3 eV) are positive.

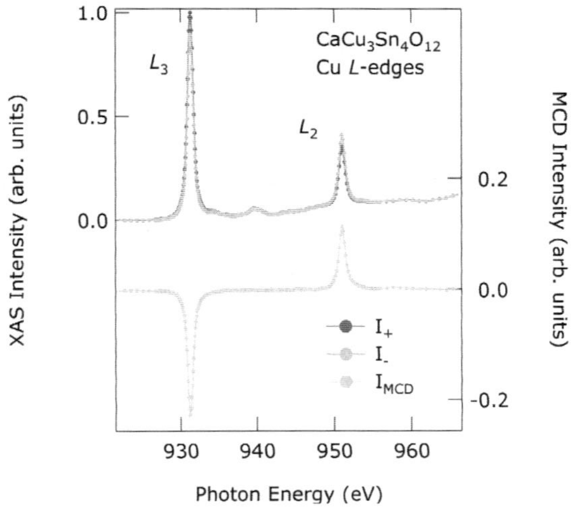

Figure 1. XAS and MCD spectra at Cu $L_{3,2}$-edges for $CaCu_3Sn_4O_{12}$ at T = 9 K. Red (I_+) and green (I_-) circles show the XAS spectra measured with external field 1.9 T applied parallel and antiparallel to the light propagation axis. The bottom line represents the MCD intensity I_{MCD}. The arbitrary unit of intensity in XAS spectrum was set to unity at 965 eV.

Figure 2 shows the XAS spectrum ((I_- + I_+)/2) at the O K-edge, which originates from an excitation of electrons from the O 1s state to the unoccupied O 2p states in the conduction band. Since the empty O 2p states hybridize with the 3d and 4sp states of cations in oxides, the observed O-XAS spectrum re ects the features of the electronic structures of the constituent transition-metal ions [9]. The spectrum from 529 to 532 eV mainly represents the nature of the Cu 3d bands (the upper Hubbard band).

The absorption spectrum at approximately 534 - 540 eV re ects the characters of the Sn 5sp and Ca 3d bands. [9, 10, 11] In the MCD spectrum also shown in Fig. 2, it is notable that

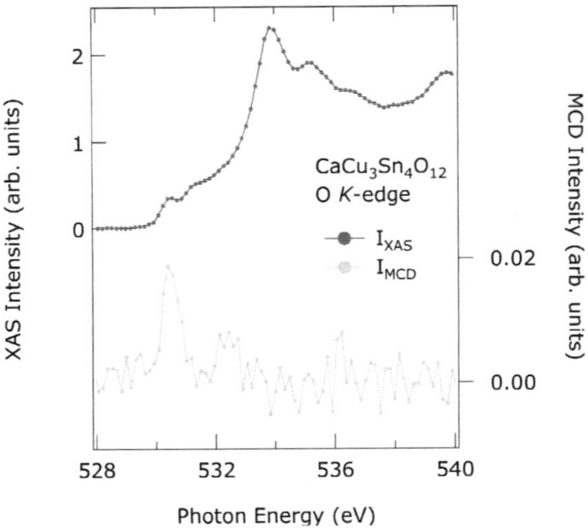

Figure 2. XAS and MCD spectra at O-K edges for $CaCu_3Sn_4O_{12}$ at T = 9 K. Red circles show the XAS spectrum I_{XAS} (= (I_- + I_+)/2) measured with external field 1.9 T applied parallel and antiparallel to the light propagation axis. The bottom line represents the MCD intensity I_{MCD}. The arbitrary unit of intensity in XAS spectrum was set to unity at 552 eV

negative signals are clearly observed around the Cu 3d region. As seen in $La_{1-x}Sr_xMnO_{3+}$, [12] the negative MCD signal at the O K-edge with intensity of a few percent of the XAS suggests the contribution of a direct transfer from the 3d orbital magnetic moment of a transition metal to the O 2p states. Thus, the result of the O K-edge MCD indicates that Cu 3d orbital magnetic moments contributed to the magnetic properties of the ferromagnetic CCSO.

We quantitatively analyze the orbital magnetic moment of Cu ions in CCSO by using the magneto-optical sum rules. [13, 14] Total magnetic moment (M_{total}) consists of orbital (M_{orb}) and spin (M_{spin}) parts and they are expressed as

$$M_{orb} = -\frac{4n_h}{3}\left(\frac{I_{L_3}+I_{L_2}}{W_{L_3}+W_{L_2}}\right)(1), \qquad M_{spin} = -n_h\left(\frac{I_{L_3}-2I_{L_2}}{W_{L_3}+W_{L_2}}\right)(2),$$

where I and W represent the spectral weights of the MCD and XAS intensities, respectively, and n_h is the number of holes in the 3d state. The boundaries between the L_3 and L_2 edges of the Cu spectra are set to 946.5 eV. Because the sample is a poly-crystal and the Cu ions in CCSO are located at the high symmetry D_{4h} sites, respectively, the magnetic dipole matrix operator is zero and thus the M_{spin} is simplified as in equation (2). From an *ab-initio* band calculation with the local spin density approximation, n_h for Cu-3d were estimated to be 0.97. [15] With these values, we can obtain M_{spin}= 0.87 ± 0.01 μ_B, M_{orb}= 0.07 ± 0.01 μ_B, and M_{total}= 0.93 ± 0.02 μ_B, per Cu ion. The orbital part of the magnetic moment for Cu is significant, consistent with the observation of the O K-edge MCD spectrum.

First, we discuss the origin of the orbital magnetic moments. Within the crystal field theory, the 3d electrons of transition metal ions in e_g orbitals under a cubic crystal field have no orbital magnetic moments [16] because the $3z^2$-r^2 and x^2-y^2 orbitals cannot be transformed into each other by rotation about any axis. The presence of orbital magnetic moments requires the spin-orbit interaction. That is to say, orbital magnetic moment is produced in the e_g state by the spin-orbit interaction between the unoccupied $|x^2\text{-}y^2{:}\downarrow\rangle$ and the occupied $|xy,yz,zx{:}\downarrow\rangle$ states and between the unoccupied $|3z^2\text{-}r^2{:}\downarrow\rangle$ and the occupied $|yz,zx{:}\downarrow\rangle$ ones. According to the perturbation theory [12, 17], orbital magnetic moments of 3d electrons are proportional to $\xi_{3d}/10Dq$ (ξ_{3d} is the spin-orbit interaction of 3d electrons.). It is determined that M_{orb} of Cu ions is 0.068 μ_B using the value of ξ_{3d} = 102 meV and $10Dq$ = 1.5 eV for Cu ions. This value is acceptably close to our estimation value of 0.07 μ_B by using sum rules. This result indicates that the orbital magnetic moments of Cu ions are originated from the spin-orbit interaction.

4. Summary

We have investigated the electronic structure of Cu ions in ferromagnetic compound CCSO by using XAS and MCD measurements around the Cu $L_{3,2}$- and O K-edges. In both edges, the signal of MCD were clearly observed. By using sum rules, we could obtained $M_{spin}= 0.87 \pm 0.01\ \mu_B$, $M_{orb}= 0.07 \pm 0.01\ \mu_B$, and $M_{total}= 0.93 \pm 0.02\ \mu_B$, per Cu ion. The orbital magnetic moments of Cu ions are significant and are originated from the spin-orbit interactions.

Acknowledgments

The XAS and MCD measurements at BL25SU (Proposal No. 2008B1001) in SPring-8 were performed with the approval of the Japan Synchrotron Radiation Research Institute (JASRI). This work was supported in part by a Grant-in-Aid for Scientific Research 19GS0207, by the Global Centers of Excellence Program "International Center for Integrated Research and Advanced Education in Materials Science," and by a grant for the Project of Integrated Research on Chemical Synthesis from the Ministry of Education, Culture, Sports, Science and Technology of Japan. The work was also supported by the Japan Science and Technology Agency's CREST program. This work was partly supported by JPSJ KAKENHI Grant Number 22540344.

References

[1] Ramirez A P, Subramanian M A, Gardel M, Blumgerg G, Li D, Vogt T and Shapiro S M 2000 *Solid State Commun.* **115** 217
[2] Zeng Z, Greenblatt M, Subramanian M A and Croft M 1999 *Phys. Rev. Lett.* **82** 3164
[3] Kobayashi W, Terasaki I, Takeya J, Tsukada I, and Ando Y 2004 *J. Phys. Soc. Jpn.* **73** 2373
[4] Tanaka S, Shimazui N, Takatsu H, Yonezawa S, and Maeno Y 2009 *J. Phys. Soc. Jpn.* **78** 024706
[5] Shimakawa Y, Shiraki H and Saito T 2008 *J. Phys. Soc. Jpn.* **77** 113702
[6] Shiraki H, Saito T, Yamada M, Tsujimoto M, Azuma M, Kurata H, Isoda S, Takano M and Shimakawa Y 2007 *Phys. Rev. B* **76** 140403
[7] Saito T, Chen W, Mizumaki M, Attfield J P and Shimakawa Y 2010 *Phys. Rev. B* **82** 024426
[8] Mizumaki M, Chen W, Saito T, Yamada I, Attfield J P and Shimakawa Y 2011 *Phys. Rev. B* **84** 094418
[9] Abatte M, Pen H, Czyzyk M T, de Groot F M F, Fuggle J C, Ma Y J, Chen C T, Sette F, Fujimori A., Ueda Y, and Kosuge K 1993 *J. Electron Spectrosc. Relat. Phenom.* **62** 185
[10] Mizumaki M, Saito T, Shiraki H and Shimakawa Y 2009 *Inorg. Chem.* **47** 8562
[11] Koide T, Shidara T, Miyauchi H, Nakjima N, Fukutani H, Fujimori A, Kawasaki S, Takano M, and Takeda Y 1999 *J. Mag. Soc. Japan* **23** 341
[12] Koide T, Miyauchi H, Okamoto J, Shidara T, Sekine T, Saitoh T, Fujimori A, Fukutani M, Takano M, and Takeda Y 2001 *Phys. Rev. Lett.* **87** 246404
[13] Thole B T, Carra P, Sette F, and van der Laan G 1992 *Phys. Rev. Lett.* **68** 1943
[14] Carra P, Thole B T, Altalleri M, and Wang X, 1993 *Phys. Rev. Lett.* **70** 694
[15] Li H, Lv S, Han D, Liu X and Meng J 2010 *Chem. Phys. Lett.* **494** 213
[16] van der Laan G 1998 *J. Phys: Condens. Matter* **10** 3239
[17] van der Laan G, Henderson C M B, Pattrick R A D, Dhesi S S, Schofield P F, Dudzik E, Vaughan D J 1999 *Phys. Rev. B* **59** 4314

Journal of Physics: Conference Series **428** (2013) 012031 doi:10.1088/1742-6596/428/1/012031

Spontaneous magnetization and magnetic susceptibility of $LaVO_3$

Hiroya Sakurai

National Institute for Materials Science, 1-1 Namiki, Tsukuba 305-0044, Japan

E-mail: sakurai.hiroya@nims.go.jp

Abstract. Five characteristic temperatures of T_M = 148 K, T_N = 142 K, T_t = 138 K, $T_f \sim$ 125 K and $T_g \sim$ 50 K were found by the measurements of the magnetization curves at various temperature. The spontaneous magnetization appears below T_M. It increases up to $M_S \simeq 2 \times 10^{-4}\ \mu_B$ at T_t and then decreases steeply below T_t, which qualitatively agrees with the temperature dependence of magnetization obtained under field-cooling (FC) condition. On the other hand, the slope of the magnetization curve, namely the magnetic susceptibility, drops below T_N, which coincides with the temperature dependence of magnetization obtained under zero-FC condition, although the magnetization curves were obtained under FC condition. The temperature dependence of the spontaneous magnetization shows a minimum at T_f and a drop at T_g although there is no anomaly in the temperature dependence of FC or ZFC magnetization.

1. Introduction

The perovskite oxide, $LaVO_3$, shows the giant diamagnetism [1, 2]. The magnetization obtained under field-cooling (FC) condition becomes largely negative when the applied field is below 0.2 T or a little higher; the M/H value at 0.02 T, for example, is about -8×10^{-3} emu/mol. The temperature dependence of magnetization under zero-FC (ZFC) condition shows the conventional behavior of an antiferromagnet with the peak at T_N = 142 K [3, 4]. On the other hand, FC data have an unusual enhancement around T_N and a sharp drop below T_t = 138 K. The canted antiferromagnetic (AF) state has been confirmed below T_t [2] and suggested between T_t and T_N [4]. Obviously the enhancement is the key to understand the giant diamagnetism. Nevertheless, there is no report on whether the spontaneous magnetization appears between T_t and T_N. The study on the magnetism of the compound is not sufficient although a large number of reports on it have been published. The purpose of this paper is to show the magnetic data, which will be useful to elucidate the magnetism in future.

2. Experiments

The powder sample of $LaVO_3$ was obtained by reduction of $LaVO_4$ in H_2 gas at 1200°C for overnight. Then, the sample was annealed in flowing Ar gas at 1200°C for overnight. The oxygen content of the final product was estimated from the weight increase after firing the sample at 600°C in air. The exact composition of the sample is $LaVO_{2.98\pm0.01}$. $LaVO_4$ was prepared from the stoichiometric mixture of La_2O_3 and V_2O_5 at 600°C for overnight and then 650°C for overnight. Magnetic measurements were performed on MPMS-XL (Quantum Design). The magnetization curves at various temperatures were measured from 7 T to −0.1 T with the intervals of 0.2 T down to 1 T and of 0.05 T below 1 T. To obtain each magnetization curve, the sample was once heated at 160 K, and the magnetic field of 7 T was applied there. Then the sample was cooled down under the field to a target temperature to start magnetization measurement. Two sets of the measurements of the magnetization curves were performed using two samples, which were taken from a single batch of the sample. They were named Run 1 and Run 2, which agree well with each other. In this paper, M, H, χ, and T represent magnetization, magnetic field, magnetic susceptibility, and temperature, respectively. Magnetization curve between 0.1 T and 1 T at each temperature was fit by the linear function of $M = M_S + \chi H$, where M_S represents spontaneous magnetization.

3. Results and Discussion

The temperature dependence of the spontaneous magnetization is shown in Fig. 1. There are four characteristic temperatures of $T_M = 148$ K, $T_t = 138$ K, $T_f \sim 125$ K, and $T_g \sim 50$ K. The spontaneous magnetization appears below T_M, and it develops up to $M_S = 2.5 \times 10^{-4}$ μ_B at T_t. Below T_t the spontaneous magnetization decreases steeply, but it starts to increase again at around T_f. The small drop of the spontaneous magnetization happens at T_g. Obviously the enhancement between T_M and T_t indicates that magnetic ordering occurs at T_M, not T_N, because the spontaneous magnetization never appear in a paramagnetic state. The M_S values above 148 K are constant at approximately 5×10^{-5} μ_B, which was likely caused by inevitable inaccuracy of the magnetic field. So, the spontaneous magnetization at T_t may be estimated to be approximately 2.0×10^{-4} μ_B. It should be pointed out that the temperature dependence of the spontaneous magnetization is concave or linear just below T_M, which is quite unusual for conventional canted antiferromagnetic ordering. Usually it is convex because it corresponds to the temperature dependence of the magnetization of each magnetic sublattice, which is the order parameter of the magnetic transition.

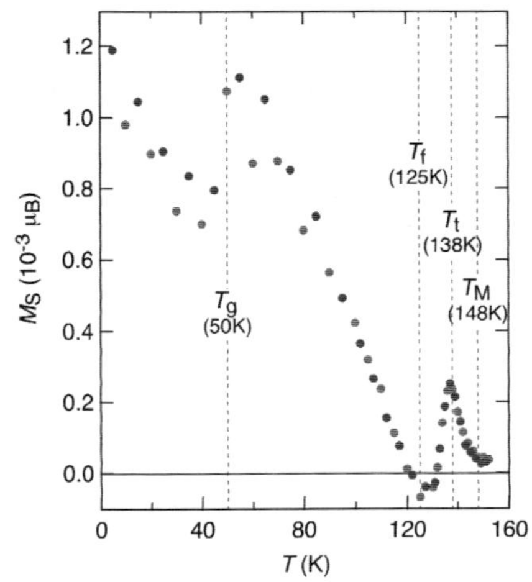

Figure 1. Temperature dependence of spontaneous magnetization. The red and blue data were obtained by Run 1 and Run 2, respectively.

The temperature dependence of magnetic susceptibility is shown in Fig. 2, together with that of magnetization under FC and ZFC conditions at $H = 0.01$ T and 1 T. It agrees well with M/H-T curves under ZFC condition, not those under FC condition, although the magnetization curves were obtained under FC condition. They all show the unusual peak at T_N; since the magnetic susceptibility here corresponds to the AC susceptibility at the lowest frequency, it should have the peak exactly at magnetic transition temperature [5].

The spontaneous magnetization is close to null at around T_f, which, naively thinking, implies that there are two kinds of magnetic moments canceling out each other around T_f. Indeed, the internal field at the V site estimated by ^{57}Fe Mössbauer spectroscopy seems to disappear at

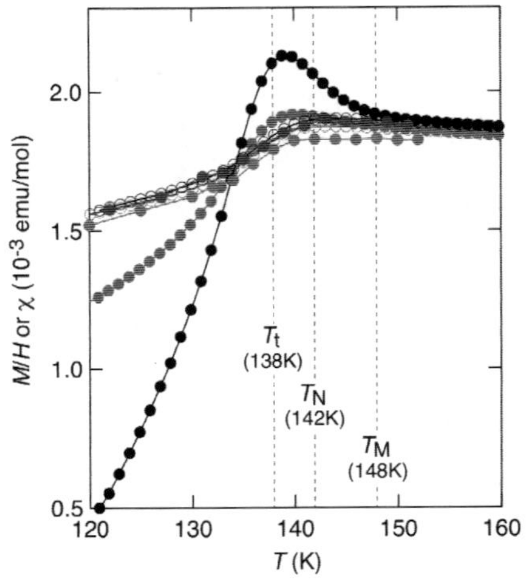

Figure 2. Magnetization or magnetic susceptibility as a function of temperature. The open and closed markers represent ZFC and FC data, respectively. The red, blue, black and gray markers represent χ obtained by Run 1 and Run 2, and M/H for 0.01 T and 1T, respectively.

around 115 K [6], which is in fair agreement with T_f considering the difference between bulk magnetization and the internal field at V site, and possible sample dependence. Of course, it can be assumed that two kinds of magnetic moments develop independently from T_M and T_t. In this case, the magnetic moments ordering at T_t are fixed in opposition to the magnetic field due to those ordering at T_M because the former is smaller in the magnitude just below T_t than the latter. The former becomes larger in the magnitude below T_f. However, it is also possible to expect that two kinds of magnetic moments order simultaneously at T_M; the smaller magnetic moments may develop quickly but the larger ones slowly. This scenario seems to be more probable considering the unusual development of M_S just below T_M. In either case, magnetic moments can flip over at around T_f when a large magnetic field is applied, and so no significant difference was observed at T_f in the magnetization between FC and ZFC data under the field. This can give the reason why there is no large difference between FC and ZFC magnetization under 7 T (See Fig. 3). The possible origins of the two kinds of magnetic moments would be the spins and the orbitals of V ions [7].

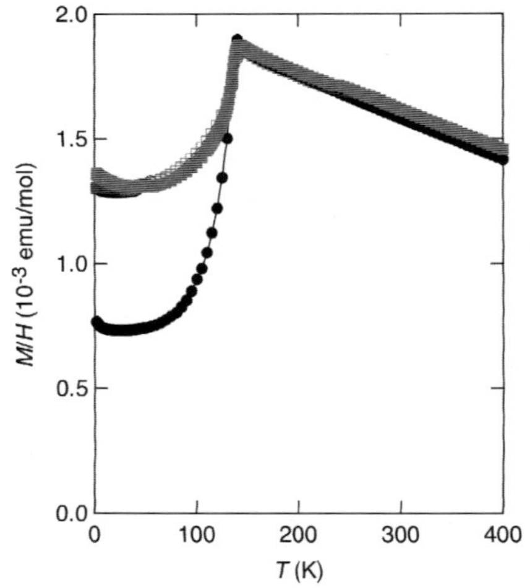

Figure 3. Temperature dependence of magnetization under 7 T (red and blue squares) and 1 T (black circles). The open and full markers represent ZFC and FC data, respectively.

The anomaly at T_g corresponds to the glassy behavior of the thermal conductivity above around T_g [7], which suggests that a part of the orbital degree of freedom survives down to T_g and orders at T_g to cause the drop of the magnetization. The magnitude of the drop at T_g

almost coincides with that at T_t, supporting the similarity between the origins of the drops. Khaliullin *et al.* suggested a partial orbital ordering and G-type spin order below the partial-orbital-ordering temperature [8, 9]. Since C-type spin order has been confirmed by neutron diffraction measurement at 4.2 K [10], the type of the spin order may change at T_g although another neutron diffraction study focused on no change in the spin structure at around T_t [11]. Because the orbital state of $LaVO_3$ could depend on thermal and field history of the sample, careful experiments would be needed to clarify the magnetism of the compound.

4. Summary

Five characteristic temperatures of T_M, T_N, T_t, T_f and T_g were found in the temperature dependence of the magnetization curve. The spontaneous magnetization appears below T_M and increases with decreasing temperature down to T_t. Below T_t it decreases steeply down to approximately zero at around T_f, at which it starts to increase again. The temperature dependence of the spontaneous magnetization shows a small drop at T_g, above which the thermal conductivity shows a glassy behavior. The magnetic susceptibility shows the peak at T_N, which is also seen in temperature dependence of magnetization under ZFC condition, although it was estimated by the magnetization curve measured under FC condition.

Acknowledgments

Acknowledgements This study was supported in part by a Grant-in-Aid for Scientific Research (A) (22246083) from MEXT and by the Japan Society for the Promotion of Science (JSPS) through its "Funding Program for World-Leading Innovative R& D on Science and Technology (FIRST Program)."

References

[1] Shirakawa N and Ishikawa M 1991 *Jpn. J. Appl. Phys.* **30** L755
[2] Mahajan A V, Johnston D C, Torgeson D R and Borsa F 1992 *Phys. Rev. B* **46** 10966
[3] Goodenough J B and Nguyen H C 1994 *C. R. Acad. Sci.* **319** 1285
[4] Nguyen H C and Goodenough J B 1995 *Phys. Rev. B* **52** 324
[5] Sakurai H, Kolodiazhnyi T, Michiue Y, Takayama-Muromachi E, Tanabe Y and Kikuchi H 2012 *Angew. Chem. Int. Ed.* **51** 6653.
[6] Yoon, S 2009 *J. Appl. Phys.* **105** 07D509
[7] Zhou J-S, Ren Y, Yan J-Q, Mitchell J F and Goodenough J B 2008 *Phys. Rev. Lett.* **100** 046401 and references therein
[8] Khaliullin G, Horsch P and Oleś A M 2001 *Phys. Rev. Lett.* **86** 3879
[9] Khaliullin G 2005 *Prog. Theor. Phys. Suppl.* **160** 155
[10] Zubkov V G, Bazuev G V and Shveĭkin 1976 *Sov. Phys. Solid State* **18** 1165
[11] Tung L D, Ivanov A, Schefer J, Lees M R, Balakrishnan G and Paul D McK 2008 *Phys. Rev. B* **78** 054416

Journal of Physics: Conference Series **428** (2013) 012032 doi:10.1088/1742-6596/428/1/012032

Magnetic properties of single crystalline $YbFe_2O_4$

K Yoshii[1], M. Mizumaki[2], K Matsumoto[3], S Mori[3], N Endo[1], H Saitoh[1], D Matsumura[1], T Kambe[4] and N Ikeda[4]

[1]Atomic Energy Agency, Sayo, Hyogo 679-5148, Japan
[2]Japan Synchrotron Radiation Institute, Sayo, Hyogo 679-5148, Japan
[3]Osaka Prefecture University, Sakai, Osaka 599-8531, Japan
[4]Okayama University, Okayama 700-8530, Japan

E-mail: yoshiike@spring8.or.jp

Abstract. We have investigated the magnetic properties of a single crystal of multiferroic $YbFe_2O_4$. Ferrimagnetic ordering of Fe spins is observed at ~250 K. The magnetization measured during field cooling becomes negative below ~10 K owing to the antiparallel coupling between the Fe and Yb sublattices. The magnetization below ~10 K is flipped by changing the applied magnetic field in a positive direction, a property which may be used to the development of magnetic devices. Magnetocaloric measurements show a broad peak of the entropy change, likely owing to a multiple magnetic transition.

1. Introduction

$YbFe_2O_4$ belongs to a class of isostructural materials RFe_2O_4 with R=Y, Ho-Lu and In [1]. Previously we have reported that this system with R=Y, Ho-Lu becomes ferroelectric below ~330 K owing to the real-space charge ordering of Fe^{2+} and Fe^{3+} on triangular lattices [2]. This origin of ferroelectricity differs from the ordinary ion-displacement mechanism. In addition, RFe_2O_4 shows ferrimagnetic ordering of Fe spins at the Néel temperature (T_N) of 230-250 K [3-5]. Hence, the system is multiferroic below T_N. Many studies of RFe_2O_4 have revealed their intriguing magnetic properties, such as huge coercivity at low temperatures [6], a competition between ferro- and antiferromagnetic states [7], spin-glass behaviour [8], and exchange bias [9]. However, full details of the magnetic properties have not been elucidated yet.

In this study, the magnetic properties of a single-crystalline sample of $YbFe_2O_4$ are investigated, as polycrystalline $YbFe_2O_4$ showed the highest T_N of 256 K among the RFe_2O_4 series [3]. Studies of RFe_2O_4 have been mainly confined to $LuFe_2O_4$, in which the R-site ion (Lu^{3+}) is nonmagnetic ($4f^{14}$) [3,6,7,8]. In contrast, reports on the properties of RFe_2O_4 with magnetic R^{3+} ions are somewhat rare [3]. Although a magnetic study of single-crystalline $YbFe_2O_4$ (Yb^{3+}; $4f^{13}$; 4.6 μ_B) was reported about 30 years ago [10], this work reports some new properties of this system, such as the magnetization reversal and the magnetocaloric effect, in which the magnetic moment of Yb^{3+} may be important.

2. Experimental

A single crystal of $YbFe_2O_4$ was grown by the floating zone method. X-ray diffraction measurements at room temperature showed that the crystal consisted of single-phase $YbFe_2O_4$ only. The magnetic measurements were carried out between 5 K and 300 K using a superconducting quantum interference device (SQUID) magnetometer (Quantum Design MPMS). The applied magnetic field was changed within ±50 kOe. The other details have been noted elsewhere [3,5].

3. Results and discussion

Figure 1a shows the magnetization-temperature (*MT*) curves measured with the applied field (H) of 1 kOe. The direction of *H* was along the *c*-axis. The deviation between field-cooled (FC) and zero-field-cooled (ZFC) curves is rooted in the ferrimagnetic ordering. The inflection of the thermoremanent magnetization (TRM) shows the Néel temperature (T_N) of ~250 K, a temperature being close to those in previous studies [3,10]. The broad peak at ~130 K suggests a complicated change of magnetic structure, as reported for $LuFe_2O_4$ [11].

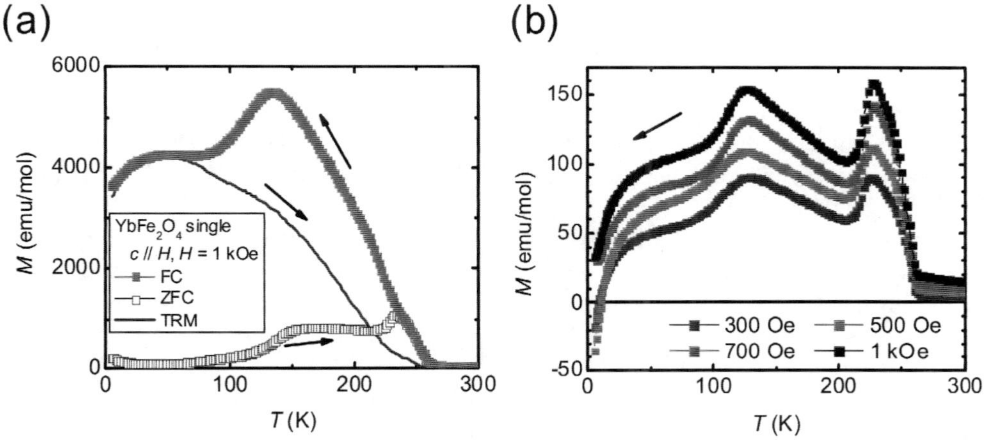

Figure 1. (a) Magnetization-temperature curves measured with the applied field of 1 kOe. The direction of the field was parallel to the *c*-axis. (b) Magnetization-temperature curves measured in the FC mode with the four applied fields. The direction of the field was parallel to the *ab*-plane.

Figure 1b shows the *MT* curves measured in the FC mode with *H* parallel to the *ab*-plane. The magnetization was much smaller (typically, ~1/10) than that in Fig. 1a. This is because of the Ising properties of RFe_2O_4, arising from the low-dimensional crystal structure [6] and the single-ion anisotropy of Fe^{2+} and Fe^{3+} [12].

We can see the interesting behaviour in Fig. 1b. That is, the magnetization is decreased below ~110 K and has a negative polarity below ~10 K for *H* lower than ~700 Oe. The same tendency is observed in the *c*//*H* condition below ~60 K (Fig. 1) [3,10], although the FC magnetization remain positive down to 5 K. The negative magnetization is plausibly brought about by the antiparallel magnetic coupling between Yb and Fe sublattices [3]; this is a similar situation to that of perovskite chromites $GdCrO_3$ and $La_{1-x}Pr_xCrO_3$ [13].

Figure 2 shows that the data could be fitted by the Curie-Weiss behaviour of Yb^{3+} moments, which have an antiparallel coupling with the Fe sublattice [13]:

$$M = M_{Fe} - \frac{C(H_I - H)}{T - \Theta} \quad (1)$$

Here, M_{Fe} represents the magnetization of the Fe sublattice. C, H_I and H stand for the Curie constant, the internal magnetic field at Yb^{3+}, and the applied magnetic field, respectively. T and Θ mean the temperature and the Weiss temperature, respectively. M_{Fe} was assumed to be independent of temperature (~94 emu/mol). The H_I and Θ values were ~1100 Oe and ~-7 K, respectively. As the negative magnetization is not observed for the *H* values greater than ~700 Oe (Fig. 1b), the obtained H_I seems to be roughly reasonable. The small absolute value of the Weiss temperature (~7 K) is

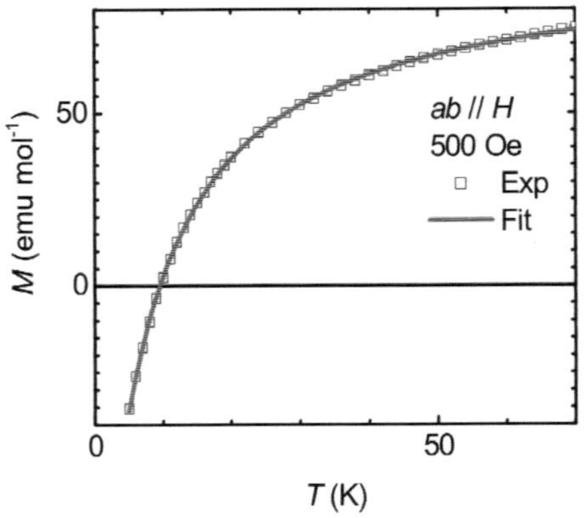

Figure 2. Curie-Weiss fit for the FC magnetization for the applied field of 500 Oe (equation (1)). Exp and Fit stand for the experimental data and the fitting curve, respectively. The details are noted in the text.

consistent with the previous result [10]. The magnetization reversal has not been observed for H parallel to the c-axis (i.e., $c//H$), even if H was lowered down to ~10 Oe (data not shown). This difference may originate from the larger M_{Fe} value (equation (1)) in the $c//H$ condition than in $ab//H$, owing to the Ising nature of RFe_2O_4.

The negative magnetization offers an interesting method of magnetization switching [14]. This is shown in Fig. 3. The sample was cooled with H=500 Oe down to 5 K. As seen in Fig. 1b, the polarity of the magnetization is negative. The magnetization measured for ~600 s was almost unchanged. Then, the applied field was increased to 1220 Oe; the magnetization was switched. This switching is repeatable by changing H between 500 and 1220 Oe. Note also that the switching is attained under H in a positive direction. This is different from the ordinary method, in which the magnetization is flipped by changing the direction of H. Hence, the present method can avoid the sample heating caused by hysteretic losses, when the magnetic field crosses zero [14].

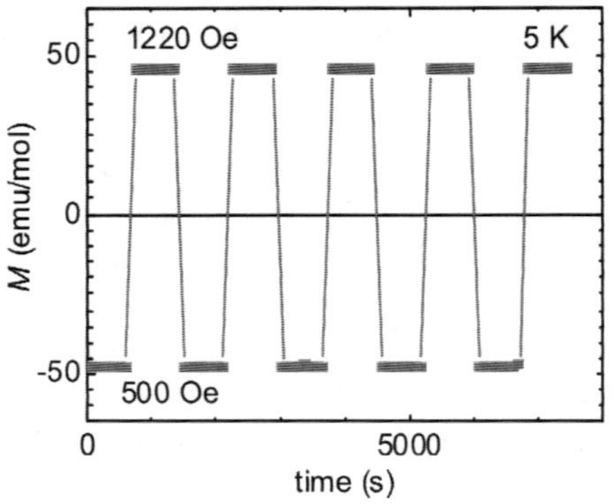

Figure 3. Magnetization switching at 5K performed after FC with 500 Oe. The applied field was changed between 500 and 1220 Oe. The details are noted in the text.

Figure 4a shows the magnetocaloric effect (MCE) of $YbFe_2O_4$. The negative values of the change of magnetic entropy ($-\Delta S$) are plotted against the temperature. ΔS is given by: [15]

$$\Delta S = \int_0^H \left(\frac{\partial M}{\partial T} \right)_H dH \qquad (2)$$

As usual, this value was calculated from magnetization-magnetic field (MH) curves [16]. The increment of the measurement temperature was 5 K. The applied field was parallel to the c-axis. The magnetic field (H) was increased from zero to 10, 20, 30, 40 and 50 kOe. Although the MCE is usually studied from the viewpoint of refrigerant applications, the sign of ΔS gives the information on the type of magnetic ordering. In ferromagnetic ordering, ΔS is negative at around the magnetic

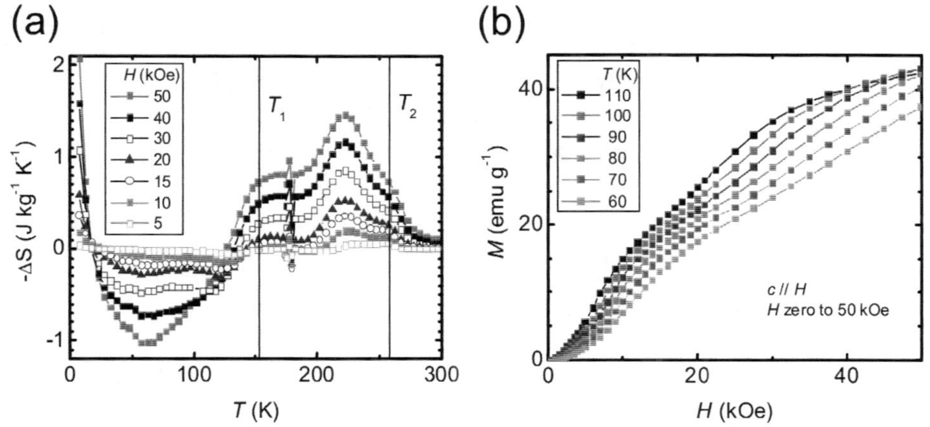

Figure 4. (a) Change of magnetic entropy (-ΔS) plotted against the temperature. The H values are defined in equation (2). (b) Magnetization-magnetic field curves at several temperatures below 110 K.

transition temperature, whereas, the value is positive for antiferromagnetism [11]. As ΔS was positive at around T_N due to the ferrimagnetic ordering, -ΔS is plotted in Fig. 4.

A peak of -ΔS appears near T_N as reported for many systems thus far [15]. The peak value is ~1 J kg^{-1} K^{-1} for H=50 kOe, and is reasonably close to that of $LuFe_2O_4$ [11]. This value is smaller than those of the large or giant MCE materials (-ΔS ~10 J kg^{-1} K^{-1}) [15]. Note that -ΔS exhibits a significantly broad peak, as found for $LuFe_2O_4$ [11]. Calculating the peak width, we have obtained the refrigerant capacity (RC) [15]. The temperatures of the hot and cold reservoirs in the refrigeration cycle are defined as the temperatures of the full width at half maximum of -ΔS. The former and latter temperatures for H=50 kOe were T_1~ 153 K and T_2~258 K, respectively. The RC values were calculated by the integration of -ΔS between T_1 and T_2. The RC was ~80 J kg^{-1}, which is ~25-50% of those of the large or giant MCE materials [15]. However, the large difference between T_1 and T_2 (~105 K) is favourable for the so-called Ericsson cycle [17]. This property is possibly owing to either a multiple magnetic transition below T_N or magnetic inhomogeneity [11], originating from a magnetic frustration on a triangular lattice. The discontinuous of -ΔS at ~190 K may be associated with a structural distortion [6], as a sharp change of -ΔS can be observed at a first-order transition [15].

The sign of -ΔS becomes negative below ~130 K. This is a different result from that of $LuFe_2O_4$, in which -ΔS is positive at all temperatures [11]. The present behaviour is rooted in a decrease in the integral term (equation (2)), because of the shrinkage of the MH curves (Fig. 4b) [9]. This shrinkage arises from the enhancement of coercivity [6,9]; we should measure the MH curves at H values greater than the highest field of the present apparatus (50 kOe). The zero value of -ΔS at ~130 K indicates a possible application to fabricate a constant temperature bath [14].

The -ΔS value increases again below ~70 K as seen in Fig. 4a. The true origin of this behaviour is unclear. However, paramagnetic Yb^{3+} moments may play some roles, considering that the magnetic entropy is decreased since Yb^{3+} spins are forced to be parallel to H by the application of large fields. A role of Yb^{3+} is also implied by the fit of the MT data (Fig. 2), in which the Curie-Weiss term (equation (1)) of Yb^{3+} is important below ~60 K. Further studies are currently in progress, and will be published in the near future.

4. Summary

We have investigated the magnetic properties of a single crystal of multiferroic $YbFe_2O_4$. Ferrimagnetic ordering of Fe spins is observed at ~250 K. The magnetization measured during field cooling becomes negative below ~10 K owing to the antiparallel coupling between the Fe and Yb sublattices. The magnetization below ~10 K is flipped by changing the applied magnetic field in a positive direction, a property which may be used to the development of magnetic devices. Magnetocaloric measurements show a broad peak of the entropy change, likely owing to a multiple magnetic transition.

Acknowledgment
This work was partially supported by a Grant-in-Aid for Scientific Research (23350071).

References

[1] Kimizuka N, Muromachi E and Siratori K 1990 *Handbook on the Physics and Chemistry of Rare Earths* vol 13 (Elsevier, Amsterdam) p 283

[2] Ikeda N, Ohsumi H, Ohwada K, Ishii K, Inami T, Murakami Y, Kakurai K, Yoshii K, Mori S, Horibe Y and Kito H 2005 *Nature* **436** 1136

[3] Yoshii K, Ikeda N, Matsuo Y, Horibe Y and Mori S 2007 *Phys. Rev.* B **76** 024423, and references therein

[4] Oka K, Azuma M, Hayashi N, Muranaka S, Narumi Y, Kindo K, Ayukawa S, Kato M, Koike Y, Shimakawa Y and Takano M 2008 *J. Phys. Soc. Jpn.* **77** 064803

[5] Yoshii K, Ikeda N, Okajima Y, Yoneda Y, Matsuo Y, Horibe Y and Mori S 2008 *Inorg. Chem.* **47** 6493

[6] Wu W, Kiryukhin V, Noh H -J, Ko K -T, Park J -H, Ratcliff W, Sharma P A, Harrison N, Choi Y J, Horibe Y, Lee S, Park S, Yi H T, Zhang C L and Cheong S –W 2008 *Phys. Rev. Lett.* **101** 137203

[7] de Groot J, Marty K, Lumsden M D, Christianson A D, Nagler S E, Adiga S, Borghols W J H, Schmalzl K, Yamani Z, Bland S R, de Souza R, Staub U, Schweika, Su W Y and Angst M 2012 *Phys. Rev. Lett.* **108** 037206

[8] Wang F, Kim J, Kim Y -J and Gu G D 2009 *Phys. Rev.* B **80** 024419

[9] Yoshii K, Ikeda N, Nishihata Y, Maeda D, Fukuyama R, Nagata T, Kano J, Kambe T, Horibe Yand Mori S 2012 *J. Phys. Soc. Jpn.* **81** 033704

[10] Kishi M, Miura S, Nakagawa Y, Kimizuka N, Shindo I and Siratori K 1982 *J. Phys. Soc. Jpn.* **51** 2801

[11] Phan M H, Frey N A, Angst M, Groot J, Sales B C, Mandrus D G and Srikanth H 2010 *Solid State Commun.* **150** 341

[12] Sugihara T, Siratori K, Shindo I and Katsura T 1978 *J. Phys. Soc. Jpn.* **45** 1191

[13] Yoshii K, and, Nakamura A 2000 *J. Solid State Chem.* **155** 447; Yoshii K 2000 *J. Solid State Chem.* **159** 204; Yoshii K, A. Nakamura A, Ishii Y and Morii Y 2001 *J. Solid State Chem.* **162** 84; Yoshii K 2011 *Appl. Phys. Lett.* **99** 142501

[14] Yusuf S M, Kumar A and Yakhmi J V 2009 *Appl. Phys. Lett.* **95** 182506

[15] Gschneidner Jr K A, Pecharsky V K and Tsokol A O 2005 *Rep. Prog. Phys.* **68** 1479

[16] Pecharsky V K and Gschneidner Jr K A 1999 *J. Appl. Phys.* **86** 565

[17] Chaturvedi A, Stefanoski S, Phan M -H, S. Nolas G S and Srikanth H 2011 *Appl. Phys. Lett* **99** 162513

Elastic Anomalies in Jahn-Teller Crystals with Competing Structural Orderings

Michael Kaplan[1], George Zimmerman[2]

[1]Department of Chemistry and Physics, Simmons College, 300 The Fenway, Boston, MA 02115
[2]Department of Physics, Boston University, 590 Commonwealth Ave., Boston, MA 02115

[1]E-mail: *mkaplan@simmons.edu*
[2]*E-mail:goz@*

Abstract. Crystals of transition metal compounds with perovskite and spinel structures are characterized by several competing electron interactions responsible for different structural and magnetic orderings. The shape of the acoustic anomaly temperature dependences allows to distinguish between them and to estimate the difference between the structural and magnetic transition critical temperatures.

1. Introduction

The interest to the crystals of the transition metal compounds with perovskite and spinel structures permanently continues to grow for a significant period of time [1]. It is related to a range of surprising discoveries of unusual properties of these materials. The anomalous properties of perovskite and spinel structure crystals are the direct consequence of the electronic orbital degeneracy or pseudo-degeneracy of the transition metal ions. The corresponding electronic states are very sensitive to the local distortions (Jahn-Teller effect [2, 3]) and to their correlation through the crystal phonons (cooperative Jahn-Teller effect [4]). This leads to the different structural transitions and the corresponding property anomalies [5, 6]. On another hand the magnetic intersite interaction in case of the local electronic orbital degeneracy results in the possibility of different unusual magnetic structures (for example, because of the electron exchange at orbital degeneracy [7]). The combination of the mentioned above interactions in Jahn-Teller

perovskite and spinel crystals causes the unordinary richness of their properties.
The real situation in these systems under discussion becomes even more interesting and complicated as the detail of the crystal structure - types of the connection between different octahedrons and tetrahedrons – could significantly affect the relative magnitude of the virtual phonon exchange responsible for the structural transitions and the magnitude of the exchange interaction at orbital degeneracy. It is often not easy to understand the primary reason of the type of structural ordering in these crystals and to attribute it to a specific virtual phonon interaction type or to an exchange magnetic interaction with the followed magnetostriction. Very often the resulting structural and magnetic orderings are the same on average (thermodynamically).
It is clear that the development of the experimental approach to the fundamental problem of the driving force in the material transformation process becomes extremely important for understanding and correct description of the phase transitions and crystal properties.
Below it is shown that such an approach could be developed basing on the analysis of the acoustical anomaly temperature dependences. The results are applicable to the Jahn-Teller crystals containing the ions with triple orbital degenerate electronic states [8-11] and in part to the MnV_2O_4 crystals the properties of which are getting a lot of attention nowadays [10, 12,13].

2. Microscopic model

We are considering a crystal with the Jahn-Teller ions in at least one of the crystal sublattices. The ground electronic state of the ions is an orbital triplet. While in this case two different types of Jahn-Teller local vibrations – e-symmetry and t-symmetry - are active in the Jahn-Teller effect, we will take into account only the e-vibrations as the interaction with them in the perovskites and spinels as a rule is the strongest. The double degeneracy of the local and corresponding crystal vibrations is responsible for the high level of degeneracy of the ordered crystal structure state caused by virtual phonon exchange. As a result many different types of packing of the tetragonally distorted structural crystal units are possible.
The most typical types of ordering in the crystals under discussion are so called ZZ-type and XY-type of orderings. In case of ZZ-type of ordering the Jahn-Teller structural units are distorted tetragonally along the c(z)-crystal axis, and all distortions are parallel, As a result of that at $T<T_s$ the

Journal of Physics: Conference Series **428** (2013) 012033 doi:10.1088/1742-6596/428/1/012033

crystal transfers from the paradistortive cubic state to a tetragonal crystal state with the elongation along the c(z)-axis and contraction in the plane perpendicular to OZ. If the XY-ordering takes place the Jahn-Teller structural units again are distorted tetragonally but at $T<T_s$ the crystal is divided in two sublattices. In both sublattices the Jahn-Teller structural units are elongated in the ab-plane of the crystal, however in one of the sublattices the elongations are parallel to the a(x)-axis, and in the another one – along the b(y)-axis. It is clear that in this case once again there is a structural transition from the cubic to the tetragonal phase. However this time in the tetragonal state the crystal is compressed along the c(z)-axis. In other words ZZ-ordering leads to a tetragonal crystal strain with $c/a>1$, and XY-ordering causes the strain with $c/a<1$. It is completely clear that if the magnitudes of the corresponding interaction constants are close the orderings compete. This is what often happens in perovskite and spinel crystal structures. Moreover, sometimes the magnetic interaction constants defining the magnetic phase transition critical temperature are also pretty close to the virtual phonon exchange constants. In this situation it is not easy to determine the major driving force of the phase transformation. But the analysis of the magnitude and of the character of acoustic anomalies helps to shed some light on the question.

The Hamiltonian of the crystal could be written as

$$H = H_{el} + H_{ph} + H_{el-ph} + H_{str} + H_{el-str} + H_{magn} \quad (1)$$

where the first two terms describe the electronic states and the free phonons of the crystal, the electron-phonon interaction is typical for the T-e-Jahn-Teller problem

$$H_{el-ph} = \sum_{m\kappa} [V^{\theta}_{m\kappa}(3L^2_{zm} - 2) + V^{\varepsilon}_{m\kappa}(L^2_{xm} - L^2_{ym})](b_\kappa + b^+_{-\kappa}) \quad (2)$$

with $V^{\theta,\varepsilon}_{m\kappa}$ as the electron-phonon constants, $L_{z,x}$ - projections of the orbital electronic moment operators defined on the ground triplet state, and b_κ, b^+_κ - phonon operators.

The magnetic interactions in the crystals under discussion are described in the simplest way reflecting the cubic symmetry of the high temperature crystal phase

$$H_{magn} = -\beta \sum_m (g_x L_x H_x + g_y L_y H_y + g_z L_z H_z) \tag{3}$$

The depending upon the homogeneous strain part of the Hamiltonian can be presented as

$$H_{str} = \frac{1}{2} C_0 \Omega (U_\theta^2 + U_\varepsilon^2) - P_\theta U_\theta - P_\varepsilon U_\varepsilon \tag{4}$$

$$H_{el-str} = -g_0 \sqrt{\frac{C_0 \Omega}{N}} \sum_m [U_\theta (3L_z^2 - 2) + U_\varepsilon (L_x^2 - L_y^2)]$$

where $C_0 \equiv C_{11} - C_{12}$ is the unrenormalized modulus of elasticity, $U_{\theta,\varepsilon}$ are the homogeneous strain components of the e-symmetry defined as

$$U_\theta \equiv 6^{-1/2} (2U_{zz} - U_{xx} - U_{yy})$$

$$U_\varepsilon \equiv 2^{-1/2} (U_{xx} - U_{yy}) \quad , \tag{5}$$

and $P_{\theta,\varepsilon}$ are the uniaxaial pressure components and g_0 is the electron-strain interaction constant.

Minimizing the Hamiltonan (1) relatively the homogeneous strain in part it could be obtained

$$U_\theta = \frac{P_\theta}{C_0 \Omega} + g_0 \sqrt{\frac{N}{C_0 \Omega}} \overline{(3L_z^2 - 2)} \tag{6}$$

From formula (6) it is clear that the orbital state ordering leads to the simultaneous strain of the crystal. Both of them are the result of the cooperative Jahn-Teller effect.

The crystal modulus is easily found from (6) as

Journal of Physics: Conference Series **428** (2013) 012033 doi:10.1088/1742-6596/428/1/012033

$$\frac{1}{C} \equiv \left(\frac{\partial U_\theta}{\partial P_\theta}\right)_{P_\theta=0} = \frac{1}{C_0\Omega} + g_0\sqrt{\frac{N}{C_0\Omega}}\overline{(3L_z^2-2)} \tag{7}$$

or

$$\frac{C}{C_0} = \frac{1}{1+g_0\sqrt{C_0\Omega N}\,\dfrac{\partial\overline{(3L_z^2-2)}}{\partial P_\theta}} \tag{8}$$

After the standard canonical transformation of the Hamiltonian (1) eliminating the linear term of the electron-phonon interaction [4], the virtual phonon exchange electron correlation is described as

$$H_{\text{int}} = -\sum_{\substack{mn\kappa \\ m\neq n}}\left[\frac{V_{m\kappa}^{\theta*}V_{n\kappa}^{\theta}}{\hbar\omega_\kappa}(3L_{zm}^2-2)(3L_{zn}^2-2)+\frac{V_{m\kappa}^{\varepsilon*}V_{n\kappa}^{\varepsilon}}{\hbar\omega_\kappa}(L_{xm}^2-L_{ym}^2)(L_{xn}^2-L_{yn}^2)+2\frac{V_{m\kappa}^{\theta*}V_{n\kappa}^{\varepsilon}}{\hbar\omega_\kappa}(3L_{zm}^2-2)(L_{xn}^2-L_{yn}^2)\right] \tag{9}$$

Due to the vibronic reduction the magnetic terms of (1) become relatively complicated. For example, the z-component of the Zeeman interaction looks like that

$$\tilde{H}_{Zeem} = -g\beta H_z\sum_m[\cos 2g_m^\varepsilon L_{zm} - \sin 2g_m^\varepsilon \tau_m] \tag{10}$$

where

$$g_m^{\theta,\varepsilon} = i\sum_\kappa\left(\frac{V_{m\kappa}^{\theta,\varepsilon*}}{\hbar\omega_\kappa}b_{-\kappa}^{+} - \frac{V_{m\kappa}^{\theta,\varepsilon}}{\hbar\omega_\kappa}b_\kappa\right) \tag{11}$$

and the electronic operator τ is represented by the matrix

$$\tau_m = \begin{pmatrix} 0 & 1 & 0 \\ 1 & 0 & 0 \\ 0 & 0 & 0 \end{pmatrix} \tag{12}$$

It is accepted that the magnetic ordering along z-direction is taking place in the crystals under discussion (the magnetic molecular field $J\gamma^2\overline{L_z}$ and the external magnetic field H_z are the largest, $\gamma \equiv \overline{\cos 2g}$ is the vibronic reduction factor).
The transformed electron correlation Hamiltonian (9) evidently (its first term) contains the ZZ-interaction. But using some transformation of the electronic operators entering (9) it could be shown that the XY-interaction is present too in the form

$$H_{\text{int}}^{xy} = -2\sum_{mn} A_{mn}^{xy}(3L_{xm}^2 - 2)(3L_{yn}^2 - 2) \tag{13}$$

where

$$A_{mn}^{xy} = 2\sum_{\kappa} \frac{|V_\kappa^\theta|^2 - \frac{1}{9}|V_\kappa^\varepsilon|^2}{\hbar\omega_\kappa} e^{i\kappa(m-n)} \tag{14}$$

In the following analysis it is accepted that both – ZZ- and XY-interactions are present together with the magnetic interactions in the crystal in agreement with possible real situation.
For a two sublattice crystal the Hamiltonian in the molecular field approximation is written as follows

$$H_m^I = -A^{I,II}(0)\overline{(3L_{xII}^2 - 2)}(3L_{ymI}^2 - 2) - B^{I,II}(0)\overline{(3L_{zII}^2 - 2)}(3L_{zmI}^2 - 2) - h_z^I\gamma L_{zmI} - P_z(3L_{zmI}^2 - 2)$$

$$H_m^{II} = -A^{II,I}(0)\overline{(3L_{yI}^2 - 2)}(3L_{xmII}^2 - 2) - B^{II,I}(0)\overline{(3L_{zI}^2 - 2)}(3L_{zmII}^2 - 2) - h_z^{II}\gamma L_{zmII} - P_z(3L_{zmII}^2 - 2) \tag{15}$$

$$h_z^{I,II} \equiv g_z\beta\gamma H_z + J(0)\gamma^2\overline{L_{zI,II}} \tag{16}$$

$$P_z \equiv g_0^E \frac{p_z}{\sqrt{C_{0E}\Omega N}} \tag{17}$$

Formulae (15)-(17) allow the calculation of the ordering parameters for the ZZ- or XY-ordering or for a magnetic ordering in the crystal. The corresponding ordering parameters are defined as $\overline{3L_{zI}^2-2} = \overline{3L_{zII}^2-2} \equiv z$, $\overline{3L_{xI}^2-2} = \overline{3L_{yII}^2-2} \equiv x$, and $\overline{L_z^I} = \overline{L_z^{II}} \equiv \overline{L_z}$. These parameters define correspondingly the strain in the sublattices and in the total crystal and the crystal magnetization [14].

The detailed analysis of the physics picture described by the developed above microscopic model was discussed in detail earlier [14]. That's why here we only are briefly reminding that the ZZ- and the XY-virtual phonon exchange interactions are competing trying to establish tetragonal c/a>1 and c/a<1 crystal distortion. Naturally, the result depends upon the relative magnitude of the interactions. However the role of the magnetic (internal or external) interaction in this situation becomes important and sometimes even crucial. Really, the magnetic interaction as it was shown in [14] are supporting the c/a<1 tetragonal crystal strain due to the crystal magnetostriction. In other words the magnetic interaction is leading to the magnetic moments parallel to the c(z)-axis and is supporting the XY-type of structural ordering and counteracts the ZZ-ordering. In the crystal where B>A (ZZ-ordering overcomes the XY-ordering), the magnetic ordering that happens below the structural phase transition can change the structure of the crystal replacing the c/a>1 at $T_N<T<T_s$ with the c/a<1 at $T<T_N$. Therefore a new type of the structural ordering of the tetragonal-tetragonal type takes place.

Of course, in this situation it is not clear in advance why this structural ordering to the state with c/a<1 takes place: because of the big magnetic interaction or because of the significant XY-interaction, or because of the combination of all these interactions.

The acoustic anomaly measurements as it is explained by the following below calculation results could shed some light on this type of questions. Plus, we hope that these calculations are helping to understand some other experimental results that before were looking contradictory.

3. Numerical calculation analysis.

The calculations of the order parameters, strains and magnetic moments were done for different combinations of the intersite electron interaction constants A, B, and J. For simplicity here only the results for one of these parameters equal to zero is discussed. We will concentrate our attention on the temperature dependence of the elastic modulus.

On the Fig.1 and 2 it is accepted that the XY-ordering only takes place (B=0). It is seen that the increase of the magnetic interaction shifts the minimum of the temperature dependence of the modulus of elasticity to higher temperature. At the same time the magnetic interaction are decreasing the relative magnitude of the acoustic anomaly. The last becomes really small when J>>A what is typical for the magnetic phase transitions accompanied by small crystal structure changes caused by magnetostriction.

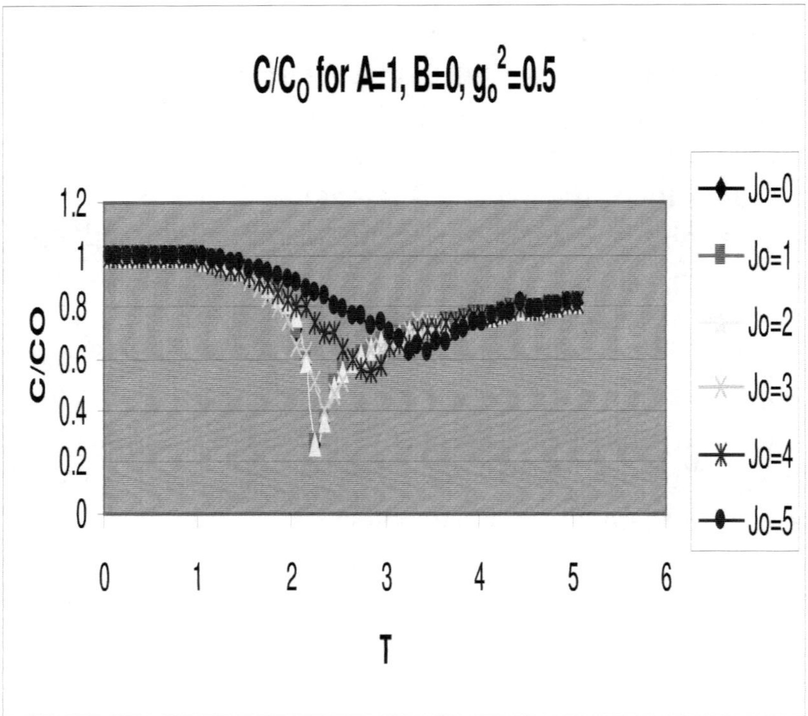

Figure 1. Temperature dependence of the modulus of elasticity at the XY-ordering for different magnitudes of the magnetic interaction J

As it is seen from the Fig.1 at the big enough magnetic interaction the acoustic anomaly is not only small, but it is additionally smeared out.

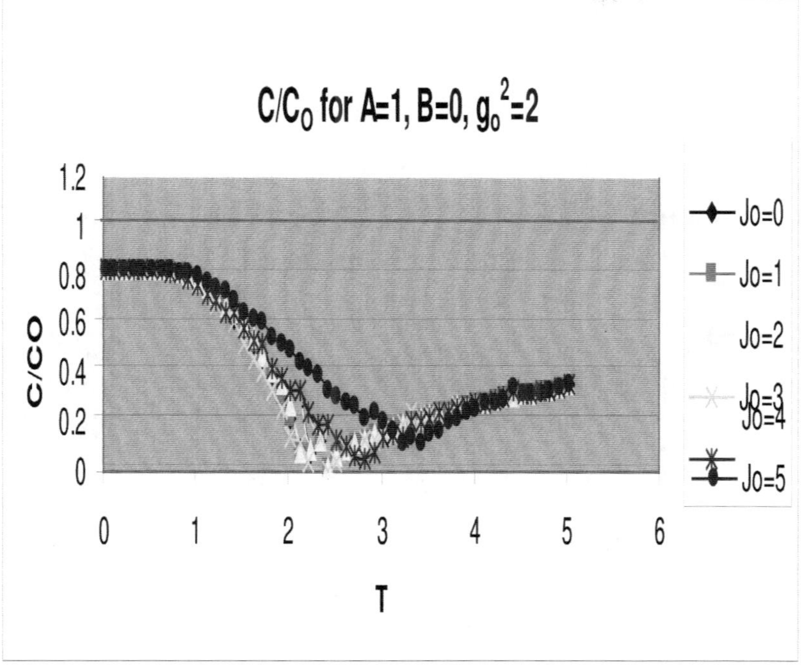

Figure 2. The same that on the Fig.1, but at the bigger electron-strain interaction constant

The increase of the electron-strain interaction constant is increasing the magnitude of the elastic anomaly (the minimum on C(T) is dipper) as always at the cooperative Jahn-Teller effect.

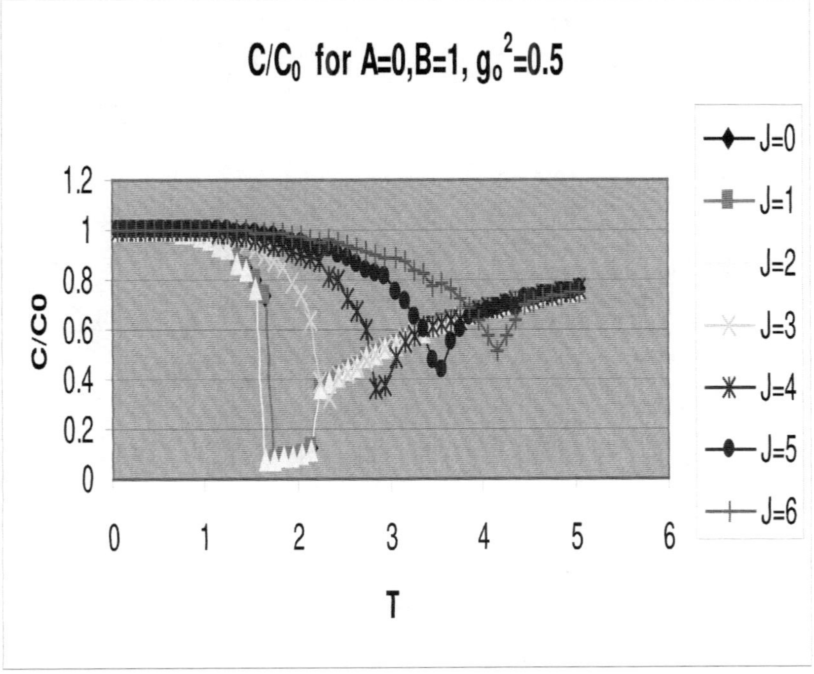

Figure 3. Temperature dependence of the modulus of elasticity at the ZZ-type of structural ordering at the different magnitudes of the magnetic interactions

On the Fig.3 the temperature dependence of the modulus of elasticity is completely different the presented on the Figs.1 and 2. If at the big enough magnetic interactions a typical smearing out and magnitude decrease of the acoustic anomaly takes place, at the magnetic interactions of the order of the B magnitude – J=1, 2 – two critical temperatures could be observed with a plato between them. At the low temperatures there is a sharp decrease of the modulus of elasticity followed by a slow increase with the temperature typically for the structural transition. However at higher temperatures the magnetic interactions prevail, they are smearing out the anomaly changing or inducing the tetragonal-tetragonal structural transition.

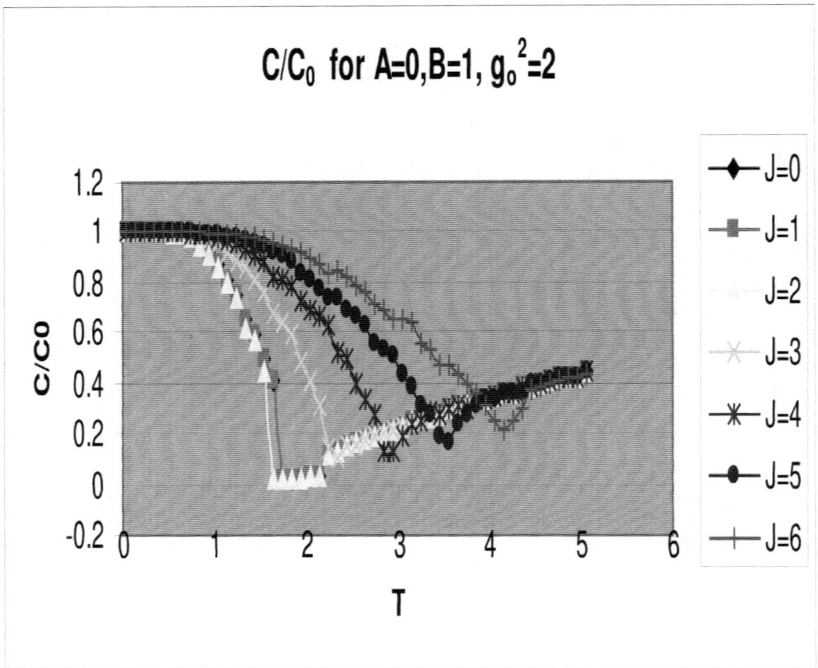

Figure 4. The same that on the Fig. 3 but at the larger electron-strain interaction constant.

Journal of Physics: Conference Series **428** (2013) 012033 doi:10.1088/1742-6596/428/1/012033

On the Fig.4 the C(T) behavior is similar to observed on the Fig.3. However due to bigger electron-strain interaction constant the minimum of C(T) is dipper and the plato between the two phase transitions is located near C=0. It means that the acoustic attenuation at this temperature interval is huge and the signal at the measurements is impossible to observe.

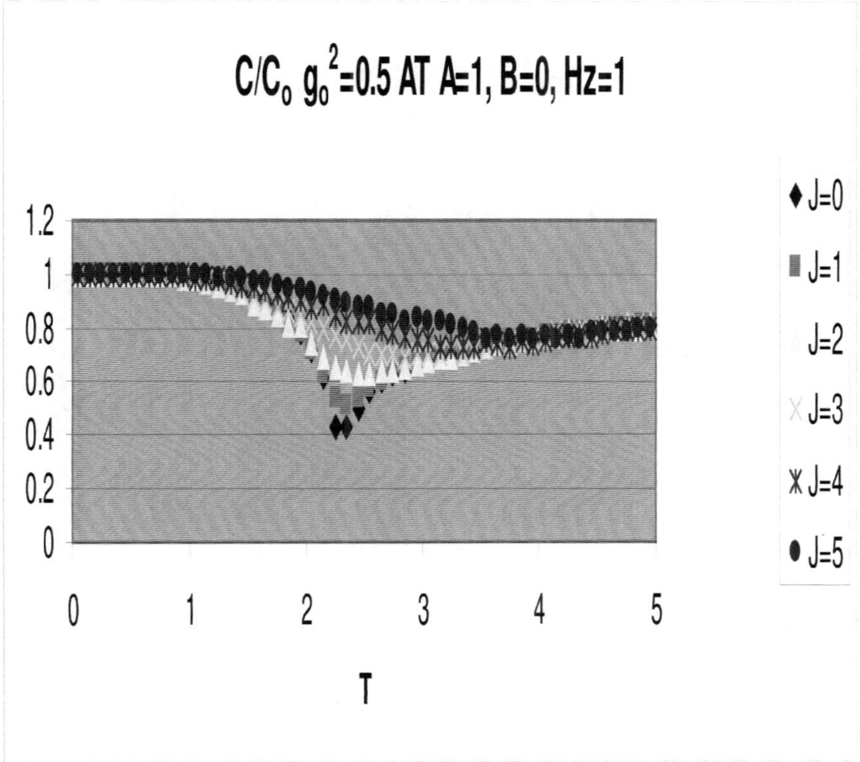

Figure 5. Influence of the external magnetic field on C(T) at the XY-type of ordering for different magnetic interaction constants J.

On the Fig.5 the temperature dependence of the modulus of elasticity is shown in the presence of the external magnetic field. The field is accelerating the decrease and the smearing out of the anomaly.

Journal of Physics: Conference Series **428** (2013) 012033 doi:10.1088/1742-6596/428/1/012033

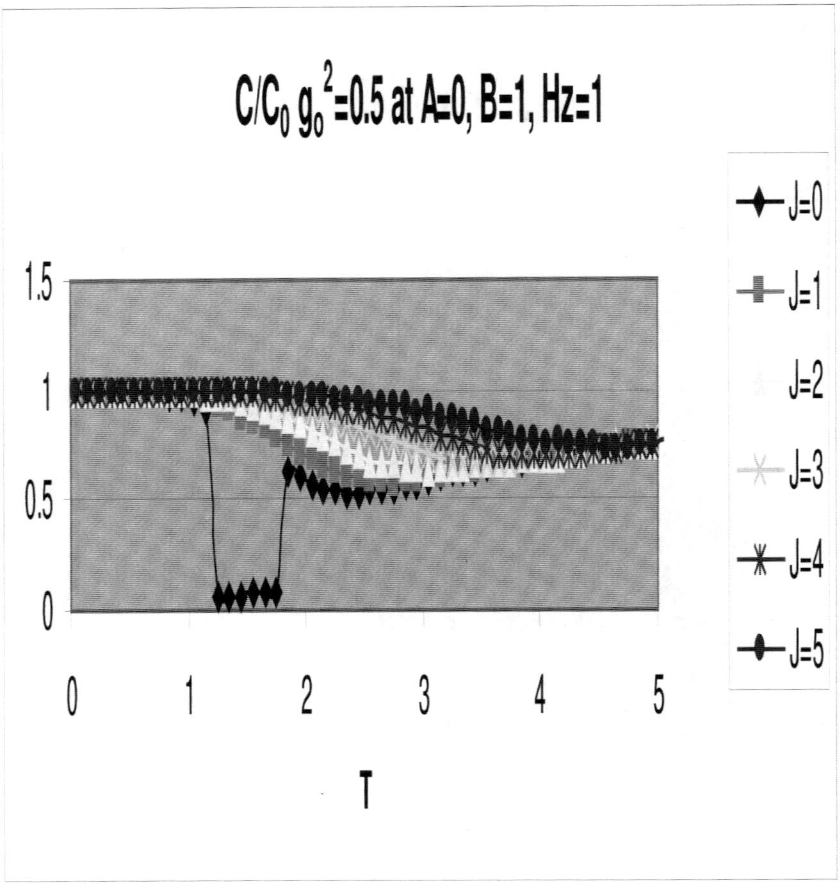

Figure 6. External magnetic field influence of the acoustic anomaly in case of the ZZ-type of structural ordering at the different magnitudes of the magnetic interaction J.

On the Fig.6 the influence of the external field on C(T) is demonstrated. The magnetic field is decreasing the ZZ-ordering critical temperature and increasing the width of the plato where the modulus of the elasticity is near its minimum. On the contrary to the Fig.4 at J=1 the plato already disappeared and the anomaly is significantly smeared out.

For example, the comparison of the calculations and ultrasonic measurements [15] in MnV2O4 crystal allows concluding that in that crystal

XY-interaction is larger than ZZ-interaction and the virtual phonon exchange of the XY-type is close in magnitude to the magnetic interaction.

Journal of Physics: Conference Series **428** (2013) 012033 doi:10.1088/1742-6596/428/1/012033

Conclusions.

In perovskite and spinel structure crystals with competing virtual phonon exchange interactions and magnetic interactions the analysis of the temperature dependences of the modulus of elasticity can help to explain the driving force of the crystal transformations and crystal properties.

References.

[1] Bersuker I 1996 Electronic Structure and Properties of Transition Metal Compounds. Introduction to the Theory (New York, Wiley)
[2] Bersuker I, Polinger V 1989 Vibronic Interaction in Molecules and Crystals (Berlin, New York, Springer)
[3] Bersuker I 2006 The Jahn-Teller Effect (Cambridge, University Press)
[4] Kaplan M, Vekhter B 1995 Cooperative Phenomena in Jahn-Teller Crystals (New York, London, Plenum Press)
[5] Kaplan M, Zimmerman G 2008 Phys. Rev. B, **77**, 104426
[6] Kaplan M, Zimmerman G 2012 Proc. APS Meeting (Boston),
[7] Kugel K, Khomskii D 1977 Usp. Phys. Nauk, **136**, 621
[8] Plumier R, Sougi M 1987 Solid State Commun. **64**, 53
[9] Plumier R, Sougi M 1989 Physica (Amsterdam), **155B**, 315
[10] Adachi K 2005 Phys. Rev. Lett. **95**, 197202
[11] Suzuki T, Katsumura M, Taniguchi K, Arima T, Katsufugi T 2007 Phys. Rev. Lett. **98**, 127203
[12] Garlea D, Jin R, Mandrus D, Roessli B, Huang Q, Miller M, Schultz A, Nagler S Phys. Rev. Lett. **100**, 066404
[13] Baek S-H, Choi K-Y, Reyes A, Kuhns P, Curro N, Ramanchandran V, Dalal N, Zhou H, Wibe C arXiv:0707.0018v1
[14] Kaplan M, Zimmerman G 2012 Vibronic Interactions and the Jahn-Teller Effect eds. Atanasov M, Daul C, Tregenna-Piggott P (Dordreht, Heidelberg, London, New York, Springer), Ch.24, 429
[15] Yanbing Luan 2011 Elastic properties of complex transition metal oxides studied by Resonant Ultrasound Spectroscopy, PhD diss. University of Tennessee

Journal of Physics: Conference Series **428** (2013) 012034 doi:10.1088/1742-6596/428/1/012034

Vibronic orbital magnetism

Michael D Kaplan

Department of Chemistry and Physics
Simmons College, 300 The Fenway
Boston, MA 02115

E-mail: mkaplan@simmons.edu

Abstract. It is shown that in systems lacking the time inversion symmetry operation the Jahn-Teller effect could be the reason of the existence of the electronic magnetic moment. In crystals with cooperative Jahn-Teller effect the last could be responsible for the vibronic magnetism – ordering of vibronic magnetic moments. The direct connection between the vibronic magnetism and piezomagnetism is discussed. If the local crystal symmetry is lacking both – the time and space inversion symmetries – the cooperative Jahn-Teller effect leads to the creation of the Jahn-Teller multiferroics.

1. **Introduction.**

Vibronic orbital magnetism is a complex of phenomena related to the orbital momentum of electrons that becomes important only if the electron-vibrational (vibronic) interaction in the system under discussion is not zero.

In general the magnetic moment of any structural crystal unit or molecule is containing two contributions coupled with electrons and nuclei correspondingly

$$\vec{M} = \vec{M}_{el} + \vec{M}_{nucl} \tag{1}$$

$$\vec{M}_{el} = -\frac{e}{2mc}\sum_i [\vec{r}_i \times \vec{p}_i] \tag{2}$$

$$\vec{M}_{nucl} = \frac{e}{2c}\sum_i [\vec{R}_i \times \vec{P}_i]\frac{Z_i}{M_i} \tag{3}$$

In the formulae (2) and (3) e is the electronic charge, m – mass of the electron, p_i is the electronic impulse conjugated to the electronic radius r_i, R_i and P_i are the radius and the impulse of a nucleus, Z_i and M_i are the effective nucleus charge and mass, i is the summation index over the structure units or molecules.

Each of the magnetic moments (2) and (3) in general may have a contribution corresponding to the fixed equilibrium positions of the nuclei and a component related to the displacement of the nuclei from equilibrium – vibronic interaction contribution.

In this presentation we will consider the electronic magnetic momentum only and only its part that becomes zero in the absence of vibronic interaction. We do not discuss here any effects and difficulties related to the finite size of crystals or crystal interfaces.

In other words we will consider M_{el} as

$$\vec{M}_{el} = -\frac{e}{2c}\sum_{ij} < \Phi_i \,|\, \vec{r} \times \vec{v} \,|\, \Phi_j > \neq 0 \tag{4}$$

and will analyze at what conditions M_{el} is not zero due to vibronic interaction, if the wave functions Φ_i in (4) are the components of an orbitally degenerate or pseudo degenerate (separated by a relatively small energy gap) electronic state.

In some sense the situation is similar to the vibronic dipole moment concept which was developed by Bersuker in [1] (see the references in [1]). The projections of the orbital momentum $\vec{L} = \vec{r}x\vec{p}$ are not zero only in the case of the broken time inversion symmetry. From the first point of view the last statement is closing the problem: only the interactions that are breaking the time inversion symmetry are responsible for the creation of the magnetic moment. However, in the presence of the broken time inversion symmetry the vibronic interactions may create new, initially

absent magnetic moment projections.
Going back to the analogy with the vibronic electric dipole moment concept it should be noted that the electric vibronic dipole moment exists in case of orbital degeneracy (not only pseudo degeneracy), if the system is lacking the space inversion center. In this situation the local distortions create an electric dipole moment as a result of Jahn-Teller (not only pseudo Jahn-Teller) effect as it was mentioned by Vekhter in [2].
Similarly to the electric dipole moment the magnetic vibronic moment can exist in case of orbital degeneracy and pseudo degeneracy if the local time inversion center is absent.
From the macroscopic (phenomenological) point of view the formation of electric and magnetic dipole moments as a result of a distortion in a system without space and time inversion symmetry correspondingly is a well known phenomenon of piezoelectricity or piezomagnetism.
It is completely clear that if the system is lacking both symmetry operations – space inversion and time inversion – the Jahn-Teller distortion creates both the electric and magnetic dipole moments. This together with the electronic structure of the unit (molecule) could be considered as a guide to a search of molecules with both – electrical and magnetic – moments. In case of crystals containing at least one such a unit sublattice the cooperative Jahn-Teller effect could lead to the formation of multiferroic compounds.
The idea of interrelation of the crystal distortions and magnetoelectricity is not new. The phenomenology of the striction mediated magnetoelectricity was suggested by Mitsek and Smolensky in [3]. The cooperative Jahn-Teller effect based microscopic theory of magnetoelectricity was considered by Kaplan and Vekhter in [4] and later developed by Kaplan in [5, 6]. However on the contrary to the first publications where the electro- and magnetostriction were in the center of the microscopic mechanism discussion, here the microscopics of piezomagnetism – vibronic magnetism – is under consideration.

2. **Microscopic model.**

Below the application of the suggested in the introduction ideas is illustrated for a following microscopic model.
A tetragonal crystal containing a sublattice of ions with the lowest electronic orbital doublet –singlet energy structure is considered. The crystal is magnetically ordered (ferromagnetism is accepted initially for simplicity), the local magnetic moments at $T<T_m$ are parallel to the crystal a-axis (it is accepted that the intersite magnetic interactions for the x- and y-components of the magnetic moments is much bigger than for the z-components).

Journal of Physics: Conference Series **428** (2013) 012034 doi:10.1088/1742-6596/428/1/012034

Among the vibronic interactions the interaction of the electrons with the E-symmetry local distortions and the corresponding monoclinic strains is considered the strongest.
After the standard in the cooperative Jahn-Teller effect theory shift transformation [4] in the second order in vibronic interaction theory of perturbation the electronic part of the crystal Hamiltonian can be presented as

$$\hat{H} = -\Delta\sum_{m}\hat{Q}_m - \sum_{mn} A_{mn}(\hat{P}_{xm}\hat{P}_{xn} + \hat{P}_{ym}\hat{P}_{yn}) - \sum_{mn} J_{mn}(\hat{S}_{xm}\hat{S}_{xn} + \hat{S}_{ym}\hat{S}_{yn}) -$$

$$-h_x\hat{S}_{xm} - h_z\hat{S}_{zm} \qquad (4)$$

In the Hamiltonian (4) Δ is the energy gap between the ground electronic doublet and the exited singlet, A and J are constants of the intersite interactions caused by virtual phonon exchange and magnetic dipole-dipole or exchange interactions correspondingly, h_x and h_z are the components of the external magnetic field. For simplicity the small vibronic reduction of the magnetic interactions is ignored. The electronic operators $\hat{Q}$, $\hat{P}$, and $\hat{S}$ on the given electronic basis are described by the following matrices

$$Q = \begin{pmatrix} 1 & 0 & 0 \\ 0 & 1 & 0 \\ 0 & 0 & 0 \end{pmatrix} \qquad P_x = \frac{1}{\sqrt{2}}\begin{pmatrix} 0 & 0 & 1 \\ 0 & 0 & -1 \\ 1 & -1 & 0 \end{pmatrix} \qquad (5)$$

$$S_x = \frac{1}{\sqrt{2}}\begin{pmatrix} 0 & 0 & 1 \\ 0 & 0 & 1 \\ 1 & 1 & 0 \end{pmatrix} \qquad S_z = \begin{pmatrix} 1 & 0 & 0 \\ 0 & -1 & 0 \\ 0 & 0 & 0 \end{pmatrix}$$

At $T<T_m$ the magnetic phase transition takes place, and $\overline{S}_{xm} = S \neq 0$, $\overline{S}_{ym} = 0$. At the lower temperature $T=T_s<T_m$ the structural phase transition from the tetragonal to the monoclinic phase takes place, and now $\overline{P}_x = P \neq 0$, $\overline{P}_y = 0$. The homogeneous monoclinic crystal strain U_{zx} is proportional to the average of the electronic operator P

$$U_{zx} \sim \overline{P}.$$

In the molecular field approximation the Hamiltonian (4) could be rewritten as

Journal of Physics: Conference Series **428** (2013) 012034 doi:10.1088/1742-6596/428/1/012034

$$\hat{H}_{mol} = -\Delta \sum_m \hat{Q}_m - AP \sum_m \hat{P}_{xm} - H_x \sum_m \hat{S}_{xm} - h_z \sum_m \hat{S}_{zm} \tag{6}$$

where

$$A \equiv \sum_n A_{mn} = \sum_{mn\kappa} \frac{V_{m\kappa}^{x*} V_{n\kappa}^{x} + V_{m\kappa}^{y*} V_{n\kappa}^{y}}{\hbar\omega_\kappa} + g_0^{E^2} \qquad H_x \equiv J\overline{S}_x + h_x \tag{7}$$

The electronic energy spectrum of the crystal is defined by the following equation

$$(E+\Delta)[A^2P^2 + H_x^2 + h_z^2 - E(E+\Delta)] == 2APH_x h_z + \Delta H_x^2 \tag{8}$$

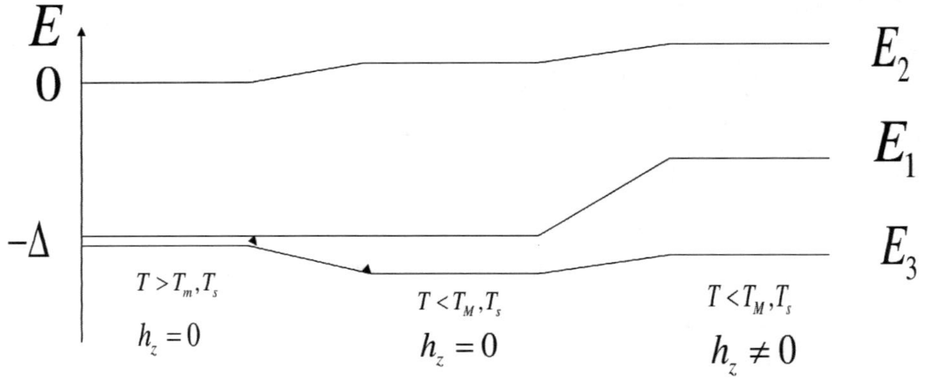

Figure 1. Schematic presentation of the crystal energy levels (8).

If the external magnetic field h_z is always small so that

$$h_z << H_x,\ AP,\ \Delta,$$

the energy levels could be found analytically in the first order in h_z of the perturbation theory approximation

$$E_1 \approx -\Delta + \frac{2APH_xH_z}{E_0^2}$$

$$E_2 \approx -\frac{\Delta}{2}(1-\sqrt{1+a}) + \frac{APH_xH_z}{E_0^2 + \frac{\Delta^2}{4}(1+\sqrt{1+a})} \qquad (9)$$

$$E_3 \approx -\frac{\Delta}{2}(1+\sqrt{1+a}) + \frac{APH_xH_z}{E_0^2 + \frac{\Delta^2}{4}(1-\sqrt{1-a})}$$

$$E_0 \equiv \sqrt{A^2P^2 + H_x^2}, \qquad a \equiv \frac{4E_0^2}{\Delta^2}$$

In the paramagnetic and paradistortive crystal phase, at $T \geq T_m > T_s$, P=0, $H_x = h_z = 0$ the energy level correspond to the accepted initially doublet-singlet structure

$$E_1 = -\Delta$$

$$E_2 = 0 \qquad (10)$$

$$E_3 = -\Delta.$$

If the crystal temperature T is below, but close to $T_m \approx T_s$, so that a<<1, $h_z << AP,\ H_x,\ \Delta$, the energy levels are described by the equations

$$E_1 \approx -\Delta + \frac{2APH_xh_z}{E_0^2}$$

$$E_2 \approx \frac{E_0^2}{\Delta} + \frac{2APH_x h_z}{3E_0^2 + \Delta^2} \tag{11}$$

$$E_3 \approx -\Delta(1+\frac{a}{2}) + \frac{2APH_x h_z}{3E_0^2}$$

From the formulae (8)-(11) it is clear that the mixture of the electronic states allowing the existence of the vibronic magnetic moment is resulting from the interference of the Jahn-Teller molecular field AP and the magnetic molecular (or/and external) field. This leads to possible specific dynamics of the systems under consideration that will be discussed later.
Using the energy spectrum it is easy to calculate the homogeneous strain and the induced by the Jahn-Teller distortion average magnetic moment $M_z = g\beta\,\bar{S}_z$. Focusing on the vibronic magnetic moment we get

$$M_z = \frac{g_z\beta}{Z_0}\frac{2A\bar{P}H_x}{E_0^2}\left(e^{\Delta/kT} + \frac{1}{3}e^{\Delta/kT(1+\frac{2E_0^2}{\Delta^2})} + \frac{E_0^2}{3E_0^2+\Delta^2}e^{-\frac{E_0^2}{kT}}\right) \tag{12}$$

$$Z_0 = e^{\frac{\Delta}{kT}}\left(1 + e^{\frac{4E_0^2}{\Delta kT}}\right) + e^{-\frac{E_0^2}{\Delta kT}} \tag{13}$$

Formula (13) demonstrates the vibronic origin of the magnetic moment

$$M_z \equiv M_{zvibr} \sim A \sim \frac{V^2}{\hbar\omega} \,. \tag{14}$$

3. Piezomagnetism, magnetic reorientation phase transitions, and multiferroicity.

The magnetic moment M_z becomes different from zero only at $T<T_S$, when the structural ordering of the crystal takes place - $\bar{P} \neq 0$. As the homogeneous U_{zx} strain is proportional to $\bar{P}$, the magnetic moment is linearly proportional to the homogeneous crystal strain – piezomagnetic effect. In other words the vibronic magnetism provides us with a microscopic mechanism of piezomagnetism with a piezomagnetic coefficient proportional to the square of the electron-phonon interaction

constant. It is clear that for this mechanism of piezomagnetism the important role belongs to the cooperative Jahn-Teller effect.
With the lowering the temperature of the structurally ordered crystal phase the spontaneous monoclinic strain is increasing and so is the vibronic magnetic moment M_z. The total magnetic crystal moment that is the vector sum of the "regular" magnetic moment $\overrightarrow{M_x}$ and the vibronic moment $\overrightarrow{M_z}$ is deviating more and more from the initial a-axis direction perpendicular to the tetragonal z-axis of the crystal. This rotation of the spontaneous crystal magnetic moment is typical for the magnetic reorientation phase transitions. But in the case under the discussion the magnetic reorientation critical temperature T_{mreor} is actually coincides with the critical T_s temperature of the structural Jahn-Teller phase transition.
The angle of the magnetic moment rotation θ is described by the formulae

$$\tan\theta = \frac{\overline{S}_z}{\overline{S}_x} = \frac{AP\Delta}{E_0^2}\frac{F_1(A,J,h_z,T)}{F_2(A,J,H_x,T)}$$

$$F_1 = e^{\frac{\Delta}{kT}}(1+\frac{1}{3}e^{\frac{2E_0^2}{\Delta kT}}) + \frac{E_0^2}{3E_0^2+\Delta^2}e^{-\frac{E_0^2}{\Delta kT}} \tag{15}$$

$$F_2 = 2e^{\frac{\Delta}{kT}(1+\frac{2E2_0}{\Delta^2})} - e^{-\frac{E_0^2}{\Delta kT}}$$

As it follows from the equation (15) the tangent of the rotation angle is linearly proportional to the Jahn-Teller molecular field A.
If the Jahn-Teller crystal structural units or molecules are lacking the local center of space inversion simultaneously with the center of time inversion, the local distortions simultaneously are creating the vibronic magnetic and electric dipole moments. In this situation the cooperative Jahn-Teller effect is responsible for generating a magnetoelectric crystal state – Jahn-Teller multiferroic. Such a situation is experimentally observed in the tetragonal $TbPO_4$ crystals where the local symmetry of the Tb^{3+} ions is D_{2d}, and the lowest electronic energy levels are a doublet and a singlet separated by a 2.2 cm^{-1} gap. The difference between this real crystal situation and the microscopic model considered above is the antiferromagnetic (not ferromagnetic as in Hamiltonian (4)) ordering. This does not change qualitatively the results of the discussion, but below the structural phase

transition the crystal is ferroelastic, antiferroelectric, antiferromagnetic, not ferromagnetic.

In the $TbPO_4$ crystal J=-1.5 K, A=3.1 K and below the critical structural transition temperature T_s=2.15 K (T_N=2.28 K) there are 6 order parameters: $\overline{S_{z\alpha}}$, $\overline{S_{x\alpha}}$, and $\overline{P_\alpha}$. At $T<T_s$ each local monoclinic distortion creates an electric dipole moment along the crystal y-axis. However in the two crystal sublattices the dipole moments are antiparallel, so that the total spontaneous electric dipole moment is zero, and the total crystal cell with two Jahn-Teller centers in it remains centrosymmetric. It is easy to see that the external electric field E along the y-axis disturbs the equivalence of the sublattices making the magnetoelectric coefficient

$$\alpha_{xy} \equiv \frac{\partial M_x}{\partial E_y} \neq 0$$

at $T_s<T<T_N$. However at these temperatures naturally $\alpha_{zy} \equiv \frac{\partial M_z}{\partial E_y} = 0$. At the same time at the temperatures below the structural phase transition both coefficients α_{xy} and α_{zy} are not zero. The results of the calculations are shown on the Figure 2.

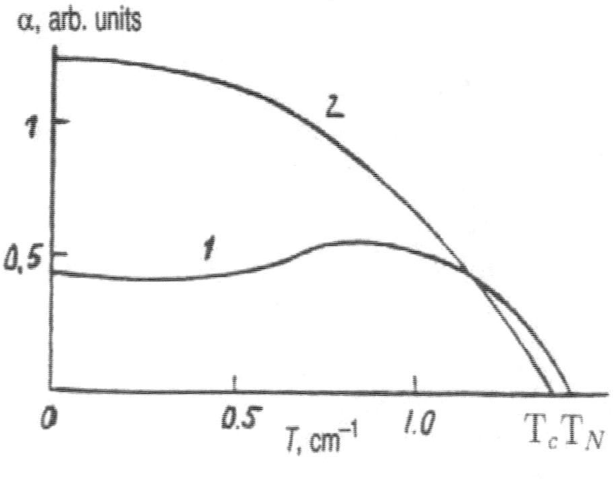

Figure 2. Magnetoelectric coefficients in $TbPO_4$ crystals as a function of the temperature. 1 - α_{xy}; 2 - α_{zy}

Journal of Physics: Conference Series **428** (2013) 012034 doi:10.1088/1742-6596/428/1/012034

The peculiarity of the Jahn-Teller multiferroics is related to the presence of the third – elastic – type of ordering additionally to magnetic and electric dipole moment orderings. As the elastic ordering is resulting from the cooperative Jahn-Teller effect, the virtual phonon exchange becomes fundamentally important for the effectiveness of the coupling between the magnetic and electric dipole moments. This leads to a very specific electron correlation in the systems under discussion. The electron correlation depends not only upon the magnetic moment correlation, but additionally upon the local distortion correlation. For example, in the considered above piezomagnetic Jahn-Teller multiferroics the electron correlation radius is the result of interference of the magnetic correlation radius and the virtual phonon exchange correlation radius. The relative contribution of the acoustic and optical phonons to the properties of the multiferroics becomes of big interest. As the experimental measurements of the electron correlation radius are possible (for example by the neutron scattering) there is a fundamentally important possibility to separate different mechanisms of multiferroicity. In part there is a hope to separate the crystal lattice mediated mutiferroicity from the one induced by helicoidal magnetic structure.

The presence of three types of ordering in the Jahn-Teller multiferroics opens a variety of possibilities to dramatically change the state of the magnetic moments by the external electric field or the state of the electric dipole moments by the external magnetic field. For simplicity it is convenient once again to consider the ferroelastic antiferroelectric antiferromagnetic Jahn-Teller crystal of $TbPO_4$ type.

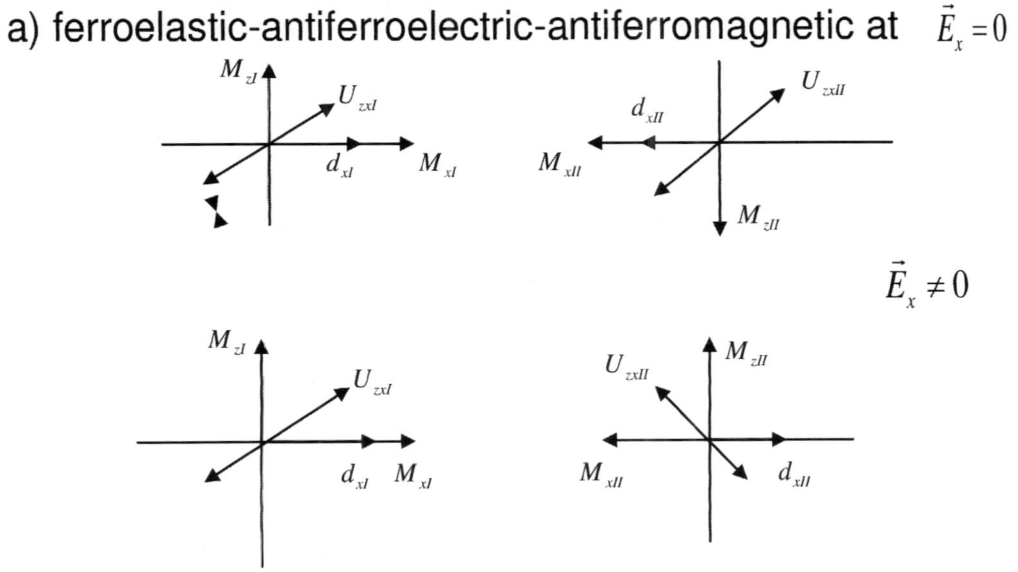

Figure 3. Jahn-Teller multiferroics in external electric field.

As it is seen from the Fig.3 at the strong enough external electric field at $T<T_s,T_N$ the electric dipole moments in one of the sublattices are overturned so that the transition from the antiferroelectric to the ferroelectric crystal state takes place. But this is accompanied by the structural transition from the ferroelastic to antiferroelastic state as in each of the sublattices the local strains and dipole moments are linearly proportional. At the same time due to piezomagnetism the overturn of the sublattice strain leads to the overturn of the vibronic magnetic moments so that the crystal is transferring from the antiferromagnetic state to the ferromagnetic one.
It is clear that in Jahn-Teller multiferroics different interesting coupled metaelastic, metaelectric, and metamagnetic phenomena could be experimentally observed [7-11].

Conclusions.

1. The mechanism of vibronic magnetism is suggested.
2. It is shown that the Jahn-Teller effect could be considered as a basic microscopic mechanism of piezomagnetism.
3. The cooperative Jahn-Teller effect can explain the magnetic reorientation phase transitions in crystals.
4. The cooperative Jahn-Teller effect provides a microscopic model for multiferroicity. The specifics of Jahn-Teller multiferroics leads to an unusual electron correlation – interference of the magnetic and virtual phonon exchange interactions - that can be observed experimentally.
5. Crystal lattice vibrations mediate the strong coupling between the electric and magnetic dipole moments.

References.

[1] Bersuker I 1966 Phys Lett **20** 589
[2] Vekhter B 1969 Proc. Second All-Union Conf. on Solid State Phys. Moscow p.49
[3] Mitsek I and Smolenskii G 1963 Sov Phys Solid State **4** 2620
[4] Kaplan M, Vekhter B 1995 Cooperative Phenomena in Jahn-Teller Crystals (New York and London, Plenum Press)
[5] Kaplan M 2006 Proc. of Int. Sugano Symposium on Left-Right Symmetry in Physics, Chemistry, and Biology, Tsukuba p.10
[6] Kaplan M 2009 The Jahn-Teller Effect, Eds. Koppel H, Yarkony D, Barentzen, (Heidelberg, Dordrecht, London, New York, Springer) p.653

[7] Kaplan M 1984 Sov. Phys. Solid State **26** 51

[8] Nojiri H, Kaneko K, Motokawa M, Hirota K, Endoh Y, Takahashi K 1999 Phys. Rev B **60** 4142

[9] Kaplan M, Zimmerman G 2008 Phys. Rev. B **77** 104426

[10] Kimura T, Goto T, Shintani H, Ishizka K, Arima T, Tokura Y 2003 Nature (London) **426** 55

[11] Kaplan M, Zimmerman G 2012 Amer. Phys. Soc. March Meeting, Portland, Session W16

Journal of Physics: Conference Series **428** (2013) 012035 doi:10.1088/1742-6596/428/1/012035

Ab initio DFT study of magneto-structural correlation of dinuclear mixed-valence Mn complexes

Shusuke Yamanaka[1], Kyohei Komi[1], Koki Ueda[1], Yasutaka Kitagawa[1], Takashi Kawakami[1], Mitsutaka Okumura[1], Haruki Nakamura[2], Kizashi Yamaguchi[3]

1. Graduate School of Science, Osaka University, Toyonaka, Osaka, Japan.

2. Protein Institute, Osaka University, Suita, Osaka, Japan.

3. Institute for nanoscience design, Osaka University, Toyonaka, Osaka, Japan.

E-mail: syama@chem.sci.osaka-u.ac.jp

Abstract. Recently, magnetism of Mn complexes has attracted much attention in relation to the metalloenzymes, in which Mn ions play important roles in the catalytic abilities. The magnetism of them are determined by structures of the Mn complexes part of the enzymes. In particular, the ferromagnetic and antiferromagnetic interactions usually compete with each other for mixed valence states. In this study, we investigated the magneto-structural correlation of a simple mixed valence complex, Mn(II)(μ-O)Mn(III)$(H_2O)_{10}$ using ab initio density functional method. Describing the landscape of magnetism of this complex for Mn-O-Mn plane, it was found that the spin-state transition occurs near the most stable structure of this complex. We consider why the spin-state transition occurs in the landscape from the viewpoint of natural orbital pictures.

1. Introduction

Recently, magnetism of Mn complexes has attracted much attention in relation to several important enzymes such as oxygen-evolving complex (OEC) in photosystem II [1-9], catalase[10-13], and superoxide dismutase (SOD)[14], in which Mn ions play important roles in the catalytic abilities of these enzymes. One important feature is that the reaction centers of these enzymes contain two or four Mn ions, of which oxidation states could vary during the reaction steps: even if we focus on two centers, there are many common oxidation states, $Mn(II)_2$, Mn(II)Mn(III), $Mn(III)_2$, Mn(III)Mn(IV), and $Mn(IV)_2$. In the models of some oxidation steps for the OEC, Mn(II)Mn(IV) and Mn(IV)Mn(V) states are also involved [4]. In the reaction centers of these enzymes, it is most likely that changes of structures and oxidation states yield Brønsted-Lowry acids and/or bases that catalyze the reactions [15,16]. Oxidation states, structures, and magnetism of Mn complexes are expected to strongly correlate with each other [17-20]. In particular, the coupling between magnetism and structures could be complicated for the mixed valence states, since double-exchange and superexchange interactions [21] compete with each other: magnetism is determined by a balance between them, which must be determined by the structure of the Mn complexes. In fact, we found that the low-lying spin-structures are nearly degenerate for the X-ray structure of OEC in previous studies [3,4], implying that correlations between magnetism and structures (magneto-structural correlations) should be taken into account for predictions or simulations of the enzyme.

Thus, we intend to investigate magneto-structural correlations of Mn complexes. For this purpose, we examined a Mn(II)(μ-O)Mn(III)$(H_2O)_{10}$ as a simple Mn complex. We calculated energy landscapes of both lowest-spin (LS) and highest-spin (HS) states for various structures of the Mn-O-Mn core. Then the landscapes of magnetic interactions are calculated, clarifying that the spin-state transition occurs near the most stable point in the LS energy surface. We analyzed the reason for the changes of stable spin states.

2. Computational details

The ab initio density functional theory (DFT) method, which always yields correct results for both stable geometries and electronic structures of manganese complexes, has not been established so far. However, we previously confirmed [22] that the B3LYP functional [23] qualitatively reproduces the magnetism of dinuclear Mn complexes, except the weak ferro and weak antiferromagnetic ones: if the absolute values of magnetic interactions of Mn complexes are more than 10 cm^{-1}, the magnetism estimated with using B3LYP were found to be consistent with their experimental values within ±25% for 16 Mn complexes we examined [22].

The B3LYP functional was reported to reproduce the experimental molecular geometries of Mn complexes when we employed LACVP* basis set [20]. Thus we employed B3LYP and LACVP* respectively for the functional and the basis set in the following all calculations. All ab initio DFT calculations presented in the next section were done with using Gaussian 09 package [24]. The UltraFine grid is used for the numerical integration scheme for DFT terms. In addition, we choose the tight convergence criteria for DFT calculations in both electronic structure calculation and geometry optimization procedures. In addition, normal frequency analyses were implemented in order to ensure that the geometries obtained are minima, not saddle points.

3. Results and discussion

We first optimized molecular geometries of Mn(II)(μ-O)Mn(III)$(H_2O)_{10}$. The optimized distances between Mn ions are 3.589 Å and 3.559 Å for high-spin (HS) and low-spin (LS) states, respectively. The angle, Mn-O-Mn are also similar, which are 150.136 and 148.483 degrees for HS and LS states respectively, indicating that there is not a significant difference in the optimize geometries between HS and LS states. This fact is reasonable from the viewpoint of Feynman's electrostatic theorem [25], which ensures that the forces and Hessians on nuclei of molecules are exerted only by other nuclei and by the electrons' charge distribution. Thus, at the first approximation, it might imply that we could reasonably optimize geometries of Mn complexes with assuming the highest-spin state of the Mn complexes, which in fact have been often employed by the precursors who have investigated the molecular structures of Mn complexes [5,11]. However, strictly speaking, the spin state of the computational targets affects on both the charge distribution and energies due to exchange and correlation effects, resulting in the difference of optimized geometries between spin states.

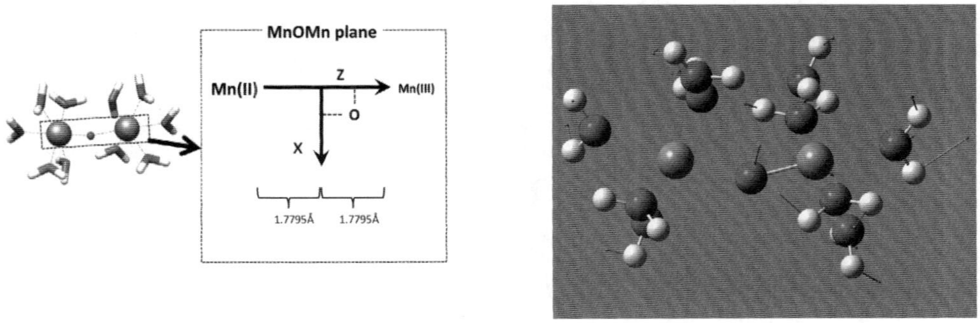

Figure 1 (A) The coordinate defined for the Mn-O-Mn plane. (B) The displacement vectors of atoms of the normal mode, of which the amplitudes of Mn ions are largest among those of all normal modes.

Thus, focusing on the lower-lying low-spin (LS) state, we illustrated the energy landscape for various Mn-O-Mn geometries around the most stable geometry of the LS state. To this end, we performed partial optimizations of Mn(II)(μ-O)Mn(III)$(H_2O)_{10}$ with fixing the X and Z coordinates of two Mn ions and the μ-oxygen dianion, which are indicated in figure 1(A). In order to plot the energy landscape, this partial optimization procedure is implemented for the various coordinates (X,Z) of the μ-oxygen dianion: X is from 0.0 to 1.0 Å at 0.05 Å intervals, and Z from 0.00 to 0.25 Å at 0.05 Å intervals. On the other hand, the interatomic distance between two Mn ions is fixed to be the value at equilibrium geometry, 3.559 Å. To validate this partially optimization, we checked the vibration amplitudes of all 93 normal modes at the equilibrium geometry.

Figure 1(B) shows the displacement vectors of a normal mode, in which the amplitudes of the Mn ions are largest among 93 normal modes. The frequency and the intensity of this mode are 203.96 cm^{-1} and 4.16, respectively. We can see from this figure that the Mn ions do not move so much, being consistent with the fact that the mass of the Mn atom is three times as heavy as that of the oxygen atom.

The calculated energy landscape is illustrated in figure 2 (A). The equilibrium point lies at (X,Z)=(0.499, 0.164) (unit:Å). At this point, the μ-oxygen dianion leans towards Mn(III), which is reasonable from the viewpoint of simple electrostatic interactions. As shown in the optimized geometry of figure 1(B), there are two hydrogen bonds between water molecules, causing the bending Mn-O-M line. In fact, the energy of the most stable point within the straight Mn-O-Mn geometries is higher than the equilibrium point by approximately 10 kcal/mol, which is approximately equal to the stabilization energies due to two hydrogen bonds. To check the behavior of magnetism for this landscape, we also calculated the energies of highest spin (HS) state with fixing the optimized geometries obtained for the LS state. The energy landscape of the high-spin state, which is colored in light red, is illustrated in figure 2(B). Comparing figure 2(B) with figure 2(A), we find that the landscapes of the HS state is quite similar to that of the low-spin state, confirming that the Feynman's electrostatic theorem holds for this system. However, it is also true that there are slight adiabatic energy differences between the LS and HS surfaces. This could be clearly shown in figure 2(C), in which we superimposed LS and HS surfaces in the same coordinate. As can be seen in this figure, there is a phase-transition line near the equilibrium point.

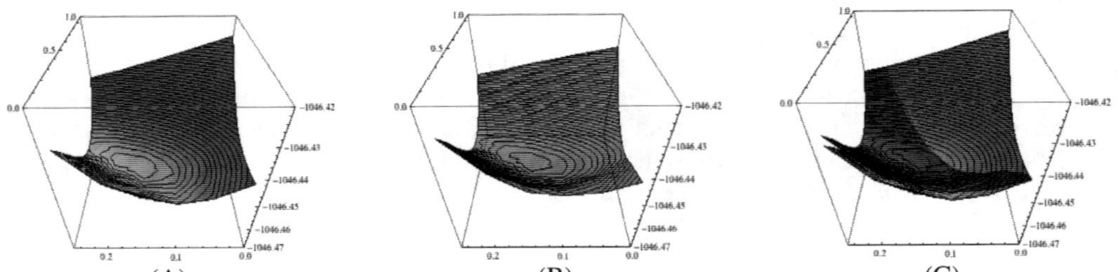

Figure 2 The energy landscape of the lowest spin state (A), and the corresponding highest spin state (B), and the superimposed landscapes of LS and HS states (C).

In order to describe the magneto-structural correlation more explicitly, we estimated the magnetic interactions using the Heisenberg model,

$$H = -2J\hat{\mathbf{S}}_1 \cdot \hat{\mathbf{S}}_2, \tag{1}$$

which is usually employed to fit magnetic susceptibility measurements to determine the magnetic interaction, J, experimentally. In order to estimate J theoretically, we use the equation,

$$J = \frac{E^{LS} - E^{HS}}{\left\langle \hat{\mathbf{S}}^2 \right\rangle^{HS} - \left\langle \hat{\mathbf{S}}^2 \right\rangle^{LS}}, \tag{2}$$

which can be derived from equation (1) with assuming that the magnitudes of spins at the Mn(II) and Mn(III) sites are equal each other [26]. Here E^X and $\left\langle \hat{\mathbf{S}}^2 \right\rangle^X$ are, respectively, energy and expectation value of square of the total spin for the X state calculated with using the B3LYP/LACVP* method. The landscape of the magnetic interaction J on the (X, Z) plane is illustrated in figure 3. Now there is an obvious line where the magnetic transition occurs.

To consider the reason why antiferromagnetic and ferromagnetic states become stable for these two regions, we picked out two optimized geometries for (X, Z) = (0.0, 0.25) and (0.5,0.0), and implemented natural orbital analysis for low-spin solutions at these two geometries. Figure 4 illustrates natural orbitals (NOs), together with occupation numbers of NOs. F1 and A1 are singly occupied molecular orbitals (SOMO) of the LS solutions for FM and AFM regions, respectively. It is obvious that the F1 localizes the left hand site, while the A1 delocalizes over two Mn sites.

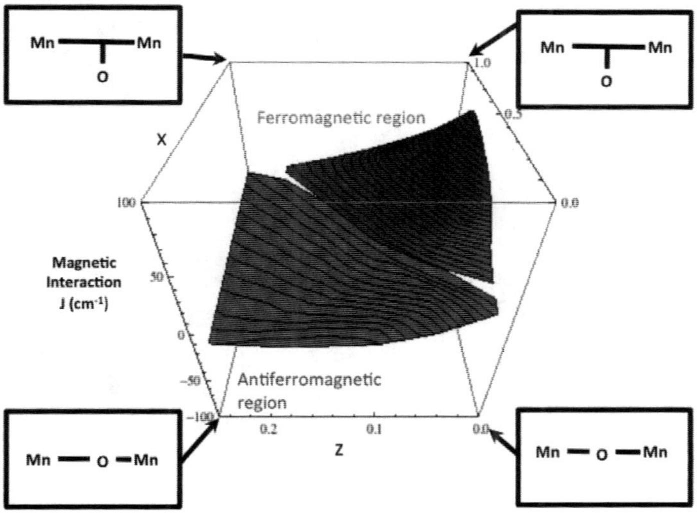

Figure 3 The landscape of magnetic interaction, J. The coordinates are those indicated in figure 1(A).

Occupation numbers (ONs) of natural orbitals, (F2)~(F5) are 1.002,1.046, 1.191, and 1.221 respectively. The corresponding ONs of (A2)-(A5) are larger than those of (F2)-(F5). Here we would like to discuss the implications of the NOs and ONs [4]. In figure 4, we only show the bonding type of natural orbitals, $\{\phi_i^b\}$, of which occupation numbers (ONs) $\{n_i^b\}$ are larger than 1.0. Correspondingly, there are the antibonding type of NOs, $\{\phi_i^a\}$, of which ONs are given by, $\{n_i^a\}$ with $n_i^a = 2.0 - n_i^b$. The relation between NOs and spin orbitals , $\{\psi_i^{\pm}\}$, is given by,

$$\psi_i^{\pm} = \cos\theta_i^{\pm}\,\phi_i^b + \sin\theta_i^{\pm}\,\phi_i^a, \tag{3}$$

where + and— indicate α and β spins, respectively. Further the occupation number of the bonding NO, ϕ_i^b, is related to the overlap integral between ψ_i^+ and ψ_i^- as follows:

$$n_i^b = \langle \psi_i^+ | \psi_i^- \rangle + 1.0. \tag{4}$$

Equation (4) implies that, as the occupation number of bonding type of NO increases, the overlap between α and β orbitals increases, resulting in the thicker superexchange path. A noteworthy point is that this holds only for active orbitals, not for closed shell orbitals. Of course, there are bonding natural orbitals, of which occupation numbers are nearly equal to 2.0, implying that the orbitals are closed-shell ones, which do not contribute to magnetism of the Mn complex. The key point to increase the overlaps is the relative position among Mn, O, and Mn ions. It is known that p orbitals at the oxygen site intermediate the transfers between two Mn ions, enhancing the orbital overlap between two Mn sites [21]. A bend of the Mn-O-Mn deteriorates this effect, resulting in small orbital overlaps. In fact, there are little amplitude at the oxygen site in (F2) and (F3), corresponding to the localization of the d orbitals at two Mn sites. This NO analysis clarifies that the landscape of the magnetic interaction of this manganese complex exemplifies the Kanamori-Goodenough rule [22,23] on the basis of ab initio calculational results.

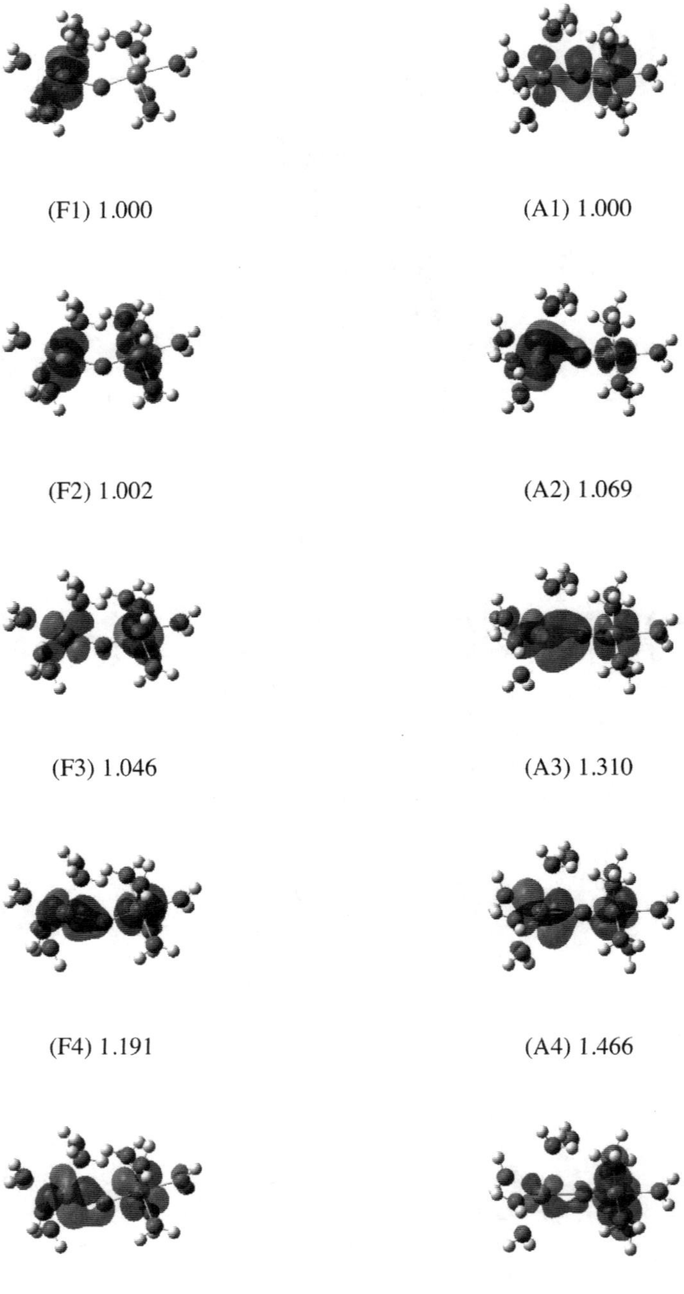

(F1) 1.000 (A1) 1.000

(F2) 1.002 (A2) 1.069

(F3) 1.046 (A3) 1.310

(F4) 1.191 (A4) 1.466

(F5) 1.221 (A5) 1.502

Figure 4 Natural orbitals (NOs) of LS solutions for (X, Z) = (0.5, 0.0) (F1~F5) and (X, Z) = (0.0,0.25) (A1~A5). The numbers indicated below the orbitals are occupation numbers of the NOs.

It might be valuable to comment upon the universality of the calculated results we showed, because it is also known that the localizability and delocalizability, and so calculated magnetism, is strongly affected by the DFT functional we employ [29]. As described above, we previously showed that the B3LYP functional often fails to describe the weak magnetic interactions for dinuclear Mn complexes, but it is also found to be reliable for magnetism of Mn complexes, of which magnitudes are larger than 10 cm^{-1}[22]. Thus, the borderline between ferro and antiferromagnetic regions could slightly be shifted if we used other functional, but we can expect that the other qualitative features of this landscape of magnetism hold for other functionals.

4. Conclusion

In this study, we examined magneto-structural correlations of a simple dinuclear Mn complex with calculating the energy landscape and the magnetic interaction landscape. It is expected that this type of landscapes provide a useful guideline to understand and/or to predict the magneto-structural correlation of Mn complexes. At the next step, we will study real synthesized complexes as well as the cores of metalloenzyems with using this approach. For instance, a typical high-valent di-μ-O complex, $[Mn(IV)_2(\mu\text{-}O)_{2\text{-}n}(\mu\text{-}OH)_n(salpn)_2]^{n+}$ (n=0,1,2) is an interesting target: the deprotonated version (n=0) exhibits the catalase activity, and the protonation to this complex deteriorates its activity [13,30]. It was also reported that the J value decreases from -92 cm^{-1} for n=0 to -48 cm^{-1} for n=1 and to -6 cm^{-1} for n=2 as the successive protonations occur. The oxidation state of the synthesized complex is only $Mn(IV)_2$ so far, but the computational study using the landscape approach with assuming any of various oxidation states, Mn(II)Mn(III), $Mn(IV)_3$ and Mn(III)Mn(IV), is possible, by which we can examine effects of oxidation, protonation, and structural changes. We expect that such typical cases exemplify the correlation among magnetism, protonation modes, and structural changes for the Mn complexes.

References

[1] Umena Y, Kawakami K, Shen J-R and Kamiya N 2011 Nature 473 55-61.
[2] Ames W, Pantazis D A, Krewald V, Cox N, Messinger J, Lubitz W and Neese F 2011 J. Am. Chem. Soc. 133 19743–19757.
[3] Kanda, K, Yamanaka S, Saito T, Umena Y, Kawakami K, Shen J-R, Kamiya N, Okumura M, Nakamura H and Yamaguchi K 2011 Chem. Phys. Lett. 506 98-103.
[4] Yamanaka S, Kanda K, Saito T, Umena Y, Kawakami K, Shen J-R, Kamiya N, Okumura M, Nakamura H and Yamaguchi K, Adv. Quantum Chem. in press.
[5] Lundberg M and Siegbahn P E 2004 Phys. Chem. Chem. Phys. **6** 4772-4780.
[6] Rivalta I, Amin M, Luber S, Vassiliev S, Pokhrel R, Umena Y, Kawakami K, Shen J-R, Kamiya N, Bruce D, Brudvig G W, Gunner M R, Batista V S 2011, Biochemistry **50** 6312–6315.
[7] Pantazis D A, Orio M, Petrenko T, Zein S, Lubitz W, Messinger J and Neese F 2009 Phys. Chem. Chem. Phys. **11** 6788-6798.
[8] Galstyan A, Robertazzi A and Knapp E W 2012 J. Am. Chem. Soc. 134 7442–7449.
[9] Yamanaka S, Kanda K, Saito T, Kitagawa Y, Kawakami T, Ehara M, Okumura M, Nakamura H, Yamaguchi K, Chapter 25 in Quantum Systems in Chemistry and Physics: Progress in Methods and Applications, Eds by Nishikawa, K. Springer Science+Business Media Dordrecht, in press.
[10] Michaud-Soret I, Jacquamet L, Debaecker-Petit N, Le Pape L, Barynin V V and Latour J-M 1998 Inorg. Chem. 37 3874-3876.
[11] Siegbahn P E M 2001 Theor Chem Acc 105 197-206.
[12] Wu A J, Penner-Hahn J E and Pecoraro V L 2004 Chem. Rev. 104 903-938.
[13] Larson E J, Riggs PJ, Penner-Han J E and Pecorro V L 1992 J. Chem. Soc. Chem. Commun. 116 102-103.
[14] Srnec M, Aquiilante F, Ryde U, and Rulisek L 2009, J. Phys. Chem. B 113 6074-6086.
[15] Lippard S J, Berg J M 1994 Principles of Bioinorganic Chemistry (Mill Valley, CA: University Science Books).
[16] Solomon E I, Scott R A, King R B Eds 2009 Computational Inorganic and Bioinorganic Chemistry. (New York:Willey).
[17] Barone V, Bencini A, Gatteschi D and Totti F 2002 Chem Eur J **8** 5019-5027.
[18] Katsuda M, Mitani M and Yoshioka Y 2012 J Biophys Chem **2** 111-126.
[19] Mitani M, Wakamatsu Y, Katsurada T and Yoshioka Y 2006 J. Phys. Chem. A **110** 13895-13913.

[20] Sproviero E M, Gascon J A, McEvoy J P, Brudvig G W and Batista V S 2006 J Inorg Biochem **100** 786–800.

[21] Mattis D C, 2006, The theory of magnetism made simple (World Scientific Publishing Co. Singapole).

[22] Yamanaka S, Kanda K, Saito T, Kitagawa Y, Kawakami T, Ehara M, Okumura M, Nakamura H, Yamaguchi K 2012 Chem. Phys. Lett. **519-520** 134-140.

[23] Becke A D 1993 J. Chem. Phys. **98** 5648-5652.

[24] Frisch MJ, Trucks GW, Schlegel HB, Scuseria GE, Robb MA, Cheeseman JR, Scalmani G, Barone V, Mennucci B, Petersson GA, Nakatsuji H, Caricato M, Li X, Hratchian HP, Izmaylov AF, Bloino J, Zheng G, Sonnenberg JL, Hada M, Ehara M, Toyota K, Fukuda R, Hasegawa J, Ishida M, Nakajima T, Honda Y, Kitao O, Nakai H, Vreven T, Montgomery J, J. A., Peralta JE, Ogliaro F, Bearpark M, Heyd JJ, Brothers E, Kudin KN, Staroverov VN, Kobayashi R, Normand J, Raghavachari K, Rendell A, Burant JC, Iyengar SS, Tomasi J, Cossi M, Rega N, Millam NJ, Klene M, Knox JE, Cross JB, Bakken V, Adamo C, Jaramillo J, Gomperts R, Stratmann RE, Yazyev O, Austin AJ, Cammi R, Pomelli C, Ochterski JW, Martin RL, Morokuma K, Zakrzewski VG, Voth GA, Salvador P, Dannenberg JJ, Dapprich S, Daniels AD, Farkas Ö, Foresman JB, Ortiz JV, Cioslowski J, Fox DJ. Gaussian 09. Revision C Wallingford CT: Gaussian, Inc. 2010.

[25] Feynmann R. 1939 Phys. Rev. 56 340-343.

[26] Yamanaka S, Kawakami T, Nagao H, Yamaguchi K 1994 Chem. Phys. Lett. **231**, 25-33.

[27] Goodenough J B 1958 J. Phys. Chem. Solids **6**, 287-297.

[28] Kanamori J 1958 J. Phys. Chem. Solids **10**, 87-98.

[29] Cohen A J, Mori-Sanchez P, Yang W, 2008 Science **321** 792-794.

[30] Baldwin M J, Stemmler T. L Xia YM, Riggs-Gelasco P J, Kirk M L, Penner-Han J E, Pecorro V L 1994 J Am Chem Soc **116** 11349-11356.

Journal of Physics: Conference Series **428** (2013) 012036 doi:10.1088/1742-6596/428/1/012036

Applicability of approximated effective Hamiltonian method to first-principle study on magnetic clusters

Hidenori Suzuki and Chikatoshi Satoko

Department of Integrated Sciences in Physics and Biology, College of Humanities and Sciences, Nihon University, 3-25-40, Sakurajosui, Setagaya-ku, Tokyo, 156-8550, Japan.

E-mail: hsuzuki@phys.chs.nihon-u.ac.jp

Abstract. In the first-principle study on the magnetic interactions in transition metal dimer complexes, we have shown that the effective Hamiltonian method can be effectively utilized (*J. Phys.: Condens. Matter* **23** 375502). To extend our method to clusters with higher nuclearity, some approximations are needed since the number of the parameters of the effective Hamiltonian is too large. In the present paper, how to decrease the number of parameters of our effective Hamiltonian and how to determine the parameters from the calculations based on the density functional theory are shown. Then, approximated Hamiltonians are examined for its applicability in the first-principle study on the magnetic interactions in transition metal clusters. From the calculations of the $Ti_2Cl_9^{3-}$ cluster whose parameters of the effective Hamiltonian are already determined, it is found that small differences caused by the symmetry of the each orbital can be neglected and not only kinetic exchange but also potential exchange terms are needed for quantitative calculation of the magnetic susceptibility. The appropriately approximated Hamiltonian reproduces the global structure of the lower energies so as to realize the nature of the magnetic interaction.

1. Introduction

First-principal studies on the magnetic interactions of transition metal complexes have been attracted much interest for long years. In the case of orbitally non-degenerate metal ions, the density functional theory (DFT) is often used to calculate the coupling constant of the isotropic Heisenberg model Hamiltonian. For example, the broken symmetry approach [1–7] and the DFT-based ligand field model method [8–11] are known as methods to obtain the coupling constant. Conversely, for orbitally degenerate metal systems, quantitative study of the magnetic properties becomes complicated because the Hamiltonian of the system must be written by not only spin operators but also operators acting within the orbital spaces.

Recently, we have investigated the magnetic interactions in the orbitally degenerate transition metal dimer complexes [12,13]. In our calculations, an effective Hamiltonian method is effectually utilized. Note that the term "effective Hamiltonian" in the present paper is used in the extensive interpretation as have been mentioned in ref. [12, 13] (In the field of the molecular magnetism, the effective Hamiltonian often means the expression of the Hamiltonian by irreducible tensor operators and spin operators, which has been applied to a kinetic exchange interaction for a pair of orbitally degenerate metal ions [14–20]). By using a set of one-electron localized orbitals as basis of the effective Hamiltonian, we can analyze the magnetic interactions on the basis of the Anderson theory [21]. In the straightforward extension of our methodology to more than trimer complexes, the number of the parameters of the effective Hamiltonian matrix will increase drastically and it becomes laborious to evaluate all parameters by DFT calculations, so that it is needed to approximate the parameters.

In the present paper, we consider how to decrease the number of parameters of the effective Hamiltonian and show how to determine the parameters using DFT calculations. Then we discuss the applicability of approximated Hamiltonians to the investigations of the magnetic interactions between transition metal complexes. For this purpose, we consider the $Ti_2Cl_9^{3-}$ cluster [12, 22, 23] whose parameters of the effective Hamiltonian are already determined in ref. [12]. Then, another simple example that orbitally non-degenerate trimer complex is also presented.

2. Effective Hamiltonian

To investigate the magnetic interactions using the effective Hamiltonian, we need to generate a set $\mathcal{L}$ of the local orbitals by the DFT calculations. However, the DFT calculations are carried out using the delocalized molecular orbitals, and it is hard to directly provide the local orbitals. Therefore, the local orbitals are defined by the linear combinations of molecular orbitals.

Here, for example, we consider the $M_2X_9^{3-}$ type cluster which contains two octahedra sharing a face so that it has D_{3h} cluster symmetry [12]. The molecular orbitals are generated in the average-of-configuration (AOC) calculation [10,11], where the AOC is the spin-restricted system in which all molecular orbitals are equally occupied. Then we obtain six molecular orbitals α_1' and α_2'' which belong to the A_1' and A_2'' terms under D_{3h} cluster symmetry, respectively, and the doubly degenerate molecular orbitals (ξ,η) and (ε,ζ) which belong to the E'' and E' terms, respectively. The set $\mathcal{L}$ of the local orbitals is composed of the following orbital functions:

$$\begin{aligned} a_1 &= \frac{1}{\sqrt{2}}\left(\alpha_1' + \alpha_2''\right), \quad x_1 = \frac{1}{\sqrt{2}}\left(\xi + \varepsilon\right), \quad y_1 = \frac{1}{\sqrt{2}}\left(\eta - \zeta\right), \\ a_2 &= \frac{1}{\sqrt{2}}\left(\alpha_1' - \alpha_2''\right), \quad x_2 = \frac{1}{\sqrt{2}}\left(-\xi + \varepsilon\right), \quad y_2 = \frac{1}{\sqrt{2}}\left(-\eta - \zeta\right), \end{aligned} \tag{1}$$

where the orbital $a_{1,2}$ belongs to the A_1 term and doubly degenerate orbitals $x_{1,2}$ and $y_{1,2}$ belong to the E term under the C_{3v} site symmetry of the cluster. Generally in the cases of the complexes with other symmetries, it is possible to represent the local orbitals by linear combinations of molecular orbitals which is generated by the AOC calculations.

It is assumed that a d-electrons in the transition metal clusters occupies one of the local orbitals $i \in \mathcal{L}$. Then, the effective Hamiltonian can be represented by the following form:

$$\mathcal{H}_{\text{eff}} = \sum_{i,j\in\mathcal{L}}\sum_{\sigma} f_{ij} c_{i,\sigma}^\dagger c_{j,\sigma} + \sum_{i,j,k,l\in\mathcal{L}}\sum_{\sigma,\sigma'} g_{ijkl} c_{i,\sigma}^\dagger c_{j,\sigma'}^\dagger c_{k,\sigma'} c_{l,\sigma} = \mathcal{H}_1 + \mathcal{H}_2, \tag{2}$$

where $c_{i,\sigma}$ and $c_{i,\sigma}^\dagger$ denote annihilation and creation operators of an electron on a local orbital i ($\in \mathcal{L}$) with a spin σ ($= \pm 1/2$). The "effective" Hamiltonian means that the exchange and correlation effects on the electron interaction are renormalized in the elements f_{ij} and g_{ijkl} of the one-body operator $\mathcal{H}_1$ and the two-body operator $\mathcal{H}_2$, respectively. Note that the elements g_{ijkl} are not two-body integrals of $1/|\boldsymbol{r}_1 - \boldsymbol{r}_2|$ because the renormalized Coulomb interaction contains a screening effect.

The number of the parameters of the above Hamiltonians is too large to extend to the complexes of higher nuclearity. In the case of the $Ti_2Cl_9^{3-}$ cluster, the effective Hamiltonian has 30 independent parameters (table 1). The notations of the Coulomb and exchange integrals are summarized in table 2, where the intra-atomic and inter-atomic Coulomb integrals are given by $j_{\phi\phi'} = \int dV_1 dV_2\, \phi_i(\boldsymbol{r}_1)\phi_i'(\boldsymbol{r}_2)\, \mathcal{H}_2\, \phi_i(\boldsymbol{r}_1)\phi_i'(\boldsymbol{r}_2)$ and $J_{\phi\phi'} = \int dV_1 dV_2\, \phi_1(\boldsymbol{r}_1)\phi_2'(\boldsymbol{r}_2)\, \mathcal{H}_2\, \phi_1(\boldsymbol{r}_1)\phi_2'(\boldsymbol{r}_2)$, respectively, and the intra-atomic and inter-atomic

Table 1. Parameters of the effective Hamiltonians.

	$\mathcal{H}_{\text{eff}}$ (ref. [12])	$\mathcal{H}_{\text{eff.1}}$	$\mathcal{H}_{\text{eff.2}}$
one-body energies	ϵ_a, ϵ_e	ϵ_a, ϵ_e	ϵ_a, ϵ_e
transfer integrals	T_{aa}, T^a_{aa} T_{ee}, T^e_{ae},T'_{ee}	t_a t_e	t_a t_e
intra-atomic Coulomb integrals	j_{aa}, j_{ee} j_{ae}, j'_{ee}	j j'	j j'
intra-atomic exchange integrals	k_{ae}, k_{ee}	k	k
inter-atomic Coulomb integrals	J_{aa}, J_{ae}, J_{ee}, $J_{ee'}$	J	J
inter-atomic exchange integrals	K_{aa}, K_{ae}, K_{ee}, $K_{ee'}$	K	0
two-electron hopping terms	F^{22}_{11}, K^{12}_{11}, K^{21}_{11}, K^{22}_{11}, K^{21}_{12} Y^{22}_{11}, X^{12}_{11}, X^{22}_{11}, X^{21}_{12}	0 0	0 0

exchange integrals are given by $k_{\phi\phi'} = \int dV_1 dV_2\, \phi_i(\boldsymbol{r}_1)\phi'_i(\boldsymbol{r}_2)\, \mathcal{H}_2\, \phi'_i(\boldsymbol{r}_1)\phi_i(\boldsymbol{r}_2)$ and $K_{\phi\phi'} = \int dV_1 dV_2\, \phi_1(\boldsymbol{r}_1)\phi'_2(\boldsymbol{r}_2)\, \mathcal{H}_2\, \phi'_2(\boldsymbol{r}_1)\phi_1(\boldsymbol{r}_2)$, respectively.

To decrease the number of the parameters of the effective Hamiltonian, we approximate the parameters on the basis of our previous result [12] as follows: (i) The two-body effects in the transfer integrals are neglected, for example, the values of $T_{aa} = t_{aa} + \int d\tau_1 \int d\tau_2 a_1(\boldsymbol{r}_1)a_1(\boldsymbol{r}_2)\mathcal{H}_2 a_2(\boldsymbol{r}_1)a_1(\boldsymbol{r}_2)$ and $T^a_{ae} = t_{aa} + \int d\tau_1 \int d\tau_2 a_1(\boldsymbol{r}_1)x_1(\boldsymbol{r}_2)\mathcal{H}_2 a_2(\boldsymbol{r}_1)x_1(\boldsymbol{r}_2)$ are equal to t_a. Note that t_a and t_e are discriminated from one-body transfer integrals t_{aa} and t_{ee} because the two-body effects will be renormalized in the evaluation procedure of the parameters. (ii) The intra-atomic integrals are reduced to three kinds of integrals, namely, intra-atomic Coulomb integral between identical orbitals ($j_{aa} = j_{ee} = j$), that of deferent orbitals ($j_{ae} = j'_{ee} = j'$) and intra-atomic exchange integral ($k_{ae} = k_{ee} = k$). This approximation corresponds to the small distortion of the C_{3v} site symmetry from the O_h symmetry. (iii) All inter-atomic Coulomb integrals are equal ($J_{aa} = J_{ae} = J_{ee} = J'_{ee} = J$) and all inter-atomic exchange integrals are equal ($K_{aa} = K_{ae} = K_{ee} = K'_{ee} = K$). The distance between two metals is enough large to neglect small differences caused by the symmetry of the each orbital. (iv) All the other integrals which are related to two-electron hopping terms, for example, $K^{12}_{11} = \int d\tau_1 \int d\tau_2 a_1(\boldsymbol{r}_1)x_1(\boldsymbol{r}_2)\mathcal{H}_2 x_2(\boldsymbol{r}_1)a_1(\boldsymbol{r}_2)$ and $Y^{22}_{11} = \int d\tau_1 \int d\tau_2 x_1(\boldsymbol{r}_1)y_1(\boldsymbol{r}_2)\mathcal{H}_2 y_2(\boldsymbol{r}_1)x_2(\boldsymbol{r}_2)$ are set to zero. Thus the approximated effective Hamiltonian $\mathcal{H}_{\text{eff.1}}$ is written in terms of the creation and annihilation operators as follows:

$$
\begin{aligned}
\mathcal{H}_{\text{eff.1}} = & \sum_{m=1,2}\sum_{\sigma=\pm1/2}\left(\epsilon_a c^\dagger_{am\sigma}c_{am\sigma} + \epsilon_e \sum_{\gamma=x,y} c^\dagger_{\gamma m\sigma}c_{\gamma m\sigma}\right) \\
& + \sum_{\sigma=1/2}\left(t_a c^\dagger_{a1\sigma}c_{a2\sigma} + t_e \sum_{\gamma=x,y} c^\dagger_{\gamma 1\sigma}c_{\gamma 2\sigma} + \text{h.c.}\right) \\
& + \sum_{m=1,2}\left((j-J)\sum_{\gamma=a,x,y} n_{\gamma_m}n_{\gamma_m} + (j'-J')\sum_{\gamma\neq\gamma'} n_{\gamma_m}n_{\gamma'_m}\right) \\
& + k\sum_{m=1,2}\sum_{\gamma\neq\gamma'} c^\dagger_{\gamma m\sigma}c^\dagger_{\gamma' m\sigma}c_{\gamma m\sigma}c_{\gamma' m\sigma} + K\sum_{\gamma,\gamma'=a,x,y}\left(c^\dagger_{\gamma 1\sigma}c^\dagger_{\gamma' 2\sigma}c_{\gamma 1\sigma}c_{\gamma' 2\sigma} + \text{h.c.}\right)
\end{aligned}
\quad (3)
$$

Table 2. Coulomb and exchange integrals between the local orbitals ϕ_k and ϕ'_l.

		ϕ'_l					
		a_1	x_1	y_1	a_2	x_2	y_2
	a_1	j_{aa}	j_{ae}, k_{ae}	j_{ae}, k_{ae}	J_{aa}, K_{aa}	J_{ae}, K_{ae}	J_{ae}, K_{ae}
	x_1	j_{ae}, k_{ae}	j_{ee}	j'_{ee}, k_{ae}	J_{ae}, K_{ae}	J_{ee}, K_{ee}	J'_{ee}, K'_{ee}
ϕ_k	y_1	j_{ae}, k_{ae}	j'_{ee}, k_{ee}	j_{ee}	J_{ae}, K_{ae}	J'_{ee}, K'_{ee}	J_{ee}, K_{ee}
	a_2	J_{aa}, K_{aa}	J_{ae}, K_{ae}	J_{ae}, K_{ae}	j_{aa}	j_{ae}, k_{ae}	j_{ae}, k_{ae}
	x_2	J_{ae}, K_{ae}	J_{ee}, K_{ee}	J'_{ee}, K'_{ee}	j_{ae}, k_{ae}	j_{ee}	j'_{ee}, k_{ee}
	y_2	J_{ae}, K_{ae}	J'_{ee}, K'_{ee}	J_{ee}, K_{ee}	j_{ae}, k_{ae}	j'_{ae}, k_{ae}	j_{ee}

where $n_{\gamma_m} = \sum_{\sigma=\pm 1/2} c^\dagger_{\gamma_m \sigma} c_{\gamma_m \sigma}$.

We also consider another approximated Hamiltonian $\mathcal{H}_{\text{eff}.2}$ which is the Hamiltonian eq.(3) with putting $K = 0$. The inter-atomic exchange integrals K correspond to the potential exchange terms in the magnetic interaction, so that only the kinetic exchange term is contained in $\mathcal{H}_{\text{eff}.2}$.

To determine the matrix elements of the effective Hamiltonian, the DFT-based ligand field model is utilized. By using the energy expectation values $E_{\text{DFT}}(\Psi_i)$ in states Ψ_i which is one of the Slater determinants (SD) of the molecular orbitals, the Coulomb and exchange integrals between two molecular orbitals are able to be determined as shown in ref. [12]. For the effective Hamiltonian without above approximation, $\mathcal{H}_{\text{eff}}$, the all parameters are explicitly related to the Coulomb and exchange integrals between the molecular orbitals. Therefore, in ref. [12], at first we have determined the Coulomb and exchange integrals between the molecular orbitals by using the least square fitting, and then we have obtained the parameters of the effective Hamiltonian. Conversely in the cases of the approximated Hamiltonians $\mathcal{H}_{\text{eff}.1}$ and $\mathcal{H}_{\text{eff}.2}$, the relations between the parameters of the effective Hamiltonians and the integrals between molecular orbitals does not hold. Therefore, the parameters of $\mathcal{H}_{\text{eff}.1}$ and $\mathcal{H}_{\text{eff}.2}$ are needed to directly determine from the least square fitting of the energy expectation values $E_{\text{DFT}}(\Psi_i)$ as follows.

By a unitary transformation U, the basis of the effective Hamiltonian composed of the local orbitals can be transformed to that of molecular orbitals. The diagonal elements of the transformed matrix, $[U\mathcal{H}_{\text{eff}.1,2}U^{-1}]_{ii}$, correspond to the energy expectation values $E_{\text{DFT}}(\Psi_i)$. From the least square fitting, namely, minimizing the following value

$$\sum_i \left\{ \left[U\mathcal{H}_{\text{eff}.1,2}U^{-1} \right]_{ii} - E_{\text{DFT}}(\Psi_i) \right\}^2, \tag{4}$$

the parameters are determined.

3. Numerical results

The following DFT calculations have been carried out with the ADF 2010.02 program package using a triple-ζ basis set with polarization function. We utilize the PBE0 functional which is hybrid 25% Hartree-Fock (HF) exchange form of PBE functional [24] proposed by Ernzerhof-Scuseria [25] and by Adamo-Barone [26].

By minimizing eq.(4), we obtain the parameters of $\mathcal{H}_{\text{eff}.1}$ and $\mathcal{H}_{\text{eff}.2}$ as shown in table 3. All values reasonably coincide with the values of the parameters of $\mathcal{H}_{\text{eff}}$ obtained in ref. [12].

Figure 1 shows the energy expectation values of the single SD's of molecular orbitals obtained from the effective Hamiltonians $\mathcal{H}_{\text{eff}}$, $\mathcal{H}_{\text{eff}.1}$ and $\mathcal{H}_{\text{eff}.2}$ with parameters in table 3. All effective Hamiltonians can be reproduce the values of $E_{\text{DFT}}(\Psi_i)$, although small disagreements appear in the result of $\mathcal{H}_{\text{eff}.1}$ and $\mathcal{H}_{\text{eff}.2}$.

The energy eigenvalues obtained by numerically diagonalization of the effective Hamiltonians are shown in figure 2. First excitation energy of $\mathcal{H}_{\text{eff}.1}$ accurately coincide with the results

Table 3. Numerical values of parameters in eV obtained from least square fit.

	$\epsilon_e - \epsilon_a$	T_{aa}	T^a_{ae}	T^e_{ae}	T_{ee}	T'_{ee}	k_{ae}	k_{ee}
ref. [12]	0.0515	−0.5083	−0.4765	0.0222	0.0240	0.0502	0.3710	0.3717
$\mathcal{H}_{\text{eff.1}}$	0.0578	$t_a = -0.3932$			$t_e = 0.0471$		$k = 0.3872$	
$\mathcal{H}_{\text{eff.2}}$	0.0578	$t_a = -0.3932$			$t_e = 0.0471$		$k = 0.4108$	

	$j_{aa} - J_{aa}$	$j_{ee} - J_{ee}$	$j_{ae} - J_{ae}$	$j'_{ee} - J'_{ee}$	K_{aa}	K_{ae}	K_{ee}	K'_{ee}
ref. [12]	3.4142	3.5530	2.5594	2.5645	0.0543	0.0104	0.0135	0.0317
$\mathcal{H}_{\text{eff.1}}$	$j - J = 3.7317$		$j' - J' = 2.6412$			$K = 0.0236$		
$\mathcal{H}_{\text{eff.2}}$	$j - J = 3.7789$		$j' - J' = 2.6884$			$K = 0$		

	F^{22}_{11}	K^{21}_{11}	K^{12}_{11}	$K^{22}_{11}(= K^{21}_{12})$	Y^{22}_{11}	X^{12}_{11}	$X^{22}_{11}(= X^{21}_{12})$
ref. [12]	0.0129	−0.0157	0.0030	0.0015	0.0042	−0.0109	0.0013
$\mathcal{H}_{\text{eff.1,2}}$	0	0	0	0	0	0	0

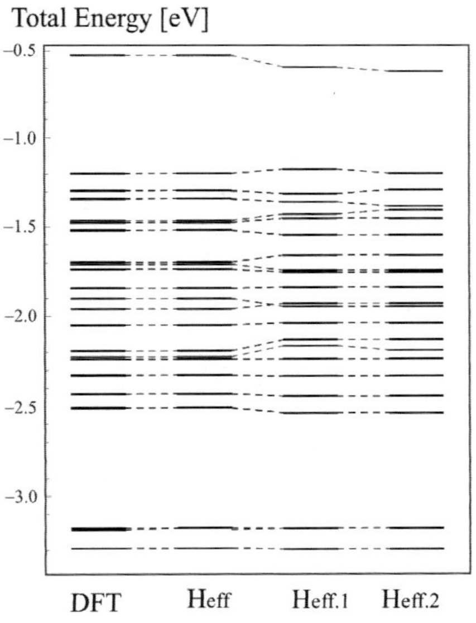

Figure 1. Energy expectation values of the single SD's of molecular orbitals calculated by DFT, and that of the systems $\mathcal{H}_{\text{eff}}$, $\mathcal{H}_{\text{eff.1}}$ and $\mathcal{H}_{\text{eff.2}}$ with the parameters given in table 3

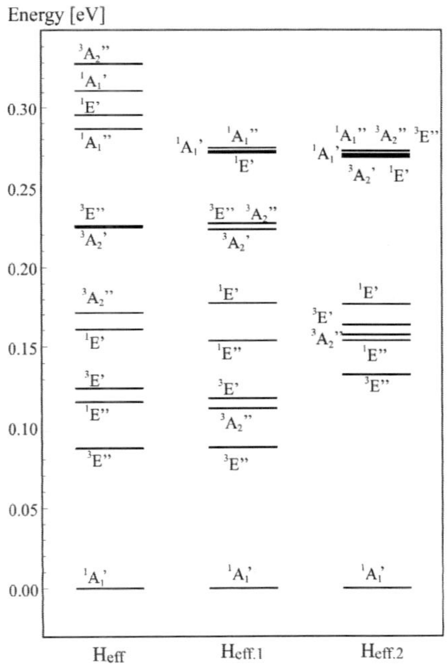

Figure 2. Energy eigenvalues of the system described by the effective Hamiltonian $\mathcal{H}_{\text{eff}}$, $\mathcal{H}_{\text{eff.1}}$ and $\mathcal{H}_{\text{eff.2}}$

of ref. [12]. As a result, the magnetic susceptibility (figure 3) is similar to that of $\mathcal{H}_{\text{eff}}$ and experimental result. At higher temperatures, both $\chi_{\parallel}(T)$ and $\chi_{\perp}(T)$ become larger than that of $\mathcal{H}_{\text{eff}}$ because the multiplet energy of the ${}^3A''_2$ term becomes lower compared with that of $\mathcal{H}_{\text{eff}}$. In the system $\mathcal{H}_{\text{eff.2}}$, the inter-atomic exchange integrals are neglected, namely, the ferromagnetic

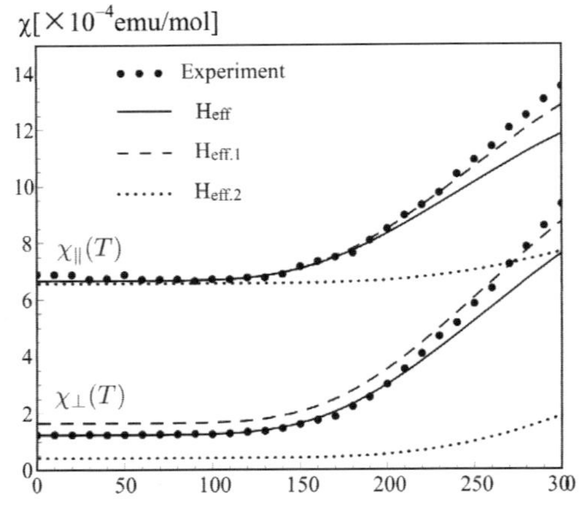

Figure 3. Temperature dependence of magnetic susceptibilities in perpendicular and parallel direction ($\chi_\perp(T)$ and $\chi_\parallel(T)$, respectively).

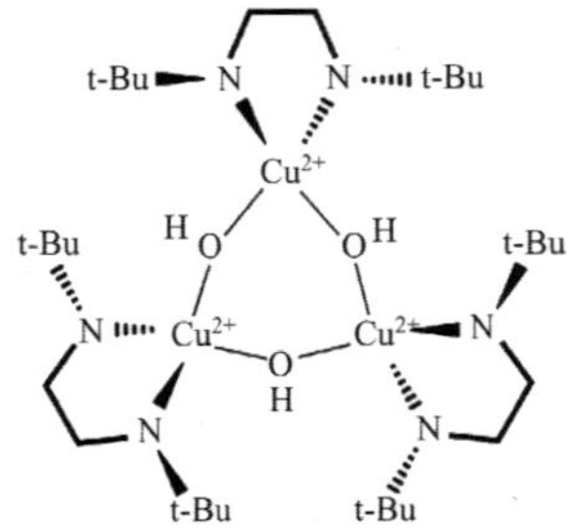

Figure 4. Structure of TrisOH.

potential exchange interaction is neglected. Therefore, the excitation energy from the singlet ground state to the triplet excitation state become larger, and thus the temperature dependences of the magnetic susceptibilities are flatten as shown in figure 3. The potential exchange term is not negligible to describe the magnetic interaction in the $Ti_2Cl_9^{3-}$ cluster because this cluster is composed of face sharing octahedra.

Here, we present another example, trinuclear copper cluster called TrisOH [27–30]. This cluster contains three Cu^{2+} ions bridged by OH^- as shown in figure 4. The ligand field from surrounding ions are so strong that the orbital degeneracy of the d-electrons in Cu^{2+} ions is resolved. The effective Hamiltonian contains too many parameters to determine its value by the least square fitting. Therefore, we consider the simplest model to describe the d-electrons in this cluster, the Hubbard model Hamiltonian: $t(c^\dagger_{1,\sigma}c_{2,\sigma} + c^\dagger_{2,\sigma}c_{3,\sigma} + c^\dagger_{3,\sigma}c_{1,\sigma} + \text{h.c.}) + U\sum_{i=1,2,3} n_i n_i$. Form the second order perturbation of the transfer integral t, the coupling constant of the spin Hamiltonian $2J_s \sum \boldsymbol{s}_i \boldsymbol{s}_j$ can be obtained as $J_s = 8t^2/U$. Assuming the three copper ions form equilateral triangle, the local orbitals a_i ($i = 1, 2, 3$) are represented by the following linear combinations of molecular orbitals ψ_i ($i = 1, 2, 3$):

$$a_1 = \frac{1}{\sqrt{3}}\psi_1 + \sqrt{\frac{2}{3}}\psi_2, \quad a_2 = \frac{1}{\sqrt{3}}\psi_1 - \frac{1}{\sqrt{6}}\psi_2 + \frac{1}{\sqrt{2}}\psi_3, \quad a_3 = \frac{1}{\sqrt{3}}\psi_1 - \frac{1}{\sqrt{6}}\psi_2 - \frac{1}{\sqrt{2}}\psi_3, \tag{5}$$

where ψ_1 and $\{\psi_2, \psi_3\}$ belong to A_1 term and E term under C_{3v} cluster symmetry, respectively.

By the least square fitting of the energy expectation values of the single SD's of molecular orbitals $E_{\text{DFT}}(\Psi_i)$, we obtain the parameters of the Hubbard model Hamiltonian as $t = 0.0972$ eV and $U = 5.212$ eV. We obtain the calculated value of the coupling constant J_s as 116 cm^{-1} which is near the experimental value 104.7 cm^{-1} [27–30]. The coupling constant is also calculated by the broken symmetry approach as $J_s = 111$ cm^{-1} [28]. In contrast to the broken symmetry approach which can be evaluated only the coupling constant, important information of t and U for the electronic structure in the metal cluster can be obtained by using our effective Hamiltonian method. This simple Hubbard model for this system does not precisely reproduce each one of the calculated energies $E_{\text{DFT}}(\Psi_i)$, but the model reproduces the global structure of the lower energies so as to realize the nature of the magnetic interaction.

4. Summary

In the present paper, the applicability of our approximated effective Hamiltonian to the first-principle study of the magnetic interactions have been examined. From the calculation of the $Ti_2Cl_9^{3-}$ cluster, it was found that small differences caused by the symmetry of the each orbital can be neglected and not only kinetic exchange but also potential exchange terms are needed for quantitative calculation of the magnetic susceptibility. The multiplet energy levels can be roughly estimated using approximated effective Hamiltonian. For the case of TrisOH, the Hubbard Hamiltonian is sufficient to obtain the coupling constant. Applications of our effective Hamiltonian method to clusters with higher nuclearity will be present in near future.

Acknowledgments

Authors are supported by the "Strategic Research Base Development" Program for Private Universities subsidized by MEXT (2009) (S0901022).

References

[1] Noodleman L 1981 *J. Chem. Phys.* **74** 5737
[2] Noodleman L, Davidson E R 1986 *Chem. Phys.* **109** 131
[3] Bencini A, Totti F, Daul C A, Doclo K, Fantucci P and Barone V 1991 *Inorg. Chem.* **36** 1391
[4] Yamagichi K, Takahara T, Fueno T and Nasu K 1987 *Jpn. J. Appl. Phys.* **26** L1362
[5] Yamaguchi K, Tsunekawa T, Toyoda Y and Fueno T 1988 *Chem. Phys. Lett.* **22** 371
[6] Yamanaka S, Kawakami T, Nagao H and Yamaguchi K 1994 *Chem. Phys. Lett.* **231** 25
[7] Onishi T, Takano Y, Kitagawa Y, Kawakami T, Toshida Y and Yamaguchi K 2001 *Plyhedron* **20** 1177
[8] Atanasov M and Daul C A 2003 *Chem. Phys. Lett.* **379** 209
[9] Atanasov M and Daul C A 2003 *Chem. Phys. Lett.* **381** 584
[10] Atanasov M, Daul C A and Rauzy C 2003 *Chem. Phys. Lett.* **367** 737
[11] Atanasov M, Daul C A and Rauzy C 2004 *Struct. Bonding (Berlin)* **106** 97
[12] Suzuki H, Delley B and Satoko C 2011 *J. Phys.: Condens. Matter* **23** 375502
[13] Suzuki H and Satoko C 2012 *Int. J. of Quant. Chem.* DOI: 10.1002/qua.24067
[14] Borrás-Almenar J J, Clemente-Juan J M, Coronado E, Palii A V and Tsukerblat B S 2001 *J. Chem. Phys.* **114** 1148
[15] Borrás-Almenar J J, Clemente-Juan J M, Coronado E, Palii A V and Tsukerblat B S 2001 *Chem. Phys.* **274** 131
[16] Borrás-Almenar J J, Clemente-Juan J M, Coronado E, Palii A V and Tsukerblat B S 2001 *Chem. Phys.* **274** 145
[17] Borrás-Almenar J J, Coronado E, Clemente-Juan J M, Palii A V and Tsukerblat B S 2003 *Polyhedron* **22** 2521
[18] Wang F, Wang B, Wang M and Chen Z 2008 *Chem. Phys. Lett.* **454** 177
[19] Palli A, Tsukerblat B S, Clemente-Juan J M and Coronado E 2010 *Int. Rev. Phys. Chem.* **29** 135
[20] Palii A, Tsukerblat B S, Klokishner S, Dunbar K R, Clemente-Juan J M and Coronado E 2011 *Chem. Soc. Rev.* **40** 3130
[21] Anderson P W 1959 *Phys. Rev.* **115** 2
[22] Briat B, Kahn O, Morgenstern-Badarau I and Rivoal J C 1981 *Inorg. Chem.* **20** 4193
[23] Leuenberger B and Güdel H U 1986 *Chem Phys. Lett.* **126** 255
[24] Perdew J P, Burke K and Ernzerhof M 1996 *Phys. Rev. Lett.* **77** 3865
[25] Ernzerhof M and Scuseria G 1999 *J. Chem. Phys.* **110** 5029
[26] Adamo C and Barone V 1999 *J. Chem. Phys.* **110** 6158
[27] Yoon J, Mirica L M, Stack T D P and Solomon E I 2004 *J. Am. Chem. Soc.* **126** 12586
[28] Mirica L M and Stack T D P 2005 *Inorg. Chem.* **44** 2131
[29] Yoon J, Mirica L M, Stack T D P and Solomon E I 2005 *J. Am. Chem. Soc.* **127** 13680
[30] Yoon J and Solomon E I 2007 *Coord. Chem. Rev.* **251** 379

Electronic and Vibronic Problems of Nanosized Mixed Valence Clusters: Advances and Challenges

Juan M. Clemente-Juan,[1] Alejandro Gaita-Ariño,[1] Juan J. Borrás-Almenar[1], Eugenio Coronado,[1] Andrew Palii,[2] Boris Tsukerblat [§3]

[1] Instituto de Ciencia Molecular, Universidad de Valencia, Polígono de la Coma, Spain
[2]Institute of Applied Physics, Academy of Sciences of Moldova, Kishinev, Moldova
[1]Chemistry Department, Ben-Gurion University of the Negev, Beer-Sheva, 84105 Israel

[§]E-mail: tsuker@bgu.ac.il

Abstract. Here we discuss the electronic and vibronic problems of mixed valency (MV) in molecular clusters which are of current interest in areas as diverse as solid-state chemistry, biochemistry, and molecular magnetism. Modern research in these areas is focused on the nanosized clusters at the border between classical and quantum scales and for this reason they are particularly difficult to study. First, we describe a general approach to the evaluation of the energy pattern of MV systems containing arbitrary number of localized spins and itinerant electrons with due account for the double exchange and other relevant interactions, like interelectronic Coulomb repulsion in instantly localized configurations. Then we present a symmetry adapted approach to the dynamic vibronic problem in large scale multimode Jahn-Teller (JT) systems, in particular, molecular MV clusters. The developed techniques are applied to the analysis of nanosized MV systems with emphasis on the 2e-reduced MV Keggin anion in which the electronic pair is delocalized over twelve sites ($\mathbf{T}_d$) giving rise to the $(^1T_2+^1E+^1A_1)\otimes(e+t_2)$ and $(^3T_1+^3T_2)\otimes(e+t_2)$ combined JT/pseudo JT problems for the spin-singlet and spin-triplet states. The delocalization of the two itinerant electrons in this remarkable systems was shown to lead to an effective spin paring. We study the energy pattern of this MV system and show that the vibronic interaction results in the ferromagnetic contribution.

1. Introduction

Modern trends in the molecular electronics and spintronics gave strong impact on the development of the synthetic, experimental and theoretical studies of large magnetic clusters and especially MV systems exhibiting double exchange [1]. These systems contain mobile itinerant electrons that provide unique opportunities for the control of quantum states through the action of external fields. The problem entire can be subdivided into two parts, namely, evaluation of the electronic levels (will be referred to as "electronic problem") and consequent consideration of the vibronic states ("vibronic problem"). For a binuclear MV system the problem of the double exchange [2] was solved by Anderson and Hasegawa [2c] and the ferromagnetic effect of the double exchange in MV dimers was elucidated. Since the solution of the double exchange problem has been essentially attributed to the case of an electron shared between pair of ions (spin cores) the results of Ref. [2c] should be generalized for more complicated high nuclearity systems [3a-c].

On the other hand, evaluation of the energy levels and JT /pseudo JT effects lead to a complicated dynamic vibronic problem [4-6]. The difficulties are aggravated for the multimode systems when the energy pattern contains several low-lying levels mixed by the active JT vibrations. This especially refers to the large scale MV systems, like biologically important systems such as iron-sulfur proteins [7], polynuclear metal-oxide clusters in which several electrons are delocalized over metal network (for example, reduced polyoxometalates (POMs) with Keggin and Wells-Dawson structures) [3,8], impurity metal ions in crystals [9-11], etc. In these cases the dimension of the truncated vibronic matrices to be diagonalized becomes very high due to sizeable electronic basis and multiple degeneracy of the excited vibrational levels.

Many efforts have been applied to get an accurate solution of the dynamic problem (see refs [3-6]). A significant progress has been achieved by the exploration of the Lie symmetries of the JT Hamiltonians [12]. Nevertheless, the vibronic Hamiltonian exhibits unitary symmetries only providing some essential restrictions, like linear vibronic coupling and special interrelations between the coupling constants which significantly reduces the area of practical applicability of these extremely elegant approaches.

The goal of this article is to describe a general solution of the electronic problem of mixed valency and a symmetry adapted approach aimed to an accurate solution of the dynamic JT/ pseudo JT problem in large MV systems. We introduce the mathematical procedure through the consideration of the MV POM with Keggin structure reduced by 2e [13]. This dodecanuclear MV system belongs to a large class of POMs, the so-called "heteropolyblues", which are attracting considerable interest in solid state chemistry, biomedical applications, catalysis and materials science in view of their unique electronic properties [14]. In terms of mixed-valency, the Keggin anion can be viewed as a high-nuclearity cluster belonging to Class II in Robin and Day classification and a complicated example of the JT system for which the dynamic problem seems to be actual for the adequate description of the non-trivial magnetic and optical (intervalence bands) properties. In this article we will focus on the elaboration of the theoretical approach, while the associated computer program with the detailed discussion of the experimental data on the Keggin anion will be published elsewhere [3d,15].

2. Electron transfer and double exchange in nanosized mixed-valence clusters

Until recently the calculation of the energy pattern of MV compounds exhibiting electron delocalization/double exchange has been restricted to comparatively simple systems comprising one or two itinerant electrons delocalized over a small number of exchange-coupled metal sites [1] (basically, dimeric and trimeric clusters). Consideration of nanoscopic MV systems require a more common approach that would allow to solve the double exchange problem for high-nuclearity MV clusters containing arbitrary number of

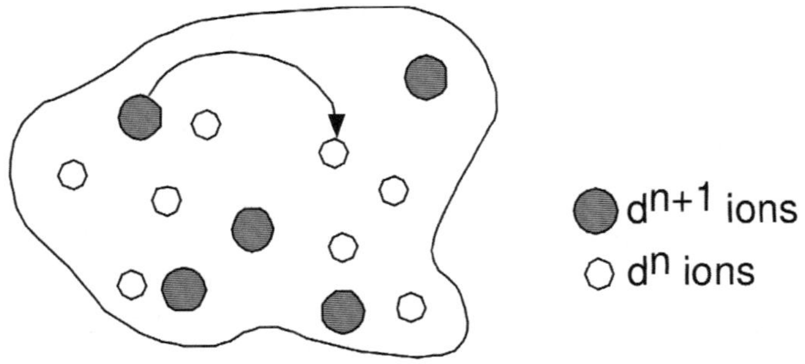

Fig.1. Scheme of a MV system containing arbitrary number of magnetic sites and itinerant electrons

localized spins and itinerant electrons [3]. The electron hopping from the site i to the site k is described by the following double exchange Hamiltonian H_{tr} (first term in Eq. (1)) supplemented by the Heisenberg type exchange interaction between localized spins (second term) and intersite Coulomb repulsion (last term):

$$
\begin{aligned}
H_e = & \sum_{i,k,\sigma} t_{ik}\left(c_{k\sigma}^{+}c_{i\sigma}+h.c.\right)+\sum_{i,j,k,l,\sigma,\sigma'} p_{ij,kl}\left(c_{l\sigma}^{+}\,c_{k\sigma'}^{+}\,c_{i\sigma'}\,c_{j\sigma}+h.c.\right)\\
& +\sum_{i,k} J_{ik}\,\boldsymbol{s}_i\,\boldsymbol{s}_k+\sum_{i,k} U_{ik}\,n_i\,n_k
\end{aligned}
\tag{1}
$$

where the operator $c_{k\sigma}^{+}\left(c_{i\sigma}\right)$ creates (annihilates) an electron on the site k (i) with the spin projection σ, t_{ik} and $p_{ij,kl}$ are the one- and two-electron transfer integrals, respectively, U_{ik} is the intersite Coulomb integrals (the distributions with the two itinerant electrons per site are supposed to be much high in energy and for this reason are neglected). Finally, J_{ik} are the exchange parameters that depend on the electronic distributions with the set of the local spins $\boldsymbol{s}_i$ defined by populations of the sites n_i. In particular, the exchange interactions between two spin cores ($s_i = s_j = s_0$), between spin cores and ions accommodating "extra" electron $s_i = s_0$, $s_j = s_0 + 1/2$ and between two centers with "extra" electrons $s_i = s_0 + 1/2$, $s_j = s_0 + 1/2$ are to be discriminated for each instantly localized electronic distribution. Using the angular momentum technique the matrix element of the double exchange operator $H_{tr}(i \rightarrow k)$ describing one-electron jump $i \rightarrow k$ (one contribution in the first term in Eq. (1)) is calculated as follows [3]:

$$
\begin{aligned}
&\left\langle\left(s'_\beta = s_\beta, \beta \neq i,k\right)\left(s'_i = s_0^r,\;\; s'_k = s_0^r + 1/2\right)\left(\tilde{S}'\right)S'\,M'_S\right|H_{tr}(i\rightarrow k)\left|\left(s_\beta, \beta\neq i,k\right)\right.\\
&\times\left.\left(s_i = s_0^r + \tfrac{1}{2},\;\; s_k = s_0^r\right)\left(\tilde{S}\right)S\,M_S\right\rangle = (-1)^{2s_0^r}\,t_{ik}\delta_{S\,S'}\delta_{M_S\,M'_S}\\
&\times Z_{i\rightarrow k}\left[\left(s'_\beta = s_\beta, \beta\neq i,k\right)\left(s_i = s_0^r + \tfrac{1}{2},\;\; s'_i = s_0^r,\;\; s_k = s_0^r, s'_k = s_0^r + \tfrac{1}{2}\right)\left(\tilde{S}\right)\left(\tilde{S}'\right)S\right],
\end{aligned}
\tag{2}
$$

where $Z_{i\rightarrow k}[\cdots]$ is the function of local spins s_i, s_k and s_i', s_k' for the initial and final electronic distributions, the corresponding sets of intermediate spins $\left(\tilde{S}\right)$ and $\left(\tilde{S}'\right)$ in the spin coupling schemes for the initial and final electronic distributions, and total spin S. It is to be noted that the function Z involves 6j-symbols rather than high order recoupling coefficients [2]. The energy levels resulting from the electronic Hamiltonian, Eq. (1), are mixed by the active JT/pseudo JT modes. In most cases of MV compounds this mixing is strong enough due to strong deformation of the ligand surrounding by the "extra" electron. In majority of cases the one-electron transfer (double exchange) is a dominant electronic interaction in MV clusters. However, in some cases, such transfer can give rise to the strongly excited states due to high intersite Coulomb repulsion energy (last term in Eq. (1)). As a result such one-electron transfer appears to be the second order contribution that is comparable (or even smaller) than that arising from the two-electron transfer (second term in Eq. (1)). The 2e reduced MV POM with Keggin structure we consider here provides an example of case when the two-electron transfer contribution proves to be important. The matrix elements of the two-electron transfer operator are expressed in terms of the product of the functions $Z_{k\rightarrow i}[\cdots]$ and $Z_{l\rightarrow j}[\cdots]$, see ref.[3] for more details.

3. Jahn-Teller Hamiltonian

The Hamiltonian of a JT (or, in general, pseudo JT) system can be represented as:

$$
H = H_e + \sum_{\nu\bar{\Gamma}\bar{\gamma}} \hbar\omega_{\nu\bar{\Gamma}\bar{\gamma}} b_{\nu\bar{\Gamma}\bar{\gamma}}^{+} b_{\nu\bar{\Gamma}\bar{\gamma}} + \left(1/\sqrt{2}\right)\sum_{\nu\bar{\Gamma}\bar{\gamma}} \upsilon_{\nu\bar{\Gamma}\bar{\gamma}} \boldsymbol{O}_{\nu\bar{\Gamma}\bar{\gamma}}\left(b_{\nu\bar{\Gamma}\bar{\gamma}}^{+} + b_{\nu\bar{\Gamma}\bar{\gamma}}\right)
\tag{3}
$$

where $b^{+}_{v\overline{\Gamma}\overline{\gamma}}$ and $b_{v\overline{\Gamma}\overline{\gamma}}$ are the creation and annihilation bosonic operators for the vibrational modes $v\overline{\Gamma}\overline{\gamma}$, ω_i are the vibrational frequencies so that the second term in Eq. (1) is the harmonic oscillator Hamiltonian. The symbol $i \equiv v\overline{\Gamma}\overline{\gamma}$ involves active vibrational irreps $\overline{\Gamma}$, $\overline{\gamma}$ numerates the basis functions of the irrep $\overline{\Gamma}$ and symbol v is introduced in order to distinguish the repeated irreps $\overline{\Gamma}$. The electronic subsystem described by the Hamiltonian H_e is supposed to consist of a set of the closely spaced electronic levels $\Gamma_1, \Gamma_2 \ldots \Gamma_t$ labeled by the irreps of the actual point group (the basis functions are marked by the symbols $\gamma_1, \gamma_2, \ldots \gamma_t$) which are mixed by the active JT/ pseudo JT vibrational modes $\overline{\Gamma}_1, \overline{\Gamma}_2 \ldots \overline{\Gamma}_f$. The third term in Eq. (1) represents the linear vibronic interaction in which $\boldsymbol{O}_i$ are the so-called vibronic matrices defined in the adopted electronic basis and υ_i are the dimensionless (in $\hbar\omega$ units) coupling parameters. Applying the creation operators to the vacuum state $(b^+)^n |0\rangle = \sqrt{n!}|n\rangle$ one can build the exited states $|n\rangle$. The computational approach will be illustrated by consideration of the MV 2e-reduced Keggin polyanion.

4. Electronic states and vibrations of the 2e-reduced Keggin polyanion

The Keggin structure [8] consists of the four edge-sharing triads of MO_6 units arranged around the heteroatom X (= B, Si, P, Co, Cu,…) by sharing corners in such a way that the system consisting of the twelve sites possesses overall tetrahedral symmetry $\mathbf{T}_d$. Fig 1b shows the idealized scheme of the metal sites in the Keggin system that consists of the four metal triangles perpendicular to the C_3 axes of the tetrahedron. The Keggin anion can be reversibly reduced by adding one or more electrons, which are delocalized over the metal network. The resulting MV systems usually belongs to Class II in the Robin and Day classification [8,16] exhibiting thus an intermediate vibronic coupling. In this

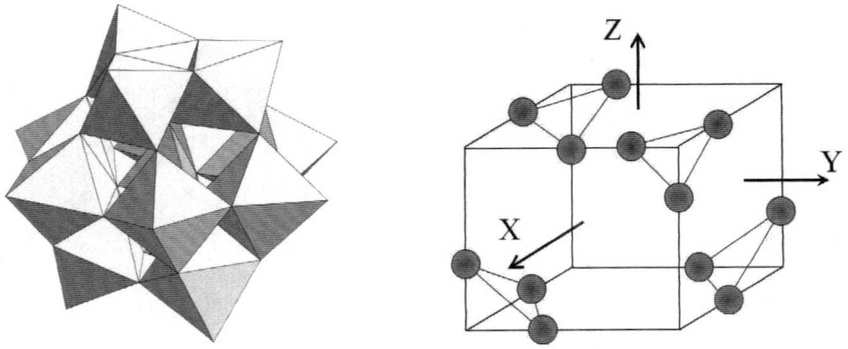

a *b*

Fig. 1. Two-electron reduced polyoxoanion with Keggin structure: network of the octahedral units (a) and scheme of the metal sites and molecular coordinate frame (b).

respect the Keggin anion with the delocalized electronic pair is of a special interest due to its unusual magnetic properties. In fact, the 2e-reduced Keggin anion was found to be antiferromagnetic that can not be explained in the framework of the conventional Heisenberg model dealing with the localized spins [8]. It was suggested [13,17] that the delocalization of the electronic pair results in the effective stabilization of the antiferromagnetic spin state of the electronic pair in the Keggin anion [8].

The interpretation of the spin pairing in the Keggin anion [13] is based on the model including delocalization of the two electrons over twelve sites (Fig.1b) as well as the Coulomb repulsion. The energy levels can be divided into five groups corresponding to the different distances between delocalized electrons in the 66 dispositions. The Coulomb repulsion tends to keep the electrons as far as possible in the ground configuration that is shown to consists of the three spin-singlets ${}^1A_1$, ${}^1E$, ${}^1T_2$ and two spin triplets ${}^3T_1$, ${}^3T_2$. The low lying energies belonging to the ground Coulomb configuration are formed by the set of states for which the electrons occupy remote sites and the two-electron jumps (transfer parameter P) keep the maximal distance between the sites of localization. The energy levels in terms of the bi-electron transfer parameter P are found as:

$$\begin{aligned} &\varepsilon\left({}^1A_1\right)=4P, \\ &\varepsilon\left({}^3T_2\right)=2P, \\ &\varepsilon\left({}^3T_1\right)=\varepsilon\left({}^1E\right)=-2P. \end{aligned} \tag{4}$$

More common consideration and a more detailed scheme of the levels with the due account for the one-electron transfer processes are given in ref. [13a] and the conditions were elucidated under which spin pairing can occur as a result of competition between the single- and double electron transfer processes and depends also on the interelectronic interaction.

It was emphasized [13c] that the conditions under which the system proves to be antiferromagnetic are strongly affected by the vibronic interactions that tends to localize the electrons and thus effectively reduce the transfer parameters. The degrees of localization prove to be different for the spin-triplet and spin-singlet states that essentially influence the magnetic properties of the reduced Keggin anion. The doubly degenerate (e) and triply degenerate vibrations (t_2) are found to be active [8a] and hence the potential surfaces are determined by the JT/pseudo JT interactions of the types of $({}^1T_2+{}^1E+{}^1A_1)\otimes(e+t_2)$ for the spin singlets and $({}^3T_1+{}^3T_2)\otimes(e+t_2)$ for the spin triplets. The adiabatic approximation [13c] provides information about the symmetry of the localized configurations of the electronic pair but it is unable to describe the magnetic characteristics and profiles of the intervalence optical bands for which the quantum structure of the excited vibronic levels is crucially important.

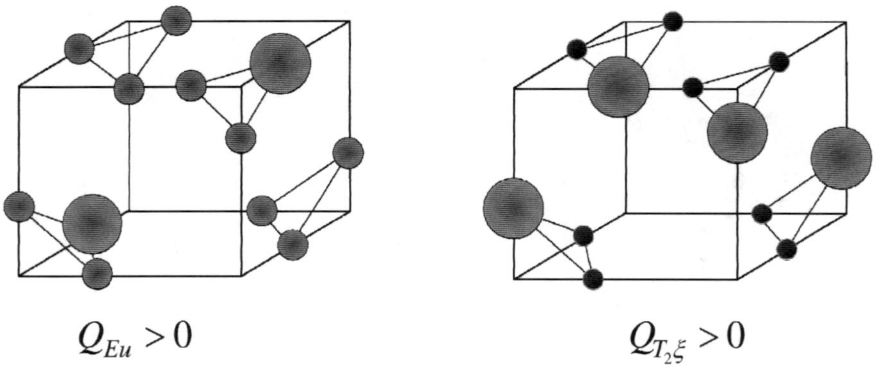

Fig. 2. Pictorial representation of the E and T_2 PKS vibrations (one component for each type) in Keggin polyoxoanion; large and small circles symbolize expanded and compressed octahedral sites, intermediate size circles denote undistorted units.

It is worthwhile to use the Piepho, Krausz and Schatz (PKS) model [16] that deals with the independent "breathing" displacements of the ions around the sites. Within the PKS model the frequencies of T_2 and E vibrations (their pictorial representation is given in Fig. 2) are equal since they simply represent the frequencies of the local "breathing" mode and therefore the electronic system is effectively coupled to a five-dimensional oscillator for which the dimension of the vibrational space $G(N)$ can be found as:

$$G(N) = (1/120)\,(N+1)\,(N+2)\,(N+3)\,(N+4)\,(N+5)\,, \tag{5}$$

where $n = n_1 + n_2 + n_3 + n_4 + n_5$ is the total number of excitation of the five-dimensional harmonic oscillator (label of the energy level), the expression under summation is the total degeneracy of the level with a given n and N is the number of the levels included in the basis. The total degeneracy of the electron-vibrational space for spin-singlets is $6G(N)$ and the same for spin-triplets, which leads to a very high dimension of the vibronic matrices. At the same time one can see that even a relatively small number of the vibronic levels involved in the evaluation results in a huge size of the matrices in the multimode systems demonstrating thus a necessity of the efficient use of symmetry.

5. Group-theoretical classification for the two- and three-dimensional oscillators

To adapt the vibrational functions to the definite irreps let us first consider the case of a two-dimensional (E) harmonic oscillator. The basis of the irrep E in $\mathbf{T}_d$ will be chosen in a standard way: $u \propto 3z^2 - r^2$, $\upsilon \propto \sqrt{3}(x^2 - y^2)$ [18]. The functions $|n_u n_\upsilon\rangle$ can be built as:

$$|n_u\, n_\upsilon\rangle = (n_u!\,n_\upsilon!)^{-1/2}\,(b_u^+)^{n_u}\,(b_\upsilon^+)^{n_\upsilon}\,|00\rangle\,. \tag{6}$$

Each level with a certain $n = n_u + n_\upsilon$ is $g = (n+1)$-fold degenerate. Before evaluation of the symmetry adapted functions it is worthwhile to classify the resulting irreps for each n and therefore to find the allowed states. Using correlation between the angular momentum representation for a two-dimensional oscillator [19] and the point group irreps [20-24], one can reach this goal. The results are summarized in Table 1.

Table 1. Group-theoretical classification of the vibrational states for E vibrations in $\mathbf{T}_d$

n	$g(n)$	vib. irreps	n	$g(n)$	vib. irreps
1	2	E	3	4	$A_1 + A_2 + E$
2	3	$A_1 + E$	4	5	$A_1 + 2E$

Table 2. Group-theoretical classification of the vibrational states for T_2 vibrations in $\mathbf{T}_d$

n	$g(n)$	vibrational irreps
0	1	A_1
1	3	T_2
2	6	A_1, E, T_2
3	10	A_1, T_1, $2\,T_2$
4	15	$2A_1$; $2E$, T_1, $2T_2$

The functions of the three-dimensional oscillator $|n_\xi n_\eta n_\zeta\rangle$ in the standard T_2 basis $\xi \propto yz, \eta \propto xz, \zeta \propto xy$ for $\mathbf{T}_d$ [18] can be built as:

$$|n_\xi n_\eta n_\zeta\rangle = (n_\xi!\,n_\eta!\,n_\zeta!)^{-1/2}\,(b_\xi^+)^{n_\xi}\,(b_\eta^+)^{n_\eta}\,(b_\zeta^+)^{n_\zeta}\,|000\rangle\,. \tag{7}$$

Each level with a given $n = n_\xi + n_\eta + n_\zeta$ is $g(n) = (1/2)(n+1)(n+2)$-fold degenerate. The set of the vibrational states with a given n can be correlated with a certain value of the angular orbital momentum l and parity $(-1)^l$ where l= 0, 2…n for even n and l= 1, 3…n for odd n [32]. Then classifying these states one obtains the results listed in Table 2.

6. Symmetry adapted vibrational basis

To create vibrational function $|n, \Gamma\gamma\rangle$ of the bidimensional vibration E with a given n let us introduce the symmetry adapted operators $\boldsymbol{T}_{\Gamma\gamma}^{(n)}$ constructed from operators $(b_u^+)^{n_u}(b_\upsilon^+)^{n_\upsilon}$ with definite $n = n_u + n_\upsilon$, where Γ is one the irreps $(\Gamma = A_1, A_2, E)$ corresponding to the set of the vibrational functions belonging the energy level n. By definition the operator $\boldsymbol{T}_{\Gamma\gamma}^{(n)}$ acts on the vacuum state $|00\rangle \equiv |n_u = 0, n_\upsilon = 0\rangle$ and creates a basis function of the type of $\Gamma\gamma$ belonging to a certain excited vibrational state n :

$$\boldsymbol{T}_{\Gamma\gamma}^{(n)}|00\rangle = |n, \Gamma\gamma\rangle . \tag{8}$$

The operators $\boldsymbol{T}_{\Gamma\gamma}^{(n)}$ realize the unitary transformation from the basis $|n_u n_\upsilon\rangle$ with a certain n to the symmetry adapted basis $|n, \nu\Gamma\gamma\rangle$ where ν is a quantum number that distinguishes the states of the system for which an irrep Γ occurs several times in the set n. We will refer $\boldsymbol{T}_{\Gamma\gamma}^{(n)}$ to as multivibronic symmetry adapted creation operators that can be considered as irreducible tensor operators of the type of $\Gamma\gamma$ in the actual point group. This allows one to build up the multivibronic operators with the use of the technique for the manipulation with the irreducible tensor operators in the point groups introduced by Sugano, Tanabe and Kamimura [18] (see refs. [20-22]). For example $\boldsymbol{T}_{\Gamma\gamma}^{(2)}$ can be found as the irreducible parts of the direct product by the use of the coupling scheme:

$$\hat{T}_{\Gamma\gamma}^{(2)} = \sum_{\gamma_1\gamma_2} b_{\Gamma_1\gamma_1}^+ b_{\Gamma_2\gamma_2}^+ \langle \Gamma_1\gamma_1\Gamma_2\gamma_2|\Gamma\gamma\rangle \equiv \{b_{\Gamma_1}^+ \otimes b_{\Gamma_2}^+\}_{\Gamma\gamma}, \quad \Gamma \in \Gamma_1 \times \Gamma_2 \tag{9}$$

The coupling (Clebsch-Gordan) coefficients $\langle \Gamma_1\gamma_1\Gamma_2\gamma_2|\Gamma\gamma\rangle$ for all point groups are given by Koster et al [21]. In the case of E- vibrations Eq. (9) gives the following result:

$$\begin{aligned} &\hat{T}_{A_1}^{(2)} = \frac{1}{\sqrt{2}}\left((b_u^+)^2 + (b_\upsilon^+)^2\right), \qquad \hat{T}_{A_2}^{(2)} = 0. \\ &\left.\hat{T}_{Eu}^{(2)} = -\frac{1}{\sqrt{2}}\left((b_u^+)^2 - (b_\upsilon^+)^2\right), \; \hat{T}_{E\upsilon}^{(2)} = \sqrt{2}\, b_u^+ b_\upsilon^+ ,\right\} E \end{aligned} \tag{10}$$

One can see that the operator $\mathbf{T}_{A_2}^{(2)}$ corresponding to the antisymmetric part $\{E \times E\}$ of the direct product $E \times E$ vanishes in the case of $n = 1$ so that the full dimension of the obtained basis is $g(n) = n + 1 = 3$. Applying operators $\mathbf{T}_{\Gamma\gamma}^{(2)}$ to the vacuum state $\mathbf{T}_{\Gamma\gamma}^{(2)}|00\rangle = |2, \Gamma\gamma\rangle$ one obtains the three-dimensional vibrational basis for $n = 2$ with $\Gamma = A_1, E$. Then one has to build the operators $\boldsymbol{T}_{\Gamma\gamma}^{(3)}$ applying the subsequent step of the coupling procedure:

$$\hat{T}^{(3)}_{\Gamma\gamma}= \left\{\left\{b^+_{\Gamma_1}\otimes b^+_{\Gamma_2}\right\}_{\Gamma_{12}}\otimes b^+_{\Gamma_3}\right\}_{\Gamma\gamma} \equiv \sum_{\gamma_{12}\gamma_3}\left\{b^+_{\Gamma_1}\otimes b^+_{\Gamma_2}\right\}_{\Gamma_{12}\gamma_{12}} b^+_{\Gamma_3\gamma_3}\left\langle\Gamma_{12}\gamma_{12}\Gamma_3\gamma_3\middle|\Gamma\gamma\right\rangle ,$$
$$\Gamma_{12}\in\Gamma_1\times\Gamma_2,\ \ \Gamma\in\Gamma_{12}\times\Gamma_3. \qquad (11)$$

In general the obtained vectors are linearly independent and therefore each step should be supplemented by the Gram-Schmidt orthogonalization within the repeated irreps that takes an arbitrary basis and generates a new orthogonal one excluding thus excessive functions. It should be noted that u-components of all E bases are automatically orthogonal to the υ-components, and, of course, the orthogonality remains between different irreps and between the functions from the sets arising from different n. Due to this fact the Gram-Schmidt orthogonalization procedure is to be applied only to the corresponding basis functions of the same irreps (let's say, to the u type functions belonging to the E bases) with the same n.

For the three-dimensional oscillator the multivibronic operators $\boldsymbol{T}^{(n)}_{\Gamma\gamma}$ are constructed from the products $\left(b^+_\xi\right)^{n_\xi}\left(b^+_\eta\right)^{n_\eta}\left(b^+_\zeta\right)^{n_\zeta}$ with a definite $n=n_\xi+n_\eta+n_\zeta$ and with Γ being one the irreps of $\mathbf{T}_d$ $\left(A_1, A_2, E, T_1, T_2\right)$ corresponding to the set of the vibrational function belonging to definite n. The general expressions for $\boldsymbol{T}^{(2)}_{\Gamma\gamma}, \boldsymbol{T}^{(3)}_{\Gamma\gamma}$, etc. can be found as in previous case. For example, one obtains the following result for the $\boldsymbol{T}^{(2)}_{\Gamma\gamma}$ operators:

$$\hat{T}^{(2)}_{A_1}=\frac{1}{\sqrt{3}}\left(\left(b^+_\xi\right)^2+\left(b^+_\eta\right)^2+\left(b^+_\zeta\right)^2\right),$$

$$\left.\begin{aligned}\hat{T}^{(2)}_{Eu}&=\frac{1}{\sqrt{6}}\left(-\left(b^+_\xi\right)^2-\left(b^+_\eta\right)^2+\left(b^+_\zeta\right)^2\right),\\ \hat{T}^{(2)}_{E\upsilon}&=\frac{1}{\sqrt{2}}\left(\left(b^+_\xi\right)^2-\left(b^+_\eta\right)^2\right),\end{aligned}\right\}E \qquad \left.\begin{aligned}\hat{T}^{(2)}_{T_2\xi}&=b^+_\eta b^+_\zeta ,\\ \hat{T}^{(2)}_{T_2\eta}&=b^+_\xi b^+_\zeta ,\\ \hat{T}^{(2)}_{T_2\zeta}&=b^+_\xi b^+_\eta .\end{aligned}\right\}T_2 \quad . \qquad (12)$$

From the direct product of $\Gamma(n=2)\times T_2$ one finds the following irreps forming the 18-dimensional space for $n=3$: $\left(A_1+E+T_2\right)\times T_2=T_2+\left(T_1+T_2\right)+\left(A_1+E+T_1+T_2\right)=A_1+E\ \ +2T_1+3T_2$. By comparing these irreps with those predicted by the group-theoretical assignation (A_1, T_1, $2T_2$) one can conclude that this 18-dimensional space contains linearly dependent vectors and therefore, beginning from this step, the Gram-Schmidt procedure of orthogonalization is to be applied. The next steps are quite similar to those so far described.

7. Symmetry adapted electron-vibrational basis

The evaluation of the vibrational basis for each type of the active modes can be extended to an arbitrary value of n by the application of the multivibronic operator of the order n to the vacuum state with the subsequent Gram-Schmidt procedure at each step n as described in [3d]. The operator $\boldsymbol{T}^{(n)}_{\Gamma\gamma}$ can be built by extending the consequent coupling procedure to the order n:

$$\hat{T}^{(n)}_{\Gamma\gamma}= \left\{\left\{\left\{b^+_{\Gamma_1}\otimes b^+_{\Gamma_2}\right\}_{\Gamma_{12}}\otimes b^+_{\Gamma_3}\right\}_{\Gamma_{123}}...\otimes b^+_{\Gamma_n}\right\}_{\Gamma\gamma} ,$$
$$\Gamma_{12}\in\Gamma_1\times\Gamma_2,\ \ \Gamma_{123}\in\Gamma_1\times\Gamma_2\times\Gamma_3\ ...,\ \ \Gamma\in\Gamma_{n-1}\times\Gamma_n . \qquad (13)$$

Then the vibrational functions for the active vibrations can be combined to get the total vibrational basis with the given symmetry. For example, for the two basis sets $\left|n_1, \nu_1\Gamma_1\gamma_1\right\rangle$ and $\left|n_2, \nu_2\Gamma_2\gamma_2\right\rangle$ one can directly use the coupling scheme, Eq.(13), within which one obtains the symmetry adapted vibrational functions $\left|n,\nu\Gamma_{\upsilon}\gamma_{\upsilon}\right\rangle (n=n_1+n_2,\Gamma_{\upsilon}\in\Gamma_1\times\Gamma_2)$. Finally, the vibrational functions $\left|n,\nu\Gamma_{\upsilon}\gamma_{\upsilon}\right\rangle$ are to be coupled to the electronic ones $\left|\alpha\Gamma_e\gamma_e,SM\right\rangle$ that are eigen-functions of the Hamiltonian H_e included in the JT/pseudo JT problem (*SM* are the quantum numbers of the full spin and its projection, α is the additional quantum number that enumerated the repeated irreps Γ_e).

This allows to achieve the final goal of the approach, namely to evaluate the basis in which the full matrix of the JT Hamiltonian is blocked according to the irreps of the point group. Finally, the approach is realized as an efficient computer program [15] that generates the blocks and evaluates required characteristics of the JT systems, like optical spectra, thermodynamic characteristics, etc. The procedure generates the basis assuming only point symmetry of the system without indication of the explicit form of the vibronic coupling, so it is not restricted by the linear terms of vibronic interactions and applicable also when the quadratic and higher order terms are taken into account. Under this condition the approaches based on high symmetries lose their advantages because the actual symmetry is reduced to the point one.

8. Dynamic Jahn-Teller problem for the 2e-reduced Keggin anion

The size of the vibronic matrices $6G(N)$ for the $(^1T_2+^1E+^1A_1)\otimes(e+t_2)$ and $(^3T_1+^3T_2)\otimes(e+t_2)$ combined JT/pseudo JT problems in MV Keggin anion is rather large, for example, for N=10, 11, 12 the sizes of the matrices are 18018, 26208 and 37128 for spin-singlets and the same dimensionalities for spin-triplets. The sizes of the matrices grow rapidly with the increase of N (Eq. (5)). Table 2 shows the sizes of the blocks (related to different irreps) of the full vibronic matrices for spin-singlet and spin-triplet states. One can see that full exploration of the symmetry essentially simplifies the vibronic problem. In fact, for N=12 the full sizes of the vibronic matrices (for S=1 and S=0) are 37128 while the maximal size of the reduced matrix proves to be only 4676 (for T_2 levels in spin-singlets and spin-triplets sets).

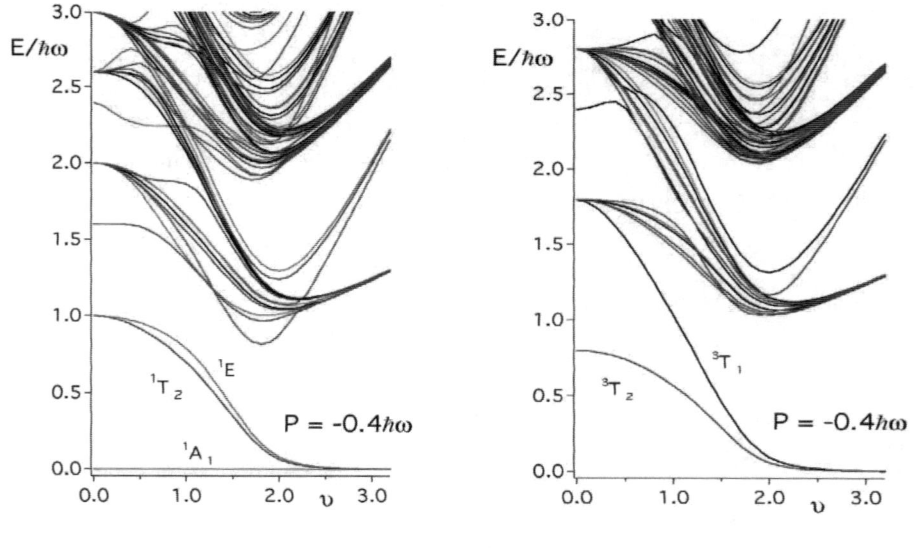

Fig. 3. Calculated vibronic levels for spin-snglet (a) and spin-triplets (b) terms belonging to the ground manifold of the 2e-reduced Keggin anion.

Fig. 3 shows a sample of the vibronic levels for spin-singlet (Fig. 3a) and spin-triplets (Fig. 3b) terms calculated for a rather narrow basis set (N=8). At P=0 one obtains the electronic levels superimposed with the unperturbed harmonic oscillator levels. The level ${}^1A_1$ which is the ground state of spin paired system, is chosen as a reference level for both patterns in Fig. 3. When the vibronic coupling

Table 3. Dimensions of the electron-vibrational spaces for the blocks of the JT/pseudo JT problems in Keggin anion.

	Spin singlets					Spin triplets				
N	A_1	A_2	E	T_1	T_2	A_1	A_2	E	T_1	T_2
10	819	728	1547	2184	2275	728	728	1456	2275	2275
11	1176	1064	2240	3192	3304	1064	1064	2128	3304	3304
12	1652	1512	3164	4536	4676	1512	1512	3024	4676	4676

increases the excited levels approach the ground level that corresponds to the multiple degenerate states in the deep wells of the adiabatic potential. Comparing Figs. 3a and 3b one can see that at the same time increase of the PKS coupling decrease the gap between the ground spin-singlet vibronic level ${}^1A_1$ (Fig. 3a) and the excited spin-triplet level ${}^3T_2$ (which is the ground level in spin-triplet pattern, Fig. 3b). This leads to the effective increase of the magnetic moment of the system that can be considered as a result of the JT/pseudo JT coupling. From the point of view of the JT theory this is a manifestation of the reduction of efficiency of the electron transfer (Ham effect).

9. Concluding remarks

We have presented a theoretical approach to obtain an accurate solution of the electronic and dynamic JT/pseudo JT problems in large scale MV systems. We have shown how to solve the double exchange problem for high-nuclearity MV clusters containing arbitrary number of itinerant electrons shared among arbitrary number of magnetic sites. The algorithm applied to the vibronic problem takes full advantage of the point symmetry arguments and thus allows to reduce the vibronic matrices to full extent. The group-theoretical assignment of the vibronic states allows to predict the symmetry labels for the vibronic levels in advance and thus to get an additional gain in the computational work. This approach is applicable to an arbitrary form of the vibronic coupling, including linear and high order terms. The approach is illustrated by consideration of the 2e-reduced MV dodecanuclear Keggin anion ($\mathbf{T}_d$) in which the electronic pair is delocalized over twelve metal sites giving rise (within PKS model) to the $({}^1T_2+{}^1E+{}^1A_1)\otimes(e+t_2)$ and $({}^3T_1+{}^3T_2)\otimes(e+t_2)$ combined JT/pseudo JT problems for the spin-singlet and spin-triplet states, correspondingly. The vibronic interaction is shown to increase the magnetic moment of such kind of systems due to partial localization of the electronic pair giving rise to reduction of the electron transfer. The results so far described are expected to open a route to a new level of accuracy in the consideration of large MV systems and to increase the size of the JT systems that can be treated.

Acknowledgments

J.M.C.J., J.J.B.A and E.C. thank Spanish MINECO (CSD2007-00010 CONSOLIDER-INGENIO in Molecular Nanoscience, MAT2007-61584, MAT2011-22785), Generalitat Valenciana (PROMETEO program), and the EU (ELFOS project and ERC Advanced Grant SPINMOL) for financial support. A.P. gratefully acknowledges the Universitat de Valencia for a visiting research grant and financial support from the Supreme Council on Science and Technological Development of Moldova. B.T. acknowledges financial support of the Israel Science Foundation (ISF, grant no. 168/09).

Journal of Physics: Conference Series **428** (2013) 012037 doi:10.1088/1742-6596/428/1/012037

References

1. (a) Borras-Almenar J.J., Clemente-Juan J.M. Coronado E., Palii A.V., Tsukerblat B.S., Magnetic Properties of Mixed-Valence Systems: Theoretical Approaches and Applications, in: *Magnetoscience-From Molecules to Materials*, eds. J. Miller, M. Drillon, Willey-VCH, 2001, pp. 155-210. (b) Tsukerblat B., Klokishner S., Palii A., Jahn-Teller Effect in Molecular Magnetism: An Overview, *Springer Series in Chemical Physics*, 2009, Vol. **97**, Ed. H. Koppel, Chapter 18, pp. 555-619.
2. (a) Zener C., 1951 *Phys. Rev.*, **82** 403. (b) Anderson P.W., Hasegawa H., 1955 *Phys. Rev.* **100** 675; (c) De Gennes, P.-G. 1960 *Phys. Rev.* **118** 141.
3. (a) Borras-Almenar J.J., Clemente J.M., Coronado E., Palii A.V., Tsukerblat B.S., 1996 *J. Chem. Phys.* **105** 6892. (b) Clemente-Juan J.M., Borrás-Almenar J.J., Coronado E., Palii A. V., Tsukerblat B., 2009 *Inorg. Chem.* **48** 4557; (c) Clemente-Juan J.M., Borrás-Almenar J.J., Coronado E., Palii A.V., Tsukerblat B.S. 2010 *J. Compt. Chem.* **31** 1321; (d) Tsukerblat B., Palii A., Clemente-Juan J.M., Gaita-Ariño A., Coronado E. 2012 *Int. J. Quant. Chem.* 112 2957.
4. Englman R., *The Jahn–Teller Effect in Molecules and Crystals*, Wiley, London 1972.
5. Bersuker I.B., Polinger V.Z., *Vibronic Interactions in Molecules and Crystals*, Springer-Verlag, Berlin 1989.
6. (a) Bersuker I.B., *The Jahn-Teller Effect*, Cambridge University Press (2006); (b) Bersuker I.B., *Chem. Rev.*, dx.doi.org/10.1021/cr300279n.
7. (a) Ding X.-Q., Bominaar E.L., Bill E., Winkler H., Trautwein A.X., Drüeke S., Chaudhuri P., Weighardt K., *J. Chem. Phys.* 1990 **92**178; (b) Gamelin D.R., Bominaar E.L., Kirk M.L., Wieghardt K., Solomon E.I. 1996 *J. Am. Chem. Soc.* **118** 8085; (c) Marks A.J., Prassides K., 1993 *New. J. Chem.*, **17** 59; (d) Marks A.J., Prassides K. 1993 *J. Chem. Phys.* **98** 4805.
8. (a) Müller A., Peters F., Pope M.T., Gatteschi D. 1998 *Chem. Rev.*, **98** 239; (b) Clemente-Juan J.M., Coronado E. 1999 *Coord. Chem. Rev.*, **193** 361; (c) Kögerler P., Tsukerblat B., Müller A. 2010 *Dalton Transactions,* **39** 21.
9. (a) Polinger V.Z., Boldirev S.I. 1986 *Phys. Stat. Sol. (b),* **137** 241; (b) Boldyrev S.I., Polinger V.Z., Bersuker I.B., *Fiz. Tverdogo Tela* 1981 (Rus) **23** 746.
10. Sakamoto N., Muramatsu S., *Phys. Rev.B* 1978 **17** 868.
11. Perlin Yu.E., Tsukerblat B.S., Optical bands and polarization dichroism of Jahn-Teller centers, in: Yu.E. Perlin, M. Wagner (Eds.), *The Dynamical Jahn-Teller Effect in Localized Systems*, vol. 7, Elsevier, Amsterdam, 1984, pp. 251-346.
12. (a) Grosso G., Martinelli L., Parravicini G.P. 1995 *Phys. Rev. B*, **51** 13033; (b) Pooler D.R. 1978 *J. Phys.A*, **11** 1045; (c) Pooler D.R. 1980 *J. Phys.A*, **13** 1029. (d) O'Brien M.C.M. 1969 *Phys.Rev.* **187** 329; (e) O'Brien M.C.M. 1971 *J. Phys.C.*, **4** 2524.
13. (a) Borras-Almenar J.J., Clemente-Juan J.M., Coronado E., Tsukerblat B.S. 1995 *Chem. Phys.* **195** 1; (b) Borras-Almenar J.J., Clemente-Juan J.M., Coronado E., Tsukerblat B.S. 1995 *Chem. Phys.* **195** 17; (c) Borras-Almenar J.J., Clemente-Juan J.M., Coronado E., Tsukerblat B.S. 1995 *Chem. Phys.* **195** 29.
14. Clemente-Juan J.M., Coronado E., Gaita-Ariño A., in *Polyoxometalate Molecular Science, NATO ASI Series 98* (Eds.: J.J. Borrás-Almenar, E. Coronado, A. Müller, M. Pope), Kluwer Acad. Publishers, Dordrecht, The Netherlands, **2003**, pp. 273
15. Clemente-Juan J.M., Palii A., Coronado E., Tsukerblat B., *J. Comp. Chem.*, to be submitted.
16. (a) Wong K.Y., Schatz P.N. 1981 *Prog. Inorg. Chem.,* **28** 369. (b) Piepho S.B., Krausz E.R., Shatz P.N. 1978 *J. Am. Chem. Soc.*, **100** 2996. (a) Piepho S.B., 1988 *J. Am. Chem. Soc.* **110** 6319; (b) Piepho S.B. 1990 *J. Am. Chem. Soc.* **112** 4197.
17. (a) Borshch S.A., Bigot B. 1993 *Chem. Phys. Lett.*, **212** 398; (b) Borshch, S.A. 1998 *Inorg. Chem.*, **37** 3116.

Journal of Physics: Conference Series **428** (2013) 012037 doi:10.1088/1742-6596/428/1/012037

18. Sugano S., Tanabe Y., Kamimura H., *Multiplets of Transition Metal Ions in Crystals*; Academic Press: London, 1970.
19. Sakurai J.J., *Modern quantum mechanics*, Addison-Wesley Publishing Company, 1994.
20. Griffith J.S. *Irreducible tensor method for molecular symmetry groups*, Dover Pub. Mineola, NY, 2006.
21. Koster G.F., Dimmok J.O., Wheeler R.G., Statz H., *Properties of the Thirty-Two Point Groups*, MIT Press, Cambridge, MA,1963.
22. Tsukerblat B., *Group Theory in Chemistry and Spectroscopy*, Dover, Mineola, NY, 2006.
23. Tsukerblat B.S., Belinsky M.I., Fainzil'berg V.E. 1987 *Sov. Sci. Rev., Chem. Rev.,* **B9** 337.
24. Landau L.D., Lifshits E.M., *Quantum mechanics: non-relativistic theory*, Pergamon Press, Oxford, NY, Benjing, Frankfurt, San Paulo, Sydney, Tokyo, Toronto, 1977.

Journal of Physics: Conference Series **428** (2013) 012038 doi:10.1088/1742-6596/428/1/012038

Fermi-Surface Topology and Superconductivity Induced by Jahn-Teller Phonons

Yuji Shiba[1] **and Takashi Hotta**[1,2]

[1] Department of Physics, Tokyo Metropolitan University, Tokyo 192-0397, Japan
[2] Advanced Science Research Center, Japan Atomic Energy Agency, Ibaraki 319-1195, Japan

E-mail: hotta@tmu.ac.jp

Abstract. We discuss emergence of superconductivity in a two-dimensional e_g-electron system coupled with Jahn-Teller phonons in the framework of the Migdal-Eliashberg theory. Here we focus on the dependence of superconducting transition temperature T_c on the Fermi-surface structure controlled by the Slater-Koster integrals of e_g-electron hopping. When the Fermi-surface structure is abruptly changed, in general, there appears the van Hove singularity in the density of states, leading to the enhancement of T_c. In addition to it, for the superconductivity induced by Jahn-Teller phonons, we also observe the increase of T_c apart from the van Hove singularity point, when the e_g-electron system exhibits disconnected Fermi surfaces. Even for the pocket-like Fermi-surface structure, we find the relatively high T_c in comparison with the case of single Fermi surface. This is understood by the fact that the pair-hopping attraction between Cooper pairs on different Fermi surfaces is enhanced by Jahn-Teller phonons.

1. Introduction

In the research field of condensed matter physics, it has been an important long-standing issue to develop materials which exhibit high superconducting transition temperature T_c. Since the discovery of high-T_c cuprate superconductors [1], it has been widely recognized that there exists a route to obtain anisotropic superconductivity with relatively high T_c originating from strong electron correlation. In particular, concerning the superconductivity characterized by singlet Cooper pair, it has been established that there occurs d-wave superconductivity due to spin fluctuations near the antiferromagnetic phase [2, 3, 4].

Recently, another possible route to obtain exotic superconductivity has attracted much attention, even though T_c is not always so high. Namely, multiband effect has been focused in correlated electron materials such as Sr_2RuO_4 [5], MgB_2 [6], $Na_{0.35}CoO_2 \cdot 1.3H_2O$ [7], and iron pnictides [8, 9]. Several kinds of heavy-electron superconductors also exhibit multi-sheets of Fermi surfaces. Immediately after the Bardeen-Cooper-Schrieffer theory [10], the multiband effect on T_c was discussed in a simple two-band electron model [11]. In recent years, multiband superconductivity has been investigated from various viewpoints [12, 13, 14, 15, 16, 17, 18, 19, 20]. In particular, the discovery of superconductivity in iron pnictides has triggered the revival of the research of multiband superconductivity [21, 22, 23, 24].

In general, in electron systems with degenerate orbitals, Jahn-Teller effect plays an important role, for instance, in manganites [25, 26], since Jahn-Teller distortions are known to lift the degeneracy in electron orbitals. The attractive interaction mediated by Jahn-Teller phonons has been analyzed in a weak-coupling limit by one of the present authors [27]. It has been found

Journal of Physics: Conference Series **428** (2013) 012038 doi:10.1088/1742-6596/428/1/012038

that T_c is significantly enhanced by the interband attraction even within the weak-coupling limit, when Jahn-Teller phonons are active.

In this paper, the effect of the Fermi-surface topology on the emergence of superconductivity is discussed in the two-dimensional e_g-electron system coupled with Jahn-Teller phonons on the basis of the strong-coupling theory. It is well known that the van Hove singularity occurs in the density of states due to the abrupt change of the Fermi-surface structure, leading to the enhancement of T_c. We find that T_c is increased for the hopping parameters apart from the van Hove singularity point, when the e_g-electron system possesses disconnected Fermi surfaces in comparison with the case of single Fermi surface. The increase of T_c is due to the large pair-hopping amplitude between different Fermi surfaces enhanced by the Jahn-Teller phonons.

2. Model and Formulation

We consider the two-dimensional square lattice and the lattice constant is taken as unity. In the unit of $\hbar = k_B = 1$, the model Hamiltonian used here is given by

$$
\begin{aligned}
H = & \sum_{\boldsymbol{k},\sigma,\gamma,\gamma'} \varepsilon_{\boldsymbol{k}\gamma\gamma'} d^{\dagger}_{\boldsymbol{k}\gamma\sigma} d_{\boldsymbol{k}\gamma'\sigma} + \sum_{\boldsymbol{q}} [g_2(a_{2\boldsymbol{q}} + a^{\dagger}_{2-\boldsymbol{q}})\tau_{x\boldsymbol{q}} + g_3(a_{3\boldsymbol{q}} + a^{\dagger}_{3-\boldsymbol{q}})\tau_{z\boldsymbol{q}}] \\
& + \sum_{\boldsymbol{q}} [\omega_2(a^{\dagger}_{2\boldsymbol{q}} a_{2\boldsymbol{q}} + 1/2) + \omega_3(a^{\dagger}_{3\boldsymbol{q}} a_{3\boldsymbol{q}} + 1/2)],
\end{aligned} \tag{1}
$$

where $d_{\boldsymbol{k}\gamma\sigma}$ denotes an annihilation operator of an e_g electron with spin σ and momentum $\boldsymbol{k}$ in the orbital γ (=a and b), a and b indicate x^2-y^2 and $3z^2-r^2$ orbitals, respectively, $\varepsilon_{\boldsymbol{k}\gamma\gamma'}$ denotes the electron energy depending on orbitals, $a_{2\boldsymbol{q}}$ and $a_{3\boldsymbol{q}}$ are annihilation operators for (x^2-y^2)- and $(3z^2-r^2)$-type Jahn-Teller phonons, respectively, g_2 and g_3 indicate electron-phonon coupling constants, $\tau_{x\boldsymbol{q}}=\sum_{\boldsymbol{k},\sigma}(d^{\dagger}_{\boldsymbol{k}+\boldsymbol{q}a\sigma} d_{\boldsymbol{k}b\sigma} + d^{\dagger}_{\boldsymbol{k}+\boldsymbol{q}b\sigma} d_{\boldsymbol{k}a\sigma})$, $\tau_{z\boldsymbol{q}}=\sum_{\boldsymbol{k},\sigma}(d^{\dagger}_{\boldsymbol{k}+\boldsymbol{q}a\sigma} d_{\boldsymbol{k}a\sigma} - d^{\dagger}_{\boldsymbol{k}+\boldsymbol{q}b\sigma} d_{\boldsymbol{k}b\sigma})$, and ω_2 and ω_3 denote Jahn-Teller phonon energies for (x^2-y^2)- and $(3z^2-r^2)$-type modes, respectively. As for the electron kinetic energies $\varepsilon_{\boldsymbol{k}\gamma\gamma'}$, they are expressed as

$$
\begin{aligned}
\varepsilon_{\boldsymbol{k}aa} &= (1/2)[3(dd\sigma) + (dd\delta)](\cos k_x + \cos k_y), \\
\varepsilon_{\boldsymbol{k}bb} &= (1/2)[(dd\sigma) + 3(dd\delta)](\cos k_x + \cos k_y), \\
\varepsilon_{\boldsymbol{k}ab} &= \varepsilon_{\boldsymbol{k}ab} = -(\sqrt{3}/2)[(dd\sigma) - (dd\delta)](\cos k_x - \cos k_y),
\end{aligned} \tag{2}
$$

where $(dd\sigma)$ and $(dd\delta)$ denote the Slater-Koster integrals.

In order to evaluate the superconducting transition temperature T_c, we solve the gap equation obtained by evaluating both normal and anomalous electron self-energies due to the electron-phonon interaction. Here we consider the spin-singlet s-wave Cooper pair with the even function of frequency. In the second-order perturbation theory in terms of g_2 and g_3, the linearized gap equation at $T = T_c$ is expressed by

$$
\phi_{\mu\nu}(i\omega_n) = T \sum_{n'} \sum_{\mu',\nu'} V_{\mu\mu',\nu\nu'}(i\omega_n - i\omega_{n'}) F_{\mu'\nu'}(i\omega_{n'}), \tag{3}
$$

where $\phi_{\mu\nu}(i\omega_n)$ is anomalous self-energy, ω_n is fermion Matsubara frequency defined by $\omega_n = (2n+1)\pi T$ with a temperature T and an integer of n, F is anomalous Green's function, and $V_{\mu\mu',\nu\nu'}(i\nu_n)$ is phonon-mediated interaction, given by

$$
\begin{aligned}
V_{aa,aa}(i\nu_n) &= V_{bb,bb}(i\nu_n) = -V_{aa,bb}(i\nu_n) = -V_{bb,aa}(i\nu_n) = g_3^2 D_3(i\nu_n), \\
V_{ab,ab}(i\nu_n) &= V_{ab,ba}(i\nu_n) = V_{ba,ab}(i\nu_n) = V_{ba,ba}(i\nu_n) = g_2^2 D_2(i\nu_n),
\end{aligned} \tag{4}
$$

and other components are zeros. Here ν_n is boson Matsubara frequency defined by $\nu_n = 2\pi T n$ and D_2 and D_3 are phonon Green's functions, respectively, given by

$$D_2(i\nu_n) = \frac{2\omega_2}{(i\nu_n)^2 - \omega_2^2}, \quad D_3(i\nu_n) = \frac{2\omega_3}{(i\nu_n)^2 - \omega_3^2}. \tag{5}$$

In the vicinity of T_c, F is given in the linearized form as

$$F_{\mu\nu}(i\omega_n) = -\sum_{\boldsymbol{k}} \sum_{\mu'\nu'} G_{\mu\mu'}(\boldsymbol{k}, i\omega_n) G_{\nu\nu'}(-\boldsymbol{k}, -i\omega_n) \phi_{\mu'\nu'}(i\omega_n), \tag{6}$$

where G denotes normal Green's function, given by

$$[G(\boldsymbol{k}, i\omega_n)]_{\mu\nu}^{-1} = [G^{(0)}(\boldsymbol{k}, i\omega_n)]_{\mu\nu}^{-1} - \Sigma_{\mu\nu}(i\omega_n). \tag{7}$$

Here G_0 is non-interacting electron Green's function, given in the matrix form as

$$G^{(0)}(\boldsymbol{k}, i\omega_n) = \begin{pmatrix} i\omega_n + \mu - \varepsilon_{\boldsymbol{k}aa} & -\varepsilon_{\boldsymbol{k}ab} \\ -\varepsilon_{\boldsymbol{k}ba} & i\omega_n + \mu - \varepsilon_{\boldsymbol{k}bb} \end{pmatrix}^{-1}, \tag{8}$$

where μ is a chemical potential determined from the relation of $n = 2T\sum_n \sum_{\boldsymbol{k}} \mathrm{Tr}G(\boldsymbol{k}, i\omega_n)$ and n is electron number per site. The normal self-energy $\Sigma_{\mu\nu}(i\omega_n)$ is given in the second-order perturbation theory in terms of g_2 and g_3 as

$$\Sigma_{\mu\nu}(i\omega_n) = -T \sum_{n'} \sum_{\mu',\nu'} \sum_{\boldsymbol{k}'} V_{\mu\mu',\nu'\nu}(i\omega_n - i\omega_{n'}) G^{(0)}_{\mu'\nu'}(\boldsymbol{k}', i\omega_{n'}). \tag{9}$$

In order to obtain T_c, first we calculate the normal self-energy Σ in eq. (9). Next we solve the gap equation eqs. (3) and (6) by using G in eq. (7). Then, we obtain T_c as a temperature at which the positive maximum eigenvalue of eq. (3) becomes unity. Throughout this paper, the energy unit t is taken as $t = |(dd\sigma)|$. Note that we set $(dd\sigma) = -t$ in this paper. For the sum on the imaginary axis, we use 131,072 Matsubara frequencies. In order to accelerate the sum of large amount of Matsubara frequencies in eqs. (3) and (9), we exploit the fast-Fourier-transformation algorithm. For the evaluation of the eigenvalue of the gap equation eq. (3), we use the power method. As for the phonon energy, we set $\omega_2/t = \omega_3/t = 0.2$. We define the non-dimensional electron-phonon coupling constants as $\lambda_2 = 2g_2^2/(t\omega_2)$ and $\lambda_3 = 2g_3^2/(t\omega_3)$, which are set as $\lambda_2 = \lambda_3 = 0.8$. Concerning the electron number density, we set $n = 1.5$ in this paper.

3. Calculated Results

In Fig. 1, we show the change of the Fermi surfaces, when we control the Slater-Koster integral $(dd\delta)/t$ from -2.0 to 2.0 for $(dd\sigma)/t = -1.0$. We note that the term of "Fermi surface" is conventionally used here, although we consider the two-dimensional system. Note also that the normal self-energy effect is included in the determination of the chemical potential μ when we fix the electron number density n as $n = 1.5$. The temperature is set as $T/t = 10^{-4}$ in Fig. 1.

For $(dd\delta) < 0$, we find a couple of Fermi surfaces, except for the case of $(dd\delta)/t = -1.0$, at which two Fermi surfaces accidentally coincide with each other, as understood from the electron dispersion of eq. (2). For the cases of $(dd\delta)/t = -1.5$ and -0.66, we observe that the Fermi surface touches the X and Y points. Namely, for such cases, the van Hove singularities will appear at the Fermi level in the density of states.

At $(dd\delta) = 0$, the center of one Fermi surface exists on the Γ point, while another has the center at the M point. Note that the nesting vectors are approximately given by $(\pi, 0)$ and

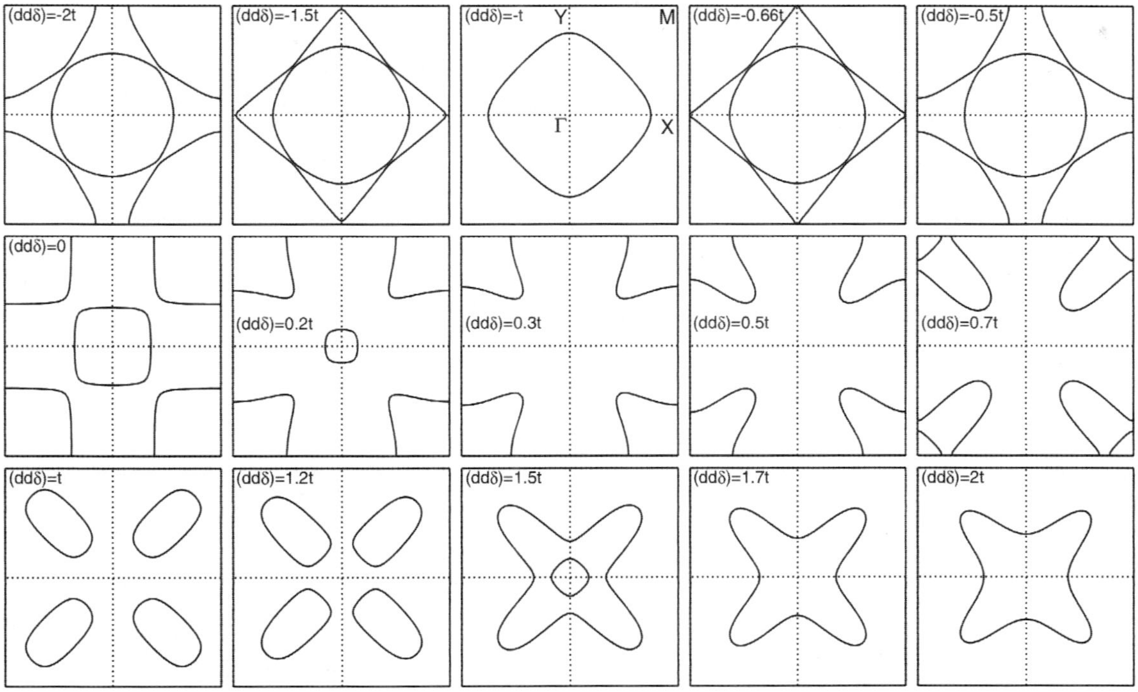

Figure 1. Variation of the Fermi surfaces for $(dd\sigma)/t = -1.0$ depicted in the first Brillouin zone of $-\pi \le k_x \le \pi$ and $-\pi \le k_y \le \pi$ with $\Gamma = (0,0)$, X=$(\pi,0)$, Y=$(0,\pi)$, and M=(π,π).

$(0,\pi)$ in this case, leading to the stabilization of the metallic phase in contrary to the cooperative Jahn-Teller distortion with the ordering vector of (π,π) [28].

When we increase the value of $(dd\delta)$ from zero, the Fermi-surface topology is the same as that of the case of $(dd\delta) = 0.0$ for a while, although the shape of Fermi surfaces is gradually changed. For $(dd\delta)/t$ between 0.2 and 0.3, the Fermi surface with the center at the Γ point disappears and we obtain single Fermi surface.

For $(dd\delta)/t \approx 0.6$, there appears another Fermi surface with the center at the M point. At $(dd\delta)/t = 1.0$, there occurs the reconstruction of the Fermi-surface structure and we find four pocket-like Fermi surfaces around the points of $(\pm\pi/2, \pm\pi/2)$ and $(\pm\pi/2, \mp\pi/2)$. This pocket-like Fermi-surface structure continues to exist for a while, but for $(dd\delta)/t \approx 1.3$, the four pocket-like Fermi surfaces are connected so as to form a star-like Fermi surface. At $(dd\delta)/t \approx 1.5$, in addition to the large-size star-like Fermi surface, we find a small-size Fermi surface with the center at the Γ point. This small-size Fermi surface disappears at $(dd\delta)/t \approx 1.6$. Then, for $(dd\delta)/t \gtrsim 1.6$, we observe the single Fermi surface.

Let us summarize the change of the number $N_{\rm F}$ of the Fermi surfaces: $N_{\rm F} = 2$ for $-2.0 \le (dd\delta)/t \lesssim 0.25$, $N_{\rm F} = 1$ for $0.25 \lesssim (dd\delta)/t \lesssim 0.6$, $N_{\rm F} = 2$ for $0.6 \lesssim (dd\delta)/t \lesssim 0.8$, $N_{\rm F} = 4$ for $0.8 \lesssim (dd\delta)/t \lesssim 1.3$, $N_{\rm F} = 2$ for $1.3 \lesssim (dd\delta)/t \lesssim 1.6$, and $N_{\rm F} = 1$ for $1.6 \lesssim (dd\delta)/t \le 2.0$. Note that we consider the extended zone when we count $N_{\rm F}$.

Now we consider the change of $T_{\rm c}$ depending on the Fermi-surface topology. In Fig. 2, we show $T_{\rm c}/t$ vs. $(dd\delta)/t$ in the range of $-2.0 \le (dd\delta)/t \le 2.0$. In the region of negative $(dd\sigma)$, we observe the overall tendency of the increase of $T_{\rm c}$ with the increase of $(dd\delta)/t$ from -2.0 to 0.0, except for the cusps at $(dd\delta)/t = -1.5$ and -0.7. As easily understood from the Fermi-surface structure, those cusps originate from the van Hove singularities. The overall tendency of the increase of $T_{\rm c}$ is explained by the increase of the density of states. When we decrease $(dd\delta)$ in

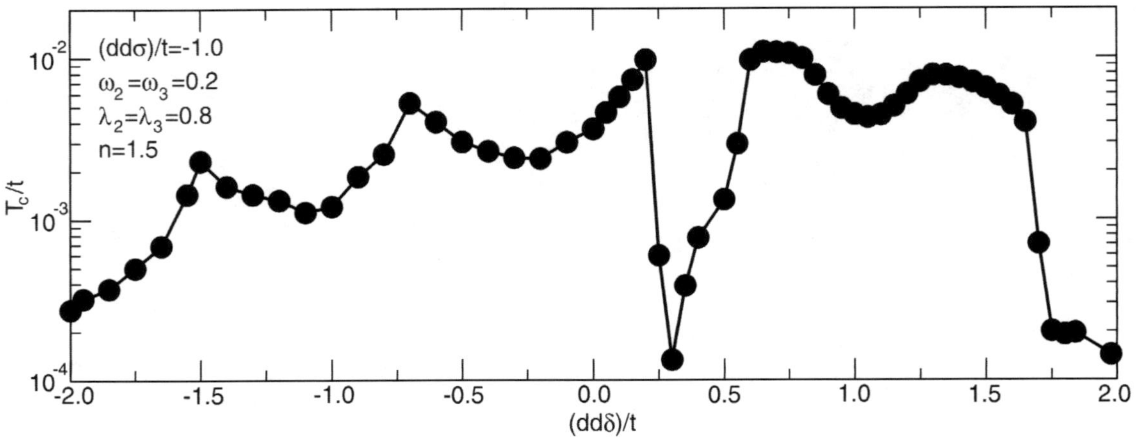

Figure 2. Superconducting transition temperature T_c/t vs. $(dd\delta)/t$ for $(dd\sigma)/t = -1.0$, $\omega_2/t = \omega_3/t = 0.2$, $\lambda_2 = \lambda_3 = 0.8$, and $n = 1.5$ in the range of $-2.0 \le (dd\delta)/t \le 2.0$.

the side of negative $(dd\delta)$, the bandwidth is monotonically increased with the increase of $|(dd\delta)|$. Since the Fermi-surface topology is not so drastically changed in the region of $(dd\delta) < 0$, as understood from the discussion of N_F, it is possible to explain qualitatively the behavior of T_c from the density of states and the van Hove singularity.

Note that T_c is continuously changed at $(dd\delta)/t = -1.0$, in which two Fermi surfaces are perfectly degenerate. In this paper, we consider only the spin-singlet s-wave solution, which has no singular behavior around at $(dd\delta)/t = -1.0$. However, when two Fermi surfaces are degenerate, in general, it is possible to consider other solutions such as odd-frequency and spin-triplet pairs. It may be interesting to pursue such exotic solutions, but we do not consider them in the present paper, since the perfectly degenerate Fermi surfaces are unrealistic.

When we increase $(dd\delta)$ from zero, T_c is monotonically increased for $0 \le (dd\delta)/t \lesssim 0.2$. This is also understood from the increase of the density of states. At $(dd\delta)/t = 0.2$, we find a cusp which originates from the van Hove singularity due to the abrupt change of the Fermi-surface topology, i.e., the disappearance of the Fermi surface with the center at the Γ point. For $0.2 \lesssim (dd\delta)/t \lesssim 0.6$, there appears the single Fermi surface and T_c is totally low, since the density of states is suppressed in this region. However, when we increase $(dd\delta)$, T_c is rapidly increased. Around at $(dd\delta)/t = 0.6$, there occurs a new Fermi surface and the Fermi-surface topology is abruptly changed again, leading to the van Hove singularity.

For $0.6 \lesssim (dd\delta)/t \lesssim 1.6$, T_c depends on the values of $(dd\delta)$, but we observe that the magnitude of T_c is kept high as $T_c/t = 0.005 \sim 0.01$, even though the Fermi-surface topology is changed in this region. Here we emphasize that we always find the plural numbers of the Fermi surfaces in the corresponding region. We imagine that the value of T_c is heavily decreased, when N_F is reduced to 1. In fact, for $(dd\delta)/t \gtrsim 1.6$ with the single Fermi surface, we observe the rapid decrease of T_c, which cannot be simply explained by the decrease of the density of states. Note that for $1.6 \lesssim (dd\delta)/t \le 2.0$, T_c is totally low and it is difficult to obtain T_c with enough precision in this region, since T_c approaches the lower limit which we can safely calculate.

The enhancement of T_c in the case of the plural numbers of the Fermi surfaces is explained by the enhancement of the pair-hopping attraction between Cooper pairs on different Fermi surfaces due to Jahn-Teller phonons [27]. This pair-hopping attraction does not work for the case of the single Fermi surface. Thus, T_c in the case of the plural numbers of the Fermi surfaces is enhanced in comparison with the case of $N_F = 1$. This enhancement mechanism still works even for the case of small-size pocket-like Fermi surfaces.

4. Discussion and Summary

In this paper, we have ignored the effect of Coulomb interaction. Here it is interesting to consider a possibility that the pair-hopping attraction due to Jahn-Teller phonons is changed to be repulsive by the effect of inter-band Coulomb interaction. If we include the breathing-mode phonons and the intra-band pair interaction is kept negative, we obtain the superconductivity even if the pair-hopping amplitude is repulsive. Note that in such a case, we may provide an alternative scenario for $s_{\pm}$-wave pairing, which has been proposed for iron pnictides [21, 22, 23, 24]. It is one of our future problems to treat both Jahn-Teller phonons and inter-band Coulomb repulsion.

In summary, we have found that in the two-band system coupled with Jahn-Teller phonons, T_c becomes relatively high for the case of the plural numbers of the Fermi surfaces. This result is understood by the pair-hopping attraction between different Fermi surfaces enhanced by Jahn-Teller phonons.

Acknowledgments

This work has been supported by a Grant-in-Aid for Scientific Research (C) (No. 24540379) of Japan Society for the Promotion of Science and a Grant-in-Aid for Scientific Research on Innovative Areas "Heavy Electrons" (No. 20102008) of The Ministry of Education, Culture, Sports, Science, and Technology, Japan. The computation in this work has been done using the facilities of the Supercomputer Center of Institute for Solid State Physics, University of Tokyo.

References

[1] J. G. Bednorz and K. A. Müller: Z. Phys. B **64** (1986) 189.
[2] T. Moriya and K. Ueda: Adv. Phys. **49** (2000) 555.
[3] T. Moriya and K. Ueda: Rep. Prog. Phys. **66** (2003) 1299.
[4] Y. Yanase, T. Jujo, T. Nomura, H. Ikeda, T. Hotta, and K. Yamada: Phys. Rep. **387** (2003) 1.
[5] Y. Maeno, H. Hashimoto, K. Yoshida, S. Nishizaki, T. Fujita, J. G. Bednorz, and F. Lichtenberg: Nature (London) **372** (1994) 532.
[6] J. Nagamatsu, N. Nakagawa, Y. Muranaka, Y. Zenitani, and J. Akimitsu: Nature (London) **410** (2001) 63.
[7] K. Takada, H. Sakurai, E. Takayama-Muromachi, F. Izumi,R. A. Dilanian, and T. Sasaki: Nature (London) **422** (2003) 53.
[8] Y. Kamihara, T. Watanabe, M. Hirano, and H. Hosono: J. Am. Chem. Soc. **130** (2008) 3296.
[9] K. Ishida, Y. Nakai, and H. Hosono: J. Phys. Soc. Jpn. **78** (2009) 062001.
[10] J. Bardeen, L. N. Cooper, and J. R. Schrieffer: Phys. Rev. **108** (1957) 1175.
[11] H. Suhl, B. T. Matthias, and L. R. Walker: Phys. Rev. Lett. **2** (1959) 552.
[12] T. Takimoto: Phys. Rev. B **62** (2000) 14641.
[13] A. Bussmann-Holder, M. Gulacsi, and A. R. Bishop: Phil. Mag. B **82** (2002) 1749.
[14] T. Nomura and K. Yamada: J. Phys. Soc. Jpn. **71** (2002) 404.
[15] T. Takimoto, T. Hotta, and K. Ueda: Phys. Rev. B **69** (2004) 104504.
[16] Y. Yanase, M. Mochizuki, and M. Ogata: J. Phys. Soc. Jpn. **74** (2005) 430.
[17] Y. Yang: Physica D **200** (2005) 60.
[18] K. Yada and H. Kontani: J. Phys. Soc. Jpn. **75** (2006) 033705.
[19] O. V. Dolgov and A. A. Golubov: Phys. Rev. B **77** (2008) 214526.
[20] C. Bersier, A. Floris, P. Cudazzo, G. Profeta, A. Sanna, F. Bernardini, M. Monni, S. Pittalis, S. Sharma, H. Glawe, A. Continenza, S. Massidda, and E. K. U. Gross: J. Phys.: Condens. Matter **21** (2009) 164209.
[21] I. I. Mazin, D. J. Singh, M. D. Johannes, and M. H. Du: Phys. Rev. Lett. **101** (2008) 057003.
[22] K. Kuroki, S. Onari, R. Arita, H. Usui, Y. Tanaka, H. Kontani, and H. Aoki: Phys. Rev. Lett. **101** (2008) 087004.
[23] H. Ikeda: J. Phys. Soc. Jpn. **77** (2008) 123707.
[24] T. Nomura: J. Phys. Soc. Jpn. **78** (2009) 034716.
[25] E. Dagotto, T. Hotta, and A. Moreo: Phys. Rep. **344** (2001) 1.
[26] T. Hotta: Rep. Prog. Phys. **69** (2006) 2061.
[27] T. Hotta: J. Phys. Soc. Jpn. **79** (2010) 023709.
[28] T. Hotta: Phys. Rev. B **67** (2003) 104428.

Journal of Physics: Conference Series **428** (2013) 012039 doi:10.1088/1742-6596/428/1/012039

Effect of electron-phonon coupling in the ARPES spectra of the tri-layer cuprate $Bi_2Sr_2Ca_2Cu_3O_{10+\delta}$

S. Ideta[1], T. Yoshida[1], M. Hashimoto[1], A. Fujimori[1], H. Anzai[2], A. Ino[2], M. Arita[3], H. Namatame[3], M. Taniguchi[2,3], K. Takashima[1], K. M. Kojima[1], S. Uchida[1]

[1]Department of Physics, University of Tokyo, Bunkyo-ku, Tokyo 113-0033, Japan
[2]Graduate School of Science, Hiroshima University, Higashi-Hiroshima 739-8526, Japan
[3]Hiroshima Synchrotron Center, Hiroshima University, Higashi-Hiroshima 739-0046, Japan

E-mail: ideta@ap.t.u-tokyo.ac.jp

Abstract. Angle-resolved photoemission spectroscopy using tunable low energy photons allows us to study the quasi-particle (QP) dispersions of the inner and outer CuO_2 planes (IP and OP) separately in the tri-layer cuprate $Bi_2Sr_2Ca_2Cu_3O_{10+\delta}$ (Bi2223). The kink energy of the OP band is $\sim$ 70 meV, as observed in various high-T_c cuprates, while that of the IP band is as large as $\sim$ 100 meV in the superconducting (SC) state. This large kink energy is attributed to the $\sim$35 meV buckling mode plus the large ($\sim$ 60 meV) SC gap of IP. The IP band also shows a weak kink feature at $\sim$ 70 meV in the SC state. The latter feature can be explained either by the 70 meV half-breathing mode or by the $\sim$ 35 meV buckling-phonon mode plus the $\sim$ 40 meV SC gap of OP if interlayer scattering of QP is involved.

1. Introduction

Attractive interaction between two electrons mediated by a virtual phonon excitation is the origin of Cooper paring in conventional superconductors. Virtual boson excitations are also likely to be the origin of Cooper paring in the high-temperature cuprate superconductors (HTSCs), but the nature of the bosons in HTSCs, phonons or magnetic excitations, has remained controversial. Because electron-boson coupling causes a change in the Fermi velocity below the boson energy, resulting in a kink in the quasi-particle (QP) dispersion curves. It has been controversial whether the kink is due to phonons [1, 2, 3, 4, 5, 6, 7] or magnetic excitations [8, 9, 10, 11, 12]. The kink in the nodal region is relatively simple because of the absence of a d-wave superconducting (SC) gap in that direction, and may allow us a relatively straightforward interpretation. According to previous theoretical work [13, 14], the nodal QP can be scattered into the antinodal region, where the SC gap is maximum Δ and, therefore, the nodal kink energy can be given by E_{kink} $= \Omega + \Delta$ in the SC state and $E_{\text{kink}} = \Omega$ in the normal state, where Ω is the energy of the boson mode and Δ is the magnitude of the SC order parameter at the anti-node.

ARPES experiment on the optimally doped double-layer $Bi_2Sr_2CaCu_2O_{8+\delta}$ (Bi2212) [15] has shown the nodal kink energy to be E_{kink} 70 meV in the SC state, which is interpreted as the sum of the boson mode energy Ω 35 meV and the anti-nodal SC gap Δ 35 meV, consistent with the relationship $E_{\text{kink}} = \Omega + \Delta$. Here, one of the likely candidates for the 35 meV mode is the out-of-plane vibration (buckling mode) of Cu and O atoms. However, an in-plane optical phonon mode (half-breathing mode), which does not strongly scatter the nodal QP into the anti-nodal region but scatters the nodal QP to another nodal region with $\Delta = 0$ [14], should have an energy 70 meV, too, according to $E_{\rm kink} = \Omega + \Delta$. Therefore, one cannot determine whether the kink at 70 meV is due to the buckling mode which scatters the nodal QP to the antinodal region where the SC gap opens, or due to the half-breathing mode which scatters the nodal QP to another nodal region, or both.

In order to study the origin of the kink, we have performed an ARPES study on the tri-layer cuprate $Bi_2Sr_2Ca_2Cu_3O_{10+\delta}$ (Bi2223), which shows the highest T_c of 110 K among the Bi-based HTSCs. The ARPES spectra of Bi2223 reveal two bands corresponding to the outer and inner CuO_2 planes (OP and IP) with different Δ's for the OP and IP [16]. We have found that the kink energies in the nodal direction are remarkably different between the IP and OP bands, which can be attributed to the different anti-nodal Δ's between IP and OP, consistent with the scenario $E_{\rm kink} = \Omega + \Delta$ [13, 14]. Furthermore, we have identified a fine structure in the kink of the IP band, which may be partially attributed to the scattering of the nodal QP from IP to OP through the excitation of a buckling phonon mode.

2. Experimental

Single crystals of optimally doped Bi2223 ($T_c = 110$ K) were grown by the traveling solvent floating zone method. ARPES experiments were carried out at beam line 9A of the Hiroshima Synchrotron Radiation Center (HiSOR), Hiroshima University, using circularly polarized light with h = 7.55, 8.4, and 11.95 eV. The total energy resolution (ΔE) was set better than 5 meV. The samples were cleaved *in situ* under an ultrahigh vacuum of 1 10 ^{11} Torr.

3. Results and Discussion

As shown in Figs. 1(a) - 1(c), the OP and IP bands are separately observed in the nodal direction by tuning the photon energy. Black dots indicate band dispersions determined by the peak positions of the momentum distribution curves (MDCs). One can see that the OP and IP bands show anomalies at 75 and 100 meV, respectively, as shown by a red arrow in the figures. In Fig. 1(d), Fermi velocities ($v_{\rm F}$'s) for the OP and IP bands are plotted as functions of hole concentration estimated from the Fermi surface area [16], and show a constant value. Here, we note that the $v_{\rm F}$ of Bi2223 has been estimated by a linear fit from the Fermi level ($E_{\rm F}$) to 30 - 40 meV below it. The $v_{\rm F}$ of Bi2212 in previous ARPES studies defined by the average group velocity in the same energy range is also plotted [17, 18], showing a nearly constant value [19]. (The doping-dependent low energy kink observed at < 20 meV [17, 18] has been ignored in the present analysis.)

In order to analyze the behaviors of the kinks more quantitatively, the real and imaginary parts of the electron self-energy, ReΣ and ImΣ, have been deduced as shown in Figs. 2(a) and 2(b), respectively. Here, ReΣ has been deduced from the difference between the observed QP dispersion and the bare band dispersion, which is assumed to be linear as shown in Fig. 1(e). The peak positions of ReΣ corresponding to the energies of the kinks for OP and IP are shown by shaded region. One finds that the peaks of ReΣ for the OP and IP bands obviously occur at different energies: for OP 75 meV, similar to that of Bi2212 [1, 2, 3, 8], while for IP as large as 100 meV. From the Kramers-Kronig relation, a kink in ReΣ is expected to manifest itself as a step in ImΣ. In fact, Fig. 2(b) shows that ImΣ's for the OP and IP bands exhibit the steepest slope around the corresponding kink energies.

The different kink energies of the OP and IP bands can be easily understood if the 35 meV boson mode scatters the nodal QP into the anti-nodal region, where the SC gaps of different magnitudes open for the OP and IP bands. According to our previous studies [16, 20], the SC gaps in the anti-nodal region are $\Delta = 40$ meV for the OP band and $\Delta = 60$ meV for the IP band. Accordingly, if the formula $E_{\rm kink} = \Omega + \Delta$ is applied, $E_{\rm kink} = \Omega + \Delta$ 35 + 40 75 (5) meV for the OP band and $E_{\rm kink} = \Omega + \Delta$ 35 + 60 95 (5) meV for the IP band, consistent

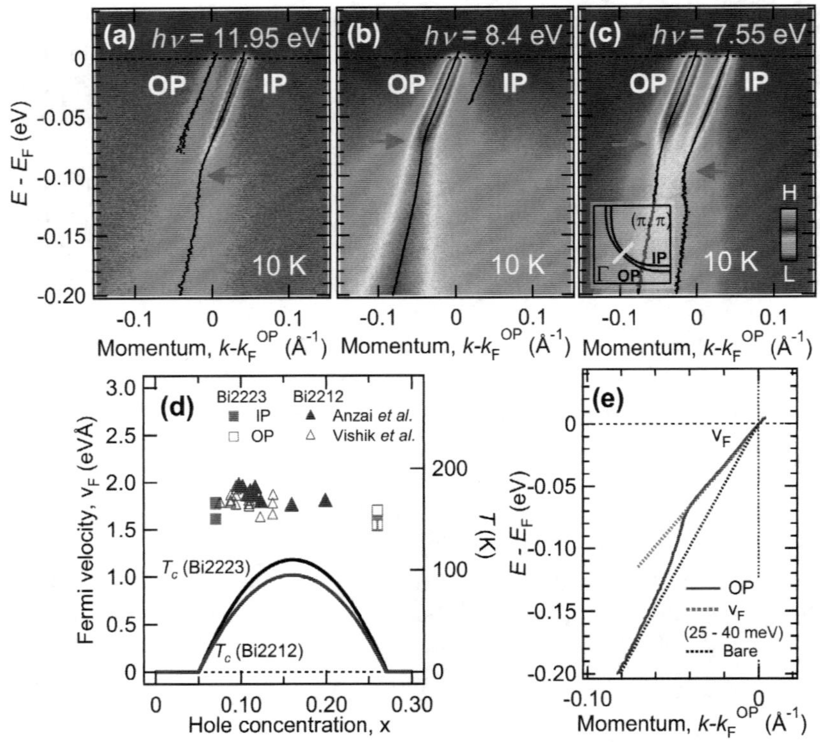

Figure 1. ARPES spectra of the tri-layer Bi2223 in the nodal direction. (a)-(c): Outer-plane (OP) and inner-plane (IP) bands in the superconducting state taken with different photon energies. Red arrows show an anomaly of the band dispersion, i.e., a kink. (d): Fermi velocities (v_F's) for the OP and IP bands plotted against hole concentration. The v_F's of Bi2212 defined by the average group velocity between E_F and 40 meV below it for various hole concentrations are also shown. (e): Dispersion of the OP band derived from panel (b). Black dotted line denotes the assumed linear bare band. Red dotted line is a linear fit to the dispersion from E_F to 25 - 40 meV below it.

with experiment, and the unusually large kink energy for the IP band can be understood. Since phonon bands are located from 12 to 72 meV according to optical measurements of Bi2223 [22], the kink energy of 100 meV observed for the IP band cannot be explained without the contribution of the large SC gap of the IP band. The half-breathing mode of energy 70 meV will also contribute to a kink at 70 - 75 meV for both the OP and IP bands though $E_{\rm kink} = \Omega + \Delta$ 70 + 0 meV, since the half-breathing mode scatters a nodal QP to another nodal region, where Δ 0. Indeed, the weak shoulder at 70 meV in the ReΣ of IP in the SC state may be attributed to the half-breathing mode. Alternatively, the 70 meV shoulder may be attributed to scattering of the nodal QP into the antinodal region of OP through the 35 meV buckling mode.

Figures 2(c) - 2(e) show the ImΣ and the MDC width at E_F of the OP and IP bands at various temperatures. Panel (e) shows that the MDC widths at E_F of the OP and IP bands increase with temperature and shows a discontinuous change around $T_c = 110$ K, like the previous experimental results on Bi2212 [3]. This means that the scattering of the nodal QP is suppressed by the opening of the SC gap. Figures 2(f) and 2(g) show ReΣ for the OP and IP bands at various temperatures. The peak positions of ReΣ do not show an abrupt change at $T_c = 110$

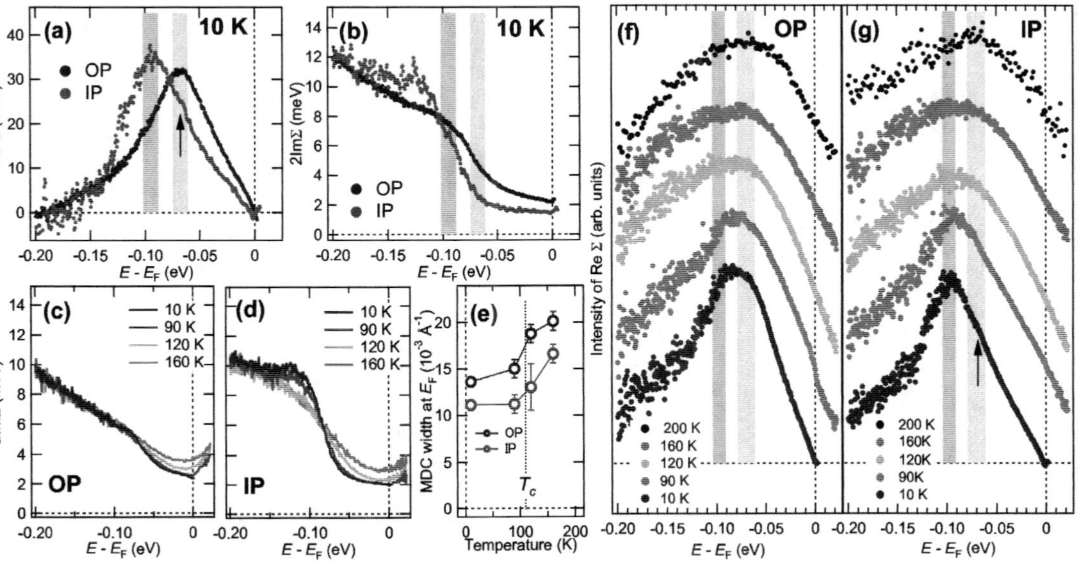

Figure 2. Real and imaginary parts (ReΣ amd ImΣ) of the self-energy Σ of Bi2223. (a),(b): ReΣ and 2ImΣ of the OP and IP bands. Shaded regions indicate the major kink energies. ReΣ has been obtained by assuming a linear bare band shown in Fig. 1(e). (c), (d): 2ImΣ for the OP and IP bands at various temperatures. (e): MDC width at E_F of the OP and IP bands at various temperatures. (f),(g): ReΣ of the OP and IP bands at various temperatures. Gray and pink shaded regions highlight the ReΣ peak positions of the OP and IP bands, respectively.

K but at $T = 200$ K, the peak position for the IP band becomes 70 meV. Because the SC gap closes at $T_c = 110$ K and the pseudogap closes at temperature T well above T_c, the change in the kink energy of the IP band would be attributed to the closure or the reduction of the pseudogap in the antinodal region. The kink energy of 70 meV at T 200 K is then attributed to the energy of the breathing mode.

As mentioned above, there are two possible explanations for the weak 70 meV shoulder in the ReΣ of IP: (i) the nodal QP of IP is scattered into the other nodal region through the excitation of the half-breathing mode of energy 70 meV, or (ii) the nodal QP of IP is scattered into the antinodal region of OP through the excitation of 35 meV phonons. If the latter scattering exist, a Cooper pair near the node of IP will be scattered into the antinodal region of OP with the virtual excitation of the 35 meV boson. This can become a mechanism which enhances T_c in multilayer cuprates [23]. As the 35 meV boson, we consider the buckling mode [24] because such a mode scatters the nodal QP to the anti-nodal one according to Refs. [14, 21].

In conclusion, we have performed an ARPES experiment on Bi2223 and found that the kink energy shows different values between IP and OP: 75 meV for the OP band and 100 meV for the IP band. This difference is attributed to the different SC gap magnitudes between OP and IP. Furthermore, we find that the ReΣ of IP shows a shoulder structure at 70 meV, which may indicate that (1) the QP of the IP is scattered by the half-breathing mode, or (2) the inter-layer scattering mediated by the buckling mode occurs.

Acknowledgments

We would like to thank Z.-X. Shen for enlightening discussion. ARPES experiments were carried out at HiSOR (Proposal No. 07-A-10, and 08-A-35). This work was supported by a Grant-in-Aid for Scientific Research, Award for Young Scientist (B), JSPS, Japan, and A3 Foresight Program "Joint Research on Novel Properties of Complex Oxides". S.I. acknowledges support from the Japan Society for the Promotion of Science.

References

[1] A. Lanzara, P. V. Bogdanov, X. J. Zhou, S. A. Kellar, D. L. Feng, E. D. Lu, T. Yoshida, H. Eisaki, A. Fujimori, K. Kishio, J.-I. Shimoyama, T. Nodak, S. Uchida, Z. Hussain, and Z.-X. Shen, Nature **412**, 510 (2001).
[2] T. Cuk, F. Baumberger, D. H. Lu, N. Ingle, X. J. Zhou, H. Eisaki, N. Kaneko, Z. Hussain, T. P. Devereaux, N. Nagaosa, and Z.-X. Shen, Phys. Rev. Lett. **93**, 117003 (2004).
[3] T. Yamasaki, K. Yamazaki, A. Ino, M. Arita, H. Namatame, M. Taniguchi, A. Fujimori, Z.-X. Shen, M. Ishikado, and S. Uchida, Phys. Rev. B **75**, 140513(R) (2007).
[4] Hae-Young Kee, Steven A. Kivelson, and G. Aeppli, Phys. Rev. Lett. **88**, 257002 (2002).
[5] T. Cuk, D. H. Lu, X. J. Zhou, Z.-X. Shen, T. P. Devereaux, and N. Nagaosa, Phys. Stat. Sol. (b) **242**, 11-29 (2005).
[6] Y. Chen, A. Iyo, W. Yang, A. Ino, M. Arita, S. Johnston, H. Eisaki, H. Namatame, M. Taniguchi, T. P. Devereaux, Z. Hussain, and Z.-X. Shen, Phys. Rev. Lett. **103**, 036403 (2009).
[7] Jinho Lee, K. Fujita, K. McElroy, J. A. Slezak, M. Wang, Y. Aiura, H. Bando, M. Ishikado, T. Masui, J.-X. Zhu, A. V. Balatsky, H. Eisaki, S. Uchida, and J. C. Davis, Nature **442**, 546 (2006).
[8] P. V. Bogdanov, A. Lanzara, S. A. Kellar, X. J. Zhou, E. D. Lu, W. J. Zheng, G. Gu, J.-I. Shimoyama, K. Kishio, H. Ikeda, R. Yoshizaki, Z. Hussain, and Z. X. Shen, Phys. Rev. Lett. **85**, 2581 (2000).
[9] H. F. Fong, B. Keimer, and P. W. Anderson, D. Reznik, F. Dogan, and I. A. Aksay, Phys. Rev. Lett. **75**, 316 (1995).
[10] J. P. Carbotte, E. Schachinger, and D. N. Basov, Nature **401**, 354 (1999).
[11] T. Dahm, V. Hinkov, S. V. Borisenko, A. A. Kordyuk, V. B. Zabolotnyy, J. Fink, B. Büchner, D. J. Scalapino, W. Hanke, and B. Keimer, Nature Physics **5**, 217 (2009).
[12] H. F. Fong, P. Bourges, Y. Sidis, L. P. Regnault, A. Ivanov, G. D. Gu, N. Koshizuka, and B. Keimer, Nature **398**, 588 (1999).
[13] A. W. Sandvik, D. J. Scalapino, and N. E. Bickers, Phys. Rev. Lett. **69**, 094523 (2004).
[14] T. P. Devereaux, T. Cuk, Z.-X. Shen, and N. Nagaosa, Phys. Rev. Lett. **93**, 117004 (2004).
[15] W. S. Lee, W. Meevasana, S. Johnston, D. H. Lu, I. M. Vishik, R. G. Moore, H. Eisaki, N. Kaneko, T. P. Devereaux, and Z. X. Shen, Phys. Rev. B **77**, 140504(R) (2008).
[16] S. Ideta, K. Takashima, M. Hashimoto, T. Yoshida, A. Fujimori, H. Anzai, T. Fujita, Y. Nakashima, A. Ino, M. Arita, H. Namatame, M. Taniguchi, K. Ono, M. Kubota, D. H. Lu, Z.-X. Shen, K. M. Kojima, and S. Uchida, Phys. Rev. Lett. **104**, 227001 (2010).
[17] I. M. Vishik, W. S. Lee, F. Schmitt, B. Moritz, T. Sasagawa, S. Uchida, K. Fujita, S. Ishida, C. Zhang, T. P. Devereaux, and Z. X. Shen, Phys. Rev. Lett. **204**, 207002 (2010).
[18] H. Anzai, A. Ino, T. Kamo, T. Fujita, M. Arita, H. Namatame, M. Taniguchi, A. Fujimori, Z.-X. Shen, M. Ishikado, and S. Uchida, Phys. Rev. Lett. **105**, 227002 (2010).
[19] X. J. Zhou, T. Yoshida, A. Lanzara, P. V. Bogdanov, S. A. Kellar, K. M. Shen, W. L. Yang, F. Ronning, T. Sasagawa, T. Kakeshita, T. Noda, H. Eisaki, S. Uchida, C. T. Lin, F. Zhou, J. W. Xiong, W. X. Ti, Z. X. Zhao, A. Fujimori, Z. Hussain, and Z.-X. Shen, Nature **423**, 398 (2003).
[20] S. Ideta, T. Yoshida, A. Fujimori, H. Anzai, T. Fujita, A. Ino, M. Arita, H. Namatame, M. Taniguchi, Z.-X. Shen, K. Takashima, K. M. Kojima, and S. Uchida, Phys. Rev. B **85**, 104515 (2012).
[21] S. Johnston and T. P. Devereaux, Phys. Rev. B **81**, 214512 (2010).
[22] A.V. Boris, D. Munzar, N. N. Kovaleva, B. Liang, C.T. Lin, A. Dubroka, A.V. Pimenov, T. Holden, B. Keimer, Y.-L. Mathis, and C. Bernhard, Phys. Rev. Lett. **89**, 277001 (2002).
[23] S. Chakravarty, H.-Y. Kee, and K. Völker, Nature **428**, 53 (2004).
[24] N. N. Kovaleva, A. V. Boris, T. Holden, C. Ulrich, B. Liang, C. T. Lin, B. Keimer, and C. Bernhard, J. L. Tallon, D. Munzar, and A. M. Stoneham, Phys. Rev. B **69**, 054511 (2004).

Journal of Physics: Conference Series **428** (2013) 012040 doi:10.1088/1742-6596/428/1/012040

Spin vortices, skyrmions and the Kosterlitz-Thouless transition in the two-dimensional antiferromagnet

Takashi Yanagisawa

Electronics and Photonics Research Institute, National Institute of Advanced Industrial Science and Technology, 1-1-1 Umezono, Tsukuba, Ibaraki 305-8568, Japan

E-mail: `t-yanagisawa@aist.go.jp`

Abstract. We investigate spin-vortex excitations in the two-dimensional antiferromagnet on the basis of the nonlinear sigma model. The model of two-dimensional Heisenberg quantum antiferromagnet is mapped onto the (2+1)D nonlinear sigma model. The 2D nonlinear sigma model has an instanton (or skyrmion) solution which describes an excitation of spin-vortex type. Quantum fluctuations of instantons are reduced to the study of the Coulomb gas, and the gas of instantons of the 2D nonlinear sigma model is in the plasma phase. We generalize this picture of instanton gas to the (2+1)D nonlinear sigma model. We show, using some approximation, that there is a Kosterlitz-Thouless transition from the plasma phase to the molecular phase as the temperature is lowered.

1. Introduction

It is important to understand the magnetic structure of two-dimensional antiferromagnetic in the light-doping region in the study of cuprate high-temperature superconductors[1]. The physics of underdoped region, which is expected to be closely related to the pseudogap state, is still not clear. The influence of doping holes on the antiferromagnetic state in the parent materials of cuprate superconductors is one of the most interesting problems in strongly correlated electron systems. It has been experimentally found that the stripe order is stabilized in the underdoped region of $La_{2\ x\ y}Nd_ySr_xCuO_4$,[4] $La_{2\ x}Sr_xCuO_4$ (LSCO)[2], and $La_{2\ x}Ba_xCuO_4$[3]. A checkerboard-like charge density modulation with a roughly $4a\quad 4a$ period (a is a lattice spacing) has also been observed by scanning tunneling microscopy experiments in $Bi_2Sr_2CaCu_2O_{8+\delta}$ (Bi-2212)[5],$Bi_2Sr_{2\ x}La_xCuO_{6+\delta}$[6], and $Ca_{2\ x}Na_xCuO_2Cl_2$ (NA-CCOC)[7]. It has been pointed out that these types of modulated structures can be understood within the framework of correlated electrons by using the variational Monte Carlo method for the two-dimensional Hubbard model[8, 9, 10].

It is expected that there appears a new state in an extremely light-doping region where the charge modulated structures such as stripes and checkerboards are instable and only spin-modulations, for example, a spin-vortex state, will be formulated. The purpose of this paper is to investigate spin-vortex excitation on the basis of the two-dimensional quantum antiferromagnet. In this paper, we use the mapping of a two-dimensional magnet onto a nonlinear sigma model. The nonlinear sigma model is a nonlinear eld theory and its renormalization properties were applied to ferromagnets in two dimensions[11, 12, 13]. The O(3) nonlinear sigma model can be interpreted as the continuum limit of an isotropic ferromagnet, and the action is

$$S = \frac{1}{2g}\int d^d x(\partial_\mu \quad)^2, \tag{1}$$

where the three-component scalar eld (x) is under the condition $\quad = 1$.

2. (2+1)D nonlinear sigma model

The two-dimensional quantum Heisenberg antiferromagnet is mapped to a (2+1) dimensional nonlinear sigma model. The (2+1)D nonlinear sigma model is[14]

$$S = \frac{1}{2g_0}\int_0^{1/k_B T} d \quad \int d^d\mathbf{r}\Big[(\nabla\varphi)^2 + \frac{1}{c_0^2}\Big(\frac{\partial\varphi}{\partial \quad}\Big)^2\Big], \tag{2}$$

where φ is a three-component scalar eld on the sphere $\varphi(\ ,\mathbf{r})^2 = 1$. The parameters g_0 and c_0 are $g_0 = a^{d\ 2}/JS^2$ and $c_0 = 2\sqrt{d}JSa$ where where J is the coupling constant of the antiferromagnetic nearest-neighbor interaction, S is the magnitude of the spin and a is the lattice spacing. In this paper we consider the two-dimensional case $d = 2$. We change the scale of by $c_0 \quad \rightarrow$, and then the action is

$$S = \frac{1}{2g}\int_0 d \quad \int d^d\mathbf{r}\Big[(\nabla\varphi)^2 + \Big(\frac{\partial\varphi}{\partial \quad}\Big)^2\Big]. \tag{3}$$

Here, we de ned in two-space dimensions $d = 2$[15],

$$g = g_0 c_0 = \frac{2\sqrt{d}}{S}\Lambda^{\ 1}, \quad = \frac{g}{t}, \quad t = \frac{k_B T}{JS^2}, \tag{4}$$

for $\Lambda = 1/a$. To be more precisely, S should be replaced by $\sqrt{S(S+1)}$. Although g has dimension in this expression, g can be dimensionless by scaling the eld variables $\mathbf{r}$ and . This model describes the Nambu-Goldstone mode, that is, the spin-wave mode of the quantum antiferromagnetic model. The renormalization of the spin-wave mode was investigated using the Wilson-Kogut renormalization method[14, 15].

In the case of hole doping, when the doping rate is small, the nonlinear sigma model is expected to be still relevant if we use the coupling constant g renormalized by the doping e ect. It is plausible that we can use the nonlinear sigma model for the extremely light-doping case.

The 2D nonlinear sigma model has an instanton solution [16, 17, 18, 19] (being interpreted as skyrmion or meron[20, 21]) which describes an excitation of spin-vortex type. Quantum fluctuations of instantons were computed and found to be reduced to the study of the Coulomb gas[22, 23, 24]. According to this study, the gas of instantons of the 2D nonlinear sigma model is in the plasma phase. This means that spin-vortex excitations are independent each other and never form dimers. This is the case for the two-dimensional classical Heisenberg model. Then, we necessarily have a question such as: is this picture still correct for the (2+1)D nonlinear sigma model? To investigate this, we generalize the instanton gas approximation to the (2+1)D nonlinear sigma model. We use the parametrization of the eld φ in terms of a single complex eld w:

$$(\varphi_1, \varphi_2, \varphi_3) = \Big(\frac{w + w}{1+|w|^2}, \ i\frac{w \quad w}{1+|w|^2}, \frac{|w|^2 \quad 1}{1+|w|^2}\Big), \tag{5}$$

where z is the complex conjugate of w. Then, using the complex variable $z = x+iy$ for $\mathbf{r} = (x, y)$, the action is written as

$$S = \frac{2}{g}\int_0 d \quad Q(\ ) + \frac{1}{g}\int_0 d \quad d^2\mathbf{r}\frac{1}{(1+|w|^2)^2}\ \ |\partial_\tau w|^2 + 4|\partial_{\bar{z}} w|^2 \ . \tag{6}$$

Here, Q is

$$Q(\ ) = \frac{1}{\ } \int d^2\mathbf{r} \frac{1}{(1+|w|^2)^2} \quad |\partial_z w|^2 \quad |\partial_{\bar{z}} w|^2 \quad . \tag{7}$$

A solution that satis es $\partial_{\bar{z}} w = 0$ is called instanton solution. A typical instanton solution is, for an integer k,

$$w = \frac{(z \quad a_1)(z \quad a_2) \quad (z \quad a_k)}{(z \quad b_1)(z \quad b_2) \quad (z \quad b_k)}. \tag{8}$$

a_i and b_i $(i = 1, \quad , k)$ indicate positions of vortex-like excitations in the antiferromagnetic spin order. This each local structure can be interpreted as a skyrmion. In general, a_i and b_i have -dependence: $a_i = a_i(\ )$, $b_i = b_i(\ )$ $(i = 1, \quad , k)$. When a_i and b_i are constants, Q is the integer quantum number equal to k[16]. We assume that $a_i(\ )$ and $b_i(\ )$ are continuous function of , then $Q(\ )$ is also a continuous function of and $Q(\ ) = k$ since Q is an integer. For this instanton solution, the action is given as

$$S = \frac{2 \quad k}{g} \quad + \frac{1}{g} \int_0 d \quad d^2\mathbf{r} \frac{1}{(1+|w|^2)^2} |\partial_\tau w|^2. \tag{9}$$

3. Fluctuation effect

Fluctuation e ect to the partition function was evaluated for the 2D nonlinear sigma model as[23]

$$Z_k = \text{const}.e^{\quad 2 \quad k/g} \frac{1}{(k!)^2} \int \prod_i da_i db_i dc c^{4k} e^{\quad V(a_i,b_i)}, \tag{10}$$

where

$$V(a_i, b_i) = \sum_{ij} \ln |a_i \quad b_j|^2 \quad \sum_{i<j} \ln |a_i \quad a_j|^2 \quad \sum_{i<j} \ln |b_i \quad b_j|^2. \tag{11}$$

Here c is the overall constant factor of w. This is the model of classical Coulomb gas interacting with the logarithmic interaction. This contribution comes from the measure of the functional integral by changing the variable w to a_i, b_i and c. This is straightforwardly generalized to the present model; the measure is given as

$$\begin{aligned} &\prod_{\tau,i} da_i(\ )db_i(\ )dc(\ ) \prod_\tau c^{4k} \prod_{i<j} |a_i \quad a_j|^2 \prod_{i<j} |b_i \quad b_j|^2 \prod_{ij} |a_i \quad b_j|^{\quad 2}. \\ = &\prod_{\tau,i} da_i(\ )db_i(\ )dc(\ ) \prod_\tau c^{4k} \exp \quad \sum_\tau \sum_{ij} \ln |a_i(\ ) \quad b_j(\ )|^2 + \sum_\tau \sum_{i<j} \ln |a_i(\ ) \quad a_j(\ )|^2 \\ &+ \sum_\tau \sum_{i<j} \ln |b_i(\ ) \quad b_j(\ )|^2 \ . \end{aligned} \tag{12}$$

This formula will be used in the following.

4. One-pair skyrmion state

Let us consider one-instanton state $w(z) = (z \quad a)/(z \quad b)$. If we set $b = \quad a \in \mathbf{R}$, for simplicity, then we obtain $\varphi = (0, 0, \quad 1)$ at $z = a$, $\varphi = (0, 0, 1)$ at $z = \quad a$ and $\varphi = (\quad 1, 0, 0)$ at $z = 0$. In the limit $|z| \to \infty$, $\varphi = (1, 0, 0)$. This indicates a pair of spin vortices that disturbs the antiferromagnetic spin order being located at $z = \quad a$. There is a spin disorder in the region that includes $z = \quad a$.

For $w = (z \quad b)/(z \quad a)$, the action in eq.(9) is given as

$$S_1 = \frac{2}{g} \quad + \frac{1}{g} \int_0 d \quad G \quad \dot{a} \quad \dot{a} + \dot{b} \quad \dot{b} \quad \dot{a} \quad \dot{b} \quad \dot{b} \quad \dot{a} \quad , \tag{13}$$

Journal of Physics: Conference Series **428** (2013) 012040 doi:10.1088/1742-6596/428/1/012040

where G is a constant that depends on $|a \quad b|$, and a and b are complex conjugates of a and b. We replace g by g/Λ so that g is dimensionless and add the fluctuation e ect to this action to obtain

$$\frac{S}{\Lambda} = \frac{2}{g} + \int_0 d \left[\frac{1}{g}G|\dot{a} \quad \dot{b}|^2 \quad \ln|a(\ ) \quad b(\ )|\right]. \tag{14}$$

This is the model of single particle in the logarithmic potential if we set $u = a \quad b$. We de ne $u = re^{i}$, then the energy is determined by the angular momentum $r^2 \dot{}$ and the potential $\ln r^2$, which is shown in Fig.1. The particle dynamics corresponds to the oscillation of the size of spin-vortex pair (Fig.2).

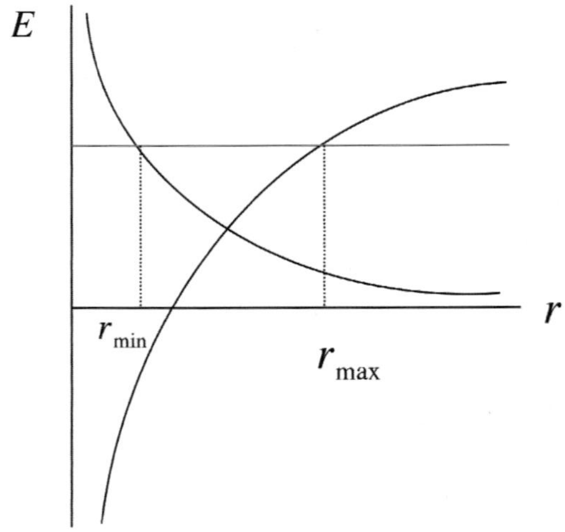

Figure 1. Energy of one-pair skyrmion state.

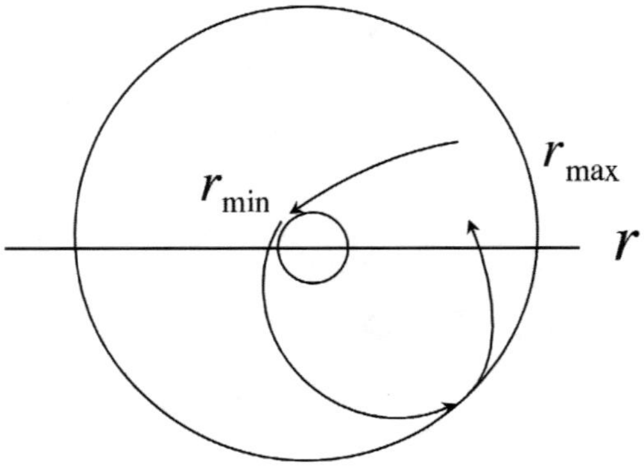

Figure 2. Oscillation of one-pair skyrmion state. This gure shows the oscillation of the distance of bounded spin-vortex pair, namely, the size of region of disordered antiferromagnetic spin.

5. Skyrmion gas

We generalize the one-pair skyrmion to many-skyrmion gas. The action is written as

$$\begin{aligned}\frac{S_k}{\Lambda} &= \frac{2 \ k}{g} + \frac{1}{g}\int_0 d \sum_{ij} G_{ij} \dot{}_i(\ ) \dot{}_j(\ ) + \int_0 d \left[\sum_{i<j} \ln|a_i(\ ) \quad a_j(\ )|^2 \right. \\ &\left. + \sum_{i<j} \ln|b_i(\ ) \quad b_j(\ )|^2 \quad \sum_{ij} \ln|a_i(\ ) \quad b_j(\ )|^2\right],\end{aligned} \tag{15}$$

where G_{ij} are constants and $\{ \ _i\}$ indicate both a_i and b_i: $(\ _1, \ _2, \quad) = (a_1, a_2, \quad , b_1, b_2, \quad)$. G_{ij} are given by integrals in terms of w and show logarithmic divergence $\ln R$ for the cuto R of the order of the system size. When we take into account only the divergent term $\ln R$, (G_{ij}) are constants. In this case, the kinetic term is simply

$$\sum_{ij} G_{ij} \dot{}_i(\ ) \dot{}_j(\ ) \propto \sum_i \dot{a}_i \quad \sum_i \dot{b}_i\Big) \left(\sum_j \dot{a}_j \quad \sum_i \dot{b}_j\right). \tag{16}$$

This means that the kinetic term is only concerned with the movement of centers of a-particles and b-particles. This term is expected to be unimportant and we neglect it. Hence, the model is reduced to the Coulomb gas model without the kinetic term.

Then we use the static approximation as a rst step. Under this approximation, the action is

$$\frac{S_k}{\Lambda} \simeq \frac{2\ \ k}{g} \quad + \quad \Big[\sum_{i<j} \ln|a_i \quad a_j|^2 + \sum_{i<j} \ln|b_i \quad b_j|^2 \quad \sum_{ij} \ln|a_i \quad b_j|^2\Big]. \tag{17}$$

This yields that the transition from the plasma phase to the molecular phase occurs at $\ = 2$[23]. Therefore, there is a Kosterlitz-Thouless transition in the (2+1)D nonlinear sigma model.

6. Summary

We have formulated skyrmion (instanton) gas picture for the two-dimensional quantum antiferromagnet on the basis of the (2+1)D nonlinear sigma model. We have shown that there is a Kosterlitz-Thouless transition from the plasma phase to the molecular phase as the temperature is lowered by using a static approximation to the instanton gas. In the molecular phase two instantons form a bound state, namely, a dipole of two spin vortices. We expect that this is related to the pairing mechanism of superconductivity in doped antiferromagnets.

References

[1] K. H. Bennemann and J. B. Ketterson Eds.: *The Physics of Superconductors* (Springer-Verlag, Berlin, 2004).
[2] T. Suzuki, T. Goto, K. Chiba, T. Shinoda, T¿ Fukase, H. Kimura, K. Yamada, M. Ohashi, and Y. Yamaguchi: Phys. Rev. B**57** (1998) 3229.
[3] M. Fujita, H. Goka, T. Adachi, Y. Koike, and K. Yamada: Physica C**426-431** (2005) 257.
[4] J. M. Tranquada, J. D. Axe, N. Ichikawa, Y. Nakamura, S. Uchida, and B. Nachumi: Phys. Rev. B**54** (1996) 7489.
[5] J. E. Ho man, E. W. Hudson, K. M. Lang, V. Madhavan, H. Eisaki, S. Uchida, and J. C. Davis: Science **295** (2002) 466.
[6] W. D. Wise, M. C. Boyer, K. Chatterjee, T. Kondo, T. Takeuchi, H. Ikuta, Y. Wang, and E. W. Hudson: Nature Phys. **4** (2008) 696.
[7] T. Hanaguri, C. Lupien, Y. Kohsaka, D. H. Lee, M. Azuma, M. Takano, H. Takagi, and J. C. Davis: Nature **430** (2004) 1001.
[8] T. Yanagisawa, S. Koike, M. Miyazaki, and K. Yamaji: J. Phys. Condens. Matter **14** (2002) 21.
[9] M. Miyazaki, T. Yanagisawa, and K. Yamaji: J. Phys. Soc. Jpn. **73** (2004) 1403.
[10] M. MIyazaki, K. Yamaji, T. Yanagisawa, and R. Kadono: J. Phys. Soc. Jpn. **78** (2009) 043706.
[11] A. M. Polyakov: Phys. Lett. B**59** (1975) 79.
[12] E. Brezin and J. Zinn-Justin: Phys. Rev. B**14** (1976) 3110.
[13] W. A. Bardeen, B. W. Lee, and E. R. Shrock: Phys. Rev. D**14** (1976) 985.
[14] S. Chakravarty, B. I. Halperin and D. R. Nelson: Phys. Rev. B**39** (1989) 2344.
[15] T. Yanagisawa: Phys. Rev. B**46** (1992) 13896.
[16] A. A. Belavin and A. M. Polyakov: JETP Lett. **22** (1975) 245.
[17] A. Jevicki: Nucl. Phys. B**127** (1977) 125.
[18] A. D'adda, M. Luscher, and P. Di Vecchia: Nucl. Phys. B**146** (1978) 63.
[19] A. M. Polyakov: *Gauge Fields and Strings* (Harwood Academic Publisher, Switerland, 1987).
[20] H. Koizumi: cond-mat/0506747.
[21] T. Morinari: Phys. Rev. B**73** (2006) 064504.
[22] I. V. Frolov and A. S. Schwarz: JETP Lett. **28** (1978) 249.
[23] V. A. Fateev, I. V. Frolov and A. S. Schwarz: Nucl. Phys. B**154** (1979) 1.
[24] B.Berg and M. Luscher: Commun. Math. Phys. **69** (1979) 57.

XXIst International Symposium on the Jahn–Teller Effect 2012 IOP Publishing
Journal of Physics: Conference Series **428** (2013) 012041 doi:10.1088/1742-6596/428/1/012041

What is the phase variable in superconductors ?: theory of superconductivity based on the spin-vortex formation

Hiroyasu Koizumi
Division of Materials Science, University of Tsukuba, Tsukuba, Ibaraki 305-8573, Japan
E-mail: koizumi@ims.tsukuba.ac.jp

Abstract. When Schrödinger solved the Schrödinger equation for the hydrogen atom, he assumed the single-valuedness of the electronic wave function. Thereafter, this assumption has been one of the fundamental postulates of quantum mechanics. When wave functions are multi-component, however, the imposing of the single-valued condition may become nontrivial.

The spin-degree-of-freedom of electron makes electronic wave functions two-component. When spin-vortices are created by the conduction electrons and they move in the self-consistent field with the spin-vortices, the twisting of the spin basis occurs; then, the imposing of the single-valued condition becomes nontrivial, and a vector potential is induced. As a consequence, the effective vector potential becomes the sum of the vector potential from the induced one and that originates from the electric current. This effective vector potential is gauge invariant and the persistent current is generated by it .

In the present work, we argue that if interactions that are omitted in the BCS reduced Hamiltonian are included, spin-vortices may be generated upon the application of a magnetic field. Then, the vector potential is induced and provides with the phase variable, θ, of the electron pair amplitude. The appearance of the spin-vortex provides with a new origin of θ; it originates from the induced gauge potential. This origin is compatible with the superselection rule for charge in contrast to the currently-accepted origin.

1. Introduction

The mechanism of the cuprate superconductivity is still not elucidated although a large amount of researches have been performed for more than 26 years. Recently, it is argued that the superconductivity without the Cooper pairs may occur in the cuprate [1, 2]. This superconductivity is based on the stable spin-vortex formation, thus, called *the spin-vortex superconductivity* [1].

If we look back the currently-accepted BCS theory, its superconducting order parameter is the electron pair amplitude given by

$$\langle \hat{\psi}_\uparrow \hat{\psi}_\downarrow \rangle = |\langle \hat{\psi}_\uparrow \hat{\psi}_\downarrow \rangle| e^{i\theta}, \tag{1}$$

where $\hat{\psi}_\sigma$ is the field operator of conduction electrons with spin σ, and $\langle \hat{O} \rangle$ denotes the thermal average of the operator $\hat{O}$ [3]. We notice that if the expectation value in Eq. (1) is calculated with fixed-particle-number states, it is zero. Since it is sensible to consider that superconductivity occurs in fixed-particle-number systems, the use of the above order parameter is for a mathematical convenience rather than for taking into account of a physical reality.

However, in the currently-accepted theory of superconductivity, θ arises from the phase difference between different charged states. This origin of θ requires the interference of different charged states, thus, is against the superselection rule for charge (SSRC); the SSRC dictates that such a phase is physically meaningless due to the nonexistence of interactions that connect different charged states [4, 5].

In the present work, we argue that the appearance of θ may be attributed to the fundamental postulate of quantum mechanics, *the single-valuedness of wave functions*, employed by Schrödinger [6]. If the interactions that are omitted in the BCS reduced Hamiltonian are included, spin-vortices will be created by the conduction electrons. Then, their wave functions have twisted spin basis since they are moving in the self-consistent field with the spin-vortices.

The twisting of the spin basis induces the following vector potential as will be explained later;

$$\mathbf{A}^{\mathrm{fic}} = \frac{c\hbar}{2q}\nabla\chi \tag{2}$$

where the charge unit q is given by $q = -e$; χ is a harmonic function that is optimized under the condition that the conduction electron wave functions are single-valued in the presence of spin-vortices [1]. Here the superscript 'fic' in $\mathbf{A}^{\mathrm{fic}}$ stands for 'fictitious' and denotes that the vector potential is not the electromagnetic field origin. Since $\nabla \times \mathbf{A}^{\mathrm{fic}} = 0$, the effect is an Aharonov-Bohm type. A remarkable point is that $\mathbf{A}^{\mathrm{fic}}$ generates *spin-vortex-induced loop currents* (SVILCs) [7]. If the spin-vortices are long-lived, the induced loop currents, i.e., SVILCs are also long-lived. The spin-vortex superconductivity is the theory of superconductivity with the SVILC as the current element [1].

The appearance of $\mathbf{A}^{\mathrm{fic}}$ in the spin-vortex superconductivity shows a marked similarity between the ferromagnetic phase and the superconducting phase. With the vector potential from the electromagnetic field $\mathbf{A}^{\mathrm{em}}$, the effective vector potential becomes the sum,

$$\mathbf{A}^{\mathrm{eff}} = \mathbf{A}^{\mathrm{em}} + \mathbf{A}^{\mathrm{fic}}. \tag{3}$$

It reminds us of the fact that a molecular field $\mathbf{B}^{\mathrm{mol}}$ exists in addition to the real magnetic field $\mathbf{B}^{\mathrm{em}}$ in ferromagnets, and the effective magnetic field is given as a sum of the two

$$\mathbf{B}^{\mathrm{eff}} = \mathbf{B}^{\mathrm{em}} + \mathbf{B}^{\mathrm{mol}}. \tag{4}$$

The occurrence of the ferromagnetic phase is attributed to the appearance of $\mathbf{B}^{\mathrm{mol}}$; the order parameter is the magnetization $\mathbf{M}$, which is calculated at $T = 0$ K as

$$\mathbf{M} = -\frac{\delta E[\mathbf{B}^{\mathrm{eff}}]}{\delta \mathbf{B}^{\mathrm{eff}}} \tag{5}$$

with taking the external mangnetic field zero. Therefore, the appearance of the ferromagnetic phase is attributed to the appearance of $\mathbf{B}^{\mathrm{fic}}$ that yields nonzero $\mathbf{M}$ even without an external magnetic field.

The similarity between Eqs.(3) and (4) suggests that in the phase in which $\mathbf{A}^{\mathrm{fic}}$ is nontrivial, a spontaneous current density $\mathbf{j}$ given by

$$\mathbf{j} = -c\frac{\delta E[\mathbf{A}^{\mathrm{eff}}]}{\delta \mathbf{A}^{\mathrm{eff}}} \tag{6}$$

may appear even without an external magnetic field. We may identify such a phase as superconducting phase.

The spin-vortex superconductivity does not require Cooper pairs; on the other hand the BCS superconductivity does. Therefore, it seems they are unrelated. However, we will show that the former may actually incorporate the latter.

Using θ, the persistent current is given by

$$\mathbf{j} = \Lambda(\mathbf{A}^{\rm em} - \frac{\hbar c}{q}\nabla\theta), \tag{7}$$

in the currently-accepted theory if the spatial variation of the magnetic field is much slower than the coherence length, where Λ is a function of the coordinate $\mathbf{r}$, and θ is an angular variable with period 2π. The observed flux quantization unit $\frac{hc}{2e}$ in superconductors is obtained by taking the charge unit q as $-2e$ in Eq.(7).

On the other hand the current density in the spin-vortex superconductivity is given by

$$\mathbf{j} = \Lambda \mathbf{A}^{\rm eff} = \Lambda(\mathbf{A}^{\rm em} + \frac{\hbar c}{2q}\nabla\chi), \tag{8}$$

where $q = -e$. The above current density also yields the observed flux quantization unit $\frac{hc}{2e}$. Whether the charge unit q is $-e$ or $-2e$ is a very important issue. The recent re-derivation of the AC Josephson frequency including the charge flow through the leads connected to the Josephson junction indicates $q = -e$ is the right one [1]; this point will be further discussed in Appendix.

The comparison of Eqs. (7) and (8) suggests the relation

$$\theta = -\chi. \tag{9}$$

In the present work, we will argue that this identification is adequate. Then, the phase of the Cooper pair amplitude is attributed to the appearance of $\mathbf{A}^{\rm fic}$. If we take this origin of θ, the current carrying state of the BCS theory becomes the one compatible with the SSRC.

2. Appearance of $\mathbf{A}^{\rm fic}$

In this section we explain the appearance of $\mathbf{A}^{\rm fic}$ using a model for the cuprate.

In the model construction, experimental results that probe the bulk properties were utilized. They indicate the appearance of significant local lattice deformations upon hole doping, and the presence of spin-configurations that generate the hourglass-shaped magnetic excitations. Since the latter is explained by the existence of spin-vortices, we construct a model with the spin-vortices [7, 8].

The minimal model for the parent antiferromagnetic insulator of the cuprate is a half-filled 2D Hubbard model with large on-site Coulomb repulsion. We include hole-lattice interaction in the Hubbard model since a number of experiments indicate that the doped holes become polarons;

$$H = -\sum_{i,j,\sigma} t_{ij} c_{i\sigma}^{\dagger} c_{j\sigma} + U\sum_{j} c_{j\uparrow}^{\dagger} c_{j\uparrow} c_{j\downarrow}^{\dagger} c_{j\downarrow} + H_{\rm hole+lattice}, \tag{10}$$

the first two terms are those of the Hubbard model, where t_{ij} is t when the sites i and j are nearest neighbors, and zero otherwise. The strong correlation condition, $0 < t \ll U$, is assumed. The third term describes the interaction between holes and underlying lattice, and also lattice vibrations.

To simplify the above Hamiltonian, we utilize the following observations: 1) the Hall coefficient measurement in a strong magnetic field indicates that the charge carriers in the magnetically-produced normal state below the superconducting transition temperatureT_c are

Journal of Physics: Conference Series **428** (2013) 012041 doi:10.1088/1742-6596/428/1/012041

electrons, not holes [9]; 2) a molecular orbital cluster calculation indicates that the energy lowering by the deformation of CuO unit is large enough to localize the hole. They indicate that holes becomes small polarons and immobile below T_c [10]. Then, the above Hamiltonian is simplified as

$$H_{\text{EHFS}} = -\sum_{i,j\in\text{acc. sites},\sigma} t_{ij} c^\dagger_{i\sigma} c_{j\sigma} + U \sum_{j\in\text{acc. sites}} c^\dagger_{j\uparrow} c_{j\uparrow} c^\dagger_{j\downarrow} c_{j\downarrow} \tag{11}$$

where EHFS stands for the 'effectively-half-filled situation'; the summation is taken only over accessible sites.

Due to the large on-site Coulomb repulsion, an effectively-half-filled situation where the number of accessible sites and that of electrons are equal, is realized; then, the ground state is well-approximated by a Slater determinant of fully-occupied lower band. Usually, the filled band does not contribute to electric current; however, the current flow by SVILCs is possible in the present model as will be explained below.

In order to obtain the wave function for the conduction electron in the self-consistent field with the spin-vortices, the following new particle operators

$$\begin{pmatrix} a_j \\ b_j \end{pmatrix} = \frac{e^{i\frac{\chi_j}{2}}}{\sqrt{2}} \begin{pmatrix} e^{i\frac{\xi_j}{2}} & e^{-i\frac{\xi_j}{2}} \\ -e^{i\frac{\xi_j}{2}} & e^{-i\frac{\xi_j}{2}} \end{pmatrix} \begin{pmatrix} c_{j\uparrow} \\ c_{j\downarrow} \end{pmatrix}, \tag{12}$$

are employed. The angular variable ξ_j describes the spin direction at site j in the x-y plane. In the cuprate, the spins are observed to lie in the CuO_2 plane [11], and we take the CuO_2 plane as the x-y plane.

The angular variable χ_j with period 2π is added; it is necessary for the single-valuedness of the above transformation matrix. Since ξ_j is an angular variable with ξ_j and $\xi_j + 2\pi$ physically equivalent, the sign-change occurs in $e^{\pm\frac{i}{2}\xi_j}$ by the phase shift $\xi_j \to \xi_j + 2\pi$; the sign-change brought about by $e^{\pm\frac{i}{2}\xi_j}$ must be compensated by the sign-change of $e^{\frac{i}{2}\chi_j}$. This condition is given using the winding numbers of χ and ξ as

$$w_\ell[\xi] + w_\ell[\chi] = \text{even number}, \quad \text{for any loop } C_\ell \tag{13}$$

where the winding number for function f around the loop C_ℓ is defined as

$$w_\ell[f] = \frac{1}{2\pi} \oint_{C_\ell} \nabla f \cdot d\mathbf{r}. \tag{14}$$

After the basis transformation in Eq. (12), H_{EHFS} becomes

$$\begin{aligned} H_{\text{EHFS}}[\nabla\chi] &= -\sum_{k,j\in\text{acc. sites}} t_{kj} e^{\frac{i}{2}\int_j^k \nabla\chi\cdot d\mathbf{r}} \left[\cos\frac{\xi_k - \xi_j}{2}(a^\dagger_k a_j + b^\dagger_k b_j) - i\sin\frac{\xi_k - \xi_j}{2}(a^\dagger_k b_j + b^\dagger_k a_j)\right] \\ &+ U \sum_{j\in\text{acc. sites}} a^\dagger_j a_j b^\dagger_j b_j \end{aligned} \tag{15}$$

A notable point in Eq. (15) is that the transfer integrals acquire phase factor $e^{\frac{i}{2}\int_j^k \nabla\chi\cdot d\mathbf{r}}$, thus, the transfer integral is modified as

$$t_{kj} \to t_{kj} e^{\frac{i}{2}\int_j^k \nabla\chi\cdot d\mathbf{r}} \tag{16}$$

This modification of the transfer integral is the one that will appear if the Peierls substitution for a magnetic field with the vector potential $\mathbf{A}^{\text{fic}}$ is performed.

We now consider the situation where ξ is given, and obtain χ that minimizes the total energy under the constraints given in Eq. (13). The optimization with respect to $\nabla\chi$ is performed under the following constraints;

$$\frac{1}{2\pi}\oint_{C_\ell} \nabla\chi \cdot d\mathbf{r} = w_\ell, \quad \ell = 1, \cdots, N_{\text{loop}}, \tag{17}$$

where w_ℓ is the winding number of χ along a loop C_ℓ, and N_{loop} is the number of independent loops.

Including the above constraints, the functional to be optimized is given by

$$F[\nabla\chi, \lambda_1, \cdots, \lambda_{N_{\text{loop}}}] = E[\mathbf{A}^{\text{fic}}] + \sum_{\ell=1}^{N_{\text{loop}}} \lambda_\ell \left(\oint_{C_\ell} \nabla\chi \cdot d\mathbf{r} - 2\pi w_\ell \right), \tag{18}$$

where $E[\nabla\chi]$ is the energy of the system for a given $\nabla\chi$, and λ_ℓ are the Lagrange multipliers.

From the stationary condition for F, we have

$$0 = \frac{\delta E[\mathbf{A}^{\text{fic}}]}{\delta \mathbf{A}^{\text{fic}}} \frac{c\hbar}{2q} + \sum_{\ell=1}^{N_{\text{loop}}} \lambda_\ell \frac{\delta}{\delta\nabla\chi} \oint_{C_\ell} \nabla\chi \cdot d\mathbf{r}. \tag{19}$$

In reality, the vector potential from the real electromagnetic field $\mathbf{A}^{\text{em}}$ must be included. Then, $E[\mathbf{A}^{\text{fic}}]$ is replaced by $E[\mathbf{A}^{\text{eff}}]$.

We note that the current density is obtained from the functional derivative of energy by the vector potential as

$$\mathbf{j}(\mathbf{x}) = -c\frac{\delta E[\mathbf{A}^{\text{eff}}]}{\delta \mathbf{A}^{\text{eff}}(\mathbf{x})}; \tag{20}$$

thus, from Eqs. (19) and (20), we have

$$\mathbf{j}(\mathbf{x}) = \frac{2q}{\hbar} \sum_{\ell=1}^{N_{\text{loop}}} \lambda_\ell \frac{\delta}{\delta\nabla\chi(\mathbf{x})} \oint_{C_\ell} \nabla\chi \cdot d\mathbf{r}. \tag{21}$$

This shows that a current is generated by the constraints. This is a novel current generation mechanism that occurs even in the filled-band situation. The current element is a loop current induced by the spin-vortex; thus, we call it *the spin-vortex-induced loop current* (SVILC). Note also that a similar calculation was done by Vergés et al. in Ref. [12]. Since they did not take care of the multi-valuedness arising from the twisting of the spin basis, they did not obtain the electric current.

3. Appearance of $\mathbf{A}^{\text{fic}}$ in the BCS current carrying state

Let us start with presenting the state vector employed in the BCS theory;

$$|\text{BCS}\rangle = \prod_{\mathbf{k}} (u_{\mathbf{k}} + v_{\mathbf{k}} c^\dagger_{\mathbf{k}\uparrow} c^\dagger_{-\mathbf{k}\downarrow})|\text{vac}\rangle, \tag{22}$$

where $u_{\mathbf{k}}$ and $v_{\mathbf{k}}$ are real variational parameters that satisfy $u_{\mathbf{k}}^2 + v_{\mathbf{k}}^2 = 1$ [13]. The BCS used this variational state vector to facilitate calculations involving the electron pair-correlation. Although it is a linear combination of states with different particle numbers, the obtained wave function can be practically regarded as that for a state with N particles since the particle number distribution has a sharp peak at the mean-value N.

Journal of Physics: Conference Series **428** (2013) 012041 doi:10.1088/1742-6596/428/1/012041

Currently, $u_{\mathbf{k}}$ and $v_{\mathbf{k}}$ are extended to complex numbers from real numbers, and their relative phases are considered physically meaningful. This extension, however, violates the SSRC [4, 5]. We stick to the BCS's original intention, and show below that what is believed to arise from the phase between $u_{\mathbf{k}}$ and $v_{\mathbf{k}}$ actually may be explained to arise from the modification of the electron pairing. The strong point of this origin is that it is in accordance with the SSRC.

In the BCS ground state the pairing occurs between two states whose annihilation operators are given by

$$\begin{aligned} c_{\mathbf{k}\uparrow} &= \sum_j \frac{1}{\sqrt{N_s}} c_{j\uparrow} e^{i\mathbf{k}\cdot\mathbf{r}_j}, \\ c_{-\mathbf{k}\downarrow} &= \sum_j \frac{1}{\sqrt{N_s}} c_{j\downarrow} e^{-i\mathbf{k}\cdot\mathbf{r}_j}, \end{aligned} \tag{23}$$

where the coordinate of the system is described as discrete lattice points; the total number of sites is denoted as N_s; $\mathbf{r}_j$ denotes the coordinate of the jth site.

Now we consider the case where an external magnetic field is applied in the z-direction. We anticipate that the state corresponding to $|\text{BCS}\rangle$ is replaced by the state with non-trivial $\mathbf{A}^{\text{fic}}$; i.e., the ground state is shifted from $|\text{BCS}\rangle$ to the new one denoted by $|\text{BCS}[\chi]\rangle$ which contains the multi-valued harmonic function χ. This shift will be caused by two interactions that are omitted in the BCS reduced Hamiltonian.

The first interaction we include is the Zeeman interaction between the magnetic field and electron spin. This interaction will create a tilting of up-spin components of electrons.

The second interaction we include is the spin-orbit interaction derived by Dirac [14]; the electron spin $\mathbf{s} = \frac{\hbar}{2}\boldsymbol{\sigma}$ is shown to interact with electric field $\mathbf{E}$ as

$$H_{\text{SO}} = \frac{\mu_{\text{B}}}{2c^2}(\mathbf{v} \times \mathbf{E}) \cdot \boldsymbol{\sigma}, \tag{24}$$

where μ_{B} is the Bohr magneton, $\mathbf{v}$ is the velocity operator; $\boldsymbol{\sigma}$ is the vector of the Pauli matrices.

At the instant when the magnetic field in the z-direction is turned on, eddy currents flow in the x-y plane. If the internal electric field $\mathbf{E}$ exists along the z-direction, the eddy current in the direction $\mathbf{v}$ and the electric field $\mathbf{E}$ create a circular effective magnetic field that acts on electron spins in the x-y plane in the direction $\mathbf{v} \times \mathbf{E}$. This is a circular effective magnetic field since the eddy current is a circular current. Thus, it will generate a circular change of spin-direction along the eddy current. As a consequence, spin-vortices will be created.

We assume that the wave function for the above situation is constructed by modifying the electron-pairing from $\{c_{\mathbf{k}\uparrow}, c_{-\mathbf{k}\downarrow}\}$ to $\{a_{\mathbf{k}}, b_{-\mathbf{k}}\}$, where $a_{\mathbf{k}}$ and $b_{-\mathbf{k}}$ are given by

$$\begin{aligned} a_{\mathbf{k}} &= \sum_j \frac{e^{i\frac{\chi_j}{2}}}{\sqrt{N_s}} \left(\sqrt{1-\varepsilon_j^2} e^{i\frac{\xi_j}{2}} c_{j\uparrow} + \varepsilon_j e^{-i\frac{\xi_j}{2}} c_{j\downarrow} \right) e^{i\mathbf{k}\cdot\mathbf{r}_j} \\ b_{-\mathbf{k}} &= \sum_j \frac{e^{i\frac{\chi_j}{2}}}{\sqrt{N_s}} e^{-i\frac{\xi_j}{2}} c_{j\downarrow} e^{-i\mathbf{k}\cdot\mathbf{r}_j}, \end{aligned} \tag{25}$$

where the phase factor $e^{i\frac{\chi}{2}}$ is added as in Eq. (12) to satisfy the single-valued condition since as before the angular variable ξ describes the spin-direction in the x-y plane, and it may give rise to singularities for the single-valued condition of wave functions.

New creation and annihilation operators satisfy the following anti-commutation relations;

$$a_{\mathbf{k}} a^\dagger_{\mathbf{k}'} + a^\dagger_{\mathbf{k}'} a_{\mathbf{k}} = b_{\mathbf{k}} b^\dagger_{\mathbf{k}'} + b^\dagger_{\mathbf{k}'} b_{\mathbf{k}} = \delta_{\mathbf{k}\mathbf{k}'}. \tag{26}$$

and

$$a_{\mathbf{k}} b^{\dagger}_{\mathbf{k}'} + b^{\dagger}_{\mathbf{k}'} a_{\mathbf{k}} = \frac{1}{N_s} \sum_j \varepsilon_j e^{i(\mathbf{k}-\mathbf{k}')\cdot \mathbf{r}_j}. \tag{27}$$

If ε_j's are non-zero in a small region along the eddy current, we may consider that the right-hand side in Eq. (27) is practically zero. If this is the case, we may ignore contributions from the right-hand side in Eq. (27) and treat them as orthogonal although states created by $a^{\dagger}_{\mathbf{k}}$ and $b^{\dagger}_{\mathbf{k}'}$ are actually not orthogonal. If the following, we assume that this is the case.

The BCS-like state vector using the $\{a_{\mathbf{k}}, b_{-\mathbf{k}}\}$ pairing is given by

$$|\mathrm{BCS}[\chi]\rangle = \prod_{\mathbf{k}} (u_{\mathbf{k}} + v_{\mathbf{k}} a^{\dagger}_{\mathbf{k}} b^{\dagger}_{-\mathbf{k}})|\mathrm{vac}\rangle. \tag{28}$$

The state $|\mathrm{BCS}[\chi]\rangle$ has local moments; the expectation value of electron spin at the jth site is calculated as

$$\begin{aligned} S^x_j &= \frac{\hbar N}{2N_s}\varepsilon_j \sqrt{1-\varepsilon_j^2}\cos\xi_j \approx \frac{\hbar N \varepsilon_j}{2N_s}\cos\xi_j \\ S^y_j &= \frac{\hbar N}{2N_s}\varepsilon_j \sqrt{1-\varepsilon_j^2}\sin\xi_j \approx \frac{\hbar N \varepsilon_j}{2N_s}\sin\xi_j \\ S^z_j &= -\frac{\hbar N \varepsilon_j^2}{2N_s} \approx 0. \end{aligned} \tag{29}$$

The above formulae indicate that ξ_j describes the angle of the direction of the spin in the x-y plane at the ith site. The circular change of ξ means the existence of spin-vortices.

It is noteworthy that paramagnetic responses have been observed in the BCS superconductors [15, 16]; the pairing shift given above agrees with this fact.

Due to the appearance of the spin-vortices, the eddy currents turn into the SVILCs that persist as long as the spin-vortices persist. The collection of these SVILCs acts as the Meissner current. For a large sample, the exclusion of the applied magnetic field by the Meissner current is always energetically favorable, thus, the appeared Meissner current will persist as long as the external magnetic field is present.

The argument above suggests that if the Cooper pair condensation occurs, the system may acquire the ability to generate $\mathbf{A}^{\mathrm{fic}}$. If this is the case, the Cooper pair condensation temperature corresponds to T_c. It is also suggested that in some materials, the superconducting phase may be realized even without the Cooper pair formation if the long-lived spin-vortex formation occurs.

The effective vector potential in the superconductor becoms $\mathbf{A}^{\mathrm{eff}}$ when $\mathbf{A}^{\mathrm{fic}}$ arises. Then, the vector potential for the magnetic field, $\mathbf{A}^{\mathrm{em}}$, in the Meissner current in the BCS theory should be replaced by $\mathbf{A}^{\mathrm{eff}}$.

If we calculate $\langle \hat{\psi}_{\uparrow} \hat{\psi}_{\downarrow} \rangle$ using $|\mathrm{BCS}[\chi]\rangle$ in Eq. (28), we obtain

$$\langle \hat{\psi}_{\uparrow} \hat{\psi}_{\downarrow} \rangle = \langle \mathrm{BCS}[\chi] | \hat{\psi}_{\uparrow} \hat{\psi}_{\downarrow} | \mathrm{BCS}[\chi] \rangle = \frac{1}{N_s} \sum_{\mathbf{k}} v_{\mathbf{k}} u_{\mathbf{k}} e^{-i\chi(\mathbf{r})}, \tag{30}$$

where the limit $\varepsilon_j \to 0$ is taken. Then, we can identify that $\theta = -\chi$.

4. Concluding remarks

The currently-accepted origin of θ assumes that the superconducting state is a coherent superposition of different charged states. Although such a treatment is against the SSRC, the observation of Josephson effects is regarded as a proof for it.

However, it was recently found that the Josephson's derivation of the AC Josephson frequency may be defective due to the lack of the inclusion of the charge flow through the leads connected to the Josephson junction [1]; the Josephson's derivation is based on the situation given in Fig. 1(a), however, the real situation is more like that given in Fig. 1(b). This point needs a serious investigation. By examining the AC Josephson problem very carefully, we may come to the conclusion that the currently accepted $q = -2e$ theory must be replaced by the $q = -e$ theory.

By the present origin of θ, the BCS theory can be incorporated into the spin-vortex superconductivity theory by identifying θ to be χ. Currently, θ is believe to arise from the modification,

$$\prod_{\mathbf{k}}(u_{\mathbf{k}} + v_{\mathbf{k}} c^{\dagger}_{\mathbf{k}\uparrow} c^{\dagger}_{-\mathbf{k}\downarrow})|\mathrm{vac}\rangle \rightarrow \prod_{\mathbf{k}}(u_{\mathbf{k}} + e^{i\theta} v_{\mathbf{k}} c^{\dagger}_{\mathbf{k}\uparrow} c^{\dagger}_{-\mathbf{k}\downarrow})|\mathrm{vac}\rangle \tag{31}$$

where the spatial variation of θ gives rise to the persistent current.

We have shown that the current carrying state is generated from the following modification,

$$\prod_{\mathbf{k}}(u_{\mathbf{k}} + v_{\mathbf{k}} c^{\dagger}_{\mathbf{k}\uparrow} c^{\dagger}_{-\mathbf{k}\downarrow})|\mathrm{vac}\rangle \rightarrow \prod_{\mathbf{k}}(u_{\mathbf{k}} + v_{\mathbf{k}} a^{\dagger}_{\mathbf{k}} b^{\dagger}_{-\mathbf{k}})|\mathrm{vac}\rangle, \tag{32}$$

where $a_{\mathbf{k}}$ and $b_{-\mathbf{k}}$ are given in Eq. (25).

The important point to remember is that the modification in Eq. (31) cannot be achieved by the known interactions [5]. On the other hand, the modification in Eq. (32) is achieve by the known interactions that are omitted in the BCS reduced Hamiltonian, i.e., the Zeeman interaction and the spin-orbit interaction of conduction electrons. They are energetically very small; however, they will introduce the singularities in ξ, thus, induces the vector potential $\mathbf{A}^{\mathrm{fic}}$.

Appendix A. The time-dependence of χ and the derivation of the AC Josephson frequency

In this appendix, we derive the AC Josephson frequency $2eV/h$ by including the charge flow through the leads connected to the junction.

Let us consider the situation where χ and ρ are only active variables for the current generation, where χ is the angular variable that appears in $\mathbf{A}^{\mathrm{fic}}$ and ρ is the electron density. Then, the Hamiltonian is given as a functional of χ and ρ as $H = H[\chi,\rho]$ [1].

The gauge invariant vector potential $\mathbf{A}^{\mathrm{eff}}$ in Eq. (3) implies that the following scalar potential is also gauge invariant

$$\varphi^{\mathrm{eff}} = \varphi^{\mathrm{em}} - \frac{\hbar}{2q}\frac{\partial \chi}{\partial t}, \tag{A.1}$$

since it is a time-component partner of $\mathbf{A}^{\mathrm{eff}}$, where φ^{em} is the scalar potential for electromagnetic field.

Then, the term of the electrostatic interaction in the Lagrangian should be given as

$$-q\int d^3x \rho \left(\varphi^{\mathrm{em}} - \frac{\hbar}{2q}\frac{\partial \chi}{\partial t}\right), \tag{A.2}$$

where $q = -e$ is electron charge.

Let us obtain the canonical momentum for χ, which we denote as p_χ. The Lagrangian L is written using χ and p_χ as

$$L = \int d^3x\, p_\chi \dot{\chi} - H. \tag{A.3}$$

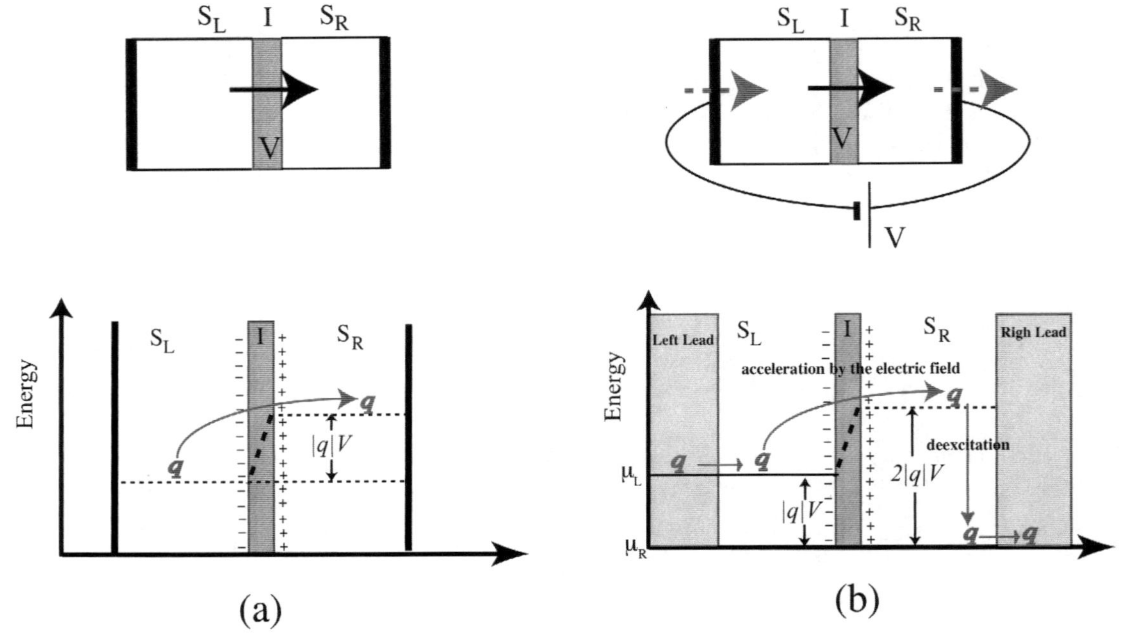

Figure 1. (a) The SIS junction considered by Josephson. The electric field in the insulator produces the voltage difference V. The Cooper-pair tunneling with unit charge $q = -2e$ across the two superconductors gives rise to the AC frequency $2eV/h$ [17]. (b) The SIS junction including the current flow through the leads connected to it. The electric field in the insulator produces the voltage difference V. The chemical potential difference eV exists between the two leads. The electron tunneling with unit charge $q = -e$ across the two superconductors is added by the energy gain due to the flow-in and flow-out through the leads. The values of the former and the latter are the same. Overall, the AC frequency is given by $2eV/h$ with the charge unit $q = -e$ [1].

The term in Eq. (A.2) implies that p_χ is given by

$$p_\chi = \frac{\delta L}{\delta \dot{\chi}} = \frac{\rho\hbar}{2}. \tag{A.4}$$

Thus, χ and $\frac{\rho\hbar}{2}$ are conjugate variables.

Then, the Hamilton's equations are given by

$$\begin{aligned} \dot{\rho} &= -\frac{2}{\hbar}\frac{\delta H}{\delta \chi} \\ \dot{\chi} &= \frac{2}{\hbar}\frac{\delta H}{\delta \rho}. \end{aligned} \tag{A.5}$$

Now we consider the SIS junction depicted in Fig. 1(b). The number of particles in the left superconductor is N_L and the right superconductor is N_R; the chemical potential on the left and right superconductors are μ_L and μ_R, respectively. The phase χ on the two superconductor are written as

$$\chi(\mathbf{r}) + \chi_L \tag{A.6}$$

and

$$\chi(\mathbf{r}) + \chi_R, \tag{A.7}$$

respectively, where χ_L and χ_R do not depend on the position in each superconductor. Actually, for the time-dependent change of the current flow to occur, $\chi(\mathbf{r})$ will change in time; however, we neglects its contribution by assuming that it is negligibly small; this also corresponds to the situation assumed by Josephson [17].

Let us assume that the link between the two superconductors is so weak that the optimized $\nabla\chi$ in S_L and S_R are not affected by the presence of the link. The conjugate variables for χ_R and χ_L are calculated using Eq. (A.4) as $\hbar N_R/2$ and $\hbar N_L/2$, respectively.

Then, the canonical equations for N_X and χ_X are obtained as

$$\begin{aligned} \dot{N}_X &= -\frac{2}{\hbar}\frac{\partial H}{\partial \chi_X} \\ \dot{\chi}_X &= \frac{2}{\hbar}\frac{\partial H}{\partial N_X}, \end{aligned} \tag{A.8}$$

where X is either L or R.

The quantum mechanical version of them are given by

$$\dot{\hat{N}}_X = \frac{i}{\hbar}[\hat{H}, \hat{N}_X], \tag{A.9}$$

$$\dot{\hat{\chi}}_X = \frac{i}{\hbar}[\hat{H}, \hat{\chi}_X]. \tag{A.10}$$

Since $\hbar\hat{N}_X/2$, and $\hat{\chi}_X$ are canonical conjugate operators, the commutation relation of them are given by

$$[\hat{N}_X, \frac{1}{2}\hat{\chi}_X] = -i. \tag{A.11}$$

By introducing the following operators,

$$\hat{C}_X = e^{-\frac{i}{2}\hat{\chi}_X}\hat{N}_X^{1/2} \tag{A.12}$$

and

$$\hat{C}_X^\dagger = \hat{N}_X^{1/2} e^{\frac{i}{2}\hat{\chi}_X}, \tag{A.13}$$

the commutation relation in Eq. (A.11) becomes,

$$[\hat{C}_X, \hat{C}_X^\dagger] = 1; \tag{A.14}$$

this shows that $\hat{C}_X$ and $\hat{C}_X^\dagger$ are the annihilation and creation operators for particles in S_X. Note that the commutation relations are used, although the particles are fermions. This is allowable since we consider not the individual particle levels, but just the number in each superconductor with the condition $N_X \gg 1$.

Then, the operator for the transfer of particles between S_R and S_L is constructed as

$$\hat{K}_J = t_{RL}\hat{C}_R^\dagger\hat{C}_L + \text{ h.c.} \tag{A.15}$$

where the hopping matrix element between S_L and S_R, t_{RL}, is given by a real parameter t_J.

Let us employ a semiclassical approximation in which $\hat{\chi}_R$, $\hat{\chi}_L$, $\hat{N}_R$, $\hat{N}_L$ are replaced by their mean values χ_R, χ_L, N_R, and N_L, respectively. Then, the semiclassical version of $\hat{K}_J$ is given by

$$K_J = E_J \cos\left(\frac{1}{2}(\chi_L - \chi_R)\right), \tag{A.16}$$

where

$$E_J = 2t_J(N_R N_L)^{1/2}. \tag{A.17}$$

If a real magnetic field exits, the hopping matrix element becomes

$$t_{RL} = t_J \to t_{RL} = t_J e^{i\frac{q}{\hbar c}\int_L^R \mathbf{A}^{\mathrm{em}}\cdot d\mathbf{r}}; \tag{A.18}$$

thus, the kinetic term in Eq. (A.16) becomes

$$K_J = E_J \cos\left(\frac{1}{2}(\chi_L - \chi_R) - \frac{q}{\hbar c}\int_L^R \mathbf{A}^{\mathrm{em}} \cdot d\mathbf{r}\right). \tag{A.19}$$

Then, the current through the junction is given by

$$J = \frac{qE_J}{\hbar}\sin\phi, \tag{A.20}$$

where ϕ is given by

$$\phi = \frac{1}{2}(\chi_L - \chi_R) - \frac{q}{\hbar c}\int_L^R \mathbf{A}^{\mathrm{em}} \cdot d\mathbf{r} = -\frac{q}{\hbar c}\int_L^R \mathbf{A}^{\mathrm{eff}} \cdot d\mathbf{r}. \tag{A.21}$$

The time-derivative of $\hbar\phi$ yields,

$$\hbar\dot{\phi} = \frac{\partial \bar{H}}{\partial N_L} - \frac{\partial \bar{H}}{\partial N_R} + q\int_L^R \mathbf{E} \cdot d\mathbf{r}, \tag{A.22}$$

where $\bar{H}$ is a part of the Hamiltonian that excludes the electrostatic potential term

$$\bar{H} = H - q\int d^3x\rho\varphi^{\mathrm{em}}. \tag{A.23}$$

The electric field $\mathbf{E}$ inside the insulator region is given by

$$\mathbf{E} = -\nabla\varphi^{\mathrm{em}} - \frac{1}{c}\frac{\partial \mathbf{A}^{\mathrm{em}}}{\partial t}. \tag{A.24}$$

In order to calculate the r.h.s in Eq. (A.22), we may use the following relation

$$\frac{\partial \bar{H}}{\partial N_L} - \frac{\partial \bar{H}}{\partial N_R} = \bar{H}(N_L + 1, N_R - 1) - \bar{H}(N_L, N_R), \tag{A.25}$$

where $\bar{H}(N_L, N_R)$ denotes the energy of the system excluding the electrostatic potential with N_L particles on the left superconductor and N_R particles on the right superconductor.

In the situation where particle flow occurs through the leads as shown in Fig. 1(b), the following two processes contribute to the r.h.s in Eq. (A.22):

(i) Particle transfer by tunneling (indicated by the black solid arrow in Fig. 1(b)). It gives rise to

$$q \int_L^R \mathbf{E} \cdot d\mathbf{r} = qV. \tag{A.26}$$

This contribution can be interpreted as *the energy gain from the acceleration during the tunneling from* S_L *to* S_R.

(ii) Particle flow through the leads (indicated by the red dotted arrows in Fig. 1(b)). The particle flows from L_L has the Fermi energy of the left lead which is equal to μ_L; the particle flows from L_R has the Fermi energy of the right lead which is equal to μ_R. Thus, we have

$$\bar{H}(N_L+1, N_R-1) - \bar{H}(N_L, N_R) = \mu_L - \mu_R = qV \tag{A.27}$$

This can be regarded as *the energy gain by the particle flow-in from* L_L *to* S_L *and the flow-out from* S_R *to* L_R.

The total contribution is the sum of the above two, i.e., $2qV$. It corresponds to the total energy gain for *the particle flow from* L_L *to* R_L *via the tunneling between* S_L *to* S_R.

Therefore, the time derivative of ϕ is given by

$$\dot{\phi} = \frac{2qV}{\hbar}. \tag{A.28}$$

This yields the observed AC frequency $\frac{2eV}{h}$ with $q = -e$.

The same frequency was obtained by taking into account only the first process in Eq. (A.26) with $q = -2e$ by Josephson and others [17]. Since the current is measured in one of the leads, it is sensible to include the effect of the charge flow through the leads. If the contribution from it is included, the AC frequency for the $q = -2e$ case becomes $\frac{4eV}{h}$, which disagrees with the experiment.

References

[1] Koizumi H 2011 *J. Supercond. Nov. Magn.* **24** 1997
[2] Hidekata R and Koizumi H 2011 *J. Supercond. Nov. Magn.* **24** 2253
[3] Anderson P W 1966 *Rev. Mod. Phys.* **38** 298
[4] Wick G C, Wightman A S and Wigner E P 1970 *Phys. Rev. D* **1** 3267
[5] Peierls R 1992 *Contemporary Phys.* **33** 221
[6] Schrödinger E 1926 *Ann. Physik* **79** 361
[7] Koizumi H 2008 *J. Phys. Soc. Jpn.* **77** 034712
[8] Koizumi H 2009 *J. Phys. Chem. A* **113** 3997
[9] Doiron-Leyraud N, Proust C, LeBoeuf D, Levallois J, Bonnemaison J B, Liang R, Bonn D A, Hardy W N and Taillerfer L 2007 *Nature* **447** 565
[10] Miyaki S, Makoshi K and Koizumi H 2008 *J. Phys. Soc. Jpn.* **77** 034702
[11] Vaknin D, Sinha S K, Moncton D E, Johnston D C and Newsam J M 1987 *Phys. Rev. Lett.* **58** 2802
[12] Vergés J A, Louis E, Lomdahl P S, Guinea F and Bishop A R 1991 *Phys Rev. B* **43** 6099
[13] Bardeen J, Cooper L N and Schrieffer J R 1957 *Phys. Rev.* **108** 1175
[14] Dirac P 1958 *Principles of Quantum Mechanics* 4th ed (Oxford: Oxford Univ. Press)
[15] Reif F 1957 *Phys. Rev.* **106** 208
[16] Androes G M and Knight W D 1959 *Phys. Rev. Lett.* **2** 386
[17] Josephson B D 1962 *Phys. Lett.* **1** 251

Journal of Physics: Conference Series **428** (2013) 012042 doi:10.1088/1742-6596/428/1/012042

Local Lattice Distortion in Superconducting Cuprates Studied by XAS

H Oyanagiand C Zhang
Research Institute of Instrumentation Frontier, AIST, Tsukuba, Ibaraki, Japan

E-mail: h.oyanagi@aist.go.jp

Abstract. *In-plane* local distortion in cuprates characterized by strong doping and temperature dependence is a signature of polaron (bipolaron). The nature of dynamic lattice response in relation to the onset of superconductivity was probed by polarized x-ray absorption spectroscopy (XAS), for pure and magnetic impurity-doped $M_xLa_{1.85}Sr_{0.15}Cu_{1-x}O_4$ (M=Mn, Ni, Co, x<0.05) single crystals grown by a TSFZ method. The results confirm that the dynamical anti Jahn-Teller (AJT) distorted domains maximize at T_d^{max}~T_c^{onset}, which is described by the disappearance of the Cu-O(1) bond alternation upon the completion of phase coherence (onset of superconductivity). In contrast to the Ni and Co dopings which quickly suppress superconductivity by magnetic pair breaking, the Mn doping shows sustainable superconductivity *i.e*., a constant T_d^{max} and T_c^{onset} upto x=0.03 with the presence of dynamical AJT distortion. The results suggest that the dynamical AJT type distortion with either Q_2 or pseudo JT symmetry extending over the basal plane could be a prerequisite for high temperature superconductivity.

1. Introduction

While the microscopic mechanism of high temperature superconductivity (HTSC) is still in mystery more than 26 years after the discovery [1], recently reported superconductivity in fluorine-doped LaFeAsO (LFAO) [2] has revived interests in the research. A conventional phonon mechanism is considered unlikely as the DFT calculations indicated a weak electron-phonon coupling, yet, the purely electronic mechanism [3] is also less likely than the case of cuprates [4] as the Hubbard *U* is not large (*U*~5eV) [5], while recent muon spin rotation experiments suggested possibility of multigap BCS-type superconductivity in $Ba_{0.6}K_{0.4}Fe_2As_2$ [6]. Although some of the iron pnictides (doped with holes) have temperature-dependent lattice anomaly described as a bond length disproportionation [7] similar to that of superconducting cuprates [8], the behaviour of local lattice is strongly dependent on the band filling, indicating a strong intimacy between local lattice and the doping-induced unconventional electronic structure, *i.e*., a pseudogap. It became therefore necessary to investigate the role of local lattice in a cuprate system with well-characterized carriers (holes) by revisiting a model system, *i.e*., $La_{1.85}Sr_{0.15}CuO_4$ (LSCO). It would be quite timely to focus on a typical cuprate and investigate the role of polarons considering the fact that high precision local probe experiments and theoretical advances (models based on specific distortion scheme) would unveil the nature of lattice effects whose "tip" is emerged in pnictides.

A polaron is formed when an electron is strongly coupled to the atoms in a crystal. Review article on a polaronic mechanism of superconductivity is elsewhere [9]. Polarons in highly correlated electron systems relate to a variety of physical properties including colossal magnetoresistance (CMR). In manganites, coherent polaron condensation is believed to be the driving mechanism of CMR phenomena [10]. Strong coupling between electrons and lattice in HTSC cuprates has been demonstrated by angle-resolved photoemission spectroscopy (ARPES) [11]. Reflecting these backgrounds, renewed interests on the role of polaron in HTSC mechanism have rapidly accumulated. In this work we revisit lattice effects in a typical cuprate $La_{2-x}Sr_xCuO_4$ (LSCO) with magnetic impurity atoms substituting copper sites [8] studied by means of polarized extended x-ray absorption fine structure (EXAFS). Previously, unusual carrier-induced local lattice anomaly found for pure LSCO was interpreted as a nanometer-scale self-organization (stripe) [12]. Here we show that the lattice effect at T_d^{max} is more catastrophic resulting in a bond alternation, *i.e.*, equal number of elongated and shortened Cu-O(1) bond length. The unusual local lattice deviation from a normal phonon behaviour is interpreted in terms of polarons [13-16]. As the magnitude of EXAFS oscillations is small, *i.e.*, in the order of 10^{-2} of the total oscillation magnitude, a highly efficient fluorescence x-ray detector is required for a statistics with 10^6 photons [17]. In this work, we have used a state-of-the-art pixel x-ray detector to obtain high-quality data and high-quality single crystals [18] grown by an in-house TSFZ furnace. As chemical doping (substitution with hetero-valent atom) itself often modifies local lattice leaving ambiguity in crystal quality, we used samples with a critical temperature controlled by uniaxial strain effect [19] and magnetic impurity doping effect [20]. Here we focus on the latter case and describe the nature of local lattice distortion.

2. Experimental

Polarized x-ray absorption spectroscopy (XAS) is advantageous as an instantaneous local probe within a coherent length of HTSC around a specific atom along a particular direction of anisotropic single crystals where the two polarization geometries (*E//ab*, *E//c*) in principle provide *in-plane* and *out-of-plane* information on the radial distribution function (RDF) of the CuO_2 plane, respectively. Difficulty in polarized EXAFS experiments on *as-grown* single crystals was overcome by a fluorescence detection method using a segmented detector that measures an emitted photon flux rather than a conventional transmitted beam intensity. Although several methods to measure polarization dependence for cuprates, our geometry (Figure 1(a)) is advantageous as a non-destructive method that covers any angle dependence between parallel and perpendicular polarization geometries by a simple rotation of the crystal around the incident beam. Segmented fluorescence monitoring was used to discriminate a signal from a background. A novel germanium 100-pixel array detector was used for collecting a purely fluorescence signal over a segmented solid angle [21]. All near-edge and EXAFS data were recorded for M:LSCO (M=Ni, Co and Mn) single crystals and polycrystalline samples at the Photon Factory. The energy and maximum electron current of storage ring were 2.5 GeV and 400–500 mA, respectively. A directly water-cooled silicon (111) double-crystal monochromator was used, covering the energy range 4–25 keV. The energy resolution was better than 2 eV at 9 keV, calibrated from the near-edge features of copper metal at the Fermi energy, E_f (8.9803 keV). A single crystal sample was attached to an aluminum holder with a strain-free glue and cooled down using a closed-cycle He refrigerator (cooling power 2 W at 20 K, stability ±0.1 K) on a high-precision goniometer (Huber 420). As a typical magnitude of normalized EXAFS oscillations is several %, each data set must have photon statistics better than 0.1% or 10^6 photons. Non-statistic (systematic) error was minimized for which segmented x-ray detection and repeated scans, typically six scans, were used. Fluorescence yield spectra for all channels of pixel array were monitored in real time and the effect of scattering was inspected.

A typical LSCO bulk single crystal grown by a traveling seed floating zone (TSFZ) method was 2 mm x 2 mm x 1 mm in dimension [8]. Using a closed cycle He refrigerator, sample temperature was controlled to a temperature error within 1 K and 0.1 K stability. Difficulty in substituting Mn with copper in LSCO is primarily a multi-phase problem; the impurity phases such as manganites stabilized at elevated temperature. Because of this difficulty, Mn substitution has been so far limited to a dilute specimen for the purpose of ESR studies. Optimizing the temperature elevation program of the TSFZ furnace led to the successful growth of $Mn_xLa_{1.85}Sr_{0.15}Cu_xO_4$ (x<0.05) [22]. The superconducting transition temperature was determined by a superconducting quantum interference device (SQUID) magnetometer (Quantum Design, MPMS) with the perpendicular applied magnetic field of 10 Oe upon field cooling.

3. Results and discussion

3.1. Carrier-induced local distortion in LSCO

LSCO is a typical HTSC cuprate with a K_2NiF_4-type structure, where copper atoms are coordinated by four *in-plane* oxygen atoms O(1) and two *out-of-plane* (apical) oxygen atoms O(2). The CuO_6 octahedron is elongated along the *c*-axis with two long (2.40 Å) and four short (1.89 Å) bonds as a result of Jahn-Teller (JT) distortion (Figure 1(a)) [22]. The geometry of our experimental setup is illustrated in Figure 1(b). The Cu K-EXAFS oscillations shown in Figure 2(a) originating from the interference of photoelectrons scattered by local atoms are sensitive to the local structure, *i.e.*, polarized radial distribution functions (RDFs) around a specific (absorbing atom) site. To analyze the relative displacement between the copper and oxygen atoms, the mean-square relative displacement $\sigma^2_{Cu\text{-}O}$ was used. In the Fourier transform (FT) shown in Figure 2(b), the *in-plane* Cu-O(1) correlation was separated. A standard FT and curve fitting based on a single scattering theory were performed for all EXAFS data taken over a wide range in temperature down to 5 K.

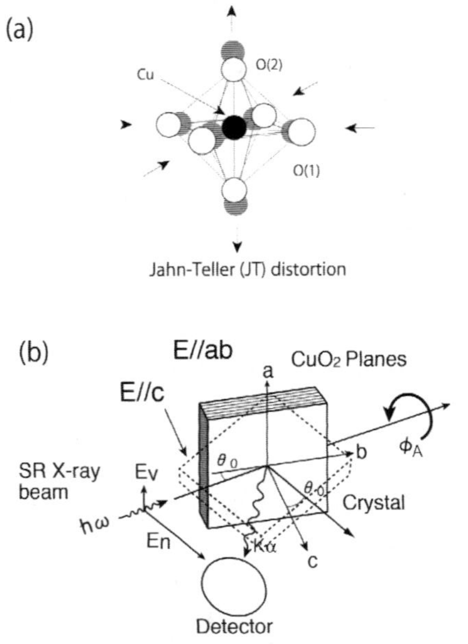

Figure 1. Jahn-Teller distortion of CuO_6 octahedral unit (a) and geometry of experimentl setup (b).

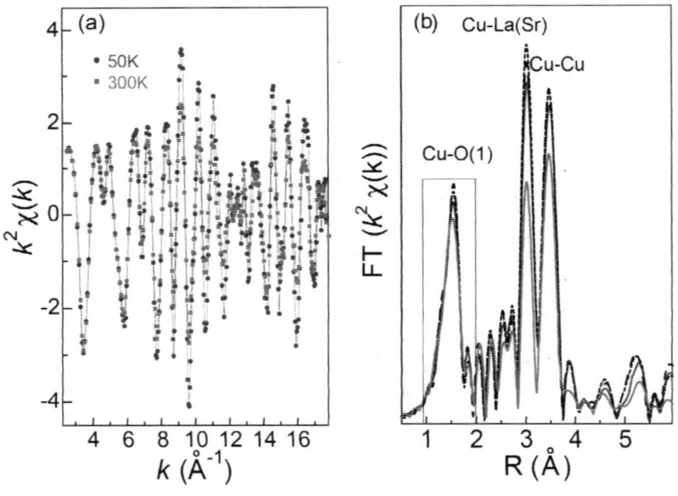

Figure 2. The normalized Cu K-EXAS oscillations for $La_{1.85}Sr_{0.15}CuO_4$ single crystal multiplied by k^2 (a) and the Fourier transform magnitude (b).

The relative displacement $\sigma^2_{Cu\text{-}O}$ is plotted in Figure 3 where open circles illustrate temperature dependence of the *in-plane* Cu-O in the optimally doped LSCO single crystal. In order to compare the superconducting and non-superconducting samples without the effect of chemical doping, careful comparison was made for the non-superconducting specimen prepared by substituting copper atom with 5% magnetic impurity (Ni), which completely suppressed superconductivity by magnetic pair breaking [8]. Reflecting a fast time scale of the probe (10^{-12} sec), the EXAFS results provide an instantaneous relative displacement of *dynamic and static* nature.

Superconducting LSCO samples show local lattice anomaly as temperature is decreased, described by: (1) an upturn $\sigma^2_{Cu\text{-}O}$ beginning at T^* (70-80 K) ~$2T_c$ (Figure 3) that maximizes at T_d^{max} ~T_c (37 K) , (2) a sharp drop upon superconductivity and (3) the continuing upturn below T_c. These anomalies are in contrast to the background temperature dependence which smoothly varies over a wide range of temperature reflecting *normal phonons* described by a non-correlated Debye model. Generally in a normal state, such a normal phononic behaviour is commonly observed if there is no structural phase transformation including a change in the Jahn-Teller distortion. An upturn (deviation from the background with a negative temperature dependence) indicates the presence of a broadened radial distribution function (RDF) either due to disorder or structural change such as a split of the scatterer atom (oxygen) positions of *static or dynamic* nature. In our case, because of a beat, the former two possibilities were differentiated. A beat occurs due to the interference of the two closely separated shells. As illustrated in Figure 4(a), a beat feature was observed in the EXAFS oscillations around k=12 Å^{-1}. Both EXAFS and FT magnitude functions were well reproduced by distortion models characterized by the two different bond lengths separated by ΔR~0.12 Å. Thus at T_d^{max}, the local structure gives rise to a bond alternation. One can find that the maximum distortion at T_d^{max} exactly coincide with the onset of superconductivity at T_c^{onset}, as illustrated in Figure 3(a) [23]. We note that our analysis of the *out-of-plane* Cu-O bond showed no anomaly, in contrast to the early works which related the anomaly to a low temperature tetragonal (LTT) deformation which causes stripe inhomogeneity [12]. The effect of anti Jahn-Teller (AJT) static displacement of apical oxygen results in the electron transport without destroying the anti ferromagnetic order [24]. We will focus on, in the following, the *dynamical* i*n-plane* lattice distortion with an anti Jahn-Teller (AJT) type displacement which indicates polaron or bipolaron are created below 1.5-$2T_c$. The present results indicate that both *static out-of-plane* and *dynamical in-plane* distortions are complementary in describing the fundamental electronic states and polaron transport in the doped CuO_2 plane, respectively.

Journal of Physics: Conference Series **428** (2013) 012042 doi:10.1088/1742-6596/428/1/012042

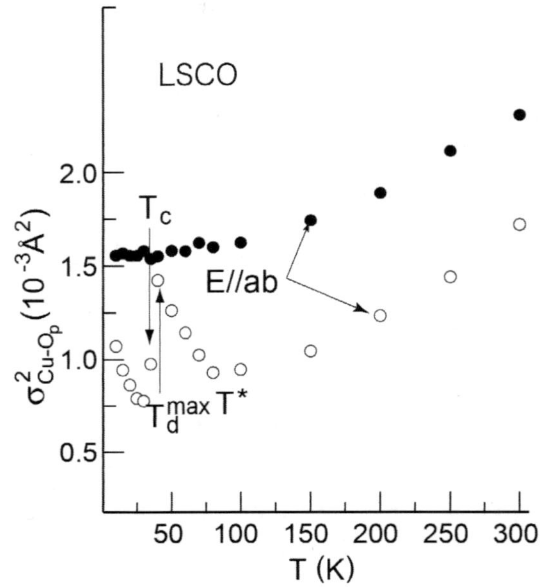

Figure 3. *E//ab* mean-square relative displacement for the Cu-O(1) bond in $La_{2-x}Sr_xCuO_4$ (x=0.15) Open and closed circles indicate the values for superconducting and non-superconducting samples of LSCO single crystals, respectively. The latter sample was prepared by substituting Cu with 5% Ni.

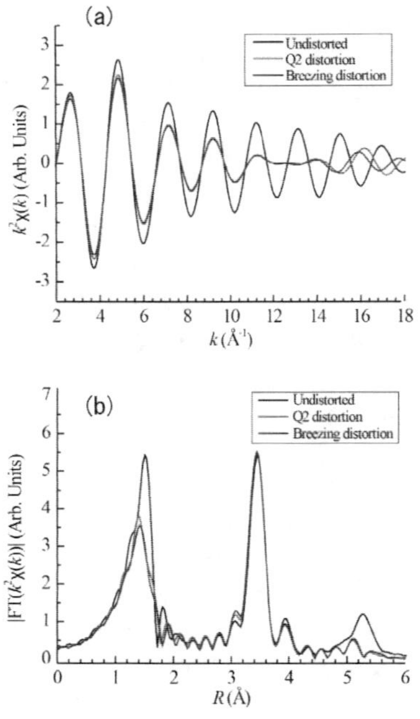

Figure 4. The first-shell E//ab Cu K-EXAS oscillations for $La_{2-x}Sr_xCuO_4$ multiplied by k^2 (a) and the Fourier transform magnitudes (b). Experimental results are compared with those of the simulation for distorted models

3.2. Local distortion in LSCO with magnetic impurities

Magnetic impurities substituting copper sites in the conducting layer of superconducting cuprates strongly affect the superconducting properties because of magnetic pair breaking described by the Abrikosov theory [25]. The *E//ab* polarized EXAFS confirmed that all dopants (Mn, Ni, Co) are incorporated in a K_2NiF_4-type structure with a chemical formula of $La_{1.85}Sr_{0.15}Cu_{1-x}M_xO_4$. Figure 5(a)-(c) summarize temperature dependence of the *in-plane* σ^2_{Cu-O} and susceptibility which shows a significant element-dependence of impurity doping. Introducing a small amount of Ni and Co impurities at the Cu site sharply suppresses both the local lattice anomaly in the Cu-O σ^2_{Cu-O} and susceptibility. With a 5% Ni or Co substitution, the lattice anomaly completely disappears associated with a collapse of superconductivity, which is not surprising since $T_d^{max} = T_c^{onset}$. In contrast, the substitution of Mn weakens the magnetic pair breaking associated with a less perturbed lattice distortion, *i.e.*, superconductivity persists as long as the lattice anomaly exists for x<0.03. Magnetic impurities have been believed to be detrimental to superconductivity based on the pair breaking theory. Such a magnetic scattering (pair breaking) has no indication of polaron (bipolaron) but at least the magnetic effect is weakened while the polaron (bipolaron) is mobile as detected by temperature-dependent local distortion durinmg which T_d^{max}=T_c^{onset}= const. which is realized only in the case of Mn doping. In Ni- and Co-doped LSCO samples, in contrast, both T_d^{max} and T_c^{onset} sharply drop as x increases, indicating a superconducting phase coherence (superfluid density) is sensitive to impurity doping.

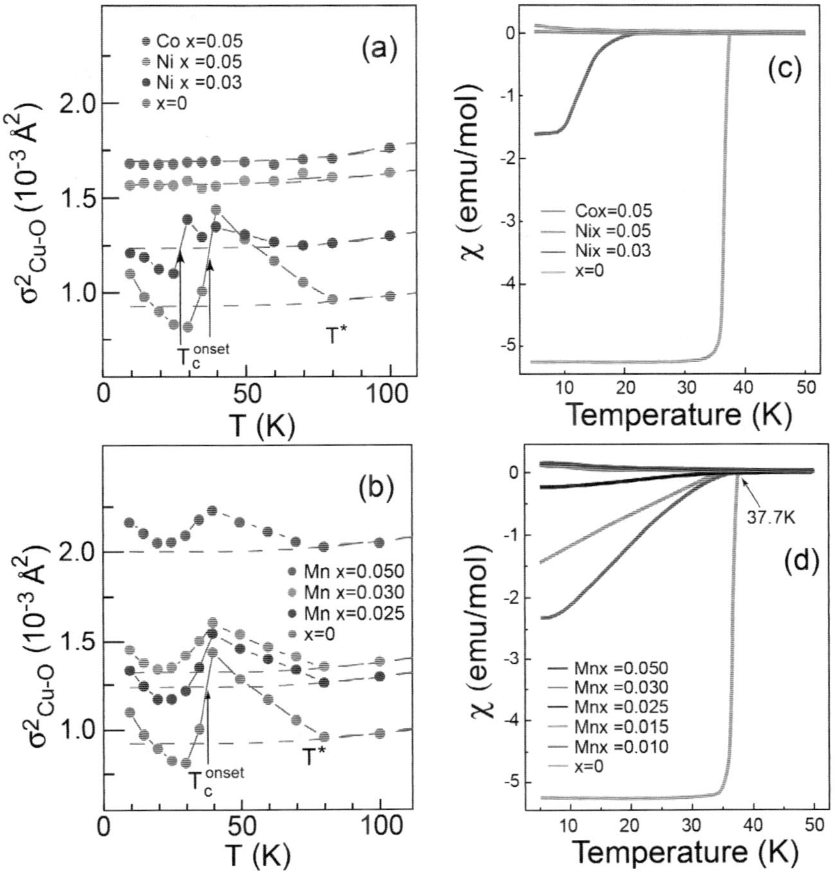

Figure 5. Temperature dependence of the mean square relative displacement of the *in-plane* Cu-O bond σ^2_{Cu-O} for LSCO single crystals with Ni, Co substitution (a) and Mn substitution (b) derived from the EXAFS data. Temperature dependence of the magnetic susceptibility of LSCO single crystals with Ni, Co substitution (c) and Mn substitution (d).

The Mn doping, however, lowers the slope of susceptibility curve and the x=0.05 specimen hardly shows superconductivity while the local lattice anomaly is present. The apparent discrepancy in the correlation between the lattice anomaly and superconductivity suggests that the Mn doping eventually causes inhomogeneous superconductivity because of the loss of coherence between the local pairs.

It is remarkable that Mn doping maintains the onset of superconductivity with the presence of AJT type distortion, unlike Ni and Co doping that immediately destroys superconductivity. Since the uniqueness of Mn doping is associated with the dynamical AJT distortion of the host lattice, it is likely that the AJT polaron (bipolaron) is a prerequisite to superconductivity. Let us examine the local structure of impurity atoms in $La_{1.85}Sr_{0.15}Cu_{1-x}M_xO_4$ (M=Mn, Ni, Co, x=0.05) single crystal samples which do not show superconductivity. The EXAFS measurement at 10 K indicated the AJT distorted structure of static nature for the case of Mn doping. Polarized Mn K-edge EXAFS spectra were analysed by the Ifeffit-1.2.9 package [26] using theoretical amplitudes and phases calculated from the crystallographic data. The *in-plane* M-O(1) distances are: 1.92 Å, 1.88 Å and 1.88 Å for M=Mn, Ni and Co, respectively. Since the Mn-O(1) distance is significantly longer than that of LSCO (~1.88 Å) while the Mn-O(2) distance is shorter, the impurity site in the CuO_2 plane is AJT distorted. In contrast, the *in-plane* Ni-O and Co-O bond lengths are close to the Cu-O(1) distance, indicating an undistorted local lattice around an impurity atom. Comparing the local distortions in the CuO_6 unit for Mn, Ni, Co doped cuprates with those parameters in perovskite manganite, nickelate, and cobaltite, the shortening of the overall M-O(2) distance (M=Mn, Ni, Co) is interpreted in terms of a conventional static AJT distortion. Moreover, the present results indicate that the *static* AJT sites (upto 0.5 %) does not critically affect the propagation of AJT-distorted domains, suggesting that polaron (bipolaron) is essential for achieving a phase coherence. In summary, a close correlation between the polaron (bipolaron) and onset of superconductivity is demonstrated.

3.3. Possible models of distortion-nature of polaron

The mean square relative displacement of the *in-plane* Cu-O bond, $\sigma^2_{Cu\text{-}O}$ in superconducting cuprates has an anomalous temperature dependence as a general feature observed in superconducting samples: (1) deviation from a normal phonon described by a simple model (negative temperature dependence), (2) a sharp drop at the a sharp drop at T_c which coincides with the inflection point of derivative resistivity variation $\partial r/\partial T$ and (3) continuing trend of negative temperature dependence. Based on the systematic studies [12] indicate that the magnitude of the drop is roughly proportional to the superfluid density. Inhomogeneous nature of cuprates is likely to relate to structural inhomogeneity consisting of undistorted and distorted domains.

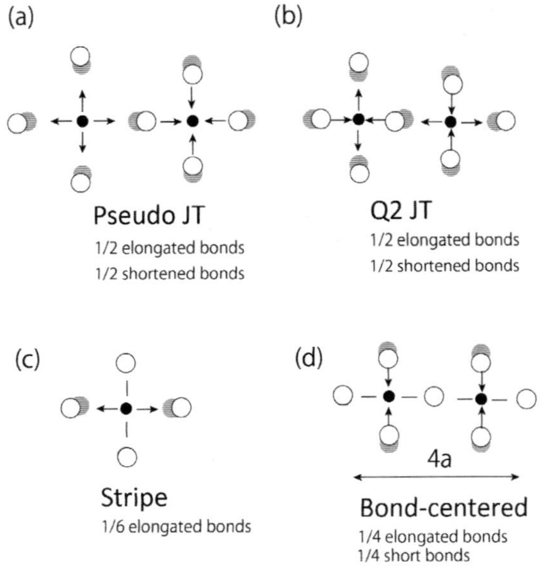

Figure 6. Schematic representation of possible distortion models; pseudo JT (breathing-like distortion) [16] (a), Q_2 JT (b) [35,15], stripe with LTT tilting of the CuO_6 unit. [12] (c), and stripe with Cu-O stretching (half breathing-like distortion) [28] (d).

The magnitude of $\sigma^2_{Cu\text{-}O}$ drop is a measure of superconducting fraction and the disappearance of inhomogeneity upon superconductivity which shows homogeneous electron states. On the other hand the observed coincidence of the onset of superconductivity with the maximum lattice distortion or fraction of distorted domain (T_d^{max}) relates to a superconducting phase coherence, for which a critical length may exist between the Josephson Junction (JJ)-coupled "metallic" domains and undistorted "insulating" domain. If the a distance between metallic domain and insulating domain exceeds this critical limit, "macroscopic" superconductivity and lattice (distortion) may loose consistence as illustrated in the case of Mn substitution. Several different models were proposed to describe the AJT-like distortion below the pseudogap temperature. Here we focus on the existence of "beat" in *k*-space at 12 $Å^{-1}$ which indicates the presence of two different bond lengths with almost equal fraction. Let us consider simple models that reproduces the experimental bond alternation.

As doping proceeds, reduced Coulomb repulsion (U) would promote integration into extended domains with a larger size. The average distance between extended domains L decreases allowing tunnelling over insulating domains [14, 27]. When L becomes greater than a critical value (z), quantum tunnelling is prohibited. For instance, suppressed tunnelling under a tensile strain decreases superfluid density, and hence the critical temperature. Near the Mn dopants in Mn:LSCO (x=0.05), non-superconducting regions are formed and superconducting domains are limited to small domains. As a distance between the superconducting domains becomes large, quantum tunneling [14, 27] or percolation [28] is prohibited and the superconductivity eventually disappears although the onset temperature remains unaffected.

If we apply the bond alternation observed at T_d^{max} as a criterion to differentiate the possible models, only the distortion models having equal number of elongated and shortened bonds remain. Possible structural models in Figure 6(a),(b) satisfy the criterion [15,16] in the limit of inhomogeneity (distortion extending all over the basal plane) while two stripe models in Figure 6(c),(d) are excluded by the criterion assuming the fraction of distorted domains 1/3 and 1/2, respectively. [12, 29]. The completion of phase coherence requires the extended AJT distortion over the crystal. It is quite interesting in possible models, the two adjacent CuO_6 units are linked each other so that the elastic energy is minimized by the complimentary distortions. This coupled CuO_6 units are familiar to a bipolaron concept. The well-known disadvantage of a JT polaron, *i.e.*, a heavy mass [30] that leads to immobility is compensated by a bipolaron [31] which does not perturb the antiferromagnetic order [32]. As the coupling of the two units is driven by elastic forces [33], the findings that both AJT models fit to the observed bond alternation seem to be reasonable. Polarons may perturb spin configurations (antiferromagnetic order) and lead to spin vortices around that grow into spin loop current [16] or their one dimensional hopping may enhance strong coupling among spin, charge and lattice [15].

Other distortion models failed to satisfy the criterion [12,29] are based on the stripes of distorted domains sandwiched by undistorted ones. Their distortions are due to either a LTT tilting of the CuO_6 units (Figure 6(c)) or a half-breezing like Cu-O bond stretching (Figure 6(d)), both of which provide the equal number of short and elongated bonds but they are diluted by the spacing of undistorted domains. Because of inhomogeneity, stripe models can not clear the criterion. The real space image of a Cu-O-Cu bond-centered electronic glass with disperse $4a_0$-wide domains of displaced oxygen atoms is observed by tunneling asymmetry imaging for $Bi_2Sr_2CaCu_2O_{8+d}$ (Bi2212) [34]. As the spatial variations in tunneling asymmetry occurs at the *in-plane* oxygen sites and the observed 2D texture pattern dispersed throughout the crystal may partially reflect the original dynamical distortion without long-range order, such patterns are taken as a frozen disorder because of a slow time scale of STM. In the model in Figure 6(d), the observed sharp contrast between a central ladder consisting of a column of oxygen atoms and the two neighboring Cu–O–Cu columns form a charge stripe [29]. Recent photoemission study on Bi2212 reported the growth of pseudogap component attributed to the preformed pair is associated with the planar modulation of electronic states observed by the STM. In this context, the polaronic lattice distortion may explain the local structural origin of pseudogap. Recent model calculations [35] indeed showed bipolarons form a charge-inhomogeneous texture pattern. The observed *in-plane* bond alternation may describe coherence of preformed pairs. More recently, the Q_2-type AJT distortion model originally based on the three-centered spin (ESR) [36] was interpreted in terms of bipolaron [37,38] which could stay one of the most likely models.

Conclusion

The carrier-induced local lattice distortion in $La_{1-x}Sr_xCuO_4$ single crystals grown by a TSFZ method was studied by a fast local probe x-ray absorption spectroscopy. We find an anomalous upturn of the mean-square relative displacement of the Cu-O(1) bond detected by the EXAFS below $T^{*}{\sim}2T_c$ strongly correlates with superconductivity. The deviation reflects an *in-plane* local lattice distortion described by an anti Jahn-Teller distortion which demonstrates the essential role of lattice, or polaron (bipolaron) in the mechanism of pairing. The observed local distortion maximizes its magnitude at $T_d^{max}{\sim}T_c^{onset}$. The presence of beat in the EXAFS oscillations at k=12 Å^{-1} indicates that the splitting of the Cu-O(1) distance into shortened and elongated bonds (a bond alternation), R_1 and R_2 where ΔR= R_1-R_2~0.12 Å. The onset of lattice distortion coincides with the opening of a pseudogap (*ca.* 80 K~$2T_c$) where an unconventional electronic state begins to grow. Upturn trend (a negative temperature dependence) of the relative displacement deviating from a normal phonon behaviour indicates the growth of dynamical AJT distorted domains. The local lattice at T_d^{max} is well described by distortion models with AJT type distortion such as Q_2 type or pseudo JT, both of which are characterized by close link of two CuO_6 units minimizing the elastic energy, which is consistent with bipolaron concept. The substitution of Mn atom into a copper site causes little perturbation of this lattice distortion, in contrast to Ni and Co substitutions which strongly suppress lattice distortion and thus superconductivity. The polarized EXAFS study on magnetic impurity doped LSCO confirmed the substitution of impurities (Mn, Ni, Co) upto 5% while susceptibility showed that Mn doping uniquely sustains the onset of superconductivity, in sharp contrast with Ni and Co which quickly destroy the superconductivity. Such an unique behavior of Mn doping is associated with the dynamic AJT type distortion retained by the host lattice. The magnetic impurity doping results indicate an intimate relation between superconductivity onset (phase coherence) and the dynamic AJT distortion of host lattice that is hardly influenced by a static distortion around impurity sites. The EXAFS data at T_d^{max} where the distorted domains maximize in population were consistent with possible models (Q_2 type AJT or pseudo JT) in the limit of inhomogeneity which could be considered as a model polaron (bipolaron) system.

Acknowledgments

The author express their thanks to A. Bussmann-Holder and H. Kamimura for fruitful discussions.

References

[1] Bednorz J G and Müller K A 1986 *Z. Phys.* B **64** 189
[2] Kamihara Y, Watanabe T, Hirano M, and Hosono H 2008 *J. Am. Chem. Soc.* **130** 3296
[3] Nakamura S, Senoh M, Nagahama S, Iwase N, Yamada T, Matsushita T, Kiyoku H and Sugimoto Y 1996 *Japan. J. Appl. Phys.* **35** L74
[4] Anderson P W 1997 *The Theory of Superconductivity in the Cuprates* (Princeton, M.A.: Princeton University Press)
[5] Malaeb W, Yoshida T, Kataoka T, Fujimori A, Kubota M, Ono K, Usui H, Kuroki K, Arita R, Aoki H, Kamihara Y, Hirano M and Hosono H 2008 *J. Phys. Soc. Jpn.* **77** 093714
[6] Hiraishi M, Kadono R, Takeshita S, Miyazaki M, Koda A, Okabe H, and Akimitsu J 2009 *J. Phys. Soc. Jpn.* **78** 023710
[7] Zhang C J, Oyanagi H, Sun Z H, Kamihara Y, and Hosono H 2008 *Phys. Rev.* B **78** 214513
[8] Zhang C J and Oyanagi H 2009 Phys. Rev. B **79** 064521
[9] Alexandrov A S 2008 *J. Supercond. Nov. Magn.* **22** 95
[10] Mannela N, Yang W L, Tanaka K, Zhou X J, Zheng H, Mitchell J F, Zaanen J, Devereaux T P, Nagaosa N, Hussain Z, and Shen Z -X. 2007 *Phys. Rev.* B **76** 233102

[11] Lanzara A *et al.* 2001 *Nature* **412** 510.
[12] Bianconi A, Saini N L, Lanzara A, Missori M, Rossetti T, Oyanagi H, Yamaguchi H, Oka K, and Ito T 1996 *Phys. Rev. Lett.* **76** 3412; Saini N L, Lanzara A, Oyanagi H, Yamaguchi H, Oka K, Ito T and Bianconi A 1997, *Phys. Rev.* B **55** 12759
[13] Mustre de Leon J, Conradson S D, Batistic I, and Bishop A R 1990, *Phys. Rev. Lett.* **65** 1675
[14] Mihailovic D 2005 *Phys. Rev. Lett.* **94** 207001-1
[15] Bussmann-Holder A and Keller H 2005 *Eur. Phys. J.* B **44** 487
[16] Koizumi H 2008 *J. Phys. Soc. Jpn.* **77** 034712
[17] Oyanagi H, Fonne C, Gutknecht D, Dressler P, Henck R., Lampert M –O., Ogawa S and Kasai K 2002 *Nucl. Instrum. & Meth.* A **513** 340
[18] Oyanagi H, Tsukada A, Naito M and Saini N L 2007 *Phys. Rev.* B **75** 024511
[19] Sato H and Naito M 1997 *Physica C* **280** 178; Locquet J –P, Perret J, Fompeyrine J, Machler E, Seo J W and Van Tendeloo G 1998, *Nature* **394** 453; Bozovic I, Logvenov G, Belca I, Narimbetov B and Sveklo I 2002 *Phys. Rev. Lett.* **89** 107001
[20] Tarascon J M, Greene L H, Barboux P, McKinnon W R, Hull G W, Orlando T P, Delin K A, Foner S, and McNiff Jr. E J 1987 *Phys. Rev.* B **36** 8393
[21] Oyanagi H, Tsukada A, Naito M, Saini N L, Lampert M –O, Gutknecht D, Dressler P, Ogawa S, Kasai K, Mohamed S and Fukano A 2006 *J. Synchrotron Radiation* **13** 314
[22] Radaelli P G, Hinks D G, Mitchell A W, Hunter B A, Wagner J L, Dabrowski B, Vandervoort K G, Viswanathan H K, and Jorgensen J D 1994, *Phys. Rev.* B **49** 4163
[23] Zhang C Z Oyanagi H, and Lee C H 2008 *Physica* C **468** 898
[24] Kamimura H and Ushio H 2012 *J. Supercond. Nov. Magn.* **25** 677; Kamimura H, Ushio H, Matsumoto S and Hamada T 2005 *Theory of copper oxide superconductors,* (Berlin: Springer)
[25] Abrikosov A A, Gork'ov L P 1961 *Sov. Phys. -JETP* **12** 1243
[26] Rehr J J, Mustre de Leon J, Zabinsky S I, and Albers R C, 1991 *J. Am. Chem. Soc.* **113** 5135
[27] Mihailovic D, Kabanov V V and Müller K A 2002 *Europhys. Lett.* **57** 254
[28] Phillips J C 2007, *Phys. Rev.* B **75** 214503
[29] Deutcher G, de Gennes P G 2007 *C.R. Physique* 8 937
[30] H¨ock K H, Nickisch H and Thomas H 1983 *Helv. Phys. Acta* **56** 237
[31] Stoneham A M and Smith L W 1990 *J. Phys.: Condens. Matter* **3** 225
[32] Hirsch J E 1993 *Phys. Rev.* B **47** 5351
[33] Kochelaev B, Safina A M, Shengelaya A S, Müller K A and Conder K 2003 *Mod. Phys. Lett.* B **17** 415
[34] Kohsaka Y, Taylor C, Fujita K, Schmidt A, Lupien C, Hanaguri T, Azuma M, Takano M, Eisaki H, Takagi H, Uchida S, and Davis J C 2007 *Science* **315** 1380
[35] Mihailovic D and Kabanov V V 2001 *Phys. Rev.* B **63** 054505.
[36] Kochelaev B J, Sichelschmidt J, Elschner B, Lemor W and Loidl A 1997 *Phys. Rev. Lett.* **79** 4274.
[37] Alexandrov A S, Kabanov V V and Mott N F 1996 *Phys. Rev. Lett.* **77** 4796
[38] Müller K A 2007, *J. Phys.: Condens. Matter* 19 251002.

On the important role of the anti-Jahn-Teller e ect in underdoped cuprate superconductors

Hiroshi Kamimura, Shunichi Matsuno[1], Takashi Mizokawa[2], Kenji Sasaoka[3], Kenji Shiraishi[4], and Hideki Ushio[5]

Tokyo University of Science, 1-3 Kagurazaka, Shinjuku-ku, Tokyo 162-8601, Japan
[1]Tokai University, Shimizu General Education Center, 3-20-1 Orido Shimizu-ku, Shizuoka 424-8610, Japan
[2]The University of Tokyo, Department of Complexity Science and Engineering, 5-1-5 Kashiwanoha, Kashiwa, Chiba 277-0882, Japan
[3]The University of Tsukuba, Graduate School of Pure and Applied Sciences, 1-1-1 Tennodai, Tsukuba 305-0006, Japan
[4]The University of Tokyo, Department of Materials Engineering, 7-3-1 Hongo, Bunkyo-ku, Tokyo 113-8656, Japan
[5]Tokyo National College of Technology, 1220-2 Kunugida-chou, Hachioji 193-0997, Japan

E-mail: kamimura@rs.kagu.tus.ac.jp

Abstract. In this paper it is shown that the "anti-Jahn-Teller effect" plays an essential role in giving rise to a small Fermi surface of Fermi pockets above T_c and d-wave superconductivity below T_c in underdoped cuprates. In the first part of the present paper, we review the latest developments of the model proposed by Kamimura and Suwa, which bears important characteristics born from the interplay of Jahn-Teller Physics and Mott Physics. It is shown that the feature of Fermi surfaces in underdoped LSCO is the Fermi pockets in the nodal region constructed by doped holes under the coexistence of a metallic state and of the local antiferromagnetic order. In the antinodal region in the momentum space, there are no Fermi surfaces. Then it is discussed that the phonon-involved mechanism based on the Kamimura-Suwa model leads to the d-wave superconductivity. In particular, it is shown that the origin of strong electron-phonon interactions in cuprates is due to the anti-Jahn-Teller effect. In the second part a recent theoretical result on the energy distribution curves (EDCs) of angle-resolved photoemission spectroscopy (ARPES) below T_c is discussed. It is shown that the feature of ARPES profiles of underdoped cuprates consists of a coherent peak in the nodal region and the real transitions of photoexcited electrons from occupied states below the Fermi level to a free-electron state above the vacuum level in the antinodal region, where the latter transitions form a broad hump. From this feature, the origin of the two distinct gaps observed by ARPES is elucidated without introducing the concept of the pseudogap. Finally, a remark is made on the phase diagram of underdoped cuprates.

1. Introduction

In 1986 Bednorz and Müller discovered high temperature superconductivity in copper oxides by doping hole carriers into La_2CuO_4 [1]. Their motivation was that higher T_c could be achieved for copper oxide materials by combining Jahn-Teller (JT) active Cu ions with the structural complexity of layer-type perovskite oxides. Undoped copper oxide La_2CuO_4 is an antiferromagnetic Mott insulator, in which an electron correlation plays an important role, as pointed out by Anderson [2]. Thus we may say that undoped cuprates are governed by Mott physics. On the other hand, the CuO_6 octahedrons in undoped La_2CuO_4 are elongated by the JT effect. To investigate the mechanism of high temperature superconductivity, it is assumed in most models that the doped holes itinerate through orbitals extending over a CuO_2 plane in the systems consisting of CuO_6 octahedrons elongated by the JT effect. Those models are called the "single-component theory", because the orbitals of hole carriers extend only over a CuO_2 plane.

In 1988, Shiraishi and coworkers [3, 4, 5] showed by first-principles calculations that the apical oxygen in the CuO_6 octahedrons tend to approach towards Cu^{2+} ions, when Sr^{2+} ions are substituted for La^{3+} ions in La_2CuO_4, in order to gain the attractive electrostatic energy in ionic crystals such as cuprates. As a result CuO_6 elongated by the JT effect shrinks with hole doping. This deformation against the JT distortion is called "anti-Jahn-Teller effect" [6]. By this effect, the energy separation between the two kinds of orbital states, which have been split originally by the JT effect, becomes smaller with hole carrier doping. These two states are the a_{1g} antibonding orbital state $|a^*_{1g}\rangle$ and b_{1g} bonding orbital state $|b_{1g}\rangle$. The a^*_{1g} antibonding orbital state is constructed by a Cu dz^2 orbital and the six surrounding oxygen p orbitals including apical O p_z-orbitals; the $|b_{1g}\rangle$ orbital is constructed by four in-plane O p_σ orbitals with a small Cu $d_{x^2-y^2}$ component parallel to the CuO_2 plane. The spatial extensions of the $|a^*_{1g}\rangle$ and $|b_{1g}\rangle$ orbitals, which are perpendicular and parallel to the CuO_2 plane, respectively, are shown in Fig. 1.

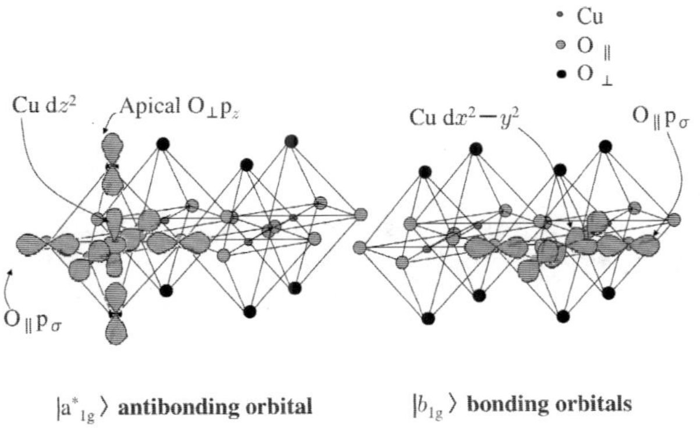

Figure 1. Spatial extension of a^*_{1g} antibonding orbital $|a^*_{1g}\rangle$ and b_{1g} bonding orbital $|b_{1g}\rangle$. **a**, a^*_{1g} antibonding orbital $|a^*_{1g}\rangle$. **b**, b_{1g} bonding orbital $|b_{1g}\rangle$.

By taking account of the anti-Jahn-Teller effect, Kamimura and Suwa proposed that one must consider these two kinds of orbital states equally in forming the metallic state of cuprates; they constructed a metallic state coexisting with the local antiferromagnetic (AF) order [7]. This model is called the "Kamimura-Suwa (K-S) model" [8]. Since the anti-Jahn-Teller effect is a central issue of Jahn-Teller physics, we may say that the K-S model bears important features originating from the interplay of Jahn-Teller physics and Mott physics. Since these two kinds of orbitals extend not only over the CuO_2 plane but also along the direction perpendicular to it, the K-S model represents a prototype of a "two-component theory", in contrast to the single-component theory.

On the basis of the K-S model, Kamimura and Ushio calculated Fermi surfaces in underdoped $La_{2-x}Sr_xCuO_4$ (LSCO) [9, 10], and showed that the coexistence of a metallic state and a local AF order results in the Fermi pockets constructed from doped holes in the nodal region. The

Journal of Physics: Conference Series **428** (2013) 012043 doi:10.1088/1742-6596/428/1/012043

appearance of Fermi pockets and small Fermi surfaces in cuprates has recently been reported by various experimental groups [11, 12, 13, 14, 15, 16].

In this paper, we first review the recent developments emerged from the K-S model. The first development is concerned with the shape of Fermi surfaces in underdoped cuprates. The second one is the occurrence of d-wave superconductivity in phonon mechanism. It is caused by the interplay of the strong electron-phonon interactions and local AF order [17, 18]. Further, in this paper we will show that the origin of strong electron-phonon interactions in cuprates is due to the anti-Jahn-Teller effect.

In the second part of the present paper, we will clarify the origin of the energy distribution curves (EDCs) of the angle-resolved photoemission spectroscopy (ARPES) profiles of cuprates below T_c observed by Tanaka and coworkers [19], based on recent theoretical calculations performed by Kamimura, Sasaoka and Ushio (abbreviated as KSU hereafter) on the basis of the K-S model [20]. We will show that the feature of the ARPES profiles consists of a coherent peak due to the d-wave superconducting density of states in the nodal region and the real transitions of electrons from the occupied states below the Fermi level to a free-electron state above the vacuum level in the antinodal region. In particular, we show theoretically that the latter transitions form a broad hump in ARPES EDCs in underdoped cuprates. From good agreement between theory and experiment, we conclude that the introduction of the phenomenological idea of the pseudogap is not necessary. Finally, in connection with the finite size of the spin-correlation length in a metallic state, we discuss the finite size effect of a metallic state on the spin-electronic structures of underdoped cuprates, and a new explanation for the phase diagram for underdoped cuprates is proposed.

The organization of the present paper is as follows: In §2, we explain about the anti-Jahn-Teller effect. Then in §3 and §4, we summarize the important features of the K-S model, which bears important features originating from the interplay of JT physics and Mott physics. In this connection, we will show that the origin of strong electron-phonon interactions in cuprates is due to the anti-Jahn-Teller effect in a sybsection of §4. In §5, on the basis of the many-body effects including energy bands obtained from the K-S model, we discuss the key features of ARPES EDCs and clarify the origin of the two-gap scenario proposed from the experimental results of Tanaka *et al* [19]. in §6, we discuss the finite size effects on the Fermi surfaces in cuprates. In §7, taking account of the finite size effect, we propose a new interpretation for the phase diagram of underdoped cuprates. We devote §8 to the conclusion and concluding remarks.

2. On the anti-Jahn-Teller e ect

As we have already described, a regular octahedron in La_2CuO_4 is elongated along the c-axis by the Jahn-Teller effect. In order to investigate whether the Jah-Teller interaction still plays a role in the local distortion of the CuO_6 octahedrons in the presence of hole carriers in ionic crystals such as cuprates, Shiraishi first performed the total energy band structure calculations of $La_{2-x}Sr_xCuO_4$ (LSCO) by varying the Cu- apical O distance in his thesis work in 1988[3]. Then Shiraishi and coworkers carried out the first-principles calculations by using the norm-conserving pseudopotential method within the local density functional formalism by varying the Sr concentration for the system of a spin-unpolarized configuration in $La_{2-x}Sr_xCuO_4$ (LSCO) with a tetragonal phase, where the doping effect of Sr is taken into account by the virtual crystal approximation [3, 4, 5].

In Fig. 2 the optimized Cu- apical O distance in the total energy calculations is plotted as a function of Sr concentration. As shown in this figure, the Cu- apical O distance decreases from 2.41 Ain the undoped case, while the Cu- inplane O distance of 1.89 Adoes not change by Sr concentration. This trend is consistent with experimental results [21]. The present result suggests that the attractive electrostatic interactions act against the Jahn-Teller intractions in the presence of hole carriers.

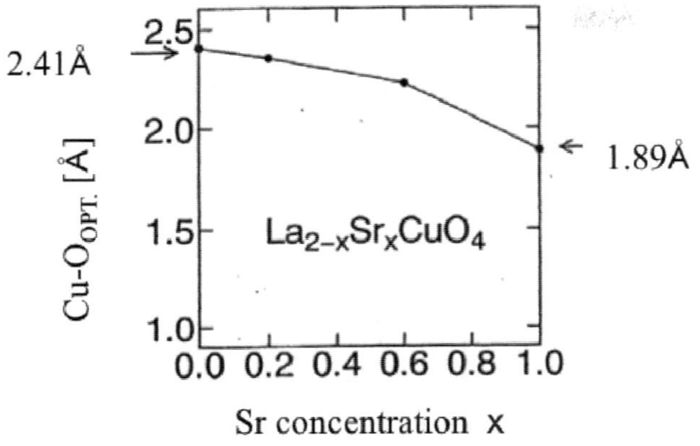

Figure 2. The optimized Cu- apical O distance as a function of Sr concentration.

3. On the K-S Model

In this section, we summarize the main features of the K-S model [7], emphasizing its important roles in underdoped cuprates due to the interplay of JT physics and Mott physics.

3.1. Key features of the K-S model

The key features of the K-S model are explained in a heuristic way using the picture of a two-story house model shown in Fig. 3. In this figure, the first story of a Cu house is occupied by Cu localized spins, which form the AF order in the spin-correlated region by the superexchange interaction. The second story in the Cu house consists of two oors due to the anti-JT effect; the lower a^*_{1g} oor and the upper b_{1g} oor. The second stories of neighboring Cu houses are connected by oxygen rooms, re ecting the hybridization of Cu d and O p orbitals. In the second story, a hole carrier with an up-spin enters the a^*_{1g} oor of the left-hand Cu house owing to Hund's coupling with a Cu localized up-spin in the first story (Hund's coupling triplet) [22, 23], as shown in the leftmost column of the figure. By the transfer interaction marked by a long arrow in the figure, the hole is transferred into the b_{1g} oor in the neighboring Cu house (second from the left) through the oxygen rooms, where a hole with up-spin forms a spin-singlet state with a localized down-spin in the second Cu house from the left (Zhang-Rice singlet) [24]. The key feature of the K-S model is that the hole carriers in the underdoped regime of LSCO form a metallic state by taking the Hund coupling triplet and the Zhang-Rice singlet alternately in the presence of a local AF order without destroying the AF order, as shown in the figure. From Fig. 3, one may understand that the characteristic feature of the K-S model is the coexistence of the AF order and a normal, metallic (or a superconducting) state in the underdoped regime. This feature of the K-S model (two-component theory) is different from that of the single-component theory.

As seen in Fig. 3, the wave functions of a hole carrier with up and down-spins have the following phase relation:

$$\Psi_{\vec{k}\downarrow}(\vec{r}) = \exp(i\vec{k}\cdot\vec{a})\Psi_{\vec{k}\uparrow}(\vec{r}). \tag{1}$$

Kamimura *et al.* have shown that this unique phase relation leads to the d-wave superconductivity [17, 18].

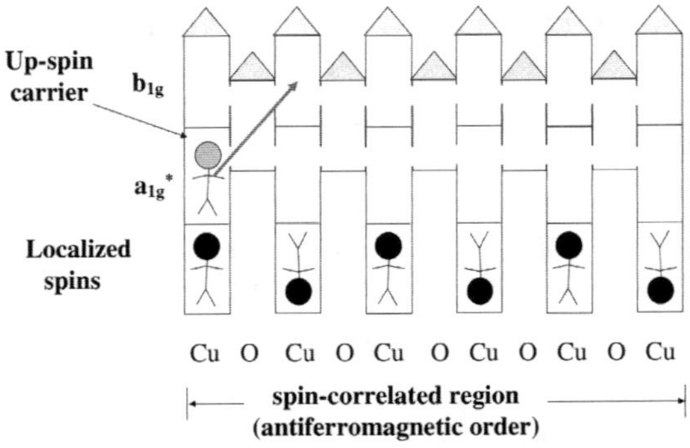

Figure 3. (Color online) Explanation of K-S model using picture of two-story house model.

3.2. Effective Hamiltonian for the K-S model

The following effective Hamiltonian is introduced to describe the K-S model according to Kamimura and Suwa [7]. It consists of four parts: the one-electron Hamiltonian H_{sing} for the a_{1g}^* and b_{1g} orbital states, the transfer interaction between neighboring CuO_6 octahedrons H_{tr}, the superexchange interaction between the Cu $d_{x^2-y^2}$ localized spins H_{AF}, and the exchange interactions between the spins of dopant holes and $d_{x^2-y^2}$ localized holes within the same CuO_6 octahedron H_{ex}. Thus, we have

$$\begin{aligned} H &= H_{\text{sing}} + H_{\text{tr}} + H_{\text{AF}} + H_{\text{ex}} \\ &= \sum_{i,m,\sigma} \varepsilon_m C_{im\sigma}^{\dagger} C_{im\sigma} \\ &\quad + \sum_{\langle i,j\rangle,m,n,\sigma} t_{mn} \left(C_{im\sigma}^{\dagger} C_{jn\sigma} + \text{h.c.} \right) \\ &\quad + J \sum_{\langle i,j\rangle} \vec{S}_i \cdot \vec{S}_j + \sum_{i,m} K_m \, \vec{s}_{i,m} \cdot \vec{S}_i \ , \end{aligned} \tag{2}$$

where ε_m ($m = a_{1g}^*$ or b_{1g}) represents the one-electron energy of the a_{1g}^* and b_{1g} orbital states, $C_{im\sigma}^{\dagger}$ and $C_{im\sigma}$ are the creation and annihilation operators of a dopant hole with spin σ in the ith CuO_6 octahedron, respectively, t_{mn} is the transfer integral of a dopant hole between the m-type and n-type orbitals of neighboring CuO_6 octahedrons, J is the superexchange interaction between spins $\vec{S}_i$ and $\vec{S}_j$ of $d_{x^2-y^2}$ localized holes in the b_{1g}^* orbital in the nearest-neighbor Cu sites i and j ($J > 0$ for AF interaction), and K_m is the exchange integral for the exchange interactions between the spin of a dopant hole, $\vec{s}_{im}$, and the $d_{x^2-y^2}$ localized spin $\vec{S}_i$ in the ith CuO_6 octahedron. There are two exchange constants, i.e. $K_{a_{1g}^*}$ and $K_{b_{1g}}$, for the Hund coupling triplet and the Zhang-Rice singlet, respectively, where $K_{a_{1g}^*} < 0$ and $K_{b_{1g}} > 0$. The appearance of the two kinds of exchange interactions in the fourth term is due to the interplay of Mott physics and JT physics. This is the key feature of the K-S model.

The electron-electron interactions between doped hole carriers are very weak for two reasons: One is the low concentration of hole carriers in the underdoped regime and the other is the wave functions of hole carriers with up and down spins in a CuO_6 octahedron occupying a_{1g}^* and b_{1g}

orbitals, respectively, as seen in eq. (1) and Fig. 1. For these reasons, we have neglected the electron-electron interactions between doped holes in the effective Hamiltonian eq. (2).

3.3. *Coexistence of a metallic state and local AF order*

By replacing the localized spins $\vec{S}_i$'s in $H_{\rm ex}$ with their average $\langle\vec{S}\rangle$ in the mean-field sense, we can calculate the change in the total energy upon moving a hole from an $\rm a_{1g}^*$ orbital state in Hund's coupling spin triplet at Cu site i to an empty $\rm b_{1g}$ orbital state in the Zhang-Rice spin singlet at the neighboring Cu site j. In the first step, the hole moves from Cu site i to infinity. The change in the total energy in the mean field approximation is equal to $\varepsilon_{\rm a_{1g}^*} + \frac{1}{4}K_{\rm a_{1g}^*}$. In the second step, the hole moves from infinity to an empty $\rm b_{1g}$ orbital state at Cu site j to form the Zhang-Rice singlet. The change in the total energy in the second step is equal to $\varepsilon_{\rm b_{1g}} - \frac{3}{4}K_{\rm b_{1g}}$. As a result, the change in the total energy by the transfer of the hole from the occupied $\rm a_{1g}^*$ orbital state at Cu site i to the empty $\rm b_{1g}$ orbital state at Cu site j is

$$\varepsilon_{\rm a_{1g}^*}^{\rm eff} - \varepsilon_{\rm b_{1g}}^{\rm eff} = \varepsilon_{\rm a_{1g}^*} + \frac{1}{4}K_{\rm a_{1g}^*} - \varepsilon_{\rm b_{1g}} + \frac{3}{4}K_{\rm b_{1g}}. \tag{3}$$

Here, $\varepsilon_{\rm a_{1g}^*}^{\rm eff}$ and $\varepsilon_{\rm b_{1g}}^{\rm eff}$ represent the effective one-electron energies of the $\rm a_{1g}^*$ and $\rm b_{1g}$ orbital states including the exchange interaction term $H_{\rm ex}$, respectively. Thus, the energy difference $(\varepsilon_{\rm a_{1g}^*}^{\rm eff} - \varepsilon_{\rm b_{1g}}^{\rm eff})$ corresponds to the energy difference between the $\rm a_{1g}^*$ and $\rm b_{1g}$ oors in the second story in Fig. 4.

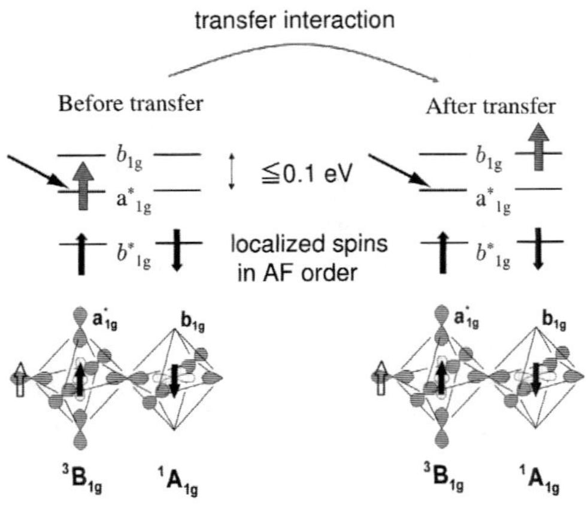

Figure 4. (Color online) Simple explanation for coexistence of metallic state and local AF order in two-component theory ($\rm a_{1g}^*$ and $\rm b_{1g}$ orbital states), by taking neighboring Cu sites i to j as an example, where localized spins are in antibonding $\rm b_{1g}^*$ orbitals. Schematic pictures of Hund's coupling triplet $^3\rm B_{1g}$ and Zhang-Rice singlet $^1\rm A_{1g}$ are also shown at the bottom. The energy in this figure is taken as the hole energy.

Now, let us estimate the energy difference $\varepsilon_{\rm a_{1g}^*}^{\rm eff}$ - $\varepsilon_{\rm b_{1g}}^{\rm eff}$ using the values of the parameters in the effective Hamiltonian eq. (2). The values of the parameters in Hamiltonian eq. (2) have been determined in the case of LSCO in ref. 6 (see also ref. 7). They are $J = 0.1$, $K_{\rm a_{1g}^*} = -2.0$, $K_{\rm b_{1g}} = 4.0$, $t_{\rm a_{1g}^*a_{1g}^*} = 0.2$, $t_{\rm b_{1g}b_{1g}} = 0.4$, $t_{\rm a_{1g}^*b_{1g}} = \sqrt{t_{\rm a_{1g}^*a_{1g}^*}t_{\rm b_{1g}b_{1g}}} \sim 0.28$, $\varepsilon_{\rm a_{1g}^*} = 0$, and $\varepsilon_{\rm b_{1g}} = 2.6$

in units of eV, where $K_{a_{1g}^*}$ and $K_{b_{1g}}$ are taken from first-principles cluster calculations for a CuO_6 octahedron in LSCO [22, 23], and the t_{mn} are obtained by band structure calculation [4, 5]. The difference in one-electron energy between the a_{1g}^* and b_{1g} orbital states in a CuO_6 octahedron for a certain x has been determined so as to reproduce the difference in the lowest state energy between Hund's coupling spin-triplet state and the Zhang-Rice spin-singlet state for the same x in LSCO calculated by Multi-Configuration Self-Consistent Field (MCSCF) cluster calculations which include the anti-JT effect [7].

Thus, the calculated $\varepsilon_{a_{1g}^*}^{\text{eff}}$ - $\varepsilon_{b_{1g}}^{\text{eff}}$ is 0.1 eV in the case of the optimum doping ($x = 0.15$). Then, by introducing the transfer interaction of $t_{a_{1g}^* b_{1g}} = 0.28eV$, a coherent metallic state in the normal phase is obtained in the presence of the local AF order for the underdoped regime. This situation is schematically shown in Fig. 4.

3.4. Features of the many-body effect including energy bands and Fermi surfaces of underdoped LSCO coexisting with the AF order

In the previous subsection, we have shown that the effective Hamiltonian eq. (2) for the K-S model can lead to a unique metallic state in the normal phase, which results in the coexistence of a superconducting state and an AF order below T_c. In 1994, Kamimura and Ushio calculated the energy bands and Fermi surfaces of underdoped LSCO in the normal phase on the basis of the effective Hamiltonian eq. (2), by treating the fourth term H_{ex} in the effective Hamiltonian eq. (2) by the mean-field approximation, that is, by replacing the localized spins $\vec{S}_i$'s with their average $\langle \vec{S} \rangle$ [9, 10]. Thus, the effect of the localized spin system was dealt with as an effective magnetic field acting on hole carriers. As a result, Kamimura and Ushio separated the localized hole-spin system in the AF order and the hole carrier system from each other, and calculated the "one-electron type" energy band for a carrier system assuming a periodic AF order. Here, "one-electron type" means the inclusion of many-body effects in the energy bands. That is, the effect of the exchange interactions between carriers and localized spins is included by the mean field approximation.

In Fig. 5, the calculated many-body effect including energy band structure for up-spin (or down-spin) doped holes in LSCO is shown for various values of the wave vector $\vec{k}$ and symmetry points in the AF Brillouin zone, where the AF Brillouin zone is adopted because of the coexistence of a metallic state and the AF order; it is shown on the left side of the figure. Here, note that the energy in this figure is taken as the electron energy but not as the hole energy. Furthermore, the Hubbard bands for localized b_{1g}^* holes, which contribute to the local AF order, are separated from this figure and do not appear in this figure.

In undoped La_2CuO_4, all the energy bands in Fig. 5 are fully occupied by electrons so that La_2CuO_4 is an AF Mott insulator, consistent with experimental results. In this respect, the present effective energy band structure is completely different from the structure of ordinary LDA energy bands [25, 26]. When Sr is doped, holes begin to occupy the top of the highest band in Fig. 4 marked by #1 at the Δ point, which corresponds to $(\pi/2a, \pi/2a, 0)$ in the AF Brillouin zone. At the onset concentration of superconductivity, the Fermi level is located slightly below the top of the #1 band at Δ, which is slightly higher than the G_1 point. Here, the G_1 point in the AF Brillouin zone lies at $(\pi/a, 0, 0)$ and corresponds to a saddle point of the van Hove singularity.

On the basis of the calculated band structure shown in Fig. 5, Kamimura and Ushio [9, 10] calculated the Fermi surface (FS) for the underdoped regime of LSCO. The calculated FS in the underdoped regime is composed of four Fermi pockets of extremely at tubes. The projected two-dimensional (2D) picture of the four Fermi pockets around the Δ point, $(\pi/2a, \pi/2a)$ and the other three equivalent points in the momentum space is shown in the antiferromagnetic Brillouin zone in Fig. 6. The total volume of the four Fermi pockets is proportional to the

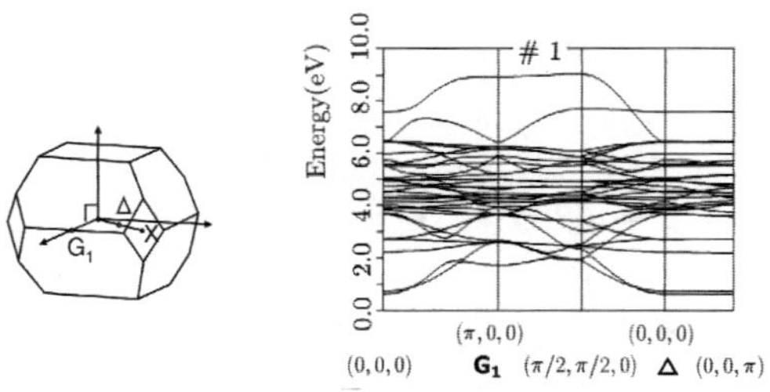

Figure 5. Many-body effect including band structure [9, 10] for up-spin (or down-spin) dopant holes in underdoped LSCO above T_c. The highest occupied band is marked by the #1 band (right) and the AF Brillouin zone (left). The Δ point corresponds to $(\pi/2a, \pi/2a, 0)$, while the G_1 point corresponds to $(\pi/a, 0, 0)$, at which a saddle-point singularity appears.

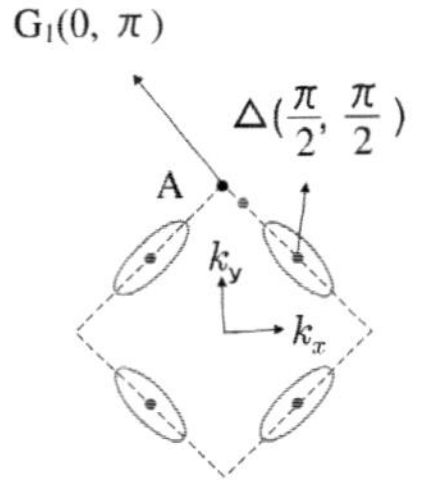

Figure 6. (Color online) Fermi surface (FS) for up-spin (or down-spin) dopant holes in underdoped LSCO above T_c. The FS consists of four Fermi pockets around the Δ point, $(\pi/2a, \pi/2a, 0)$, and the other three equivalent points (the nodal region) in momentum space. The figure shows the two-dimensionally projected Fermi pockets. A represents one of the electronic states.

concentration of the doped hole carriers. Thus, the feature of Fermi pockets constructed from the doped holes shown in Fig. 6 is consistent with Luttinger's theorem in the presence of AF order [27].

In 1996 and 1997, respectively, Mason *et al.* [28] and Yamada *et al.* [29] independently reported the magnetic coherence effects on the metallic and superconducting states in underdoped LSCO, by neutron inelastic scattering measurements. Since then, a number of papers suggesting the coexistence of local AF order and superconductivity in cuprates as a result of neutron and NMR experiments have been published [30, 31, 32, 33, 34, 35].

The Fermi surface structure in Fig. 6 is completely different from that of the single-component theory, in which the FS is large. Recently, Meng *et al.* have reported the existence of the Fermi pocket structure in the ARPES measurements of underdoped $Bi_2Sr_{2-x}La_xCuO_{6+}$ (La-Bi2201) [11]. Their results are clear experimental evidence of our Fermi pocket structure for underdoped LSCO predicted by Kamimura and Ushio in 1994 [9].

In 1997, Anisimov *et al.* calculated the energy band structure of the ordered alloy $La_2Li_{0.5}Cu_{0.5}O_4$ by the LDA+U method [36], and they showed that a fairly modest reduction in the apical Cu-O bond length is sufficient to stabilize Hund's coupling spin triplet state with dopant holes in both b_{1g} and a^*_{1g} orbitals. Their calculated result supports the K-S model.

4. High T_c superconductivity — Appearance of d-wave superconductivity in K-S model —

In high T_c superconductivity, d-wave superconductivity is observed in π-junction experiment for hole-doped YBCO by Wollmam *et al.* [37] and by Kirtley *et al.* [38]. Based on the K-S model we explain the mechanism of d-wave superconductivity by phonon mechanism, following the theory developed by Kamimura and coworkers [17, 18].

4.1. Mechanism of d-wave superconductivity

Before we discuss the mechanism of high temperature superconductivity, we would like to point out that the coexistence of a metallic state and Af order in the K-S model is due to the kinetic-energy lowering mechanism. In connection with the observed finite size of a spin-correlated region of the AF order [28, 29], Hamada and his coworkers [39] and Kamimura and Hamada [40] determined the ground state of the K-S model in a two-dimensional (2D) square lattice system with 16 (4 × 4) localized spins by the exact diagonalization method. They clarified that, in the presence of hole-carriers, the localized spins in a spin-correlated region tend to form an AF order rather than a random spin-singlet state like the resonance valence bond (RVB) state[2]. In this way they clarified that the hole-carriers can lower the kinetic energy of hole-carriers in the spin-correlated region by taking the two kinds of orbitals alternately in the lattice of AF order. This mechanism of lowering the kinetic energy of the hole-carriers has led to the coexistence of a metallic state and the local AF order, which is the key-point of the K-S model. Recently a similar mechanism has been discussed for superconductivity by Wrobel *et al.*[41].

Now in this section we will explain the mechanism of d-wave superconductivity in cuprates based on the K-S model: The interplay among the phonon mechanism, the local AF order and the characteristic feature of the K-S model that carrier holes occupy different orbitals with different symmetry in neighbouring sites leads to the d-wave pairing mechanism. First, we will point out that the phase relation between the wave functions of a hole carrier with up and down spins plays an important role in creating the d-wave superconductivity. As a result the matrix elements of electron-phonon interactions have a spin-dependent property. Let us start showing a sketch of the wavefunctions of a hole carrier with up- and down-spins in the tight binding scheme in Fig. 7(a) and (b), respectively. As seen in Fig. 7, the wavefunction of a down-spin carrier is displaced from that of an up-spin carrier by the vector $\vec{a}$ representing the Cu-O-Cu distance. Thus, the wavefunctions of a hole carrier with up- and down-spins have the following phase relation in K-S model, as already shown in equation (1):

$$\psi_{\vec{k}\downarrow}(\vec{r}) = \exp(i\vec{k}\cdot\vec{a})\psi_{\vec{k}\uparrow}(\vec{r}) \tag{4}$$

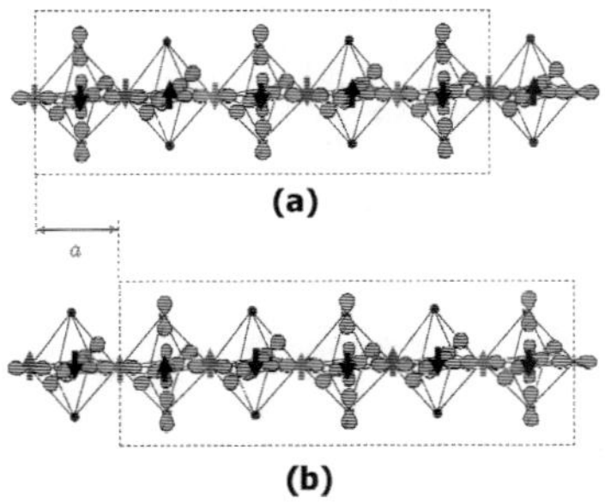

Figure 7. (Color online) A wave function for down-spin hole **(a)** and that for up-spin hole **(b)**. They coincide with each other, if one of them is displaced by vector $\vec{a}$ ($\vec{a}$ is Cu-O-Cu distance).

Now let us define the matrix elements of electron-phonon interactions for up- and down-spin carriers between hole-states $\vec{k}$ and $\vec{k}$' scattered by a phonon with wave vector $\vec{q}$, $V_{\uparrow}(\vec{k},\vec{k}',\vec{q})$ and $V_{\vec{k}\downarrow}(\vec{k},\vec{k}',\vec{q})$, respectively. By using the phase relation (1), that is equation (4), these matrix elements have the following spin-dependent property:

$$V_{\uparrow}(\vec{k},\vec{k}',\vec{q}) = \exp(i\vec{K}\cdot\vec{a})V_{\downarrow}(\vec{k},\vec{k}',\vec{q}) \tag{5}$$

where $\vec{K} = \vec{k} - \vec{k}' - \vec{q}$ is a reciprocal lattice vector in the AF Brillouin zone and $\vec{a}$ is a Cu-O-Cu distance. In ordinary BCS case, $\exp(i\vec{K}\cdot\vec{a})$ is always $+1$, and thus the spectral function is always positive leading to the s-wave superconductivity. In K-S model, however, $\exp(i\vec{K}\cdot\vec{a})$ takes the value of $+1$ or -1, because of $\vec{K} = (n\pi/a, m\pi/a, 0)$ with $n+m =$ even. This is a characteristic feature of the K-S model. By using the phase relation eq.(4), the effective interactions between a pair of holes at $(\vec{k}\uparrow, -\vec{k}\downarrow)$ and that at $(\vec{k}'\uparrow, -\vec{k}'\downarrow)$ via a phonon of wave vector $\vec{q}$ is expressed as

$$\begin{aligned} V_{\uparrow}(\vec{k},\vec{k}',\vec{q})\cdot V_{\downarrow}(\vec{k},\vec{k}',\vec{q}) &= \exp(i\vec{K}\cdot\vec{a})|V_{\uparrow}(\vec{k},\vec{k}',\vec{q})|^2 \\ &= \begin{cases} +1 & \text{for } n = \text{even(normal-scattering)} \\ -1 & \text{for } n = \text{odd(umklapp-scattering)}. \end{cases} \end{aligned} \tag{6}$$

In the K-S model one should notice that the process of scattering from $\vec{k}$ to $\vec{k}$' by a phonon of wave vector $\vec{q}$ includes the two sub-processes of normal- and umklapp-scatterings. Suppose that a hole occupies the state A in one sections of Fermi pockets in Fig. 8. Then a hole is scattered by phonon with wavevector $\vec{q}$ from A to state C$'$ on a Fermi pocket(see Fig. 8). Since state C$'$ is equivalent to state B by the translation of a reciprocal lattice vector $(-\vec{Q}_1) - \vec{Q}_2$, the effective interaction from A to B is attractive according to eq.(6) with $n = 2$. On the other

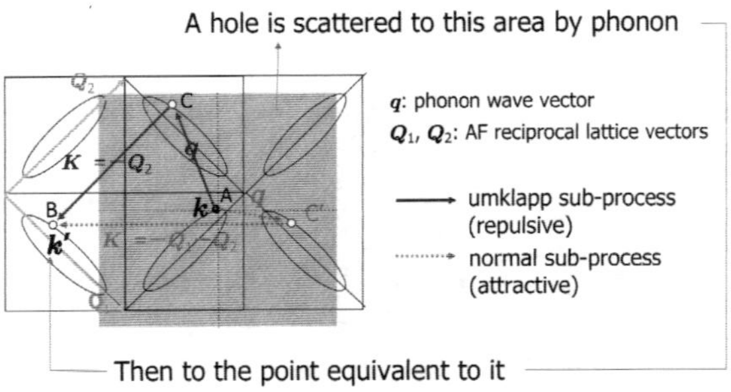

Figure 8. (Color online)The process of scattering from $\vec{k}$ to $\vec{k}$'. It includes normal subprocess and umklapp subprocess.

hand, when a hole is scattered from state A to C on Fermi pockets in the shaded area in Fig. 8, we notice that state C is equivalent to state B by the translation of a reciprocal lattice vector $(-\vec{Q}_2)$. Thus the effective interaction is repulsive by eq.(6) with $n = m = 1$. Since state C$'$ is inside the AF Brillouin zone which can be placed in the first Brillouin zone by the translation of AF-reciprocal lattice vector $\vec{K} = n\vec{Q}_1 + m\vec{Q}_2$ with $n + m$ being an even number while state C is outside the AF Brillouin zone, one may say that a scattering process from A to B via C$'$ is a normal scattering while a scattering process from A to B via C is umklapp scattering. Thus the total effective interactions between hole-states $\vec{k}$ and $\vec{k}$' is expressed by the sum of normal term and umklapp term as follows,

$$[V_\uparrow(\vec{k},\vec{k}',\vec{q}) \cdot V_\downarrow(\vec{k},\vec{k}',\vec{q})]_{\rm total} = V_\uparrow(\vec{k},\vec{k}',\vec{q})_{\rm normal} \cdot V_\downarrow(\vec{k},\vec{k}',\vec{q})_{\rm normal} + V_\uparrow(\vec{k},\vec{k}',\vec{q})_{\rm umklapp} \cdot V_\downarrow(\vec{k},\vec{k}',\vec{q})_{\rm umklapp} \quad (7)$$

Normal and umklapp scatterings have different sign, so that if the normal term is larger, the effective interaction is attractive while if the umklapp term is larger, the effective interaction is repulsive.

As an example of calculated results, in Fig. 9 we show the results of the $\vec{k}$ and $\vec{k}'$ dependence of the electron-phonon spectral function $\alpha^2 F_{\uparrow\downarrow}(\Omega,\vec{k},\vec{k}')$ with spin-singlet calculated by Kamimura and coworkers for one of out-of-plane modes, A_{1g} mode (Fig. 9 **(a)**) in LSCO with tetragonal symmetry [17], where $F_{\uparrow\downarrow}(\Omega,\vec{k},\vec{k}')$ is the momentum-dependent density of phonon states, and α^2 is the square of the electron-phonon coupling constant.

In the right column of Fig. 9 we can see that the momentum-dependent spectral function varies by taking values with + and − sign in each section of ① to ④ in Fig. 9 **(c)**, when the wave vector $\vec{k}'$ changes from the section ① of Fermi pocket to the section ④ while $\vec{k}$ is fixed in Fig. 9 **(b)**. This is clearly a d-wave behavior. By using this spectral function, we can obtain the $\vec{k}'$ dependence of a gap function. The obtained d-component gap function $\Delta(\vec{k})$ varies as a function of $(\cos(k_x a) - \cos(k_y a))$ and the wavefunction of a Cooper pair has a spatial extension of $d_{x^2-y^2}$ symmetry in the momentum space.

The calculated result for d-wave spectral function is shown as a function of phonon energy Ω in Fig. 10**(a)**. Out-of-plain phonon modes, an example of which is shown in Fig. 10**(b)**,

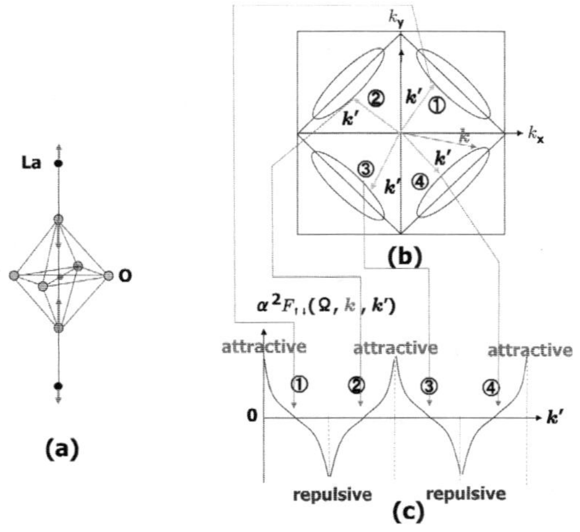

Figure 9. (Color online)An example of phonon modes which contribute to the formation of Cooper pairs (**a**). Changing $\vec{k}$' as shown in (**b**), the spectral function varies as shown in (**c**).

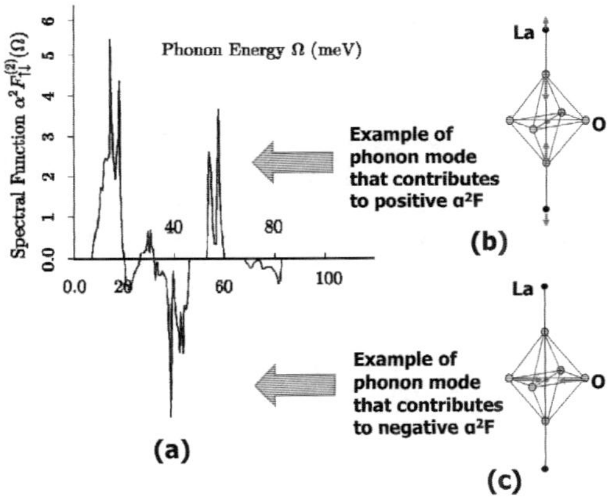

Figure 10. (Color online)The calculated d-wave component of electron-phonon spectral function for LSCO as a function of phonon frequency Ω.

contribute to the positive part of spectral function. The positive part of spectral function contributes to superconductivity. In-plain phonon modes, an example of which is shown in Fig. 10(**c**), contributes to the negative part of spectral function. The negative part of spectral function does not contribute to superconductivity.

4.2. The electron-phonon coupling constant for d-wave pairing

Kamimura, Ushio, Matsuno and Hamada calculated the electron-phonon coupling constant λ_d for LSCO by integrating the positive part of the calculated d-wave spectral function shown in Fig. 10(**a**), $\alpha^2F^{(2)}_{\uparrow\downarrow}(\Omega)$, over the phonon frequency Ω [6]. The value of λ_d thus calculated for the optimum doping of LSCO ($x = 0.15$) is 1.96. This indicates that cuprates are strong coupling superconductors. In this paper we will show that the strong coupling of superconductivity in cuprates is due to the anti-Jahn-Teller effect.

According to the K-S model, there are two contributions to the electron-phonon interactions. One contribution is due to the change of transfer interactions by atomic displacements. The other is due to the change of onsite energies by atomic displacements. The appearance of this term is due to the anti-Jahn-Teller effect. It is characteristic of the K-S model. By the latter contribution, the electron-phonon interactions are very strong in cuprates.

Kamimura and Eto calculated the energy difference in the lowest state between Hund's coupling triplet $^3B_{1g}$ and Zhang-Rice singlet $^1A_{1g}$ as a function of Sr concentration, by taking account of the anti-Jahn-Teller effect [22, 23]. In Fig. 11, the calculated energy difference between the lowest energies of $^1A_{1g}$ and $^3B_{1g}$ is shown as a function of the apical O-Cu distance c, where a is the inplane O-Cu distance, and c decreases with increasing the Sr concentration by the anti-Jahn-Teller effect. From the slope in the dependence of the energy difference on c, we can estimate the change of onsite energies by atomic displacements. This value is large so that the electron-phonon coupling in cuprates is large. This important result was pointed out by Kamimura, Matsuno, Suwa and Ushio for the first time in 1996. [17]

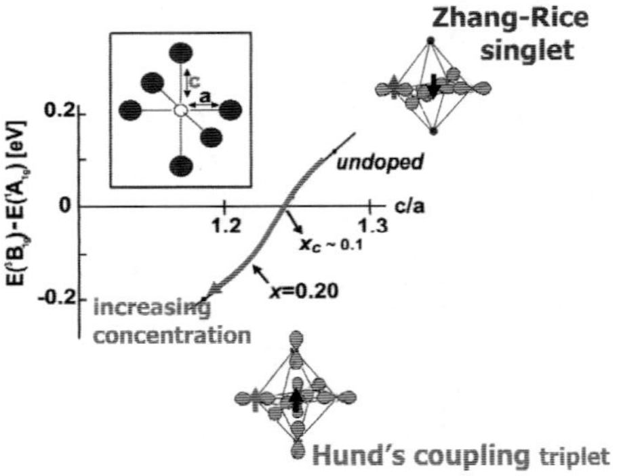

Figure 11. (Color online) The calculated energy difference in the lowest state between Hund's coupling triplet and Zhang-Rice singlet as a function of Sr concentration in LSCO, where the distance between apical O and Cu, c, decreases with increase of Sr concentration, by Kamimura and Eto [22, 23]. The energy difference between the two multiplets is small.

4.3. Finite size effects on superconductivity

Before calculating T_c, we would like to point out that a mean free path ℓ_0 in a metallic region where K-S model stands up all right is much expanded from the spin-correlated region λ_s by the spin uctuation effect in the 2D localized spin system.

The spin-correlated region λ_s is shown by a red circle in Fig. 12. On a boundary, localized spins are frustrated. The spin- ip time in a frustration region, $\tau_s = \hbar/J$, is 10^{-14} sec while the traveling time of a hole over λ_s, $\tau_s = \lambda_s/v_F$, is 10^{-13} sec, where v_F is the Fermi velocity of a hole carrier. Thus the spin- ip time is shorter than the traveling time of a hole carrier over λ_s

Before a hole moves rightward out of spin-correlated region, several spin- ips occur and spin-correlated region moves rightward. So that, a hole can move coherently beyond spin-correlated region.

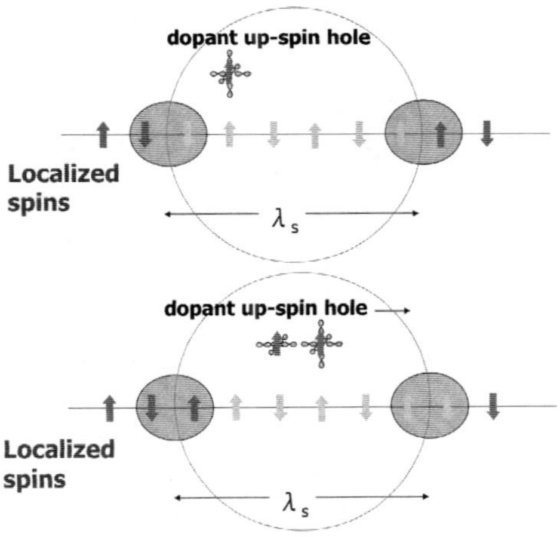

Figure 12. (Color online) A spin-correlated region and a dopant hole. On a boundary, localized spins are frustrated, so that a metallic region expanded from the spin-correlated region.

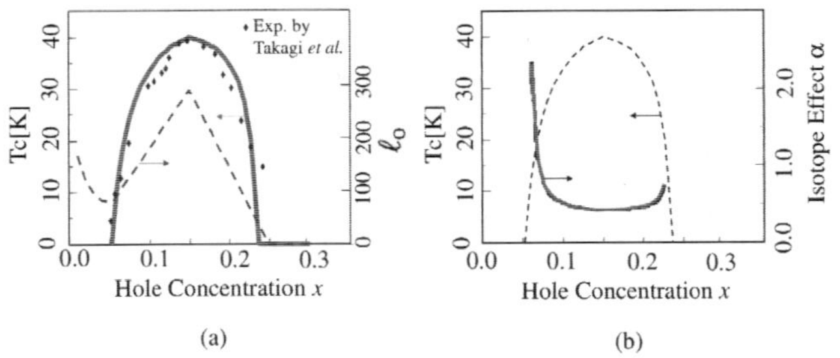

Figure 13. (Color online) The calculated results on the hole-concentration dependence of $T_c(x)$ **(a)** and isotope effect **(b)** for LSCO.

4.4. The hole-concentration dependence of T_c *and of the isotope effect calculated by Kamimura et al*

In Fig. 13 we show the hole-concentration dependences of T_c and the isotope effect α for LSCO calculated by Kamimura *et al* [18]. In this figure the calculated results are compared with experimental results . In doing so they introduced one disposal parameter, ℓ_0 . This represents the mean free path of a hole carrier. One may consider it as a size of a metallic state. Kamimura *et al* determined it so as to reproduce the highest value of T_c at the optimum doping in LSCO. The mean free path thus determined for the optimum doping of LSCO is 300A. Then we notice good agreement between the theoretical results by the K-S model and experimental results by Takagi *et al* [42] in Fig. 13**(a)**. In this figure the sharp increase of DOS at the onset of superconductivity gives rise to a sharp increase of T_c. The breakdown of Cooper pairs by the finite size gives rise to a sharp decrease of T_c.

The isotope effect on T_c in LSCO depends on the hole concentration critically, as seen in Fig. 13**(b)**. The present calculated result shows that it is remarkably large near the onset concentration of superconductivity. This is consistent with the experimental results reported by Bishop and coworkers[43] and Weyeneth and Müller[44].

5. Calculation of ARPES Spectra Based on the K-S Model and the Conclusion of the Absence of a Pseudogap

Recently, considerable attention has been paid to the phenomenological idea of the pseudogap. When a portion of the Fermi surface in cuprates was not observed in the ARPES experiments, the idea of the pseudogap was proposed as a type of gap for truncating the Fermi surface (FS) in a single-particle spectrum [45, 46]. The disconnected segments of a Fermi surface are called the "Fermi arc" [14, 46, 47]. Further ARPES experiments showed that such a pseudogap develops below a temperature denoted T^*, which depends on the hole concentration x in the underdoped regime of cuprates; thus, we write $T^*(x)$ hereafter. $T^*(x)$ decreases with increasing hole concentration x and disappears at a certain concentration x_o in the overdoped region [48]. In this section, on the basis of the K-S model, we clarify the origins of the pseudogap and $T^*(x)$.

5.1. Calculation of the photoemission intensity and clarification of the origin of the observed two distinct gaps

Below T_c, the hole carriers in the Fermi pockets shown in Fig. 6 form Cooper pairs, contributing to the formation of a superconducting state, and a superconducting gap appears across the Fermi level. This feature is consistent with Uemura *et al.*'s plot [49]. In Fig. 14(a), the d-wave node below T_c predicted by the K-S model [17, 18] is schematically shown as dots, and the d-wave superconducting density of states is schematically shown in Fig. 14(b). Here, note that

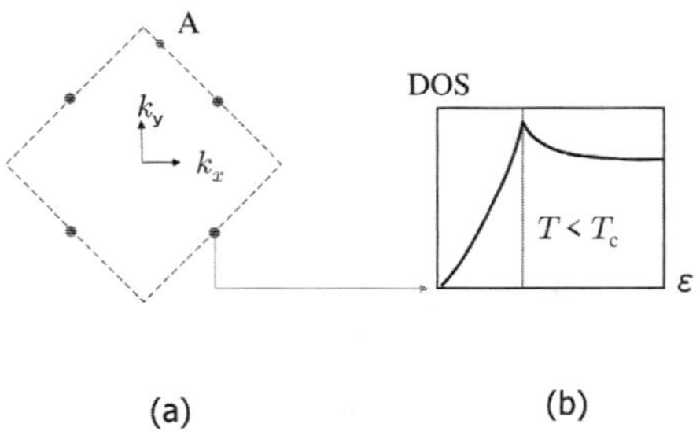

Figure 14. (Color online) Sketch of d-wave superconductivity in K-S model below T_c. (a) Change of Fermi pockets in nodal region to d-wave nodes below T_c. (b) d-Wave superconducting density of states.

the AF order still coexists with a superconducting state below T_c so that we can use the same AF Brillouin zone as that shown in Fig. 6.

On the other hand, in the antinodal region, the states occupied by electrons that do not participate in the formation of superconductivity still exist below T_c. As an example of such states, the state A is shown in Fig. 14(a), and the state corresponding to A above T_c is also shown in Fig. 6. Then, real transitions of electrons from the occupied states, say, the state A, below the Fermi level ε_F in the #1 energy band in Fig. 5 to a free-electron state above the vacuum level occur by photoexcitation both above and below T_c around the G_1 point $(\pi/a, 0, 0)$ and other equivalent points in momentum space. These transitions appear in the antinodal region in momentum space.

Such a transition is shown in Fig. 15. Let us consider the case in which an electron in the occupied state A with energy ε_i and momentum $\vec{k}_i$ in the #1 energy band below ε_F is excited to a free-electron state with the bottom of the energy dispersion at $(\varepsilon_F - \varepsilon_o)$ in a crystal by a photon with energy $h\nu$, where we use the suffix i to emphasize the initial state of the transition in the crystal. Thus, the final state of the transition with energy ε_f in the crystal is expressed as

$$\varepsilon_f = \frac{\hbar^2}{2m}(k_\parallel^2 + k_\perp^2) + \varepsilon_F - \varepsilon_o, \tag{8}$$

where $k_\parallel$ and $k_\perp$ are the momenta of the photoexcited electron parallel and perpendicular to the crystal surface, respectively. As shown in Fig. 14, the energy conservation for this excitation process from the initial state $|i\rangle$ to the final state $|f\rangle$ in the crystal is expressed as $h\nu = \varepsilon_f - \varepsilon_i$ for a photon of energy $h\nu$.

When an electron is ejected into the vacuum level of the crystal by a photon with energy $\hbar\nu$, it acquires kinetic energy. Through ARPES experiments, we measure the kinetic energy of photoelectrons emitted in vacuum. We define the kinetic energy of such photoelectrons emitted in vacuum as ε_{kin}, where $\varepsilon_{kin} = (\hbar^2/2m)(K_\parallel^2 + K_\perp^2)$.

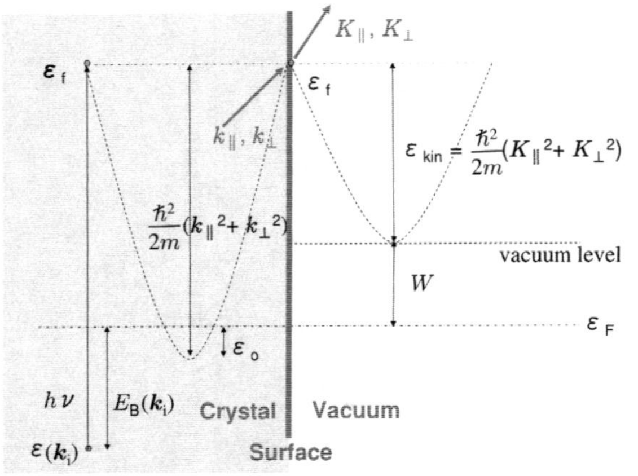

Figure 15. (Color online) Energetics of photoemission process.

On the other hand, the binding energy of the electron in the initial state, E_B, is introduced as a new variable instead of $\varepsilon(k_i)$. E_B is defined as

$$E_B = -\varepsilon(k_i) + \varepsilon_F. \tag{9}$$

The following equation also holds for ε_f:

$$\varepsilon_f = \varepsilon_{kin} + W + \varepsilon_F, \tag{10}$$

where W is the work function of the crystal (see Fig. 15).

In the ARPES experiment, when a photoelectron is emitted from a crystal in vacuum through a surface, it is assumed that the momentum parallel to the surface is conserved: $k_{\parallel} = K_{\parallel}$. Now, note that the #1 band in Fig. 5 has been calculated by the mean field approximation for the fourth term in the Hamiltonian (2). The important consequence of this approximation is that, having taken into account the strong spin exchange interaction in the mean field approximation, the probability of removing an electron in the state with momentum $\vec{k_i}$ and energy ε_i in the #1 energy band to the free-electron state in vacuum can be treated in a framework similar to that for single-particle photoexcitation.

Recently Kamimura, Sasaoka and Ushio (KSU)[20] calculated the EDCs in the ARPES experiments corresponding to the transition from the occupied states in the many-body effect including energy band in Fig. 5 to the free-electron band can be calculated using the formula for the photoemission intensity $I(\vec{k}, \omega)$:

$$I(\vec{k}, \omega) = \left|M_{\vec{k}}\right|^2 A(\vec{k}, \omega)\rho_f(\varepsilon_{kin}). \tag{11}$$

Here, $A(\vec{k}, \omega)$ is the spectral function that gives the probability of removing or adding an electron at $(\vec{k}, \omega)$, where ω is the electron energy relative to the Fermi level. It is related to the imaginary part of the one-electron Green's function; $A(\vec{k}, \omega) = -(1/\pi)\mathrm{Im}G(\vec{k}.\omega)$. Furthermore, $\rho_f(\omega)$ is the density of final states and $\left|M_{\vec{k}}\right|^2$ is the squared one-electron transition matrix element. It is clear from Fig. 15 that $A(\vec{k},\omega)$ gives the highest probability when $\hbar\omega$ is equal to $(h\nu - \varepsilon_{\text{kin}} - W)$, where $h\nu = \varepsilon(k_{\text{f}})$ - $\varepsilon(k_{\text{i}})$.

By taking account of the lifetime effects due to the finite size of a metallic state, the deviation from the mean field approximation, and other factors, the spectral function $A(\vec{k},\omega)$ is given by

$$A(\vec{k},\omega) = (1/\pi)\frac{\delta}{[\hbar\omega - (\varepsilon_{\text{F}} - \varepsilon(\vec{k_{\text{i}}}))]^2 + \delta^2}, \tag{12}$$

where δ denotes the lifetime effects, and the momentum dependence in ε_{i} is expressed as $\varepsilon(\vec{k_{\text{i}}})$ explicitly.

The density of final states $\rho_{\text{f}}(\varepsilon_{\text{kin}})$ in the EDCs is defined from the dispersion of the momentum of a photoexcited electron perpendicular to the crystal surface in the crystal, $k_{\perp}$, as,

$$\rho_{\text{f}}(\varepsilon_{\text{kin}}) = \left|\frac{\mathrm{d}k_{\perp}}{\mathrm{d}\varepsilon_{\text{kin}}}\right|. \tag{13}$$

Using eqs. (4) and (6) with the conservation of momentum of a photoexcited electron parallel to the crystal surface, $k_{\|} = K_{\|}$, the density of final states is obtained as,

$$\rho_{\text{f}}(\varepsilon_{\text{kin}}) = \frac{1}{2\sqrt{(\hbar^2/2m)\left(\varepsilon_{\text{kin}} + V - (\hbar^2/2m)k_{\|}^2\right)}}, \tag{14}$$

where $V = W + \varepsilon_{\text{o}}$ is the inner potential.

Since much of the ARPES EDC data is expressed as a function of the binding energy E_{B}, we express eqs. (8) and (10) in terms of E_{B}. For this purpose, we first insert eq. (5) into eq. (8), and simultaneously replace $\hbar\omega$ in eq. (8) by $(h\nu - \varepsilon_{\text{kin}} - W)$. As a result, eq. (8) can be written as,

$$A(\vec{k},\omega) = (1/\pi)\frac{\delta}{[(h\nu - \varepsilon_{\text{kin}} - W) - E_{\text{B}}]^2 + \delta^2}. \tag{15}$$

Furthermore, using the expression for the inner potential, $V = W + \varepsilon_{\text{o}}$, and the energy conservation relation in Fig. 15 given as

$$h\nu = \varepsilon_{\text{kin}} + W + E_{\text{B}}, \tag{16}$$

eq. (10) can be expressed as

$$\rho_{\text{f}}(\varepsilon_{\text{kin}}) = \frac{1}{2\sqrt{(\hbar^2/2m)\left(h\nu - E_{\text{B}} + \varepsilon_{\text{o}} - (\hbar^2/2m)k_{i\|}^2\right)}}, \tag{17}$$

where $k_{i\|}$ is the component of $\vec{k}_i$ parallel to the crystal surface.

Using eqs. (7), (12), and (13), we have calculated the photoemission intensity $I(\vec{k},\omega)$ as a function of $E_{\text{B}}(\vec{k_{\text{i}}})$. In performing the numerical calculations, we have considered that the photon energy $(h\nu)$ range in synchrotron radiation experiments is 10 to 100 eV and the kinetic energy range of the photoelectron is also 10 to 100 eV. Since the width of the energy dispersion of the #1 energy band in Fig. 5 is about 1 eV, we notice that the E_{B} range is up to 1 eV. For δ, whose inverse gives a measure of the lifetime broadening in the #1 band, we assume 100 meV on the basis of the discussion in the subsequent section.

In this context, we choose 15 eV for $h\nu$, 10 eV for $\varepsilon_{\rm kin}$, 3.5 eV for W [50], and 4.5 eV for the inner potential $V = \varepsilon_{\rm o} + W$ in the present numerical calculations. As regards the ith point in $\hbar^2 k_{i\|}^2/2m$, we choose the center of the antinodal region, i.e., the G_1 point, or the edge of the AF Brillouin zone in Fig. 5, because the antinodal region is narrow around the G_1 point, so that $\hbar^2 k_{i\|}^2/2m$ does not change much upon varying the ith point. By adopting the empty lattice test for the free-electron energy bands, we estimate $\hbar^2 k_{i\|}^2/2m$ to be 3 eV for $i = G_1$.

The calculated $I(\vec{k}, \omega)$ with the values of the above parameters is shown as a function of E_B in Fig. 16(a). Since $\varepsilon_{\rm kin}$ is very large, a divergent point in the density of final states $\rho_{\rm f}(\omega)$ appears at a large $E_{\rm B}$. Thus, the photoemission intensity $I(\vec{k}, \omega)$ shows a feature of a broad hump, re ecting a peak in the spectral function $A(\vec{k}, \omega)$ given by eq. (11), as seen in Fig. 16(a). This trend is consistent with the experimental results of the ARPES spectra of underdoped Bi2212 samples below $T_{\rm c}$ in the antinodal region by Tanaka *et al.* [19], although the shape of the broad hump is slightly different.

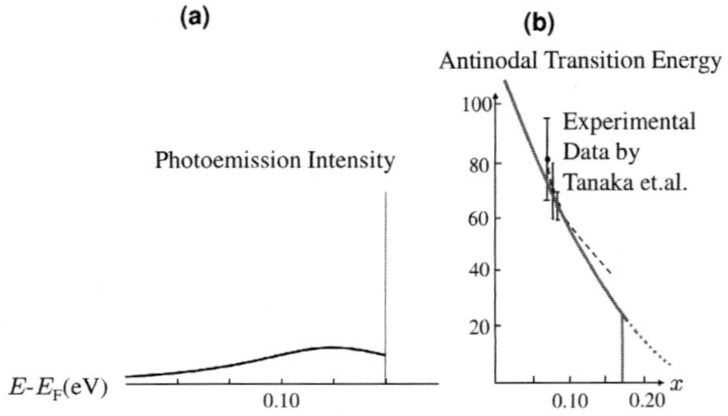

Figure 16. (Color online) Calculated ARPES spectra of LSCO and their comparison with experimental results of Bi2212. (a) Calculated photoemission intensity as a function of binding energy $E_B(= E - E_{\rm F})$. (b) Calculated energy difference $|\varepsilon({\rm G}_1) - \varepsilon_{\rm F}(x)|$ (antinodal transition energy) as a function of hole concentration x at $T = 0$K. Experimental results of Tanaka *et al.* [19] are shown by dots.

From the ARPES spectra in the nodal region shown in Fig. 14(b), which was predicted from the d-wave superconductivity due to the K-S model, [17] and those in the antinodal region shown in Fig. 16(a), we can conclude that the features of ARPES spectra below $T_{\rm c}$ are theoretically as follows: ARPES spectra consist of a coherent peak due to the superconducting density of states that appears in the nodal region around the Δ point and a broad hump that appears in the antinodal region, which corresponds to the real transition of electrons from the occupied states below the Fermi level $\varepsilon_{\rm F}$ to a free-electron state above the vacuum level. These theoretical results of ARPES EDCs are similar to the experimental ones reported by Tanaka *et al.* for Bi2212 [19], where the experimental results revealed two distinct energy gaps in the nodal and antinodal regions exhibiting different doping dependences. Thus, we designate the broad hump in the antinodal region as an "antinodal transition", where $\varepsilon_{\rm F}$ varies with the hole concentration x, so that we write $\varepsilon_{\rm F}(x)$ hereafter.

Furthermore, Tanaka *et al.* reported the doping dependence of the position of the hump, which is determined from the second derivative of the spectra in the antinodal region around the G_1 point for the three underdoped samples in Bi2212. We compare this experimental result of Bi2212 with the calculated doping dependence of the binding energies E_B at the G_1 point for LSCO, which correspond to the energy difference $|\varepsilon(\mathrm{G}_1) - \varepsilon_\mathrm{F}(x)|$. We call this energy difference the "antinodal transition energy".

Since the shape of the density of states (DOS) for the highest conduction band does not depend on the type of cuprate material, we can compare in Fig. 16(b) the calculated doping dependence of the antinodal transition energy for LSCO (solid lines) with the experimental results of the antinodal gap of Bi2212 in ref.16, which are shown as dots in the figure. As seen in Fig. 16(b), the agreement between the theory and the experiment is remarkably good. From this quantitative agreement, we can conclude that among the observed two gaps below T_c, the gap associated with the antinodal regime corresponds to the real transitions of electrons from the occupied states below the Fermi level to a free-electron state above the vacuum level, while the other gap associated with the near-nodal regime corresponds to the superconducting gap created on Fermi pockets. From the excellent agreement between the present theoretical results and the experimental results of Tanaka *et al.*, we can conclude that the real transitions of electrons from occupied states below the Fermi level to a free-electron state above the vacuum level by photoexcitation appear in the antinodal region in underdoped cuprates so that the introduction of a pseudogap is not necessary.

5.2. Physical meaning of $T^*(x)$ and the temperature dependence of ARPES spectra

In calculating the temperature dependence of the antinodal transition energy, recently Kamimura, Sasaoka and Ushio (KSU)[20] clarified the physical meaning of $T^*(x)$. In this subsection we review the recent results obtained by KSU. When a hole concentration x is fixed at a certain value in the underdoped region and the temperature increases beyond T_c, in the normal phase, the local AF order constructed by superexchange interaction in a CuO_2 plane is destroyed by thermal agitation, and thus a phase showing the coexistence of a metallic state with the Fermi pockets and the local AF order diminishes gradually. As a result, an electronic phase consisting of a large Fermi surface (FS) without the AF order is mixed with a phase of the K-S model. Finally, at a certain temperature, a uniform phase consisting of the electronic phase consisting of a large FS without the AF order will appear in the underdoped regime. This temperature is defined as $T^*(x)$. Thus, the phase of the Fermi pockets coexisting with the local AF order in the K-S model holds only below $T^*(x)$. KSU designated the phase of the Fermi pockets in the K-S model as the "small FS" phase and the electronic phase consisting of a large FS without the AF order as the "large FS" phase. Hereafter, the former and latter are abbreviated as the SF and LF phases, respectively. In this context, one may consider that a phase below $T^*(x)$ is a mixed phase of the SF and LF phases in the underdoped regime; thus, $T^*(x)$ represents a crossover from the mixed phase to the LF phase.

To calculate $T^*(x)$ on the basis of the K-S model, one must take account of the effect of thermal agitation in the system of Cu localized spins in the AF order (the first story in Fig. 3). However, such calculation is possible only for a finite system, as Hamada and coworkers have shown [39, 40]. In this context, KSU calculated $T^*(x)$ approximately, neglecting the effect of thermal agitation in the system of Cu localized spins.

For this purpose, they have introduced a quantity that defines the difference between the free energies of pure LF and SF phases:

$$\Delta F(T,x) \equiv F_\mathrm{LF}(T,x) - F_\mathrm{SF}(T,x), \tag{18}$$

where $F_\mathrm{LF}(T,x)$ and $F_\mathrm{SF}(T,x)$ are the free energies of the LF and SF phases, respectively. Here the free energy $F(T, x)$ is defined as

$$F(T,x) \;=\; E(T,x) - TS(T,x), \tag{19}$$

where $E(T, x)$ and $S(T, x)$ are the internal energy and entropy of each phase, respectively. These quantities are calculated from

$$E(T,x) \;=\; \int_{-\infty}^{\infty} \varepsilon\rho(\varepsilon) f(\varepsilon,\mu(x))\mathrm{d}\varepsilon, \tag{20}$$

and

$$S(T,x) \;=\; -k_{\mathrm{B}} \int_{-\infty}^{\infty} \Big[f(\varepsilon,\mu(x)) \ln f(\varepsilon,\mu(x)) + \{1 - f(\varepsilon,\mu(x))\} \ln\{1 - f(\varepsilon,\mu(x))\}\Big] \rho(\varepsilon)\, d\varepsilon\,, \tag{21}$$

where $\mu(x)$ is the chemical potential of each phase, $\rho(\varepsilon)$ is the DOS for each phase, and $f(\varepsilon, \mu(x))$ is the Fermi distribution function at energy ε and chemical potential $\mu(x)$. Then, $\tilde{T}^*(x)$ is defined by

$$\Delta F(\tilde{T}^*(x), x) = 0. \tag{22}$$

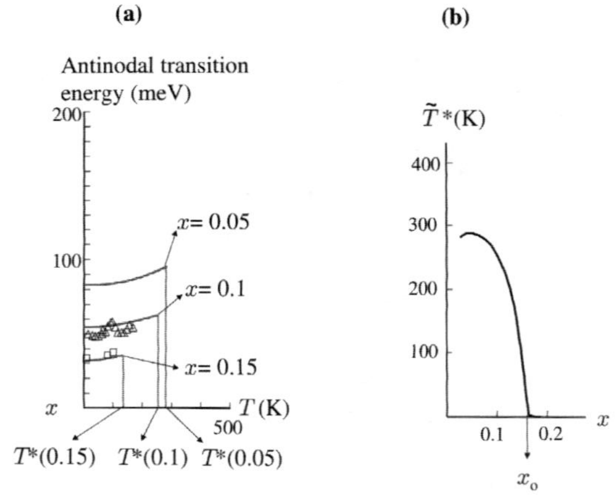

Figure 17. (Color online) (a) Calculated temperature dependence of antinodal transition energy for $x = 0.05$, 0.1, and 0.15. Experimental data obtained by Norman *et al.* [51] (triangles) and Lee *et al.* [52] (squares) are shown. (b) Calculated result of $\tilde{T}^*(x)$ as a function of the hole concentration x

Kamimura *et al.* calculated the electronic entropies for the SF and LF phases of LSCO [8]. According to their results, the difference in electronic entropy between the SF and LF phases increases with increasing hole concentration x in the underdoped regime. Using this result, KSU calculated $\tilde{T}^*(x)$ from eq. (22) as a function of x, instead of $T^*(x)$. In doing so, they introduced two parameters, $\tilde{T}^*(x = 0.05)$ and x_o, where x_o is the critical concentration that satisfies $\tilde{T}^*(x_o)$=0. Here, $\tilde{T}^*(x = 0.05)$ represents a quantity related to the energy difference between the phase of the doped AF insulator and the LF phase at the onset concentration of the metal- insulator transition ($x = 0.05$). $\tilde{T}^*(x = 0.05)$ is chosen to be 300K. On the other hand, KSU explained the physical meaning of $\tilde{T}^*(x_o) = 0$ as follows: When the hole concentration exceeds the optimum doping level ($x = 0.15$) for LSCO and enters a slightly overdoped region, the local AF order via superexchange interaction in a CuO_2 plane is destroyed by an excess of hole carriers. Thus, the K-S model does not hold at a certain concentration x_o in the overdoped region, and hence the small FS in the K-S model changes to a large FS. Thus, $\tilde{T}^*(x)$ vanishes at x_o. From the analysis of various experimental results, they chose $x_o = 0.17$ for LSCO [53, 54]. The calculated result of $\tilde{T}^*(x)$ with $\tilde{T}^*(x = 0.05) = 300$K and $x_o = 0.17$ is shown as a function of x in Fig. 17**b**.

From the above-mentioned result, KSU have suggested that the area below $\tilde{T}^*(x)$ in the underdoped regime represents the region in which the normal (metallic) phase above T_c and the superconducting phase below T_c coexist with the local AF order. In a real system, a region of a mixed phase consisting of the SF and LF phases appears between $\tilde{T}^*(x)$ and $T^*(x)$ owing to the dynamical interaction of the fourth term in the effective Hamiltonian (2). Thus, $T^*(x)$ always appears above $\tilde{T}^*(x)$ [55].

Under this circumstance, it is clear that the antinodal transition energy defined by $|\varepsilon(G_1) - \varepsilon_F(x)|$ appears at temperatures below $T^*(x)$ and vanishes at $T^*(x)$. By using $\tilde{T}^*(x)$ instead of $T^*(x)$, KSU have calculated the temperature dependence of the antinodal transition energy using eqs. (18)-(21). The calculated results for three concentrations, i.e., $x = 0.05$, 0.10, and 0.15, of LSCO in the underdoped-to-optimallydoped region are shown in Fig. 17(a) as functions of temperature, where $\tilde{T}^*(0.05) = 300$K is used. As seen in the figure, the antinodal transition energy increases slightly with temperature up to $\tilde{T}^*(x)$ and vanishes suddenly at $\tilde{T}^*(x)$. These calculated results are compared with the experimental results of the underdoped sample of Bi2212[51, 52], which are indicated by triangles and squares in Fig. 17(a). As seen in the figure, the agreement between the theory and the experiment is remarkably good.

6. Spatially Inhomogeneous Distribution of Fermi Pocket States and Large Fermi Surface States due to the Finite Size E ects

6.1. *Finite size effects of metallic state on Fermi surface*

According to the results of neutron inelastic scattering experiments by Mason *et al.* [28] and Yamada *et al.* [29], the AF spin-correlation length λ_s in the underdoped region of LSCO is finite. In the underdoped regime of LSCO, it increases as the Sr concentration increases from $x = 0.05$ in LSCO, the onset of superconductivity, and reaches about 50A or more at the optimum doping level ($x = 0.15$). In this subsection, we discuss the effects of the finite size of the AF spin-correlation length on the structure of the Fermi pockets shown in Fig. 6. According to the K-S model in Fig. 3, in the spin-correlated region a doped hole in the underdoped regime of LSCO can itinerate coherently by taking the a^*_{1g} and b_{1g} orbitals alternately in the presence of the local AF order without destroying the AF order.

In the case of a finite spin-correlated region, one may think that there are frustrated spins at the boundary between the spin-correlated region of the AF order and the region of the "resonating valence bond" (RVB) state proposed by Anderson [2] without hole carriers. Here, the frustrated spins mean that the localized spins at the boundary are not in the AF order, but directed parallel to each other. Suppose that one of the frustrated spins in a parallel direction at the boundary changes its direction from parallel to antiparallel by the uctuation effect in the 2D Heisenberg AF spin system during the time of τ_s defined by $\tau_s \equiv \hbar/J$, where J is the superexchange interaction (~0.1 eV). At the time of τ_s, on the other hand, hole carriers at the Fermi level can move with the Fermi velocity inside the spin-correlated region of the AF order. The traveling time of a doped hole at the Fermi level over an area of the spin-correlation length is given by $\tau_F \equiv \lambda_s/v_F$, where v_F is the Fermi velocity of a doped hole at the Fermi level. In the case of underdoped LSCO, τ_s is 6×10^{-15} s. Since v_F is estimated to be 2.4×10^4 m/s from the dispersion of the #1 band in Fig. 5, τ_F is 2×10^{-13} s for the underdoped region of $x = 0.10$ to $x = 0.15$ in LSCO, where for the spin-correlation length λ_s at $x = 0.15$, we have chosen 50 A[28, 29]. Thus, τ_F becomes much longer than τ_s. As a result, the frustrated spins on the boundary change their directions from parallel to antiparallel before a hole carrier in the spin-correlated region of the AF order reaches the boundary. Thus, a metallic state for a doped hole becomes much wider than the observed spin-correlation length by the passing of a doped hole through the boundary without spin scattering. In this way, a metallic state is surrounded by RVB states and the spatial distribution of the metallic states is inhomogeneous.

Finally, we explain why we have chosen 100 meV for δ in calculating the photoemission intensity shown in Fig. 16(a). For example, the initial state of photoexcitation in ARPES near the G_1 point is either a component of Hund's coupling triplet $^3B_{1g}$ or Zhang-Rice singlet $^1A_{1g}$ shown in Fig. 4 in the #1 band [10]. Thus, if the local AF order between neighboring Cu sites in Fig. 4 is destroyed, the calculated result in Fig. 16(a) may not be valid. This is the reason why δ is the same order of magnitude as the inverse of τ_s, that is, the superexchange interaction J (~0.1 eV).

6.2. Origin of the coexistence of a local AF order and a metallic state: The kinetic-energy-driven mechanism

Concerning the finite system of cuprates, Hamada *et al.* [39] and Kamimura and Hamada [40] attempted to determine the ground state of the effective Hamiltonian (2) for the K-S model by carrying out the exact diagonalization of the Hamiltonian (2) using the Lanczos method for a 2D square lattice system with 16 (4 × 4) localized spins with one and two doped holes, respectively, as already mentioned in §4.. According to their results, in the presence of hole carriers the localized spins in a spin-correlated region tend to form an AF order rather than a random spin-singlet state, and thus hole carriers can lower the kinetic energy by itinerating in the lattice of the AF order (the first story in Fig. 3). This is the mechanism leading to the coexistence of a metallic state and a local AF order in the K-S model.

Generally, a hole-carrier in the spin-correlated region of the AF order can propagate through the boundary of the spin-correlated region with the above-mentioned mechanism of the K-S model; hence, the region of a metallic state coexisting with the AF order becomes much wider than the observed spin-correlated region. In fact, Kamimura *et al.* estimated the length of the metallic region at the optimum doping level of LSCO to be about 300 A from the T_c at the optimum doping level [18].

7. Remark on a Phase Diagram for Underdoped Cuprates

From the calculated results shown in Fig. 17(b), we would like to comment on the T vs x phase diagram for cuprates shown in Fig. 18, for which it has been said that a pseudogap state exists below the temperature $T^*(x)$. According to our calculations in previous sections, the SF phase constructed from Fermi pockets appears in the presence of the local AF order below $T^*(x)$ in the underdoped region. However, when the temperature increases at a fixed concentration in the underdoped regime, the AF order is destroyed gradually with increasing temperature, and thus the K-S model does not hold slightly below $T^*(x)$. On the other hand, when the hole concentration increases at a fixed temperature, the AF order is destroyed by overdoped holes. Thus, the K-S model does not hold at a certain hole concentration. The thermal effect and excess hole density effect cause a mixing of the SF and LF phases, as explained in the previous section.

In this context, we would like to point out that the area below $T^*(x)$ and above T_c in Fig. 18 is not a pure phase but an inhomogeneous distribution of Fermi pocket states and large FS states.

Thus, we can say that, when the temperature increases from T_c at a fixed hole concentration in the underdoped region, the population showing Fermi pockets decreases while that showing large FS increases with increasing temperature. Finally, when the temperature exceeds $T^*(x)$, a uniform LF phase appears as a metallic state. Thus, we may say that $T^*(x)$ is a crossover line from the inhomogeneous mixed phase of Fermi pockets and large FS to the LF phase.

This result can explain the strange temperature evolution of a Fermi arc observed by Norman *et al.* [46] and Kanigel *et al* [48]. Furthermore, in the superconducting phase below T_c and $T^*(x)$ in Fig. 18, an s-wave component of superconductivity originating from the LF phase may be mixed with the d-wave superconductivity. Such a mixing effect was experimentally reported by Müller [56]. In this context, we conclude that $T^*(x)$ represents a crossover from the SF phase to the LF phase rather than a phase boundary between the pseudogap phase and a metal [57].

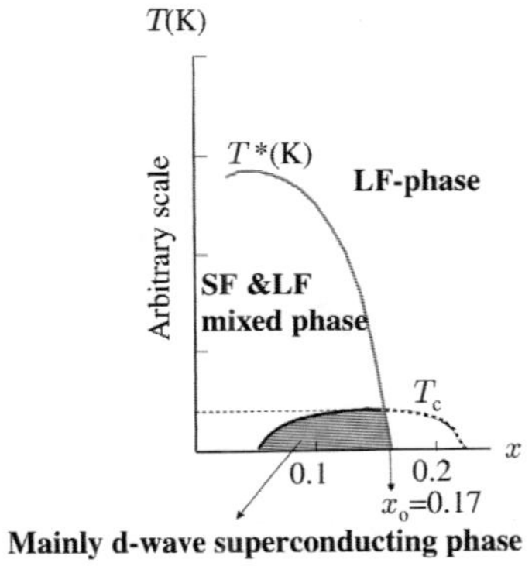

Figure 18. (Color online) New explanation for phase diagram of LSCO. Since $T^*(x)$, which is higher than $\tilde{T}^*(x)$, has not been calculated, its numerical values are not shown on the vertical axis.

8. Conclusions and Concluding Remarks

In this study we have shown on the basis of the K-S model how the interplay of the anti-Jahn-Teller effect in the JT physics and the electron correlation in the Mott physics plays an important role in determining the superconducting state as well as the metallic state of underdoped cuprates. In connection with this interplay, the following important results have been obtained: It has been clarified on the basis of the K-S model that the concept of the pseudogap discussed theoretically[58, 59] and reported by ARPES, STM, and tunneling experiments below $T^*(x)$ in underdoped cuprates [47, 48, 60] can be explained by the occurrence of Fermi pockets in the underdoped region without the pseudogap hypothesis. We have shown that the appearance of the broad hump observed in the antinodal region in ARPES can be explained by the real transitions of photo-excited electrons from the occupied states in the highest conduction band in the antinodal region to a free-electron state above the vacuum level. Furthermore, the present results have supported the idea of KSU that the physical meaning of T^* represents a crossover line from an inhomogeneous phase consisting of Fermi pockets and large FS in the normal state to a phase consisting of a large FS.

Journal of Physics: Conference Series **428** (2013) 012043 doi:10.1088/1742-6596/428/1/012043

Acknowledgments

We would like to thank Dr. Wei-Shen Lee, and Profs. Atsushi Fujimori, Nobuaki Miyakawa, Naurang Saini, Tomohiko Saitoh, Hideaki Sakata, and Xingjiang Zhou for valuable discussions on ARPES experimental results. One of us (H.K.) would like to thank Professor Dr. Alex Müller and Dr. Jaw-Shen Tsai of RIKEN for valuable comments on the present work, and Professor Yoshihisa Yamamoto of National Institute of Information, Director of FIRST "Quantum Information Processing" project for his kind encouragements. This work was partially supported by FIRST "Quantum Information Processing.

References

[1] J. G. Bednorz and K. A. Müller: Z. Phys. B **64** (1986) 189.
[2] P. W. Anderson: Science **235** (1987) 1196.
[3] K. Shiraishi: Dr. of Science Thesis, Dept. of Physics, Graduate School of Science, University of Tokyo, Tokyo, (1988). K. Shiraishi and H. Kamimura, Butsuri **44** (1989) 188 (Physical Society of Japan) (in Japanese).
[4] N. Shima, K. Shiraishi, T. Nakayama, A. Oshiyama, and H. Kamimura: in Proc. 1st Int. Conf. Electronic Materials in New Materials and New Physical Phenomena for Electronics of the 21st Century, ed. T. Sugano, R. R. H. Chang, H. Kamimura, I. Hayashi, and T. Kamiya (Materials Research Society, Pittsburgh, 1989) p.51.
[5] A. Oshiyama, N. Shima, T. Nakayama, K. Shiraishi and H. Kamimura: in *Mechanism of High Temperature Superconductivity. Springer Series in Materials Science* ed. H. Kamimura and A. Oshiyama (Springer, Berlin, Heidelberg, 1989) **vol. 11**, p.111.
[6] H. Kamimura, H. Ushio, S. Matsuno and T. Hamada: *Theory of Copper Oxide Superconductors* (Springer, Berlin, Heidelberg, 2005).
[7] H. Kamimura and Y. Suwa: J. Phys. Soc. Jpn. **62** (1993) 3368.
[8] H. Kamimura, T. Hamada and H. Ushio: Phys. Rev. B **66** (2002) 054504.
[9] H. Kamimura and H. Ushio: Solid State Commun. **91** (1994) 97.
[10] H. Ushio and H. Kamimura: J. Phys. Soc. Jpn. **64** (1995) 2585.
[11] J. Meng, G. Liu, W. Zhang, L. Zhao, H. Liu, X. Jia, D. Mu, S. Liu, X. Dong, J. Zhang, W. Lu, G. Wang. Y. Zhou, Y. Zhu, X. Wang, Z. Xu, C. Chen, and X. J. Zhou: Nature **462** (2009) 335.
[12] N. Doiron-Leyraud, C. Proust, D. LeBoeuf, J. Levallois, J.-B. Bonnemaison, R. Liang, D. A. Bonn, W. H. Hardy, and L. Taillefer: Nature **447** (2007) 565.
[13] A. F. Bangura, J. D. Fletcher, A. Carrington, J. Levallois, M. Nardone, B. Vignolle, P. J. Heard, N. Doiron-Leyraud, D. LeBoeuf, L. Taillefer, S. Adachi, C. Proust, and N. E. Hussey: Phys. Rev. Lett. **100** (2008) 047004.
[14] T. Yoshida, X. J. Zhou, M. Nakamura, S. A. Keller, P. V. Bogdanov, E. D. Lu, A. Lanzara, Z. Hussain, A. Ino, A. Fujimori, H. Eisaki, Z.-X Shen, T. Kakeshita and S. Uchida: Phys. Rev. Lett. **91** (2003) 027001.
[15] T. Yoshida, X. J. Zhou, K. Tanaka, W. L. Yang, Z. Hussain, Z.-X. Shen, A. Fujimori, S. Sahrakorpi, M. Lindroos, R. S. Markiewicz, A. Bansi, S. Komiya, Y. Ando, H. Eisaki, T. Kakeshita, and S. Uchida: Phys. Rev. B **74** (2006) 224510.
[16] S. Charkravarty and H.-Y. Kee: Proc. National Academy of Sciences of USA **105** (2008) 8835.
[17] H. Kamimura, S. Matsuno, Y. Suwa, and H. Ushio: Phys. Rev. Lett **77** (1996) 723.
[18] H. Kamimura, H. Hamada, S. Matsuno, and H. Ushio: J. Supercond. **15** (2002) 379.
[19] K. Tanaka, W. S. Lee, D. H. Lu, A. Fujimori, T. Fujii, Risdiana, I. Terasaki, D. J. Scalapino, T. P. Devereaux, Z. Hussain, and Z.-X. Shen: Science **314** (2006) 1910.
[20] H. Kamimura, K. Sasaoka, and H. Ushio: J. Phys. Soc. Jpn. **80** (2011) 114715.
[21] T. Egami, B.H. Toby, S.J.L Billinge, Chr Janot, J.D. Jorgensen, D.G. Hinks, M.A. Subramanian, M.K. Crawford, W.E. Fargeth, and E.M. McCarron, in *High Temperature Superconductivity* ed. J. Azhkenazi and G. Vezzoli (Plenum Press, New York, 1992) p. 389, and related references therein.
[22] H. Kamimura and M. Eto: J. Phys. Soc. Jpn. **59** (1990) 3053.
[23] M. Eto and H. Kamimura: J. Phys. Soc. Jpn. **60** (1991) 2311.
[24] F.C. Zhang and T.M. Rice: Phys. Rev. B **37** (1988) 3759.
[25] See, for example, L. F. Mattheiss: Phys. Rev. Lett. **58** (1987) 1028.
[26] See also J. Yu, A. J. Freeman and J.-H. Xu: Phys. Rev. Lett. **58** (1987) 1035.
[27] J. M. Luttinger and J. C. Ward: Phys. Rev. **118** (1960) 1417.
[28] T. Mason, A. Schroder, G. Aeppli, H. A. Mook and S. M. Haydon: Phys. Rev. Lett. **77** (1996) 1604.
[29] K. Yamada, C. H. Lee, J. Wada, K. Kurahashi, H. Kimura, Y. Endoh, S. Hosoya, G. Shirane, R. J. Birgeneau and M. A. Kastner: J. Supercond. **10** (1997) 343.

[30] K. Yamada, C. H. Lee, K. Kurahashi, J. Wada, S. Wakimoto, S. Ueki, H. Kimura, and Y. Endoh: Phys. Rev. B **57** (1998) 6165.
[31] Y.-J. Kao, Q. Si, and K. Levin: Phys. Rev. B **61** (2000) R11898.
[32] N. B. Christensen, D. F. McMorrow, H. M. Ronnow, B. Lake, S. M. Hayden, G. Aeppli, T. G. Perring, M. Mangkorntong, M. Nohara, and H. Takagi: Phys. Rev. Lett. **93** (2004) 147002.
[33] J. M. Tranquada: Nature **429** (2004) 534.
[34] S. M. Haydon, H. A. Mook, P. Dai, T. G. Perring and F. Dogan: Nature **429** (2004) 531.
[35] H. Mukuda, M. Abe, Y. Araki, Y. Kitaoka, K. Tokiwa, T. Watanabe, A. Iyo, H. Kito, and Y. Tanaka: Phys. Rev. Lett. **96** (2006) 087001.
[36] V. L. Anisimov, S. Yu Ezhov and T. H. Rice: Phys. Rev. B **55** (1997) 12829.
[37] D. A. Wollman, D. J. Van Harlingen, J. Giapintzakis, and D. M. Ginsberg: Phys. Rev. Lett **74**, 787 (1995).
[38] J. R. Kirtley, C.C. Tsuei, J.Z. Sun, C.C. Chi, Lock See Yu-Jahnee, A. Gupta, M. Rupp, and M.R. Ketchen, Nature **373**, 225 (1995).
[39] T. Hamada, K. Ishida, H. Kamimura and Y. Suwa, J. Phys. Soc. Jpn. **70**, 2033 (2001).
[40] H. Kamimura and T. Hamada, in *Models and Method of High-T_c Superconductivity: Some Frontal Aspects* **Volume 2**, J.K. Srivastava and S.M. Rao, Eds. (Nova Science Publishes, New York, 2003) Chapter 2. pp. 9-71
[41] P. Wrobel, R. Eder and P. Furde, J.Phys.: Condens. Matt. **15**, 6599 (2003).
[42] H. Takagi, T. Ido, S. Ishibashi, M. Uota, S. Uchida, and Y. Tokura: Phys. Rev. B **40** (1989) 2254.
[43] A.R. Bishop, A. Bussmann-Holder,M. Cardona, O.V. Dolgov, A. Furrer, H. Kamimura, H. Keller, R. Khasanov, R.K. Kremer, D. Manske, K.A. Mueller, and A. Simon, J. Supercond. Novel Magnetism **20**, 393 (2007).
[44] S. Weyeneth and K. A. Müller, J. Supercond. Novel Magnetism **24**, 1235 (2011)
[45] D. S. Marshall, D. S. Dessau, A. G. Loeser, C.-H. Park, A. Y. Matsuura, J. N. Eckstein, I. Bozovic, P. Fournier, A. Kapitulnik, W. E. Spicer, and Z.-X. Shen: Phys. Rev. Lett. (1996) **76**.
[46] M. R. Norman, H. Ding, M. Randeriat, J. C. Campuzano, T. Yokoya, T. Takeuchi, T. Takahashi, T. Mochiku, K. Kadowaki, P. Guptasarma, and D. G. Hinks: Nature **392** (1998) 157.
[47] M. R. Norman, A. Kanigel, M. Randeria, U. Chatterjee, and J. C. Campuzano: Phys. Rev. B **76** (2007) 174501 and related references therein.
[48] A. Kanigel, M. R. Norman, M. Randeria, U. Chatterjee, S. Souma, A. Kaminski, H. M. Fretwell, S. Rosenkranz, M. Shi, T. Sato, T. Takahashi, Z. Z. Li, H. Raffy, K. Kadowaki, D. Hinks, L. Ozyuzer and J. C. Campuzano: Nat. Phys. **2** (2006) 447.
[49] Y. J. Uemura, L. P. Le, G. M. Luke, B. J. Sternlieb, W. D. Wu, J. H. Brewer, T. M. Riseman, C. L. Seaman, M. B. Maple, M. Ishikawa, D. G. Hinks, J. D. Jorgensen, G. Saito, and H. Yamochi: Phys. Rev. Lett. **66** (1991) 2665.
[50] For the value of the work function for cuprates, see, for example, G. Kinoda, T. Hasegawa, S. Najao, T. Hamaguri, K. Kitazawa, K. Shimizu, J. Shimoyama, and K. Kishio: Phys. Rev. B **67** (2003) 224509.
[51] M. R. Norman, M. Randeria, H. Ding and J. C. Campuzano: Phys. Rev. B 57 (1998) R11093.
[52] W. S. Lee, I. M. Vishik, K. Tanaka, D. H. Lu, T. Sasagawa, N. Nagaosa, T. P. Devereaux, Z. Hussain, and Z.-X. Shen: Nature **450** (2007) 81.
[53] T. Nakano, M. Oda, C. Manabe, N. Momono, Y. Miura and M. Ido: Phys. Rev. B **49** (1994) 16000.
[54] J. R. Cooper and J. W. Loram: J. Phys. I. France **6** (1996) 2237.
[55] See, for example, N. Miyakawa, P. Guptasarma, J. F. Zasadzinski, D. G. Hinks, and K. E. Gray: Phys. Rev. Lett. **80** (1998) 157.
[56] K. A. Müller: Phil. Mag. Lett. **82** (2002) 279.
[57] K. McElroy: Nat. Phys. **2** (2006) 441.
[58] See J. Schmalian, D. Pines and B. Stojkovic: Phys. Rev. Lett. **80** (1998) 3839, and related references therein.
[59] K.-Y. Yang, T. M. Rice and F.-C. Zhang: Phys. Rev. B **73** (2006) 174501, and related references therein.
[60] Ch. Renner, B. Revaz, J.-Y. Genoud, K. Kadowaki, and O. Fischer: Phys. Rev. Lett. **80** (1998) 149.

XXIst International Symposium on the Jahn–Teller Effect 2012 IOP Publishing
Journal of Physics: Conference Series **428** (2013) 012044 doi:10.1088/1742-6596/428/1/012044

Some unique superconductive Properties of Cuprates

K. A. Müller

Physics Institute, University of Zurich, CH 8057, Switzerland

Copper oxides are the only materials that show transition temperatures, T_c, above the boiling point of liquid nitrogen, with a maximum T_c^m of 162 K under pressure. Their structure is layered, with one to several CuO_2 planes, and upon hole doping, their transition temperature follows a dome-shaped curve with a maximum at T_c^m. In the underdoped regime, i.e., below T_c^m, a pseudogap T* is found, with T* always being larger than T_c, a property unique to the copper oxides [1]. In the superconducting state, Cooper pairs (two holes with antiparallel spins) are formed that exhibit coherence lengths on the order of a lattice distance in the CuO_2 plane and one order of magnitude less perpendicular to it. Their macroscopic wave function is parallel to the CuO_2 plane near 100% d at their surface, but only 75% d and 25 % s in the bulk, and near 100% s perpendicular to the plane in YBCO. There are two gaps with the same T_c [2]. As function of doping, the oxygen isotope effect is novel and can be quantitatively accounted for by a two-band vibronic theory [3] near T_c^m, and underdoped below it till $T_c = 0$ with by a formula valid for (bi)polarons [4]. These cuprates are intrinsically heterogeneous in a dynamic way. In terms of quasiparticles, Jahn-Teller bipolarons are present at low doping, and aggregate upon cooling [1], so that probably ramified clusters and/or stripes are formed, leading over to a more Fermi-liquid-type behavior at large carrier concentrations above T_c^m.

This is a video lecture. The video is in the supplementary data.

[1] For an overview see: K.A. Müller, J. Phys: Condens. Matter 19, 251002 (2007).
[2] R. Khasanov, A. Shengelaya et al. Phys. Rev. Lett. 98, 0570007 (2007).
[3] H. Keller, A. Bussmann-Holder, and K.A. Müller, Materials Today 11, 38 (2008).
[4] S Weyeneth and K.A. Müller, J. Supercond. Nov. Magn. 24, 1235, (2011).

AUTHOR INDEX

AUTHOR INDEX

Institute of Physics
Dirac House, Temple Back
Bristol BS1 6BE UK

CURRAN ASSOCIATES INC
proceedings.com